Isolation of Rabaul

HISTORY OF U.S. MARINE CORPS OPERATIONS IN WORLD WAR II

VOLUME II

by USMC

HENRY I. SHAW, JR.

MAJOR DOUGLAS T. KANE, USMC

Historical Branch, G–3 Division, Headquarters, U.S. Marine Corps

1963

Other Volumes in this Series

I

LtCol Frank O. Hough, Maj Verle E. Ludwig, and Henry I. Shaw, Jr.

Pearl Harbor to Guadalcanal

Library of Congress Catalog Card No. 58-60002

For sale by the Superintendent of Documents, U.S. Government Printing Office
Washington, D.C., 20402 - Price $5.50

Foreword

This book, the second in a projected five-volume series, continues the comprehensive history of Marine Corps operations in World War II. The story of individual campaigns, once told in separate detail in preliminary monographs, has been largely rewritten and woven together to show events in proper proportion to each other and in correct perspective to the war as a whole. New material, particularly from Japanese sources, which has become available in profusion since the writing of the monographs, has been included to provide fresh insight into the Marine Corps' contribution to the final victory in the Pacific.

The period covered in these pages was a time of transition in the fighting when the Allied offensive gradually shifted into high gear after a grinding start at Guadalcanal. As the situation changed, the make-up of the Fleet Marine Force changed, too. We passed through the era of hit and run and through the time for defensive strategy. Our raider and parachute battalions were absorbed in regular infantry units, the seacoast batteries of our defense battalions became field artillery, and our air squadrons were re-equipped with newer and deadlier planes.

In the converging drives that made the Japanese fortress Rabaul their goal—one under Navy command and the other under Army leadership—Marines played a significant part well out of proportion to their numbers. In those days, as in these, the use of trained amphibious troops in a naval campaign overloaded the scale in our favor.

As one hard-won success followed another in the Solomons and on New Guinea, a progression of airfields wrested from island jungles gave us the means to emasculate Rabaul. While the enemy garrison waited helplessly for an assault that never came, we seized encircling bases that choked the life out of a once-potent stronghold.

Once the front lines passed by Rabaul, other island battles seized the headlines—battles of the great two-pronged advance on Japan, which was made possible in large part by the victories of 1943 in the Southwest Pacific. For thousands of Americans, Australians, and New Zealanders, however, the campaign against Rabaul never ended until the last day of the war. In this unheralded epilogue of blockade and harassment, Marine air units took the lead just as they had in the all-out aerial battle that preceded.

The outstanding aspect of all the operations covered in this volume, one evident in every section of the narrative, was the spirit of cooperation between different services and national forces. No finer example exists in recent history of the awesome combined power of distinct military forces pursuing a common goal.

DAVID M. SHOUP
GENERAL, U.S. MARINE CORPS
COMMANDANT OF THE MARINE CORPS

Reviewed and approved
16 May 1963

Preface

The Allied campaign to reduce Rabaul was not an uninterrupted series of flawless operations. It had, like most human enterprises, a share of mistakes to match its successes. Since we learn by both errors and accomplishments, the lessons, good and bad, absorbed during the fighting on New Georgia, Bougainville, and New Britain were priceless in value. They undoubtedly saved the lives of many Marines who went on to take part in the Central Pacific drive that culminated in the battle for Okinawa.

Our purpose in publishing this operational history in durable form is to make the Marine Corps record permanently available for study by military personnel and the general public as well as by serious scholars of military history. We have made a conscious effort to be objective in our treatment of the actions of Marines and of the men of other services who fought at their side. We have tried to write with understanding about our former enemies and in this effort have received invaluable help from the Japanese themselves. Few peoples so militant and unyielding in war have been as dispassionate and analytical about their actions in peace.

This volume was planned and outlined by Colonel Charles W. Harrison, former Head, Historical Branch, G–3 Division, Headquarters, U.S. Marine Corps, working in conjunction with Mr. Henry I. Shaw, Jr., the senior historian on the World War II historical project. Major Douglas T. Kane wrote the narratives of the New Georgia and Bougainville operations, using much of the research material gathered for the monographs prepared by Major John N. Rentz, *Marines in the Central Solomons* and *Bougainville and the Northern Solomons*. The remainder of the narrative was written by Mr. Shaw, who in treating the story of operations at Cape Gloucester and Talasea drew upon the research data assembled for the monograph, *The Campaign on New Britain*, by Lieutenant Colonel Frank O. Hough and Major John A. Crown. The appendices concerning casualties, command and staff, and chronology were prepared by Mr. Benis M. Frank. Colonel Harrison, Major Gerald Fink, Colonel William M. Miller, Major John H. Johnstone, and Colonel Thomas G. Roe, successive heads of the Historical Branch, did most of the final editing of the manuscript. The book was completed under the direction of Colonel Joseph F. Wagner, Jr., present head of the branch.

A number of leading participants in the actions described have commented on preliminary drafts of pertinent portions of the book. Their valuable assistance is gratefully acknowledged. Several senior officers, in particular General

Alexander A. Vandegrift, General Lemuel C. Shepherd, Jr., and Vice Admiral Daniel E. Barbey, made valuable additions to their written comments during personal interviews. General Vandegrift, in addition, made his private correspondence with senior commanders in the Pacific available for use and attribution.

Special thanks are due to the historical agencies of the other services for their critical readings of the draft chapters of this book. Outstanding among the many official historians who measurably assisted the authors were: Dr. John Miller, Jr., Deputy Chief Historian, Office of the Chief of Military History, Department of the Army; Mr. Dean C. Allard, Head, Operational Archives Branch, Naval History Division, Navy Department; and Dr. Robert F. Futrell, Historian, U.S. Air Force Historical Division, Research Studies Institute, Air University, Maxwell Air Force Base.

Chief Warrant Officer Patrick R. Brewer and his successor as Historical Branch Administrative and Production Officer, Second Lieutenant D'Arcy E. Grisier, ably handled the many exacting duties involved in processing the volume from first drafts through final printed form. The many preliminary typescripts and the painstaking task of typing the final manuscript for the printer were done by Mrs. Miriam R. Smallwood. Much of the meticulous work demanded by the index was done by Mrs. Smallwood, Miss Mary E. Walker, and Miss Kay P. Sue.

The maps were drafted by Chief Warrant Officer Brewer and Corporal Robert F. Stibil. Official Defense Department photographs have been used throughout the text.

R. E. CUSHMAN, Jr.
MAJOR GENERAL, U.S. MARINE CORPS
ASSISTANT CHIEF OF STAFF, G-2

Contents

PART I STRATEGIC SITUATION—SPRING 1943

PART II TOENAILS OPERATION

PART III NORTHERN SOLOMONS OPERATIONS

PART IV THE NEW BRITAIN CAMPAIGN

PART V MARINE AIR AGAINST RABAUL

PART VI CONCLUSION

APPENDICES

ILLUSTRATIONS

ILLUSTRATIONS—Continued PAGE

MAPS

PART I

Strategic Situation—Spring 1943

2

Setting the Stage

World War II had the dubious distinction of being the first truly global conflict. The Allied and the Axis Powers clashed on a dozen widely separated fronts and a thousand different battlefields. Six years, lacking only 26 days, passed between the fateful dawn when Nazi tanks rumbled across the Polish border and the solemn moment when the *Enola Gay* released its bomb load over ground zero at Hiroshima. The United States was in this war from the beginning, perhaps not as an active belligerent, but certainly as an open and material supporter of its friends and allies.[1]

Germany was tagged "the predominant member of the Axis Powers" and the Atlantic and European area "the decisive theatre" eight months before the Japanese struck at Pearl Harbor.[2] The stark fact of that surprise attack and its resulting havoc did not alter the basic decision made by the responsible American military and naval chiefs to give priority of men, equipment, and supplies to the campaign against Germany. Their analysis of the situation boiled down to the simple conclusion that Germany was more dangerous to the United States than Japan.

The "Germany-first" decision was made in terms of overall war potential, not solely in terms of fighting men. Indeed, the sobering succession of Allied reverses in the Pacific during the early days of 1942 gave ample evidence of the formidable fighting qualities of Japanese soldiers and seamen. Japan was no pushover; her defeat would require years of all-out effort. However slim the allotment of resources to the Allied troops that faced the Japanese, constant military pressure had to be maintained. Casualties and costs would soar if ever the enemy was allowed time to consolidate his hold on the strategic islands, to dig in and construct defenses in depth.

The United States had the primary responsibility for halting the Japanese advance south and east through the Pacific. The fact that the battleground included thousands of open miles of the world's largest ocean added immeasurably to the logistic problem involved and made mandatory the assignment of amphibious-trained troops to the fighting. In such a situation, the Marine Corps, which had argued and coaxed, sweated and struggled, to develop workable amphibious techniques in the 20's and 30's, soon proved the worth of its findings and training.

[1] See Parts I and II of Volume I of this series for an examination of the extent to which the U.S. was prepared for and participating in World War II prior to 7 December 1941.

[2] Para 13, ABC-1, dtd 23Mar41, quoted in Navy Basic War Plan—Rainbow No. 5, dtd 26May41. The "Rainbow" plans outlined possible courses of action in the event of a multi-nation war, the term deriving from the custom of giving color names (Japan was Orange) to war plans involving one major enemy. Rainbow-5 was the basic American war plan at the time of Pearl Harbor.

A Marine occupied a unique position among American servicemen during World War II. While his country battled a coalition of enemies, and most of his countrymen in arms were fighting halfway across the globe from him, the Marine trained to meet only one enemy—Japan. As the war moved inexorably onward, the men who flocked to join the Corps in unprecedented numbers were literally and consciously signing up to fight the Japanese. This orientation toward a single enemy and towards one theater, the Pacific, colored every Marine's life in and out of battle and had an incalculable but undeniably beneficial effect on the combat efficiency of the Fleet Marine Force (FMF).

A glance back over the first year highlights of the Pacific war will set the stage for the stirring events to follow—for the story of the Marine Corps' vital part in the all-out Allied shift to the offensive.

THE FIRST YEAR OF THE PACIFIC WAR [1]

The homespun philosophy of America furnishes an apt saying that described Japan's plight in World War II: "she bit off more than she could chew." Not only did the Japanese militarists grossly underestimate the staying power and counterpunching ability of the United States and its allies; they also failed to make a realistic appraisal of their own nation's capabilities. Compounding their original error of starting the war, the enemy leaders in-
invincibility of their fleets and armies.
dulged in some wishful thinking about the

Certainly the Japanese had cause to view their parade of early victories with chauvinistic pride. There were only a few moments during the first half year of fighting when the Allies were not faced with the alternatives of retreat or defeat. But even then, for every outpost like Guam or Hong Kong where token garrisons had no choice but to lay down their arms, there was a Wake or Bataan where a desperate last-ditch defense was fought. True, the Japanese prevailed on all fronts, but the bitter nature of the fighting should have furnished a clue to the spirit of the defenders and the certainty of retaliation.

In Tokyo, the staff members of *Imperial General Headquarters* ignored or misread the warning signs. Japan had caught the Allies off balance and ill-prepared; she had taken all of her original objectives and held the "Southern Resources Area," the Netherlands Indies and Malaya, in a tight grip. Ostensibly, she now had the means to make herself self-sufficient, and she needed every bit of time and every man she could muster to consolidate her hold on her prize. Her next logical move, and the one called for in original war plans, was to strengthen defenses. A line along which she would make her stand had been picked out: a long, looping arc that ran south from the Kuriles through Wake to the Marshalls and Gilberts and then west to include the Bismarck Archipelago, Timor, Java, Sumatra, Malaya, and Burma. The defense of this perimeter was probably a task beyond Japan's resources, even with the help of the newly seized territories. At the war's end, one

[1] Unless otherwise noted, the material in this section is derived from: USSBS (Pac), NavAnalysisDiv, *The Campaigns of the Pacific War* (Washington: GPO, 1946), hereafter USSBS *Campaigns;* USSBS (Pac), JapIntelSec, G-2, *Japanese Military and Naval Intelligence Division* (Washington: GPO, Apr46); *The War Reports of General of the Army George C. Marshall—General of the Army H. H. Arnold—Fleet Admiral Ernest J. King* (Philadelphia and New York: J. B. Lippincott Company, 1947).

senior Japanese officer described this perimeter as "just about the limit, the maximum limit of our capability." [4]

The natural clairvoyance of hindsight similarly aided a number of enemy officers to recognize the fact that Japan had overextended herself by early spring of 1942. At that time, however, the headquarters faction that had authored the original ambitious war plan was still in the saddle and their aggressive philosophy prevailed. Orders went out from Tokyo to continue the advance, to seize further positions that would shield the initial perimeter. It was this decision more than any other taken by *Imperial Headquarters* during the course of the war that hastened the downfall of the Japanese Empire. In less than a year's time, enemy forces were reeling back all across the Pacific, and the reserves that would have bolstered the original perimeter were dissipated in a fruitless effort to continue the offensive.

The new expansionist plans called for the occupation of strategic islands, suitable for air and naval base development, in the North, Central, and South Pacific. The grand prize sought was Midway; it was hoped that a thrust there would bring out the American fleet for a decisive engagement. Closely linked to this projected attack was the movement of an occupation force into the Aleutians to seize Kiska, Attu, and Adak Islands. The two operations would be conducted simultaneously, and both enemy supporting fleets would be available to combine against the American ships. In the south, the objective was to strengthen the Japanese position in the Bismarcks and on New Guinea. Plans were laid to take Port Moresby in southeastern New Guinea and to move outpost garrisons into the Solomons. After the successful conclusion of the Midway operation, the Japanese planned to move against New Caledonia, Fiji, and Samoa, and sever Australia's lifeline to the States.

The enemy timetable for expansion listed the seizure of Port Moresby for early May, followed in a month's time by the attack on Midway. In both cases the carefully selected occupation troops never got a chance to set foot on their objectives. Seen in retrospect, the issue was decided at sea, and the decision was final.

On 7–8 May in the Coral Sea, an American carrier task force intercepted the invasion fleet bound for Port Moresby and was successful in turning it back. In "the first major engagement in naval history in which surface ships did not exchange a single shot," [5] carrier aircraft inflicted all the damage. Each side lost a carrier, each had one severely damaged, but the honors of the field fell justly to the American pilots who forced the Japanese to withdraw. The Port Moresby operation was put off until July, but the outcome of the Battle of Midway ensured a permanent postponement. (See Map I, Map Section.)

Midway could hardly have been called a surprise target. The intelligence available to Admiral Chester W. Nimitz, Commander in Chief, Pacific Fleet (CinCPac), regarding where and when the enemy would strike next was conclusive. When the Japanese carrier attack force approached within launching distance of the atoll on 4 June, it ran into a whirlwind of

[4] USSBS (Pac), NavAnalysisDiv, *Interrogations of Japanese Officials*, 2 vols (Washington: GPO, 1946), Interrogation No. 393, FAdm Osami Nagano, IJN, II, p. 353, hereafter USSBS, *Interrogation* with relevant number and name.

[5] King, *War Reports*, op. cit., p. 523.

American planes. Nimitz had brought up all his available carriers, had added long-range bombers staging from Hawaiian fields, and had given the Midway garrison's Marine Aircraft Group 22 (MAG-22) new planes to meet the enemy threat. The result of these preparations was electrifying; all four of the Japanese carriers were sent to the bottom and the invasion force streaked back for the relative safety of home waters. The Battle of Midway was a disaster from which the Japanese naval air arm never recovered. The battle has frequently been termed *the* decisive engagement of the war in the Pacific and its results were certainly far reaching. The severe and sudden cut in enemy carrier strength put a crimp in all plans for further offensive action.[5]

The immediate reaction of Admiral Isoroku Yamamoto, Commander in Chief of the *Combined Fleet*, to the news of his Midway losses was to recall the Aleutian occupation forces. Then, almost immediately, he reversed himself and ordered the operation to continue but with the modification that only the two westernmost targets, Kiska and Attu, would be seized. Perhaps Adak Island was too close to the U.S. base at Dutch Harbor for comfort. Although Yamamoto's exact reasoning in ordering the operation to continue is not known, it is probable that he gave a great deal of weight to the fact that more American territory would be occupied, a definite boost to Japanese morale that would be needed if the truth of the Midway battle leaked out. On 7 June, occupation troops landed on the two bleak islands, there to stay until the Allies could spare the men, supplies, and equipment which were needed to drive them out. Although there was considerable public alarm in the States, especially the Pacific Northwest, over the presence of Japanese in the Aleutians, actually the new enemy bases were not much of a threat. The rugged island chain, cursed with more than a fair share of the world's miserable weather, was no avenue for conquest.

Midway's results went far to redress the balance of naval strength in the Pacific and to give the Allied leaders a chance to launch a limited offensive. The logical target area was the South Pacific, where the Japanese, despite their Coral Sea misadventure, were still planning to take Port Moresby and were continuing their encroachment into the Solomons. The enemy field headquarters for this two-pronged approach to the Australia-United States supply route was Rabaul on New Britain, a prize whose capture dominated Allied planning. But Rabaul was far too ambitious an objective for the summer of 1942, when almost any offensive effort severely strained available resources.

The calculated risk of the first offensive—a "shoe-string" operation—was made at Guadalcanal, a hitherto obscure jungle-clad island in the lower Solomons. The Japanese first moved into the area in April, when they occupied tiny Tulagi and set up a seaplane base and anchorage in the fine natural harbor between that island and neighboring Florida. A stretch of some 20 miles of open water, which was soon to earn the grim name of "Iron Bottom Sound," separated Tulagi from Guadalcanal. The larger island was one of the few places in the Solomons where

[5] See Part V of Volume I of this series for details of the Marine participation in the Midway battle.

terrain favored rapid airfield development, and the Japanese, soon after Midway, began to clear ground and construct a fighter strip along its northern coastal plain.

Guadalcanal's airfield and Tulagi's harbor became prime objectives once Washington okayed the opening offensive in late June. In contrast to the months of meticulous planning that characterized later amphibious operations, this first effort, code-named WATCHTOWER, was surrounded by an aura of haste. The unit picked to do the job was the one most likely to be successful, one which had more of the requisite amphibious training and indoctrination than any other at this stage of the war—the 1st Marine Division (Reinforced). The division was in the process of completing a move to New Zealand, its rear echelon still at sea, when warning orders were received designating it the WATCHTOWER assault force. In less than a month, the division had changed its orientation from routine training to preparation for jungle fighting, had prepared its tactical plans in light of the scanty information available on enemy and terrain, and had unloaded its ships and then reloaded them for combat. A rendezvous was made at sea in the Fiji rehearsal area with the convoy of the 2d Marines, which had been sent out from San Diego to take the place of the 7th Marines, one of the division's regular regiments detailed to Samoa's garrison.

On 7 August, assault elements of the 1st Division landed on Guadalcanal and moved inland according to plan without meeting any opposition. Simultaneously, Marines stormed ashore on Tulagi and its neighboring islets, where the landings were opposed violently. Several days of hard fighting were needed to secure Tulagi's harbor, but when this first battle was over the scene of ground action shifted to Guadalcanal. There, engineers worked feverishly to put the partially completed airstrip in shape to receive friendly fighters. And the Marine defenders desperately needed aerial reinforcement, in fact any kind of reinforcement that they could get, for the Japanese reaction to the Guadalcanal landing was swift and savage.

For six hectic months, during which it often seemed that WATCHTOWER would prove a fiasco, the 1st Division and an all-too-slowly swelling number of Army and Marine reinforcements stood off a series of sharp enemy counterattacks. The Japanese poured thousands of crack troops into the jungles that closed on the Marine perimeter, but never were able to put ashore enough men and equipment at one time to overcome the garrison. From the captured airfield (Henderson Field), a weird and wonderful composite force of Navy, Army, Marine, and New Zealand planes fought the Japanese to a standstill in the air and immeasurably strengthened the Allied hand at sea by attacking enemy transport and surface bombardment groups as they steamed from bases in the upper Solomons to Guadalcanal.

Although Allied naval forces lost heavily in the series of sea battles that were fought for control of Solomons' waters, the American and Australian ships kept coming back on station. The Japanese admirals strove mightily to seize the advantage when it was theirs, but the opportunity faded. By the end of November, enemy losses had increased so sharply

691-360 O—63——2

that capital ships were no longer risked in Iron Bottom Sound.[7]

When the anniversary of Pearl Harbor rolled around, the Japanese situation on Guadalcanal was desperate. A steady parade of men, ships, and planes had been committed to drive out the Americans and every effort had failed. Even the firebrands in *Imperial General Headquarters* were now convinced that Japan had overreached herself. By the year's end, the decision had been made to evacuate Guadalcanal and orders were sent out to consolidate positions on the original perimeter.[8]

GUADALCANAL AND PAPUA[9]

By the time of the Guadalcanal landing the Japanese held effective control of all the Pacific islands they had invaded but one—New Guinea. In March of 1942, the enemy had occupied positions along the northeast coast of the enormous island at Lae and Salamaua, and their local naval superiority gave them the means of moving in wherever else they wished along this virtually undefended coast. Allied air, operating from carriers or staging from Australia through Port Moresby, was the principal deterrent to further Japanese encroachment. When, during the Battle of the Coral Sea, the *Port Moresby Invasion Force* was forced to turn back to Rabaul, the obvious capability of the enemy to attack again prompted the Allies to make a countermove to ward off this threat. In June and July, Australian ground units and fighter squadrons supported by American engineers and antiaircraft artillery moved to Milne Bay on the eastern tip of New Guinea to build and hold an air base that would cover Port Moresby's exposed flank.

The Japanese thwarted a further Allied advance planned for early August when they landed their own troops near Buna Mission on 22 July. Buna was the northern terminus of the Kokoda Trail, a difficult 150-mile route over the Owen Stanley Mountains to Port Moresby. The superior enemy landing force soon fought its way through the light Australian defenses and reached Kokoda village, about 30 miles inland, where it held up. This first move by the Japanese into Papua, the Australian territory which comprised most of the eastern part of New Guinea, was essentially a reconnaissance in force to test the feasibility of an overland drive on Port Moresby. Thousands of enemy reinforcements arrived from Rabaul in August to strengthen the Buna position and add weight to the proposed attack. By 26 August the Japanese were ready, and they jumped off from Kokoda in a determined assault that quickly overpowered the few

---

[7] The story of the Navy's bitter struggle for control of the Guadalcanal waters is well told in Samuel E. Morison, *The Struggle for Guadalcanal, August 1942–February 1943—History of United States Naval Operations in World War II*, v. V (Boston: Little, Brown and Company, 1949).

[8] MilHistSec, G–2, FEC, Japanese Monograph No. 45, IGHQ Army High Command Record, Mid-1941–Aug45, 2d rev. 1952 (OCMH), p. 67, hereafter *IGHQ Army Record*.

[9] Unless otherwise noted, the material in this section is derived from: *IGHQ Army Record;* Part VI of Volume I of this series; John Miller, Jr., *Guadalcanal: The First Offensive—The War in the Pacific—United States Army in World War II* (Washington: HistDiv, DA, 1949); Samuel Milner, *Victory in Papua—The War in the Pacific—United States Army in World War II* (Washington: OCMH, DA, 1957); Morison, *Struggle for Guadalcanal, op. cit.;* USSBS, *Campaigns.*

Australians who tried to block their advance. The problem of supporting these defending troops was a logistician's nightmare, but it was a nightmare that the Japanese inherited as the distance from the front line to their base at Buna increased.

The enemy troops attacking along the Kokoda Trail were operating with minimal air cover, in fact the Allied air forces were doing their best to cut them off completely from Buna and to sever Buna's supply lines from Rabaul. These Japanese were now making an isolated effort since the secondary operation planned to complement the overland drive had miscarried.

Originally, the enemy operation plan had called for the seizure of Samarai Island, off the eastern tip of New Guinea, as a seaplane base and staging area for an amphibious assault on Port Moresby, timed to coincide with the Kokoda Trail approach. When reconnaissance planes discovered the Allied activity at Milne Bay, the target was shifted to this new base. The Japanese, in a move characteristic of their actions in this period, underestimated their opposition and assigned a grossly inadequate landing force for the operation. On 25 August, about a thousand enemy troops from Kavieng began landing in the bay and immediately made contact with the Australians. A reinforcement of 500 men came in on the 29th, but by that time they were only enough to fill the gaps in the ranks of the first unit. The Milne defense force, a reinforced brigade almost 10,000 strong, first blunted, then smashed the Japanese attack. The dazed survivors were evacuated on the nights of 4–5 September, victims of an Australian victory that did much to hearten Allied morale.

The failure at Milne Bay, coupled with similar disasters on Guadalcanal, prompted *Imperial General Headquarters* to check the overland advance on Port Moresby and concentrate its efforts on achieving success in the lower Solomons. The Japanese troops on the Kokoda Trail had reached a point so close to Port Moresby that "they could see the lights of the city,"[10] but it is doubtful if they could have ever reached their objective. An outpouring of Allied troops from Australia into Port Moresby had strengthened the position to the point that preparations were underway to mount an offensive when the enemy fell back with the Australians hot on their heels. Throughout October the pressure was increased until the Japanese position had contracted to a perimeter defense of Buna and Gona (a native village about seven miles north of Buna Mission).

The Australian 7th Division and the American 32d Infantry Division closed on the perimeter. The Australians came overland for the most part, the majority of the Americans by air and sea. The fighting was bitter and protracted in jungle terrain even worse than that encountered by the Marines on Guadalcanal and against a deeply dug-in enemy who had to be gouged out of his bunkers. Gona fell to the Australians on 9 December and Buna Mission to the Americans on 2 January; the last organized resistance was overcome on the 22d, six months to the day after the Japanese had landed in Papua.

On the same day that the Australians drove the Japanese out of Gona, the 1st

---

[10] Interrogation of Gen Hitoshi Imamura and LtGen Rimpei Kato, IJA, in USSBS (Pac), NavAnalysisDiv, *The Allied Campaign Against Rabaul* (Washington: GPO, 1946), p. 89.

Marine Division was officially relieved on Guadalcanal, its mission completed. The tide of battle had swept full course to the Allied favor, and strong Army and Marine forces of the XIV Corps were now capable of annihilating the remaining Japanese. When evacuation orders were received from Tokyo, however, the Japanese Navy in a series of high-speed night runs managed to bring off about 13,000 men from the island. On 9 February, Guadalcanal was cleared of enemy units and the campaign was ended. American losses in dead and wounded by ground action were close to 6,500, but more than 23,000 enemy lay dead in the jungles around Henderson Field, victims of combat and disease. The loss of additional thousands of enemy sailors and pilots, hundreds of planes, and more than a score of warships and transports increased the wastage of Japanese strength that marked the fruitless effort to retake Guadalcanal.

With the victories in Papua and on Guadalcanal, the Allies had flung down the gauntlet. The Japanese had to accept the challenge; they had lost the initiative.

JAPANESE STRATEGY [11]

The original impetus for the Japanese move into the Solomons and Eastern New Guinea came from enemy naval officers who felt "that a broad area would have to be occupied in order to secure Rabaul." [12] Although the Navy promoted the advance, the Army accepted the concept readily enough, and both services began to develop outlying bases which would cover the approaches to New Britain. When the Allies struck at Guadalcanal, the Japanese Navy "was willing to stake everything on a decisive fight" [13] to regain the island and turn back the offensive thrust. Army leaders, interested mainly in the war on the Asian mainland and in the spoils of the Netherlands Indies, woke up too late to the realities of the Guadalcanal campaign.

Two months passed before realistic estimates of the strength of Henderson Field's defenders began to figure in enemy reinforcement plans. By the time the Japanese were ready to commit enough men to retake Guadalcanal, the chance for them to reach the island in decisive numbers had passed. The Allies were able to choke off most attempts, and the shattered units that did reach shore were seldom in shape to mount a sustained attack. The situation called for a reevaluation of Japanese strategic objectives in the light of Allied capabilities.

At the year's end, military planners in Tokyo, acting on the discouraging reports from the field, projected accurately the course of Allied action for the next months, pointing out that:

> . . . the enemy plans to attack Rabaul since it is the operational base for Army, Navy, and Air Forces. The enemy will try to accomplish this task in the Solomon Is. Area by driving our units off Guadalcanal Is. and advancing northward on the Solomon Is. In the Eastern New Guinea Area, the enemy will secure the Buna Vicinity and attack the Lae and Salamaua Areas from

---

[11] Unless otherwise noted, the material in this section is derived from: MilHistSec, G-2, FEC, Japanese Monograph No. 35, SE AreaOpsRec, Seventeenth Army Ops—Part II (OCMH), hereafter *Seventeenth Army Ops—II;* MilHistSec, G-2, FEC, Japanese Monograph No. 99, SE Area NavOps—Part II (OCMH), hereafter *SE Area NavOps—II; IGHQ Army Record;* USSBS, *Campaigns.*

[12] USSBS, *Interrogation* No. 503, VAdm Shigeru Fukudome, IJN, II, pp. 524–525.

[13] *Ibid.*, p. 526.

> the sea. After penetrating Dampier [Vitiaz] Strait, they will attack Rabaul in joint operations with forces on the Solomon Is. After this, planning to attack the Philippine Is., they will continue operations along the northern coast of New Guinea.[14]

On 3 January, the text of the "Army-Navy Central Agreement on South Pacific Area Operations" was radioed to Rabaul; it laid down Tokyo's newly approved strategy. Although expressed in the bombastic language characteristic of the spirit of the offensive permeating Japanese military documents, the "Agreement" was in fact the outline of a defensive pattern. Key points, mainly airfields and anchorages, were to be occupied or strengthened in the North and Central Solomons and in Eastern New Guinea after the first order of business, the evacuation of the troops on Guadalcanal, had been completed. Some of the names that were to figure prominently in the war news—Lae, Salamaua, Wewak on New Guinea; New Georgia, Bougainville, and Buka in the Solomons—were emphasized in the allotment of defensive sectors. The Japanese Army and Navy had divided the responsibility for base defense along service as well as geographic lines, a factor that was to have considerable influence on the conduct of the fighting.

The enemy naval planners, running true to form, wanted to get the main defenses in the Solomons as far away from their major base at Rabaul as possible. The Army authorities, made cautious by the outcome of the attempt to reinforce Guadalcanal over a long, exposed supply route, were willing to move only major forces into the Northern Solomons. As the Army already had primary responsibility for ground defense of the Bismarcks and New Guinea, the additional task of conducting the defense of Bougainville, Buka, Choiseul, and the Shortland-Treasury Islands was considerable. Since the Navy wanted the New Georgia Group and Santa Isabel included in the defended area, it received operational responsibility for these islands and their garrisons. Land-based naval air squadrons were to operate primarily in the Solomons and Bismarcks, while most Army air units were assigned to the defense of the New Guinea area. The *Combined Fleet*, its main strength concentrated at Truk, stood ready to engage any Allied striking force moving north through the Solomons or west from Hawaii.

One of the fundamental differences between the Japanese and the Allied conduct of the war in the Pacific was pointed up by the high command setup established in the "Agreement." There was no area commander appointed with authority to exercise final control of all defensive measures; consequently, there was no joint staff with the function of preparing and executing an overall defense plan. Instead, the senior Army and Navy commanders in the field were responsible directly to their respective headquarters in Tokyo.[15] This duality of command was a feature of the Japanese military system, and to a great extent it also existed in Tokyo at the heart of the enemy war effort. *Imperial General Headquarters* was only the term used to connote the co-equal existence of the general staffs of the two services. Any order tabbed as coming from the *Headquarters*

[14] *IGHQ Army Record*, p. 71. Although this record was assembled after the war, Japanese defensive actions agreed with the quoted estimate and it very probably represents contemporary thought.

[15] Imamura-Kato Interrogation, *op. cit.*, p. 88.

was simply an Army-Navy agreement. In operation, this system could mean, as one Japanese admiral expressed it, that:

> . . . as far as questions of Army operations are concerned, if the Chief of the Army General Staff says we will do this, that is the end of it; and so far as the Navy operations are concerned, if the Chief of the Naval General Staff says we will do this, that fixes it. . . .[16]

Obviously, decisions involving inter-service operations had to be made; stalemate was unacceptable, but the opportunity for unnecessary delay and uncoordinated unilateral action was inherent in the system.

Fortunately for the Japanese, the two commanders at Rabaul got along well together and were determined to cooperate. The single aim of both General Hitoshi Imamura and Vice Admiral Jinichi Kusaka was to hold their portion of Japan's defenses with all the men and material at their disposal. Imamura's command, the *Eighth Area Army*, comprised the *Seventeenth Army* in the Bismarcks and Solomons and the *Eighteenth Army* defending Eastern New Guinea. Both were supported by the *6th Air Division*. Kusaka, as Commander of the *Southeast Area Fleet*, controlled the land-based planes of the *Eleventh Air Fleet* and the ships and ground units of the *Eighth Fleet* which were strung out from New Guinea to New Georgia. Both men expected that the next Allied targets would be found in the area under their control. The choice of the time, place, and strength of those attacks was made, however, by planners in Brisbane, Noumea, Pearl Harbor, and Washington.

[16] USSBS, *Interrogation* No. 379, Adm Mitsumasa Yonai, IJN, II, p. 328.

ALLIED STRATEGY [17]

In World War II the military fortunes of Great Britain and the United States were so closely enmeshed that it was imperative that a workable inter-Allied command system be developed both in the field and at the national level. Meeting in Washington five days after the attack on Pearl Harbor, the political and military leaders of the two major Western powers agreed to adhere to the principle of unity of command in the various theatres of operations. The same basic decision was reached in regard to the higher direction of the military effort of the two nations and of the numerous other Allied powers that they perforce represented. While the concept of a single commander who would control all national forces committed was accepted for limited areas and specific operations, there was no inclination to trust overall command to one man, if indeed such a superman existed. The chosen instrument for the direction of what might

[17] Unless otherwise noted, the material in this section is derived from: ComSoPac WarDs, Jan–Mar43 (COA, NHD), hereafter *ComSoPac WarDs* with appropriate months; CinCSWPA Plan for the Seizure of the New Britain-New Ireland-New Guinea Areas—ELKTON, dtd 28Feb43 (COA, NHD); FAdm Ernest J. King and Cdr Walter M. Whitehill, *Fleet Admiral King—A Naval Record* (New York: W. W. Norton & Company, 1952), hereafter King and Whitehill, *King's Naval Record;* Richard M. Leighton and Robert W. Coakley, *Global Logistics and Strategy 1940–1943—The War Department—United States Army in World War II* (Washington: OCMH, DA, 1955); Samuel E. Morison, *Breaking the Bismarcks Barrier, 22 July 1942–1 May 1944—History of United States Naval Operations in World War II*, v. VI (Boston: Little, Brown and Company, 1950), hereafter Morison, *Breaking the Bismarcks Barrier, 22 July 1942–1 May Roosevelt and Hopkins—An Intimate History* (New York: Harper and Brothers, 1948).

best be called the Western war effort was the Combined Chiefs of Staff (CCS); its membership, the chiefs of the land, sea, and air services of Great Britain and the United States.

Washington was selected as the site of the new headquarters and Field Marshal Sir John Dill, as the senior on-the-spot representative of the British chiefs, was permanently stationed in the American capital with an executive staff. In order to represent adequately the military views of the United States in CCS discussions, it was necessary that the American chiefs meet regularly and air the problems of their respective services. In short order, a series of inter-service staff agencies came into being to support the deliberations of the American chiefs, and a flexible working organization, the Joint Chiefs of Staff (JCS), became the right hand of the President, acting as Commander in Chief.[18]

Admiral Ernest J. King, Chief of Naval Operations and Commander in Chief of the United States Fleet, was the naval representative in the JCS. The Army was represented by two officers, its Chief of Staff, General George C. Marshall, and its senior airman, General Henry H. Arnold, whose opposite numbers on the British Chiefs were the heads of the Imperial General Staff and the Royal Air Force. Through most of the war there was a fourth member of the JCS, Admiral William D. Leahy, who acted as Chief of Staff to the President.

The Combined Chiefs, working directly with Churchill and Roosevelt, established spheres of strategic responsibility best suited to national interests and capabilities. In mid-April, the United States was given responsibility for directing operations in the Pacific from the mainland of Asia to the shores of the Western Hemisphere. This decision had the effect of placing all Allied troops and materiel alloted to the Pacific under control of the Joint Chiefs and of the two men they selected for command.

The JCS divided the Pacific into two areas of command responsibility, one including Australia, the Netherlands Indies, and the Philippines and the other the rest of the ocean and its widely scattered islands. To head the relatively compact Southwest Pacific Area, where most operations could be conducted under cover of land-based air, the JCS chose the colorful commander of the defense of the Philippines—General Douglas MacArthur. The appointment of MacArthur, made with the assent of the Australian government, was announced on 18 April 1942 after the general was spirited out of beleaguered Corregidor; his new title was Supreme Commander, Southwest Pacific Area (CinCSWPA). For Commander in Chief, Pacific Ocean Areas (CinCPOA) the logical choice was Admiral Nimitz; his concurrent command of the Pacific Fleet as CinCPac recognized that the war in his area would be essentially a naval one.

The initial boundary line between the SWPA and POA included all of the Solomon Islands in MacArthur's command; however, the fact that Nimitz' forces were going to mount the first offensive at Guadalcanal made a shift of the line westward a matter of practicality. The new boundary just missed the Russell Islands, ran north to the Equator, turned west to 130° north longitude, then north and west

[18] As a result of a British suggestion at the ARCADIA Conference (23Dec41–14Jan42), the term "combined" was applied thereafter to collaboration between two or more nations, while "joint" was reserved for collaboration between two or more services of the same nation.

again to include the Philippines in the SWPA. The line hugged the tortuous Indochinese, Thai, and Malayan coastlines to Singapore and then cut south between Sumatra and Java to divide the American area of responsibility from the India-Burma sphere of operations, which came under the British Chiefs of Staff.

The JCS issued a directive on 2 July 1942 to govern offensive operations in the Southwest Pacific, setting forth a concept that included three tasks: 1) the seizure and occupation of the Santa Cruz Islands, Tulagi, and adjacent areas; 2) the seizure and occupation of the rest of the Solomons and the northeast coast of New Guinea; and 3) the seizure and occupation of Rabaul and surrounding positions. A subordinate command, the South Pacific Area, was established under Admiral Nimitz and charged with responsibility for executing Task One—the Guadalcanal operation. The post of Commander, South Pacific (ComSoPac) was held first by Vice Admiral Robert L. Ghormley and then by Vice Admiral William F. Halsey. Task One was completed under Halsey with the evacuation of Guadalcanal by the Japanese, but neither Nimitz nor MacArthur considered that he had available the forces or supplies necessary to initiate Task Two immediately. The relatively few Australian and American infantry divisions assigned to the Southwest Pacific were either committed to forward garrisons, still forming and training, or badly in need of rest and rehabilitation as a result of hard campaigning.[19]

[19] CinCPac msg to ComInCh, dtd 8Dec42, Subj: Future Ops in the Solomons Sea Area (COA, NHD); CinCSWPA msg C-251 to CofSA, dtd 27Jan43 (WW II RecsDiv, FRC Alex).

Equally as important, though hardly as well publicized as the feats of the fighting troops and ships, were the accomplishments of the service and supply agencies furnishing logistic support to the combat operations. The South and Southwest Pacific are certainly not areas that would be voluntarily chosen for amphibious campaigns. When the fighting started, there was almost a total lack of ports and bases suitable for support of large scale operations. In a surprisingly short time, however, islands like Espiritu Santo and Efate in the New Hebrides and New Caledonia sprouted vast compounds of supplies, tank farms for fuel storage, and a host of vital maintenance, repair, and service facilities. Hardly had the smoke and dust of battle settled before Tulagi was turned over to the engineers, base personnel, and defense troops who quickly converted it into an essential advance naval base. Guadalcanal in its turn underwent extensive development as the Japanese were driven off. A full stride forward in terms of the 2 July JCS directive could be taken only after an adequate stockpile of military materiel had accumulated in the forward dumps and depots of an expansible logistic network.

A good part of the supply and manpower difficulties of the Pacific commanders were traceable directly to the favored apportionment given to the European and North African theaters of operations. The basic war policy of the Western Allies was affirmed by the Combined Chiefs in January at the Casablanca Conference where their outline of action for 1943 emphasized again the primacy of the defeat of Germany. First priority of resources was allotted to the campaign to wipe out the U-Boat threat in the Atlantic; the occupation of Sicily, a stepped-up bomber offensive against Germany, and "the sending of the greatest volume of supplies possible" to Russia were among the other priority programs. Offensive operations in the Pacific were to be kept within limits that would not jeopardize the chance for a decisive blow against Germany.[20] In their report to the President and Prime Minister, the CCS indicated a number of prospective lines of action in the Pacific, including an advance west from Midway toward the Marianas and Carolines and a drive north from Samoa into the Marshalls. Implicit in these projections of possible offensive action was the successful completion of a campaign to capture or neutralize Rabaul.[21]

In early February, Admiral Halsey was queried by King on his reaction to an operation to seize the Gilbert and Ellice Islands using South Pacific forces. Halsey strongly recommended against it, preferring instead to continue pressure in the Solomons. Admiral Nimitz supported Halsey's opinion, but asked if South Pacific operations could be depended upon to pin down the Japanese Fleet. On 17 February, ComSoPac replied that he believed "that the best way to pin down the Japanese Fleet is to threaten Rabaul," and went on to indicate that he intended to occupy the Russell Islands inside of a week and move into the New Georgia Group "as soon as possible." [22] He soon set early April as his target date for the New Georgia operation, but a re-evaluation of Pacific strategy forced a revision of his plans.

[20] CCS 155/1, dtd 19Jan43, Subj: Conduct of the War in 1943 (COA, NHD).

[21] CCS 170/2, dtd 23Jan43, SYMBOL—Final Rept to the President and Prime Minister (COA, NHD).

[22] *ComSoPac Feb43 WarD*, p. 33.

Under terms of the JCS directive of 2 July 1942, General MacArthur had been given responsibility for strategic direction of all operations against Rabaul, including those undertaken by South Pacific forces after completion of Task One. On 28 February 1943, his staff completed a plan (code-named ELKTON) that reflected MacArthur's conviction that the Japanese were now much stronger in the Southwest Pacific than they had been the previous summer. The situation prompted him to submit a new concept of operations calling for a more deliberate advance than had once been contemplated and a substantial increase in all categories of forces.

Under ELKTON, the command position of Admiral Halsey as ComSoPac was an unusual one. The operations contemplated in the Solomons would of necessity get their logistic support from SoPac bases and be executed in the main by SoPac forces. Naval officers were strongly of the opinion that these forces should remain under command of Halsey, but did not question the need for MacArthur to continue to give strategic direction to the overall campaign against Rabaul. Halsey's plan to attack New Georgia in April, tentatively approved by Nimitz, clashed with the sequence of operations thought necessary by SWPA planners. The upshot of the submission of ELKTON to the JCS was that a Pacific Military Conference of representatives of SWPA and POA was called together in Washington to resolve differences and to try to find the additional troops and equipment that MacArthur thought necessary.

En route to Washington, MacArthur's representatives, headed by his chief of staff, Major General Richard K. Sutherland, stopped at Noumea to talk with Halsey and hear his plan for New Georgia.

They then flew on to Pearl Harbor where in a round of conferences with Nimitz' staff they learned the views of that commander on ELKTON. On 10 March, the conferees arrived in Washington to begin two weeks of discussion in an atmosphere where the requests from the Pacific could be best assessed against the world-wide commitments of the United States.

The sequence of operations called for in the ELKTON Plan listed the capture of airdromes on the Huon Peninsula of Eastern New Guinea as a necessary preliminary move to closing in on Rabaul. Bomber squadrons operating from fields in the Lae-Salamaua-Finschhafen area would then control the Vitiaz (Dampier) Strait and could neutralize the Japanese strongpoints at Kavieng, New Ireland, and on New Britain, Buka, and Bougainville. With this assistance from SWPA air, SoPac forces would seize and occupy positions in the New Georgia Group. Next would come a simultaneous drive on western New Britain from New Guinea and on Bougainville from the lower Solomons. The two-pronged attack would then converge in the capture of Kavieng, or if the situation seemed favorable, the last step, the capture of Rabaul, would be attempted directly.

General Sutherland and Major General Millard F. Harmon, commander of Army forces in Halsey's area, agreed that in order to accomplish ELKTON as outlined, all the men, ships, and planes asked for would have to be made available.[23] There was no chance that this would be done, since the JCS was already engaged in a reexamination of the resources available for all the strategic undertakings decided on at Casablanca. It was now apparent that there just was not enough to go around to give full coverage to every scheme; forces requested for ELKTON would have to be cut drastically.[24]

The requirements of the heavy bomber offensive against Germany changed one aspect of ELKTON immediately. The planned aerial interdiction of Japanese rearward bases from the Huon Peninsula depended on more long range planes reaching the Southwest Pacific. Since these planes could not be made available, airbases closer to Solomons' objectives within range of medium bombers would have to be taken. Woodlark and Kiriwina Islands in the Solomon Sea east of Papua were agreed upon as suitable objectives. Despite this modification of the ELKTON concept, General Sutherland still considered that the Huon Peninsula operations would have to precede all others; on the other hand, Halsey's Chief of Staff, Captain Miles R. Browning, USN, maintained that the seizure of Woodlark and Kiriwina would allow Halsey to make his move into New Georgia without waiting for the capture of Huon airfields. The varying points of view were presented to the JCS for decision.[25]

The solution arrived at by the JCS was workable and retained elements of both the unity of command concept and that of cooperative action. Subject to the checkrein authority of the JCS, General MacArthur was given overall control of the campaign. Admiral Halsey would have direct command of operations in the Solo-

[23] 4th PMC meeting, dtd 15Mar43, Anx A (COA, NHD).

[24] JCS 238, Memo by the JSP, dtd 16Mar43 (COA, NHD); JSSC 11, Surv of the Present StratSit, dtd 22Mar43 (COA, NHD).

[25] Minutes of the JCS 68th meeting, dtd 21 Mar43 (COA, NHD).

mons within the scope of MacArthur's general directives. Any Pacific Ocean Area forces not specifically approved by the JCS for inclusion in task forces engaged in ELKTON operations would remain under Admiral Nimitz.

On 28 March 1943, the Joint Chiefs issued a new directive that cancelled that of 2 July 1942 and outlined the new scheme of operations for the campaign against Rabaul. The schedule of tasks now called for establishment of airfields on Woodlark and Kiriwina Islands, to be followed by seizure of bases on Huon Peninsula concurrently with Halsey's move into New Georgia. Western New Britain and southern Bougainville were the next steps toward the goal of Rabaul. The purpose of these operations was set down as "the ultimate seizure of the Bismarck Archipelago." [26]

[26] JCS 238/5/D, dtd 28Mar43, Directive—Plan for the Seizure of the Solomons Islands-New Guinea-New Britain-New Ireland Area (COA, NHD).

The Opening Moves

AREA OF CONFLICT[1]

Prior to the outbreak of war, the strategic area centered on Rabaul was a slow-paced frontier of Western civilization. Economic development of the Bismarck Archipelago, the Solomon Islands, and Eastern New Guinea was pretty well limited to the cultivation of coconut palms for copra. The coconut plantations, together with a scattering of trading posts, missions, and government stations, housed the relative handful of non-native inhabitants. The islands had little in the way of climate or terrain to attract tourists or anyone else without a surpassing good reason for visiting them. For the most part the area remained as it had been when the first European explorers visited it in the middle of the sixteenth century.

More than 300 years passed before a Western nation thought it worthwhile to lay claim to any of the islands. Then Germany, as part of her belated attempt to build a colonial empire, followed her traders and missionaries into Northeast New Guinea, the Bismarcks, and the Northern Solomons, proclaiming them the protectorate of Kaiser-Wilhelmsland in 1884. Britain countered by establishing her own protectorate over the Southern Solomons and by annexing the rest of Eastern New Guinea. As the Territory of Papua, with a capital at Port Moresby, this latter area was turned over to Australia in 1905. At the outbreak of World War I, the Australians occupied Kaiser-Wilhelmsland and kept it, under a League of Nations mandate following the peace, as the Territory of New Guinea. The mandate capital was established at Rabaul,[2] and the territory divided into government districts of Northeast New Guinea, New Britain, New Ireland, Manus (the Admiralty and Northwest Islands), and Kieta (Buka and Bougainville). South of Bougainville, the Solomons, including the Santa Cruz Islands, formed the British Solomon Islands Protectorate, administered by a resident commissioner at Tulagi.

Each island in the Rabaul strategic sphere, with the exception of a few outlying atolls, has a basic similarity of appearance which holds true regardless of size. High hills and mountains crowd the interior, sending out precipitous spurs and ridges to the coasts. A matted jungle canopy of giant trees covers all but the highest peaks, and the sun touches the ground only along the banks of the numerous streams

[1] Unless otherwise noted, the material in this section is derived from: MID, WD, Surv of the Solomons Islands (S30-677), 2 vols, dtd 15Mar43, hereafter *Solomons Survey;* MIS, WD, Surv of NE New Guinea and Papua (S30-678), dtd 15Jul43; MID, WD, Surv of Bismarck Archipelago (S30-675), 2 vols, dtd 5Oct43, hereafter *Bismarcks Survey.*

[2] The capital of the New Guinea Territory was transferred to Lae in Northeast New Guinea on 1 December 1941 because of the danger of volcanic activity in the Rabaul area.

that slice the slopes. Where fire or water forces a temporary clearing, the vine and bush barrier of second growth springs up to add to the difficulties of transit. Along the shores most of the low-lying ground is choked with rank second growth, and vast stretches of fetid mangrove swamp mark the mouths of streams and rivers. Easily accessible and well-drained land is at a premium and on most such sites coconut plantations had been established. Years are required to grow the trees and constant attention is needed to prevent the encroachment of jungle. These plantations, together with the few significant reaches of grassland scattered throughout the larger islands, were the potential airfields that figured so prominently in Southwest Pacific planning.

Along with tropical forest and rugged hills, the area shares a common climate—hot, humid, and unhealthy. There is a rainy season around December when the northwest monsoon blows, but the "dry months" of the southeast trade winds, April through October, are wringing wet by temperate zone standards. Although the amount of precipitation varies considerably according to locale, an average rainfall of more than 200 inches in the uplands and 100–150 inches along the coasts is not unusual. The islands lie in the only latitudes in the world where evaporation is greater over land than water.[3] The temperature of the moisture-saturated air stays in a steady range of 75–90 degrees the year around. Constant high heat and humidity sap a man's strength and make him prey to a wide variety of tropical diseases.

[3] G. T. Trewartha, *An Introduction to Climate* (New York: McGraw-Hill Book Company, 1954), p. 112.

Strange though it may seem, this uninviting area has well over a million inhabitants. The majority are Melanesians, the dominant race in the islands northeast of Australia. Primitive in habit and appearance, these people have dark brown, almost black, skins, small but solidly-built bodies, and frizzled, upstanding mops of hair. The natives of Papua belong to a related but separate race, shorter, darker, and more Negroid in aspect. In the atolls around the periphery of the area there are a few thousand Polynesians, tall, fair, and fine-featured members of the race that occupies the Central Pacific islands. Mixed marriages among these peoples are not uncommon; the Melanesians themselves are thought to be the product of a merging of Papuan and Polynesian strains.

Tribalism is the way of life in the islands; there is no native national spirit or tradition. The frankly paternalistic British and Australian administrations respect the tribal organization and govern through the local chiefs. Almost all the natives live in small villages, their outlook limited to what they can see, feel, or hear. Village garden plots, temporarily wrested from the jungle, grow only enough taro, yams, and sweet potatoes for local needs; fruit and fish supplement an otherwise monotonous and starchy diet. Although Christian missionaries have been moderately successful in gaining converts, the basic religion of these simple people is a natural animism. The diversity of dialects is so great that the traders' jargon of pidgin English is the only universally understood language. By Western standards, the natives live a severely limited life, but this simple existence has the sanction of centuries.

In general, a view of life outside the village is sought and seen only by the laborers who work the coconut plantations and the relatively few natives who serve in the government or police. These men, especially the "police boys," are capable of great personal loyalty to those who can understand and lead them. The evidence for this statement is easily found in the existence of the spy system that operated behind Japanese lines in the Southwest Pacific.

Years before the enemy invaded the New Guinea Territory and drove south toward Port Moresby and Tulagi, the probability of hostile approach through the island screen had been foreseen by the Australians. In 1919, their Navy began to set up a network of observers along the sparsely settled northeastern shores of the continent. As compact and reliable radios were developed, the observer system spread northward into the islands where strategically located officials and planters were recruited and trained to send coded reports of enemy movements. Many of these veteran islanders, famed as the coastwatchers, remained behind when the Japanese advanced, and from vantage points deep in the midst of enemy-controlled waters, sometimes even from enemy-held islands, fed a steady stream of valuable intelligence into Allied hands. The natives who stayed with the coastwatchers were in many cases their eyes and ears in the enemy camp. Though the opportunity for betrayal was great, it was seldom seized.[4]

[4] Cdr Eric A. Feldt, RAN, *The Coastwatchers* (New York & Melbourne: Oxford University Press, 1946), p. 4*ff*, hereafter Feldt, *The Coastwatchers*.

While the success of the coastwatching system was a tribute to human courage and resourcefulness, it was equally an acknowledgment of the complex geographic factors making it possible. Each island and island group that figured in the Allied drive on Rabaul has its own peculiar character, and its detailed description is part of the narrative of the operations that concern it. A general sketch of the whole strategic area is needed, however, to set in mind the relationship of these islands to each other.

On the map, New Guinea, the world's second largest island, dominates the sea north of Australia. More than 900,000 natives live in the scattered villages of Papua and Northeast New Guinea, an area roughly the size of California. Lofty mountains, some ranging well above 13,000 feet, form a spine for the Papuan Peninsula which juts out into the Coral Sea. In the bulging midsection of the island near the border of Dutch New Guinea, thousands of square miles of soggy ground and tangled swampland spread out along the wandering courses of torpid rivers coming down out of the highlands. The Huon Peninsula, which harbors near its base and southern flank the airfield sites so prominent in the ELKTON planning, thrusts east toward New Britain, less than 50 miles away across Vitiaz (Dampier) Strait. (See Map I, Map Section.)

The tip of western New Britain, Cape Gloucester, has enough low grassland near the coast to allow airfield development. Thus, from the Cape and from the Huon Peninsula, directly opposite on New Guinea, control could be easily maintained over Dampier Strait, the only entry into the Solomon Sea from the northwest. New

Britain, an elongated and crescent-shaped island, 370 miles long and 40–50 miles wide, is heavily forested and has the usual prominent jumble of mountains and hills ridging its interior. Midway along the coasts, Talasea in the north and Gasmata in the south offer way-point airdrome sites for a drive on Rabaul, which "has by all odds the best natural harbor and base for military operations in the entire New Guinea-New Britain-Solomon Islands area."[5]

Curving to the northwest from the waters off Rabaul's Blanche Bay, scimitar-like New Ireland parallels New Guinea's coastline 300 miles away and closes one side of the Bismarck Sea. The airfields and harbor at Kavieng on the slim island's northern point made the small colonial town a prime strategic objective. The Bismarck Sea is outlined by a staggered arc of islands which swings north from New Hanover off Kavieng to the Saint Matthias group, then west to the Admiralties and on to the atolls known as the Northwest Islands, which dip south toward New Guinea. The native population of the whole area of the Bismarck Archipelago is approximately 150,000, the largest number by far living on New Britain and New Ireland.

Planes based at Rabaul and at airfields on Buka or Bougainville can effectively close off the passage between New Ireland and Buka, the second major gateway to the Solomon Sea. Politically speaking, these two northern islands are part of the New Guinea Territory; geographically, they are one with the rest of the Solomons. The principal islands of the Solomons constitute a double mountain chain running northwest and southeast for about 700 miles between the Bismarcks and the New Hebrides; the width of this central grouping is 100 miles. Several offshoot islands well away from the main chain—Ontong Java atoll to the northeast, Rennell due south of Guadalcanal, and the Santa Cruz group to the southeast—are also considered part of the Solomons.

In all the islands there are some 165,000 natives living in a total land area equivalent to that of West Virginia. The terrain fits the general pattern of the whole strategic area—jungle and hills extremely difficult to traverse which tend to localize land combat and put a premium on air and sea power. The major islands of the northeast chain, Choiseul, Santa Isabel, and Malaita, have few natural military objectives, and the same lack characterizes San Cristobal, the southernmost of the southwest chain. The other large islands of the Solomons, Guadalcanal, New Georgia, and Bougainville, have sizable harbors and airfield sites sufficient to make them logical stopping points in a deliberate advance on Rabaul. Each of these major island objectives has several smaller islands nearby which also class as potential targets: Tulagi and Florida are coupled with Guadalcanal; Vella Lavella and Kolombangara with New Georgia; and Buka and the Shortland-Treasury Islands with Bougainville.

Typical of the smaller island groups in the Solomons is the Russells, which lie 30–35 miles northwest of Cape Esperance on Guadalcanal. The accident of location rather than any considerable strategic advantage made them Admiral Halsey's first objective after Guadalcanal was secured.

[5] *Bismarcks Survey*, p. 1.

SIMPSON HARBOR AND RABAUL *appear in a composite aerial photograph taken during an Allied air raid prior to the Bismarck Sea battle. (USAF F23272AC)*

MARINES OF THE 3D RAIDER BATTALION *land from rubber boats on the beach at Pavuvu to take part in the seizure of the Russells. (USMC 54468)*

OCCUPATION OF THE RUSSELLS [6]

The Russell group consists of two main islands fringed by a scattering of lesser islets. Pavuvu, the larger island, is very irregular in shape and no more than ten miles across at its widest point. The low land along the shoreline of many of its coves and bays is clear of undergrowth and lined with coconut trees, but these plantings only edge the jungle and mark the steep rise towards the hills of the interior. On the north coast, several deep water bays provide sheltered anchorages which will accomodate large ocean-going vessels. Only a narrow channel separates Pavuvu from its smaller neighbor to the east, Banika, which has unusual terrain for the Solomons. Except in its southwestern portion, where hills rise to 400 feet, the island is low and rolling and suitable for military development. Banika's north coast is cleft by Renard Sound, a deep inlet that provides access to the low ground. (See Map 1.)

In January 1943, when it was evident that the Japanese were losing their fight to regain control of Guadalcanal, the possibility of moving forward to the Russells was given serious consideration at ComSoPac headquarters. To Halsey's staff, the island group seemed a desirable objective and one that could be taken and held with the limited resources available in the South Pacific. A presentation of this concept was made to Admiral Nimitz when he visited Noumea for conferences with Halsey on 23 January. The Pacific Fleet commander gave oral approval to the idea and before the month was out he gave specific authorization for the operation.

The Japanese had not occupied the Russells when they moved into the lower Solomons, but once the decision was made to pull out of Guadalcanal about 300 troops were sent to tiny Baisen Island off Pavuvu's northwest tip, Pepesala Point, to set up a barge-staging base.[7] The enemy unit left when its job was done; the withdrawal was reported on 11 February by a coastwatcher who had been landed in the islands earlier to keep tab on Japanese activity. The prospect of an unopposed landing was cheering to Allied planners, but it resulted in no reduction in assault troop strength for the proposed operation. A healthy respect for Japanese offensive capabilities kept the figure high.

South Pacific Area planners felt that a further "attempt on the part of the enemy to reestablish himself on Guadalcanal was a distinct possibility," and that if this happened the reaction to Allied occupation of the Russells would be violent.[8] The landing force allotted under these circumstances had to be strong enough to

[6] Unless otherwise noted, the material in this section is derived from: *ComSoPac Jan–Apr43 WarDs*; ComPhibSoPac Rept of Occupation of the Russell Islands (CLEANSLATEOp) 21Feb–17Apr43, dtd 21Apr43; 43d InfDiv FO No. 2, dtd 15Feb43; 3d RdrBn Rept of the Russell Islands (CLEANSLATEOp), dtd 9Apr43; Russell Islands Det. 11th DefBn Jnl; LtCol E. S. Watson, G–3, 43d InfDiv, "Movement of a Task Force by Small Landing Craft," dtd 17Apr43; *Solomons Survey;* Morrison, *Breaking the Bismarcks Barrier*. Documents not otherwise identified are located in the Russell Islands Area Operations File of the Archives, Historical Branch, G–3 Division, Headquarters Marine Corps.

[7] HistSec, G–2, FEC Japanese Monograph No. 48, SE Area NavOps—Part I, n.d. (OCMH), p. 61, hereafter *SE Area NavOps—I*.

[8] CGUSAFISPA 1st Ind, dtd 1May43, to 43d InfDiv Summary of the Occupation of the Russell Islands, dtd 17Apr43, in Watson Rept, *op. cit.*

sustain a major counterattack. A further consideration in determining the size of the force was the belief that the assault troops would be favorably disposed to take part in future operations against New Georgia.

On 7 February, Halsey's directive for CLEANSLATE, the Russells operation, was issued. Named to overall command was Rear Admiral Richmond K. Turner, who headed South Pacific Amphibious Forces. The 43d Infantry Division, less its 172d Regimental Combat Team (RCT), was designated the principal component of the CLEANSLATE occupation force. Major reinforcing units were the Marine 3d Raider Battalion, antiaircraft elements of the Marine 11th Defense Battalion, half of the 35th Naval Construction Battalion, and Acorn 3.[9] Once the acorn unit had an airfield in operation on Banika, MAG-21, then en route to the South Pacific, would move in its fighter squadrons for intercept and escort missions..

The assembly of the CLEANSLATE task force was hard to detect. No ship larger than a destroyer was assigned to transport troops or supplies; most of the workload fell to newly arrived landing ships and craft getting their first offensive test in the South Pacific. While larger vessels brought the 43d Division from New Caledonia to Guadalcanal, the movement was made in normal-sized convoys. Japanese planes located and unsuccessfully attacked one of these convoys near San Cristobal on the 17th, but the enemy pilots saw nothing about the transports to indicate that they were anything more than another reinforcement-replacement group headed for the Guadalcanal garrison. Ships arriving off Koli Point, staging area for the operation, unloaded immediately and cleared the vicinity. Only a cluster of innocuous small vessels, mainly LCTs (landing craft, tank) and LCMs (landing craft, medium), and a screen of destroyers stood by for the run to the Russells.

The projected D-Day for planning purposes was 21 February; four days before, when it was evident that the operation was proceeding on schedule, Admiral Halsey confirmed this date. Late on the 19th loading out began, LCTs first, followed by the smaller craft, and topped off by the destroyer types. Only the APDs (high speed transports) were equipped to hoist on board landing craft, and the destroyers and destroyer mine-sweepers assigned as transports each took a quartet of small boats under tow. Near midnight on 20 February, the strange flotilla got underway. Destroyers were in the van, throttled down to the speed of a dozen squat LCTs that followed in trace, with the rear brought up by a tug-drawn barge loaded with ammunition and barbed wire.

The support given CLEANSLATE was impressive. Bombers from SWPA hit Japanese rearward bases in the Northern Solomons and Bismarcks. Aerial cover over the target and interdiction missions against enemy installations in the Central Solomons were flown by squadrons from Henderson Field, temporarily reinforced by the *Saratoga's* air group. Nearly every combat ship in Halsey's command put to sea, ready to meet a Japanese surface attack; four cruisers and four destroyers steamed up New Georgia Sound (aptly nicknamed "The Slot") as a covering

[9] An acorn was a naval unit designed to construct, operate, and maintain an advanced landplane and seaplane base and to provide facilities for operations.

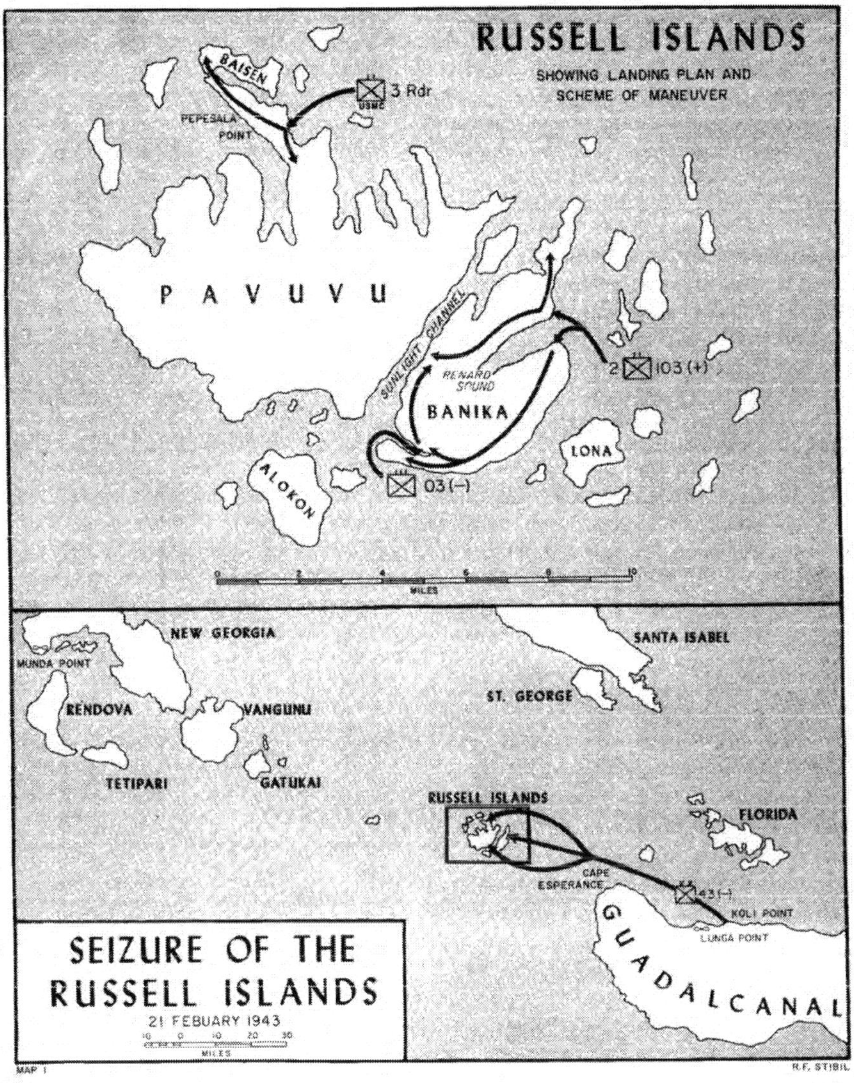
RUSSELL ISLANDS
SHOWING LANDING PLAN AND
SCHEME OF MANEUVER
BAISEN
PEPESALA POINT
3 Rdr
USMC
PAVUVU
SUNLIGHT CHANNEL
RENARD SOUND
BANIKA
2 103 (+)
03 (–)
LONA
ALOKON
MILES
NEW GEORGIA
MUNDA POINT
RENDOVA
VANGUNU
TETIPARI
GATUKAI
SANTA ISABEL
ST. GEORGE
RUSSELL ISLANDS
FLORIDA
CAPE ESPERANCE
43 (–)
KOLI POINT
LUNGA POINT
GUADALCANAL
SEIZURE OF THE RUSSELL ISLANDS
21 FEBUARY 1943
MILES
MAP 1
R.F. STIBIL

force. The precautions paid off; no enemy scout plane or vessel spotted the task force and the landing was made without opposition.

Reconnaissance parties sent to the Russells several days before the landing had selected suitable beach exits, gun sites, and camp and dump areas. On the morning of the 21st, the Army battalions landing on the two beaches of Banika and the Marine raiders going ashore on Pavuvu's Pepesala Point were met by guides who led them to pre-selected positions. One hour after the waves of assault troops landed from their destroyer transports, the LCTs nosed ashore and began unloading. The first echelon of antiaircraft guns and crews of the 11th Defense Battalion were in position on Banika by noon. The raiders and infantrymen were dug in to meet a counterlanding attempt before nightfall.

Marine antiaircraft gun crews and the Army field artillerymen whom they had retrained for the air defense job[10] were the only ones to see action during the ensuing weeks. On 6 March, the first Japanese attack occurred when a dozen fighters and bombers made a low-level strike on the main islands. Early warning radar was not yet in operation and the first enemy bombing and strafing run caused some casualties before the antiaircraft defenses were manned; at least two enemy raiders were shot down. Sporadic air attacks followed this first effort, but none were of serious import.

The Marine elements of the original CLEANSLATE landing force were only temporarily assigned to the operation. The 11th Defense detachment was used only until the 10th Defense Battalion arrived in the Russells; on 15 March, the new unit began taking over the 11th's battery positions. The changeover was completed by 17 March, and the detachment's gun crews returned to Port Purvis on Florida Island the following night. The 3d Raider Battalion pulled out on 20 March and returned to Espiritu Santo; 43d Division units occupied the raiders' defensive positions.

The withdrawal of the Marines was about the only rearward movement of troops from the Russells during this period. Each day after D-Day, LCTs loaded at Guadalcanal and under cover of darkness made the run to the new forward base; succeeding echelons of Turner's task force arrived at the islands for 50 nights running. By the end of February over 9,000 men were ashore and by 18 April, when responsibility for logistic support and defense of the Russells passed to the commanding general at Guadalcanal, 16,066 men and 48,517 tons of supplies had been brought in by the LCT shuttle. Banika now boasted an operating airfield for MAG-21's three fighter squadrons, a motor torpedo boat base, extensive base defense installations, and the start of a considerable supply handling capacity.

Admiral King in Washington was somewhat dubious of the value of putting so much into the Russells, but Halsey defended his policy as necessary for the protection of the new air base. As far as ComSoPac was concerned, CLEANSLATE was merely the completion of the first stage of his move toward New Georgia, and the troops and supplies stationed there were earmarked for the continued advance up the Solomons chain.

[10] Maj Joseph L. Winecoff ltr to CO, 11th DefBn, dtd 22Feb43.

BATTLE OF THE BISMARCK SEA [11]

While ComSoPac was consolidating his hold on the new Russells airdrome, an event occurred in the Southwest Pacific Area that emphasized dramatically the importance of land-based air in the campaign against Rabaul. On 2–3 March, a Japanese troop convoy headed for Lae at the base of the Huon Peninsula was engaged in a running fight by Australian and American squadrons based on New Guinea. The results of this Battle of the Bismarck Sea were so significant that General MacArthur stated:

> We have achieved a victory of such completeness as to assume the proportions of a major disaster to the enemy. Our decisive success can not fail to have most important results on the enemy's strategic and tactical plans. His campaign, for the time being at least, is completely disrupted.[12]

The convoy was an attempt initiated by *Eighth Area Army* to strengthen its defenses in the Lae-Salamaua area and to insure continued control of both shores of Dampier Strait. Eight transports, varying in size from 500 to 6,800 tons, and eight escort destroyers made the run. On board were approximately 6,000 soldiers of the *51st Division* and 400 replacements for Special Naval Landing Force units.[13] A canopy of fighter planes, both Army and Navy, was provided for overhead protection.

The enemy convoy cleared Rabaul on 1 March, steaming at slow speed along the northern coast of New Britain, partially hidden by lowering skies which made observation difficult. A reconnaissance bomber of MacArthur's Fifth Air Force sighted the ships and escorting fighters, however, and radioed in its find; a flight of B–17s was unable to locate the target when the weather closed. The following morning the convoy was spotted again and this time the Flying Fortresses found their quarry, broke through the screen of planes and antiaircraft fire, and sank one transport, the *51st Division* command ship. Survivors were transferred to two destroyers, which separated from the main convoy and steamed ahead to Lae. Australian and American planes continued to seek out the ships throughout the day, but the continued bad weather helped to foil these attacks and further damage was minor.

Despite the aerial harassment, the Japanese adhered to their original sailing schedule, which was to bring them off Lae about 1700 on 3 March. By early morning of the 3d, the convoy was well within Huon Gulf and also well within range of Allied airbases on New Guinea; for all practical purposes the enemy had sailed into a trap. MacArthur's air commander, Major General George C. Kenney, sprang it with a coordinated attack led by low-level fighter-bombers specially practiced in

[11] Unless otherwise noted, the material in this section is derived from: HistSec, G–2, FEC, Japanese Monograph No. 32, SE Area AirOps Nov42–Apr44, n.d. (located at OCMH), hereafter *SE Area AirOps; SE Area NavOps—II;* Wesley F. Craven and James L. Cate (eds), *The Pacific: Guadalcanal to Saipan, August 1942 to July 1944—The Army Air Forces in World War II*, v. IV (Chicago: University of Chicago, 1950), hereafter Craven and Cate, *Guadalcanal to Saipan; Morison, Breaking the Bismarcks Barrier.*

[12] Quoted in MajGen Charles A. Willoughby and John Chamberlin, *MacArthur 1941–1951* (New York: McGraw-Hill Book Co., 1954), p. 112.

[13] USSBS, *Campaigns*, p. 174.

anti-shipping strikes, backed up by medium- and high-level bombers and an escort of fighters. The resulting melee was disastrous for the Japanese. Every transport was burning and in a sinking condition in less than half an hour; two destroyers were sunk and a third heavily damaged. The attacks continued until a fourth destroyer, engaged in rescue operations, was hit that afternoon. Motor torpedo boats reached the scene during the night and finished off one of the cripples, and the next morning the bombers completed the score. After rescuing 3,800 men, four destroyers, all that was left of the original convoy, made it back to Rabaul; fewer than 900 men got through to Lae.

The overwhelming Allied success had, as General MacArthur observed, important results. Not only did the Japanese fail to get a substantial reinforcement through to the Huon Peninsula area, but the transport losses forced them to abandon large-scale reinforcement attempts altogether. Dampier Strait did not belong to the Allies yet, but Kenney's fliers made it clear that the Japanese had no clear title either. Supplies and men slated for enemy garrisons in Eastern New Guinea or the Solomons—for any base within effective range of Allied planes—were now moved forward by destroyers, whose high speed, excellent maneuverability, and antiaircraft guns gave them a measure of protection, or by small craft hugging the island coasts. The Japanese had been decisively defeated, but a battle is not the war, and the sorry record of their defending aircraft prompted an all-out effort to restore at least a parity of airpower in the Rabaul strategic area.

JAPANESE "I" OPERATION [14]

Following the Bismarck Sea debacle, Japanese scout planes reported increased activity in Papua and the lower Solomons. Allied troop and material strength was clearly increasing, and all intelligence pointed to the imminence of offensive operations. Allied air raids and antishipping strikes seriously disrupted enemy defensive preparations and curtailed the movement of reinforcements to forward bases. The Japanese decided that a strong counter-stroke was needed to blunt the Allied air spearhead and to gain a respite for their own defense build-up. Tokyo assigned the task, designated *I Go* ("I" Operation), to Admiral Yamamoto and his *Combined Fleet*.

In order to bolster the strength of Rabaul's *Eleventh Air Fleet*, Yamamoto ordered forward from Truk the planes and pilots from four of the carriers of his *Third Fleet*. On 3 April, the admiral himself flew to Rabaul to take personal command. The combined force available for *I Go* was at least 182 fighters, 81 dive bombers, and 72 medium-range land bombers, plus a few torpedo planes. The 15-day operation was planned to proceed in two phases, the first incorporating a strike against the Solomons and the second, attacks on Allied positions in Papua.

The busy cluster of ships, large and small, in the vicinity of Tulagi and Guadalcanal was the initial *I Go* target. Near noon of 7 April, 67 enemy Val dive bombers with an escort of 110 Zeke fighters

[14] Unless otherwise noted, material in this section is derived from: HistSec, G-2, FEC, Japanese Monograph No. 122, SE Area NavAirOps—Part III, Nov42–Jun43, n.d. (OCMH), hereafter *SE Area NavAirOps—III; SE Area NavOps—II;* Morison, *Breaking the Bismarcks Barrier.*

took off from staging airfields on Buka and Bougainville to make the attack. Their departure was duly noted and reported to Henderson Field by coastwatchers on Bougainville; at 1400, radar in the Russells picked up the oncoming flights and 76 interceptors, Guadalcanal's typical joint-service mixture, scrambled and tangled with the Zeke escort over The Slot. The Japanese bombers, hiding behind a blanket of heavy black clouds that covered Indispensible Strait between Malaita and Florida, headed for Tulagi. Almost all the ships were out of Tulagi harbor when the raiders struck; only a fleet oiler and a New Zealand corvette were caught in the confined waters. Both were sunk. The attack continued against the rapidly maneuvering vessels in Iron Bottom Sound, but ship and shore antiaircraft fire kept the Vals high and the bombing inaccurate. The destroyer *Aaron Ward*, attempting to protect an LST that had become the focus of enemy attention, was seriously damaged; she later sank under tow. The bombers caused no other significant damage and drew off soon after they loosed their loads. The Zekes scored just as lightly as the Vals, accounting for only seven planes, all of them Marine. The welter of conflicting claims for enemy aircraft shot down was winnowed to an estimate that less than 25 Japanese were lost,[15] a figure that argued well with the highest official enemy report of 24 planes downed.[16]

Aside from the relatively light damage to Allied ships and aircraft, the 7 April attack had one other tangible result for the Japanese. It enabled them to slip reinforcements into Kolombangara by destroyer transport while ComSoPac concentrated his air strength at Guadalcanal to meet further attacks. Similar reinforcement efforts were executed for Western New Britain and the Huon Peninsula under cover of the trio of attacks on New Guinea targets that made up the second phase of *I Go*.

On 11 April, 94 *Third Fleet* carrier planes attacked shipping in Oro Bay, 20 miles southeast of Buna. Fifty Allied fighters fought them off but not before Japanese Val pilots had sunk one merchantman, beached another, and damaged an Australian minesweeper. On the following day, 174 Japanese naval planes, including 43 medium bombers, made a mass attack on the airfields surrounding Port Moresby. While defending fighters took on the Zeke escort, the bombers plowed up the airstrips, but otherwise did little damage. The third *I Go* raid was made on ships and airfields at Milne Bay on 14 April. The Japanese again attacked in overwhelming force, 188 planes; only 24 Australian fighters were available to meet them, but all ships were forewarned and underway, firing their antiaircraft guns to make the enemy pilots shear off. The major result of the attack was the sinking of a merchantman and some minor damage to other shipping. In all three attacks, 5 Allied planes were lost, and the Japanese admitted the loss of 21 aircraft.

On 16 April, Admiral Yamamoto called off the "I" Operation, ordering the remaining *Third Fleet* planes back to Truk. He had been completely misled by the glowing reports of his pilots into believing that *I Go* had been a tremendous success. The total damage claim for the four raids was

---

15 CinCPac, Ops in the POA for Apr43 (COA, NHD), p. 14.

16 Morison, *Breaking the Bismarcks Barrier*, p. 124, indicates that Japanese postwar records confirm the loss of only 12 Vals and 9 Zekes.

staggering: 1 cruiser, 2 destroyers, and 25 assorted transports and cargo vessels sunk, with heavy damage to 2 more transports and several smaller vessels; 134 planes shot out of the air (including 39 probables). Matched against these totals was the actual loss of 1 destroyer, 1 corvette, 2 merchantmen, and less than 20 aircraft. It would seem that the *Third Fleet* pilots had adopted the penchant for reporting "gross exaggeration of damage inflicted" that was rampant in the ranks of the *Eleventh Air Fleet*.[17] Whatever the explanation for pilot error, be it willful exaggeration or wishful thinking, the premature ending of *I Go* without any significant results was chilling to Japanese hopes of delaying Allied offensive preparations.

ALLIED RAIDS[18]

The prime targets of Halsey's pressure tactics in the early months of 1943 were the enemy airfields at Munda on New Georgia and Vila on Kolombangara. The tempo of air raids against these bases increased steadily as Allied strength mounted. Coupled with these air strikes was a limited program of naval bombardment made possible by the fact that SoPac planes and ships had wrested control of the waters immediately north of Guadalcanal from the Japanese. The Japanese could and did risk their warships within range of Henderson Field's bombers, but the need had to be great as in the evacuation of Guadalcanal. The chance for a showdown sea battle still brightened the hopes of enemy naval officers, but there was little desire any longer to seek this critical fight in the confined waters of the lower Solomons. Halsey's cruisers and destroyers had time to get some experience at shore bombardment, thus exposing enemy garrisons to a bit of the bitter medicine dished out by Japanese naval gunners in the darkest days of the struggle for Guadalcanal.

Rear Admiral Walden L. Ainsworth took a cruiser-destroyer force against Munda on 5 January and again on the 24th brought his bombardment group up to New Georgia, this time to shell the field at Vila. On both occasions, fires started by naval gunfire lit up the night and the results of the bombardment were at least spectacular. Follow-up attacks by Allied air caused more damage, according to the Japanese, but neither bombing nor shelling had any lasting effect on the progress of the enemy airbase construction. Admiral Ainsworth noted that, while these air-sea attacks might render the fields unusable for critical periods of time, "the only real answer is to take the field away from them."[19]

Admiral Halsey was in complete agreement with this sentiment; he expected little more from his attack pattern in the Central Solomons than the harassment and delay he achieved. The Japanese became expert at filling in the craters in the runways and dug-in their scattered dumps and shelters to minimize the effect of the Allied raids. Life was mighty unpleasant under the constant round of attacks, however, and the portents were hardly encouraging for the success of Japanese arms.

Another of Ainsworth's shore bombardment groups was underway for New Georgia after dark on 5 March when he received

[17] USSBS, *Interrogation* No. 601, Cdr Ryoske Nomura, IJN, II, p. 532.

[18] Unless otherwise noted, the material in this section is derived from: *SE Area NavOps—II;* Morison, *Breaking the Bismarcks Barrier*.

[19] CTF 67 Rept of Vila-Stanmore Bombardment 22-24Jan43, dtd 28Jan43 (COA, NHD).

word from Guadalcanal that two enemy ships had left Bougainville headed south earlier in the day. Guadalcanal's seaplane scouts spotted the ships, two destroyers that had just delivered supplies to the Vila base steaming north on a return course. Fire control radar screens pinpointed the location of the targets for ships' batteries, and the enemy vessels were buried in a deluge of shells. Both quickly sank; only 49 crewmen survived to tell their harrowing tale to the garrison on Kolombangara.[29]

The cruiser-destroyer force was slated to hit Munda and Vila again on the night of 7–8 April and was already out of port when Admiral Yamamoto launched his "I" Operation with the assault on Tulagi. The Japanese planes failed to sight the bombardment ships, but the prospect of further enemy attacks prompted Halsey to call off the mission and concentrate his forces. The temporary slackening-off of surface and air raids made necessary by this assemblage of power may have been instrumental in convincing Yamamoto that *I Go* was a success. The enemy admiral had no time to discover his mistake.

On 18 April, two days after Yamamoto ended *I Go*, he left Rabaul with his staff on an inspection trip to Bougainville. As a result of a message intercept, Allied intelligence knew the itinerary of the inspection party and a killer group of Army long-range fighters from Henderson Field met the Japanese planes over their destination, Buin airfield. The execution was swift and sure; a few moments after the interceptors attacked, the staff transports crashed in flames and the enemy's most famous naval commander was dead. Yamamoto died primarily because Nimitz' staff evaluated him as the best man the Japanese had to command their *Combined Fleet;* had he been less competent, less of an inspiration to enemy morale, he might well have lived. His death dealt a telling blow to the spirit of the defenders in the Solomons, and furnished grim warning of the downward course of Japanese fortunes.

---

[29] USSBS, *Interrogation* No. 138, LCdr Horishi Tokuno, IJN, I, p. 142; Morison, *Breaking the Bismarcks Barrier*, p. 110, indicates that other sources show there were 174 survivors.

Order of Battle

FLEET MARINE FORCE [1]

By 30 April 1943, the Fleet Marine Force in the Pacific had reached formidable strength in comparison to the few battalions and squadrons that had been its aggregate at the outbreak of war. Over 110,000 Marines and sailors were serving in three divisions, three air wings, and a wide variety of supporting units positioned at Allied bases along a broad, sweeping arc from Midway to Australia. The majority of combat troops were located in the South Pacific under Admiral Halsey's command, where the highest Marine ground echelon was Major General Clayton B. Vogel's I Marine Amphibious Corps (IMAC). The senior Marine aviator, Major General Ralph J. Mitchell, wore two hats as commander of a newly established area headquarters, Marine Aircraft, South Pacific (MASP), and of its principal operating component, the 1st Marine Aircraft Wing (1st MAW). Neither IMAC nor MASP had any substantial tactical function; both commands were organized primarily to serve as administrative and logistical headquarters.

From his command post at Noumea, General Vogel controlled the 2d and 3d Marine Divisions, then in training in New Zealand, as well as a strong body of supporting troops either attached to the divisions or encamped in New Caledonia, the lower Solomons, and the New Hebrides. General Mitchell's units, all temporarily assigned to the 1st Wing, were stationed at airfields from New Zealand to the Russells. Guadalcanal was the focal point of air activity as a steady rotation of squadrons was effected to maintain maximum combat efficiency in the forward areas. Also part of MASP was Headquarters Squadron of the 2d MAW, newly arrived in New Zealand to prepare for a command role in future operations.

In addition to the troops assigned to IMAC and MASP, there was still another sizeable body of FMF units in the South Pacific—those units which were part of the garrisons of American and British Samoa, Wallis Island, and Funafuti in the Ellice Group. American bases on these islands were all included in Major General Charles F. B. Price's Samoan Defense Command. For ground defense, Price

[1] Unless otherwise noted, the material in this section is derived from: CMC AnRept to SecNav for the Fiscal Year Ending 30Jun43; "Historical Outline of the Development of Fleet Marine Force, Pacific, 1941–1950 (Preliminary)," MS official history written at FMFPac Hq about 1951; Kenneth W. Condit, Gerald Diamond, and Edwin T. Turnbladh, *Marine Corps Ground Training in World War II* (Washington: HistBr, G–3, HQMC, 1956); 1stLts Robert A. Aurthur and Kenneth Cohlmia, *The Third Marine Division* (Washington: Infantry Journal Press, 1948), hereafter Aurthur and Cohlmia, *3d MarDivHist*; C. W. Proehl (ed.), *The Fourth Marine Division in World War II* (Washington: Infantry Journal Press, 1946); Robert Sherrod, *History of Marine Corps Aviation in World War II* (Washington: Combat Forces Press, 1952), hereafter Sherrod, *MarAirHist*.

had two rifle regiments, one (3d Marines) under orders to join the 3d Division, and four defense battalions. In special combat training centers were two replacement battalions learning the fundamentals of jungle warfare.[2] Price also had operational control of the squadrons of Marine Aircraft Group 13 (MAG–13), which was administratively part of the 4th Marine Base Defense Aircraft Wing (4th MBDW).

The remaining squadrons of the 4th Wing were stationed in the Central Pacific, on Oahu, and at the outpost islands, Midway, Johnston, and Palmyra, that guarded the approaches to the Pacific Fleet's main base. Ground garrisons for these outposts were furnished by Marine defense battalions administered from a headquarters at Pearl Harbor. The remaining major unit of the FMF in the Pacific, the 1st Marine Division, was in Australia assigned to General MacArthur's command, and just beginning to feel fit again after its ordeal on Guadalcanal.

There was no single headquarters, operational or administrative, for all FMF organizations in the Pacific, although Marine air units did have an administrative headquarters on Oahu—Marine Air Wings, Pacific under Major General Ross E. Rowell. Senior ground commanders, like Vogel and Price, had to consult the Commandant directly on many organizational, administrative, and logistical matters that could well have been handled by a type command at the fleet level. As the FMF grew in size, and its components' missions in complexity, the lack of a higher Marine headquarters to support and coordinate the activities of the air-ground team was to be felt more acutely. The lessons to be learned in the fighting in the Solomons and Bismarcks and on the atolls of the Central Pacific would have to be absorbed before such a headquarters was established.

Most Fleet Marine Force activity in the States was concentrated in a complex of neighboring bases on each coast. In the east, the major ground training center was Camp Lejeune at New River, North Carolina, a site incorporating thousands of acres of tangled, stream-cut forest backing 11 miles of dune-topped beaches. The sprawling Marine Corps Air Station at Cherry Point, less than 40 miles north of Lejeune, controlled a number of smaller airfields scattered throughout the Carolinas. On the west coast, most ground training was carried on at Camp Elliott, a relatively small area just outside San Diego, or at Camp Pendleton, which stretched north from Oceanside for 18 miles along the coastal highway—a vast area of rolling hills, steep-sided canyons and arroyos, and frequent thickets as dense as tropical jungle. A network of air stations and auxiliary fields, the largest being El Toro near Los Angeles, housed the squadrons training for Pacific duty.

These bases, like the Marine Corps itself, were feverishly building at the same time they performed their function of readying men for combat. The 1st Marine Division developed the New River area for amphibious training, and when it shipped out in April 1942 it left behind cadres which

[2] In the fall of 1942, it was decided to season the Marine replacement battalions organized on the east coast of the U.S. in Samoa, where they could receive advanced combat training under climatic conditions and over terrain matching the battle area. Beginning with the 1st Replacement Battalion, which arrived 17 December 1942, seven battalions were trained before the high incidence of filariasis forced a discontinuance of the program in July 1943.

formed the nucleus of the 3d Marines, organized in June. In like manner, the 2d Marine Division, which gave Camps Elliott and Pendleton their baptism as combat training areas, furnished the cadres for most of the units of the 3d Division, which was activated at Elliott on 16 September 1942. The 4th Marine Division was not scheduled for formal activation until August 1943, but its major components were in being by midyear, again by the process of building on a skeleton of veteran officers and enlisted men.

On the air side, the picture of experienced cadres forming the core of new units was much the same as with ground organizations. In contrast to the division and the regiment, however, the Marine aircraft wing and group were essentially task forces shaped to the job at hand and constantly changing their make-up. The 1st MAW, for example, joined a number of squadrons of the 2d Wing during the air battles over Guadalcanal, while the 2d MAW operated largely as a training command in the States. When the 2d Wing left California for the South Pacific in January 1943, its training functions were taken over by Marine Fleet Air, West Coast—a subordinate command of Marine Air Wings, Pacific, in Hawaii. Additional squadrons tentatively assigned to the wings already overseas were in training at every Marine air base in California in 1943.

On the east coast, the 3d MAW was activated in November and its component units grew up with the new airfields then building. Nearly a year's forming and training time was needed before the first of the wing's squadrons was combat ready.[3]

[3] See Appendix E for a location and strength breakdown on the FMF on 30 April 1943.

The overall growth of the Marine Corps matched the rapid swelling of the ranks of the FMF. Although the lion's share of new officers and men ended up in FMF units, thousands of Marines were needed for sea duty, guard assignments, and the supporting establishment. Beginning in February 1943, a steady stream of young women entered the Corps to free men for combat by taking over a host of administrative and technical jobs in non-FMF units. Their performance of duty as Marines "proved highly successful in every way."[4] The enlisted strength of the Marine Corps rose from 222,871 at the start of 1943[5] to 287,621 within six months; on 30 June the number of officers had reached 21,384. Projected total strength for the end of the year was more than 355,000 officers and men,[6] a far cry from the 66,000-man Marine Corps that existed on 7 December 1941.[7]

The second year of fighting saw a cherished tradition of the Marine Corps, its all-volunteer composition, become a war casualty. A Presidential executive order of 5 December 1942 put an end to voluntary enlistment of men of draft age in any of the services. The intent of the directive was to give manpower planners in Washington a greater measure of qualitative control over the influx of men into each service in keeping with the quantitative control already exercised through a quota system. Starting with the intake of February 1943, the recruit depots at

[4] LtGen Keller E. Rockey ltr to CMC, dtd 6Nov55.

[5] M-1 Sec OpDiary 7Dec41–31Dec44, dtd 20Mar45, p. 8.

[6] *Ibid.*

[7] For a location and strength breakdown of the Marine Corps on the eve of WW II, see Volume I, Part I, Chapter 5 of this series.

Parris Island and San Diego saw only a sprinkling of men (mainly draft-exempt 17-year-olds) who did not come in through the Selective Service System. It was still possible, however, for many draftees who anticipated their call-up to enter the service of their choice. The Commandant, Lieutenant General Thomas Holcomb, assigned liaison officers to state governors and draft boards to encourage the deferment of those men who wanted to be Marines until they could fit into the Corps' quota.[8] This program, which was quite successful, resulted in the seeming paradox that most of the draftees in Marine uniforms were still volunteers, in fact if not in name.

The intangible but clearly evident atmosphere of a volunteer outfit was retained by the Marine Corps throughout the war. This spirit was especially evident in the units of the Corps' striking arm—the Fleet Marine Force—where officers and men alike were intolerant of anyone attempting to get by with a marginal performance. The prevailing attitude was that every man had asked to be a Marine and no complaints were expected when the going got a little rough. Each Marine assigned to a unit earmarked for the impending Central Solomons operations seemed quietly determined to equal, even if he could not better, the fighting record of his fellows on Guadalcanal.

THE BATTLE LINES ARE DRAWN[9]

ComSoPac anchored his ELKTON attack against enemy positions in the Solomons on a trio of islands, New Caledonia, Espiritu Santo, and Efate. On each there grew up a complex network of port installations, air bases, supply depots, and salvage and repair facilities geared to operate at a pace that meshed well with Halsey's aggressive offensive philosophy. Like the tactical task forces which actually closed combat with the Japanese, the logistic organizations formed an integrated whole in which the various services cooperated to solve supply and support problems. All units were under orders to "consider themselves as part of the same team rather than Navy, Army, or Marine services in a separate and independent sense. . . ."[10]

The hustling bases in the New Hebrides and at Noumea fed a growing stream of supplies forward to the Guadalcanal area to meet the immediate needs of the garrison and to build a stockpile for future operations. In combat training camps scattered throughout the South Pacific, the interservice exchange and cooperation characteristic of the logistic agencies was repeated. A sense of impending action was high; there was a distinct "get the job done" atmosphere.

At this stage of the war—spring of 1943—no Allied position in either the South or Southwest Pacific could yet be considered a "safe" rear area. As a consequence, large ground garrisons, kept

---

[8] Rockey ltr, *op. cit.*

[9] Unless otherwise noted, the material in this section is derived from: RAdm Charles M. Cooke memo for Pacific Conferees, "Availability of Naval Forces," in notes of 3d PMC Meeting, dtd 13Mar43 (COA, NHD); Army-Navy Central Agreement on SE Area Ops, dtd 22Mar43 in IGHQ NavDirective No. 213, dtd 25Mar43 (OCMH), hereafter *IGHQ Agreement of 22Mar43; Seventeenth Army Ops—II; SE Area NavOps—II*: RAdm Worrall R. Carter, *Beans, Bullets and Black Oil* (Washington: NHD, ND, 1953); Morison, *Breaking the Bismarcks Barrier*.

[10] Carter, *Beans, Bullets and Black Oil, op. cit.*, p. 46.

strong in tribute to Japan's offensive capabilities, were immobilized at key points well away from the prospective center of conflict: *e.g.*, Samoa, Fiji, Tonga. Adding to this drain of offensive strength was the slow recovery of battle-tested units from the debilitating effects of sustained jungle warfare. The troops available for an offensive, therefore, were quite limited in view of the considerable job at hand. In all, MacArthur and Halsey could count on having only 14 divisions, both veteran and untried, ready for offensive action by mid-year.[11] Of this total the SoPac share was six divisions, four Army and two Marine.

Although the manpower squeeze brought on by the shipping demands of the two-front war set a low ceiling on Pacific ground forces, Allied plane and ship strength were on the upswing. American war production made the difference. Over 2,000 combat aircraft would be available for the campaign against Rabaul,[12] a fair match for anything the Japanese could put up against them. At sea, the Pacific Fleet was rapidly approaching a position of absolute superiority over the Japanese as new ships of every type, including the carriers and landing craft vital to amphibious operations, reported to CinCPac for duty.[13] The naval elements of Halsey's and MacArthur's area commands, now designated Third Fleet and Seventh Fleet respectively,[14] could be reinforced from Nimitz' mobile striking force as strategic requirements dictated.

Japanese preparations to meet the offensive that they knew was pending in the Solomons began to take shape concurrently with the evacuation of Guadalcanal. The *Southeast Area Fleet* set naval defense troops to building bases on New Georgia, Kolombangara, and Santa Isabel. In March, the first of several reinforcing units from the *Eighth Area Army* was added to the naval forces in the New Georgia Group. On Buka and Bougainville in the Northern Solomons, the *6th Division* was moved in from Truk to provide the bulk of the garrison.

A steady build-up of defenses, with troops and supplies brought in by barge and destroyer, took place despite the incessant and telling attacks of Allied planes and submarines. Enemy air stayed north of New Georgia except for occasional raids on Guadalcanal; enemy combat ships stuck close to Truk and Rabaul waiting for the opportune moment to strike.

Defense of the Solomons took second place in Japanese plans to measures for continued retention of the Lae-Salamaua region of New Guinea. On 22 March, the Army and Navy staffs in Tokyo agreed on a new directive for operations in the Rabaul strategic area, replacing the one that had governed during the Guadalcanal withdrawal. The new order spelled out the primacy of defensive efforts in New Guinea, but its general tenor was the same as that of its predecessor. In emphatic language, the senior commanders in the field, General Imamura and Admiral Kusaka, were enjoined to hold all the positions that their troops then occupied.

Although the Japanese retained a dual command structure in Rabaul under the

---

[11] CCS 239/1, dtd 23May43, Subj: Ops in the Pacific and Far East in 1943–44 (COA, NHD).

[12] *Ibid.*

[13] King and Whitehill, *King's Naval Record*, pp. 491–495.

[14] Admiral King established a numbered fleet system on 15 March 1943 with all fleets in the Pacific having odd numbers, those in the Atlantic even.

22 March directive, Imamura and Kusaka were told to cooperate closely and elements of both services were ordered to "literally operate as one unit." [15] In the field, the senior Army or Navy ground commander in an area would take charge of the operations of troops of both services. Until the first Allied assault force attacked, those operations consisted, in the main, of constructing defensive positions skillfully wedded to the terrain. Although Japanese soldiers and sailors were deeply imbued with an offensive spirit, they seemed to have a special affinity for defensive fighting where the pick and shovel often rated equal with the rifle. On New Georgia, South Pacific forces were due to get their first real taste of the burrowing, grudging, step-by-step advance that characterized the later stages of the Pacific War.

[15] *IGHQ Agreement of 22Mar43.*

PART II

TOENAILS Operation

891-360 O—63——4

Objective: New Georgia

BACKGROUND OF MUNDA [1]

Occupation of the Russells, following closely on the heels of the Guadalcanal victory, seemed to whet the appetite of Allied forces in the South Pacific for more action, more show-downs with the Japanese. In Admiral Halsey's New Caledonia headquarters, optimism and enthusiasm ran high. Singleness of purpose and a spirit of camaraderie united all representatives on ComSoPac's staff; and, charged by Halsey's impatience to get on with the war, his staff busied itself planning for the next major offensive in the Solomons. The objective: seizure of the New Georgia Group.[2] (See Map II, Map Section.)

[1] Unless otherwise noted, the material in this section is derived from: Allied GeographicalSec, GHQ, SWPA, Study of New Georgia Group—Terrain Study No. 54, dtd 26Mar43; IntelSec, SoPacFor, ObjectiveRept 25-13, New Georgia Group, dtd 15Feb43; SoPacFor PhotoInterpretationU Repts Nos. 37-39, 42, 43, and 47, 24Nov-17Dec43; Morison, *Breaking the Bismarcks Barrier;* Maj John N. Rentz, *Marines in the Central Solomons* (Washington: HistBr, G-3, HQMC, 1952), hereafter Rentz, *Marines in the Central Solomons.* Documents not otherwise identified in this part are located in the following files of the Archives, Historical Branch, G-3 Division, Headquarters Marine Corps: Aviation; Monograph and Comment; New Georgia Area Operations; Publications: Unit Historical Reports.

[2] In the succeeding chapters, the term "New Georgia Group" will refer to the entire island group. The term "New Georgia" will refer only to the island of that name.

A compact maze of islands separated by shallow, coral-fouled lagoons or narrow reaches of open water, the New Georgia Group lies on a northwest-southeast axis between Bougainville, 110 miles to the northwest, and Guadalcanal, 180 miles to the southeast. Nearly 150 miles long and 40 miles wide, it comprises 12 major islands outlined by many smaller islands and formidable reefs. A dense, forbidding jungle growth covers the rugged terrain and accents the abruptly rising, conical mountains which mark the volcanic origin of the group.

Largest island in the group is its namesake, New Georgia. It is hugged closely on the north by the islands of Wana Wana, Arundel, and Baanga and guarded to the southeast by Vangunu and Gatukai. Standing off to the south are Rendova and Tetipari, with two islands—Vella Lavella and Ganongga—in a line to the northwest. Gizo Island chains Vella Lavella to Wana Wana and blocks the southern end of Vella Gulf. Completing the New Georgia Group is the circular, 5,450-foot mountain peak, Kolombangara, which juts out of the sea between Vella Gulf and Kula Gulf, only a few miles northwest of Arundel.

The group centers on New Georgia. A tortuous, misshapen mass with a spiny ridge of peaks, it lies pointing north in a big inverted V, 45 miles in length and 20 miles wide. Its southern coastline is bordered for nearly 20 miles by Roviana

Lagoon, a coral-laced and treacherous stretch of water varying from one to three miles in width. Only small boats can safely trace a channel between the narrow openings in the reef and around the shallow bars of the lagoon. Viru Harbor, southeast of Roviana, is one of the few easy-access points on the southeast coast. A land-locked anchorage, it has an unobstructed but zigzag channel.

The entire east and northeast side of the island is reef-lined, with Marova and Grassi Lagoons bordering that coast almost as Roviana does on the south. As the coastline turns south at Visuvisu Point—apex of the V—Kula Gulf swells directly into three deep-water anchorages formed by jungle rivers rising in the mountains on the north coast. Rice Anchorage and Enogai Inlet are short and mangrove-lined with deep forest crowding the shores; Bairoko Harbor, deeper and longer, is partially blocked by reefs but is the best anchorage along the gulf. Past Bairoko, Hathorn Sound connects the gulf with the passage through to the south, Diamond Narrows. Only 224 feet wide at its narrowest point and 432 feet across at its widest, the Narrows separates New Georgia from Arundel and is the northern entrance to Roviana Lagoon. Its twisting channel is navigable, however, only by small boats.

The dank, oppressive nature of the island characterized even the life of the New Georgia natives. Theirs was a scrubby existence from small gardens, native fruits, some fishing, and occasional trading. One-time headhunters, they became aggressive sailors who moved from point to point through the lagoons by canoe, avoiding the rugged inland travel. As a result, they were, in 1943, excellent guides to the coastlines of the islands but almost completely ignorant of the interiors.

In November 1942, while still contesting possession of Henderson Field on Guadalcanal, the Japanese sought another airfield which would bring their fighter planes within shorter striking distance of the southern Solomons. They found it at Munda Point on New Georgia, about two-thirds of the way from Rabaul to Guadalcanal. It was a natural selection. Munda Point was relatively flat and could be reached from the sea only through one narrow break in its barrier reef, which was risky even for shallow-draft ships at high tide, or through several openings in the string of islets locking Roviana Lagoon to the island. An overland approach required an arduous jungle trek either from river inlets 10 miles to the north or from points to the east in Roviana. The position of the proposed airfield made an ally of the entire island, utilizing in protection all the reefs and islets which ringed New Georgia and the matted canopy of jungle growth which covered it.

The Japanese came to Munda in force on 13 November 1942. Their transports stopped off Munda reef late that day and, by early morning of the 14th, troops completed debarking by small boats. The occupation unit immediately sent out armed patrols to "subjugate" the natives and inform them of the Japanese intentions. Kolombangara, Rendova, Vangunu, and surrounding smaller islands were visited and quickly put under control. The construction of an airfield began with the arrival of additional troops and engineers on 21 November.

Coastwatcher Donald G. Kennedy at Segi Plantation on the extreme southeastern tip of New Georgia was one of the first to hear of the occupation. In October, a month previous, when the Japanese first reconnoitered New Georgia, Kennedy organized a band of natives to help him defend his post. When they informed him of the Munda landing, he sent Harry Wickham, a half-native co-worker, to Rendova to watch Munda and report on the progress of the airfield.

Wickham's report of Japanese activity at Munda was investigated immediately by Allied air reconnaissance. The first report, on 24 November, was negative. Photographs clearly showed a plantation area, a small cluster of buildings at Munda, and a similar cluster of buildings at Kokengola Mission north of Munda. There was no activity which could be classed as enemy, no evidence of airfield construction. Allied planes bombed the area anyway. It was a gesture of confidence in Kennedy and Wickham.

Then photo interpreters picked up interest. New buildings began to show up in later photo strips, and a strange white line appeared beneath the plantation trees. On 3 December, SoPac interpreters announced their discovery: a possible landing strip under construction. Two distinct strips, 125 feet wide and about 1,000 feet apart in a direct line with each other, were visible in the prints. One strip was about 175 feet long, the other about 200 feet. Natural camouflage, it was decided, partially shielded the construction. Two days later the field was 2,000 feet long. No trees had been cut down, but piles of either loose earth or coral appeared beneath each tree. New buildings, obviously control towers, had been built adjacent to the field. On 9 December, photos showed the field nearly clear, the trees apparently pulled up and taken away, and the holes filled in with coral. The Japanese, alerted by the continued interest of Allied planes over Munda, had abandoned further camouflage attempts.[3]

By 17 December, after only a month at Munda and despite multiple bombing raids, the enemy had an operational airstrip 4,700 feet long. A series of revetments and a turn-around loop eventually finished the field. An advance echelon of 24 aircraft was moved to Munda upon its completion, but all were destroyed or badly damaged by bombing raids within a week after arrival. Thereafter, the Japanese used the field mainly for servicing planes after raids on Guadalcanal and the Russells, and few pilots dared Allied bombings to tarry at Munda very long. Repair of the strip was easy; bulldozers quickly filled in the holes. Despite the rain of bombs and occasional shellings, the field was never out of operation longer than 48 hours.[4]

New Georgia, the Allies had decided, would be the target of the next offensive in the South Pacific. Munda airfield was the bull's-eye. As a military prize, it held the enemy's hopes for a re-entry into the lower Solomons and the Allied hopes for another step towards Rabaul.

---

[3] One of the popular but unverified stories about the camouflage of Munda field is: "The Japanese had spun a web of wire cables between the tops of the palm trees. The trunks were then cut out from under the branches which remained suspended exactly in place, held by the cables." Capt Walter Karig, USNR, and Cdr Eric Purdon, USNR, *Battle Report, Pacific War: Middle Phase* (New York: Rinehart and Company, Inc., 1947), p. 201.

[4] USSBS, *Interrogations* No. 195, LCdr S. Yunoki, IJN, I, p. 192.

NEW GEORGIA RECONNAISSANCE[5]

Halsey had intended to be in New Georgia by mid-April. His planning date scrapped by the JCS and his offensive tied to construction of airfields at Kiriwina and Woodlark, the admiral waited for the go-ahead signal. While waiting, he sent reconnaissance patrols probing the Central Solomons.

Guadalcanal land operations had been plagued by a dearth of information on terrain and topography. New Georgia was likewise unmapped and hydrographic charts were badly out of date. Since aerial photography revealed only thick jungle growth, actual physical scouting by trained men was the only answer. A combat reconnaissance school with experienced Marine and Army personnel and selected coastwatchers as instructors was organized at Guadalcanal, and about 100 men were trained and formed into scouting teams. Halsey found their reports invaluable, and beginning with ELKTON planning, "never made a forward move without their help."[6]

First terrain information on New Georgia had been received from a patrol of six Marines and a ComSoPac staff officer that had prowled Roviana Lagoon and the Munda area in late February, contacting coastwatchers, scouting and mapping trails, and selecting possible landing beaches. Their report helped the admiral reach a decision on hitting the Central Solomons and gave SoPac planners the information for tentative strategy.[7]

On 21 March, a group of Marine scouts drawn from the raider battalions and graduates of the combat reconnaissance school landed by PBY (Catalina flying boat) at Segi Plantation.[8] With Kennedy's natives as guides, the group split into patrols and set out to scout possible landing beaches, landmarks, and motor torpedo boat (MTB) anchorages. Traveling by canoe at night and observing during daylight hours, the patrols checked travel time from point to point, took bearings on channels, scouted enemy dispositions and installations, and sketched crude maps to help fill in the scanty information already available. One group had the mission of "collecting information about the Viru garrison, armament and accessibility to the area, both by way of direct attack up the harbor cliffs and by inland native trails through the jungle,"[9] which marked

[5] Unless otherwise noted, the material in this section is derived from: *ComSoPac Apr–May42 WarDs;* ComPhibFor, SoPacFor WarD, 17–30Jun43; Col William F. Coleman, "Amphibious Recon Patrols," *Marine Corps Gazette,* v. 27, no. 12 (Dec43); Sgt Frank X. Tolbert, "Advance Man," *Leatherneck,* v. 28, no. 3 (Mar45); Feldt, *The Coastwatchers;* Rentz, *Marines in the Central Solomons.*

[6] FAdm William F. Halsey and LCdr Julian Bryan, III, *Admiral Halsey's Story* (New York: Whittlesey House, 1947), p. 158, hereafter Halsey and Bryan, *Halsey's Story.*

[7] A member of that first patrol said that Halsey, after hearing the reports on New Georgia, declared, "Well, gentlemen, we're going to hit that place. I don't know when or how, but we're going to hit it." Maj Clay A. Boyd interview by Maj John N. Rentz, dtd 16Feb51.

[8] The senior member of this patrol group later commented: "I never heard of the 'combat reconnaissance school' and know that I and the other two members of the patrol from the 3d Raider Bn. didn't graduate from it." Col Michael S. Currin ltr to Head, HistBr, G–3, HQMC, dtd 11Oct60, hereafter *Currin ltr.*

[9] Maj Roy D. Batterton, "You Fight by the Book," *Marine Corps Gazette,* v. 33, no. 7 (Jul49), hereafter Batterton, "You Fight by the Book."

it as a possible target in the assault. Other patrols ranged from Roviana Lagoon to Arundel and Kolombangara, along the northern shore of New Georgia from Enogai Inlet to Marova Lagoon, and around the coast of Vangunu. Another patrol contacted the Rendova coastwatcher, Harry Wickham.[10] The missions were virtually the same: to bring back all possible data on the enemy and terrain.

At this early date in the spring of 1943, tentative invasion plans envisioned a divisional landing at Segi Plantation followed by a sweep overland to capture Munda field. The patrol reports confirmed the growing suspicions of the ComSoPac war plans staff: Segi's beaches would not accommodate a large landing force, and a sizable body of troops could not move through untracked jungle to Munda with any hope of success. Another method of attack would have to be developed.

The patrols continued to shuttle back to New Georgia for more information. Coastwatchers A. R. Evans on Kolombangara, Dick Horton and Harry Wickham at Rendova, and Kennedy at Segi played hosts to furtive guests who slipped in by native canoes from submarines, fast destroyers, or PBYs. The patrols searched openings in the barrier reef of Roviana, checked overland trails from Rice Anchorage on the north coast to Zanana Beach on the south in Roviana, and looked for easy access to Munda field. In this connection, Wickham—who had lived on New Georgia most of his life—"was particularly valuable."[11]

The reports on Munda were discouraging. Hathorn Sound had no beaches and shallow landing craft could pass safely only halfway through Diamond Narrows.[12] LSTs might possibly skirt the west shore of Baanga Island to get to Munda, but it would be a hazardous, obstacle-lined trip. Crossing the reef at Munda bar was another risk. Soundings indicated that the opening, through continued coral deposits, had become more shallow and restricted than admittedly outdated reports indicated. A direct assault over Munda bar, the closest entrance to Munda, was patently the most dangerous course and held the least chance of success.

Final assault plans were a concession to the terrain. They provided for landings off-shore from Munda, followed by a troop buildup on New Georgia and then a strong attack on the airfield from all sides. The last reconnaissance patrols went into the New Georgia Group on 13 June. Landing at Segi they took off in log canoes for the four landing spots finally selected: Rendova, Rice Anchorage, Viru Harbor, and Wickham Anchorage. Teams of Marine Corps, Army, and Navy officers studied the designated beaches and sought artillery positions, observation posts, water points, bivouac areas, and interior trails. Some of the patrols skirted Japanese defenses, noting the strength and habits of the enemy, before striking inland for terrain information. When the teams paddled back to Segi, some of the members stayed behind with natives to

[10] *Currin ltr.*

[11] Feldt, *The Coastwatchers*, p. 149.

[12] In the later stages of the campaign, some LSTs made it through the Narrows with 25th Infantry Division troops and equipment on board, but the division's chief of staff remembers the trip as "a tight squeeze w/fast tidal flow—no picnic." MajGen William W. Dick, Jr. USA, ltr to ACofS, G-3, HQMC, dtd 31Oct60, including comments by MajGen David H. Buchanan, USA.

guide the landing parties to the beaches with lights flashed from the shore.

AWAITING ASSAULT[13]

Munda assumed a new role in enemy strategy during the spring of 1943. Instead of the proposed springboard for recapture of Guadalcanal, it became a keystone in Japan's decision to build up the Lae-Salamaua defense line while maintaining the Solomons as delaying positions.

Japanese engineers, after rushing Munda into completion, hurried to Kolombangara to construct another field at Vila Plantation on the southeast shore. Here they did not attempt concealment. The task went ahead despite almost daily bombings and occasional naval bombardments. The enemy now had two strips from which they could stage attacks against Allied positions on Guadalcanal and the Russells; but the air over The Slot was a two-way street, and most of the traffic was from Henderson Field.

Buildup of troop strength in the Vila-Munda area was steady but slow. Air supremacy was still contested, but the initiative was with the Allies. Japanese plans for reinforcing the Central Solomons were slowed by the continual harassment from planes of Commander, Aircraft Solomons (ComAirSols), and the enemy was eventually reduced to scheduling troop transfers "from the end of the month to the beginning of the following month to take advantage of the new moon."[14] Then, too, the transport losses in the Battle of the Bismarck Sea on 3 March and the steadily mounting attrition of naval craft from air attacks was slowly sapping Japanese sea power.

By the end of April, land defenses in the Central Solomons had been strengthened with Army and Navy troops, and additional reinforcements were standing by in the Buin-Shortland area for further transportation. The *8th Combined Special Naval Landing Force (CSNLF)*, which included the *Kure 6th Special Naval Landing Force* and the *Yokosuka 7th SNLF* was the Navy's contribution to the defense of Vila-Munda.[15] After Japan lost the initiative in the Pacific, these amphibious assault troops were changed to defense forces. Named for the naval base at which the unit was formed, an SNLF generally

---

[13] Unless otherwise noted, the material in this section is derived from: MilHistSec, G-2, FEC, Japanese Monograph No. 34, Seventeenth Army Ops—Part I (OCMH), hereafter *Seventeenth Army Ops—I*; *SE Area NavOps—II*; CIC, SoPacFor Item No. 635, New Georgia Area DefButai SecretO No. 16, dtd 23Mar43, Item No. 647, Kolombangara Island DefO A No. 5, dtd 2Jul43, Item No. 672, Outline of Disposition of SE Det, dtd 20Jun43, Item No. 690, Kolombangara Island DefTaiO, dtd 1Mar43, Item No. 702, New Georgia DefOpO "A" No. 8, c. late Jun43, Item No. 711, Seventeenth ArmyO No. 244, dtd 27Apr43, Item No. 753, SE DetHq IntelRec No. 2 (Middle June Rept), dtd 24Jun43; SoPacFor POW Interrogation Repts 105 and 106, dtd 9Oct43, and 138 and 140, dtd 24Nov43; USAFISPA OB G-2 Rept No. 27, 17-24Jul43; Rentz, *Marines in the Central Solomons*.

[14] *Seventeenth Army Ops—I*, p. 6.

[15] The total ordnance of the *8th CSNLF* included: 8 140mm coast guns; 8 120mm coast guns; 16 80mm coast guns; 4 120mm AA guns; 8 75mm AA guns; 12 40mm AA guns; 2 75mm mountain (artillery) guns; 2 70mm howitzers; 40 13mm AA machine guns; 38 heavy machine guns; 102 light machine guns. MilIntelDiv, WarDept, Handbook on Japanese Military Forces (TM-E 30-480), dtd 15Sep44, pp. 76, 78; GHQ, SWPA, MilIntelSec, Organization of the Japanese Ground Forces, dtd 22Dec44, p. 299.

included: a headquarters unit; two rifle companies; a heavy weapons company with howitzer, antitank, and machine gun units; an antiaircraft company; a heavy gun or seacoast defense unit; and medical, signal, supply, and engineer troops.

The *Yokosuka 7th* landed at Kolombangara on 23 February with 1,807 men, and was followed on 9 March by the *Kure 6th* with 2,038 men. This unit went into positions between Bairoko and Enogai and around the airfield at Munda. Rear Admiral Minoru Ota, commanding the *8th CSNLF*, assumed responsibility for the defense of the New Georgia sector.

By prior agreement between the *Seventeenth Army* and the *Eighth Fleet*, Army and Navy strength in the Central Solomons was to be about even. After sending the *8th CSNLF* to the New Georgia area, the Navy was determined to hold the Army to its end of the bargain. Following a number of conferences, the Army reinforcements began arriving in late March. The original force at Munda consisted of two companies from the *2d Battalion, 229th Regiment* of the *38th Division* with two antiaircraft battalions for protection for the naval base construction troops. Kolombangara was garrisoned early in 1943 with troops from the *51st Division* including an infantry battalion, an artillery detachment, and engineer and air defense units.

The remainder of the *229th Regiment* at Buin, with supporting troops, began to filter into New Georgia late in April and the *51st Division* troops on Kolombangara were relieved. The *229th* moved to the Munda airfield area, and a battalion from the *13th Infantry Regiment, 6th Division*, took over actual defense of Kolombangara. As opportunities arose, the Japanese moved in more troops.

Guadalcanal had been a well-learned—albeit painful—lesson. The reinforcement of the Vila-Munda area reflected Japan's new strategy:

> Our fundamental policy was to bring the desired number of troops into strategic key points before the enemy offensive, in spite of manifold difficulties; and in event of an enemy offensive, to prevent our supply transportation from being hampered; to throw in our entire sea, land, and air strength at the first sign of an enemy landing to engage it in decisive combat; and to secure completely the strategic key positions linking the Central Solomons, Lae, and Salamaua, which formed our national defense boundary on the southeastern front.[18]

On New Georgia, the Japanese prepared defenses for all eventualities. Munda Point and the airfield vicinity bristled with antiaircraft and artillery weapons. The enemy did not discount the threat of a direct assault over Munda bar and sited some of their armament to cover that approach; but the bulk of the weapons pointed north toward Bairoko—from which an overland attack might come—and toward Laiana Beach on Roviana Lagoon—where an attack seemed logical. The Japanese believed, however, that the next Allied objective was to be Kolombangara in an attempt to attack Munda from the rear; so Vila likewise was prepared to repulse any assault. Increased Allied air activity, the presence of a great number of troop transports in the Guadalcanal area, and increased reconnaissance convinced the enemy that an attack was imminent. Their intelligence reports of about 50 cargo-type airplanes at Henderson Field also prompted speculation on the possibility of airborne operations against Vila-Munda.

---

[18] *SE Area NavOps—II*, p. 4.

With both Army and Navy troops occupying identical Central Solomons positions, a more unified command arrangement was sought. Admiral Ota, the senior commander in the area, had been responsible for both Army and Navy land defenses in the Vila-Munda area. On 2 May, however, *Imperial Headquarters* directed that a command post be established in New Georgia, and on 31 May, Major General Noboru Sasaki of the *38th Division* arrived at Kolombangara to head the new *Southeast Detachment*, a joint Army-Navy defense force. Administratively attached to the *Seventeenth Army* but under the operational command of the *Eighth Fleet*, General Sasaki was assigned responsibility for all land defenses in the New Georgia sector and command of all Army troops in the area. Admiral Ota, still in command of Navy troops, was directed to give him fullest cooperation. It was a command structure which crisscrossed Army and Navy channels, but with Sasaki's assignment spelled out, and with Ota's cooperation assured, a unified force was established.

By late June, as Japan waited for an Allied thrust she believed was coming, the defensive positions in the New Georgia Group were set. To obtain greater coordination, General Sasaki divided his defense area into three zones of responsibility: the Central (Munda); the Western (Kolombangara); and the Eastern (Viru-Wickham). The task of defending Munda Point he gave to Colonel Genjiro Hirata and the *229th Regiment*, augmented by two batteries of the *10th Independent Mountain Artillery Regiment*. Air defense would be provided by the *15th Air Defense Unit* which combined the *41st Field Antiaircraft Battalion* (less one battery), the *31st Independent Field Antiaircraft Company*, the *27th Field Machine Cannon Company*, and the *3d Field Searchlight Battalion* (less one battery). One company of the *229th Regiment* was dispatched to Rendova.

To aid Sasaki in the defense of the airfield, Admiral Ota established three seacoast artillery batteries at Munda with 140mm, 120mm, and 80mm guns. Also based there was an antiaircraft machine gun company of the *Kure 6th SNLF*, the *21st Antiaircraft Company* and the *17th* and *131st Pioneers* (labor troops). Ota also sent a rifle company from the *Kure 6th* to Rendova. The remainder of the *Kure 6th*, under Commander Saburo Okumura, was to defend the Bairoko Harbor area. Kolombangara's defense was entrusted to a battalion of the *13th Regiment*, reinforced by a battery of the *10th Independent Mountain Artillery*. Air defense of Vila airstrip rested with the *58th Field Antiaircraft Battalion* (less one battery), the *22d* and the *23d Field Machine Cannon Companies*, and a searchlight battery. The main detachments of the *Yokosuka 7th SNLF* and the *19th Pioneers* were also based on Kolombangara.

Viru Harbor was garrisoned by the *4th Company* of the *229th*, less one platoon which went on to Wickham Anchorage to augment a seacoast defense battery from the *Kure 6th*. To complete the defensive picture, lookout platoons were scattered about the coastline of New Georgia and on some of the small adjacent islets to act as security detachments.

In all, as Sasaki's reinforcement and defense plans raced right down to the wire with Allied offensive preparations, the Japanese had about 5,000 Navy and 5,500

Army troops in the New Georgia-Kolombangara area. Although the *8th CSNLF* was not combat tested, the *229th Regiment* and the *13th Regiment* were another matter. The *229th* had participated in the capture of Hong Kong before taking part in the occupation of Java. Committed to combat again, the regiment had one battalion nearly annihilated on New Guinea and another battalion suffered heavy casualties at Guadalcanal. Reinforced by fresh troops at Rabaul and Bougainville, the survivors had been formed into new battalions to join the *2d Battalion* at New Georgia. The elements of the *13th Regiment*, before being sent to Kolombangara, were part of the *6th Division* which garrisoned the Northern Solomons. One of Japan's oldest divisions, the *6th* was likewise hardened by combat in China before being sent to the Solomon Islands.

SEA OFFENSIVES [17]

Expanding sea and air offensives by the Allies in the late spring of 1943 had a definite bearing on Japan's outlook toward her defenses in the South Pacific. Widening the scope of the war, a large-scale bombing attack in mid-May plastered the Japanese-held atoll of Wake in the Central Pacific. This strike followed a landing in the Aleutians on 11 May by U.S. Army troops covered by naval forces. The enemy believed that the counter-landings in the North Pacific were a direct threat to the Home Islands, and plans for the southeast area were immediately curtailed. About 20 per cent of the troops earmarked for the Solomons and New Guinea were shifted to the northeast area; and Admiral Mineichi Koga, successor to Admiral Yamamoto, pulled his *Combined Fleet* headquarters out of Truk and moved to Tokyo so that he could better control operations throughout the Pacific. His main fleet units, however, remained at Truk.

True to his promise to the JCS at the time of the 28 March directive, Admiral Halsey kept the pressure on the Japanese in the Central Solomons. Under the pounding of bombs and sea bombardments, the Vila-Munda area never had the opportunity to develop past its use as a refueling point for enemy planes. The Allied strikes scored few casualties among the Munda defenders, relatively secure in underground defenses near the airfield, but kept enemy engineers busy repairing the cratered runways. The attacks lowered morale, however, by keeping the Japanese "sleepless and fatigued," [18] and occasional hits were scored on fuel and supply dumps. Prior to May, Munda and Vila had taken nearly 120 bombing raids, and four major naval bombardments had rained shells on the two airfields.

The Tokyo Express—fast destroyers carrying troops and supplies to the New Georgia Group—still steamed on. The Allies found they could not possibly cover all avenues of supply, and that to halt the traffic entirely would require more planes

[17] Unless otherwise noted, the material in this section is derived from: CinCPac Ops in POA, May43, dtd 15Aug43; CinCPac Ops in POA, Jun43, dtd 6Sep43; ONI, *Combat Narratives, Solomon Islands Campaign: IX—Bombardments of Munda and Vila-Stanmore, January–May 1943* (Washington, 1944), hereinafter cited as ONI, *Combat Narratives IX: SE Area NavOps—II;* Morison, *Breaking the Bismarcks Barrier;* Miller, *Reduction of Rabaul;* Sherrod, *MarAirHist.*

[18] USSBS, *Interrogation* No. 224, Cdr Yasumi Doi, IJN, I, p. 210.

and ships than South Pacific forces could muster at this stage of the war. On 6 May, however, the express runs were abruptly, if only temporarily, disrupted. Rear Admiral Walden L. Ainsworth, heading a Third Fleet task force of three cruisers, five destroyers, and three converted destroyer-mine layers, steamed up the gap between Gizo and Wana Wana Islands into the Vella Gulf. As Ainsworth's cruisers and four destroyers blocked the northern entrance to Vella Gulf, the three mine layers escorted by a radar-equipped destroyer laid three rows of mines across the straits between Kolombangara and Gizo. Then the entire force turned for home bases at Guadalcanal. (See Map II, Map Section.)

Dividends were almost immediate. The next night, four Japanese destroyers slipped into Blackett Straits with Vila as their destination. They never reached it. The trap was sprung. Blundering into the mine field, one ship went down almost immediately; two others were badly damaged. The fourth ship stood by to pick up survivors. And that's the way Allied planes, somewhat delayed by adverse weather, found them the next day. The two damaged ships were sunk by bombs, but the fourth ship, heavily bombed and strafed, managed to limp back toward Bougainville. Gleeful coastwatchers radioed the box score to Guadalcanal.

Heartened by the success, the Third Fleet planned another surprise. This time, Vila would be shelled as the northern entrance to Kula Gulf was mined. On the night of 13 May, Admiral Ainsworth led a force of three cruisers and five destroyers in firing runs past Vila, steaming in from the north, while a destroyer and three fast mine layers planted mines off the east coast of Kolombangara. As each of Ainsworth's ships completed her run past Vila, she turned and pumped heavy fire into the Bairoko and Enogai-Rice Anchorage areas.

At the same time, a force of one cruiser and three destroyers plastered Munda on the opposite side of New Georgia. The airfield had not been included in the original bombardment plans, but a last-hour switch in orders—accomplished by dispatch and a message drop from planes to the ships designated—had added that stronghold. Vila was hit by a total of 2,895 six-inch and 4,340 five-inch shells, Munda by 970 six-inch and 1,648 five-inch. The operation was covered by an air strike in the Northern Solomons and additional fighter planes flew cover and reported bombardment results.

The mine-laying did not produce the earlier results. It slowed the Japanese supply chain by forcing it to be more cautious, but it did not halt it. The bombardment was a bigger disappointment. Less than 12 hours after the last shell had been fired, a flight of 26 Japanese fighters staged from Munda-Vila was chasing the attack force back to Guadalcanal. Coastwatchers radioed the warning; 102 Allied aircraft formed a welcoming committee. Seventeen enemy planes were reportedly shot down; 16 of them were claimed by Marine fighters. Five Allied planes and three pilots were lost in the action. The bombardment was the last scheduled before the actual invasion; the results, it was apparent, were not worth the price. A harassing bombardment, CinCPac later advised, was not justified when "all ships were subjected to the hazard of enemy MTB and SS [submarine] attacks with no prospect of equal opportunity to damage the enemy."[19]

Air activity increased during June. Airfields in the Russells gave the Allies a shorter range to targets in the Northern and Central Solomons, as well as providing another launching area for getting planes into the air to repel attacks. Attracted by the concentration of shipping in the Guadalcanal area, the Japanese tried a new one-two punch of heavy flights of fighters followed by large numbers of bombers, but three major strikes on 7, 12, and 16 June resulted in staggering losses. The enemy had hoped to break even in fighter tolls, which would then give their bombers opportunity to attack unmolested. The maneuver boomeranged. Each time, ComAirSols was able to meet the threat with from 105 to 118 aircraft and in the three strikes, a total of 152 enemy airplanes was claimed. The Allies lost 21.

PREPARING TO STRIKE[20]

The assault of New Georgia, viewed in optimism contagious at the time, seemed an easy assignment despite the inaccessability of Munda. Reconnaissance had virtually pinpointed Japanese strong points, and the combat effectiveness of the Vila-Munda airfields had been reduced considerably by the Allied pounding. Intelligence sources, which later proved remarkably accurate, estimated that there were only about 3,000 Japanese at Munda, with another 500 troops at Bairoko and a detachment of 300 men at Wickham Anchorage and about 100 more at Viru Harbor. The bulk of the forces, estimated at 5,000 to 7,000 troops, was on Kolombangara, together with an additional 3,000 laborers.

Japan's reinforcement ability from points in the Northern Solomons was noted, but there was no ready estimate of the numbers available for quick assignment to combat. Her sea strength in the Solomons was believed to be 6 destroyers, 5 submarines, and 12 transports, with a cruiser, 5 destroyers, 7 submarines, and 25 attack transports at Rabaul. Japanese air strength was put at 89 land-based aircraft in the Solomons with another 262 at Rabaul. While troop estimates were near the actual enemy totals, ComSoPac guesses on air and sea numbers of the enemy were low. The entire *Eighth Fleet* was in the Shortlands area, while a part of the *Combined Fleet* at Truk was committed to lend assistance in Southeast Area operations. The Japanese Navy had 169 land-based planes available for combat from a total of nearly 300 deployed in the Bismarck Archipelago and the Northern Solomons. The *Eighth Area Army* had about 180 aircraft attached directly to it; however, most of these were supporting operations against the Allies in New Guinea.

The target had been marked. Early in June, Admiral Halsey published his orders for the seizure and occupation of New Georgia. The improbable code name TOENAILS masked Halsey's part in the CARTWHEEL offensive. The missions: capture Wickham Anchorage and Viru Harbor as small-craft staging areas; seize Segi Plantation as a possible airfield site; seize Rendova as a base for the neutralization of Munda by artillery fire. Orders for the actual assault of Munda airfield

[19] ONI, *Combat Narratives IX*, p. 74.

[20] Unless otherwise noted, the material in this section is derived from: ComSoPac OPlan 14-43, dtd 3Jun43; ComThirdFlt OPlan 12-43, dtd 5Jun43; CTF 31 OpO A8-43, dtd 4Jun43; NGOF FO No. 1, dtd 16Jun43 and No. 2, dtd 24Jun43; Rentz, *Marines in the Central Solomons*.

would be issued by ComSoPac after the successful completion of the first phase of TOENAILS.

Task units of the Third Fleet were assigned covering missions which would insure success of the operation by blocking any enemy force attempting to disrupt the landings with a counteroffensive. While one force of destroyers and cruisers moved in to mine the main sea channels around the Shortland Islands, another heavier force of battleships and destroyers was to stage a bombardment of Japanese strong points in the Northern Solomons and Shortlands. Air units of the South Pacific Air Command, under Vice Admiral Aubrey W. Fitch, were assigned strikes against shipping in the Shortlands area and bombing missions on airfields on Bougainville. Carrier air groups were to intercept any enemy ships or aircraft heading for the New Georgia Group. SoPac submarines were to range into the Northern Solomons for interception and early warning of any Japanese force, and destroyer units would provide close-in support for the transport groups engaged in the actual landing operations. Thus, with Admiral Halsey's forces guarding the northern and eastern approaches to New Georgia, and General MacArthur's operations in New Guinea shielding the western flank, the assault forces could proceed with the seizure of TOENAILS objectives.

The Army's 43d Infantry Division, part of Major General Oscar W. Griswold's XIV Corps, was named as the assault and occupation troops. The 2d Marine Aircraft Wing headquarters under Brigadier General Francis P. Mulcahy was assigned to direct tactical air support over the target during the operation. Rear Admiral Turner, commander of amphibious forces in the South Pacific, was given overall responsibility for New Georgia operations. Detailed planning for the actual seizure and occupation of the objectives outlined in Halsey's broad plans would be Turner's job.

To accomplish the TOENAILS missions, Turner divided his command into two units. He would personally direct the larger Western Force in the main landing at Rendova and would be responsible for movement of troops and supplies to the objective and for their protection. The Eastern Force, under the direction of Rear Admiral George H. Fort, would seize Viru, Segi Plantation, and Wickham Anchorage. Admiral Fort would be responsible for movement to these targets and for embarking troops and supplies from the Russells for subsequent operations.

Admiral Turner's ground commander, Major General John H. Hester, headed the New Georgia Occupation Force (NGOF). Its combat units consisted of Hester's own 43d Infantry Division, including the 172d and 169th Regiments and one battalion of the 103d Regiment; the Marine 9th Defense Battalion; the 136th Field Artillery Battalion from the 37th Infantry Division; the 24th Naval Construction Battalion (NCB); Company O of the Marine 4th Raider Battalion; the 1st Commando, Fiji Guerrillas;[21] and assigned service troops.

Fort's Eastern Force would include the 103d RCT (less the battalion with Hester); Companies N, P, and Q from the 4th Raider Battalion; elements of the 70th Coast Artillery (Antiaircraft) Battalion; parts of the 20th NCB; and service units.

---

[21] Central Office of Information, *Among Those Present* (London, 1946), pp. 53–56.

The landing force would be headed by Colonel Daniel H. Hundley, commanding the 103d RCT. Selected as ready reserve for the operation was the 1st Marine Raider Regiment (less the 2d, 3d, and 4th Battalions), commanded by Colonel Harry B. Liversedge. The Army's 37th Infantry Division (less the 129th RCT and most of the 148th RCT) would be in general reserve at Guadalcanal, ready to move on five-days' notice.

Execution of the assigned tasks looked easy. Turner's original concept was to seize the southern end of New Georgia simultaneously with Rendova. Artillery based on Rendova and offshore smaller islands would soften Munda field while the buildup of assault forces began. Four days later, it was planned, Munda would be attacked through Roviana Lagoon and over Munda Bar, while Bairoko would be struck either from the Russells or by a force hitting overland from Roviana Lagoon. This maneuver would block reinforcements for the airfield. Capture of Munda would then trigger the next shore-to-shore jump to Kolombangara, the last phase of Operation TOENAILS.

These were the first plans. ComSoPac orders stressed their successful completion with a minimum of forces. It could not be foreseen at the initial planning conferences that, before Munda could be captured and the New Georgia Group occupied, elements of four infantry divisions would be committed and extensive changes in plans would be required. The problems mounted early. Laiana Beach on Roviana Lagoon east of Munda was heavily defended, although the best landing area. The channel through Roviana, scouted from canoes, was too shallow for LCMs. The islands near Rendova originally considered for artillery positions were not within effective 105mm howitzer range of Munda airfield. And coastwatcher reports indicated that the enemy—despite Allied efforts—was slipping reinforcements into the Vila-Munda area. Further, a reconnaissance team reported that a strike at Bairoko from Roviana was impossible within the time limits planned.

The solutions plagued Turner's staff. Zanana Beach, about 5,500 yards east of Laiana, was smaller but virtually undefended, the scouts reported. While it would hold only a few landing craft, the Piraka River mouth 1,000 yards farther east could permit beaching of additional boats. Hester decided on Zanana as his landing beach, and Turner gave his approval. Reaching Zanana would be a problem, however. Landing boats would have to slip through narrow, coral-choked Onaiavisi Entrance that threaded between the small offshore islands and then follow a twisting channel to the beaches. The selection of Zanana was based as much on its undefended nature as on its capability of being reached by LCMs. It had, however, one apparent drawback. The attacking troops would be unloaded at a considerable distance from their objective. (See Map 5.)

The planning problems were unexpectedly magnified by an emergency. An urgent call for assistance by Coastwatcher Kennedy at Segi resulted in the premature commitment of two Marine raider companies and two companies from the 103d RCT on 21 June. Admiral Turner made the decision. The speedup in schedule upset previous planning, but it was deemed necessary. It required a change in basic strategy, a shuffling of troops, a change in the transport plans—and some around-the-clock supply duty by the Marine 4th Base Depot in the Russells—but the decision retained possession of Segi for the Allies until the actual New Georgia invasion.[22]

General Hester, who would direct the operations ashore, continually faced thorny problems. To deal with the mounting complexities, he delegated the planning for the Rendova landings to a 43d Division staff headed by his assistant division commander, Brigadier General Leonard F. Wing. A second staff, the NGOF staff, completed the New Georgia attack planning. Hester retained command of both staffs. The final assault plans evolved from the best solutions to a multiplicity of problems. In the scheme of maneuver, part of the Western Force would hit at Munda through Onaiavisi Entrance with two regiments landing at Zanana and pivoting to the west in an overland attack with one flank resting on the lagoon. This force, designated the Southern Landing Force, would be commanded initially by General Wing.

Liversedge's raiders—now titled the Northern Landing Group—would strike directly at Bairoko from Kula Gulf. This would be coordinated with the landings at Zanana and would block reinforcements to Munda. It was not expected that the Munda forces would attempt to reinforce the Bairoko defenders. This half of a pincer movement faced one handicap; the area was not as well scouted as that of Roviana Lagoon. The Hester plan of attack envisioned a short campaign during which the Japanese would be caught between a hammering force from the south and a holding force in the north. Thus, the enemy would be pushed back towards an area where, ringed by Allied troops, they could be pounded into submission by aircraft and Rendova-based artillery. To insure success, additional 43d Division artillery (the 192d and the 103d Field Artillery Battalions) was added to the NGOF.

D-Day assignments were set. Troops of the 172d Infantry would seize two small islands guarding the approaches to Rendova and then establish a beachhead on Rendova itself. Through the secured passage, Hester and Wing would funnel the rest of the landing force, with the 103d Infantry given the task of expanding the beachhead and mopping up the island, reported to be lightly defended. Simultaneously, two companies of the 169th Infantry would land on islets flanking Onaiavisi Entrance and a detachment of Fiji guerrillas and Marine raiders would mark the channel with buoys to Zanana Beach and the nearby Piraka River. Four days later, the 172d Infantry would make the Rendova-Zanana move and establish a beachhead for the landing the following day of the 169th Infantry from the Russells. The 169th was to move inland to the north of the 172d, then face to the west. This would put two regiments abreast, ready to launch an attack from a line of departure along the Barike River, some 2,000 yards closer to Munda. Artillery on the offshore islands and Rendova would support the attack. Five days later, it was planned, the 3d Battalion of the 103d Infantry and the eight tanks of the Marine 9th Defense Battalion would cross Munda bar for the final, direct assault on Munda airfield.

[22] The capture of Segi is related in the following chapter.

MARINE CORPS SUPPORT[23]

Marine units which were to participate in the seizure of Munda were fulfilling a number of tasks and training missions prior to the operation. The actual job of pushing the enemy from New Georgia belonged mainly to General Hester's 43d Division; contributions to the campaign by the Marine Corps would be in support of the main effort. The 9th Defense Battalion was given a dual mission of making enemy positions on Munda untenable by artillery fire and of providing antiaircraft protection for the landing forces. The 1st Marine Raider Regiment, at first intended as a reserve element, was thrust into an active role with its mission of wedging a block on Dragons Peninsula between the Munda defenders and reinforcements at Bairoko Harbor.

Colonel Liversedge's raiders were a cocky, confident group which prided itself on being a volunteer unit within a volunteer Corps. Carrying only 60mm mortars and light machine guns as supporting weapons, each battalion was generally organized with four rifle companies, an engineer and demolition platoon, and a headquarters company. Smaller in authorized strength than the regular Marine infantry battalion, the actual strength of the raider battalions varied between 700 and 950. Specially trained for jungle fighting, amphibious raids, and behind-the-lines guerrilla action, the raiders had participated in the Tulagi assault, a hit-and-run raid at Makin Island in the Gilberts, the defense of Midway, and jungle warfare on Guadalcanal. These Marines thus brought to the New Georgia campaign considerable combat experience plus the conviction that the fighting ahead would follow no orderly lines of battle. The vexing problems presented by the jungle in maintaining communications and supply would demand the utmost in courage, ingenuity, and stamina; but the raiders felt up to the task. They were firm in the belief that these difficulties, inextricably complicated by the terrain and enervating climate, could be overcome by their tough physical training, combat experience, and high morale.

At the time of consolidation of the four battalions under one command on 15 March 1943, the raiders were scattered throughout the South Pacific with regimental headquarters and the 2d and 4th Battalions at Espiritu Santo, the 1st at Noumea, and the 3d in the Russells. Upon assignment to the TOENAILS operation, the regiment (less the 2d and 3d Battalions) moved to Guadalcanal, arriving there the first week in June. Here the raiders had only a few days to go over their orders, iron out organizational kinks, and practice as a single unit before the 4th Battalion was abruptly assigned to Kennedy's assistance.

[23] Unless otherwise noted, the material in this section is derived from: 1st MarRdrRegt WarD 15Mar–30Sep43, dtd 6Oct43; 9th DefBn RecofOps 1–28Jun43, dtd 3Oct43; 9th DefBn Rept on AA Ops 18Jun–18Sep43, hereafter *9th DefBn AA Ops;* 9th DefBn NarrativeHist 1Feb42–14Apr44, dtd 2May44; 4th BaseDep OrgHist 1Apr–31Jul43, dtd 24Aug43; HistSec G–2, SoPacBaseComd, MS Hist of the New Georgia Campaign, 2 vols., c. 1947 (OCMH), hereafter *New Georgia Campaign;* LtCol Wilbur J. McNenney, Observers Rept New Georgia Ops to CG, IMAC, dtd 17Jul43; LtCol Wright C. Taylor ltr to CMC, dtd 4Mar52; Col Archie E. O'Neil ltr to CMC, c. 1Mar51; LtCol Robert C. Hiatt ltr to CMC, c. 26Feb52; Maj Cyril E. Emrich ltr to CO, 10th DefBn, dtd 1Jul43; Maj John L. Zimmerman, *The Guadalcanal Campaign* (Washington: HistDiv, HQMC, 1949); Rentz, *Marines in the Central Solomons;* Sherrod, *MarAirHist.*

The 9th Defense Battalion commanded by Lieutenant Colonel William J. Scheyer, had particular reason to be proud of its assignment in the TOENAILS operation. With a quick conversion of its seacoast batteries to field artillery units, the 9th would be in an offensive role against the Japanese at Munda and the prospect pleased the entire battalion. One of 14 such highly specialized defense forces scattered from Cuba to New Zealand, the 9th was providing antiaircraft protection for Guadalcanal forces when picked for the New Georgia offensive. Activated early in 1942, the 9th trained extensively in Cuba before arriving at Guadalcanal on 30 November 1942. The battalion was in defensive action almost immediately, and its 90mm batteries bagged a total of 12 enemy aircraft in the following months.

Organizational changes had to be made, however, to get the unit ready for its part in the capture of Munda. In 21 days, the seacoast batteries, augmented by 145 new men, were trained in field artillery fire direction methods and had test-fired newly arrived 155mm pieces. The change from seacoast sights to field artillery sights and different fire commands was only part of the problem, though. As one battalion officer reported:

> Our problem was not one of training but one of obtaining the necessary equipment and ammunition so that a relative calibration could be fired to obtain some idea as to the relative velocity errors of the new weapons in order to mass their fires. We were plagued throughout the operation with this equipment and ammunition problem. When the ammunition did arrive from Noumea, there were 19 different powder lots in a shipment of 25 rounds. Obviously, calibrations could not be conducted with propellents of different powder lots and about all that was accomplished was test firing of the weapons so that the men could be familiar with them.[24]

The battalion, with an assist from its relieving Army unit, the 70th Coast Artillery (Antiaircraft) Battalion, picked up new 90mm antiaircraft guns equipped with power rammers and remote control equipment in exchange for the old guns which were left in position. In addition, power-operated mounts were placed on spare 20mm guns, increasing speed and efficiency over the standard mounts which were pedal-operated. The 9th also borrowed 12 amphibian tractors from the 3d Marine Division, and Griswold's XIV Corps exchanged new trucks and jeeps for old. As the 9th readied itself for its mission, its armament included a platoon of 8 light tanks, 8 155mm guns, 12 90mm guns, 16 40mm guns, 28 20mm guns, and 35 .50 caliber antiaircraft machine guns.

Relieved of its defensive role on Guadalcanal on 17 June, the battalion spent the remaining time in familiarization firing of weapons, gun drills which included reconnaissance, selection, and occupation of positions, and practice landings. Gunners and loaders from the antiaircraft batteries turned riflemen to give the tankmen practice in tank-infantry tactics. The amphibian tractors were test-loaded until a loading arrangement was obtained which would provide enough 40mm, 20mm, and .50 caliber ammunition for all three types of antiaircraft guns to go into action immediately upon landing. The 9th also took advantage of a liberal interpretation of its orders to get more ammunition for the 90mm batteries. Loading orders specified three units of fire were to be carried. Since an Army unit of fire for the 90mm guns was 125 rounds and a Marine unit of

---

[24] Hiatt ltr, *op. cit.*

fire 300 rounds, the 9th interpreted the orders to mean Marine Corps units of fire and carried the extra ammunition. Despite some misdirected trucks and some confusion as to unmarked dock areas, the eager 9th was aboard ship and waiting hours before the scheduled departure.[25]

In time, elements of the 10th and the 11th Defense Battalions would be called upon to augment the 9th in its mission at Rendova and Munda, but until placed on alert, they continued to assist in the defense of Guadalcanal and the Russells. A fourth unit, the 4th Defense Battalion, which had been in the New Hebrides before going to New Zealand, was soon to be recalled to Guadalcanal for participation in the final phase of the campaign in the Central Solomons. The employment of these battalions as offensive elements instead of defense forces illustrated the change in the character of the war.

Although not carried on the orders as part of the New Georgia Occupation Force, another Marine Corps element was to provide invaluable support to the operation. This was the 4th Base Depot, a supply organization which had been activated at Noumea on 1 April 1943 as the direct result of a logistics logjam in the South Pacific. Prior to the New Georgia operation, the Army had responsibility for unloading all supplies, but as the size of forces in the area grew, the inadequate and limited facilities and the understaffed corps of laborers in the Pacific were strained to maintain a smooth and uninterrupted flow of necessary supplies. Despite the Army's best efforts, the result was a confused backlog of equipment and supplies at New Caledonia and Guadalcanal which almost sidelined the New Georgia operation.

Shipping to the lower Solomons, except for vital aircraft engines and spare parts, motor transport spare parts, rations, and medical supplies, was curtailed for a time, and all other goods were routed to Noumea for transshipping to Guadalcanal on call. Supplies necessary for the New Georgia operation were then plucked from the stockpiles at Noumea and assembled at Guadalcanal. Other war materials were directed to the Southwest Pacific forces, added to the growing dumps in the New Hebrides, or stored in New Zealand.

The 4th Base Depot, under the command of Colonel George F. Stockes, and with personnel gleaned from the 1st, 2d, and 3d Base Depots and the Marine 12th Replacement Battalion, moved with 61 officers and 1,367 men to Guadalcanal to help relieve the congestion. Placed under the command of the XIV Corps, it was ordered by Griswold to relieve the service elements of the 43d Infantry Division in the Russells, and to bring order out of the general confusion. The 4th Base Depot was then to receive and store all supplies for the New Georgia operation and the Russells garrison; maintain a 60-day level of supplies for TOENAILS forces; and handle and load aboard ships all supplies as called for by the 43d Division and supporting troops on New Georgia.

The assignment was insurance that logistical problems would not slow the attack. It was a timely move. Shortly after the 4th Base Depot began working on the jumbled stockpiles of material, the initial phase of TOENAILS began with the Segi Plantation occupation, and the Marines were called upon for supply as-

[25] Emrich ltr, *op. cit.* In order to avoid similar misunderstandings and to facilitate logistical planning, Nimitz' headquarters subsequently published a CinCPac order listing units of fire for all types of weapons.

sistance. By the time the main operations started at Rendova, the depot had the necessary material ready for forward movement, and in the following months it funneled a steady stream of lumber, cement, ammunition, rations (including fresh fruit and meats), clothing, tires, spare parts, gasoline, lubricants, sand bags, tents, engineer equipment, post exchange items, and many other types of supplies into New Georgia.

For Marine Corps aviation units, establishment of an exact date for the start of the New Georgia campaign is difficult. The conflict for air superiority was constant and continuing, not boundaried by beachheads or D-Days. The struggle for undisputed possession of the lower Solomons phased directly into the New Georgia campaign, and it is hard to differentiate between the squadrons which supported the consolidation of the Solomons and those which directly took part in the capture of Munda airfield. In any event, most Marine squadrons then based at Guadalcanal or in the Russells participated in both campaigns, either in part or in whole.

Rear Admiral Marc A. Mitscher, as ComAirSols, had an Allied force of 627 planes with which to support operations in the Central Solomons. It was a composite of Marine, Army, Navy, and New Zealand aircraft, and included 290 fighters, 94 scout bombers, 75 torpedo bombers, 48 heavy bombers, 26 medium bombers, 30 flying boats, 24 seaplanes, and a miscellany of 40 search, rescue, and transport planes.[26]

[26] ComAirPac to ComSoPac ltr ser 00517 of 4Jun43, quoted in *New Georgia Campaign*, p. 60.

Although plans for garrisoning New Georgia were still in the tentative stage, a number of Marine squadrons were to be based at Munda airfield following its capture and would become an integral part of the New Georgia Air Force. Prior to the campaign, however, this term was a paper designation for a forward echelon of the 2d Marine Aircraft Wing, attached to the NGOF; its commanding officer, General Mulcahy, would "exercise operational control of aircraft in flight assigned to air cover and support missions in the New Georgia area."[27] Requests for air support strikes would be made to liaison parties with each landing force, and General Mulcahy as ComAir New Georgia would approve, disapprove, or modify. It was, in effect, a fighter-bomber direction center for both air defense and direct support missions. Control of the assigned aircraft would pass to ComAir New Georgia when the planes took off from their home fields.

Available for such tactical air support missions as would be assigned them in the months ahead were seven Marine fighter and four scout bomber squadrons, backed up by three utility squadrons and a photo reconnaissance detachment. For the most part, though, the role of the Marine squadrons in the seizure of Munda is part of the bigger story of how Allied air strength reduced the Japanese stronghold at Rabaul to impotency. This will be related in Part V of this volume.[28]

[27] NGOF FO No. 1, *op. cit.*

[28] Stationed in the Solomons in June 1943 were VMF-112, -121, -122, -123, -124, -213, and -221, with several more squadrons due to arrive as replacements later; VMSB-132, -143, -144, and -234; VMJ-152, -153, and -253; and a photographic detachment from VMD-154.

ELKTON Underway

WOODLARK-KIRIWINA[1]

The planned moves of the Allied forces in the Central Solomons—Papuan area in the summer of 1943 resembled pieces of a jigsaw puzzle. Each operation in itself did not represent a serious threat to the enemy's defense line, but, as part of a bigger picture, each was important and contributed to the success of all. The pieces fitting together formed a pattern of coordinated, steady advance.

D-Day (30 June) for ELKTON was practically a planning date only. ComSoPac operations began at Segi on 21 June; and Woodlark-Kiriwina landings two days later opened the action by Southwest Pacific forces, well in advance of the date set. The near-concurrent start was a coincidence; a two-pronged attack by Halsey and MacArthur had been postponed three times before 30 June as a mutual D-Day was accepted. A number of factors forced the delay, chief among which was the scarcity of amphibious troops required by the missions of ELKTON. The 43d Division was the early choice as the New Georgia assault force, and that unit was scheduled for extensive ship-to-shore training prior to the operation. In the Southwest Pacific, an entire new command—the VII Amphibious Force (VII PhibFor)—was activated to assemble and train the needed troops.

Marine Corps divisions, whose specialty was such amphibious movements, were not available for assignment to CARTWHEEL operations. Two divisions were undergoing rehabilitation and training; a third was not yet combat-ready; and a fourth was still forming in the States. The result was that Marine raider and defense battalions were at a high premium to augment available Army units for the twin operations of TOENAILS in New Georgia and CHRONICLE at Woodlark-Kiriwina.

A tentative lineup of forces for the planned attacks was made in April. Admiral Halsey made a quick trip to Brisbane on the 18th to meet the general under whom he would be operating, and he and MacArthur quickly came to an agreement based upon mutual respect. MacArthur

[1] Unless otherwise noted, the material in this section is derived from: VII PhibFor, SWPA, ComdHist 10Jan43–23Dec45, n.d.; MIS, WD, Survey of North East New Guinea and Papua (S30-678), dtd 15Jul43; 12th DefBn WarD, 30Jun–31Jul43, dtd 2Aug43; OCE, USAFPac, *Airfield and Base Development—Engineers in the South West Pacific, 1941–1945*, v. VI (Washington, 1951), *Engineer Supply—Engineers in the South West Pacific, 1941–1945*, v. VII (Washington, 1950), and *Critique—Engineers in the South West Pacific, 1941–1945*, v. VIII (Washington, 1950); Jeter A. Isely and Philip A. Crowl, *The U.S. Marines and Amphibious War* (Princeton: Princeton University Press, 1951), hereafter Isely and Crowl, *Marines and Amphibious War;* George C. Kenney, *General Kenney Reports* (New York: Duell, Sloan and Pearce, 1949), hereafter Kenney, *Reports*; Morison, *Breaking the Bismarcks Barrier*; Halsey and Bryan, *Halsey's Story*.

needed some help in his amphibious venture; Halsey offered it. He ordered his Noumea headquarters to assign the 20th NCB (Acorn 5) to Brisbane and to select one combat-ready RCT plus one Marine defense battalion for further transfer to SWPA. Assignment of the Marine unit was easy; the 12th Defense Battalion had arrived in Pearl Harbor in early January and was awaiting further transfer. But the many needs of the expanding South Pacific defense area had left few Army regiments without active assignments. It was finally decided, after a musical-chair shuffle of troops, that the 112th Cavalry (dismounted) on New Caledonia would join the 12th Defense Battalion, Acorn 5, and other naval base and service units in a transfer to SWPA. Here they would serve as the Woodlark defense force. Lieutenant General Walter Krueger's Sixth Army troops would garrison Kiriwina.

MacArthur's targets, Woodlark and Kiriwina Islands, lay in the Coral Sea off the southeastern shore of New Guinea, about 60 miles north and east of the D'Entrecasteaux Islands. Kiriwina, in the Trobriand Group, is about 125 miles directly south of New Britain; Woodlark is about 200 miles southwest of Bougainville. Their designation as future airfield sites to support operations in both New Guinea and the Solomons sent Army engineers scrambling over them to obtain beach and terrain information to supplement native reports and aerial photography. The reconnaissance teams were wary, but prior information was correct—the Japanese had not occupied the islands. (See Map 2.)

Kiriwina, shaped like a bent toadstool, was ringed by an extensive coral reef broken by only a few narrow openings for shallow-draft boats. Twenty-five miles in length, and from two to eight miles wide, the island held about 7,500 natives, had a sub-surface coral base which would support an airstrip, and had many good trails for jeep roads. But there were no good beaches. Woodlark, about 100 miles southeast of Kiriwina, was nearly 44 miles long and from 10 to 20 miles in width. Curved in shape, it held a number of good anchorages tucked within the protected shorter arc. The beaches, however, ran inland only a few hundred yards before bumping into a coral cliff. Sparsely settled, Woodlark was covered with a thick jungle growth and dotted with large outcroppings of coral.

Together, these islands could provide bases for fighter escorts of Lieutenant General George C. Kenney's Allied Air Forces hitting at New Guinea, New Britain, and New Ireland, and for SoPac strikes against the Northern Solomons in subsequent operations. Their capture, the JCS had decided earlier, would provide the first test of the newly formed VII Amphibious Force.

This force had come into being under the direction of Rear Admiral Daniel E. Barbey, who opened his headquarters at Brisbane in mid-January 1943. By April, it was apparent that the task of forming and training an amphibious force was far more difficult than had been supposed at first. An assortment of United States and Australian ships formed the transport division, and Sixth Army troops, recuperating from the hard fighting in the Buna-Gona campaign, were trained in amphibious operations. Practice landings which were sandwiched between troop lifts to New Guinea were never realistic. Few troops, ships, or pieces of heavy

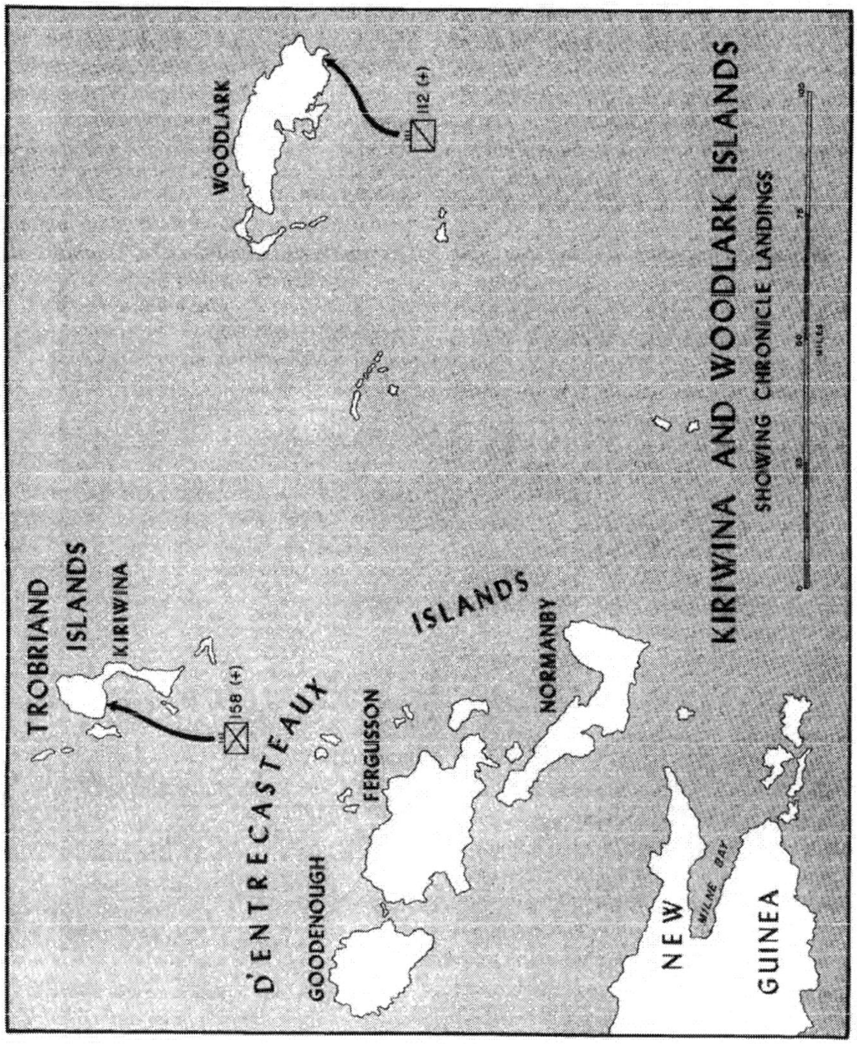

MAP 2

R.F. STIBIL

equipment could be spared from that operation for practice purposes. With an operational deadline pressing, Admiral Barbey scoured the Southwest Pacific for more ships. Some new LSTs were assigned him; others he borrowed from ComSoPac. The USS *Rigel*, a repair ship with none of the desired command facilities, was pressed into service as a flagship.

MacArthur, in his first conference with Halsey, had tentatively set 15 May as D-Day for the combined operation. Late in April, MacArthur announced that he could not meet this date and directed its postponement to 1 June. It was later changed to 15 June as logistical and shipping problems piled up in the Pacific. On 26 May, the general proposed the 30th of June as D-Day and requested ComSoPac concurrence. This date, MacArthur pointed out, would also coincide with landings by other SWPA forces at Nassau Bay on New Guinea, about 10 miles south of Salamaua. Halsey agreed.

The CHRONICLE forces assembled, Kiriwina's garrison (code-named BY-PRODUCT) at Milne Bay on New Guinea and Woodlark's garrison (code-named LEATHERBACK) at Townsville, Australia. On 21 June, nine days ahead of schedule, the advance echelon of the 112th Cavalry, with heavy bulldozers and operators from the 20th NCB, set off for Woodlark. The next night, troops and equipment were landed. The speedup resulted because the troops were ready, there would be no enemy to oppose the landing, and Barbey's transports would need the extra time to carry two landing forces to their destinations. Kiriwina's advance echelon was landed on the nights of the 23d and the 25th, the last group landing across the reef over a coral causeway 300 yards long and 7 feet high which had been built by combat engineers and natives.

The main landing of the Kiriwina force, which included the 158th RCT, the 46th Engineer Combat Company, and antiaircraft artillery and service troops, was made on the night of the 29th according to the ELKTON schedule. Additional Woodlark advance echelons had been landed on the nights of the 25th and 26th, with the main landing of support elements coming on the 30th, also as scheduled. Woodlark's garrison, in addition to the troops transferred from ComSoPac, included the 404th Engineer Combat Company as well as other service and ordnance troops.

Enemy opposition was neither expected nor received, although a fighter cover of General Kenney's forces provided assurance of success. The landings at Woodlark proceeded smoothly throughout. With a better area in which to land and with experience gained in a last minute rehearsal, the LEATHERBACK force went ashore with a minimum of effort. The Kiriwina operation, however, left much to be desired. Lack of prior training and insufficient equipment, complicated by poor landing areas, contributed to the confusion. In addition, the island's coral circlet made resupply of the island difficult. Regardless of these handicaps, VII PhibFor carried 12,100 troops to Woodlark and 4,700 to Kiriwina without a casualty, while a total of 42,900 tons of supplies and equipment were unloaded without loss of a ship or landing boat.

For the Marine 12th Defense Battalion, the Woodlark landing was anticlimactic. Organized in San Diego in August 1942 under the command of Colonel William H. Harrison, the battalion trained extensively and test-fired all its armament before moving to Pearl Harbor and further combat training. The battalion joined the LEATHERBACK force in Australia prior to the operation. Two 90mm antiaircraft batteries went ashore from LCIs on 30 June and the remaining batteries and groups followed them ashore during the next 12 days. The first two 90mm batteries were ready to fire by 1300 on 1 July, and the other units were in firing positions in equally short order once ashore. But the opportunity for combat firing seldom came. It was not until 27 July that a solitary Japanese plane, after making several false attempts, hurried over Woodlark to drop five small bombs. There was no damage, and the plane escaped. After that, only occasional alerts were noted in the 12th Defense Battalion's log. Kiriwina, however, was bombed several times during construction with some damage to equipment and installations and some casualties to the BY-PRODUCT troops.

Construction of the airstrip on Woodlark progressed speedily; the Kiriwina field was slowed by heavy rains and the fact that much of the heavy equipment had seen too much prior service and was deadlined for repair within a few days. On 14 July, Woodlark was declared operational with a strip 150 feet wide and 3,000 feet long available. The first fighter squadron from South Pacific forces arrived on 23 July. The runway at Kiriwina was operational in late July, and on 18 August, a Fifth Air Force fighter squadron arrived on station. Kiriwina staged its first strike against enemy forces on New Guinea in late August, and later was a base for a Fifth Air Force fighter group.

No Allied strike was ever staged from Woodlark's strip, and South Pacific aircraft commanders lost interest almost as soon as it was completed. In fact, after the capture of Munda, Woodlark was turned over to the Fifth Air Force. Kiriwina remained for a time as a fighter plane base, but later the war moved northward toward the Bismarck Archipelago and the Admiralties and left both fields far behind. However, the Woodlark-Kiriwina operation gave needed experience to a new amphibious force and provided a protective buffer to the New Georgia operation which was concurrently underway.

OCCUPATION OF SEGI AND SEIZURE OF VIRU [2]

The man who was to call the shots in the defense of Munda airfield unknowingly tripped the alarm which set the ELKTON plans into action. Major General Sasaki, in his area headquarters at Kolombangara, was irked at Coastwatcher Kennedy at Segi Plantation near Viru Harbor, and—after months of tolerating Kennedy's presence—determined to get rid of him. Sasaki had good reasons: Kennedy's station was the center of resistance on New Georgia, and his air raid warning activities had contributed greatly to the lack of success of Japanese strikes against Guadalcanal. On 17 June, Sasaki sent re-

[2] Unless otherwise noted, the material in this section is derived from: *New Georgia Campaign;* TF 31 ltr to holders of EasternFor LoadingO 1–43, dtd 22Jun43; TG 31-3 OpO AL 10–43, dtd 21Jun43; TG 31.3 LoadingOs 1–4, dtd 16Jun–7Jul43; 4th RdrBn SAR 10Jun–10Jul43, dtd 14Sep43; 4th RdrBn WarD, 26Feb–31Aug43, hereafter *4th RdrBn WarD; SE Area NavOps–II; Seventeenth Army Ops–I;* Col Michael S. Currin ltr to CMC, dtd 8Feb51; LtCol Anthony Walker ltr to CMC, dtd 23Feb51; Batterton, "You Fight by the Book;" Feldt, *The Coastwatchers;* ONI, *Combat Narratives, Solomon Islands Campaign: X—Operations in the New Georgia, 21Jun–5Aug43* (Washington, 1944), hereafter ONI, *Combat Narratives X;* Rentz, *Marines in the Central Solomons.*

inforcements to the Viru Harbor garrison with orders "to pacify that area."[3] (See Map 3.)

Prior to Sasaki's decision to reinforce Viru, Segi Plantation on the southeast coast of New Georgia had been important only to the Allies. Segi was an ideal entryway into the island. Amphibious patrols had landed here, and the plantation had been a haven for many downed aviators. For the new advance, the Allies planned to build an airstrip here, but Kennedy reported on 18 June that he would not be able to hold this position if he did not get help in a hurry. The Japanese were closing in on him.

Admiral Turner ordered an immediate occupation of Segi. If Kennedy said he needed help, he was to be taken at his word. This determined New Zealander, the former District Officer for Santa Isabel Island across The Slot from New Georgia, was no alarmist. He had moved to Segi Point after the Japanese occupied the Solomons and there he had been completely surrounded by enemy garrisons. But he had held on, and his position 160 miles northwest of Lunga Point had fitted in admirably with the system of air raid warnings. His reports on Japanese flights meant that their arrival over Guadalcanal could be forecast within a minute or two. Kennedy had told the natives of New Georgia that Britain was not going to give up these islands, and the success of the Allies at Guadalcanal and Tulagi gave convincing evidence of this. He continued to live almost openly in the plantation house at Segi. There were no trails leading to his station, and the approach along the beach could be watched. But Kennedy and his natives had been forced to ambush Japanese parties to keep the the position secret. Some enemy had escaped Kennedy's attacks, however, and Sasaki had issued the order which made Kennedy the most wanted man on New Georgia.

Already at Viru Harbor was the *4th Company* of the *1st Battalion, 229th Regiment*, plus a few assorted naval personnel from the *Kure 6th* and *Yokosuka 7th SNLF*, a 3-inch coastal gun, four 80mm guns, eight dual-purpose guns, and a varying number of landing craft. To augment the Viru garrison, Major Masao Hara was to take another infantry company and a machine gun platoon from his *1st Battalion* and comb southeastern New Georgia for the coastwatcher's hideout.[4] When this force came close, Kennedy faded into the hills and radioed:

> Strong enemy patrol has approached very close, and by their numbers and movement, it is believed they will attack. Urgently suggest force be sent to defend Segi.[5]

The message reached Turner at Koli Point, Guadalcanal, during the night of 18–19 June, and the admiral decided to send a force to Segi at once rather than wait until 30 June, the D-Day established by ELKTON plans. Fortunately, the admiral had combat units ready. The destroyer-transports *Dent*, *Waters*, *Schley*, and *Crosby* were standing by in Guadalcanal waters for the operations against New Georgia, and Lieutenant Colonel Mi-

[3] CIC SoPac Item No. 786, 8th CSNLF Operational Rad and TgOs, translated 17Dec43.

[4] *Ibid.*

[5] 4th RdrBn SAR, *op. cit.*, p. 9. Kennedy had good reason for his suspicions. A diary later taken from the body of Second Lieutenant Harumasa Adachi at Viru Harbor indicated that the Japanese had discovered Kennedy's hideout and that an attack was being planned. ICPOA Item No. 598, Translation of Captured Japanese Document, dtd 6Jul43.

chael S. Currin's 4th Marine Raider Battalion, which included personnel who had been to New Georgia on prelanding reconnaissance missions, was also completely combat-ready. With these ships and men, Turner could mount out a force to protect Kennedy and also thrust a toe in the Central Solomons door that the Japanese were trying to slam shut.

Currin's battalion (less Companies N and Q, scheduled to attack Vangunu Island on 30 June) went on board the *Dent* and *Waters* on 20 June for a night run to Kennedy's aid. This force was followed the next night by Companies A and D of the Army's 103d Infantry Regiment. Initially, these units would defend Segi, and then carry out the planned attack on Viru Harbor on 30 June as scheduled. With the exception of raider Company O, previously detached to duty with Turner's Western Force but now returned to Currin, these units were part of Admiral Fort's Eastern Force and were scheduled for use in this area of New Georgia. Thus the landing at Segi on 21 June, which set off the CARTWHEEL operations, amounted only to stepping up the timetable.

All was not smooth sailing for the *Dent* and *Waters*. The natural obstacles which had contributed to Kennedy's security at Segi Point were hazards for these ships. There is deep, sheltered water off Segi, but the channels to this anchorage were uncharted, dismissed on the charts as "foul ground." There are so many reefs and coral outcroppings in these waters that Vangunu appears to be almost a part of the larger island of New Georgia. There is no suitable route to Segi from the north, and only the natives and a few local pilots were acquainted with the passages to the south. Even with a local pilot sighting on Kennedy's bonfire signal on the beach, the transports scraped bottom and rode over reefs. At 0530 on 21 June, the Marines went over the side and into ships' boats for the landing, and by 1030, all supplies had been brought ashore and the transports were picking their way through the coral heads and reefs for a speedy return to Guadalcanal. Currin immediately established defensive positions and sent out patrols, but there was no contact with the enemy. At 0600 on the following day, the two Army companies plus an airfield survey party from Acorn 7 came ashore from the *Schley* and *Crosby*.

Kennedy was grateful that these troops had come to his rescue, but both his pioneer spirit and his scouting routine were pinched by this population influx. For peace and quiet, and to re-establish a schedule of unrestricted movements for his native scouts, he moved across the narrow channel to Vangunu Island. Currin kept contact with the coastwatcher, and, with natives provided by Kennedy, sent out patrols to determine the most suitable means of approach to Viru Harbor. At the same time, Seabees began converting Segi's uneven and muddy terrain into an airstrip. With bulldozers and power shovels, working at night under floodlights, the men had an airstrip ready for limited operations as a fighter base by 10 July. It was the intentions of ComSoPac to have the field capable of servicing 20 planes an hour at first, and then—by 25 September—of basing about 60 light bombers.[6]

---

[6] ComSoPac ltr ser 00534, dtd 10Jun43, Subj: "Proposed Master Plan for Construction of Airfields and Seaplane Bases in the South Pacific Area, Guadalcanal Island, Koli Point Section," quoted in *New Georgia Campaign*.

BURIAL CEREMONY *at Viru Harbor honors Marines of the 4th Raider Battalion killed in the first American offensive on New Georgia.* (USMC 57581)

155MM GUNS *of the 9th Defense Battalion on board an LCT off Rendova's beach as they move to a new firing position.* (USMC 60658)

The Allies had plans for Viru Harbor, too. This small, landlocked cove 35 miles from Munda was to be developed into a minor naval base for small craft. The best anchorage on the New Georgia coast, it had an entrance 300 yards wide and 800 yards long, outlined on both sides by coral cliffs. The inner harbor widened, and was fed by three small rivers, the Mango, Tita, and Viru. Previous amphibious patrols had reported the bulk of the Viru defenders to be located on the high headlands on the west side of the harbor at the village Tetemara, with another detachment at Tombe, a village facing Tetemara across the channel. But intelligence reports on the size of the Viru garrison conflicted. Early estimates had ranged from 20 to 100 men; an early-June reconnaissance patrol revised these figures to 200 enemy troops. (See Map 3.)

As Companies A and D of the 103d set up a defense against any further attempts by the Japanese to wipe out Kennedy's station, raider amphibious reconnaissance teams concentrated their attention on finding the most suitable route to Viru. Several times they narrowly missed bumping into Japanese patrols or sentries as the Marines examined a number of small river inlets searching for a beach which would exit to an overland route to the rear of Viru Harbor. While Currin's raiders scouted the area between Segi and the proposed landing site, a member of the staff, Captain Foster C. LaHue, slipped by native canoe through the bay to Hele Islands in Blanche Channel to meet the *Schley* and receive Admiral Fort's orders for the Marines' attack on Viru.

Currin had hoped to land during the night of 27 June at Regi, a village just seven miles from Viru Harbor and considerably west of Segi Plantation. From here his force could move overland to a point east of the Viru River, and there split for attacks down both sides of the inlet to seize the village of Tombe on the east bank and Tetemara on the west. Fort's orders, however, directed only Company P to land on the 28th at Nono village, just a few miles west of Kennedy's station. Currin was then to strike through the jungle to attack Tetemara at 0700 on 30 June, and capture the seacoast guns reported to be in Tetemara. The APDs *Kilty* and *Crosby* would then sail into the harbor and put ashore a 355-man occupation force consisting of Company B of the 103d; one-half of Company D, 20th NCB; Battery E (less one platoon) of the 70th Coast Artillery (Antiaircraft) Battalion; and a naval base unit.

Additional paragraphs of the order gave details concerning the proposed seizure of Wickham Anchorage and the development of Segi Point, but contained no instructions for Company O of the Raiders and that portion of the 4th Raider Battalion headquarters already at Segi.[7] At 1600 that afternoon, Colonel Currin radioed Admiral Fort for permission to land at Regi, to use Company O as well as Company P, and to begin the operation on 27 June rather than 28 June. The raider commander had spent 20 days in this area with amphibious patrols during March and April, and he estimated that even if he started a day earlier he would be hard-pressed to make the D-Day of 30 June at Tombe and Tetemara. An overland trek would mean tortuous trails over ridges, rivers, and swamps, and the hiking distance was considerably more than map

[7] TG 31.3 OpO AL-10, dtd 21Jun43; TF 31 ltr FE25/L1 over 0013b, dtd 22Jun43; TG 31.3 LoadingO 1-43, dtd 16Jun43.

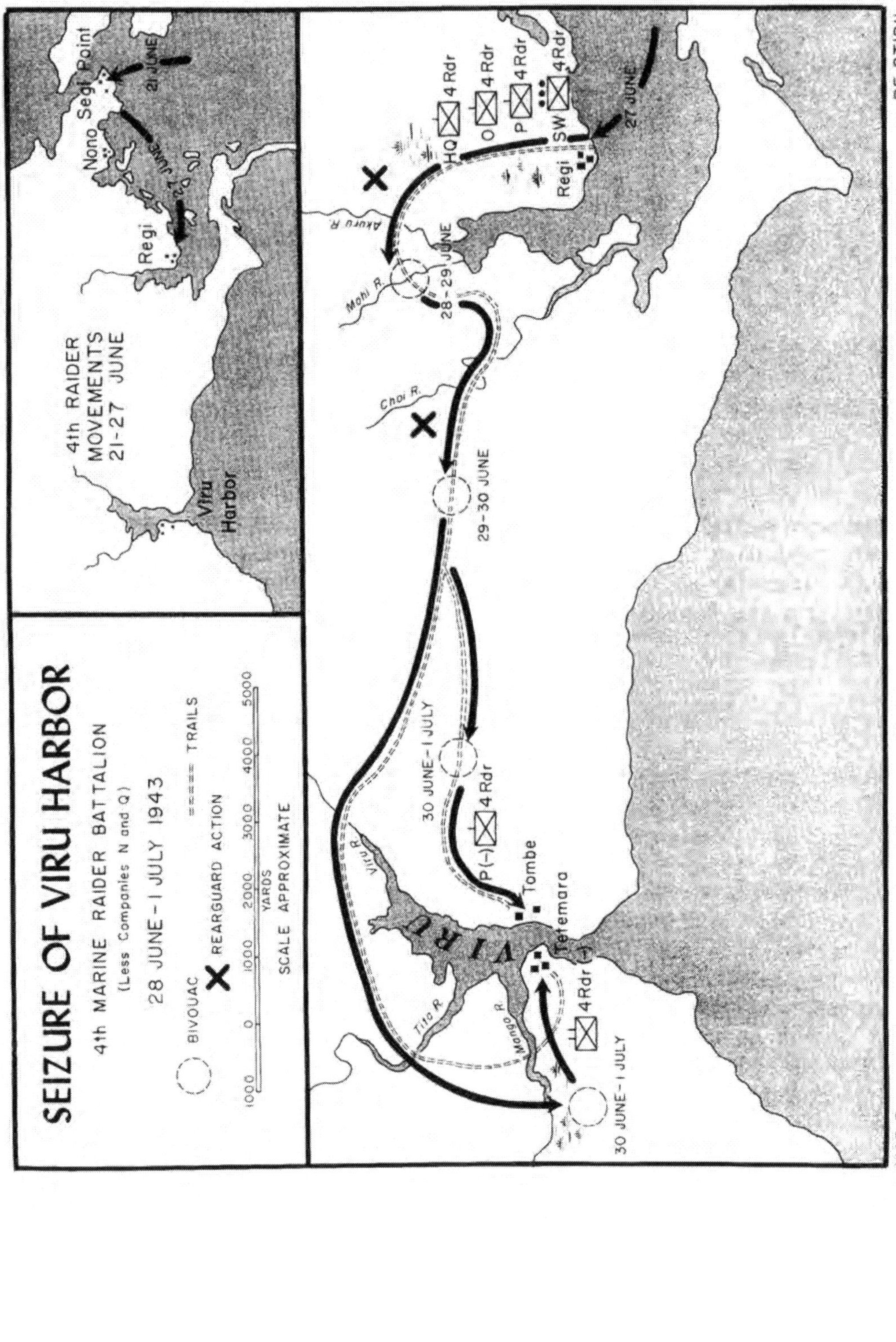
SEIZURE OF VIRU HARBOR
4th MARINE RAIDER BATTALION
(Less Companies N and Q)
28 JUNE - 1 JULY 1943
BIVOUAC
TRAILS
REARGUARD ACTION
1000 0 1000 2000 3000 4000 5000
YARDS
SCALE APPROXIMATE
4th RAIDER MOVEMENTS 21-27 JUNE
Viru Harbor
Regi
Nono
Segi Point
21 JUNE
27 JUNE
HQ 4 Rdr
O 4 Rdr
P 4 Rdr
SW 4 Rdr
28-29 JUNE
29-30 JUNE
30 JUNE-1 JULY
P(-) 4 Rdr
Tombe
Tetemara
Mohi R.
Akuru R.
Choi R.
Viru R.
Tita R.
Mango R.
VIRU
RF STIBIL

miles. Besides, the distance in miles was not a realistic indication of the problems the Marines faced in the thick jungle. Currin knew the job would be much too tough for a single company. The enemy situation had changed since Admiral Fort's plans were made, and there was now a larger Japanese force in the Viru area with patrols active at Nono. Currin felt that if his men landed in their rubber boats at Nono they would be "sitting ducks" for the Japanese.[8] Within an hour and a half, Admiral Fort had radioed his approval of the modified plan.

At 2000 on 27 June, the Marines boarded their rubber boats and started paddling the eight miles to Regi, Currin and his staff leading the way in two large Melanesian war canoes. As one Marine described the trip:

> It was a weird moonless night with black rubber boats on black water slipping silently through the many islands of Panga Bay. The trip was uneventful except for one scare. It came just before reaching Regi, while lying offshore waiting for word from native scouts who had gone ahead to make certain no Japanese were in the village. Due to the sudden appearance of a half moon which began to cast a sickly reflection, a small island appeared to be an enemy destroyer.[9]

The scouts came back with an "all clear," and, by 0100, all hands were ashore, and the rubber boats were being towed back to Segi by natives in the war canoes. At dawn, the battalion followed the scouts into the jungle with Company O in the advance guard followed by the headquarters group, and with Company P furnishing the rear guard.

There were many signs of Japanese patrols, but they indicated small scouting parties rather than forces large enough to offer determined opposition. Currin instructed his Marines to meet Japanese harassment with forces no larger than absolutely necessary so that the main column could continue to advance. The Marines would have to fight against time if they were to reach Viru Harbor on schedule and silence the Japanese coastal guns before Admiral Fort's landing forces entered the harbor.

The Marines' battle against the New Georgia jungle began just outside Regi where the force encountered a mangrove swamp two miles wide. There was no suitable way to skirt this obstacle, so the column struck out through it. The first enemy contact was a five-man patrol that came in off a side trail and apparently surprised itself as well as the Marines by stumbling into the rear party of the raider battalion. The 3d Platoon of Company P killed four of these men in a brief skirmish before resuming the march with the rest of the column. At 1115 another enemy group hit the rear guard, and five Marines at the end of the column were cut off from the main body as Company P drove these Japanese off with rifle and machine gun fire. The five men, evading the Japanese but unable to catch up with the column, returned to the landing site and paddled back to Segi in a native canoe they found on the beach.

In all, the force made about six miles the first day. The terrain grew more difficult as the Marines moved deeper inland, and the advance became more of an up-and-down climb than a march. The raiders bivouacked in a tight perimeter, ate their K rations, and huddled under their ponchos throughout the rainy night. Realizing that the slow going would keep him from making his assault on schedule,

[8] *Currin ltr.*

[9] Batterton, "You Fight by the Book," p. 16.

Currin sent two native runners back to Segi with a message for Admiral Fort that the raiders would be a day late in reaching Viru. Kennedy had trouble contacting the Russell Islands, and when this message got through, the landing force was already underway toward Viru Harbor.

On the second day, Currin's force covered seven miles of the difficult terrain, and was forced to make three crossings of the meandering Choi River, now swollen and swift from the heavy tropical rains. At about 1400, just as the rear guard completed its first crossing of the Choi, it drew fire from 30 to 40 Japanese in positions on the right flank of the advance. Captain Anthony Walker, commanding Company P, sent First Lieutenant Devillo W. Brown with a reinforced platoon of 60 men to deal with this enemy force. The Marines located the enemy dug in on the crest of a hill some 300 yards from the trail. The raiders wasted no time. With one squad in position for covering fire, the other two squads went up the ridge by infiltration, firing as they climbed. Eighteen enemy dead were found, but five raiders had been killed and another wounded in the attack.

Burying their dead and carrying the wounded man, Brown's men pushed on to catch Currin's column. The battalion crossed the Choi River again, skirted a large swamp, and then halted for the night just after crossing the Choi for the third time. There Lieutenant Brown and his platoon rejoined the battalion. Colonel Currin tried to report his position to Guadalcanal, but his radio failed him. The battalion commander could only hope that the message he had sent via runner and Kennedy during the previous night would keep the transports from sailing into range of the enemy's 3-inch coastal gun before the raiders could get to Viru and silence that weapon.

As Currin's force moved out on the morning of 30 June with a full day's march remaining between it and its objective, Commander Stanley Leith's Viru Occupation Unit in the *Hopkins*, *Kilty*, and *Crosby* edged toward Viru Harbor and the Japanese gun which Currin's force was to have silenced. Leith, however, had received roundabout word that Currin was going to be a day late in his attack at Viru. Remaining close by in case the Marines were in trouble, the APDs at 0730 came within range of the enemy's 3-inch gun, and the shells began splashing all around the ships.

Leith withdrew to the harbor mouth, where he steamed back and forth until 1000. Then, with Admiral Fort's approval, he withdrew from Viru area and the next day put the landing force ashore at Nono. These troops, under the command of Captain Raymond E. Kinch of Company B, 103d Infantry, would go overland to Viru, as Currin was doing. From Viru, Major Hara reported to General Sasaki at Munda Point that he had repulsed an attempted American landing.[10]

Early on 30 June, the raider battalion reached the trail fork from which one route extended south toward Tombe. Currin had planned to send one platoon against Tombe. The enemy opposition of the previous days, however, and the fact

[10] Hara was, it seems, vague as to his orders. Sasaki, with American troops landing at Rendova, had previously ordered Hara to return to Munda. CIC SoPacFor Item No. 702, New Georgia DefOpO "A" No. 11, dtd 30Jun43.

that enemy patrols working the jungle between Viru and Segi Point could reinforce this village more quickly than they could Tetemara, prompted Currin to increase the size of the force attacking the east side of the harbor. Two platoons from Company P (Lieutenants Brown and Robert J. Popelka) with Captain Walker in command were assigned this mission. The attack at Tombe would be made independently of the assault at Tetemara.

Currin moved on toward Tetemara with a smaller force than he had originally intended. For the men with Currin, this day's march was the worst yet. They met no enemy, but by mid-morning had to ford the Viru River and then struggle through mountainous terrain—rugged jungle ridges along the course of the Tita River which they crossed later in the day. Going was slow for the men weighted down with arms, equipment, and ammunition, and there was but an hour of daylight remaining when they came out of the bush on the bank of the Mango River. Fifty yards wide, deep and swift, the Mango was a formidable obstacle. But the Marines clasped hands and moved out, the human chain snaking the force across the river.

Beyond the Mango, the Marines were caught by darkness and a mangrove swamp. Water, knee-to-waist-deep, hiding twisted, snakelike roots under the surface, trapped the raiders. In a matter of minutes, the column was stalled as men fought to keep contact with each other. However, "tree-light" — phosphorescent wood from dead logs and trees—was provided by the native guides and re-established contact. With each man carrying a piece of this dimly glowing wood, and guiding on the piece carried by the man ahead, the column closed up and moved out. Four hours later, the Marines were out of the swamp and facing the last half-mile of steep slope to the rear of Tetemara Village. Weary raiders struggled up the slick and muddy trail, falling exhausted at the top of the ridge after crawling on their hands and knees the last 100 yards of nearly vertical slope.

On the east side of Viru, Walker's force bivouacked a short distance from Tombe, and at 0900 on the morning of 1 July launched its attack. The surprise assault killed 13 Japanese, scattered the remainder of the small garrison, and carried the position at no loss to the raiders. The firing aroused the enemy across the harbor at Tetemara. When they rushed out in the open, they were bombed and strafed by six planes from VMSB–132 and VB–11. The strike had been requested by the air liaison party at Segi and approved by General Mulcahy in his headquarters at Rendova. It was the first strike logged in the new records of ComAir New Georgia.[11]

Currin's force, moving along the high ground overlooking Tetemara, heard the explosions and firing during the air strike, but the jungle screened the planes from view.[12] Fifteen minutes later, Currin attacked the village. With First Lieutenant Raymond L. Luckel's Company O in the lead, the raiders moved down the slope, then fanned out in an attempt to confine the Japanese to an area bordered by the harbor and the sea. Luckel's machine guns were attached to his assault platoons, and with the help of this additional firepower the advance continued slowly. A few out-

[11] ComAirSols StrikeComd WarD, 2Apr–25Jul43; ComAir New Georgia SAR, 29Jun–13Aug43.

[12] *Currin ltr.*

guard positions were overrun before the Marines were forced to halt under steady fire from the enemy's main line of defense.

Advance was slow and sporadic, with long periods of silence broken abruptly by a series of short, sharp fire fights lasting only a few minutes each. In an hour, the Marines had gained about 100 yards. Deciding that a buildup for an envelopment around his left flank was developing, Luckel committed his 3d Platoon to that flank, and the advance continued. By 1305, the Marines had reached a low crest of ground from which the terrain sloped away toward Tetemara.

The bottled-up enemy, realizing their predicament, began withdrawing toward the northeast with much frantic yelling. Anticipating a *banzai* charge in an attempt to break through the Marine's leit flank, Currin dispatched his slim battalion reserve of the 3d Platoon and two sections of machine guns from Company P to the aid of Company O. The reinforcements arrived just in time. In a matter of minutes, the hopeless rush of the enemy was broken, and the Marines began to move forward against spotty resistance. The 3-inch gun was captured, Tetemara occupied, and the few remaining Japanese flushed out of caves and jungle hiding places. Currin's force counted 48 enemy dead, and captured, in addition to the 3-inch gun, the four 80mm guns and eight dual-purpose guns of the Viru garrison, as well as 16 machine guns, food, clothing, ammunition, and small-boat supplies. Eight Marines were killed in the attack.

Even while the fighting was in progress, three LCTs sailed into the harbor with gasoline, oil, and ammunition for the proposed naval base. They remained safely offshore until Tetemara was secured, and then came in to drop their ramps and unload. Three days later, on 4 July, Company B of the 103d Infantry struggled into Tombe after an enervating march overland from Nono. On 10 July, after a new garrison force came in to hold and develop the Viru area, the raiders returned to their old camp at Guadalcanal.[13] Seizure of Viru had cost the battalion 13 killed and 15 wounded out of an original force of 375 officers and men.

Major Hara's Viru garrison force lost a total of 61 killed and an estimated 100 wounded in the defense of Tombe and Tetemara. Another estimated 170 escaped into the jungle. Hara's force, under orders from the *Southeast Detachment* to return to Munda, marched over the rugged jungle trails and reached the airfield about 19 July, just in time to take part in the final defense of that area.[14]

SECURING VANGUNU [15]

Another side show to the main New Georgia landing in the Munda area was

[13] Because of the length of the Viru Harbor operation, the 4th Raider Battalion missed being available for the Rice Anchorage landing of Liversedge's 1st Marine Raider Regiment. The 4th Raiders' place at Rice Anchorage was taken by the 3d Battalion of the 145th Infantry. Currin's battalion, after a short rest at Guadalcanal, joined the Liversedge force on 18 July.

[14] *SE Area NavOps—II*, p. 34; CIC SoPacFor Item No. 702, New Georgia Defense OpO A No. 44, dtd 19Jul43.

[15] Unless otherwise noted, the material in this section is derived from: *New Georgia Campaign*; TF 31 ltr, dtd 22Jun43, *op. cit.*; TG 31.3 OpO AL 10–43, *op. cit.*; TG 31.3 LoadingOs 1–4, *op. cit.*; 4th RdrBn SAR, *op. cit.*; Capt James E. Brown ltr to CMC, dtd 6Mar51; Col Lester E. Brown, USA, ltr to Maj John N. Rentz, dtd 19Mar52; RAdm George H. Fort ltr to Maj John N. Rentz, dtd 30Jan52; Maj Earle O. Snell, Jr., ltr to CMC, dtd 16Feb51; ONI, *Combat Narratives X*; Rentz, *Marines in the Central Solomons*.

the taking of Vangunu Island for the purpose of establishing a base along the supply route between the lower Solomons and the main target area. Pre-landing reconnaissance revealed that the island would not be suitable for airfield construction as planned earlier. It could be taken with a relatively small force, however, because it was not heavily defended. Thus, it would be an economical prize for the Allies with the promise of a useful way station at Wickham Anchorage, a sheltered harbor tucked behind coral reefs between Vangunu and neighboring Gatukai Island to the east. (See Map 4.)

An amphibious scouting party sent to Vangunu in mid-June radioed Admiral Turner's headquarters on the 20th, confirming reports that the Japanese had not reinforced the island and that beaches at Oloana Bay on the south side of the island could accommodate the landing of a reinforced battalion. Admiral Fort was then directed to occupy the island with a small force on 30 June. His D-Day landing would not be a complete surprise. Japanese sentries spotted the amphibious patrol and reported "enemy surface forces" in the Wickham area; all units were cautioned to be on the alert.[16]

As his landing force, Fort selected Lieutenant Colonel Lester E. Brown's 2d Battalion, 103d Infantry Regiment; Battery B (90mm), 70th Coast Artillery (Antiaircraft) Battalion; and half of the 20th NCB. To augment Brown's soldiers, Admiral Fort also assigned that portion of the 4th Marine Raider Battalion which had not gone to Segi Point and Viru Harbor under Lieutenant Colonel Currin. The raider battalion's executive officer, Major James R. Clark, commanded these units which included Company N (Captain Earle O. Snell, Jr.), Company Q (Captain William L. Flake), a demolitions platoon, and a headquarters detachment. For added firepower, Battery B (105mm howitzers) of the 152d Field Artillery Battalion, and a special weapons group from Battery E (40mm and .50 caliber antiaircraft guns) of the 70th Coast Artillery Battalion were added to the Wickham Anchorage force.

The plan called for the Marines to land before dawn at Oloana Bay from APDs *Schley* and *McKean*, contact the scouting party still on Vangunu, and establish a beachhead. A first echelon of Army troops would land over this beach 30 minutes later from seven LCIs, followed by a second and final echelon of Army troops landing from seven LSTs at 1000. From Oloana Bay, Brown's force would move inland to widen the beachhead line while Company E, 103d Infantry, reinforced with the battalion's 81mm mortars, skirted along the beach eastward toward Vura Bay, reported as the main enemy base. Native scouts operating from the base of coastwatcher Kennedy near Sergi Point had reported that there were approximately 100 Japanese at this point.

After a rendezvous at Purvis Bay off Florida Island, Fort's transports sailed north beyond the Russell Islands and reached the debarkation area off Oloana Bay at 0230 on 30 June. The scouts had placed markers on the beach and were showing a signal light, but the ships arrived in the midst of such a heavy downpour that these aids could not be spotted. High winds put a bothersome chop on the sea, and the APDs and landing craft pitched and tossed as the Marines groped

[16] CIC SoPacFor Item No. 641, 8th CSNLF RadLog, 4–22June43.

their way over the side to prepare for the "blind" landing. The best radar in the task force was an old model in Admiral Fort's flagship, the destroyer-mine sweeper *Trever*, but it was not able to fix the position of the force accurately in relation to the beach.

Admiral Fort called off the landing until the weather cleared, or until dawn when the beach could be seen, but the APDs either did not receive or misunderstood these orders and went ahead with the landing. The whole operation became a classic example of how not to land troops on a hostile beach. At 0345, while in the midst of debarkation, the APD commanders decided that their ships had not been correctly positioned, so they moved 1,000 yards to the east and continued the operation. The move added to the confusion, since it forced the landing craft to cross paths with the seven LCIs, resulting in thorough dispersion of the landing craft just as they were heading toward the beach. Regaining contact proved impossible and the coxswains had to do the best they could on their own. Not having been given anything but a general course to the beach, they landed in widely separated spots along seven miles of the Vangunu coast. Six boats, but no men, were lost in the pounding surf.[17]

Two boats carrying the 1st and 2d Platoons of Company Q managed to stay together, but they headed in the wrong direction and finally grounded on a reef approximately seven miles west of Oloana Bay. The craft which contained Second Lieutenant James E. Brown's 1st Platoon managed to clear the reef, but in doing so lost its rudder. Marines tied buckets to the ends of lines and then guided the boat by trailing these buckets astern and pulling on the lines like reins. The boat carrying Second Lieutenant Eric S. Holmgrain's 2d Platoon broached to in the surf on the reef. Holmgrain and his men waded and swam nearly two miles to shore and set up a local defense until dawn. Brown remained just off shore with his platoon boated. The next morning, Holmgrain hiked his platoon along the beach toward Oloana Bay while Brown steered along the coast with his buckets. The *McKean* spotted this craft and sent out a sound boat in which Brown finished his trip.

The other scattered Marines met no opposition and were able to regroup at Oloana Bay. There the first waves of soldiers landed in calmer seas at 0700, followed by the remainder of the force within an hour. The amphibious scouts reported to Colonel Brown that the Japanese garrison was not at Vura as expected, but was instead occupying Kaeruka, a small village about 1,000 yards northeast of Vura Bay on another coastal indentation. Colonel Brown immediately issued new orders, designating the mouth of the Kaeruka River as the objective, and just before 0800, the drenched force moved out. Company E retained its original mission of capturing Vura village. There this force would hold up and prepare to give mortar support to the other units attacking

---

[17] Admiral Fort commented on this passage; "The chief lesson here is *to obey orders!* The APDs acknowledged for this order, and I was amazed to learn later that they had landed the Marines or that the Marines had agreed to land under circumstances that would insure . . . failure. The Army troops in the LCIs obeyed the order—landed in perfect order as planned without any difficulty whatever. The Marines straggling ashore hours later would not have been of much help had there been much resistance." VAdm George H. Fort comments on draft MS, dtd 3Oct60.

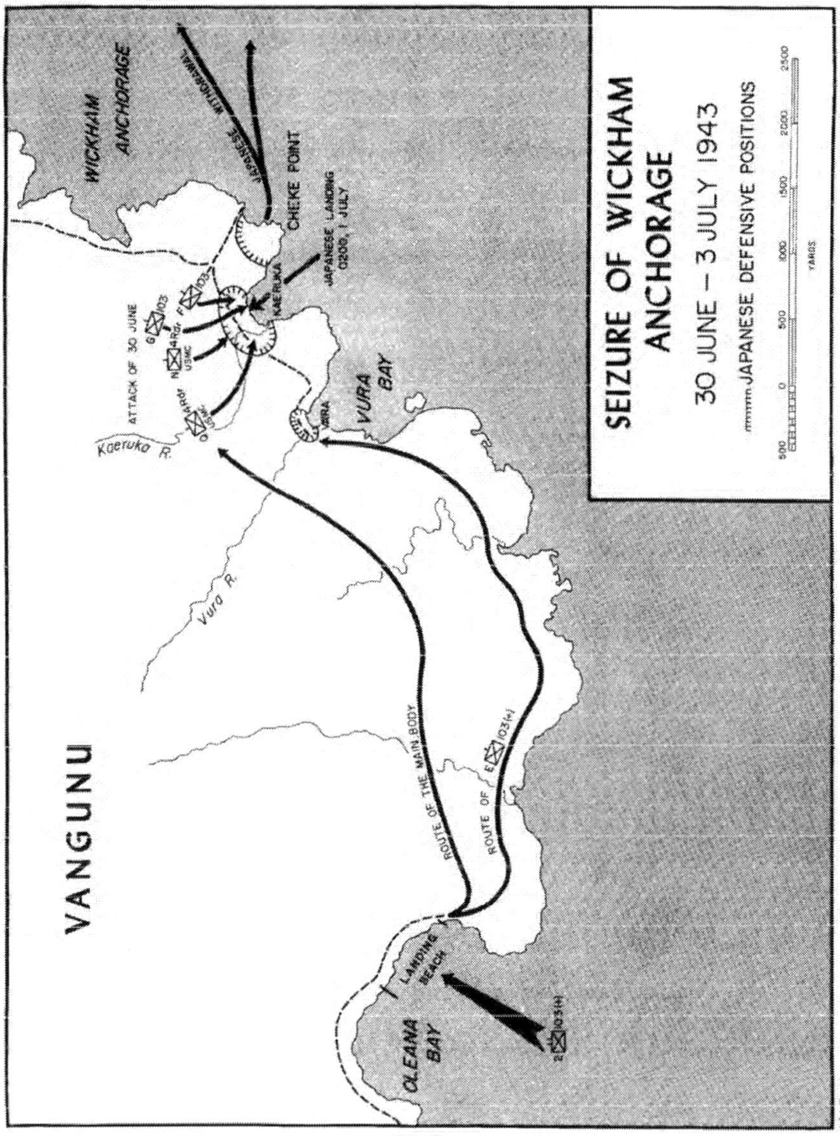
SEIZURE OF WICKHAM ANCHORAGE
30 JUNE – 3 JULY 1943
JAPANESE DEFENSIVE POSITIONS
500 0 500 1000 1500 2000 2500
YARDS
VANGUNU
WICKHAM ANCHORAGE
CHEKE POINT
JAPANESE LANDING 0200, 1 JULY
KAERUKA
VURA BAY
VURA
Kaeruka R.
Vura R.
ATTACK OF 30 JUNE
ROUTE OF THE MAIN BODY
ROUTE OF
LANDING BEACH
OLEANA BAY
JAPANESE WITHDRAWAL
R.F STIBIL

MAP 4

Kaeruka. Companies F and G of the 103d, along with Marine units and eight native guides, would swing inland along a coast-watcher trail, which it was believed had not been discovered by the Japanese, and assemble on high ground some seven miles from the beachhead. This hilly terrain would give the attackers an attack line of departure just east of the Kaeruka River and 700 yards from the village where the Japanese were camped. Artillerymen and Seabees would hold the beachhead at Oloana.

At Vura village, Company E met 16 enemy armed with two light machine guns, but the mortars quickly knocked out this opposition. The company then prepared to support the other attacking force which had to deal with a more difficult march and stronger enemy defenses.

The driving rain had turned the coast-watcher trail to slick mud and the Vura and Kaeruka Rivers into shoulder-deep torrents. Swimmers strung ropes across these streams and the Marines and soldiers then managed to cross, each man pulling himself along the ropes. Brown's force finally reached its line of departure at 1320. By this time all scattered Marine units, including the two platoons which grounded on the reef seven miles from the landing beach, had rejoined their parent companies, much to the gratification of Colonel Brown:

> This in itself was a considerable feat because some of the landing boats had gone ashore far down the coast . . . and the Marines were all heavily laden with weapons and ammunition.[18]

The attack jumped off without preparatory fires. The rain had put all radios out of commission, and Colonel Brown could not contact either Vura for mortar fire or Oloana Bay for artillery support. The Marines, commanded by Major Clark, and the soldiers moved south from their line of departure at 1405. On the right, Company Q (raiders) guided on the meandering Kaeruka River with orders to cross the river farther south to turn the left flank of the enemy. Company N (raiders) in the center drove straight towards the Japanese; and on the left, Company F of the 103d Infantry moved to position for a partial envelopment of the Japanese right. The 103d's Company G, in reserve, stood ready to exploit any weakness in the Japanese defenses and to protect the American flanks.

Off Company Q's right, the Kaeruka River made a 300-yard loop to the east before turning south again to flow 300 or 400 yards into the sea. This long bend in the river partially enclosed the Japanese camp on the coast and made the stream, in effect, a major obstacle facing the Marine companies. Fifteen minutes after the Marines moved from their line of departure, Company Q began to draw fire from enemy riflemen hidden in trees and camouflaged spider traps.[19] As the Marines deployed, they met heavier fire from Japanese positions across the river. At 1445, Major Clark told his raider companies to cross the river, reduce the opposition, and then attack the main enemy positions.

Marines of Company Q struggled down the slippery bank of the river, crossed over and climbed the other side. But the Japanese concentrated so much rifle and machine gun fire on the crossing site that only

[18] L. E. Brown ltr, *op. cit.*

[19] Individual foxholes of kneeling or standing depth covered by "lids" camouflaged to conceal the position from observation even at short range.

one squad of Company N managed to cross before the attack was called off. Contact between the soldiers and Marines was now broken, and while the two Marine companies attempted to tie together, patrols were sent out to re-establish contact with the Army companies.

On the left, soldiers of Company F tried to envelop the right of several Japanese machine guns which they encountered shortly after starting the attack. This maneuver further broke contact between the soldiers and Company N. Colonel Brown then sent Company G to fill the gap. The reserve company moved almost directly south toward the beach meeting only scattered opposition. Although both flanks of the American advance had lost contact with the center unit, this unhandy tactical maneuver split the Japanese force. As Company G moved through the gap between Companies N and F and reached the beach, it placed itself squarely in the enemy rear, and the Japanese opposing the Marines and Company F gave way in disorder.

Resistance in front of Company Q faded, and Company N moved up quickly through the jungle to exploit the confusion of the enemy and drive them to the southwest. The Marine companies, one on each side of the river, then pressed on to the beach below the village of Kaeruka. The soldiers of Company F also reduced the opposition facing them, reaching the beach shortly thereafter. Twelve Marines were killed in the action and 21 wounded; Army casualties numbered 10 killed and 22 wounded. One hundred and twenty Japanese dead were counted.

As Colonel Brown displaced his command post forward from the line of departure, Major Clark established a perimeter defense along the beach east of the river. Company G, the raider demolition platoon, and Company Q dug in facing seaward. Company N tied in on Company Q's right flank extending inland along the east bank of the Kaeruka River. Soldiers of Company F closed the perimeter with a line which faced inland. Patrols were set out to mop up any bypassed enemy, but darkness forced these men back to the perimeter before any contact could be made. Intermittent enemy mortar fire exploded inside the perimeter and along the beach during the early part of the night, and Japanese machine guns harassed the northern portions of the perimeter, but no attempt at penetration was made.

At about 0200, the American force hit a jackpot. Defenders reported three enemy barges approaching the beach. This was evidently a supply run, bringing food and reinforcements to a Japanese garrison which no longer existed. As the barges moved out of the darkness toward the beach area near the junction of the demolitions platoon and Company G, they met an overwhelming reception. A concerted burst of rifle and machine gun fire set the three barges foundering and drifting out of control. The Japanese called out, evidently believing they were being fired upon by friends, and for a time did not return the fire. The barges continued to drift toward the beach, and Company Q added rifle grenades to the small-arms fire that the other units along the beach directed into the landing craft. The Japanese finally returned the fire, and a few enemy soldiers jumped overboard and splashed ashore. The demolitions platoon killed these with hand grenades. One barge sank offshore and the others broached to in the surf. The fight was over in half an hour, and 109 of an estimated 120 Japanese were dead.[20] Two Marines and one soldier were killed.

The choice of beach defense, made in half darkness, without the aid of maps, was particularly fortunate. Unknown to Major Clark, the beach was the only possible landing point in the entire area; and an unsuspecting enemy had picked that night to resupply and reinforce his garrison. The Japanese "had walked blindly into a hornet's nest. For the Marines, it was like filling an inside straight."[21]

The next morning (1 July) patrols searched for the remnants of the Japanese garrison and learned that the main group of Japanese survivors was digging in at Cheke Point, a bulge of shoreline about 500 yards east of the Kaeruka River. Despite the success of the defense the night before, Colonel Brown marched his force back toward Vuru village since he considered it a more suitable defensive area and his troops could be more easily supplied. Moreover, Cheke Point was readily identifiable from the air, and artillery fire from Oloana Bay coupled with air strikes could neutralize the position with considerable saving of lives. A few Japanese harassed the column with long-range fire from machine guns and a 37mm gun, inflicting some casualties, but the enemy made no organized counterattack.

From a new perimeter at Vura, Brown sent out patrols and organized a coordinated attack on Cheke Point. By this time, with the help of Seabees, the 105mm howitzers of the 152d Field Artillery Battalion were in position, and after registering on Cheke Point, fired all day 2 July. In the afternoon, Admiral Fort's flagship added naval gunfire to the pounding. On the morning of the 3d, 18 SBDs from ComAir New Georgia staged a strike on Cheke Point while Brown's force moved forward. Kaeruka was reoccupied without a shot and even Cheke Point was taken with little opposition since the bombardments had forced the Japanese to evacuate the area. Brown's attack killed seven Japanese and destroyed supply and ammunition dumps which had been overlooked earlier.

On 4 July, the Marines, detached from Colonel Brown's force, went back to Oloana Bay on LCIs. There they rested until 8 July when they were sent across to Gatukai Island east of Vangunu to look for some 50 to 100 enemy troops which natives reported were holding that small island. The Marines patrolled Gatukai for two days but did not locate any Japanese, although bivouac areas attested to recent occupation. After one night back at Oloana Bay, the Marines returned to Guadalcanal on 12 July to rejoin the remainder of the battalion which Colonel Currin had just brought back from the operation at Segi Point and Viru Harbor. Casualties within Major Clark's original force of 18 officers and 350 men totaled 14 dead and 26 wounded.

[20] Five of 11 Japanese who survived this encounter were later killed in the Kaeruka area. Six others who escaped made their way along the coasts of Vangunu and New Georgia to Rice Anchorage on northern New Georgia. There they later met the same demolitions platoon in another action, in which five were killed and one captured. *4th RdrBn WarD.*

[21] Rentz, *Marines in the Central Solomons*, p. 50. Another possible explanation is that the Japanese were collecting scattered outposts. CIC SoPacFor Item 702, New Georgia DefOpO A No. 11, dtd 30 June states: "The Wickham Butai in large landing barges will collect the lookouts from the West Harbor and North Harbor and the Barike Butai and return to Munda around the north coast of Vangunu and New Georgia Islands."

BEACHHEAD AT RENDOVA[22]

Allied occupation of Segi on 21 June provided the clincher to a Japanese error in judgment: The New Georgia defenders were set to repel an invasion; night and day Allied radio traffic and reports of a troop and transport buildup in the Guadalcanal area had convinced them that an attack was imminent. To intercept such a move, the Japanese assembled their air attack forces at Buin and deployed to defend the Central Solomons. Then, when the occupation of Segi occurred without further immediate buildup, the enemy was positive that the operation was only a limited infiltration, and that the abrupt decline of radio traffic meant a curtailment of further plans. On 26 June the air fleet was ordered to return to Rabaul.

This strategy backfired. A Japanese submarine spotted Turner's task force near Gatukai on the night of the 29th, but, before the Japanese could determine the significance of the submarine's report, Vila and Buin were rocked by naval bombardments and Turner's force was dropping anchor in Rendova Bay. The attacks at Vila and Buin were a diversion only, planned to place striking units in position to protect the Rendova landing. The bombardments were conducted in a driving rain which shielded the retirement of the cruisers and destroyers. Unfortunately, the poor weather also cancelled a Fifth Air Force strike at Rabaul which was supposed to cripple further the Japanese potential to lash back at the New Georgia landings.

The Rendova operation cast the Japanese command in the role of a poor second-guesser:

> The next enemy counter-offensive was estimated in various ways and a series of measures was taken to meet the situation. However, we hardly anticipated that the enemy would first occupy the small islands across from Munda at the time of the invasion of Munda and that they would proceed with their operations under the cover of heavy guns on these islands. Therefore, the landing on Rendova Island completely baffled our forces.[23]

For the Rendova-bound attackers, the movement from Guadalcanal was uneventful. The task force streamed north from Koli Point at 1545 on the 29th. Screened by destroyers, the six transports and two APDs sailed in a double column north of the Russells before turning west and then northwest again to head up Blanche Channel. Shortly before dawn, the same weather front which shielded the bombard-

[22] Unless otherwise noted, the material in this section is derived from: CinCPac Ops in the POA, Jun43, dtd 6Sep43; InteiSec, SoPacFor Objective Rept 25–13, New Georgia Gru, dtd 15Feb43; SoPacFor PhotoInterpretationU Rept No. 41, dtd 1Dec42, and Rept No. 67, dtd 26Dec42; *New Georgia Campaign;* 3d MarDiv Observers Rept Rendova and Munda (Col George W. McHenry, LtCol John T. L. D. Gabbert, LtCol J. M. Smith, Jr.), dtd 9Sep43; 9th DefBn Rept to ComMarDefGruSols, dtd 24Jun43; 9th DefBn Rept of Ops, dtd 2May44, hereafter *9th DefBn OpsRept*; 9th DefBn Informal CbtRept New Georgia Campaign, dtd 9Sep43; 9th DefBn Narrative Hist 1Feb42–14Apr44, dtd 2May44; *9th DefBn AA Ops;* 9th DefBn OpO 5–43, dtd 24Jun43; *SE Area NavOps—I; SE Area NavOps—II;* Maj Wilson F. Humphreys ltr to CO, 14th DefBn, dtd 25Jul43; LtCol Francis M. McAlister ltr to CG, IMAC, dtd 7Aug43; LtCol Wilbur J. McNenney Rept on New Georgia Ops to CG, IMAC, dtd 17Jul43; LtCol Henry H. Reichner Jr., ltr to CMC, dtd 27Feb51; Capt Michael Taylor ltr to CMC, dtd 9Feb51; LtCol Wright C. Taylor ltr to CMC, dtd 4Mar52; ONI, *Combat Narratives X;* Rentz, *Marines in the Central Solomons;* Sherrod, *MarAirHist.*

[23] *SE Area NavOps—II*, p. 26.

ment forces covered the invasion fleet, and troops aboard the transports had only hazy glimpses of the rain-drenched volcanic peak which identified Rendova. (See Map II, Map Section.)

The invasion site was on the north end of Rendova, a haunch-shaped island nearly 20 miles long and 8 miles wide. Mountainous and densely wooded, Rendova was a fitting counterpart to the other islands in the group. Few of the marshy beaches along its otherwise irregular, steep coastline could be used as landing sites, and most of the shoreline was fouled by coral patches. The island's best anchorage was Rendova Harbor, a cove three-fourths of a mile wide and one and one-half miles long, protected by a barrier of three small islands. The cove had two deep-water entrances and pre-landing reconnaissance teams had designated the eastern of the two as the funnel for the ship-to-shore landing movement.

While the escorting destroyers took up their screening positions—the *Jenkins*, *Gwin*, *Radford*, *Buchanan*, and *Farenholt* echeloned at 1,000 yard intervals to the northwest and the *McCalla*, *Ralph Talbot*, and *Woodworth* blocking Blanche Channel to the east—the transports began unloading. Troops from the *McCawley*, *Algorab*, *Libra*, and *President Adams* were to land on the east beach of the cove; the troops from the *President Hayes* and *President Jackson* would go ashore on the west beach.

At 0640, only minutes after arriving at debarking stations, the transports had landing craft lowered and headed toward the beach 3,000 yards away. Through the slot between the offshore islets, the boated troops could see Rendova Mountain and the relatively flat area of the Lever Brothers plantation at its foot which was the landing area. There was some momentary confusion as the flood of boats hit the entrance; but quickly forming into columns, the landing craft plowed on toward shore. As the boats scraped to a halt, disorganized enemy machine gun and rifle fire from the plantation area greeted the disembarking troops.

This was the first indication that plans had been fouled-up. Companies C and G of the 172d Infantry, scout troops called "Barracudas," were supposed to have landed from the APDs *Dent* and *Waters* an hour earlier, secured the beachhead, and then provided a covering force for the first wave of troops. The enemy resistance, obviously, was evidence that the beachhead had not been taken. The Barracudas, missing the rain-obscured beacon fires, had drifted some 10 miles down the Rendova coast before landing. Then, realizing that they had missed the designated beach, they reembarked and headed upshore toward the cove. They arrived in time to land unopposed over a beach secured by soldiers of the 103d, Seabees of the 24th NCB, and Marines of the 9th Defense Battalion.

The amphibious maneuver was not a classic. The beachhead had not been expanded beyond 15 yards or so, and in this confined space, soldiers, Seabees, and Marines milled about in the midst of a growing mountain of supplies. Only the first wave had been coordinated. After that, eager coxswains rushed back to the transports for additional loads, and the ship-to-shore movement became an uncontrolled race. To add to the confusion, an occasional enemy machine gunner would spray the landing area from the interior of the plantation, drawing in return a flurry of uncontrolled shots from riflemen on shore and automatic weapons from the landing craft. Eventually, combat patrols were organized, and soldiers of the 103d and 172d began to push inland in skirmish lines, flushing snipers and hidden machine gun nests.

The landing area had not been defended in any great force.[24] Sporadic and desultory fire had been the enemy's only resistance to the invasion. Although warned earlier of the possibility of a landing, the Rendova garrison had gone back to sleep and awoke to find an invasion fleet in its front yard. Too late to defend the harbor in force, most of the garrison fled to the hills to escape later to Munda by canoe. Wet batteries silenced the enemy radios, and contact with Munda could not be made. The first warning the airfield defenders had of the invasion across the channel was a message flashed by lights and flares from a lookout station on a promontory south of the harbor. After inflicting nine casualties, including a face wound to the commander of the 172d Infantry, Colonel David N. M. Ross, the Japanese defenders finally fled the plantation area, leaving 65 dead behind.

The end of ground resistance marked the end of enemy efforts to dislodge the invasion force until the air fleets at Rabaul could get into action. General Sasaki at Munda could offer only slight opposition. Intermittent shellfire began to register in the transport area and around the destroyers shortly after the invasion was launched, but only the *Gwin* was hit. Two veterans of Guadalcanal sea action, the *Buchanan* and the *Farenholt*, took up the challenge and fired shells back at suspected points, meanwhile changing direction and speed so that the Japanese batteries could not register on them. It was estimated that seven guns were silenced, but the destroyer screen and the Japanese continued the sporadic exchange throughout the unloading activities.

It was a frustrating experience for the Munda defenders:

> Because of insufficient preparations and installations, our naval guns could not engage the enemy. Because of the range, the mountain guns were not able to fire against the enemy. Therefore, our unit was in the predicament that the enemy landed in broad daylight while our unit watched helplessly.[25]

While soldiers of the 103d and 172d pressed inland against spotty resistance, the establishment of a base of operations began in the continual downpour of rain. Unenthusiastic infantrymen were organized as working parties to sort and disperse the jumbled piles of ammunition, rations, lubricants, and other materials. The mushrooming dumps of supplies accentuated the fact that an insufficient beach control party and working party had been provided, and that too high a priority had been given to barracks bags,

[24] The original Rendova garrison consisted of 150 members of the *7th Company, 229th Regiment* augmented by 76 men from a *Kure 6th SNLF* signal detachment. In mid-June, 18 unarmed engineers moved to Rendova under orders to complete a torpedo boat base before the end of the month. Total defenders: 224. CIC SoPacFor Item No. 702, New Georgia DefOpO "A" No. 8, late Jun43; USAFISPA G-2 POW Interrogation Rept No. 105, dtd 4Aug43; CIC SoPacFor Item No. 632, 1st New Georgia Area ButaiO No. 3, dtd 5Aug43. The garrison had been weakened by malaria and other sicknesses, however, and probably numbered about 140.

[25] *Seventeenth Army Ops—I*, p. 15. Evidently Sasaki's naval guns and artillery, although unable to register accurately on the invasion fleet, fired at random hoping for a lucky hit.

officers' locker boxes, tents, chairs, and other personal comfort items.[26]

As the unloading continued, cargo-staging areas turned into seas of mud through which trucks churned and skidded. The road through the plantation area soon became a quagmire which caught and held all wheeled vehicles. Tractors were required to extricate them. Culverts, which had been judged strong enough to support heavy traffic, crumbled under the weight of loaded trucks and increased the difficulty of movement. Finally, only the wide-tread prime movers, the amphibian tractors of the 9th Defense Battalion, and the bigger tractors of the Seabees could plow through the mud. All other vehicles stalled, and infantrymen had to hand-carry most supplies to designated dumps, bivouac areas, and gun positions.

Tank lighters were unloaded by soldiers wading through 50 feet of knee-deep water. Later unloading proceeded faster after bulldozers pushed ramps of coral out to the lighters. Cargo was finally shunted to offshore islands in an effort to relieve the congestion, and, with virtually every truck mired down, a message was sent to the ships to delay sending in more vehicles. The mud, however, had convinced observers that future scouting of landing beaches would include engineer as well as tactical reconnaissance.

With the landing well underway, 32 fighter planes from ComAirSols appeared overhead, and troops on the beach and sailors on the ships breathed easier. Their concern was well-founded. Because of the poor weather, General Kenney's Fifth Air Force had been able to hit Rabaul with only 25 bombers in the 5 days prior to the landing, and the Japanese were still able to launch a powerful counterpunch through the air from Rabaul, Buin, Ballale, and Kahili airfields. It was not long in coming. The *Eleventh Air Fleet* at Rabaul dispatched a strike of 26 medium bombers and 8 carrier bombers shortly after dawn. Picking up a fighter escort of 72 planes at Bougainville, the flights swept down on Rendova. Intercepted by the Allied fighter cover, the enemy formations were forced away from the landing area, but in their reckless attempts to strike a crippling blow to the invasion, the Japanese lost 18 bombers and 31 fighters. Two hours of valuable unloading time, however, had been lost by the ships maneuvering to escape the enemy bombing runs.

At 1505, with all the troops unloaded and most of the supplies on the beach, Admiral Turner decided that the attack force had stretched its luck long enough and ordered the return to Guadalcanal. As the ships headed down Blanche Channel, a flight of about 50 Japanese fighters and torpedo bombers swung in over Munda Point and started bombing and strafing runs. The *Farenholt* dodged two torpedos before being bumped by a third—a dud; the *McCalla* was bracketed front and rear while a third torpedo plunged under the ship. The *McCawley* was not as lucky. A solid hit amidships opened a gaping hole, and Turner's flagship came to a dead halt. The admiral transferred his flag to the *Farenholt*, and the *Libra* took the *McCawley* under tow. After sur-

[26] Marine observers, in reports on the Rendova operation, were unanimous in reporting that too much personal gear was unloaded the first day, that it contributed to the confusion on the beach, and that infantrymen were fatigued sorting and carrying it through the mud.

viving another attack by 15 dive bombers, the *McCawley* continued to settle and was abandoned. That night, three more torpedos slammed into the transport and it sank. Believed the victim of an enemy submarine, the *McCawley* actually was sunk by an American MTB which had mistaken her for an enemy ship.

The day's air action cost the Japanese heavily. Determined to stop the invasion, the *Eleventh Air Fleet* flooded the skies with every type of plane available. Despite the waiting interceptors of ComAirSols, the Japanese plunged recklessly toward Rendova. Fighter protection for the bombers was insufficient, however, and each attack resulted in scores of flaming crashes. Claimed kills in the one morning and two afternoon raids totaled 101 enemy planes; Marine squadrons (VMF-121, -122, -213, and -221) reported downing 58 of them. The Allies lost 17 planes, but 8 pilots were fished out of the water by PBYs and torpedo boats. In addition, ComAirSols hit Vila with 16 torpedo bombers and 12 scout bombers in a morning strike, and then bombed Munda with an afternoon strike by 25 medium bombers, 18 scout bombers, and 18 torpedo bombers. These attacks further crippled Vila and Munda, and forced the Japanese to contest the Rendova landing without any close-in points for rearming and refueling.

The same false optimism which had given Admiral Yamamoto a distorted picture of the success of the April *I Go* operation prevailed, though, and surviving enemy pilots reported that they had sunk 2 destroyers and 1 cruiser, damaged 8 transports, set 2 destroyers afire, and downed 50 planes. Their own losses they set at 17 attack bombers and 13 fighters. Despite the seeming top-heavy score reported by the Japanese, they ruefully admitted that ". . . due to tenacious interference by enemy planes, a decisive blow could not be struck against the enemy landing convoy." [27]

That night, the Japanese hastily tried to assemble a strong raiding force in the Shortlands area for a counterlanding on Rendova, but only five of the destroyers made contact at the rendezvous area. Moving south around Vella Lavella, the force arrived off Rendova at about 0130 on 1 July. Ironically, the same rain squalls which resulted in more mud ashore reduced visibility to such an extent that the Japanese ships could not determine the debarkation point and were forced to withdraw.

The abortive naval raid climaxed a confusing day of action that saw many elements of the landing force fill roles never laid out for them in operation plans. Typically, Marines of the 9th Defense Battalion who went ashore early on 30 June to provide antiaircraft protection for the beachhead found themselves instead taking part in its seizure. The unexpected role as infantry was handled competently, and often eagerly, by the Marine gunners.

Prior to the operation, Colonel Scheyer had divided his battalion into four task groups. The special weapons group (Lieutenant Colonel Wright C. Taylor) was to land on 30 June and position eight 40mm weapons on one of the offshore islands, Kokorana. The 20mm guns and .50 caliber machine guns were to be used on Kokorana and Rendova for beach defense and protection for the antiaircraft weapons. The 90mm group, under the direction of Major Mark S. Adams, was to land one battery on 30 June on Kokorana

[27] *SE Area NavOps—II*, p. 29.

for immediate antiaircraft protection, with another two batteries to be landed and emplaced on Rendova on 1 July. Lieutenant Colonel Archie E. O'Neil, in command of the 155mm artillery group, was to land his big guns on the 1st and 2d of July to deliver neutralization fire on Munda airfield positions and to support the eventual assault on the airfield. The tank platoon, under First Lieutenant Robert W. Blake, was to land in later echelons and wait on Rendova for commitment in the final push on Munda.

Initial resistance by the Japanese did not delay execution of the 9th's missions. Quickly organizing the advance parties into combat patrols, the Marines secured Kokorana before starting the job of clearing firing areas for the 90mm battery. Some assistance in unloading was given by Seabees and late-arriving Barracudas. On the east beach of Rendova, Marines seeking possible gun positions frequently found themselves ahead of the front lines engaged in flushing snipers. One patrol of the 9th wiped out a machine gun nest during such a reconnaissance. For the 9th Defense Battalion, this was the first close contact with the enemy, and many Marines took the opportunity to turn infantrymen and help secure the island.

While the beach perimeter was being expanded, Marines selected spots for future battery positions, command posts, fire direction centers, and observation posts. Telephone lines were strung, and fields of fire for the big guns cleared by blasting down palm trees. By the end of the first day ashore, the advance elements of the 9th Defense Battalion were bivouacked on Rendova's beach and along the plantation road. Battery E of the 9th (90mm guns) was in position on Kokorana, and had fired its first shots against a low-flying enemy fighter at 1645. Twelve 40mm guns, eight 20mm guns, and eighteen .50 caliber machine guns were set up along the beaches on both islands, bolstering the defense positions. Only one small hitch had delayed the quick installation of the 90mm battery on Kokorana. The gun director was missing, and members of the battalion had to rummage through scattered piles of material on Rendova's shore until they found it.

The next day, 1 July, troops and supplies in the second echelon of the Western Landing Force began to arrive, and the four LSTs and five LCIs encountered the same unloading problems that had plagued the assault troops. The ships had to approach the island at slow speed, inching along through the shallow water until grounded by mud at considerable distance from shore. Vehicles which attempted to churn through to the beach became bogged and had to be rescued by tractors. The weight of heavy artillery pieces, towed ashore by tractors in tandem, further ruined the road along the beach, and, after the guns were manhandled into position, traffic of any kind over the road was impossible.

While the rain poured on, almost without cessation, most of the personnel ashore were pressed into service again as beach working parties to carry rations, fuel, ammunition, communication gear, and other supplies from the jumbled piles on the beach to dumps inland. Attempts to gain some measure of traction for vehicles in the soft underfooting met with failure. Seabees tried to corduroy the former road with 12-foot coconut logs, but the logs and steel matting they used soon sank under the mud. In addition, areas believed suitable for gun positions or bivouac areas became swamps, and dispersion of troops was almost impossible. Soldiers and Marines who attempted to dig foxholes morosely watched their efforts become sunken baths.

Despite the difficulties caused by the rain, by the end of the second day ashore, two guns of Battery A of the Marines' 155mm gun group were in permanent positions on Rendova and had test-fired several rounds at Munda. Battery B of the same group was ashore in a temporary position but had not fired. In addition, two 90mm batteries were in place and all special weapons dug-in nearby for protection. The Marines, unable to dig habitable positions in the mud, built above-ground shelters with coconut logs and sandbags.

Army artillerymen, taking positions on Kokorana, found the island had a solid coral subsurface that held the 155mm howitzers without difficulty. Moreover, since the island was open on the north side, no fields of fire had to be cleared. Soldiers of the 192d Field Artillery Battalion pushed their guns into position, took general aim at Munda some 13,000 yards distant, and began firing registration shots late the second day. While the Army artillerymen and Marines struggled with their heavy guns, ammunition, generators, and radar units, combat patrols of the 43d Division secured over half of Rendova. The stage was being set for the move to New Georgia.

Air activity during the second day was limited. The ComAirSols fighter cover over Rendova intercepted and fought off only one attempted enemy attack. The covering fighters also mounted guard over a strike by 28 torpedo and scout bombers at Vila, which further reduced that field to a nonoperational status. Before returning to Guadalcanal, each fighter plane worked over Munda defenses, strafing possible bivouac areas. General Mulcahy, assuming an active role in the operation, scheduled and directed the strike which helped American forces rout the enemy at Viru Harbor.

The third day ashore, 2 July, promised to be just as wet as the previous two days. While the 103d and 172d Infantry prepared for the move to New Georgia, the Marine 155mm guns and the Army 155mm howitzers continued firing registration missions on Munda airfield. Direct observation was used, with spotters clinging precariously to perches atop palm trees. As yet, no artillery fire control maps were available, so only area targets were selected. The 192d Field Artillery and the Marine group fired with impunity; fears that the Japanese could retaliate with counterbattery fire proved unfounded.

It was at this point, shortly after 1330, that the Japanese air commander at Rabaul, Admiral Kusaka, finally had his inning. His timing was perfect. The ComAirSols fighter cover had reluctantly been withdrawn under threat of bad weather, and the Japanese bombers arrived only a few minutes after the Allied fighters departed. An early-warning radar unit was temporarily out of operation, while its generator was drained of diesel oil mistakenly used in place of white gasoline.

The Japanese flight, variously estimated at from 18 to 25 medium bombers, swung in over the east side of Rendova Mountain, catching the troops in the open on the beach. A bombing pattern that stitched the beachhead from one end to the other quickly dispelled any illusion that these might be friendly planes. There was time only for a shouted, "Condition Red," before troops frantically sought cover. But

GUN CREW *of Battery C, 9th Defense Battalion is revealed in the flash of a 90mm shell fired at Japanese night raiders striking at New Georgia.* (SC 185876)

many were caught in the open, an extra dividend to the attractive target of ships, equipment, and supplies jammed into a restricted area. Many of the bomb salvos hit ration and fuel dumps; others exploded ammunition dumps. Highest casualties occurred among the Seabees concentrated on a promontory off the beach. A dynamite dump there was hit, its blast adding to the casualties of the bombing. The peninsula was promptly dubbed "Suicide Point." Further, the clearing station of the 43d Division was hit, which reduced the amount of assistance which could be given. Most of the victims were rushed to ships in the bay for treatment of wounds.

Because of the confusion, early estimates of the number of dead and wounded varied widely. Some men were reported missing, either killed by exploding ammunition or direct hits, or, more probably, removed to ships and hurried to Guadalcanal for treatment. In all, 64 men were killed and another 89 wounded. Seabees in the boat pool and soldiers in the 43d Division bivouac areas sustained the heaviest casualties. In spite of the congestion, damage to materiel on the beach was relatively light. Besides the ammunition and fuel dumps hit, two of the 155mm guns were scarred by bomb fragments, two 40mm guns were damaged, and three amphibian tractors were holed. All were repairable, though, with the exception of one of the tractors.

The attack's success was the result of many factors. For one thing, Army radar units had gone out of commission shortly after landing, and although a Marine radar unit had been landed on 1 July, it was this one that was being drained of diesel fuel. Also, on the day previous, the troops had believed a flight of American medium bombers to be enemy planes and had scrambled for cover. This day they believed the enemy planes to be the same mediums back on station. A third factor was lack of dispersion. Shelters had been dug along the beach, but the troops were now busy handling other materiel, and had not provided other protection. But as a result of the raid, the area became dotted with foxholes—deep foxholes.

By 3 July, the routine of operations ashore was established. Troops of the 43d Division began the shuttle to Zanana Beach on New Georgia, and the big guns of the 9th Defense Battalion and the 192d Field Artillery picked at Munda's defenses, seeking for a hidden strong point, a bivouac area, or a supply or ammunition dump. A 130-foot coast artillery observation tower of 1½-inch angle iron made spotting easier than viewing from a swaying palm tree. Erected on high ground about 200 yards back of General Hester's command post at the foot of Rendova Mountain, the tower provided a central point from which Marine and Army spotters could radio corrections to the artillery fire direction center and then observe the strike of the shells on Munda airfield and its bordering hills across the channel, and on the nearby islands off New Georgia's shore. In time, a system was developed whereby films dropped near the tower by photographic planes were immediately picked up, developed, and then studied for assessment of damage to Munda defenses.

On the night of 3 July, the enemy attempted to follow up its devasting strike of the 2d with an attack from the sea. A Japanese naval force suddenly appeared offshore and spattered the Rendova beachhead area with a bombardment

691-360 O—63——7

which did little or no damage. Allied destroyers and torpedo boats forced the enemy ships to withdraw hastily without accomplishing the hoped-for crippling blow to the invasion troops.

As following echelons of the Western Landing Force unloaded on 4 July, a desperate Japanese command at Rabaul tried once more to knock the invasion force off Rendova. Since the air attack on 2 July represented the only measure of success in their efforts so far, the Japanese repeated the act. The cast and the script remained the same, except for the final curtain. This time the Japanese found themselves holding the wrong end of a Fourth of July Roman candle. From a force of more than 100 planes trying to press home on attack through the ring of Allied interceptor planes, only 16 bombers were able to swing over Rendova Mountain in the repeat performance. But this time, alerted by sound locators and radar, the 9th Defense Battalion antiaircraft batteries were ready, and 12 of the 16 bombers and an escorting fighter were knocked down in flames. The 90mm guns expended a total of only 88 rounds, a feat which the Marines jubilantly proclaimed a record for rounds per plane.

This attack on Rendova was the last daylight assault on the island of any size made by the Japanese air fleet. From this point on, the attacks were made at night. Although the ComAirSols fighter cover still maintained a vigil over Rendova, the focus of the air war shifted to New Georgia as the troops shuttled from the beachhead at Rendova to the beachhead at Zanana. There the second phase of Operation TOENAILS was to begin.

Munda Victory

ASHORE AT ZANANA BEACH [1]

The Allied landings in the Central Solomons and the New Guinea area caused Japanese planners some anxious moments. Plainly, the situation called for prompt action to relieve the pressure on the first defensive lines of Japan's war-flung empire, but the question was: Where should the major effort be directed? To date, all attempts to repulse the landings had proved futile, and prospects for future success didn't look too promising, either. Mindful of earlier basic plans to retain the Central Solomons while holding out in New Guinea, the Japanese commanders at Rabaul scheduled a conference for 4 July to reach a decision.

To General Sasaki and Admiral Ota, ruefully watching the Rendova operations from a well-protected headquarters on Kokengola Hill at Munda airfield, the situation was a bit more pressing and a lot more personal. From observation it was apparent that the troops across the channel had come to stay and were building up for an offensive in strength. When 155mm guns and howitzers began to register on the airfield, the pattern of the campaign became all too clear. Munda was going to need reinforcements, and quickly, if it was to be held.

The two commanders reported their appraisal of the situation, and then took steps to strengthen the airfield defenses as best they could with the troops available. In a series of orders signed jointly by Sasaki and Ota, all eastern New Georgia lookout detachments were recalled on 30 June, and two recently arrived 140mm guns and two smaller mountain guns were ordered rushed overland from Bairoko. In addition, a reserve force, the *12th Company, 229th Regiment*, was alerted to move from Kolombangara to New Georgia.

As the Allied buildup on Rendova continued, however, these defensive measures began to look woefully weak, so the remainder of the *3d Battalion, 229th Regiment*, was ordered to Munda's aid. By Sasaki's own estimate, all defenses must be ready by dusk on 3 July. Meanwhile, the combined Army and SNLF units were exhorted to "maintain alerted conditions throughout the night and guard against enemy landings; if the enemy commences

[1] Unless otherwise noted, the material in this section is derived from: *ComSoPac Jul43 WarD;* ComThirdFlt Narrative Rept, SoPac Campaign, dtd 3Sep44, hereafter cited as *ThirdFlt Rept;* NGOF (XIV Corps) Narrative Account of the Campaigns in the New Georgia Group, B.S.I., n.d., hereafter *NGOF Account;* History of 43d Infantry Division, World War II, 24Feb41–Jun44, n.d., hereafter *43d InfDivHist;* 43d InfDiv FO Nos. 1–17, 16Jun–25Aug43; *9th DefBn OpsRept;* Combat Infantry, Part Eight: New Georgia, dtd 31Aug44, hereafter *New Georgia Combat; New Georgia Campaign;* ONI, *Combat Narrative X;* Rentz, *Marines in the Central Solomons;* Joseph A. Zimmer, *The History of the 43d Infantry Division, 1941–1945* (Baton Rouge, La.: Army and Navy Publishing Co., 1947), hereafter Zimmer, *43d's History*.

to land, destroy them at the water's edge." On 2 July, the command relationship was changed. Sasaki, as the senior officer, "in response to the conditions in this area," assumed sole command of all New Georgia garrisons. Admiral Ota, relieved of his landing forces, was assigned control of Army and Navy barge and shipping units in the area.[2]

The actual invasion of western New Georgia was not the direct assault on Munda airfield which Sasaki and Ota believed was coming. Instead, in a landing on 30 June which actually preceded the Rendova assault by several hours, soldiers of Companies A and B, 169th Infantry scrambled ashore on the islands that guarded the Onaiavisi Entrance to Roviana Lagoon. Lashed by heavy rain squalls and hampered by the darkness, the soldiers nevertheless managed to make contact with a waiting pre-D-Day amphibious patrol and native scouts. The landing was unopposed, but not uneventful. The mine sweeper *Zane*, which had been used as a transport, went aground on a small island just inside the entrance, and lay exposed as a telltale marker. Her helpless state and the landing area were ignored, however, by Japanese planes striking at the Rendova landing. The ocean tug *Rail*, summoned from Guadalcanal, pulled the *Zane* off the reef late that afternoon.

After securing the entrance islands, the soldiers began the move to Zanana Beach on the shore line of New Georgia. Earlier plans had called for Company O of the 4th Marine Raider Battalion to act as scouts for this phase of the operation, but with the raiders still at Segi and Viru, reconnaissance teams from the Rendova forces were organized. These were later augmented by a company of Fijian and Tonganese scouts, who were aggressive and skilled jungle fighters.[3] (See Map 5.)

The patrols moved into the area between Zanana and the Barike River, marking water points, trails, coastal roads, possible artillery positions, and all avenues of approach to Munda. They were also ordered to probe Japanese defenses between the airfield and Bairoko Harbor, and to report all barge activity observed. One of the first radioed messages from the patrols reported a successful ambush of a Japanese group and that uniform markings on a dead enemy rifleman indicated that he had been a member of the *229th Regiment*. The ambushed Japanese had been part of the *5th Company, 2d Battalion*, which had been ordered to investigate the Onaiavisi Entrance landings and "drive out the enemy who has landed there and make the area secure."[4] Later the *5th Company* was told to resist stubbornly against this new phase of landings and fight to the last at their present positions. These instructions set the pattern for Japanese resistance in New Georgia.

General Hester received Admiral Halsey's approval to proceed with the New Georgia phase of TOENAILS on 2 July. That night, elements of the 172d's 1st Battalion began the move from Rendova to Zanana Beach. The troop transfer was

[2] CIC SoPacFor Item No. 702, dtd 13Sep43, New Georgia DefOpO A Nos. 11-15, 30Jun-1Jul43.

[3] R. A. Howlett, *The History of the Fiji Military Forces, 1939-1945* (Christchurch, N.Z.: Whitecombe and Toombs, Ltd., 1948), p. 5.

[4] CIC SoPacFor Item No. 705, dtd 9Sep43, New Georgia Defense Butai, 2d Bn Order A No. 141, dtd 30Jun43.

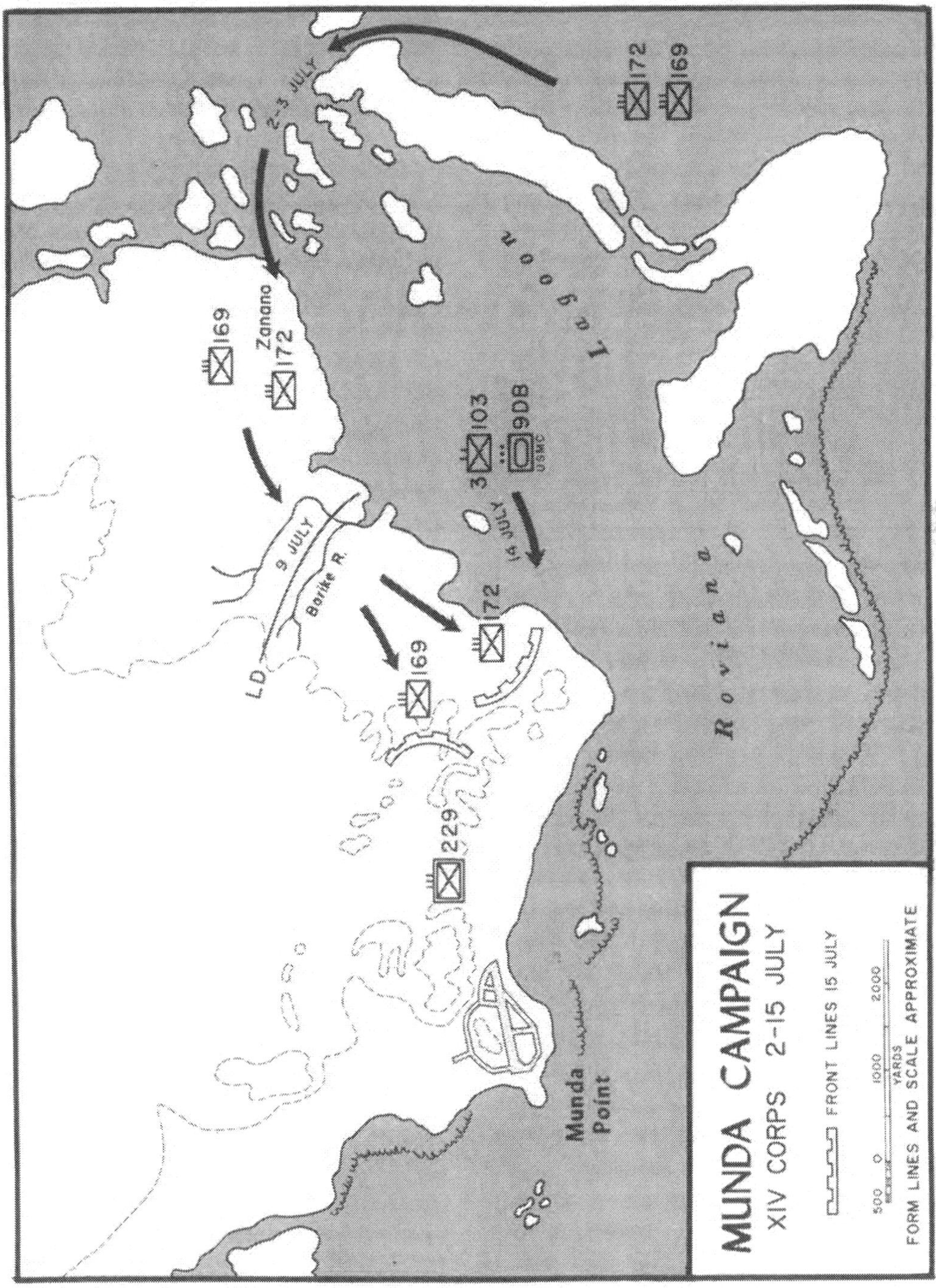
MUNDA CAMPAIGN
XIV CORPS 2-15 JULY
FRONT LINES 15 JULY
500 0 1000 2000
YARDS
FORM LINES AND SCALE APPROXIMATE
Munda Point
Roviana Lagoon
Zanano
Barike R.
9 JULY
LD
2-3 JULY
14 JULY
169
172
103
3
9DB
USMC
72
229

MAP 5

made in landing craft, which towed additional rubber boats carrying soldiers. Torpedo boats furnished an escort across Blanche Channel, and, at Onaiavisi Entrance, native guides in canoes took over and directed the landing craft through the lagoon to the beachhead. The following day, 3 July, Brigadier General Wing established the 43d Division's forward command post (CP) on New Georgia. A 52-man detail from the 9th Defense Battalion's special weapons group arrived on 4 July and immediately emplaced four 40-mm guns for antiaircraft protection. Four .50 caliber machine guns were sited to protect the antiaircraft positions and to add depth to the firepower of the soldiers.

The Japanese air attack of 4 July at Rendova managed to make targets out of most of the troops that were to participate in the push on Munda airfield. The 172d was still in the process of shuttling troops to Zanana Beach; and the fifth echelon of the NGOF, the remainder of the 169th Infantry and the 136th Field Artillery Battalion carried in 14 LCIs and 4 LSTs, had just arrived at Rendova Harbor. The 169th had remained in the Russells as division reserve during the early part of the operation, and the 136th was detached from the 37th Division on Guadalcanal. The air attack hit as the 169th and 136th were debarking. Unloading activities were abruptly abandoned. Luckily, no ships were hit. But for the new arrivals, the bombing attack following a sea-tossed trip from the Russells was a rough welcome to New Georgia.

Transferring their equipment and supplies to small craft from the Rendova boat pool, the soldiers began the movement to Zanana almost immediately. The 155mm howitzers of the 136th were unloaded on one of the islands guarding Onaiavisi Entrance and positioned to provide artillery support to the troops attacking Munda. Other heavy weapons, the 105mm howitzers of the 169th and 103d Field Artillery Battalions were also emplaced on the off-lying islands for additional fire support. By dusk of 5 July, the 172d and the 169th Infantry were ashore on New Georgia, ready to begin the march toward the line of departure along the Barike River. A secondary landing, early on the morning of 5 July by the Northern Landing Group (NLG), commanded by Marine Colonel Liversedge, established a beachhead at Rice Anchorage on the north coast of New Georgia to threaten Sasaki's forces from that direction.[5]

On the 6th, the 172d moved west toward the Barike. Little opposition was encountered. The next day, however, as the 169th Infantry began its move to positions north of the 172d, determined enemy opposition decisively stalled the entire regiment. Stopped short of the Barike, the 169th went into bivouac.

Accounts of the action during the night of 6 July combine fact and fancy. Reports that Japanese riflemen had infiltrated the loose perimeter set up by the 169th's leading battalion caused a panic among the soldiers. Although the regiment had been on Guadalcanal and the Russells prior to New Georgia, the troops evidently were not prepared for jungle combat at night. Soldiers reported the next morning that enemy infiltrators threw grenades, screamed, whistled, shouted invective, and jumped into foxholes to bayo-

[5] The account of the Rice Anchorage landing is related in the following chapter.

net the occupants. After a wild night of grenade bursts, shooting, and screaming, however, no enemy dead were found in the perimeter when dawn came and the soldiers were able to look around. But NGOF casualties were numerous.

The action on the night of 6 July, which started a wave of near hysteria among the troops, seriously impaired the combat efficiency of the 169th Infantry. Despite many later aggressive and determined attacks, the 169th's initial failures along the Barike River were attributed to an apparent lack of combat conditioning and training.[6]

Regardless of speculation as to whether such night attacks were wholly real or in part imagined, there was no denying the end results—the loss of many front-line troops through actual wounds and war neurosis. Later all regiments in the attack were subjected to this type of enemy tactics. In defense against such raids, 43d Division soldiers adopted a policy of joint foxholes for two or more men protected by trip wires with noise makers attached. In addition, a rigid fire plan was adopted which prohibited promiscuous shooting and movement at night and allowed only the outside perimeter to fire or use grenades. These defensive measures restored discipline and stability.

After delaying most of the morning of 7 July in reorganization, the 169th resumed its push toward the Barike. Again the regiment was stopped almost immediately by aggressive enemy resistance. Although the 169th managed to overcome this first enemy opposition, the soldiers had to fight another lengthy action before reaching the low hills east of the river. The 172d, in its zone of action, had pushed to the Barike without too much trouble. When it became apparent that the 169th could not reach the Barike River in time to begin the attack on 8 July as planned, General Hester—with Halsey's approval—ordered the operation delayed one day. The NGOF commander also cancelled that part of his plan that called for a direct assault on the airfield over Munda bar by a battalion of the 103d Infantry with Marine 9th Defense Battalion tanks in support. Mounting evidence that the Japanese held the area in great strength dimmed the prospects for the success of such a thrust.

After another night of infiltrators' attacks, during which soldiers crouched sleepless in foxholes, the advance was resumed the next morning. The 172d moved fairly easily along a coastal trail in a column of battalions. The 169th, struggling through the jungle with an open flank screened only by the Fiji scout company, was echeloned to the right rear. A heavy concentration of mortar and artillery fire on the Japanese position to the immediate front of the 169th broke resistance there; and, aided by a flanking attack by the 172d hitting from the left, the 169th was able to push ahead. Late in the afternoon of the 8th, the fatigued 169th struggled into position on line with the 172d to start the drive toward Munda the following morning.

[6] Commenting on this phase of the campaign, Admiral Halsey said: "The [169th] regiment sent 300 men back to Guadalcanal as 'war nerves' casualties after one day's fighting. General Harmon met them there, promptly returned 300 of them to the combat zone. . . ." Halsey and Bryan, *Halsey's Story*, p. 161. See also Miller, *Reduction of Rabaul*, pp. 108–109, for a description of this action.

NGOF IN ATTACK: ZANANA TO LAIANA[7]

Booming salvos from four destroyers at 0512 on the 9th of July signalled the start of the NGOF attack. The one-hour naval bombardment, which dumped 2,344 five-inch shells on positions in the rear of the enemy lines, was followed by a cannonade by all artillery battalions of the NGOF. The shelling combined the fires of two 155mm howitzer battalions, one 155mm gun battalion, and two 105mm howitzer battalions. In all, the Munda-Barike area was battered by 5,800 rounds of high explosives. Enemy defensive positions, lines of communication, bivouac areas, and command posts were blasted for one hour before the fires were shifted to the area to be assaulted by the ground troops. As artillery lifted, 52 Navy and Marine torpedo bombers and 36 scout bombers struck, dropping high explosive and fragmentation bombs on the area. At 0900, heartened by this extreme concentration of firepower, the 43d Division started its attack toward the NGOF objective—Munda airfield.

After clearing the initial Japanese resistance, the advancing soldiers encountered only snipers and small outposts. Progress, however, was slow. Each new enemy opposition forced deployment and attack. Hidden snipers, pinning down the advance units, held up the regiments for hours. Every step forward was a struggle against a determined enemy and multiple jungle obstacles—dense, vine-choked underbrush, steep ridges, numerous swamps, constant and enervating heat, and almost incessant torrents of rain.

The only maps provided the attacking force were sketches based on aerial photos. The drawings outlined jungle areas with conventional symbols which did not reveal the intricate, abrupt mass of hills, ridges, and swamps—jumbled without pattern—that lay under the thick jungle canopy. Contour lines on the maps were based on scouting reports, and, as 43d Division soldiers discovered, were usually in error. The ridges and hills, bending and twisting in all directions, forced the attacking units to move in one direction, then another. As a result, by the end of the second day of the attack, both regiments had become intermingled and were attacking in virtually a single column. The initial frontage of 1,300 yards had collapsed to almost one-half that distance. In addition, the lines of communication and supply were now stretched over two miles through the jungle from Zanana Beach, an extension exceedingly vulnerable to counterattack from the north, or right, flank.

For the 169th, the advance had been particularly harrowing. Given a zone of action that forced them to cross the meandering Barike River a number of times, the soldiers slowly pressed forward over the steep ridges and through the deep swamps in the upper river region. Fatigued from the initial struggle through the jungle from Zanana, and continually harassed at night by enemy soldiers probing at the exposed right flank, the 169th was a dispirited outfit. After such a

[7] Unless otherwise noted, the material in this section is derived from: *ThirdFlt Rept;* CTF 31 SAR, Rept of Bombardment of Munda Point, dtd 9Aug43; ComAir New Georgia SAR, 29Jun–13 Aug43, dtd *c.* 1Jan44; *New Georgia Combat; New Georgia Campaign; NGOF Account; 43d InfDivHist*; 9th DefBn Rept of TkOps in the New Georgia Area, dtd 24Aug43, hereafter *9th DefBn TkOps;* ONI, *Combat Narratives X;* Rentz, *Marines in the Central Solomons;* Zimmer, *43d's History.*

disappointing start, the regiment mustered only lethargic attacks against enemy opposition. Wounded soldiers and combat fatigue cases wandered back along the trail to Zanana, draining the front lines of needed strength and creating a serious evacuation problem. Additionally, with the regiment so strung out, troops were needed to carry food, water, and ammunition to the attackers as well as help evacuate the wounded, tasks which further sapped the fighting strength of the outfit.

The pattern of enemy resistance developed by the end of the second day of attack, 10 July, plainly indicated that the Japanese were holding a barrier position in the high ground east of Munda airfield which they would defend in strength. The NGOF offensive—grinding against this line of mutually supporting fortifications of logs and coral, strongly defended by automatic weapons, mortars, and artillery—faltered.

As the NGOF struggled against the jungle and a tenacious enemy, engineers attempted to established a supply route to the front lines by hewing a jeep road out of the matted underbrush. Native guides pointed out a trail which took advantage of as much high ground as possible, but most of the route had to follow the marshy banks of the Barike River and in some instances ran parallel to the front lines. Bridging of the Barike was accomplished in several spots by trestles made of felled timber. Even while constructing the road in the rear of the front lines, however, the engineers were under almost constant attack from bypassed snipers and wandering squads of enemy. Bulldozer operators were a prime target, and engineer casualties mounted as the road clearing proceeded. Metal shields were eventually welded to the tractors to protect the 'dozer operators. Since no heavy graders were available, the jeep road could not be ditched or crowned, and any traffic over the road after a rainstorm usually meant extensive road repairs.[8]

With the need for a closer reinforcing and resupply point made obvious by conditions to the rear of the NGOF front, Hester's staff focused attention on Laiana Beach. Rejected earlier as a landing site because it was deemed too heavily defended and too inaccessible for quick resupply, Laiana now appeared to be the answer to NGOF logistic problems. The beach was some 5,000 yards closer to Munda, and its possession would shorten supply, evacuation, and reinforcement lines as well as put fresh attacking troops considerably closer to the main objective. On 11 July, General Hester ordered the 172d to disengage from the frontal assault and pivot southwest in an attack toward the coast line to secure Laiana Beach. At the same time, the NGOF commander alerted the 3d Battalion, 103d Infantry and the tank platoon of the 9th Defense Battalion to be ready to leave Rendova for Laiana as soon as the 172d reached the coast.

Though the 172d was only a short distance northeast of the beach when directed to attack, the area was not secured until 13 July. Despite near-constant artillery assistance which shredded and blasted the jungle covering from defenses on the sharp hills between the 172d and Laiana, the enemy clung stubbornly to his positions. Repeated air strikes failed to dent the defenses, and the Japanese, apparently aware of NGOF intentions, rained mortar and artillery fire between the 172d and its

[8] *McAllister ltr.*

AVENGER TORPEDO BOMBERS *wing toward New Georgia on 9 July 1943 to strike Munda airfield in support of the 43d Division's attack.* (*USMC 57685*)

MARINE LIGHT TANK, *accompanied by Army infantrymen, moves through the jungle toward the front lines on New Georgia.* (*SC 395877*)

objective. Marine tanks and the 103d Infantry Battalion, scheduled to land on the 12th, were held back. The 172d reached Laiana on the 13th, and, on the following day, landing craft and tank lighters carried the reinforcements ashore. Artillery smoke shells covered the landing activities. Although the infantry hit the shore line without incident, enemy 75mm guns hidden in the jungle fired random shots at the lighters. No hits were scored, and all tanks were put ashore without damage. From his headquarters at Munda, General Sasaki observed the smoke screening this new development; but in his orders for the 14th of July, he erroneously reported that 70 large barges had attempted to land but had been repulsed with the loss of 15 of the barges.[9]

While the 172d held the new beachhead area and waited for the 169th to close the gap between the two regiments and come abreast, the Marine tanks and 3/103 moved into division reserve. A special weapons detail from the 9th Defense Battalion accompanied the infantry to Laiana and set up 40mm, 20mm, and .50 caliber antiaircraft weapons for protection against Japanese strafing and bombing attacks.

In the 169th's zone, strong mortar and artillery fires were placed on Japanese defensive positions in an effort to reestablish forward movement, but the enemy resistance continued. At this time, the regiment—tired and understrength—was opposed by a determined, dug-in enemy to the front and continually harassed by snipers and infiltrators in the rear areas. On the 11th, the 169th's commanding officer and his staff were relieved by Colonel Temple Holland and a staff from the 145th Infantry, 37th Division. The new regimental commander postponed further attacks by the 169th until the next day so that he might have time to reorganize his command.

A new push by the 169th on the 12th, following a rolling artillery barrage, failed to gain ground, however, and a return was made to the line of departure. The following morning, 1,000-pound bombs dropped by 12 scout bombers of ComAir New Georgia further hammered the defenses holding up the 169th's progress. Pilots returning from the strike noted that the target area marked by smoke shells was 600 yards east of the grid coordinates given in the air mission request, an indication of the difficulties the 169th was experiencing in locating its position on the ground. The whole regiment was committed to the attack after the air strike, but only the 3d Battalion on the left managed to gain ground. Successful in seizing the crest of a small knoll about 600 yards to the front, the battalion hung grimly to its position and repelled several strong counterattacks. During the next two days, the 3d Battalion took 101 casualties, dead and wounded. Despite strong enemy pressure, the infantrymen held their position. Barrages fired by supporting artillery units boxed the front and flanks of the salient, and discouraged the development of a large-scale Japanese counterattack.

In an effort to aid the beleaguered 3d Battalion, the 1st Battalion attacked on the 15th toward a dominating rise of ground about 400 yards to its right front. When opposition failed to develop, the attackers clambered to the top of the ridge, only to find deserted pillboxes, abandoned fox-

[9] CIC SoPacFor Item No. 702, dtd 13Sep43, New Georgia DefOpO A No. 36, dtd 14Jul43.

holes, and empty trenches. The Japanese defenders had finally withdrawn.

The victory lifted the spirits of the entire regiment, but more heartening was a glimpse of the NGOF's ultimate objective—Munda airfield. On its coral white runways and taxiways some three miles away could be seen wrecked and burned enemy planes. With new vigor, the 169th took over the enemy positions and prepared to defend the newly won ridgeline.

COUNTERATTACK PREPARATIONS [10]

While General Hester's NGOF fought its way from the Barike to Laiana, General Sasaki's defenders were operating on the simple strategy of trading space for time. Considerably outnumbered, the *229th Regiment* and *8th CSNLF* had nevertheless forced the invading American division to move slowly and cautiously. Sasaki's defensive lines had reduced the NGOF invasion to a groping, stumbling advance—much in contrast to the swift, hard-hitting operation envisaged earlier by the Americans. The Japanese played for time during which reinforcements could arrive.

The plight of the Munda defenders had received immediate attention. General Imamura, commanding the *Eighth Area Army* at Rabaul, on 3 July ordered the New Georgia defense augmented by the remainder of the *13th Regiment* as well as by additional antitank, mountain artillery, engineer, and medical units. In addition, the rear echelons of the *229th Regiment*, which were still in the Shortlands area, were ordered to join their parent unit. A number of large landing barges were also dispatched to New Georgia. Most of the fresh troops were to stop at Kolombangara, but the elements of the *229th*, the antitank units, and most of the engineers were to go directly to Munda.[11] In all, Imamura ordered about 3,000 troops from the Shortlands-Faisi area to the New Georgia Group. More reinforcements were to follow. The joint Army-Navy conference at Rabaul, on 4 July, cemented the understanding between the *Eighth Area Army* and the *Southeast Area Fleet* that the main sea and air effort would be directed against the Central Solomons while the troops already on New Guinea would hold out without additional help for the time being.

Imamura's promised reinforcements started to New Georgia on schedule, but the transports bumped into an Allied destroyer force lurking in Kula Gulf and turned back to the Shortlands to await a better time. The next night, 5–6 July, the transports sailed again, and, although part of the force was ambushed by Allied ships, the Japanese managed to land about 850 troops on Kolombangara.[12] On New Georgia, General Sasaki shoved all available *229th Regiment*, *8th CSNLF*, and *38th Division* support troops into the defense of the airfield in an attempt to hold out as long as possible. His line of forti-

[10] Unless otherwise noted, the material in this section is derived from: *New Georgia Campaign;* CIC SoPacFor Item No. 702, dtd 13Sep43, New Georgia DefOpO A Nos. 11–50, 30Jun–23Jul43; *SE Area NavOps—I; SE Area NavOps—II;* Rentz, *Marines in the Central Solomons.*

[11] CIC SoPacFor Item No. 740, dtd 23Sep43, Go Area (Eighth Area Army) OpO No. 35, dtd 3Jul43, and Item No. 838, dtd 11Nov43, Oki (Seventeenth Army) Group OpO No. 270, dtd 4Jul43.

[12] This sea encounter, known as the Battle of Kula Gulf, will be related in connection with the Rice Anchorage landing in the following chapter.

fications, spiked with seacoast and dual-purpose guns, ringed the coastline of Munda Point for some 6,500 yards and then swung inland from Roviana Lagoon for almost 3,000 yards. As NGOF troops were to find out, it was a formidable area to crack.

Sasaki's tactics in the defense of the terrain between the Barike River and his main positions around the airfield were to counterattack continually in the hopes of offsetting any gain which the NGOF might make. Skillfully deploying the forces available, his field commanders ordered one company to hold and threaten a flank of the Allied line while other units slipped to the rear of the attackers to raid and cut communications. This infiltration had the calculated two-fold effect of creating casualties and demoralizing the attacking force. In instances where it became necessary to hold a particular strong point, an ambush squad with orders to fight to the death was left in position.

While part of the Munda defense force wrestled with the advancing Allied units, other engineers and soldiers feverishly built pillboxes, dug trenches, and cleared lanes of fire in defensive lines to the rear. Each time the Japanese gave ground, they fell back to another strong position. Well-camouflaged and protected, the barrier of mutually-supporting positions allowed Sasaki's troops to contest any advance stubbornly. The terrain was an ally, since it hid the Japanese defenses and forced the Allied attackers to battle against the jungle and enemy troops simultaneously. Sasaki had another advantage, too. He was close to Bougainville and the Shortlands, and although reinforcements—mainly machine gun, antitank, and artillery units—dribbled into New Georgia in an unsteady stream, his strength remained nearly constant. Troops from Kolombangara, transported to Munda by barges during the night, were at the front lines the next day.

With the Allied lines inching slowly toward Munda, the Japanese were aware that the only means of re-establishing any type of order in New Georgia depended upon a strong counterattack. Weighing the time element against the danger, the Japanese decided on a delaying action in the Munda area while a counterattacking force struck through the upper Barike River region. As reinforcements arrived at Kolombangara, this counteroffensive was kept in mind. The ground attack would be staged simultaneously with a sea campaign, which would cut Allied supply lines while the air fleets pounded the Allied lines and rear areas on New Georgia.[13]

The *13th Regiment*, which had moved in parcels from the Shortlands, was selected to straighten the lines in New Georgia. On 8 July, Colonel Tomonari was alerted to send the *2d Battalion* to Bairoko Harbor to help Commander Saburo Okumura's *Kure 6th SNLF* defend that area from another but smaller Allied landing force. At the same time, Tomonari was to relinquish command of Kolombangara's defenses to the commander of the *Yokosuka 7th SNLF* and with the remaining two battalions of the *13th Regiment* advance to Munda for the new attack.[14] Okumura, at Bairoko, was to cover the *13th*'s advance from Kolombangara and then defend the Bairoko area without further assistance. Sasaki's orders

---

[13] CIC SoPacFor Item No. 730, dtd 23Sep43. SE Area ForOpO No. 10, dtd 18Jul43.

[14] CIC SoPacFor Item No. 676, dtd 30Aug43. Kolombangara DefOpO No. 6, dtd 8Jul43.

to Tomonari were for the counterattacking force to move to a bivouac area on a plantation about five miles north of Munda. The *13th* was to remain there until Sasaki deemed that the time was opportune for the attack.

To ensure that the operation would go smoothly, Sasaki established a liaison post at the plantation area and then sent a guide to meet Tomonari at Bairoko. Plans proceeded without a hitch as the first echelon of about 1,300 men moved by barge to Bairoko on 9 July. On the 11th, another 1,200 troops moved across Kula Gulf and a further 1,200 men made the cross-channel journey by barge on the night of the 12th. The movements were postponed several days by naval action in the gulf, but just as soon as they were able to make the crossing, all units of Colonel Tomonari's attacking force, mainly the *1st* and *2d Battalions*, assembled at Bairoko.

In moving into the bivouac area, Tomonari's force abruptly ran into a trail block set up by part of Colonel Liversedge's Northern Landing Group. In a brief but sharp encounter, the American force scattered the *13th Regiment's* leading elements, and reported to Liversedge that a large movement of Japanese reinforcements had been prevented from reaching Munda. Actually, Tomonari had broken off the engagement so as not to disclose the impending counterattack. Instead of staying to slug it out with the NLG, Tomonari withdrew his two battalions, and Sasaki's guides then led the Japanese soldiers toward Munda over another trail. By the morning of the 13th, Tomonari's main elements were at the plantation assembly area.

With two regiments now in position to oppose the landing force hitting toward Munda on the south, Sasaki was confident of his ability to reclaim the initiative. Some of his optimism could have been used by his superiors, however, because Army-Navy disagreements were stalling the progress of further help in the airfield's defense. The Navy, seeking the commitment of an additional Army division in New Georgia, wanted reassurance that Navy installations in Bougainville, the Shortlands, and Rabaul would be protected. The Navy suggested a possible 2,000 troops for the Rice Anchorage area, 3,000 more for Munda airfield, another 2,000 to take over the Roviana Lagoon islands, and an additional 4,000 to be used as an attacking force.

The Army turned thumbs down. The *Eighth Area Army* had no intention of further reinforcing the New Georgia area. To Army planners, there was no way in which the war situation could be altered, and, as a matter of fact, a reappraisal of the situation had convinced them that Bougainville could not be held long if the Allies attacked there. While this difference of opinion existed, General Sasaki would have to make do with the *Southeast Detachment Forces* already at hand and those few scattered rear echelon and support troops which destroyer-transports could rush to Kolombangara for barge transfer to New Georgia.

MARINE TANKS VS. PILLBOXES [15]

The occupation force's struggle to advance on New Georgia was anxiously

[15] Unless otherwise noted, the material in this section is derived from: *ComSoPac Jul43 WarD; ThirdFlt Rept; 43d InfDivHist; NGOF Account; 9th DefBn TkOps; New Georgia Campaign*: Halsey and Bryan, *Halsey's Story*; Rentz, *Marines in the Central Solomons*; Zimmer, *43d's History*.

watched by the remainder of the NGOF on Rendova and the barrier islands. Artillerymen, executing fire missions, noted that front lines did not move forward. Landing craft coxswains, returning from supply runs to Zanana and Laiana beaches, brought back reports of the fighting and distorted stories of the Japanese infiltration raids. All NGOF units knew that the 172d was stalled in the hills west of Laiana and that the 169th was understrength and fatigued by the struggle through the jungle. Despite the continual and intense pounding by three 155mm and three 105mm gun and howitzer battalions, which seemed to have leveled all aboveground installations, the enemy still seemed as strong as ever and apparently as disposed to continue the fight. Air strikes, which included as many as 70 planes, bombed the enemy defenses without apparent results except to strip foliage from the jungle.

Realization that more Allied troops would be required had come early in the campaign. On 6 July, General Hester had requested, and had been granted, the use of the 148th Infantry (less one battalion with the NLG) as division reserve. The 145th Infantry (also less one battalion with the NLG) was additionally attached to Hester's NGOF. Both regiments were alerted for possible commitment to combat and, prior to 14 July, were moved to Rendova where they would be readily available.

With the addition of two regiments as NGOF reserve, a needed change in the command structure became more apparent. For some time, observers had believed that General Hester's 43d Division staff, split between the two tasks of directing a division in combat and a larger occupation force in a campaign, had been unequal to the job. Moreover, on the 13th, General Griswold of the XIV Corps had some disquieting reports for Admiral Halsey and General Harmon:

> From an observer viewpoint, things are going badly. Forty-three division about to fold up. My opinion is that they will never take Munda. Enemy resistance to date not great. My advice is to set up twenty-fifth division to act with what is left of thirty-seventh division if this operation is to be successful.[16]

Halsey, on 9 July, had directed Harmon to name a corps commander to take command of all ground troops on New Georgia. Now, after Griswold's first-hand report from the front lines, Halsey told Harmon to take whatever steps he thought necessary to straighten out the situation. Griswold and his XIV Corps staff was ordered to assume command of the NGOF and Hester was returned to the command solely of the 43d Division.[17] All ground forces, including those of the 37th Division, now in the NGOF, as well as the 161st Regiment from the 25th Division, were assigned to Griswold's command. The new NGOF leader, requesting a few days for reorganization, promised a prompt, coordinated attack. The command change was effective at midnight, 14 July, a date which happened to coincide with the long-planned relief of Rear Ad-

[16] Griswold disp to Harmon, dtd 13Jul43, quoted in *New Georgia Campaign*, p. III-39.

[17] Among reasons he later cited for recommending the shift in command, General Harmon noted that Admiral Turner "was inclined more and more to take active control of land operations." Turner disagreed strongly with Harmon's recommendation on Hester's relief as NGOF commander, but Harmon convinced Halsey of the necessity for this change. Miller, *Reduction of Rabaul*, pp. 123-124.

miral Turner by Rear Admiral Theodore S. Wilkinson as Commander, III Amphibious Force. Turner returned to Pearl Harbor to take command of amphibious forces in the Central Pacific.

The addition of tanks and a fresh battalion of infantry to the forces at Laiana beach buoyed the hopes of the NGOF that the impetus of the attack could be resumed. The tank platoon of the 9th Marine Defense Battalion had landed on Rendova with its parent unit, but had not been required for seizure of the island. The tanks later moved to Zanana Beach to support an engineer mission shortly after the NGOF began its attack. The marshy ground in the vicinity of the Barike balked attempts to use armor in support of infantry operations, however, so the eight tanks were withheld from action until Laiana was taken. Here, it was reported, the ground was more firm and could support armored operations.

Forward movement of the 172d Infantry in the Laiana area had virtually ceased when the Marine armor arrived. The enemy's defensive line, a series of pillboxes dug into the hill mass rising just forward of the American lines, stubbornly resisted attack. Infantrymen attempting to push ahead were driven back by fierce machine gun fire from the camouflaged positions. In the hopes that a coordinated tank-infantry thrust could crack the defenses, an attack was planned for 15 July.

On the morning of the 15th, three tanks reported to the 2d Battalion, 172d on the left, while another trio of tanks moved toward the 3d Battalion on the right. Tangled underbrush hid stumps and logs that hampered attempts to get into position, and the drivers had to back and turn the machines constantly to move ahead. In the left zone, the first opposition, which came from a log and coral emplacement, was promptly knocked out by 37mm high explosive rounds and machine gun fire. Two grass bivouac shelters were peppered with canister rounds [18] and machine gun fire, and six to eight dead enemy were reported in each by the 172d's infantrymen following the machines.

Further progress was stopped, however, by enemy machine gun and rifle fire which began to pour from other camouflaged positions. The infantrymen sought cover. The Marine tanks, without infantry support, were forced to resort to a deadly game of blind man's bluff. Hit from one direction, the tanks wheeled—only to receive fire from another quarter. By alternating canister with high explosive rounds, the tankers stripped camouflage from emplacements and then blasted each bunker as it was uncovered. Enemy soldiers attempting to flee the positions were killed by machine guns. Opposition gradually ceased, and the infantrymen moved forward. The advance marked the first significant gain in several days.

In the right zone, the other three tanks were also blasting hidden positions which supporting infantrymen marked with tracer bullets. At one time the tanks were under fire from five hidden bunkers and dugouts. Combat was so close in the thick, hilly jungle that in several instances the muzzles of the 37mm guns could not be depressed enough to engage the enemy positions. Continually drummed upon by small-arms fire, and blasted repeatedly by grenade and mortar bursts, the armor withdrew after clearing the enemy from one hill. The 3d Battalion immediately

[18] Short-range 37mm ammunition similar to an over-sized shotgun shell.

occupied the positions and set up defenses. The only casualty suffered by the Marines in the engagement was one driver injured when a hidden log jammed its way through a floor hatch.

On the following day, three tanks with six infantrymen following each machine moved around the base of the hill taken by the 3d Battalion and pushed through the heavy jungle toward the next hill. The tanks raked the underbrush with fire and then pumped explosive shells into the enemy positions. A number of pillboxes, dugouts, and enemy shelters were knocked out. Only rifle and automatic weapons fire opposed the advance, and the infantrymen quickly moved forward. In the 2d Battalion zone on the left flank, defenses on the coast were outflanked by the tanks, which maneuvered along the shore line firing at the blind sides and rear of the bunkers. After nearly 200 yards of progress, the tankmen discovered they were without infantry support and returned to the lines. A second attack was stalled by heavy mortar fire which drove the supporting infantrymen back to their foxholes.

Unprotected by infantry, the tanks kept firing to the front and sides to keep enemy soldiers from attacking. Heavy jungle growth limited visibility to only a few yards and restricted maneuver of the machines. While trying to disengage from the battle, the tanks were rocked by heavy explosions, apparently from magnetic antitank grenades tossed against the machines by enemy soldiers hidden in the dense thicket all about the armor. The rear machine was blasted twice, and each of the other two tanks was damaged slightly by similar explosions. Swiveling and turning, the tanks fired at every movement in the brush, and, by sweeping the jungle with canister and machine gun fire, managed to break clear and crawl back toward friendly lines.

That night, the 3d Battalion, 103d Infantry relieved 2/172 in the left zone and another coordinated tank-infantry attack was scheduled. Working all night, 16–17 July, the Marines had five tanks available for combat. By prior agreement, 30 infantrymen were to accompany each machine and the tanks were not to move unless soldiers supported them. The day's attack had hardly begun, however, before stiff enemy opposition developed. Machine gun and rifle fire spewed from a number of concealed positions, and bullets ricocheted among the infantrymen following the armor. Soldiers, returning the fire, attempted to locate the emplacements so that the tanks' 37mm guns could be directed against the enemy.

As the tanks maneuvered toward the enemy defenses, the lead machine was suddenly sprayed with flame thrower fuel by a Japanese in a camouflaged position. The fuel did not ignite, and the enemy soldier was quickly killed. In such close combat, however, even nearby infantrymen could not protect the tanks from hidden enemy soldiers who suddenly appeared to toss magnetic grenades on the tanks. The third machine, hit by such a missile, took a gaping hole near the hull. Two crewmen were wounded. A hasty look behind them convinced the Marines that the infantrymen had fallen behind, and that protection was gone. Covering each other by fire, the tanks moved back with one of the undamaged vehicles towing the disabled machine.

Although no long gains had been made in the three-day attack, the commitment

691-360 O—63——8

of armor on the extreme left flank of the NGOF front had helped wedge an opening into Sasaki's defenses. A line of pillboxes stretching from Laiana beach northwest for more than 400 yards had been breached. Typical of the defenses was a cluster of seven pillboxes which covered a frontage of only 150 yards, each position defending and supporting the next. Overhead and frontal protection consisted of two thicknesses of coconut logs and three feet of coral. Skillfully camouflaged, with narrow firing slits, the bunkers were virtually a part of the terrain and surrounding jungle.

TOMONARI REPULSED[19]

The Japanese counterattack hit just as the NGOF paused to consolidate its gains, restore contact and communication, and effect a reorganization and reinforcement. Through coincidence or superior combat intelligence, General Sasaki committed the *13th Regiment* at a time when its appearance would provide the greatest shock effect. (See Map 6.)

Following its arrival at Bairoko and the move to the plantation area, the *Tomonari Force* scattered in small groups to reassemble north of the Barike River area. Sasaki's orders to Tomonari were:

> The 13th Regiment will immediately maneuver in the area of the upper reaches of the Barike River; seek out the flank and rear of the main body of the enemy who landed on the beach east of the Barike River and attack, annihilating them on the coast.[20]

To accomplish this task, Colonel Tomonari was to take over the defensive positions in the designated area and establish a base from which attacks could be staged. Colonel Hirata's *229th*, with as much strength as possible, was to coordinate with the *13th* and attack the American left flank.

Despite Sasaki's precautions, however, the *Tomonari Force* was observed moving toward the Barike. On 17 July, the 43d Division Reconnaissance Troop, screening the open right flank of the NGOF, reported that a large body of enemy, numbering from 200 to 300 men, had been observed moving toward the rear of the NGOF. One platoon of the troop attempted to ambush this force but was overrun. Sasaki's admonitions to keep contact notwithstanding, communication between the *Tomonari Force* and the *229th* was broken, and the two counterattacks were never synchronized. On the right flank of Sasaki's units, the *3d Battalion, 229th* was kept off balance by the tank-infantry attacks of the 172d. Farther north, the 169th was in a commanding position and was able to call down artillery fire on any observed group of enemy infantry, and thus effectively forestalled any threat of a push through the center of the line. Only the attack from the upper Barike materialized.

Shortly after dark on the 17th, enemy troops hit almost simultaneously at the rear area and beach installations of the 43d Division. Soldiers helping to evacuate wounded were themselves cut down.

---

[19] Unless otherwise noted, the material in this section is derived from: *New Georgia Campaign; NGOF Account; 43d InfDivHist;* 9th DefBn WarD, Jul43; *SE Area NavOps—I; SE Area NavOps—II;* ONI, *Combat Narratives X;* Rentz, *Marines in the Central Solomons;* Zimmer, *43d's History.*

[20] CIC SoPacFor Item No. 702, dtd 13Sep43, New Georgia DefOpO No. 35, dtd 13Jul43.

In a series of sharp skirmishes, Japanese infiltrators struck at the medical collecting station, the engineer bivouac area, the 43d Division CP, and the beach defenses. For a short time, the fate of the command post was held in one thin telephone line. Although most lines were cut, contact with the artillery units on the adjacent islands was still open over one line, and support was urgently requested. Accurate and destructive artillery fire that virtually ringed the command post was the quick reply. In several instances, concentrations within 150 yards of the CP were requested and received. In a matter of moments, the *Tomonari Force* was scattered, and although the CP area was under attack all during the night, repeated concentrations falling almost within Allied positions kept any large-scale assault from developing.

In the beachhead area, Army service units, the 172d's antitank company, and the 9th Defense Battalion's antiaircraft detachment were also hit. A Marine patrol, investigating the CP situation, returned to report that a body of enemy infantry of near battalion strength was moving between the CP and the beach. Reclaiming two .30 caliber machine guns from an Army supply dump by piecing together parts from a number of guns, half of the 52-man Marine detachment went forward to set up an ambush ahead of the advancing Japanese, while the other half remained behind to man the antiaircraft defenses. The ambush stopped the first enemy attack, and, after the Marines fell back to the beach defenses, the attack was not renewed. The reason was apparent the next morning. Two Marines who volunteered to remain behind at the ambush had effectively stopped the counterattack by repulsing four attempts. Only one of the two Marines survived the attack, which left 18 enemy dead littered about the guns.

The night of 17 July virtually ended all Japanese attempts to regain the initiative. The *Tomonari Force*, in small groups, appeared from time to time in various areas, raiding and infiltrating, but was not an effective threat. Up to the time of the resumption of the NGOF attack, Sasaki still harbored hopes that he could collect his scattered forces for another attempt, but the rapidly-accelerating Allied buildup nullified all his efforts.

CORPS REORGANIZATION AND ATTACK [21]

A number of Army units were close at hand for ready reinforcement of the NGOF lines. These were promptly ordered to New Georgia when the Japanese

[21] Unless otherwise noted, the material in this section is derived from: *New Georgia Campaign; New Georgia Combat; NGOF Account;* ComAirSols StrikeComd WarD, 2Apr–25Jul43; ComAir New Georgia SAR, 29Jun–13Aug43; 37th InfDiv Rept of Ops in the Munda Campaign, dtd 25Aug43; 37th InfDiv AR 22Jul–5Aug43, n.d.; 37th InfDiv Jnl 22Jul–5Aug43; *43d InfDivHist;* Col Stuart A. Baxter Rept of Ops of the 148th Inf (-3d Bn) in New Georgia 18Jul–5Aug43, n.d.; NarrativeRept CbtActivities of 1/148, dtd 13Sep43; *9th DefBn TkOps;* 10th DefBn AR of TkPlat in New Georgia Campaign, dtd 3Sep43; 11th DefBn WarD, Aug–Sep43, hereafter *11th DefBn WarD; SE Area NavOps—I; SE Area NavOps—II;* MajGen Oscar W. Griswold Rept to LtGen Leslie J. McNair, dtd 21Sep43; Robert F. Karolevitz, ed., *The 25th Division and World War II* (Baton Rouge, La.: Army and Navy Publishing Co., 1947), hereafter Karolevitz, *25th InfDivHist;* Stanley A. Frankel, *The 37th Infantry Division in World War II* (Washington; Infantry Journal Press, 1947), hereafter Frankel, *37th InfDivHist;* Rentz, *Marines in the Central Solomons;* Zimmer, *43d's History.*

counterattacked. The 148th Infantry was on Kokorana when the emergency alerted that unit at 0100 on the 18th; the 1st Battalion, dispatched immediately, came ashore at Zanana fully expecting to find the beach area in enemy hands and the 43d Division CP wiped out. By this time, however, the serious threat had passed and when the regiment was assembled, it began moving to the front lines. Although an advance party was hit by remnants of the *13th Infantry*, the 148th pushed forward aggressively, cleared the opposition, and moved into the rear area of the 169th by nightfall of the 18th.

The 145th Regiment, which already had one battalion in place as reserve for the 43d Division, reached the rear of the 169th lines on the 20th. Upon the arrival the same day of Major General Robert S. Beightler, the 37th Infantry Division assumed responsibility for the sector and the 169th Infantry was relieved. Colonel Holland, who had directed the 169th in its capture of the hills overlooking Munda, returned to command of the 145th. The 169th's 1st and 2d Battalions, tired and badly depleted, departed for Rendova for a needed rest. The 3d Battalion remained on New Georgia as 43d Division reserve.

The arrival of other units also strengthened the NGOF lines. The 161st Infantry, detached from the 25th Division on Guadalcanal, debarked on the 21st. Attached to the 37th Division, the regiment moved into bivouac on the division's right flank. The remainder of the 103d Regiment joined the 3d Battalion on New Georgia on the 21st and 22d of July, and, from that point on, the 103d (less the 1st Battalion still at Segi) fought as a regiment. Additional antiaircraft protection against the periodic Japanese air raids on New Georgia and Rendova was provided by a detachment of 4 officers and 140 men from the Marine 11th Defense Battalion. Alerted early in the campaign for possible commitment, a 90mm battery, augmented by four 40mm guns and four .50 caliber machine guns, was sent to Kokorana Island from Guadalcanal on 18 July.

During the period 18–24 July, while the NGOF swelled in size as fresh regiments poured in, the front lines of the New Georgia Force remained static. At this time, the main positions of the NGOF traced an irregular pattern through the hilly jungle in a northwest direction from Laiana Beach to the steep hills guarding the northern approach to Munda. Into this 4,000-yard front, still about three miles from Munda, General Griswold moved the two divisions with orders to continue the attack on the 25th. In the southern sector, General Hester's 43d Division had the 103d Infantry (Lieutenant Colonel Lester E. Brown) anchored to the coast with the 172d Infantry (Colonel Ross) on the right. In the 37th Division's zone of attack on the north, General Beightler had placed the 145th Infantry (Colonel Holland) on the left flank and the 148th Infantry (Colonel Stuart A. Baxter) on the extreme right flank with the added mission of protecting the right flank and rear of the NGOF. The 161st Infantry (Colonel James M. Dalton) was assigned as the interior unit between the 145th and the 148th. To insure a rapid advance, the frontline units were directed by General Griswold to bypass all strong points, leaving these for the reserve units to eliminate.

Combat action during the period in which the NGOF reorganized and rested was limited. As each front-line unit moved into place, patrols sought to determine the disposition and strength of the Japanese units to the front. Occasionally, scattered bands of *13th Regiment's* soldiers were encountered, and a number of confused, short skirmishes resulted. Casualties to both sides were light.

The NGOF had one advantage. The ground fighting had been relatively free of air interference, and most of the bombing attacks were by friendly planes on rear area enemy defenses. The Japanese had attempted but failed in several attempts to locate the NGOF front lines for a bombing and strafing attack. Segi, Wickham, and Viru, however, were visited regularly by nocturnal aircraft which the troops—conforming to South Pacific custom—tagged with the euphemisms of "One-Bomb Bill" or "Washing-machine Charlie." Most of the Japanese air attempts, though, appeared to be aimed at Rendova where the bulk of supplies was stockpiled. An alert air cover, helped by antiaircraft batteries, kept enemy planes at a wary distance.

Air support missions requested by General Mulcahy as ComAir New Georgia were generally directed at the easily identifiable targets around Munda field. Close air support for troops fighting in dense jungle had proven impractical with target designation so difficult. Air-ground coordination, struggling against the handicaps of visibility and communications, was not helped by the inaccurate operation maps. Even though gridded, the photo-mosaics were not precise enough for such close work, where a slight error might result in heavy NGOF losses. Then, too, in the fighting where daily progress was measured in 200- or 300-yard gains, the troops were reluctant to withdraw for an air strike. Soldiers reported that when they had pulled back to provide a zone of safety for air strikes or artillery and mortar preparations, the enemy simply moved forward into the abandoned area and waited for the bombing or artillery to lift before moving back into their original positions in time to defend against the expected ground attack.

Requested support missions were flown by Strike Command, ComAirSols. The New Georgia support was in addition to the repeated bombing and strafing strikes at enemy shipping and airfields at Kahili, Ballale, Vila, Enogai-Bairoko, and Bougainville. The planes flew cover for task groups and friendly shipping as well. During the period 30 June to 25 July, the start of the corps offensive in New Georgia, the Strike Command squadrons flew 156 missions involving 3,119 sorties. In addition to more than four million pounds of explosives dropped on enemy installations, the ComAirSols planes counted 24 enemy ships sunk and another 28 damaged. A total of 428 fighter planes and 136 bombers were reported as destroyed by ComAirSols pilots. Strike Command losses in the Central Solomons during the period were 80 planes.

The final push on Munda promised the hardest fighting of the campaign. Between the NGOF and its objective were more than 4,500 yards of low but steep hills, irregular and broken, densely covered with tropical rain forest, and laced with enemy defenses. Reports of the patrols and observation of bunkers already taken indicated that the enemy soldiers were dug in and covered by low, two-level camouflaged coral and log emplacements with deadly interlocking fields of fire. Trenches bulwarked by coconut logs connected the bunkers. NGOF soldiers were well aware that the enemy would have to be routed from these positions and that resistance until death was standard practice. Further, the soldiers knew that the enemy often abandoned one bunker to man another, and then, after the first bunker had been overrun, returned to defend it again. An area gained in attack during one day had to be cleared of infiltrators the following day.

Prior to the 25 July attack by the NGOF, an attempt was made by Marine tanks to crack the hill complex south of Laiana Beach and bring the 43d Division units on a line with the 37th Division. Withdrawn from further engagements in that sector after the 17 July attack, the 9th Defense Battalion tanks were sent into action again on the 24th. An artillery preparation prior to 0700 pounded a 100-yard zone in front of the lines before the armor moved out from the lines of the left battalion of the 172d Infantry. Repulsed by a strongly defended position in that sector, the Marine tanks tried again from the adjoining battalion of the 103d Infantry on the left. Although several pillboxes were knocked out, the tanks were forced to withdraw after one machine was blinded by hits on the periscope. Two other machines sputtered with engine trouble caused by low-octane fuel and overheating. The withdrawal was made under fire, the disabled machine under tow by another.

Another point of tenacious defense was met by the 161st Infantry. Dalton's regiment, attempting to move up to the line of departure, was told that only two pillboxes were to his immediate front. A reinforced platoon, making the initial attack, knocked out the two pillboxes but then uncovered another network of fortifications. A strong company was sent into the area. Two more pillboxes were knocked out, but 12 more were uncovered. At this point, the regiment moved in and knocked out these strong points before discovering more pillboxes. At last, with the 25 July attack impending, the regiment bypassed the fortifications and moved up to the line of departure. But before the pocketed strong point was reduced, "it took the combined efforts of two battalions, 3,000 rounds of 81mm mortar fire, the use of tanks, and the passage of seven day's time." [22]

As General Griswold's NGOF poised for the final make-or-break assault on Munda, his adversary was forced to face the contest with a dwindling stack of chips. XIV Corps intelligence officers estimated that General Sasaki had lost about 2,000 troops, including 1,318 counted dead, of the more than 4,500 which he had available earlier.[23] His biggest gamble had failed—matched and beaten by a larger reserve. The *13th Regiment* had now filtered back toward Munda to take up defensive positions to the northeast. The main units of the *229th Regiment*, which had so bitterly contested the advance of the NGOF from the Barike, had taken steadily mounting casualties. Nearly cut off from the rest of the command by the pressure of the NGOF attack, the *229th* took up final positions in the Munda hills, the battalions and companies considerably intermingled. General Sasaki, hoping to avoid some of the pounding aimed at Kokengola Hill, moved his headquarters from the airfield to the plantation north of it.

[22] 37th InfDiv Ops, *op. cit.*, p. 4.

[23] USAFISPA IntelRept No. 27, dtd 24Jul43.

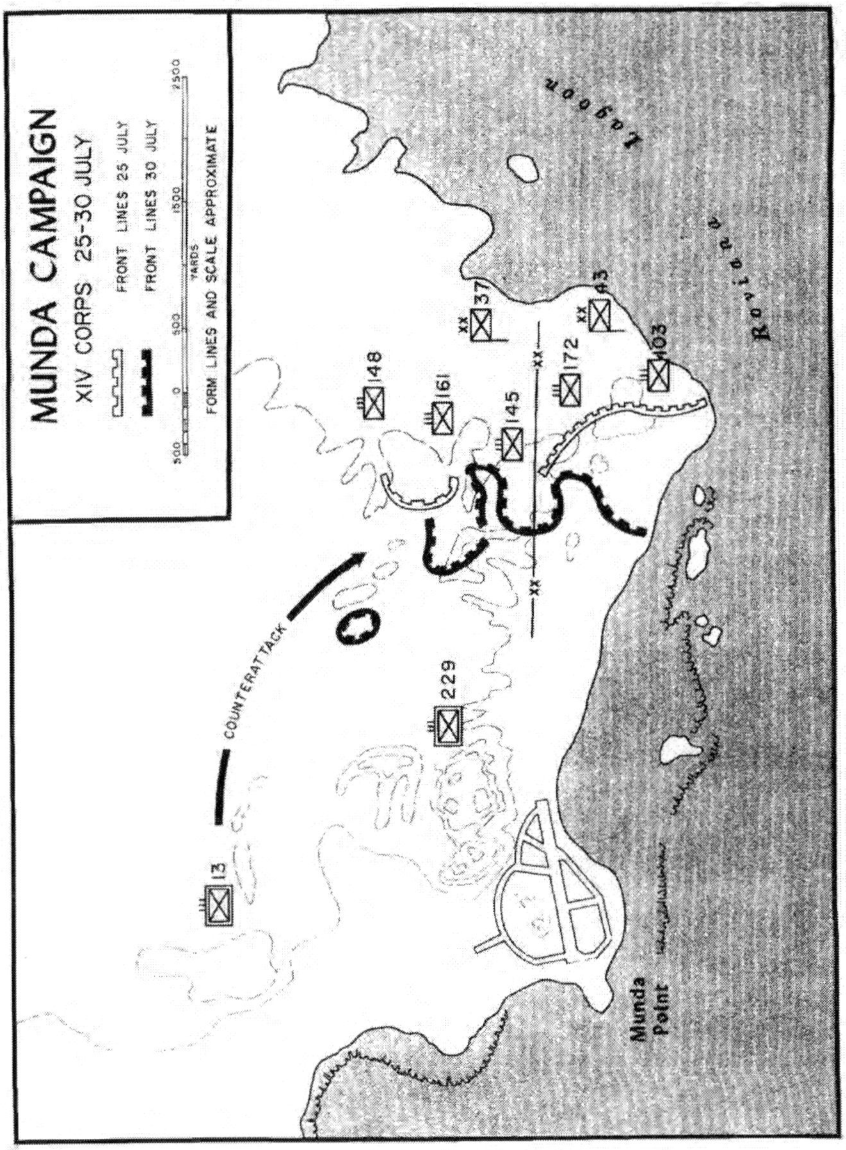
MUNDA CAMPAIGN
XIV CORPS 25-30 JULY
FRONT LINES 25 JULY
FRONT LINES 30 JULY
500 0 500 1500 2500
YARDS
FORM LINES AND SCALE APPROXIMATE
COUNTERATTACK
13
229
148
161
145
172
37
43
103
Munda Point
Roviana Lagoon

MAP 6

With the worsening situation in New Georgia came new realization and uneasiness that Japanese positions in Bougainville would be as quickly overrun. A seaplane carrier protected by five destroyers, trying to reach that island on 22 July, was attacked by a force of 16 dive bombers, 18 torpedo bombers (all from VMTB-143), and 16 heavy bombers which stopped the reinforcement effort cold. Only 189 men out of 618 Army personnel aboard the carrier survived. Also lost were 22 tanks, heavy equipment, guns, fuel, and ammunition destined for the Central Solomons defenders. The destroyers, however, managed to land some troops.

Sasaki continued to hope for reinforcements, but the Allied clamp on Kula Gulf was too tight. The only major unit to reach New Georgia was the understrength *230th Regiment*, a remnant from the Guadalcanal withdrawal. Only about 400 men reached Munda, and these were put into the final defense around Kokengola Hill. The pincers movement of the NGOF and the concentrated shelling and bombing counted toward making the Central Solomons situation doubtful, but the blockade of Kula Gulf by Allied destroyer forces, torpedo boats, and night and day air patrols was perhaps the telling factor. "Consequently," the enemy was forced to admit, "the fate of the Munda sector became a matter of time." [24]

General Sasaki, a realist, confessed that the Allies had complete material superiority and that a sustained push by the NGOF would collapse his command. Although he was envious of his opponents' artillery, communication, and large landing boats, he was critical of the NGOF soldier—who, he said, advanced slowly, failing to take advantage of his strength and equipment:

> They awaited the results of several days' bombardments before a squad advanced. Positions were constructed and then strength increased. When we counterattacked at close quarters, they immediately retreated and with their main strength in the rear engaged our pursuing troops with rapid fire. The infantry did not attack in strength, but gradually forced a gap and then infiltrated. Despite the cover provided by tank firepower, the infantry would not come to grips with us and charge. The tanks were slow but were movable pillboxes which could stop and neutralize our fire.[25]

The defense of the airfield had also depleted Sasaki's forces. The Japanese soldier, fatigued and muddy, was forced to fight in some instances on only one rice ball a day. Kept irritated and sleepless by the constant bombardment, the Munda defender was gaunt, weary, and hungry—but still determined. Despite the hardships, morale was high and the Japanese soldier was "prepared to die in honor, if necessary." [26]

The NGOF attack, now corps-size, opened on 25 July when five-inch shells rained upon the Munda area from seven destroyers. At 0630, heavy bombers began dropping 500-pound bombs and followed up with a rain of 120- and 300-pounders. Next came flights of torpedo bombers and scout bombers which dropped 2,000-pound and 1,000-pound bombs. In all, 171 planes took part in the saturation bombing of the area paralleling the entire front lines. Special attention was given to defensive positions in the

[24] *SE Area NavOps—II*, p. 32.

[25] CIC SoPacFor Item No. 877, dtd 2Dec43, SE DetCondRept to Seventeenth Army CofS, late Jul43.

[26] CIC SoPacFor Item No. 1026, dtd 8Feb44, Translated Enemy Diary.

hills near the lagoon and the heavily defended strong point in the center of the Japanese defensive line, which the NGOF troops called Horseshoe Mountain because of its U-shaped appearance. Bibolo Hill, guarding Munda, was also worked over. (See Map 6.)

As the attack began, Japanese air units attempted to retaliate. At 0930, a flight of from 60 to 70 enemy fighters bore down on New Georgia, but the air cover provided by ComAirSols held off the attack. Additional Allied fighter planes, hastily scrambled from Segi's newly completed airstrip, arrived in time to discourage a second attempt by the enemy planes.

NGOF artillery, firing parallel to the front lines, lashed the area to be attacked; and, with this awesome display of firepower to pave the way, the NGOF regiments began to move forward. One Japanese soldier, astounded by the volume of shelling, wondered, "Are they intending to smash Munda with naval and heavy artillery?"[27] In the 43d Division sector, the 9th Defense Battalion tanks were called to rescue troops of 3/103 held up by a strong point. Aided by a flanking movement of the 172d's 2d Battalion, the tanks slashed through the rear of the enemy positions facing the 103d, and the Japanese hastily abandoned their positions to flee toward the next line of hills. Elements of the 103d then pushed toward the relatively clear plantation area between Laiana and Munda. The advance was about 500 yards. The 3d Battalion of the 169th then moved out of reserve positions to fill the gap between the 103d and the 172d.

The main effort of the first day's attack was made in the 37th Division zone. The 145th Infantry, the left flank unit, held its positions in order to straighten the NGOF lines, while the 161st and 148th pressed the attack. Stiff resistance from the defenders of Horseshoe Mountain held the 161st to a slight gain, but the 148th easily advanced about 600 yards against occasional fire from small outposts. By nightfall, the NGOF had pressed itself against the Japanese front lines.

Marine tanks were in support of both divisions the following day. A newly arrived weapon making a first appearance in the fighting, the flame thrower, was combined with tanks from the 9th Defense Battalion to crack a belt of 74 pillboxes on a 600-yard front which faced the 103d and 172d regiments. The day's attack put the 43d Division well into the rear of the Laiana defenses. Farther north, the 145th continued to hold fast while the 161st attempted to crack the resistance to the front. A fresh Marine tank platoon, six of the machines from the 10th Defense Battalion, was committed to action in an attempt to clear the Horseshoe Mountain defenses.

After a five-hour struggle against the thick jungle and steep terrain, a total of 14 pillboxes had been demolished. The tanks, crashing through a thick underbrush tangled by fallen logs and stumps, finally located the enemy fortifications near a large clearing. Infantry support, however, was often pinned down by murderous enemy fire, and the tanks were forced to twist and turn, pivot and backtrack, to keep enemy riflemen from assaulting the machines with magnetic grenades. Three tanks were knocked out and abandoned before the Marine tankers could disengage from the furious fighting. The strong point remained, however, only partially silenced. That night, close-in artillery fire ringed the abandoned tanks so

[27] USAFIPSA IntelRept No. 40, dtd 26Oct43.

that enemy soldiers could not use them as pillboxes.

On the far right, Colonel Baxter's 148th Infantry continued to drive ahead against only slight resistance, advancing another 800 yards the second day. The move, however, put the 37th Division far ahead of the 43d Division. To straighten the lines, the next attack effort would be directed against the enemy in the south. If the 103d and 172d could press past the open south side of the Horseshoe Mountain defenses, the penetration might relieve the pressure on the central portion of the NGOF line.

Marine tanks were to spearhead the 43d Division attack in the south on the 27th, but the advance had hardly started before the lead tank was blasted by an antitank gun. Confusion resulted. The first tank, with casualties among the crew, stalled. As it started again and attempted to back up, it rammed the second tank. A third tank was hit immediately by antitank fire. As a fourth and fifth machine moved up, one was blasted by magnetic mines and the other, after raking the jungle with machine gun fire, was also disabled by a grenade. All machines, however, by mutual fire assistance, managed to limp back to friendly lines. But the day's attack virtually ended the combat efficiency of the 9th Defense Battalion tank platoon. Of the eight machines brought ashore, five had been disabled that day, a sixth had been disabled previously, and two others were under repair. Four tanks were reported deadlined permanently. In addition, the platoon had a number of drivers and crewmen killed or wounded.

Progress along the line on the 27th had been slight, for two localized strong points continued to hold up the advance. The 43d Division still faced a rugged defensive area in the south which repeated tank-infantry assaults had failed to dent, and the 37th Division was hung up against the Horseshoe Mountain line, kingpin of Sasaki's resistance. To XIV Corps observers, it was plain that the capture of either strong point would result in the downfall of the other.

On 28 July, 3/103 followed four Marine tanks into attack on the coast area after a 30-minute mortar and artillery preparation. The attack proved to be the finest example of tank-infantry tactics of the campaign. With the machines guarded and supported by the infantry, the battalion advanced in a series of spurts. For the first time, the tanks were operating over relatively flat and open terrain with dry footing. Enemy opposition began to falter, then dwindled rapidly, as the attackers rushed ahead. Even three direct hits by antitank guns on the lead tank failed to stop the attack. The enemy gun emplacement was overrun a few moments later. Completely routing the enemy in a 500-yard advance, the infantrymen took up defensive positions while the tanks continued to range ahead. One tank was hit, but managed to limp back to the lines. The day's advance had completely broken the Japanese defenses in the south.

In the north, the 161st jumped off in an attack without prior artillery preparation and caught the enemy unawares. In a brief skirmish, the 161st occupied a ridge which had held up the advance for two days. At this time, the attention of the NGOF was suddenly drawn to the right flank where the 148th had abruptly found itself in trouble. As Colonel Baxter ruefully admitted later: "Don't forget, being *too* aggressive can often get you into as much hot water as doing nothing."[28]

Baxter's regiment, pushing ahead against weak and scattered opposition, had reached the Munda-Bairoko trail, but in so doing had opened a hole between the 148th and 161st. With two battalions in the attack, the 148th had been unable to plug the gap, and, as at the Barike River earlier, alert Japanese soldiers quickly infiltrated. That night, the rear supply dump of the 148th was under determined attack by an enemy force of considerable size. Support troops managed to beat off a three-sided enemy assault by using ration boxes and supply cartons as barricades, much in the manner of frontier wagon trains under attack by Indians. Elements of the 148th, which had reached as far as Bibolo Hill west of the airfield to confirm indications that the enemy was abandoning that front, now rushed back to the defense of the supply dump. In this instance, the 148th virtually had to fight its way to the rear as about 250 Japanese in small bands with machine guns and mortars, probably remnants of the *Tomonari Force*, harassed the unit for three nights. The 148th reached the supply dump and established contact with the 161st before turning about to resume the attack toward the northern part of Bibolo Hill.

Although the 43d Division, now under the command of Major General John R. Hodge who had relieved General Hester, continued to push forward along the coast in rapidly increasing gains, the center of the NGOF continued to be snagged on the enemy defenses on Horseshoe Mountain. First break in the barrier came on 30 July when the 172d attacked and occupied a small ridge complex southeast of the main defenses. The following day, 31 July, the 169th attacked and completed the reduction of the southern anchor of the Japanese strong point. The advance, however, still failed to break the Horseshoe defenses.

On 1 August, the 43d Division punched through to the outer taxiway of Munda airfield. The move put the Allied force almost in the rear of Sasaki's last strong point, and enemy resistance on Horseshoe Mountain suddenly dissolved. The airfield defenders had at last succumbed to the steady pressure of the NGOF.

The withdrawal had been ordered after the *New Georgia Defense Force* had become steadily weakened by lack of ammunition, food, and additional troops. Although a few destroyers managed to make Kolombangara, practically all Japanese transportation and supply lines had been strangled. On 29 July, an officer courier of the *Eighth Fleet* had arrived at Munda to relay to Sasaki the order to fall back to the line of hills ringing Munda for a last-ditch stand. The airfield was to be defended even at the price of Kolombangara. Reinforcements would come. Following instructions, Sasaki pulled what scattered elements he could find back to his last defense. As the campaign drew to a close, his line was held by the *229th Regiment* on the south part of Bibolo Hill with the undermanned *230th Regiment* on Kokengola Hill. On the extreme left flank were units of Tomonari's *13th Regiment*.[29] Remnants of the *8th CSNLF* were combined with Army units for a last-ditch stand.

At the close of the fighting on 2 August, the 43d Division was perched on the last

[28] Baxter Rept, *op. cit.*, p. 16.

[29] SoPacFor POW InterrogationRept No. 140, dtd 24Nov43.

low row of hills overlooking Munda airfield, and the 37th Division was gradually tightening the lines around the northern part of the airfield. The following day, Hodge's troops captured the southern part of Bibolo Hill while the 37th Division moved cautiously but swiftly through isolated pillbox areas northwest of the field. The 148th, reaching the Munda-Bairoko trail once more, ambushed a large force of enemy fleeing the area. (See Map 7.)

As the two divisions resumed the attack on 4 August, the only opposition facing the 43d Division came from Kokengola Hill in the middle of the airfield. While a rain of artillery and mortar shells blasted the hill, Marine tanks from the 10th and 11th Defense Battalions roamed about the airfield, flushing snipers and blasting rubble-hidden fortifications. The tanks from the 11th Defense Battalion had been hurriedly dispatched to take part in the assault of the airfield after the 9th Battalion's tanks had been deadlined. Alerted on Tulagi since 30 June, the Marine tankers reached New Georgia on 3 August, just in time to join the final attack.

North of Munda, while the 145th mopped up the last shreds of opposition, the 161st and 148th Regiments plunged rapidly through to Diamond Narrows. In that final drive, the 37th Division soldiers staged a slashing, stabbing charge that overwhelmed all outposts. That night, the last shots fired were those sent after Japanese trying to swim to islands across the Narrows.

The following day, 5 August, tanks of the 10th and 11th Defense Battalions—accompanied as a courtesy gesture by the sole remaining operational tank of the 9th Defense Battalion—made five sorties over the airfield. The only fire received was from Kokengola Hill, and this the Marine tanks quickly squelched with 37mm rounds. At 1410, the airfield was officially declared secured, and Allied troops took over the enemy fortifications ringing the war prize which had taken more than a month of bitter combat to obtain. Along the blasted and cratered runways were hulks of 30 enemy airplanes, some still in revetments. All were stripped of armament and instruments. None would ever fly again. Japanese supplies, including tasty tinned foods, beer, *sake*, and rice gave triumphant infantrymen a change from the weary routine of combat rations.

Beach defenses were strengthened the next day, and grimy soldiers bathed, washed clothes, and rested from the tough grind of battle. Patrols, ranging far to the north, reported no opposition. The patrols' only result was the capture of one forlorn Japanese soldier, whom one officer described as typical of the enemy who were thwarted in their attempts to hold their precious airfield: "Injured, tired, sick, no food, dirty torn clothes, little ammunition and a battered rusty rifle."[30] For both victor and vanquished, the campaign had been hard.

The fall of Munda almost coincided with another disaster which heaped additional misery upon the Japanese. In a belated and ill-fated attempt to help Sasaki hold the Central Solomons, the *Seventeenth Army* at Bougainville organized two well-equipped infantry battalions, bolstered by the addition of artillery and automatic weapons. The troops were taken from the *6th* and *38th Divisions*. The reinforcement unit started for New Georgia on the night of 6 August in four destroyers. As the ships steamed through the

[30] 1/148 Rept, *op. cit.*

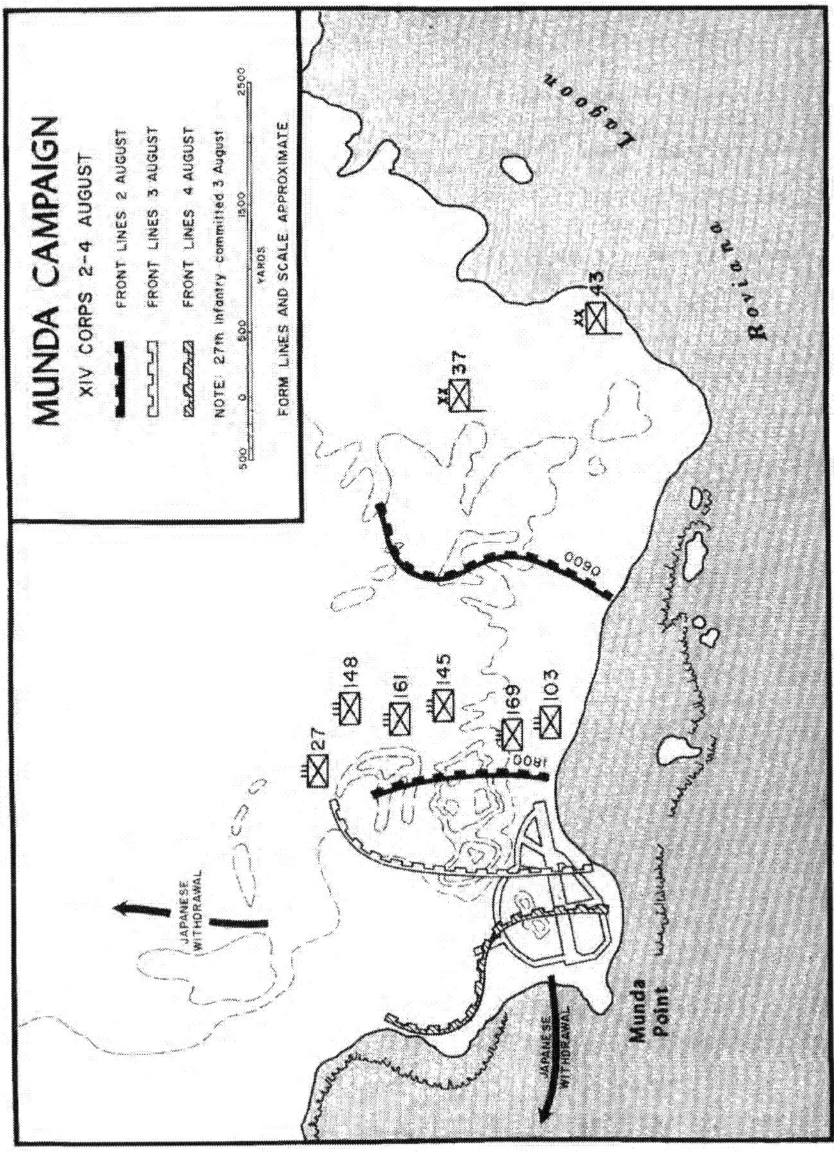
MUNDA CAMPAIGN
XIV CORPS 2–4 AUGUST
FRONT LINES 2 AUGUST
FRONT LINES 3 AUGUST
FRONT LINES 4 AUGUST
NOTE: 27th Infantry committed 3 August
500 0 500 1500 2500
YARDS
FORM LINES AND SCALE APPROXIMATE
JAPANESE WITHDRAWAL
JAPANESE WITHDRAWAL
27
148
161
145
169
103
37
43
0600
1800
Munda Point
Roviana
Lagoon

MAP 7

north entrance of Vella Gulf trying to make Kolombangara, an ambush set by an Allied force of six destroyers (Commander Frederick Moosbrugger) struck suddenly. In a matter of moments, three of the Japanese destroyers were in flames and sinking. The ambush in Vella Gulf resulted in the loss of 820 Army troops and 700 crew members in a single stroke. It was the last attempt by the Japanese to reinforce the Central Solomons.

Munda's capture was marked by the commitment of the 27th Infantry from Major General J. Lawton Collins' 25th Division. Augmented by division support troops, the regiment joined the NGOF on 2 August and took over the mission of guarding supply and communication lines along the 37th Division's right flank. After Munda was taken, the 161st Infantry reverted to 25th Division control and joined the 27th Infantry in a new push toward Kula Gulf.

With hardly a pause at the airfield, the two regiments pivoted north to complete the rout of all enemy forces in the area between Diamond Narrows and Bairoko Harbor. Only spotty resistance was encountered, for increased barge activity revealed that the Japanese were feverishly trying to evacuate the scattered remnants of the New Georgia garrison. After two weeks of locating and eliminating Japanese positions north of Munda, the 27th Infantry declared its zone secured. The 161st, meanwhile, had advanced toward Bairoko after knocking out enemy strong points on two jungle peaks. The final ground action on New Georgia came on 25 August, when the 161st Infantry combined with Liversedge's force to attack the harbor area from three sides—only to find that the Japanese had just completed evacuation of the area. All organized enemy resistance on the island was ended.

RENDOVA: FINAL PHASE [31]

During the period that NGOF soldiers slogged their way through jungle mud on the way to the airfield, the Rendova force settled into a routine of firing artillery missions and combatting enemy air raids. After the initial units of General Hester's force departed for New Georgia, the harbor at Rendova became the focal point for all reinforcements, supplies, and equipment moving into the Central Solomons.

During July, daily transport shuttles from the rear echelons on Guadalcanal poured a total of 25,556 Army, 1,547 Navy, and 1,645 Marine troops into Rendova for eventual commitment in New Georgia. Additionally, the beaches at Rendova and its offshore islands became piled high with rations, oil and lubricants, ammunition, vehicles, and other freight, all of which found its way to the NGOF.

This bustling point of entry—with troops unloading and stockpiles of material lining the beaches—was a tempting target to the Japanese. The Rendova air patrol of 32 fighter planes constantly flying an umbrella over the island drained the resources of ComAirSols, but, at the same time, was a successful deterrent to enemy attacks. During the New Georgia campaign, only three enemy hits were

---

[31] Unless otherwise noted, the material in this section is derived from: 9th DefBn Rept to ComMarDefGruSols, dtd 5Jul43; *9th DefBn OpRept; 9th DefBn AA Ops.;* 9th DefBn Informal CbtRept New Georgia Campaign, dtd 9Sep43; 9th DefBn Narrative Hist 1Feb42–14Feb44, dtd 2May44; 155mm Gun Gru, 9th DefBn Work Sheets, 18Jun–21Jul43, dtd 13Oct 43; *11th DefBn WarD;* ONI, *Combat Narratives X.*

registered on ships in the harbor by bombers or torpedo bombers, and only one horizontal bombing attack was able to close on Rendova during the daylight hours when the fighter umbrella was on station.

Playing a major role in the defense of the harbor, the 90mm batteries and the Special Weapons Group of the 9th Defense Battalion shot down a total of 24 enemy planes during the month of July. For the Marine antiaircraft crews, the defense of Rendova was virtually an around-the-clock operation which was a deadly contest of skill between enemy and defender. The Japanese tried all methods of attack, including hitting the target area with planes from various directions and altitudes simultaneously. Since large areas of the search radar screens were blocked by mountains on New Georgia, this approach route became the favorite of the Japanese pilots. Warnings for attacks from this direction were so short as to be almost useless, so Marines were forced to keep at least one 90mm battery manned continually with fire control radars constantly in operation. The Marines found that early in the campaign the enemy pilots dropped their bomb loads as soon as they were fired upon or pinpointed by searchlights. Later attacks, however, were pressed home with determination, and only well-directed shooting deterred them.

Marines also had a prominent part in the artillery support of the NGOF. After registering on Munda field prior to the NGOF overland attack, the Marine 155mm guns began a systematic leveling of all known enemy installations and bivouac areas. Since the exact location of the NGOF front lines was ill-defined most of the time, the Marine group left the close-support firing missions to Army 105mm units which were much nearer to the combat. The Marine guns were directed instead against rear installations, supply and reinforcement routes, and targets of opportunity.

Most of the firing missions were requested by NGOF headquarters with corrections directed by aerial observers or spotters at the 43d Division observation post. The Marine group had notable success interdicting supply dumps, bivouac areas, and enemy positions in the immediate vicinity of Munda field. Cooperation between air spotters from the 192d Field Artillery Battalion and the 155mm Group of the 9th Defense Battalion reached such a high state of efficiency that missions were fired with a minimum of time and adjustment. The Marines were occasionally rewarded by the sight of towering columns of smoke, indicating that a supply or ammunition dump had been hit.

Ammunition problems plagued the 155-mm batteries. On the 13th of July, just as the NGOF stalled against General Sasaki's defenses, an ammunition restriction was placed on the Marine batteries and the number of rounds expended dropped abruptly. After four days of limited firing, all shooting was stopped entirely while the NGOF reorganized in New Georgia. The only mission fired during this interval was on 20 July in answer to an emergency request to keep Japanese troops from moving back into an area which had been shelled and neutralized previously. The ammunition limitation resulted from powder becoming wet and unserviceable in containers broken from much handling. Further compounding the difficulties was the fact that during the period of ammunition scarcity, 11 miscellaneous lots of powder were used which resulted in varying initial velocities. Marines could only guess from one shot to another whether the shell would be over the target or fall short. When the powder situation was remedied and the 43d and 37th Divisions began the final drive for Munda, the Marine gunners, now experienced field artillerymen, returned to firing accurate missions.

After the fall of Munda, the 9th Defense Battalion began the move to New Georgia to help defend the newly won prize. Antiaircraft batteries were placed around the airfield and 155mm gun positions established on offshore islands and at Diamond Narrows. The 9th was relieved on Rendova by the Marine 11th Defense Battalion, which moved to that island from Guadalcanal to take part in the final stages of the Central Solomons fighting.

Although the capture of Munda was essentially an Army operation and the number of Marines participating was proportionately small, the contributions of the Marine Corps tanks, artillery, and antiaircraft units were essential to the success of the operation. Their exploits are an integral part of the story of the campaign. A handful of Marine tanks spearheaded most of the successful attacks; and even though handicapped by the rugged terrain, the armored vehicles were usually the factor which tipped the balance to the Americans' favor. Victory at Munda was won by inter-service teamwork—one of the frequent examples of coordinated Army, Navy, and Marine Corps effort in World War II.

The Dragons Peninsula Campaign

RICE ANCHORAGE[1]

Munda's eventual capture was a triumph over initial frustration and failure. Admittedly, the campaign to take the airfield had been costly and time-consuming. But while the spotlight was focused on the New Georgia Occupation Force as it fought its way from Zanana Beach to the airstrip, another tense struggle was waged simultaneously in the northern part of the island in which the jungle combat was as bitter and as deadly. The results were much less conclusive. From the initial ship-to-shore movement of the Northern Landing Group through the following six weeks of fighting, this phase of the New Georgia campaign contributed as much to the feeling of disappointment and futility as the first Munda attacks.

Early plans of the NGOF called for Colonel Harry B. Liversedge's 1st Marine Raider Regiment (less two battalions) to be a ready reserve. When intelligence reported a garrison of some 500 enemy troops with coast defense guns at Bairoko Harbor, the Kula Gulf landing was written into the attack order. Of prime concern to the Allied planners was the road connecting Bairoko with Munda airfield. Scarcely more than an improved jungle trail, the road was nevertheless a vital link between Munda and Vila, the main source of Japanese reinforcements and supplies. Bairoko Harbor was the knot which tied the overland route to the Kula Gulf barge system. An Allied ground force between Munda and Bairoko Harbor would have the double-barreled effect of cutting off the flow of enemy supplies and reinforcements to Munda as well as keeping the airfield forces and the Bairoko garrison from reinforcing one another.

[1] Unless otherwise noted, the material in this section is derived from: ComInCh Rept of SoPac Action, 2Dec42–31Jan44, hereafter *ComInCh Rept of SoPac Action;* WesternForO No. A11–43, dtd 28Jun43; TG 36.1 OpO No. 10–43, dtd 1Jul43; *New Georgia Campaign; NGOF Account;* 1st RdrRegt SAR 4Jul–29Aug43, dtd 6Oct43, hereafter *1st RdrRegt SAR;* 1st RdrRegt R–2 Est of Situation, c. 23Jun43; 1st RdrRegt R–2 Jnl 2Jan–31Aug43, hereafter *1st RdrRegt Jnl;* 1st RdrRegt MsgFile 5Jul–28Aug43, hereafter *1st RdrRegt MsgFile;* 1st RdrRegt PtlRepts 23Jul–14Aug43; 1st RdrBn WarD, 20Jun–29Aug 43, dtd 14Sept43, hereafter *1st RdrBn WarD;* 3/148 URepts, 7Jul–3Aug43, hereafter *3/148 Rept; SE Area NavOps—I; SE Area NavOps—II;* Maj Clay A. Boyd interview by HistDiv, HQMC, dtd 16Feb51; Col Samuel B. Griffith, II, ltr to DirPubInfo, HQMC, dtd 12Feb51; Col Samuel B. Griffith, II, ltr to Maj John N. Rentz, dtd 12Feb51, hereafter *Griffith ltr;* Col Samuel B. Griffith, II, ltr to Col Eustace R. Smoak, dtd 3Mar52; Lt Robert B. Pape ltr to CMC, dtd 22Feb51; LtCol William D. Stevenson ltr to CMC, dtd 22Feb51, hereafter *Stevenson ltr;* Adm Richmond K. Turner ltr to CMC, dtd 22Feb51; Col Samuel B. Griffith, II, "Corry's Boys," *Marine Corps Gazette*, v. 36, no. 3 (Mar52) and "Action at Enogai," *Marine Corps Gazette*, v. 38, no. 3 (Mar54); TSgts Frank J. McDevitt and Murrey Marder, "Capture of Enogai," *Marine Corps Gazette*, v. 27, no. 9 (Sep43); Morison, *Breaking the Bismarcks Barrier;* ONI, *Combat Narratives X;* Rentz, *Marines in the Central Solomons.*

691-360 O—63——9

Factors involved in risking a secondary attack north of the airfield had been carefully considered before a decision to land at Rice Anchorage at the mouth of the Pundakona (Wharton) River in Kula Gulf had been reached. Two areas—the Pundakona and the Piraka River in Roviana Lagoon—were scouted before the former was selected. Admiral Turner's staff reasoned that a landing from Roviana Lagoon would be unopposed but that the resultant overland trek would be excessively slow, fatiguing, and difficult to resupply. Further, this landing would not bring the enemy under immediate attack. Despite the native trails crossing the island, a large force could not travel fast enough through the jungle to give assistance to the expected rapid seizure of Munda.

On the other hand, a landing at Rice Anchorage would likewise be unopposed, and the enemy could be taken under attack almost immediately. This would force the Japanese into one of three courses of action: withdrawal to either Munda or Vila, a counterattack in strength, or an attempt at defending the Bairoko Harbor area. The latter course, it was believed, would be the logical enemy reaction to such a threat to the Munda-Vila link. Defense by the enemy at Bairoko would keep that garrison from reinforcing Munda. Though the disadvantages of making a landing on a narrow, confined beach on the Pundakona River nearly outweighed the advantages, the Rice Anchorage attack held the most hope for success in dividing the Munda-Bairoko forces. (See Map 8.)

Liversedge's group, augmented by the 3d Battalion, 148th Infantry, was given a multiple mission in NGOF orders. After landing at Rice, the Northern Landing Group was to move overland to the southwest, capturing or killing any enemy forces encountered in the Bairoko and Enogai Inlet area. After establishing road blocks across all roads leading from Bairoko to Munda, the NLG was to advance along the Munda-Bairoko trail as far as possible to prevent any enemy supplies or reinforcements reaching Munda, and also to block any withdrawal from that area. Contact with the right (north) flank of the 169th Infantry was to be maintained by Liversedge's command.

The Marine-Army force had only a limited knowledge of the terrain between Rice and Enogai Inlet and practically no information on Dragons Peninsula, the area between Bairoko and Enogai. For one thing, no oblique angle aerial photographs of the area were available. This type of aerial intelligence was particularly desirable, since jungle terrain photographed from high altitudes directly overhead rarely revealed anything of tactical value. In addition, the peninsula had not been scouted. The New Georgia guides had been reluctant to enter this area, fearing treachery because of vague rumors that the natives of this area were hostile to men from Roviana Lagoon.

Most of the SoPac reconnaissance patrols had been more concerned with Munda where the main effort of the NGOF was to be made. Those few patrols which ventured into the vicinity of Enogai Inlet were forced to turn back by close brushes with Japanese patrols. Only the long, narrow Leland Lagoon which borders the north shore of Dragons Peninsula had been patrolled, and this had been done in canoes. As a result, the dark stretches of jungle between Bairoko and Enogai were still an unknown area.

With the date of the landing set for 4 July, a one-day postponement was granted to allow another 37th Division unit, the 3d Battalion of the 145th Infantry, to join Liversedge's force. Unexpectedly, the 4th Raider Battalion was still engaged in the Viru Harbor attack and could not be withdrawn in time to join the NLG.

It was a lightly armed force. The only weapons carried, other than individual arms and light machine guns, were the 60mm mortars of the raiders and the 81mm mortars and heavy machine guns of the Army battalions. Noticeably lacking in artillery support, the NLG expected to have air power available upon request.

Shortly after midnight, 5 July, a covering bombardment of Kolombangara and Bairoko by a cruiser-destroyer force began on schedule. Prompt retaliatory fire from enemy shore batteries at Enogai surprised the task force, however, because the presence of large guns at Enogai as well as Bairoko had not been reported. In a matter of moments, part of the covering fires was shifted to these new targets and the bombardment continued. The destroyer *Strong* was the only task force casualty; it was hit at 0046, not by shellfire but by a torpedo fired by a Japanese destroyer running along Kolombangara's northeast shore.[2] The ship sank fast, but most of the crew was saved.

The actual landing of the Liversedge group started about 0130 in the midst of a torrential downpour and sporadic shellfire. For a short time the success of the amphibious venture seemed in serious doubt. Rice Anchorage could not be located in the darkness and rain. The transport group slowed and waited—uncomfortably remembering warnings of a lurking enemy submarine force—while one destroyer with a sweep radar probed ahead, seeking the anchorage. After a short delay, the Pundakona River mouth was located and the transport group moved into debarkation positions. As Marines and soldiers clambered into landing craft alongside the APDs, enemy star shells glimmered through the rainy darkness and shells splashed among the transports. This fire the raiders shrugged off with the comment, "erratic and inaccurate,"[3] but it was disconcerting, too.

A shallow bar blocking the entrance to Rice Anchorage further delayed the landing operation. The landing craft, each towing 10 men on a rubber raft, were forced to return to the transports to lighten loads before crossing the reef. Some of the rations were unloaded before the boats returned for a second try. Scouting reports had termed the beach as "small." The raiders found this almost an understatement to describe the narrow stretch of landing area hacked out of the jungle on the south side of the river, about 500 yards upstream from the anchorage. While four boats at a time beached to unload troops and supplies, the other boats jammed the river mouth or idled in Rice Anchorage waiting for a turn to unload. The black night obligingly curtained the milling confusion.

Ashore, drenched Marines and soldiers stumbled about the confined beach, slipping in the mud and tripping over hidden banyan roots. Since enemy shellfire ranged overhead to hit about 2,000 yards

[2] Most sources credit this successful torpedoing to enemy destroyers which were fleeing the task force; however, one official postwar source assigns credit to a submarine. NavHistDiv, Off of the CNO, ND, *United States Naval Chronology in World War II* (Washington, 1955), p. 53.

[3] *1st RdrBn WarD*, p. 1.

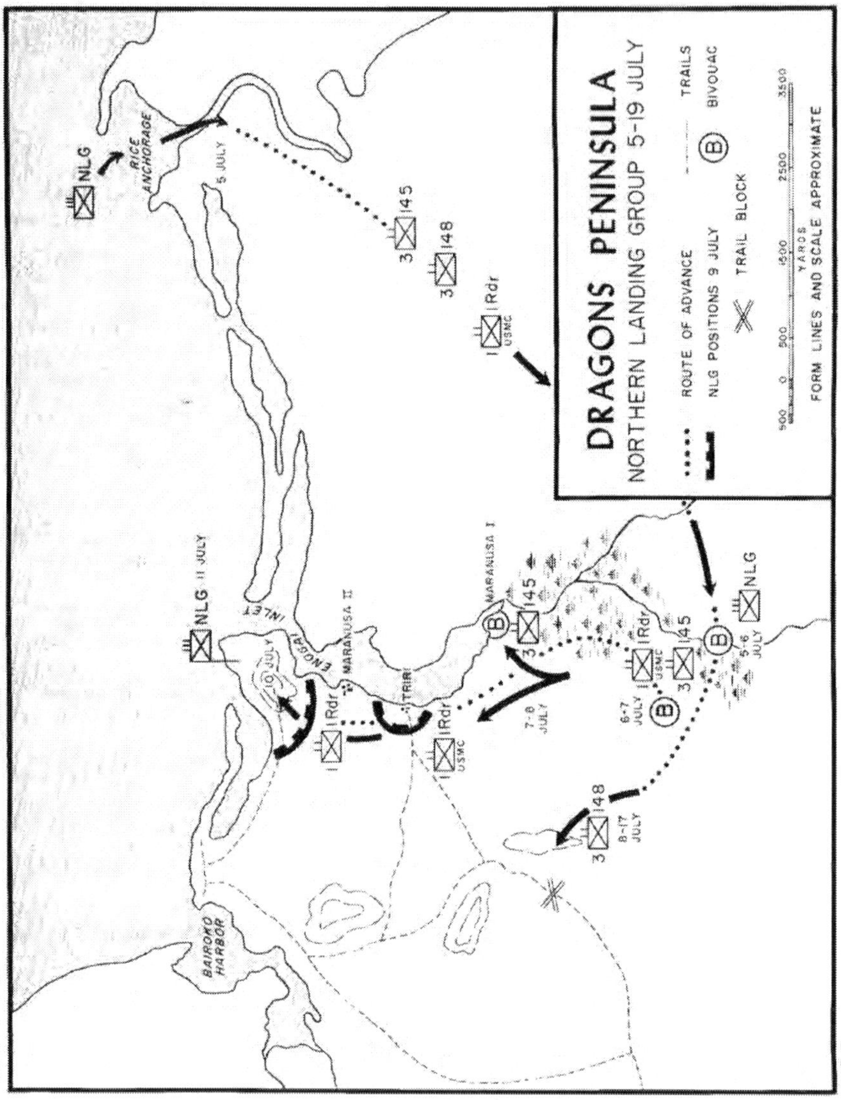

DRAGONS PENINSULA
NORTHERN LANDING GROUP 5-19 JULY
ROUTE OF ADVANCE
NLG POSITIONS 9 JULY
TRAIL BLOCK
TRAILS
BIVOUAC
YARDS
FORM LINES AND SCALE APPROXIMATE
RICE ANCHORAGE
ENOGAI INLET
MARANUSA II
MARANUSA I
TRIRI
BAIROKO HARBOR
NLG
5 JULY
10 JULY
11 JULY
5-6 JULY
6-7 JULY
7-8 JULY
8-17 JULY

MAP 8

farther up the coast, Liversedge's officers decided that the Japanese at Enogai were not aware of the exact location of the landing and risked the use of hooded flashlights. Thereafter, the unloading and reorganization proceeded more smoothly. Near dawn, with almost all supplies ashore, Colonel Liversedge broke radio silence with one uncoded word, "Scram." The anxious APDs and destroyers, unhappily expecting enemy retaliation at any moment, quickly turned and headed back for the southern Solomons. The landing, although delayed, had been accomplished without serious mishap. One unit, an Army company, was taken to the wrong landing area; it went ashore farther north along the coast. The company rejoined the main body later in the day.

The NLG had been welcomed ashore by a mixed greeting committee. Heading a large group of native guides and carriers—who were obviously frightened and bewildered by the sudden influx of so many white men to their island—were an Australian coastwatcher, Flight Lieutenant J. A. Corrigan, and a Marine raider patrol leader, Captain Clay A. Boyd of Liversedge's regiment. Corrigan had been on the island for some time, radioing reports of enemy activity in Kula Gulf and recruiting a labor force of nearly 200 natives. Small, wiry men with powerful arm, back, and leg muscles, the native carriers were to receive one Australian shilling, a stick of trade tobacco, and two bowls of rice and tea each day for carrying ammunition and rations for the NLG. A few spoke pidgin English, a jargon of simple words which bridged the language barrier. Colorful in cotton "lap laps" wrapped around their waists, they were intensely loyal to the coastwatchers.

Boyd had made several scouting trips to New Georgia. The last time, in mid-June, he and his men had remained with coastwatcher Harry Wickham to direct the landings at Onaiavisi Entrance and Zanana Beach before cutting across the island to link up with Liversedge. After the arrival of the NLG at Rice Anchorage, he resumed command of his company in the 1st Raider Battalion.

On one of his earlier trips, Boyd had scouted a trail leading from Rice to Enogia, and Corrigan's natives had then chopped a parallel trail on each side of this track. After the NLG stacked all excess ammunition, equipment, rations, and blanket rolls in assembly areas prepared by the natives, the march to Enogai started over these three trails. Companies A and B of the 1st Raider Battalion (Lieutenant Colonel Samuel B. Griffith, II) were assigned to move along the left trail (southern) with the demolitions platoon of the raiders heading along the right (northern) path. Thus protected on each flank, the main elements of the NLG started along the center trail with the remaining Marine companies leading and the two Army battalions following. Two companies, M and I, of Lieutenant Colonel George G. Freer's 3/145 with a medical detachment, communications unit, and the antitank platoon remained behind to guard the supply dump.

Scouting reports had termed the Rice-Enogai area an open jungle with small, broken hills and few swamps. Rain-soaked Marines and soldiers, struggling over the sharp, irregular slopes made treacherous by the mud and hidden roots, could not agree. The meager trails, hardly more than narrow defiles gashed through the thick, sodden jungle, were trapped with sprawling banyan roots slick with green moss, coral outcroppings, vines, and underbrush. The rain continued unabated.

The Army battalions, carrying heavier weapons and more ammunition and gear than the lightly equipped raiders, were forced either to travel at a slower pace or to stop to establish another supply dump. The soldiers, choosing to keep going, pushed on but dropped steadily behind. The leading NLG units, heading deeper into the New Georgia jungle on a course roughly south from Rice, reached the Giza Giza River late in the afternoon and set up a perimeter defense. Shortly after dark, all units were in bivouac on both sides of the Giza Giza. The NLG estimated that it had traveled eight miles during the first day. Actually, progress had been only about five miles, but undoubtedly the hardship of jungle travel had helped give every indication of greater distance. (See Map 8.)

That night, men of the Northern Landing Group listened to the distant sounds of a naval battle in Kula Gulf. A U.S. cruiser-destroyer force had intercepted a group of 10 Japanese destroyers, 7 of them transporting reinforcements. In a short but violent action, the U.S. force lost the light cruiser *Helena*. The Japanese lost two destroyers but managed to land 850 troops at Vila.

At daybreak on 6 July, the NLG stirred from its wet bivouac and resumed winding its way through the dripping jungle toward Enogai. The trails chopped by Corrigan's natives ended abruptly at the river, and the Marines were forced to slash their way through the mangrove swamp lying between the Giza Giza and the Tamakau Rivers. Rain continued to drizzle through the jungle canopy. The battalions became one thin, straggling line snaking its way through the swamp on an indistinct trail. The rains drowned the radio equipment, and communication wire laid along the trail was grounded as the protective covering peeled off in the hands of the infantrymen who used the wires as guidelines. Runners carrying field messages kept Liversedge in contact with his base at Rice.

The NLG had divided into two segments early that morning. Lieutenant Colonel Delbert E. Schultz had been directed to take his 3d Battalion, 148th Infantry, along another trail to the southwest to cut the Munda-Bairoko road and establish a block there. The remainder of the NLG continued toward the Tamakau River. Captain Boyd, leading Marine Company D, reached the river shortly before noon. Much to his dismay, the small stream he had scouted early in June was now a raging, flooded torrent. The raiders and soldiers paused while equipment was carried across or ferried on rafts made from branches and ponchos. Then the infantrymen began crossing the river, single-file, over a fallen tree which bridged the swollen stream. A rope stretched from bank to bank provided an unsteady guideline, and strong swimmers fished from the river those individuals who were unfortunate enough to slip and plunge into the water.

The crossing delayed the NLG until late in the afternoon. While intermittent rains continued, the Liversedge force bivouacked for the night of 6–7 July in the midst of a swamp. Muddy and tired, the raiders and soldiers swallowed canned rations and huddled in wet ponchos under banyan roots, waiting for dawn.

Late that night, in answer to a plea from NGOF headquarters, Liversedge broke radio silence to give a progress report. A listening watch had been set up at all halts, but the NLG commander had not used the radio in the hope that his cross-country march was still a secret. Although Liversedge carried medium-powered radios (TBX), contact with Hester some 20 miles away was made with difficulty. Such communication problems were to seriously handicap NLG operations. A high-powered radio, deck-loaded on one of the APDs, had not been unloaded during the anxious landing operations and was now back at Guadalcanal — a logistics oversight which was to be regretted many times.

First contact with the enemy came shortly before noon on the 7th of July. Liversedge's wet and hungry men struggled out of the swamp early in the morning and moved along a tortuous ridge paralleling the west bank of Enogai Inlet. The delay caused by the Tamakau and the swamps was emphasized when the sounds of an air strike at Enogai were heard. This had been the designated day for Liversedge's assault of that strongpoint. After moving through the native village of Maranusa I without incident, the point platoon of the NLG suddenly encountered seven Japanese riflemen. Surprise to both forces was apparent, but the Marines recovered first. In a brief fight, two of the enemy were killed before the rest fled. Uniforms identified the dead as members of the SNLF, probably from the *Kure 6th* at Enogai.

Realizing that the fight had warned the Enogai garrison of an attack through the back door, Liversedge increased the speed of the advance. Griffith was directed to take his raider battalion forward as quickly as possible to take advantage of any remaining element of surprise and to screen the advance of the rest of the force. The next brush with the enemy came as suddenly as the first. The demolitions platoon, meeting a strong enemy patrol, withdrew slightly to high ground and engaged the Japanese in a hot fire fight. Boyd's Company D then flanked the enemy and killed 10 before the Japanese fled. The brief fight cost the raiders three killed and four wounded. By nightfall, Griffith's Marines had occupied the native village of Triri on Enogai Inlet. Liversedge's CP was set up at Maranusa I with the NLG reserve units, Companies K and L of the 145th Infantry. Hasty perimeters were placed around each village.

The absence of defensive works at Triri further convinced the Marines that the Japanese at Enogai had not been expecting an attack from the direction of the inlet. The only item of value found at Triri was a detailed enemy map which pinpointed the location of four 140mm guns at Enogai. As Griffith's battalion prepared hasty defensive positions, the document was rushed to Liversedge at Maranusa I. The NLG commander immediately radioed for an air strike to knock out these weapons, but his message failed to raise a response from either NGOF headquarters or the 43d Division. An Army radio station at Viru, hearing the request for a relay, accepted the message for transmission to ComAir New Georgia.

Early on the morning of 8 July, Griffith hurried two platoons down divergent paths north and west from Triri to ambush any enemy probing attacks. The Marines on the west trail scored first. A Japanese patrol of near-company strength, blundering along the trail without advance security, walked into the trap.

Premature firing, however, spoiled any surprise effect and the enemy withdrew without difficulty. Within a few minutes, a full-scale counterattack had been directed at the Marine ambushing party, and Griffith rushed Boyd's Company D forward to help hold the trail. In the meantime, Colonel Liversedge picked up his command post and the two Army companies and rushed to Triri to be closer to the conflict.

The fight continued for three hours, the close jungle terrain handicapping the observation and maneuvering of both forces. Company C (Captain John P. Salmon), moving forward to relieve Company D under fire, broke the deadlock with a 60mm mortar barrage and continuous machine gun fire. As the raiders moved forward, the enemy disengaged and fled down the trail. Fifty dead Japanese were left littered about the scene of the fight.

The Marines did not pursue. Enogai was the first objective. While the Army companies took over the defense of Triri, the raider battalion hastily reorganized and resumed the march toward Enogai along the north trail where the second ambush force had set up. The trail, however, ended abruptly in an impassable swamp. Reluctantly, after considerable time had been spent in trying to find an acceptable trail to Enogai, the battalion commander ordered the return to Triri for another start the following day.

Meanwhile, the Japanese force had reinforced and reorganized for another attack on Triri. Late on the afternoon of the 8th, an estimated 400 Japanese struck quickly at the left flank of the thin perimeter established by the two Army companies. The lines of Company K of the 145th slowly began to give way under the continuing pressure of the enemy assaults. Company L, on the right, received only scattered sniper fire. The demolitions platoon of the raider battalion, which had remained behind with Liversedge's CP, rushed to assist Company K in its defense just as Griffith's battalion returned. On orders from Liversedge for a quick counterattack, Griffith directed First Lieutenant Robert Kennedy's platoon from Company B to circle back and hit the left flank and rear of the Japanese. Kennedy's countermove completely surprised and crushed the enemy's left flank. The Japanese fled once more. Another 20 enemy dead were left behind. Company K, which had three soldiers wounded, estimated that 75 additional Japanese had been killed in the attempted breakthrough. Kennedy's platoon suffered no casualties.

CAPTURE OF ENOGAI[4]

After a quiet night at Triri, the Marines again started toward Enogai the following morning. A radio team with a TBX and headquarters personnel of the raider regiment remained behind with the Army companies, but Liversedge moved out with Griffith's battalion. The raiders had more luck this day. A good trail, apparently unknown to the Enogai garrison, was discovered and rapid progress was made by the Marines. Sounds of an air strike at Enogai indicated that the request for the destruction of the enemy guns there was

---

[4] Unless otherwise noted, the material in this section is derived from: *NGOF Account; New Georgia Campaign; 1st RdrRegt MsgFile; 1st RdrRegt SAR; 1st RdrRegt Jnl; 1st RdrBn WarD; Griffith ltr; Stevenson ltr;* Pape ltr, *op. cit.;* Boyd interview, *op. cit.;* Rentz, *Marines in the Central Solomons;* Griffith, "Action at Enogai," *op. cit.;* McDevitt and Marder, "Capture of Enogai," *op. cit.*

CASUALTIES *from the fighting on Dragons Peninsula are readied for evacuation from Enogai by PBY.* (*USMC 182121*)

COLUMN OF MARINE RAIDERS *crosses a jungle stream near Enogai during the active patrolling in August.* (*USMC 60166*)

being executed. Shortly before noon, Leland Lagoon was sighted and the Marine battalion turned east toward the enemy defenses at Enogai Point. After several hours of cautious approach, the raiders were halted by the stutter of two light machine guns. The Marines paused for battle orders. As they waited, the volume of enemy fire picked up. The Enogai defense line was being reinforced.

The attack was made without mortar preparation—Company A (Captain Thomas A. Mullahey) with its left flank resting on the lagoon, Salmon's Company C in the center, and Company B (Captain Edwin B. Wheeler) on the right flank. Boyd's Company D was held in reserve. The frontal assault, made with grenades and machine guns, was beaten back. With the jungle daylight fast closing into deep twilight, Liversedge called off the assault. Griffith was told to hold in place and resume the attack the following morning.

The Marines' defensive positions, facing commanding ground, were not to Liversedge's liking, but the NLG commander wanted to keep the pressure on the enemy during the night and decided to risk a Japanese counterattack. The gamble paid off. The night passed without incident except for the sudden crash of a huge, bomb-weakened banyan tree in the command post area which crushed one raider, injured three others, and completely smashed the command's TBX.

Breakfast on the morning of the 10th was not a problem for the raiders who had not eaten since the morning of the 9th. There was no food. After a few quiet orders from Griffith, the 1st Raider Battalion renewed the attack. Wheeler's Company B on the right front reported no opposition and moved forward rapidly. Companies A and C, as expected, however, were hit by intense fire from rifles and automatic weapons. The two companies paused for a 60mm mortar barrage to soften the enemy line before plunging on. Company B, at last meeting strong defensive fire, raced through a small native village on the inlet's shore south of Enogai. Dead enemy were sprawled throughout the village. A number of machine guns were taken and turned about to put more fire on the fleeing Japanese. The breakthrough put raiders almost in the rear of the enemy lines. Opposition facing Company C in the center abruptly faltered, then scattered.

As enemy resistance began to crumble, the raider attack gained momentum. Behind a withering fire of automatic weapons and machine guns, the raiders moved through Enogai. Mortarmen, in positions on the high ground overlooking the village, dropped 60mm mortar shells along the shoreline of Kula Gulf, trapping the village defenders between two fires. Stragglers, attempting to swim across Leland Lagoon, were machine gunned by the raiders. By early afternoon, the coast defense positions were in raider hands, and only two small pockets of enemy resistance remained. These the Marines contained, postponing mopping-up operations until the next day. Late that afternoon, Company L of the 145th struggled into Enogai, each soldier carrying rations, bandoleers of ammunition, and three extra canteens of water. Without food for more than 30 hours, the raiders had been reduced to catching drinking water in ponchos during the intermittent rains.

The food was part of an air drop which the rear headquarters at Triri had received early on the morning of the 10th. Liversedge had requested the drop the previous day. The original three-day supply of rations carried ashore at Rice Anchorage had been stretched over five days, and fresh water was also scarce. Wounded were fed wormy rice which had been found at Triri. The situation had become tense—so serious, in fact, that the Marines were far more concerned with the prospect of continued diminished rations than they were with the threat of having another enemy garrison in their rear at Bairoko.

Anxiety increased when the planes appeared over Triri on schedule but could not locate the purple smoke grenades marking the NLG positions. An air liaison officer finally made contact with the planes and directed the air drop. Parachutes drifted down, and soldiers and Marines dodged the welcome "bombing" to collect the bulky packages. The first containers opened held only mortar shells, and the troops howled their disappointment. K-rations and chocolate bars soon followed, however. An immediate relief party was organized to carry supplies and water to Griffith's battalion, then hotly engaged at Enogai.

That night the Marines dined on K-rations and Japanese canned fish, rice, and *sake*. The captured enemy rations were liberally seasoned with soy sauce found in several large barrels. Articles of Japanese uniforms were used to replace the muddy and tattered Marine uniforms. The evening passed without further activity, the Marines resting easily behind a perimeter defense anchored on Leland Lagoon on the right flank and Enogai Inlet on the left. The defenses faced toward Bairoko. During the night, Japanese barges were heard in Kula Gulf and the raiders scrambled for positions from which to repel an enemy counterlanding. The Japanese barges, however, were only seeking to evacuate stragglers from the sandspit between Leland Lagoon and Kula Gulf.

The following morning, mop-up operations began with Companies A and D moving quickly through the two remaining points of opposition, although Company D was hard-hit initially. Only a few Japanese were flushed by the other patrols, and these the Marines killed quickly. The 1st Raiders now owned all of Enogai Point between Leland Lagoon and the inlet. Japanese casualties were estimated at 350. The raiders, in moving from Triri, had lost 47 killed in action and 74 wounded. Four others were missing and presumed dead. The wounded were placed in aid stations housed in the thatched huts at Enogai.

The four 140mm naval guns, three .50 caliber antiaircraft guns, and numerous machine guns, rifles, and small mortars were captured, in addition to large stocks of ammunition, food, clothing, two tractors, and a searchlight. Allied bombardments and bombings had not materially damaged any of the Enogai installations.

The Japanese retaliated quickly on the morning of the 11th with a bombing attack which lasted for more than an hour and left the Marines with 3 more men killed and 15 wounded. Three American PBYs were called in that afternoon to evacuate the more seriously injured, and, after landing at Rice Anchorage, the big flying boats taxied along the shoreline to Enogai where the wounded were loaded aboard from rubber rafts. Shortly before takeoff, the PBYs were bombed and strafed by two enemy floatplanes. The Marines on shore fired everything they had at the attackers, including small arms and captured weapons, but the Japanese went unscathed and the PBYs hastily departed for Guadalcanal. On the same afternoon, headquarters personnel of Liversedge's CP arrived at Enogai and, at 2100, seven landing craft from Rice made the initial supply run into the inlet.

TRAIL BLOCK ACTION[5]

After splitting with Liversedge's main force early on the morning of 6 July, Lieutenant Colonel Schultz started his 3d Battalion, 148th Infantry down a trail which his sketch map showed would put him in position to intercept Japanese traffic over the Munda-Bairoko trail and establish the road block which Liversedge had directed. The Army battalion was hardly on its way down the new trail when one of Corrigan's native guides—looking at Schultz' map—insisted that the map was wrong. The Army commander, relaying this information to Liversedge by field message, reported that he was going to press on in the hope that the trail would cross the Munda-Bairoko trail at some point.

The soldiers moved down the inland trail without undue difficulty; the ground was more rolling and less swampy than in the coastal area. Crossing the Tamakau proved no problem farther upstream, and, late on the afternoon of 7 July, Schultz informed Liversedge that he had reached a trail junction which he believed to be the main Munda-Bairoko trail and that a block would be established the following morning. Footprints on the trail, evidence of recent use, convinced Schultz that he had indeed reached his objective. He also requested that rations be carried to him, and reported that the native carriers had become apprehensive and had returned to Rice.

The next morning, 8 July, Schultz set up his road block. Company I defended the approach from the north and Company K the approach from the south. Company L filled in a thin perimeter between I and K.

The first enemy contact was made shortly after 1300 when a squad of Japanese, sighted coming down the trail from the north, was taken under fire. The fight was brief, the enemy quickly fleeing back toward Bairoko. Two hours later a full-scale attack by 40 to 50 Japanese hit Company I's outposts, driving them back into the perimeter, but the enemy did not penetrate the battalion's defenses. The afternoon's engagement cost the Japanese about 7 killed and 15 to 20 wounded. One American was killed and three others wounded.

After a quiet night, Schultz sent patrols forward on each trail in an attempt to locate the enemy. No contact was made in either direction, although an abandoned enemy bivouac area was discovered about two miles down the Munda fork of the trail. Schultz also tried to contact the 169th Regiment, by this time supposed to be well on its way to Munda field. Unknown to Schultz, neither force was in position to make contact. That night, after listening to reports from his patrols, Schultz reported to Liversedge that he believed himself to be about six miles north of Munda.

Early on the morning of 10 July, the battalion was hit on the right flank by about 50 Japanese and then on the left flank by a larger force of about 80 men. Both probing attacks were repulsed, the

[5] Unless otherwise noted, the material in this section is derived from: *NGOF Account*; *New Georgia Campaign*; *1st RdrRegt SAR*; *1st RdrRegt Jnl*; *1st RdrRegt MsgFile*; *3/148 Rept*; Rentz, *Marines in the Central Solomons*.

Japanese losing 14 killed in the two skirmishes. After a number of similar searching attacks, the Japanese suddenly unleashed a strong attack on the right flank at the junction of Companies I and L. The Army positions were quickly overrun, but an orderly withdrawal was made. The Japanese force, estimated at more than two companies, quickly occupied a small ridge and set up a number of automatic weapons and heavy machine guns.

Company K hurriedly organized a strong counterattack with the battalion's reserves, but the enemy's hold on the rise remained intact. An 81mm mortar barrage, which Schultz directed to be placed along the ridge, kept the enemy from continuing the attack further. The following day, 11 July, Company K attacked again toward the ridge, but was driven back. A later attempt by the same company was also repulsed. Casualties, however, in both attacks were few. That night, Company K was hit in return by a bayonet charge. The *banzai* attack was beaten back with only three soldiers being wounded.

Schultz' force, by now just as ill-fed and unkempt as Griffith's battalion, was on ⅓ rations. The food problem had become more acute on the afternoon of the 11th when Company I of the 145th Infantry arrived from Triri to reinforce Schultz' battalion. A food drop that same afternoon had been greatly disappointing. As Schultz had predicted in an early report to Liversedge, the jungle prevented aircraft from spotting either flares or colored panels. Consequently, the air drop was wide of the mark. Schultz' men, engaged closely with the enemy, could recover only a few of the packages, and these contained mostly mortar shells. Most of the ammunition was found to be outdated and of the wrong caliber, and nearly all the rations were spoiled. Little of either could be used.

The next morning, Company I of the 145th moved up to the rear of the 148th's positions and then lunged forward toward the ridgeline, following a heavy machine gun and mortar preparation. The position had been abandoned. The absence of any dead or wounded enemy indicated that the withdrawal had been effected during the night. The positions around the trail block were restored, and, with the arrival of some natives with rations from Triri, the situation began to look brighter. Defense of this area had cost Schultz 11 killed and 31 wounded. Japanese casualties were estimated at 150.[6]

ENOGAI: 12–19 JULY[7]

Another attempt by the Japanese to reinforce Vila and Munda through Kula Gulf was partially blocked shortly after midnight on 12–13 July. An Allied force of 3 cruisers and 10 destroyers ambushed 4 enemy transports escorted by several destroyers and a light cruiser. Enemy torpedos damaged two U.S. cruisers, the *Honolulu* and *St. Louis*, and the New Zea-

[6] Trail block casualties reported by Schultz in 148th Infantry's After Action Report, quoted in *New Georgia Campaign* p. V–21, totaled 11 killed and 29 wounded with 250 estimated Japanese casualties. The figures given here are from the original day-to-day reports from the trail block.

[7] Unless otherwise noted, the material in this section is derived from: *NGOF Account; New Georgia Campaign; 1st RdrRegt SAR; 1st RdrRegt Jnl; 1st RdrRegt MsgFile; 1st RdrBn WarD; 4th RdrBn Jul43 WarD; SE Area NavOps—I; SE Area NavOps—II;* ONI, *Combat Narratives X;* Morison, *Breaking the Bismarcks Barrier;* Rentz, *Marines in the Central Solomons.*

land cruiser *Leander*. The U.S. destroyer *Gwin* was sunk, and two others were damaged slightly in a collision. The Japanese lost only one cruiser in the engagement, and managed to land 1,200 troops on Kolombangara. The battle, however, persuaded the Japanese to abandon further attempts to run the gantlet of Kula Gulf. Thereafter, the enemy resorted to attempts to sneak barges through the waters west of Kolombangara. The battle also lessened the threat of a counterlanding against Liversedge's force.

At Enogai, the possibility of such an enemy attempt had been considered and the defenses of the captured village strengthened and extended. Marines strung captured barbed wire from Enogai Inlet across the point to Leland Lagoon and constructed defensive positions behind this line, but the Japanese did not attempt to regain the area. Enemy bombing attacks, too, became less frequent.

Enogai became the new NLG command post. Liversedge directed that supplies at Rice Anchorage be moved to the new CP, and, with the exception of a small detail, 3/145 moved to Triri. Rice then became a relay point where APDs anchored to unload supplies into landing craft. The busy small boats then skirted the shoreline to Enogai, carrying supplies to the NLG and evacuating wounded on the return trip.

For some time, Liversedge had been concerned about his tactical situation. His original orders had given him the dual mission of capturing or destroying Japanese in the Bairoko-Enogai area while blocking the Munda-Bairoko trail, but the distance between his command post at Enogai and Schultz' trail block was too great for effective control. A new landing area on the upper reaches of Enogai Inlet made resupply and evacuation of the trail block easier by eliminating much of the overland hike, but the combined boat trip and march still took considerable time. Moreover, General Hester on 9 July had insisted that Liversedge keep his battalions within supporting distance of each other. So, as soon as Enogai had been captured and a defensive perimeter established at Triri, the Marine colonel turned his attention to the trail block where Schultz' battalion had suddenly found itself facing first a determined enemy of considerable strength and then no enemy at all.

Following the withdrawal of enemy forces from the trail block area on 12 July, no further Japanese troops had been encountered. Combat patrols, hitting along the Munda-Bairoko trail in both directions, failed to make contact. With Munda under heavy attack, this seemed surprising since it appeared logical that the Japanese would make some attempt to reinforce the airfield. Disturbed by the reports from the trail block, Liversedge sent his operations officer Lieutenant Colonel Joseph J. McCaffery, to check Schultz' position. McCaffery left Triri on the morning of the 13th accompanied by part of the regimental staff and the 145th's Company K.

He later radioed Liversedge that the situation at the trail block was "okay," and that the defense of the trail was tight and not split as had been reported. Rations were needed badly, since natives could not carry enough supplies to support the augmented trail block force and the front lines could not be weakened to supply carriers. An air drop was requested.

By this time, however, Liversedge was already en route to the trail block for a personal reconnaissance. The NLG commander left Enogai with a small patrol on the 15th of July and, after bivouac on the Triri trail, joined McCaffery and Schultz early on the morning of the 16th. One day at the defensive position was enough to convince Liversedge that the trail block should be abandoned. Schultz' battalion, unable to contact the 169th and at a considerable distance by boat and foot from the supply base at Enogai, was definitely out on a shaky tactical limb. Moreover, 3/148 was in a weakened condition, and many soldiers were ill from eating contaminated food. Their ability to ward off a sustained attack was questionable. Resupply was a problem, too: very little of the rations dropped were recovered. The intended purpose of the trail block seemed to have been served:

> The presence of our force at the road block since 8 July had materially assisted in the capture of Enogai by holding enemy forces at Bairoko in position and preventing them from reinforcing their Enogai garrison. It further established the fact that the enemy was not using the Bairoko-Munda trail as a supply route.[8]

On the morning of the 17th, executing Colonel Liversedge's orders, Schultz directed his battalion to abandon the trail block, and the two companies of the 145th Regiment and 3/148 retraced the path to Triri. There the soldiers changed clothes, bathed, and ate a good meal after nearly two weeks in the jungle. Their rest was to be short-lived, though.

At Enogai, the Marines, now rested and well-supplied, had been actively patrolling the trails toward Bairoko. Enemy contacts after the capture of Enogai had been limited to an occasional brush between opposing patrols, which resulted in brief fire fights with few casualties to either side. The raiders lost one killed and one wounded during the period 13–17 July. Japanese planes, however, continued to make Rice Anchorage and Enogai a favored target. Each night enemy floatplanes droned over the NLG positions to drop bombs from altitudes of about 500 feet. No damage was inflicted, and no casualties resulted.

Patrol reports definitely established the fact that the Japanese intended to defend Bairoko Harbor. Several patrols reported glimpses of Japanese working parties constructing emplacements and digging trenches east of the harbor. Two-man scouting teams, attempting to get as close to Bairoko as possible, returned with the information that the high ground east of the Japanese positions had not been occupied by the enemy and that two good trails leading to this area had been found. The scouts reported that a battalion could reach this position in two and one-half hours. There was still no reliable estimate of the size of the defending force at Bairoko, however.

Upon his return from the trail block on the 17th, Liversedge was greeted with the news that Lieutenant Colonel Currin's 4th Raider Battalion would arrive the next day to augment the NLG. The NLG commander had requested this reinforcement shortly after the capture of Enogai. Major William D. Stevenson, the regiment's communication officer, had hitchhiked a ride on one of the PBYs carrying casualties out of Enogai on 11 July and had gone to Guadalcanal to relay Liversedge's request personally to Admiral Turner. Ten-

[8] *1st RdrRegt SAR*, p. 4. The Army's official history of this campaign notes in regard to this contemporary judgment that "knowledge gained after the event indicates that none of these beliefs was warranted." Miller, *Reduction of Rabaul*, p. 104.

tative approval for the reinforcement was given. After conferring with Currin, Stevenson returned to Enogai with supplies and mail.[9]

Early on the morning of the 18th, four APDs anchored off Enogai Point and the 4th Marine Raiders debarked, bringing additional supplies and ammunition with them. Liversedge, who had expected a full battalion, was taken aback when Currin reported his battalion nearly 200 men understrength. The captures of Viru Harbor and Vangunu, as well as recurring malaria, had taken their toll. Liversedge put the NLG sick and wounded aboard the APDs to return to Guadalcanal and turned his attention toward the seizure of Bairoko Harbor. The orders were issued late that afternoon, after a conference with his battalion commanders at the Enogai CP.

Approach to Bairoko was to be made by two columns. Two full-strength companies (B and D) of the 1st Raider Battalion and the four companies of the 4th Raider Battalion were to make the main effort, advancing along the south shore of Leland Lagoon straight toward Bairoko and the north flank of the Japanese positions. Schultz' battalion was to move from Triri toward Bairoko to hit the south flank of the Japanese positions. Freer's 3/145 was to remain in reserve at Triri and Enogai. The departure time was set for 0730, with an air strike scheduled for 0900 to precede the actual attack on the harbor defenses.

As soon as Liversedge's orders had been given, Schultz and Freer returned to Triri, and Currin and Griffith began a last reconnaissance. A reinforced platoon from Wheeler's Company B under the command of Second Lieutenant William J. Christie moved down the sandspit between Leland Lagoon and Kula Gulf to get into position for the morning's attack and to protect the seaward flank. At 1600, an air strike by 18 scout bombers and 19 torpedo bombers pounded the east side of Bairoko Harbor while 8 mediums strafed Japanese supply dumps and bivouac area. The strike marked the fourth time since 15 July that Bairoko had been worked over by ComAirSols planes.

That night Enogai was rocked in return by enemy bombing and strafing attacks that lasted nearly seven hours. Ten Marines were wounded. The NLG wondered: Had the enemy accurately guessed the date for the NLG attack or were the Japanese just giving as good as they had received in the air attacks of the previous days? If the former, enemy intelligence work had been much better than the NLG's.

Although the Liversedge force knew the general location and nature of the Japanese defenses at Bairoko, there was a disturbing lack of intelligence about the size of the Japanese garrison. The pre-landing estimate had been about 500 enemy at the harbor. The 350 Japanese encountered and killed at Triri and Enogai were identified as members of the *Kure 6th SNLF*. Schultz' attackers at the road block had not been identified, but were believed to have been from the Bairoko garrison. The NLG concluded—wrongly—that only about two reinforced companies held Bairoko.

At the time of the Rice Anchorage landing, Enogai was lightly defended by a detachment from Commander Okumura's *Kure 6th SNLF*. When Liversedge's force split on the second day, the Japa-

[9] *Stevenson ltr.*

nese believed that two regiments were attacking Dragons Peninsula and ordered half of the *2d Battalion, 13th Regiment* from Vila to Okumura's assistance. The reinforcements included a machine gun company. The new troops were to have been rushed to Enogai to defend the coast defense guns but the move was made too late. By the time the *2d Battalion* units reached Bairoko, Enogai had been captured. When Liversedge's intentions to continue the attack toward Bairoko became more evident, more reinforcements were rushed to the harbor. These included several companies of the *2d Battalion, 45th Regiment* and the *8th Battery* of the *6th Field Artillery Regiment* which had recently arrived from Bougainville.

Since contact with the enemy had been negligible after the capture of Enogai, the NLG had no basis for comparison of strength and were not aware of the added enemy capability to defend Bairoko. Patrols did not aggressively test the Japanese defenses; in fact, no probing attacks against the outposts guarding Bairoko were attempted. The only enemy prisoner taken during this period was a badly burned pilot, rescued from an offshore island and immediately evacuated. In effect, the NLG was facing an unknown quantity in its attack against the harbor.

"I HAVE COMMITTED THE WORKS"[10]

The approach to Bairoko by the raiders began over trails and terrain now familiar through much patrolling. Wheeler's Company B led the approach march with Company D (now commanded by First Lieutenant Frank A. Kemp, Captain Boyd having been evacuated with malaria), the demolitions platoon, Currin's 4th Raider Battalion, and the regimental command post following in column. The two companies in the 1st Battalion had been brought up to near-full strength for the attack by taking men from Companies A and C. These understrength companies remained behind with the 145's Company L at Enogai. (See Map 9.)

As the NLG file moved through the dripping jungle, scrambling over sharp coral rocks and climbing low but steep hills and ridges, the Marines waited to hear the first sounds of bombing and strafing which would indicate that the 0900 air strike on Bairoko's defenses was being executed. The raiders waited in vain—there would be no strike.

Unknown to Liversedge, his request was apparently made too late. The support strikes by ComAirSols for 20 July were already scheduled and the planes allotted. The NLG commander, however, did not know this. Considerable difficulty was encountered in transmitting the message on the afternoon of the 19th, but the message was finally cleared. Scheduling of the strike was not confirmed, Liversedge's communication officer recalls:

> Acknowledgment was requested, as I remember, but this acknowledgment did not come until night. It was actually nothing more than an acknowledgment of the re-

[10] Unless otherwise noted, the material in this section is derived from: *New Georgia Campaign; NGOF Account;* ComAirSols WarD, Jul43; ComAir New Georgia SAR, 29Jun–13Aug43; *1st RdrRegt SAR; 1st RdrRegt Jnl; 1st RdrRegt MsgFile;* 1st RdrRegt PtlRepts, *op. cit.; 4th RdrBn Jul43 WarD; Griffith ltr;* LtCol Anthony Walker ltr to CMC, dtd 23Feb51; LtCol Edwin B. Wheeler ltr to CMC, dtd 20Mar52; ONI, *Combat Narratives X;* Rentz, *Marines in the Central Solomons.*

> ceipt of the message by the staff officer on duty at the headquarters addressed.[11]

Without air support, the odds for success in capturing Bairoko lengthened considerably. Disappointed but determined, the two Marine battalions kept moving forward.

The first shots came shortly after 1015. A Japanese outpost opened fire on the NLG column, and Wheeler and Kemp quickly deployed their companies into attack formation. The outpost was overrun. Without pause, the raiders continued forward, feeling their way through the tangled jungle. At 1040, Griffith informed Liversedge by message that he was deployed and pushing forward against several machine guns.

Five minutes later, the raiders were in a violent, all-out battle. A sudden eruption of intense and accurate fire from close range raged at them. The Marine attackers were pinned down, closely pressed against banyan roots, logs, and coral outcroppings, unable to move against the withering fire from automatic weapons and machine guns which raked the jungle. Recovering quickly, the Marines returned the fire, the battle racket becoming louder. As the intensity of the firing increased, the din was punctured by hoarse shouts and curses as the Marines tried to maneuver against the murderous fire pouring from the jungle facing them.

Confronting the raiders was a series of log and coral bunkers dug into the rising ground under banyan roots, and well camouflaged with palm fronds and branches. The ridge ahead blazed with fire from these low fortifications. Similar to those encountered by the NGOF in its approach to Munda, the emplacements supported each other with lanes of interlocking fire. Further protection was furnished by Japanese soldiers in trees overhead who sniped at the Marines with Nambu (.25 caliber) light machine guns. Okumura had prepared his defenses well.

The pitched battle went on, both sides firing at a rapid rate. Wheeler's company, with its right flank near the end of the lagoon, was unable to move forward and could not make contact with Christie's platoon on the sandspit. Heavy firing across the lagoon indicated that Christie, too, was engaged. Kemp's company, on the left, finally regained fire superiority, however, and began to inch forward in an attempt to take high ground to the front. As casualties began to mount in both companies, Griffith moved his sole reserve unit—the demolitions platoon under Marine Gunner Angus R. Goss—to the left flank for protection from attacks from that direction. At 1105, after 20 minutes of furious combat, Griffith reported to Liversedge: "Harry: I have committed the

---

[11] *Stevenson ltr.* All NLG records, both Marine and Army, indicate that such a strike was expected. ComAirSols and ComAir New Georgia records, however, do not reveal any notation of the request. This particular incident, which illustrates NLG liaison difficulties with higher echelons, remains unresolved. Evidently a staff officer at ComAirSols, adhering to a policy that air support requests had to be received before 1600 on the day prior to the date of execution, took no action on the request. The XIV Corps G–3 Journal of 19 July contains a message from Liversedge, sent at 2235, 18 July, requesting a 12-plane strike on the 19th and a "large strike to stand by for July 20 A M and SBD's to stand by for immediate call remainder of day." Corps headquarters replied that a "large strike standby" for the 20th was "impracticable." Quoted in Miller, *Reduction of Rabaul*, p. 130*n*.

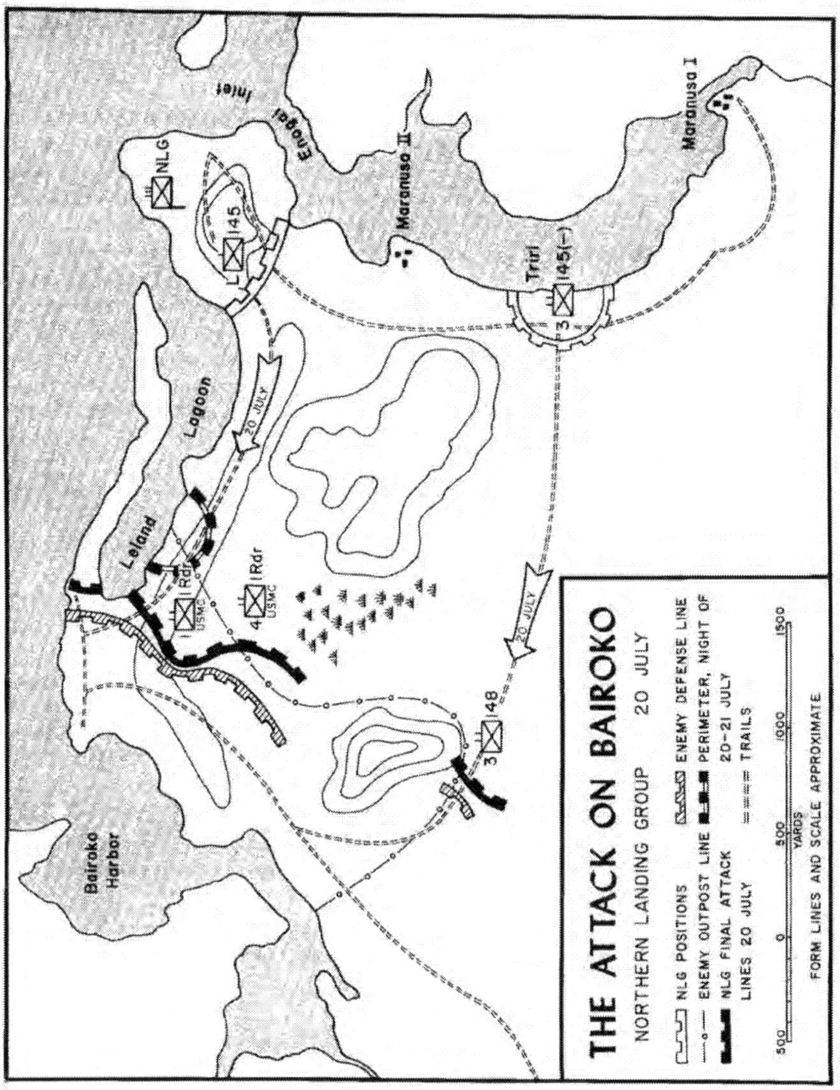
THE ATTACK ON BAIROKO
NORTHERN LANDING GROUP 20 JULY
NLG POSITIONS
ENEMY OUTPOST LINE
NLG FINAL ATTACK LINES 20 JULY
ENEMY DEFENSE LINE
PERIMETER, NIGHT OF 20-21 JULY
TRAILS
500 0 500 1000 1500
YARDS
FORM LINES AND SCALE APPROXIMATE
Bairoko Harbor
Leland
Lagoon
Enogai
Inlet
Maranusa II
Maranusa I
Triri
20 JULY
20 JULY
NLG
145
3 145(-)
3 148
4 Rdr USMC
1 Rdr
1 Rdr USMC

MAP 9

works. . . . Movement forward continues. Sam."[12]

By noon, the first line of enemy resistance crumbled, broken under the relentless pressure of the raider units. Unable to use the 60mm mortars because of the jungle canopy, lacking the new flame-thrower weapons, and without air or artillery support, the Marines breached Okumura's defensive line by knocking out first one pillbox and then another by demolitions and overwhelming small-arms fire. But losses were heavy and progress was slow.

Shortly after noon, with the 1st Battalion clearly needing quick assistance, Liversedge committed Currin's battalion to the fight. Company P (Captain Walker) was in close support behind Griffith's battalion, and thus able to move quickly into the line. Kemp's Company D, which had moved steadily ahead despite numerous casualties, was receiving heavy fire on its left flank and Walker now attacked toward this opposition. Goss' demolition platoon, in turn, circled through the rear of the 1st Battalion to take up a new position on Kemp's right flank to bridge the gap between Company D and Company B.

Walker's fresh company, under orders to attack southwest to the shores of the inlet before turning north to hit the enemy's right flank, was barely able to move forward before criss-crossing fire from both right and left flanks held it back. While Walker scouted his front lines to determine the location of the machine guns facing him, Captain Snell moved his Company N into position behind Walker's unit to refuse the left flank and support Walker's attack. The battle continued in full fury all along the line, the raider gains measured a yard at a time. Contact with the platoon on the sandspit still had not been made. Christie's unit, facing a marshy swamp backed by a strong line of Japanese fortifications, could not advance. Seven enemy machine guns, pouring a deadly fusilade over the swamp and along the shores of the sandspit, resisted every attempt at forward movement.

In the next two hours, the raider attack slowly punched through two different defensive lines, uncovering a number of bunkers on the reverse slopes. Company D, riddled with casualties by the heavy and continuous enemy fire, scrambled to the top of a ridge line which overlooked the harbor at Bairoko, about 500 yards away. But between the raiders and their objective lay another series of formidable fortifications. Hoping to cement Kemp's position on the commanding terrain, Liversedge directed First Lieutenant Raymond L. Luckel's Company O into the gap between Company D on the ridge and Company P. Both companies had been hit hard by several machine guns in this area, and Luckel's company was ordered to silence these weapons. As Company O lunged forward, the maneuver reduced fire on Company P and Company N. Walker and Snell then moved their companies forward to take a small ridgeline to the left front.

At this time, the NLG front lines arched in a wide U pointed towards the harbor with Company D as the leading unit. On the left flank, Currin had three companies, bent around to the southwest. Griffith's two companies and the demolitions platoon, on the right, had managed to move nearly to the end of the lagoon, but a slight gap still existed between the bat-

[12] *1st RdrRegt MsgFile*, Griffith to Liversedge, dtd 20Jul43.

talion and the lagoon's shoreline. Liversedge, in an attempt to plug this gap and try once more to contact Christie, moved First Lieutenant Leonard W. Alford with a reinforced platoon from Company O to this flank. Alford's platoon made a spirited attack, but the volume of enemy fire prevented movement beyond that of Wheeler's company. The move, however, tied Christie's platoon closer to the main NLG line.

At 1445, sporadic but accurate mortar fire from enemy positions on the inlet suddenly changed into an intense barrage that shook the attacking lines. The Marines, without weapons for counterbattery fire, could only press closer into their shallow positions behind scant cover on the ridge lines and try to weather the pounding. Estimated to be 90mm rounds, the shells inflicted further casualties, mainly from tree bursts overhead. The barrage was immediately followed by a screaming counterattack. Kemp's company, bearing the brunt of the enemy charge, was pinned between searing fire from the front and the mortar shelling. Withdrawing to the first ridge taken, Kemp organized a counterattack of his own, and with a badly depleted company stormed back to his old position in a sudden rush. The quick conquest was the first visible crack in the Japanese defenses. Marines reported the enemy fleeing, many of them without weapons. Griffith sent a hasty note to Liversedge, advising the NLG commander that the addition of just one company (L of 3/145) would take Bairoko by night. The Japanese, Griffith believed, were on the run, but casualties were heavy among the raiders and reinforcements would be needed.

Unfortunately, there were no ready reserve units. Nothing had been heard from the Army battalion which was supposed to hit the south flank of the enemy, but sounds of firing from that direction indicated that Schultz was engaged. Freer's battalion, scattered between Rice, Triri, and Enogai was not in position to help, even if those bases could have been left unguarded. Company L at Enogai had been ordered to the front lines with ammunition, rations, and blood plasma at 1400, but had not yet arrived. Liversedge would have to take the Japanese position with the troops already at hand.

Following Company D's return to its former position, the 4th Battalion found movement easier, and Companies N and P managed to move forward in the face of stiffening fire to extend the NLG lines more to the southwest. But the move was costly: both companies received heavy casualties. Company Q (Captain Lincoln N. Holdzkom), the sole remaining company as yet uncommitted, moved up to the rear of the other three 4th Battalion units to be in position for an attack when directed.

By 1600, the Japanese had been pushed, still defiant and dangerous, into an area on the Bairoko Harbor headlands about 300 yards wide and 800 yards long. Their back to the sea, the enemy defenders kept up a sustained and murderous machine gun and mortar fire that showed few signs of slackening. In an effort to strike one last, conclusive blow, Liversedge ordered Company Q into the lines. Holdzkom's company moved around the left flank of Company N in an attack straight into the teeth of heavy enemy fire. Action along the rest of the front line dwindled as the fury of the attack on the south flank increased. Now all combat units had been committed; only the demolitions platoon of Currin's battalion remained as security for the command posts in case of an enemy breakthrough. Wheeler's Company B, requesting reinforcements for a last attack, was told that no help was available.

The outcome of Liversedge's last attempt to take his objective was not long in doubt. Despite the vigor of Company Q's attack, the overwhelming fire of the enemy won. Badly depleted in a matter of moments, Company Q was forced to retire. Repulsed, the company reeled back, virtually noneffective through its losses. The tactical situation had been opportune for one last heavy punch to knock out the enemy defenders, but without artillery, air support, or other heavy weapons, the raider battalions could not deliver it.

During the early part of the Marines' attack, Colonel Liversedge heard nothing from Schultz, who was supposed to have hit the enemy's other flank. From his command post just behind the raiders' front lines, the NLG commander tried to contact Schultz by telephone to order registration of the battalion's 81mm mortars on the harbor's defenses. The wires, however, were dead, apparently grounded somewhere in the relay linking Liversedge to Enogai and then to Schultz. And, in this crucial moment, the TBXs carried by the raider regiment failed to reach even the short distance back to Enogai. Chagrined by the absence of contact with 3/148, and desperately needing assistance in his bid to capture Bairoko, Liversedge, at 1345, directed McCaffery to take a small patrol and try to contact Schultz as soon as possible. For the operations officer, this entailed a rugged trip to Enogai, then a boat ride to Triri, and a subsequent march to Schultz' position.

The first word Liversedge had from Schultz, a field message from Enogai at about 1500, was not encouraging:

> Harry: Steve [Stevenson] has contact with Dutch. Dutch has been hit 3,400 yards from Triri. Steve told Dutch to keep pushing and try to connect with our outfit. Artillery fire is falling between Rice and Triri. LaHue.[13]

Schultz' battalion had departed Triri on schedule that morning, Company K leading the column down the Triri-Bairoko trail. Progress was slow, the heavy machine guns and mortars carried by the soldiers adding to the difficulty of movement over the slippery jungle terrain. By noon, the battalion had reached a point on the trail where enemy positions had been reported, but the Japanese had apparently evacuated the area. The only enemy contact was a glimpse of a Japanese patrol of about 15 men moving hurriedly down the trail ahead of the column, but no shots were fired. Shortly afterwards, however, the chatter of an enemy light machine gun sent the column off the trail. Several probing attacks were made to determine the enemy position, and at 1515 Schultz sent a message to Enogai for relay to Liversedge:

> Light Horse Harry: Have met Nips about 3½ miles down trail. Have not yet hit Munda-Bairoko trail. Strength of enemy undetermined, but know they have four automatic weapons. We are attacking. Will keep you informed as situation develops. They hold high ground to our front. Dutch Del.[14]

Schultz then moved his companies into attack formation and ordered a mortar barrage on the Japanese positions. The

---

[13] *Ibid.*, Capt Foster C. LaHue to Liversedge, dtd 20Jul43.

[14] *Ibid.*, Schultz to Liversedge, dtd 20Jul43. Liversedge, a former Olympic athlete, was widely known by the nickname used in the message. Lieutenant Colonel Schultz signed most of his messages with the name used here, evidently a coupling of a nickname and the contraction of his given name, Delbert.

pre-attack bombardment was to start at 1600. Enemy strength, Schultz decided, was about one company. Shortly after the mortar barrage began, Liversedge was able to contact Schultz directly by telephone and advise him of the situation that faced the Marines on the right flank. Schultz must establish contact with the main positions at Bairoko—and soon—Liversedge told him, or the attack on Bairoko would fail.

The Army commander, not knowing whether his present attack would succeed, reported that he did not think it possible that contact with the Marine units could be made before nightfall. McCaffery, who had reached Schultz after the attack had been started, could only urge that Schultz push forward as rapidly as possible. The battalion's attack carried forward only a few hundred yards before stiffening enemy resistance stopped the advance. Schultz then ordered his men to dig in and hold the ground taken. He had, he figured, reached a position from which he could launch an attack the following morning.

For Liversedge, Schultz' failure to attack aggressively on the left flank was the final blow in a series of sharp disappointments. To his front, the battle din had subsided into an uneasy calm broken occasionally by the stutter of a machine gun or the sharp report of a rifle. While both forces—the Japanese compressed into a corner and the Marines clinging tenaciously and tiredly to shell-pocked ridges won through sacrifice and courage—waited for the next move, Liversedge asked Griffith to reconnoiter the front lines and report what action could be taken. Griffith's recommendation: withdraw.

> By this time the Raiders (1st and 4th) had nearly 250 casualties, or about 30 percent of the force. We had another 150 men tied up getting them evacuated to aid stations and to Enogai. There was nothing to do but pull back to reorganize, re-equip, get some rest, try to get something to cope with the Jap 90mm mortars, and get the wounded out.
>
> The decision to pull back was made by Harry the Horse on recommendation from me after I had talked to Currin and his and my company commanders and had made a personal reconnaissance of the front. Harry had a mission and was understandably loath to abandon it. The final determining factor was the Japanese capability to reinforce from Vila Stanmore during the night by barge. We were already up against a stone wall, low on ammunition and out of water, and had a responsibility to 200 wounded men. In any case, reorganization was a paramount requirement. I feel that the decision to withdraw was entirely sound and the only sensible one to have made.[15]

Victory had been close. At 1630 Griffith had joined Kemp on his hard-won ridgeline overlooking Bairoko. The harbor was about 300 yards away—but still unattainable. For more than seven hours, the raiders had been in continuous attack, trading punch for punch with the enemy and had almost won. Exhausted and nearly out of ammunition, with almost as many men wounded as were still fighting, the raiders could only retire, carrying their dead and wounded. The positions won through courage and indomitable will could not be held during the night because there were no other troops ready to pick up the fight. Regretfully, Liversedge ordered the withdrawal of his forces.

The retirement began shortly after 1700. First to leave were the litter cases, about 90 in number. Marines from the battalion and regimental headquarters companies carried the wounded off the ridgeline in crude stretchers made from folded pon-

---

[15] *Griffith ltr.*

chos and tree branches. The walking wounded followed, a thin stream of lurching, bloody men who had remained in the fight despite injuries. While Companies N and P held the main positions, Company Q pulled back. Companies O and D disengaged next. Despite a continued spatter of enemy mortar and machine gun fire, the retirement was orderly, Marines assisting the wounded and each other whenever necessary. As they moved back, the men salvaged weapons and ammunition which had been dropped in the fight.

As the abrupt jungle darkness closed in, the rest of the raider companies disengaged to retire to the high ground east of the end of the lagoon. A rough defensive perimeter was set up both flanks resting on the lagoon. Company L of 3/145, which arrived at 1800 with badly needed medical supplies and water, also moved into the defensive line. Christie's platoon, pulled back a short distance on the sandspit, blocked a possible enemy counterattack from that direction.

After seeing 80 walking wounded start the long and tortuous night march back to Enogai, the Marines settled down into an uneasy rest in their shallow foxholes. That night Liversedge made another request for air support. To forestall any swift counterattack by the Bairoko defenders, the NLG commander asked that the area between the NLG perimeter and the harbor be worked over by a bombing and strafing attack the next morning. Liversedge then concluded his request with: "You are covering our withdrawal." [18]

The night of 20–21 July passed with only one enemy attack to test the hasty perimeter. A light Japanese force attempted to penetrate the defenses on the west flank, but was repulsed by Companies B and D in a sharp fight that wounded nine more Marines and killed another. Four dead Japanese were found the next morning.

At dawn on the 21st, another group of walking wounded started toward Enogai where three PBYs waited. The main body of the NLG followed, the Marines carrying the more seriously wounded men on stretchers. Shortly after the grueling march began, a group of Corrigan's natives appeared to take over the stretcher bearing. Progress was slow and exhausting as the natives and Marines, burdened with extra weapons and packs, labored over the rough terrain. A stop was made every 200 yards to rest the wounded and the carriers. The main body of troops had gone about halfway to Enogai when the Marines were met by Company I, 3/145, which had hurried from Triri to take over the rear guard. The rough march was further eased when a number of the wounded were transferred to landing craft about halfway down Leland Lagoon. After that, the march speeded and by 1400 all troops were within the defensive perimeter at Enogai. Christie's platoon, which retired down the spit, also arrived without incident.

Schultz, who had been surprised at the abrupt change of events, had kept his soldiers on the alert for a morning attack if a switch in orders came. When the order for withdrawal was repeated, Schultz turned his battalion around and within several hours was back at Triri.

During the march toward Enogai, the Marines had been heartened by the sounds of continuous bombing and strafing attacks

[18] *Stevenson ltr.*

at Bairoko. Although Liversedge's request for air support the night previous had been received at 2244, well past the required deadline for such requests, the ComAir New Georgia headquarters apparently read the appeal in the NLG message and the request was passed to ComAirSols. Every available plane, including some outmoded scout planes, was diverted to attack the enemy positions at Bairoko. The strikes began at 0950 on the 21st and lasted until 1710, long after the raiders had reached the base at Enogai. In all, 90 scout bombers, 84 torpedo bombers, 22 medium bombers, and 54 fighter planes took part in the continuous air attack. A total of 135 tons of bombs were dropped on enemy positions, and strafing attacks by the mediums started a number of fires in supply dumps and bivouac areas. The only resistance by the Japanese was a flight of 17 fighters which attempted to intercept the last flight of medium bombers, but was driven off by the Allied fighter cover.

Evacuation of the wounded from Enogai continued despite attempts by Japanese planes to strafe the big, lumbering PBYs which landed in Enogai Inlet. The interruptions delayed, but did not halt, the removal of wounded for hospitalization at Guadalcanal. With all the troops in bivouac at Triri or Enogai, a sobering count of wounded and dead was made. The 1st Battalion with two companies in the attack had lost 17 killed and 63 wounded. Currin's battalion counted 29 dead and 137 wounded. In the action along the trail south of Bairoko, Schultz lost 3 killed and 10 wounded.[17]

The raiders had faced an estimated 30 machine guns in coral and log emplacements, cleverly camouflaged with narrow, hard-to-detect firing slits. Only 33 enemy dead had been counted during the day-long attack, but the evidence of much blood in the bunkers which had been reduced indicated that the Japanese casualties had been considerably higher.

The following day, 22 July, Liversedge received orders from Griswold to remain at Enogai and Rice Anchorage. Active patrolling was to be continued, and the NGOF was to be apprised of any hostile troop movement from Bairoko to Munda. Evidently, no further attempt to take the well-fortified harbor would be made for a while. With these orders, the conflict on Dragons Peninsula settled down to a state of cautious but active watchfulness.

Occasional fire fights flared as opposing patrols bumped into each other, but close contact between the two forces was infrequent. The Japanese reclaimed the high ground overlooking Bairoko and reconstructed their fortifications. Evidently hoping to keep the NLG off balance, the enemy harassed the Enogai positions nightly with bombing attacks by one or more planes. Some nights the number of such attacks or alerts reached as high as seven. The Allies, meanwhile, pounded Bairoko with short-range shelling from three destroyers on 24 July and bombed the harbor defenses on 23 and 29 July and 2 August. For the most part, however, the operation reverted to a routine of enervating patrolling and air raid alerts. Of particular benefit was a rest camp established by Corrigan's natives near Rice Anchorage where Marines were able to relax for three days away from the weary monotony of patrols and air raids.

---

[17] *1st RdrRegt SAR; 1st RdrBn WD; 4th RdrBn Jul43 WD; 3/148 Rept.*

END OF A CAMPAIGN [18]

The virtual stalemate on Dragons Peninsula ended on 2 August. The XIV Corps, poised for a last headlong break-through to Munda field, directed the NLG to rush another blocking force between Munda and Bairoko to trap any retreating enemy. After a hurried night conference with his battalion commanders, Liversedge ordered Schultz' battalion on a quick march down the Munda-Bairoko trail from Triri. The 4th Raider Battalion, at Rice, returned to reserve positions at Enogai and Triri. Schultz' battalion, leaving Triri on the 3d, moved quickly past its old positions abandoned on 17 July to another trail junction farther southwest. Here he established a road block. On 5 August, as Munda fell, Liversedge joined him with a reinforcing group (Companies I and K) from the 145th Infantry and a reinforced platoon from each of the two raider battalions. The first enemy contact came on 7 August when a patrol from Schultz' battalion encountered Japanese building a defensive position and killed seven of them.

Contact between the forces capturing Munda and Liversedge's command was made on 9 August when a patrol from the 1st Battalion, 27th Infantry, commanded by Lieutenant Colonel Joseph F. Ryneska, appeared at Schultz' road block.

The following day, 10 August, on Griswold's orders, operational control of the NLG passed to the 25th Division. Control of Schultz' battalion passed to the 27th Infantry, and Ryneska's battalion joined the NLG in Schultz' place. Leaving the road block position to be defended by Ryneska's outfit, Liversedge and his Marine-Army force returned to Triri and Enogai. There the Marines had been actively patrolling to determine if the enemy was preparing to make another determined stand at Bairoko. Heavy barge traffic, however, and lack of aggressive resistance indicated that Bairoko was being evacuated. Meanwhile, the nightly enemy air raids continued with practically the same results as before: "No casualties, no damage, no sleep." [19]

On 9 August, a light antiaircraft battery from the 11th Defense Battalion arrived at Enogai. The 50 Marines with 40mm antiaircraft guns and .50 caliber machine guns were a welcome addition to the base's defense. The first night that the battery was in action, the 40mm guns scored a hit on a surprised Japanese plane which hurried away trailing smoke. The gleeful Marines scored the hit as a "probable." Thereafter, the nightly enemy raiders climbed considerably higher; and as the altitude increased, the accuracy of the bombing decreased.

The final assault on Bairoko was made on 24 August after two regiments of the 25th Division (161st and 27th) had pushed inexorably toward the last Japanese stronghold. In the late afternoon of 24 August, Ryneska—whose battalion had advanced steadily toward the objective on the Munda-Bairoko trail—sent a message to Liversedge that he was one hour's

[18] Unless otherwise noted, the material in this section is derived from: *New Georgia Campaign*; *NGOF Account*; *1st RdrRegt SAR*; *1st RdrRegt Jnl*; *1st RdrRegt MsgFile*; *1st RdrBn WarD*; 4th RdrBn WarD, Aug43; *11th DefBn WarD*; 11th DefBn Quarterly AA Rept, dtd 15Sep43; Maj Marvin D. Girardeau, USA, ltr to CMC, dtd 6Feb57; *Griffith ltr*; Rentz, *Marines in the Central Solomons*.

[19] *1st RdrRegt Jnl*.

march from the southern end of the harbor and that he was going into Bairoko on the following morning "come hell or high water."[20] Ryneska's message was followed by another message from the 3d Battalion, 145th (now commanded by Major Marvin D. Girardeau) which had advanced from Enogai over the raiders' route of 20 July. A company from that battalion reported that it had entered Bairoko without opposition. The harbor had been evacuated. Composite raider companies, formed from the effective members of each battalion, were in reserve at Triri and Enogai but were not needed for the final phases. The long fight for Bairoko was over.

On the 28th, General Collins, commanding the 25th Division, arrived at Enogai and after an appraisal of the situation ordered the Marines withdrawn. That night and early on the 29th of August, the raiders went aboard APDs. By 1130 on the 30th, the raiders were back at Guadalcanal. The last entry in the 1st Raider Regiment Journal, at midnight of 31 August 1943, is significantly eloquent: "1st Marine Raider Regiment relaxes (bunks, movies, beer, chow)."[21]

The Marine raider battalions which returned to Guadalcanal were a pale shadow of the two units which had originally been assigned to the NLG. Malnutrition, unavoidably poor sanitary conditions, exposure, fatigue, and continued loss of sleep and malaria had taken their toll. Battle casualties had been unexpectedly high—25 percent of the total command of the 1st Battalion, 27 percent of the 4th. Griffith's battalion had lost 5 officers killed and 9 wounded, with 69 enlisted men killed and 130 wounded. Currin's battalion, in three operations (Viru, Wickham, and Bairoko) had 2 officers killed and 8 wounded, 52 enlisted men killed and another 160 wounded. Of the 521 men remaining in the 1st Battalion, only 245 were judged effective by battalion medical officers. Only 154 Marines out of the 412 officers and men in the 4th Battalion could be classed as effective. The doctors concluded that further commitment to combat at this time was impossible:

> Not more than fifty percent of the present personnel would be able to move out on a march without extreme exhaustion and of these, the undermining of physical and nervous stamina has been so great as to render none of them capable of exerting sixty percent of their usual offensive effectiveness.[22]

CONCLUSIONS

The contributions of the NLG to the eventual success of the New Georgia campaign appear slight in a post-operational review. The trail block, as originally situated, lost all surprise value and usefulness after one engagement. The Japanese did not contest its presence further, and simply moved reinforcements to Munda over another route. As later reconnaissance proved, the actual location of the trail block should have been another 1,200 yards farther southwest at the junction of the main Munda-Bairoko trail.

Liversedge's force, in attacks on Enogai and Bairoko, inflicted a large number of casualties on the enemy and forced the Japanese to commit additional troops to the Dragons Peninsula area — troops which the enemy could have used to advantage in the defense of Munda. This,

[20] *1st RdrRegt MsgFile*, Ryneska to Liversedge, dtd 24Aug43.

[21] *1st RdrRegt Jnl.*

[22] InformalRept, Bn Surgeons to CO, 1st Rdr Regt, dtd 8Aug43.

perhaps, was the principal benefit derived from the NLG's operations at Enogai and Bairoko.

The failure of the attack on Bairoko can be ascribed to the burden of handicaps under which the NLG labored—lack of intelligence, poor communications, the vital need for supporting air and artillery, and insufficient support from higher echelons. Each handicap, in its turn, contributed to the eventual failure.

Operational planning was handicapped by the failure of the NGOF in making maps, mosaics, and aerial photographs available to the NLG prior to the landing. Other than the operational mosaic, the Liversedge force received only one high-level stereographic set of prints of Bairoko, which revealed nothing. And, as Liversedge later pointed out, no provision was made for the NLG to receive further intelligence.[23]

Realistic estimates as to enemy strength and reinforcement capabilities were lacking. On a par with the assumption that Munda would be captured in a matter of days was the equally poor reasoning that the Japanese would not stoutly defend against an attack on their major port of entry into New Georgia. Pre-attack patrolling by the Marine and Army battalions was extensive but, as Liversedge admitted, not aggressive enough to force the enemy to reveal the added strength of the Bairoko defenses.

The serious disadvantage imposed by communication failures in the dripping jungle balked the operation constantly. Contact with NGOF headquarters at Rendova was difficult, and NLG messages usually had to be relayed by a variety of stations, including those at Segi and Guadalcanal. Not all the communications woes were equipment failures, however. In some instances, transmission of messages was refused. After Enogai was captured, Liversedge reported, permission to transmit three urgent messages to the NGOF was not granted, and the NLG was directed to clear the message with another station, unknown to the NLG. The urgent messages to the NGOF were finally cleared after 15 hours of waiting.[24]

The attack on Bairoko, started and continued without air bombardment, the only supporting weapon available to the NLG, raises questions which existing records do not answer. Since his request for air preparation on the objective had apparently been rejected and there was no assurance that another request would be honored, Liversedge undoubtedly believed that a higher echelon had deemed air support unnecessary for the attack. As the next day, 21 July, was to prove, however, air support — and lots of it — was available. The only restriction, apparently, was that requests had to reach the headquarters of ComAirSols on Guadalcanal before the end of the working day.

Another question unanswered was the complete absence of any supporting artillery. Although it would have been impossible to pull artillery pieces through the jungle from Rice to Enogai, there seems to be no reason why artillery could not have been unloaded at Enogai after that village was captured. It is believed that one battalion of 105mm howitzers could have been spared from the many battalions then at Munda. Based at Enogai, these guns would have made a vast difference in the attack on Bairoko.

---

[23] *1st RdrRegt SAR*, p. 17.

[24] *Ibid.*, pp. 12–13.

Naval gunfire support, as known later in the war, was at this point in mid-1943 still in the exploratory stages; "reliable, fool-proof communications and the development of gunnery techniques for the delivery of accurate, indirect fire from afloat onto unseen targets"[25] ashore had not been fully worked out yet. As before, records do not indicate the reasons why Allied planners waited until after the repulse at Bairoko to plaster that enemy point with air and naval bombardments.

Although the Marine battalions were forced to admit failure in taking the assigned objective of Bairoko, the seven-hour attack by men armed with only grenades, rifles, and light machine guns[26] against an enemy of near equal numerical strength barricaded in heavily fortified bunkers stands as one of the finest examples of personal courage in Marine annals. It is to the raiders' credit that victory over these overwhelming odds was at one point very nearly in their grasp. Whether the harbor could have been taken by more aggressive action by the 3d Battalion, 148th Infantry is pure conjecture. The records indicate that action of the left flank was not coordinated with the raider attack, and that apparently the urgency of the situation was not realized by Schultz. Why the 3d Battalion, 145th Infantry was never used except as a support force and not committed to combat is another question which was unanswered in reports of the action. Equally puzzling is the fact that the Army battalion's 81mm mortars were not employed to support the raiders' attack.

In any event, an evaluation of the Dragons Peninsula campaign does not discredit the troops and their leaders who fought there. Rather than being remembered for failure, the Dragons Peninsula operation and the attack on Bairoko in particular are a testimonial to the personal courage of the Northern Landing Group, which achieved at least partial success, although almost hopelessly handicapped by innumerable shortcomings in the initial planning and in the support subsequently received. Faulty intelligence which underestimated the enemy, faulty task organization which neglected the inclusion of required fighting elements, and something less than full support by higher headquarters are the main shortcomings which analysis reveals.

---

[25] Col Robert D. Heinl, Jr., ltr to Head, HistBr, G-3, HQMC, dtd 3May62.

[26] Major General Robert S. Beightler, who commanded the 37th Infantry Division at New Georgia, noted that he had "personally urged" Colonel Liversedge "to adequately equip the Marine battalions with heavy automatic weapons" before they left Guadalcanal. MajGen Robert S. Beightler, USA, ltr to ACofS, G-3, HQMC, dtd 15Dec60, hereafter *Beightler ltr*.

End of a Campaign

BAANGA AND ARUNDEL[1]

In a matter of days after its seizure by the NGOF, the airfield at Munda—shell-cratered, with stripped and fire-blackened palm stumps outlining the runways—was converted to an Allied base for further operations in the Central Solomons. Almost as soon as enemy resistance around the airfield was ended, the busy bulldozers of the Navy's construction battalions were smoothing the coral landing strips and repairing revetments for use by ComAirSols planes. As the 25th Division turned north to follow the enemy's withdrawal toward Bairoko, the 43d Division took over defense of the airfield and began mop-up operations on the offshore islands.

Separated from New Georgia by only a few yards of shallow water, Baanga Island north of Munda Point was a ready-made sanctuary for Japanese fleeing the bigger island. As such, the densely-wooded appendage was a stepping stone along the Japanese route of retreat. The original island garrison had been small—about 100 Army and Navy troops—but the general exodus from Munda swelled the population. Tag ends and remnants of Munda's defenders fled to the island either to go overland toward Arundel or await evacuation by barge. (See Map II, Map Section.)

On 11 August, as the 43d Division widened its cleanup efforts around the airfield, a patrol confirmed reports of Japanese activity on Baanga. The following day, a company-sized unit moved by landing craft to the island. As the soldiers disembarked, a withering fire from the jungle felled about half of the force and forced its withdrawal. Two days later, while an artillery barrage from 155mm guns hastily emplaced at Munda paved the way, two battalions of the 169th made an unopposed dawn landing on the shore opposite the site of the ill-fated assault of the 12th. As the infantrymen moved inland, crossing the island from east to west, resistance stiffened. An estimated 400 Japanese manned a strong line of hastily-built fortifications blocking the advance.

On 16 August, two battalions of the 172d Regiment went to Baanga to reinforce the attack. As more artillery units (including the 155mm gun batteries of the 9th Marine Defense Battalion) moved into po-

[1] Unless otherwise noted, the material in this section is derived from: CinCPac-CinCPOA Rept of Ops, POA, for Sep43, dtd 17Dec43; *ComInCh Rept of SoPac Action; ComSoPac Aug–Sep43 WarDs; NGOF Account; 43d InfDivHist;* 9th DefBn WarD, Aug–Sep43; *11th DefBn WarD; 9th DefBn Tank Ops;* 10th DefBn Rept on Action of the TkPlat in the New Georgia Campaign, dtd 3Sep43; 43d InfDivRept to CG, XIV Corps, Employment of Tanks in Arundel Ops, dtd 23Sep43; *SE Area NavOps—II; Seventeenth Army Ops—I;* JICPOA Item No. 1973, Translation of Captured Japanese Document, dtd 22Nov43; CIC SoPacFor Item Nos. 799 and 814, dtd 26Oct43, Translation of Captured Japanese Documents; *New Georgia Campaign;* Karolevitz, *25th InfDivHist;* Rentz, *Marines in the Central Solomons;* Zimmer, *43d's History.*

sition at Munda and on the offshore islands, and systematically knocked out every known enemy gun emplacement, resistance dwindled. Increased barge traffic on the night of 19 August indicated that the Japanese were withdrawing. The following day, the southern part of the island was quickly occupied, and two battalions then moved north along opposite coastlines. Only scattered stragglers were encountered; the enemy had abandoned Baanga. The 43d Division lost 52 men killed and 110 wounded in the week-long battle.[2]

Contact with the Japanese was reestablished on Arundel. One of the smallest of the major islands in the group and virtually unoccupied by the Japanese except as a barge staging base, Arundel was within easy distance of both New Georgia and Kolombangara. Its eastern shore bordered Hathorn Sound and its northern fringe of narrow reef islands was just 1,200 yards from Kolombangara—a strategic position which became increasingly important to both forces. For the Japanese, the island was an important outpost to Kolombangara and an invaluable evacuation point. The NGOF wanted the island because Arundel in Allied hands would bring Vila airfield within range of artillery. (See Map 10.)

On 27 August, troops from the 172d Infantry crossed Diamond Narrows from New Georgia and landed unopposed on the southeastern tip. After securing the southern part of the island, the landing force split into two reinforced companies to begin extended patrol action north along the east and west coastlines of Arundel.

As on New Georgia, the dense jungle and large mangrove swamps made travel difficult. First enemy contact was made by the east shore patrol on 1 September south of Stima Lagoon. Pushing on, the patrol fought its way through brief skirmishes and delaying actions without trouble. To help cut off the retreating enemy, the 2d Battalion of the 172d established a beachhead near the lagoon and reinforced the eastern patrol. Meanwhile, the 1st Battalion moved by LCMs through Wana Wana Lagoon to link up with the western patrol which had reached Bustling Point on the northwest coast without so much as seeing an enemy soldier. The beachhead on that coast was then expanded to include the extreme western end of Bomboe Peninsula.

When the 2d Battalion's attack near Stima Lagoon on 5 September was abruptly halted by fierce enemy resistance, the 3d Battalion was landed to reinforce the effort. Neither battalion, however, was able to penetrate the enemy's strong line of defense which included mine fields and booby traps as well as many machine guns. Artillery fire from Kolombangara supported the defense. The arrival of the 1st Battalion from Bustling Point, where a battalion of the 169th had assumed responsibility for the western beachhead, placed the entire 172d Infantry on the east coast and paved the way for the commitment of the 27th Regiment (25th Division) on Bomboe Peninsula. Two batteries of 155mm howitzers and a 4.2-inch chemical mortar company also landed at Bustling Point to support the 27th Regiment on that coast, while NGOF artillery on New Georgia emplaced on the shores of Hathorn Sound delivered counterbattery fire on Kolombangara to support the 172d's attack on the east coast. Of the two infantry regiments, however, only the 27th Infantry was relatively fresh, although its

[2] Miller, *Reduction of Rabaul*, p. 172.

rifle companies were seriously understrength and its men "well seeded with malaria."[3] The 172d had been through nearly two months of arduous fighting and was badly understrength.

While troops from the 169th held the Bustling Point area, the 27th Infantry on 12 September opened a drive east along the length of Bomboe Peninsula. The leading battalion, restricted to a narrow strip of island only 400 yards wide and unable to make a flanking attack, could only grind straight ahead when it ran into stiff opposition. Small gains with mounting casualties were the inevitable results.

As the front lines inched abreast of Sagekarasa Island, which parallels Bomboe Peninsula, a second battalion swam and waded across a lagoon to establish another front on that island. Unable to erase the beachhead in a series of screaming counterattacks that night, the Japanese then hurriedly evacuated their barge base on the extreme western tip of the island. Sounds of barge traffic each night, however, indicated that the enemy still had other bases on Stima Peninsula which could be used to resupply and reinforce the Arundel defenders.

By dusk of 14 September, the two battalions of the 27th were in secure positions astride Sagekarasa Island and Bomboe Peninsula while the 172d Infantry pressed slowly northward along the east coast. In the gap between, stragglers from the *229th* and a battalion from Tomonari's *13th Regiment* fought determinedly to hold Stima Peninsula and a corner of Arundel's northeastern coast.

[3] MajGen William W. Dick, Jr., USA, ltr to ACofS, G-3, HQMC, dtd 31Oct60, including comments by MajGen David H. Buchanan, USA.

On the night of 14–15 September, the remaining battalions of the *13th Regiment* on Kolombangara were loaded on barges for transfer to Arundel to begin a counteroffensive which was supposed to regain the initiative in the Central Solomons. Undaunted by the loss of Colonel Tomonari and two battalion commanders who were killed by American artillery fire as their barge beached on the Arundel coast, the Japanese unleashed a near-fanatical attempt to break out of the perimeter. The desperation thrust failed. The 172d and 27th, reinforced quickly on line by troops from the 169th, contained the attack although the battle was touch-and-go for some time. As the attack subsided, the Japanese reverted once more to delaying tactics to preserve their thin foothold on Arundel. The repulse decided the Japanese upon withdrawal from Arundel and eventual evacuation of the Central Solomons.

The counterattack, however, resulted in Marine Corps tanks joining the 43d Division. Alerted earlier for possible commitment, the tank platoons of the 9th, 10th, and 11th Defense Battalions moved their remaining 13 serviceable machines by LCM from Munda to Bomboe Peninsula on the 16th. While the tanks of the 9th and 10th went into bivouac, five tanks of the 11th Defense Battalion moved up to help the 27th Regiment in the Bomboe Peninsula area. The armored attack on 17 September took the Japanese by complete surprise. The heavy jungle rains apparently drowned the noise of the tanks clanking into attack position. Moving forward in two waves with infantrymen following, the Marine tanks crunched through the enemy defenses before abruptly turning to the left in a flanking

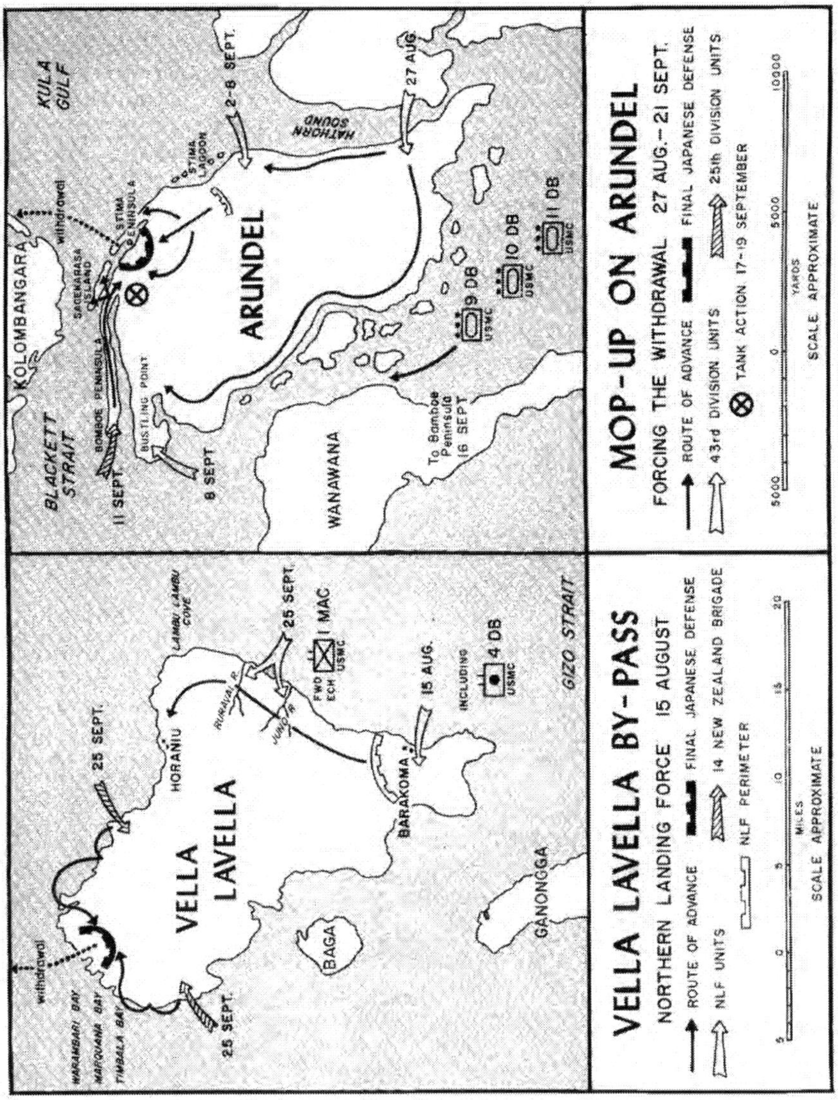
VELLA LAVELLA BY-PASS
NORTHERN LANDING FORCE 15 AUGUST
ROUTE OF ADVANCE
FINAL JAPANESE DEFENSE
NLF UNITS
14 NEW ZEALAND BRIGADE
NLF PERIMETER
MILES
SCALE APPROXIMATE
VELLA LAVELLA
HORANIU
BARAKOMA
BAGA
GANONGGA
GIZO STRAIT
LAMBU LAMBU COVE
WARAMBARI BAY
MARQUANA BAY
TIMBALA BAY
withdrawal
25 SEPT.
15 AUG.
I MAC
FWD ECH USMC
4 DB
INCLUDING
MOP-UP ON ARUNDEL
FORCING THE WITHDRAWAL 27 AUG.–21 SEPT.
43rd DIVISION UNITS
25th DIVISION UNITS
TANK ACTION 17-19 SEPTEMBER
YARDS
ARUNDEL
KOLOMBANGARA
BLACKETT STRAIT
KULA GULF
HATHORN SOUND
WANAWANA
BAMBOE PENINSULA
BUSTLING POINT
SAGEKARASA ISLAND
STIMA PENINSULA
STIMA LAGOON
To Bamboe Peninsula 16 SEPT
9 DB
10 DB
11 DB
USMC
11 SEPT.
8 SEPT.
2–8 SEPT.
27 AUG.

MAP 10

maneuver to complete the rout of enemy in that sector. Infantry units advanced about 500 yards in the attack. The following day, however, as four tanks and an infantry company jumped forward in another assault, the enemy suddenly opened point-blank fire with 37mm antitank guns. Two of the 11th Defense Battalion tanks were knocked out of action, but quick and effective covering fire by the infantry allowed the tank crewmen to escape. The attack stalled.

On 19 September, the remaining effective tanks—two from the 9th, four of the 10th—joined those of the 11th. Lined up in two ranks virtually tread to tread they started toward the enemy lines. The rear rank covered the front with fire. Concentrated blasts of 37 mm canister rounds and bursts of machine gun fire from the leading tanks withered the jungle ahead, stripping foliage from the enemy positions and hewing out an avenue of attack. Behind this shield of firepower, the infantry advanced rapidly. Afterwards described by 27th Infantry officers as one of the finest examples of tank-infantry coordination they had seen, the attack moved quickly and steadily forward.

This fearsome mass assault, coupled with the Japanese decision to quit Arundel, settled the fight for the island. That night, despite near-continuous artillery and mortar barrages, Japanese barges began evacuating the bulk of Arundel's defenders. While enemy artillery fire from Kolombangara kept the two American regiments from closing in, the remainder of the *13th Regiment* was withdrawn the next night. On 21 September, with only a few overlooked stragglers to contend with, the NGOF declared Arundel secured.

Instead of being a routine mopping-up job, the fight for Arundel had unexpectedly developed into a major operation which required the principal elements of three infantry regiments as well as armored and artillery support. Japanese losses in three weeks of fighting were 345 counted dead, although the enemy must have lost considerably more. Countless shallow graves dotted Arundel's northern coast, and the lagoons and Blackett Strait yielded many other bodies of enemy dead who had been killed in evacuation attempts or had drowned attempting to swim to Kolombangara.

Allied losses for the island's capture were relatively light, 44 killed and 256 wounded. Army observers credited the timely support of Marine Corps tanks for abruptly terminating the campaign and preventing the loss of additional Allied lives.

VELLA LAVELLA *

With Munda taken and the Allied drive slowly turning toward Vila airfield, the Japanese in mid-August had every right to expect that the decisive battle in the Central Solomons would be fought on the big, volcanic island of Kolombangara. But Admiral Halsey, a former Naval Academy halfback, knew the value of an end run in warfare as well as in football. Ten days after Munda was captured, the Allies skirted the strongly defended positions prepared by the enemy on Kolombangara and hit at lightly-held Vella Lavella.

---

* Unless otherwise noted, the material in this section is derived from: CinCPac Rept. Sep43, *op. cit.*; CinCPac–CinCPOA Rept of Ops, Oct43, dtd 20Jan44; *CominCh Rept of SoPac Action*; *ComSoPac Aug–Sep43 WarDs*; Com III PhibFor

The decision to switch targets was made a month earlier. On 12 July, just six days after asking for Admiral Turner's plans for Kolombangara, Halsey changed his mind and directed that this island be sidestepped and Vella Lavella taken instead. By this time it was obvious to the staff of ComSoPac that Munda was not going to be taken as quickly as estimated and that the island of Kolombangara, with nearly 10,000 entrenched defenders, would be even harder to take. Further, Vila airfield was reported to be poorly drained and poorly situated. If a better airfield site could be found, the chance to land virtually unopposed at Vella Lavella would be a much sounder tactical move.[5]

A reconnaissance team which scouted the island in late July returned to report that the southern end of the island near Barakoma drained sufficiently well to enable construction of an airfield there, and that there were adequate beaches, bivouac areas, and MTB anchorages in the area. Vella Lavella, the patrol reported, differed little from New Georgia. A dense jungle of tangled creepers and huge trees covered the island from coastline to the low but sharp mountain peaks in the interior. One of the most developed islands in the group before the war, Vella Lavella's European-type buildings included a hospital, several missions, and a leprosarium. (See Map 10.)

Coastwatchers on the island added to the report. Only about 250 Japanese were estimated to be occupying the northern part of the island where Vella Lavella's irregular coastline provided many coves for protection for barges shuttling between Kolombangara and Bougainville. The natives on the island had remained friendly to the Allies, and were well organized. They had, in fact, aided the many survivors of the USS *Helena* who had managed to swim to the island and had assisted in their evacuation by fast APDs on 16 July.[6]

On 11 August, orders were issued by ComSoPac for the seizure of Vella Lavella by Admiral Wilkinson's Task Force 31. The forces on New Georgia were directed to continue the cleanup operations in the Munda area and to interdict Vila airfield by artillery fire. The Allies had decided that enemy troop concentrations on Kolombangara did not necessitate an attack, that neutralization of the island would be as effective as occupation and not as costly in terms of troop casualties or supplies.

AR, 16–19Aug43, dtd 20Dec43; Com III PhibFor Rept of Occupation of Vella Lavella, 12Aug–3Sept43, dtd 20Sep43; CTF 31 OpO A12–43, dtd 11Aug43; *New Georgia Campaign*; *NGOF Account*; NLF FO No. 1, dtd 11Aug43; *Seventeenth Army Ops—I*; *SE Area NavOps—II*; Frankel, *37th InfDivHist*; O. A. Gillespie, *The Official History of New Zealand in the Second World War—The Pacific* (Wellington: War History Branch, Department of Internal Affairs, 1957), hereafter Gillespie, *New Zealand History*; Halsey and Bryan, *Halsey's Story*; Karolevitz, *25th InfDivHist*; Morison, *Breaking the Bismarcks Barrier*; ONI, *Combat Narratives, The Solomon Islands Campaign: XI—Kolombangara and Vella Lavella 6 August–7 October 1943* (Washington, 1944), hereafter ONI, *Combat Narratives XI*; Rentz, *Marines in the Central Solomons*; USSBS, *Campaigns*.

[5] This is not the first instance of successful bypass strategy. The amphibious force which landed in the Aleutians in May 1943 took Attu before forcing the evacuation by the enemy of strongly-held Kiska Island.

[6] U.S. ships picked up 93 men and 11 officers at one point and 59 men and 2 officers at another point on the island. Natives had protected and fed both groups. The rescue of these survivors from an enemy-held island in enemy waters was a distinct boost to morale to all South Pacific forces.

Further, soggy Vila airfield was no longer deemed worthy of capture.

The Northern Landing Force (NLF),[7] organized to attack and occupy Vella Lavella, comprised the Army's 35th Regimental Combat Team which included the 64th Field Artillery Battalion, the 58th Naval Construction Battalion, and the Marine 4th Defense Battalion, as well as additional Army and Navy support units. Brigadier General Robert B. McClure, the 25th Division's assistant commander, was named to head this organization.

Embarkation of major units began at Guadalcanal on 12 August. That same night, an advance force landed near Barakoma to mark channels and landing beaches and to select bivouac areas and defensive positions. After being forced to fight their way to shore, however, through fire from a motley collection of survivors from sunken barges, the reconnaissance group hurriedly requested reinforcements. The next night a infantry company landed to help them.

The main landing force departed Guadalcanal on 14 August on a split-second, staggered schedule. The slowest transport group, LSTs, started first and was passed later by the faster APDs. In this manner, the transports which had departed Gaudalcanal in reverse order arrived off Vella Lavella in the proper order and at the right time.

Debarkation of troops began at dawn on 15 August, the APDs unloading quickly in one hour. The first snag in the invasion schedule occurred when it was discovered that the beach could accommodate only 8 of the following 12 LCIs. The LSTs, which arrived later at the correct time, were forced to stand offshore waiting to unload. Limited beach areas had resulted in the very delay and exposure which it was hoped the staggered schedule would prevent. There was, however, no enemy opposition ashore. As the beachhead widened, soldiers reported scattered Japanese troops fleeing northward.

Shortly before 0800, just as the LCIs were in the unloading stage, the first of four frantic Japanese air attacks struck. After making one pass at the protective destroyer screen standing offshore, the enemy bombers and fighters turned their attack on the LCIs and LSTs, evidently figuring that the smaller transports carried the bulk of invasion troops and supplies. All four attacks were driven off by alert planes from ComAirSols and the fierce antiaircraft fire from the task force destroyers.

The fighter cover came from Munda airfield, which had begun operations only the day before. As a dividend for having won an airfield closer than Segi or Guadalcanal, the Allies were able to keep an umbrella over the beachhead most of the day. Despite the presence of this air cover, however, the Japanese persisted in sporadic attacks, striking from different altitudes and directions. The results were negligible. None of the ships in the convoy were damaged, and during the day more than 4,600 troops and 2,300 tons of equipment and supplies were unloaded at Barakoma. Twelve men were killed and 40 wounded in the day's attacks. That night, as the convoy withdrew slowly down Gizo Straits, the ships fought off repeated torpedo attacks. Enemy floatplanes kept the area lit with flares.

[7] This was a new echelon of the NGOF and is not to be confused with the Northern Landing Group commanded by Colonel Liversedge at Enogai and Bairoko.

The successful jump from Munda to Vella Lavella asserted Allied domination in the Central Solomons. Failing to repulse the landing, officers of the *Eighth Fleet* and the *Seventeenth Army* hastily called a conference to consider making a counterlanding on the island. One battalion was all that could be spared, it was decided. This proposal was promptly squelched by *Eighth Area Army*. Such a move would require at least two brigades, the higher headquarters decided; and, in view of the existing difficulties in reinforcing and resupplying other Central Solomons garrisons, the idea was better forgotten. The only help the stragglers on Vella Lavella received was reinforcement on 19 August by 290 Army and 100 Navy personnel.

The NLF beachhead expanded rapidly. Within the first 20 days of the operation, 6,505 troops, 1,097 tons of rations, 843 tons of gasoline and oil, 2,247 tons of ammunition, 547 vehicles, and 1,011 tons of other classes of supplies were landed. Shipping was to have been unloaded during night hours, but the attacks on the convoys in the narrow confines of Gizo Strait changed that schedule. After 18 August, the convoys arrived and departed Barakoma during daylight hours, protected during the unloading and passage through Gizo Strait by Allied planes from Munda.

There was little opposition to the advance of the NLF. By 18 August, the three battalions of the 35th Infantry had established a firm defensive perimeter across the southern end of the island. Behind this protective barrier, airfield construction began immediately. The Marine 4th Defense Battalion provided antiaircraft and seacoast defense.

As the fight for Vella Lavella progressed, the 35th began driving the enemy before it. Toward the end of August, increased resistance was met on the east coast near Lambu Lambu, and it was 15 September before the regiment's assault battalions broke through the Japanese defenses to overrun the barge base at Horaniu on the northeastern coast. The enemy, however, escaped and fled north.

At this point, the 14th Brigade of the New Zealand 3d Division landed at Barakoma with two infantry battalions, the 35th and 37th, as the main units. In a reshuffle of command, Major General H. E. Barrowclough of the 3d Division was named as commanding general of all Allied forces on Vella Lavella. With the arrival of a third New Zealand battalion, the 30th, the American frontline troops were relieved. On 25 September, the colorful New Zealanders—the majority of whom disdained the use of steel helmets to wear their distinctive flat visored field hats—began their attack.

While the 35th Battalion leapfrogged around the west side of the island in a series of landings, the 37th Battalion began moving by landing craft up the northeastern coast, making landings at various points to cut off the fleeing Japanese. When cornered, the enemy soldiers fought stubbornly and fiercely for survival, but it was apparent that they were not under a single command or organized into a single unit.

By early October, the New Zealanders were in position to put the squeeze on the Japanese, who had been backed into a jutting piece of land between Marquana and Watambari bays. The two battalions of the 14th Brigade made contact and joined for the final push to crowd the enemy into the sea. A Japanese prisoner reported that the tired and hungry enemy force was willing to surrender, but that Japanese officers would not permit it.

On the night of 6–7 October, Allied troops heard voices and the sound of barges scraping coral, but the supporting fires were ordered too late. The next morning, only littered stocks of Japanese equipment and supplies were scattered over the peninsula. The enemy cornered on Vella Lavella—589 by Japanese accounts—had been successfully withdrawn.

The New Zealanders had estimated that the Vella Lavella campaign would end in less than two weeks. The conclusion came several days early—one of the few successful timetables in the Central Solomons fighting. The 14th Brigade casualties totaled 32 killed and 32 wounded. Japanese losses for the defense of the entire island were about 250.

The price for an island of considerable strategic and operational value was not exorbitant. Allied casualties for the seven weeks of fighting were less than 150 killed, most of these in air attacks. The practice of bypassing a strong point to hit at a weaker point somewhere else was now established. Future Pacific operations followed the strategy initiated in the Aleutians and used with success in the Central Solomons.

MARINES AT VELLA LAVELLA[8]

Success of the Rendova beachhead had proven the value of a Marine defense battalion in a landing effort. When the task organization to seize Vella Lavella was planned, the inclusion of a similar unit seemed logical. Closest and most available was the 4th Defense Battalion, then on Guadalcanal. Organized at Parris Island in 1940, the battalion was stationed at Efate, New Hebrides, before transfer to New Zealand, then Guadalcanal. Its organization was similar to other Marine defense battalions—155mm seacoast artillery group, 90mm antiaircraft group, a special weapons group of 40mm, 20mm, and .50 caliber weapons, and a tank platoon.

By nightfall of the first day ashore at Barakoma, about two-thirds of the light antiaircraft weapons were in temporary firing positions. Other guns of Major McDonald I. Shuford's special weapons group were kept set up on two LSTs in the harbor, an innovation which increased the firepower of the beachhead. The addition was effective. A total of five enemy planes were claimed by the 4th the first day.

During the next six days, other echelons of the battalion arrived and moved into positions to defend the beach. Only the air defense units of the battalion got into action, however. The 155mm gun groups, which moved ashore shortly after the original landings, were in coastal defense positions ready to fire within a few days, but the need never arose. The tank platoon

[8] Unless otherwise noted, the material in this section is derived from: *NGOF Account;* 4th DefBn WarD, Aug–Oct43; 4th DefBn SAR, dtd 9Jan44; CO, Corps Troops and Staging Area, IMAC, Narrative Rept of Ops in Vella Lavella, dtd 30Jan44; Fwd Ech, Corps Troops, IMAC, OpO No. 1–43, dtd 23Sep43; Col John H. Cook, Jr., ltr to CMC, dtd 2Mar52; Col McDonald I. Shuford ltr to CMC, dtd 4Mar52; LtCol Donald M. Schmuck ltr to CMC, dtd 28Apr52; LtCol Charles T. Hodges ltr to CMC, dtd 21Mar52; LtCol Carl M. Johnson ltr to CMC, dtd 2Mar52; *New Georgia Campaign;* Rentz, *Marines in the Central Solomons.*

which landed on 21 August was never committed to action.

After the initial landings, the Japanese bombing attempts dwindled in frequency and ferocity. During the early part of the operation, the enemy attacks were pressed home with fanatical fury and many reckless planes were knocked spinning into Vella Gulf. Later the Japanese became more cautious, and fewer mass assaults were attempted. Since any activity at Barakoma was readily discernible from Kolombangara, the arrival of Allied ships was generally followed closely by a strike by a conglomerate force of enemy bombers, fighters, and float planes. Most of the attacks were less than vigorous, however, as the Japanese pilots soon gained a healthy respect for ComAirSols planes and the accurate shooting of the 4th Defense Battalion. By late August, those few enemy planes that did attack usually did not get close enough to bomb accurately.

During the Vella Lavella operation, 15 August to 6 October, the 4th Defense Battalion compiled an enviable accuracy record. During 121 different air attacks aimed at the island, the Marine antiaircraft gun crews knocked down the following: 90mm gun group—20 planes; 40mm batteries—10 planes; 20mm batteries—5 planes; the .50 caliber weapons of the special weapons groups—4 planes; and the .50 caliber weapons of the seacoast artillery group—3 planes. The total: 42.

Other Marines, not part of the NLF, also took part in the Vella Lavella operation. After the 35th RCT moved past Horaniu, establishment of a Marine advance staging point on the island was ordered. Planning for the Bougainville operation was already underway, and the I Marine Amphibious Corps wanted a base closer than Guadalcanal to the Northern Solomons. On 17 September, the new Commanding General, IMAC, Major General Charles D. Barrett, who had taken command of the corps on the 15th, named Major Donald M. Schmuck to head the proposed Corps Forward Staging Area, Vella Lavella. The task organization included elements of the Marine 4th Base Depot, a motor transport company, a special weapons battery, a communication team, part of the Navy's 77th Seabees, as well as two provisional infantry companies from the 3d Marine Division. All told, the forward echelon of Corps Troops included 28 officers and 850 men.

The task force was to land at two points: Juno river and Ruravai beach, on Vella's east coast. Part of the organization was to begin the establishment of a base camp while the combat elements provided local security. Hastily organized, the forward echelon made one practice landing at Guadalcanal before proceeding to Vella Lavella. On 25 September, troops went ashore by landing craft at Juno river while LSTs beached some three miles north at Ruravai beach.

Unloading at both points proceeded without incident until about 1115, when 15 Japanese bombers and about 20 fighters swept over. After one brief sideswipe at the destroyer screen offshore, the enemy planes turned toward Ruravai. Some 40mm and .50 caliber antiaircraft weapons had been hastily set up on the beach, and these opened with a steady fire that was accurate and effective. Three bombers were downed and a fourth damaged. Two of the doomed bombers, however, managed to complete their bombing runs before crashing in the jungle. The other planes continued to bomb and strafe. One 40mm crew and gun was destroyed by a direct hit and a second crew knocked out of action. Volunteers quickly manned the second gun and continued the fire.

As the bombing attack ended, Allied fighter cover appeared to clear the sky in a series of running dogfights. But the landing area at Ruravai was a shambles. Exploding ammunition continued to wreak havoc. Casualties and damage to supplies were high. One LST had been sunk outright, others had been damaged. A total of 32 men on the beach had been killed and another 58 wounded.

The Japanese did not let up. Each day brought a number of pressing air attacks. Despite frequent interruptions, the construction of roads, LST beaching areas, and base installations continued. The work was further handicapped by wandering bands of enemy stragglers, which necessitated active combat patrols as well as increased guards at all construction projects. Progress, however, was fairly rapid.

On 1 October, as the second echelon of corps troops (including the 2d Parachute Battalion), arrived, the Japanese struck another heavy blow. Four air attacks during the day resulted in further damage and more casualties. One LST was sunk and another damaged. The Japanese lost only one plane. Convinced at last of the futility of trying to land men and supplies over a beach inadequately protected against air attacks, IMAC then directed all further echelons and supplies to be unloaded at Barakoma under the protection of the guns of the 4th Marine Defense Battalion. The supplies were then trucked to Ruravai, where wide dispersal and increased aircraft defense measures ensured fewer losses.

After surviving a number of such severe air strikes during the next week (while Barakoma was studiously avoided by the Japanese), the Corps Staging Area was replaced on 8 October by the newly arrived Vella Lavella Advance Base Command. Some troops were returned to their parent organizations; others remained at the base under the new command. Ruravai was seldom used for its intended purpose, since most ships preferred the loading facilities at Barakoma. Later, however, the sawmills and hospital of the base command proved valuable during the Bougainville campaign by providing timbers for bridging and ready medical facilities for seriously wounded men. Its construction had proved costly, however. In the two weeks at Vella Lavella, the Forward Echelon had lost 17 men killed and 132 wounded during the many air attacks.

JAPANESE WITHDRAWAL[9]

As the campaign in the Central Solomons drew closer to its inevitable end, the Japanese efforts during August and September became those of near-desperation. The Allied attack on Vella Lavella, which effectively shunted the enemy forces at Kolombangara to the sidelines of the war, had the added effect of nearly isolating Japanese garrisons from the main sources of supplies and reinforcements in the Northern Solomons. Aggressive action by Allied destroyer squadrons tightened the blockade. Camouflaged enemy barges, trying to keep the supply lanes open by sneaking along the coves and breaks of island coastlines were hounded and ha-

[9] Unless otherwise noted, the material in this section is derived from: CinCPac Repts, Sep-Oct43, *op. cit.; ComInCh Rept of SoPac Action; ComSoPac Oct43 WarD; New Georgia Campaign; Seventeenth Army Ops—I; SE Area NavOps—II;* ONI, *Combat Narratives X;* Rentz, *Marines in the Central Solomons;* USSBS, *Campaigns.*

NEW ZEALAND TROOPS *of the 14th Brigade land on Vella Lavella to relieve American soldiers battling the Japanese.* (*SC 184437*)

MUNDA AIRFIELD *after its capture and reconstruction, as seen from the control tower atop Kokengola Hill.* (*SC 233548*)

rassed by the vigilant MTBs and the black-hulled Catalina flying boats ("Black Cats") which prowled the waters of Vella and Kula Gulfs. Nearly stymied in their barge supply attempts, the Japanese finally resorted to supplying garrisons by floatplanes and submarines.

These inadequate measures and a careful second look at the strategic situation forced the enemy to make the only decision possible: general evacuation of all forces from the Central Solomons. The operation began with the removal of troops from the seaplane base at Reketa on Santa Isabel Island in early September. An Allied patrol, landed from an MTB on 3 September, verified the absence of enemy troops. Quantities of rations and ammunition found on shore indicated that the withdrawal had been hurried.

After scattered outposts on Gizo and Ganongga Islands returned to Kolombangara on the 19th and 23d of September, the only remaining enemy troops were the small force defending Arundel, a sizeable body of troops on Kolombangara, and the stragglers back-tracking along the coast of Vella Lavella—about 12,000 troops in all, by Japanese estimates.

Weighing two factors—the direction of the Allied effort and the capability of the *13th Regiment* on Arundel to conduct a delaying action—the Japanese scheduled the withdrawal for late September during a moonless quarter. The northern coast of Kolombangara was designated as the evacuation point. Landing craft from the Buin area would ferry troops across The Slot to Choiseul for further transfer to Bougainville. Sick and wounded would be evacuated by fast destroyers.

The Japanese schedule began none too early. By 27 September the fighter airfield on Vella Lavella was operational although not yet completed, and enemy troops on Kolombangara were caught in a vise between ComAirSols planes at Munda and Barakoma. In addition, Allied 155mm guns and howitzers emplaced on New Georgia's northern coast were pounding a steady tattoo on Kolombangara's defenses.

The effect of waning moonlight — plus the increased barge activity—was not lost on the Allies. By late September it became evident that all Japanese activity was directed toward withdrawal. Immediately, all available Third Fleet destroyer squadrons rushed with protecting cruisers into interception duty in Vella and Kula Gulfs.

The planned withdrawal began, but was disrupted many times by the sudden appearance of Allied planes and ships. On the night of 28 September, the Japanese managed to load 11 destroyers with 2,115 sick and wounded for a quick sprint to safety at Bougainville. Despite the Allied interference and considerable loss of small craft and men, the Japanese relayed another 5,400 men by landing barges to Choiseul during the next few dark nights and an additional 4,000 men were picked up by six destroyers. In the squally weather and murky darkness of the period, the Allied destroyers were hard-pressed to keep track of all enemy activity. In a number of instances, the destroyers had to choose between steaming toward targets which radar contacts indicated as small craft or heading towards reported enemy destroyer forces. Sometimes contact could not be made with either target. Allied ships, however, reported a total of 15 barges sunk on the nights of 29 and 30 September.

During the night of 1–2 October, all available Allied destroyers steamed through The Slot seeking the main Japanese evacuation attempt. Few contacts were made in the pitch darkness. About 20 of the 35 barges encountered were reported sunk. The following night the Allied ships again attempted contact with the Japanese but could not close to firing range. Aware that the enemy destroyers were acting as obvious decoys to lure the attackers away from the barge routes, the Allied ships abandoned the chase and returned to The Slot to sink another 20 barges.

Further enemy evacuation attempts were negligible, and the Allies reasoned that the withdrawal had been completed. A patrol landed on Kolombangara on 4 October and confirmed the belief that the Japanese had, indeed, successfully completed evacuation of all troops. Jumbled piles of supplies and ammunition attested to the fact that the enemy had been content to escape with just their lives. The withdrawal, the Japanese reported later, was about 80 percent successful, the only losses being 29 small craft and 66 men.

The final evacuation attempt was made on 6 October from Vella Lavella. A sizeable enemy surface force was reported leaving Rabaul in two echelons, and three U.S. destroyers moved to intercept the enemy. Another Allied force also hurried to the scene. Contact was made in high seas and a driving rain. In a fierce battle which lasted less than 12 minutes, the United States lost one destroyer to enemy torpedoes and two other destroyers were badly damaged. The Japanese lost only one of nine destroyers, and during the battle the transports managed to evacuate the troops stranded almost within the grasp of New Zealand forces.

The removal of troops from Vella Lavella ended the Japanese occupation of the New Georgia Group. The loss of the islands themselves was not vital, but the expenditure of time and effort and the resultant loss of lives, planes, and ships was a reverse from which the Japanese never recovered. There could only be a guess as to the number of casualties to the enemy in the various bombings, sea actions, and land battles. Postwar estimates placed the number at around 2,733 enemy dead,[10] but this does not account for the many more who died in air attacks, barge sinkings, and ship sinkings. In any event, the units evacuated after the costly defense of the New Georgia Group were riddled shells of their former selves, and few ever appeared again as complete units in the Japanese order of battle.

More than three months of combat had been costly to the New Georgia Occupation Force, too. Casualties to the many units of the NGOF totaled 972 men killed and 3,873 wounded. In addition, 122 died of battle wounds later, and another 23 were declared missing in action. Marine Corps units, other than the 1st Marine Raider Regiment, lost 55 killed and 207 wounded.[11]

CONCLUSIONS

New Georgia lacked the drama of the early months of Guadalcanal and the awesome scope of later battles in the Central

[10] *NGOF Account*, p. 29. The figure reported here includes the 250 enemy dead at Vella Lavella.

[11] *Ibid.*, p. 29; Rentz, *Marines in the Central Solomons*, p. 174. These figures do not include non-battle casualties.

Pacific. Instead, it was characterized by a considerable amount of fumbling, inconclusive combat; and the final triumph was marred by the fact that a number of command changes were required to insure the victory. There were few tactical or strategic successes and the personal hardships of a rigorous jungle campaign were only underscored by the planning failures. And too, the original optimistic timetable of Operation TOENAILS later became an embarrassing subject. For these reasons, a postwar resumé of the battle for the Central Solomons pales in comparison with accounts of later and greater Allied conquests.

The primary benefit of the occupation of the New Georgia Group was the advancement of Allied air power another 200 miles closer to Rabaul. The fields at Munda, Segi Point, and Barakoma provided ComAirSols planes with three additional bases from which to stage raids on the Japanese strongholds in the Northern Solomons and to intercept quickly any retaliatory raids aimed by the enemy at the main Allied dispositions on Guadalcanal and in the Russells. Behind this protective buffer, relatively free from enemy interference, the Allies were able to mass additional troops and materiel for future operations. This extended cover also gave Allied shipping near immunity from attack in southern waters. Although most fleet activities continued to be staged from Guadalcanal, the many small harbors and inlets in the New Georgia Group provided valuable anchorages and refueling points for smaller surface craft.

The capture of the Central Solomons also afforded the Allies the undisputed initiative to set the location and time for the next attack. The simple maneuver of bypassing Kolombangara's defenses won for the Allied forces the advantage of selecting the next vulnerable point in the enemy's supply, communication, and reinforcement lines. The Japanese, guarding an empire overextended through earlier easy conquests, could now only wait and guess where the next blow would fall. The New Georgia campaign presented the Japanese in his true light—an enemy of formidable fighting tenacity, but not one of overwhelming superiority. His skill at conducting night evacuation operations, demonstrated at Guadalcanal and confirmed at New Georgia, could not be denied, however. Both withdrawals had been made practically under the guns of the Allied fleet.

On the Allied side, the campaign furthered the complete integration of effort by all arms of service—air, sea, and ground. Seizure of the Central Solomons was a victory by combined forces—and none could say who played the dominant role. Each force depended upon the next, and all knew moments of tragedy and witnessed acts of heroism. The New Georgia battles marked a long step forward in the technique of employment of combined arms.

There were valuable lessons learned in the campaign, too—lessons which were put to use during the many months to follow. As a result of the New Georgia operation, future campaigns were based on a more realistic estimate of the amount of men and time required to wrest a heavily defended objective from a tenacious enemy. Another lesson well learned was that a command staff cannot divide itself to cover both the planning and administrative support for a campaign as well as the active direction of a division in combat. After New Georgia, a top-level staff was superimposed over the combat echelons to plan and direct operations.

On a lower level, the tactics, armament, and equipment of individual units were found basically sound. As a result of campaign critiques, a number of worthwhile equipment improvements were fostered, particularly in communications where the biggest lack was a light and easily transported radio set. From the successful operation of Marine Corps light tanks over jungle terrain came a number of recommendations which improved tactics, communication, and fire coordination of the bigger and more potent machines, which were included in the task organization for future jungle operations. The battle against the enemy's bunker-type defenses on New Georgia also pointed up the desirability of tank-mounted flame throwers. Experimental portable models used in the fight for Munda had proved invaluable in reducing enemy pillboxes. Increased dependency upon this newly developed weapon was one direct result of its limited use in the Central Solomons fighting.

Throughout the entire campaign, the improvement in amphibious landing techniques and practices was rapid and discernible. Despite seeming confusion, large numbers of troops and mountains of supplies were quickly deposited on island shores, and rapid buildup of men and material continued despite enemy interference. One contributing factor was the increased availability of the ships needed for such island-to-island operation—LCIs, LSTs, LSDs, and the workhorse LCMs. By the end of the Central Solomons campaign, two years of war production was beginning to make itself felt. Equipment and ships were arriving in bigger numbers. The efficiency of these ships and craft was, in part, a reflection on the soundness of Marine Corps amphibious doctrines—vindication for the early and continued insistence by the Marine Corps on their development and improvement.[12]

[12] For the story of the Marine Corps part in the development of landing craft, see Chapter 3, Part I in Volume I of this series.

PART III

Northern Solomons Operations

CHAPTER 1

Continuing the Pressure

STRATEGIC REQUIREMENTS[1]

There was little time for extended rest for Allied sea, air, and ground forces after the final Central Solomons action in October. As early as the previous March, a decision had been made that the Northern Solomons would be the target next after New Georgia; and by the time Munda field was operational under Allied control, plans for the seizure of a beachhead in the Shortlands - Choiseul - Bougainville area were in the final stages. In effect, ComSoPac operations of September and October were the last act to the successful completion of the Central Solomons campaign as well as the overture to the forthcoming Bougainville attack.

The importance of Rabaul had not diminished during the long period involved in taking New Georgia, but the events of the last six months had put the Japanese stronghold on New Britain in a new light. In 1942 there had been no doubt that Rabaul would have to be physically eliminated to insure an inviolate hold on the Southwest-South Pacific area. By mid-1943, there was a growing realization that the enemy air and naval base might not have to be erased by force, that neutralization would serve the Allied cause as well. Concurrent with this shift in thinking was a proposal that the main Allied effort be directed through the scattered islands of the Central Pacific instead of through the larger land masses of New Guinea and the Philippines.

Both concepts had many high-ranking proponents. The divergent views resulted in some open disagreement among the many strong-willed commanders and staff officers responsible for the Allies' operational strategy, but the eventual solution was born of the imagination and experience of all, and there was no further dispute once the course of action had been charted. The actual decision to strangle Rabaul by air instead of capturing it was made by the Combined Chiefs of Staff upon recommendation of the JCS. The CCS conference at Quebec in August also directed that the advance through the

[1] Unless otherwise noted, the material in this section is derived from: CinCPac–CinCPOA WarDs, Aug through Nov43 (COA, NHD); *ComSoPac Jun–Nov43 WarDs; ThirdFlt NarrRept;* IIIPhibFor AR, Seizure and Occupation of Northern Empress Augusta Bay, Bougainville, 1–13Nov43, dtd 3Dec43 (Bougainville AreaOpFile, HistBr, HQMC) hereafter *IIIPhibFor AR;* IMAC AR, Phase I, Sec A, Rept on Bougainville Operation, dtd 21Mar44 (Bougainville AreaOpFile, HistBr, HQMC) hereafter *IMAC AR–I;* Maj John N. Rentz, *Bougainville and the Northern Solomons,* (Washington: HistSec, DivInfo, HQMC, 1948) hereafter Rentz, *Bougainville and the Northern Solomons;* Morison, *Breaking the Bismarcks Barrier;* Isely and Crowl, *Marines and Amphibious War;* Halsey and Bryan, *Halsey's Story;* King and Whitehill, *King's Naval Record;* Miller, *Reduction of Rabaul;* Hist Div, HQMC, "The Bougainville Operation," MS, ca. Feb45 (HistBr, HQMC), hereafter *HistDiv Acct.*

691–360 O—63——12

Southwest-South Pacific by General MacArthur and Admiral Halsey was to continue while Admiral Nimitz aimed a new offensive along the Central Pacific axis. The idea of two campaigns was an effective compromise. Although some realignment of forces was necessary, an extensive shuffle of troops or shipping from either theater would not be required; and a coordinated attack along two fronts would have the advantage of keeping Japanese defenses off balance and committed over a wide area.

Throughout this evaluation period, General MacArthur held fast to the original ELKTON concept. On the 4th of September, the VIIPhibFor (Admiral Barbey's command) landed SoWestPac troops on the Huon Peninsula of New Guinea to set the stage for eventual passage of MacArthur's forces through the Vitiaz-Dampier Straits. Success of the venture, though, depended upon insurance in the form of air bases within fighter plane distance of Rabaul. Thus, MacArthur's continued surge toward the capture of Salamaua, Lae, and Finschhafen virtually dictated establishment of Allied air facilities in the Northern Solomons by November or December of 1943. Plans for a landing in the Bougainville vicinity, temporarily shelved while other strategic concepts were being examined, were once again restored to the status of a full-scale operation by MacArthur's insistence.

The exact location for such an undertaking had been the cause for considerable discussion and reconnaissance by Halsey's staff. Several islands had been proposed as targets but closer examination eliminated them. The island of Choiseul was little regarded as a major landing site because of two factors: it was not within fighter escort range of New Britain; and MacArthur was of the opinion that an assault on Choiseul did not directly threaten Rabaul. Buka, the small island appendage to northern Bougainville, was too far from Allied air bases for a landing there to be protected adequately. Initial plans to seize the southern end of Bougainville were canceled because the airfields at Kahili and Kara were too strongly protected to attack with the forces then available to Halsey. With the drawn-out campaign on Munda still fresh in his mind, ComSoPac was reluctant to mix with the Japanese in a prolonged struggle that would take too many lives and too much time without paying immediate dividends.

After some deliberation, Halsey proposed that the SoPac forces seize the Shortland Islands (Ballale and Faisi) as airfield sites and then interdict Kahili and Kara with artillery fire as the Rendova forces did at Munda. This move would put Allied air support within fighter range of Rabaul. MacArthur, willing to settle for any action which would help him realize his expressed ambition to return to the Philippines, approved this concept. But later reconnaissance revealed that the Shortlands had no beaches big enough or good enough over which an amphibious assault could be staged, and that airfield sites were limited.

Halsey's top-echelon planners, abandoning the Shortlands idea,[2] on 6 September advanced another plan to seize the Treas-

---

[2] The major factors governing the abandonment of the Shortlands as primary targets were their distance from Rabaul, "a little far . . . for some of our short-legged planes," and the strong feeling among the planners that "we had to spend too much to get too little." LtGen Field Harris ltr to ACofS, G-3, HQMC, dtd 27Oct60.

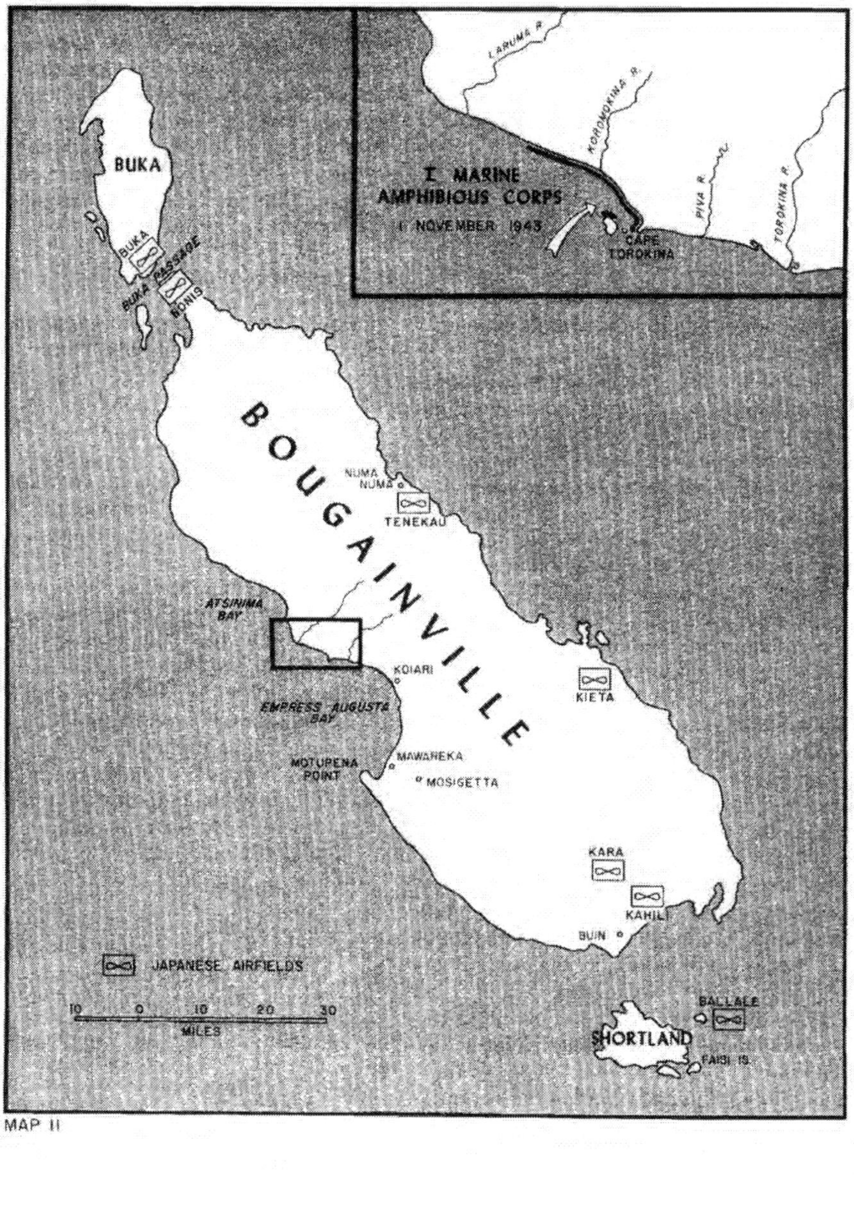
I MARINE
AMPHIBIOUS CORPS
1 NOVEMBER 1943
LARUMA R.
KOROMOKINA R.
PIVA R.
TOROKINA R.
CAPE
TOROKINA
BUKA
BUKA
BUKA PASSAGE
BONIS
BOUGAINVILLE
NUMA
NUMA
TENEKAU
ATSINIMA
BAY
KOIARI
EMPRESS AUGUSTA
BAY
KIETA
MOTUPENA
POINT
MAWAREKA
MOSIGETTA
KARA
KAHILI
BUIN
JAPANESE AIRFIELDS
10 0 10 20 30
MILES
BALLALE
SHORTLAND
FAISI IS.

MAP II

ury Islands and Choiseul as radar and PT bases and jumping-off points for further operations against Bougainville. MacArthur, however, had concluded that the original intentions of ELKTON called for a landing directly on Bougainville and that the interests of the JCS-approved plan could best be executed by an early operation (within the next few months) against Bougainville. MacArthur indicated that any target Halsey selected would be acceptable, but that a decision must be reached soon.

Kieta, on the northeastern coast of Bougainville, was a protected anchorage with an airfield close by; but an assault on Kieta involved long approaches by sea—and the Allied shipping shortage was critical. Kieta was too close to Choiseul, moreover, and that island would have to be attacked as a preliminary measure to protect the landing on Bougainville. Empress Augusta Bay, on the west coast, at first glance had little to recommend it as a landing site. The beaches along the bay were exposed to wind and waves and held no sheltered anchorage for the amphibious shipping required for such an operation. The terrain inland was known to be swampy, heavily timbered, and with few routes of communication. On the other hand, the enemy had apparently dismissed this area as a probable landing spot and the bay was only lightly defended. As late as 17 September, the Allies stood at a figurative crossroads, undecided about which fork in the road to take.

Five days later, on 22 September, Halsey announced a decision that canceled all previous plans and alerted his forces for one of two alternate courses of action: seize and hold the Treasury Islands and northern Empress Augusta Bay area on Bougainville as airfield sites; or, as a second course, seize and hold the Treasury Islands and the Choiseul Bay area as advance radar points, torpedo boat anchorages, and a staging base for landing craft before moving on to construct an airfield on the east coast of Bougainville later in the year. The final decision depended upon last-minute reconnaissance efforts.

SPOTLIGHT ON BOUGAINVILLE [3]

For the Japanese, the conviction that Bougainville was the ultimate Allied objective in the Northern Solomons was hardly a random guess. Lying in a position athwart the northern entrance to The Slot, Bougainville's big bulk dominated the rest of the Solomons chain. By virtue of this ideal geographic location, the island served the Japanese as an advanced supply and refueling base for most of the sea and air operations against the Allies at Guadalcanal and in the Central Solomons. Here, too, were staged the infantry replacements destined for combat or garrison duty on other South Pacific islands. No less than six major airfields and a number of naval operating bases were established by the Japanese on Buka, Bougainville, and the Shortlands to help guard the outer defenses of the airfields and Simpson Harbor at Rabaul. Easily supported by air bases on New Britain, New Ireland,

---

[3] Unless otherwise noted, the material in this section is derived from: *ThirdFlt NarrRept*; *IIIPhibFor AR*; *IMAC AR–I*; *HistDiv Acct*; 3d MarDiv Terrain InfoRept, n.d. (Bougainville AreaOpFile, HistBr, HQMC); MilHistSec, G–2, FEC Japanese Monograph No. 100, SE Area NavOps, Part III (OCMH), hereafter *SE Area NavOps—III*; Rentz, *Bougainville and the Northern Solomons*; Morison, *Breaking the Bismarcks Barrier*; Miller, *Reduction of Rabaul*; Feldt, *The Coastwatchers.*

and Truk—points outside the range of most Allied land-based planes—Bougainville was also under the umbrella of naval support from Rabaul and Truk. In short, the Allies needed Bougainville for further operations against Rabaul; the Japanese needed Bougainville to defend Rabaul. Viewed from either camp, the island was a priority possession.

Bougainville was never a part of the British Solomon Islands protectorate. German expansion had claimed the island in 1899, some 130 years after its belated discovery by the French explorer whose name it bears. Mandated to Australia after World War I, the island still was influenced by German missionaries, for the Bougainville natives in 1943 were recognized as hostile to the Allies and considerably more friendly to the Japanese.[4] Some 40,000 natives were on the island, gathered mainly in small villages of less than 200. Energetic and industrious, they provided an adequate labor force for the large plantations which, before the Japanese occupation, were owned by European companies.[5]

The island is the largest of the Solomons group. Nearly 30 miles in width, Bougainville is divided down the center by mountain ranges which extend nearly the entire 125-mile length of the island. The northern Emperor range is capped by the distinctive 10,171-foot peak, Mt. Balbi. The second range, Crown Prince, is less rugged and settles gradually into a broad, widening plain at the southern end of the island. This area has the best anchorages and the biggest plantations. The two coastlines, east and west, are markedly different. The eastern side of the mountain ranges slopes to a fairly open plain with good beach areas. The western side, however, is deeply etched by rushing mountain streams which carry silt into a swampy, alluvial plain bordering the coastline. The characteristic result is a series of deep valleys ending in swamps and sand bars cut by meandering waterways and sluggish rivers of varying depths. This soggy fringe area, covered with tall marsh grass and bamboo-like growth, is trapped between mountains and the sea by a grey-black beach strip which seldom exceeds 15 yards in width.

The island interior is enveloped by a dense rain forest and choking jungle growth which combine with the rugged mountain ranges to discourage overland exploration. A number of good trails traverse the more populated areas in the south and east, but only a few native tracks venture across the inhospitable interior. One path cuts across the mountains from Numa Numa to Empress Augusta Bay where it connects with the East-West trail. This path joins the western villages of Mawareka and Mosigetta to Buin in the south. Fairly wide and cleared, the East-West trail skirts the coastal swamps and can be traveled most of the year.

Although the Japanese had occupied the island since March 1942, only those facilities necessary to maintain the war in the Solomons had been constructed initially. In time, four airfields were in operation, two at each end of the island, and additional troops were stationed on Bougainville. Scant attention, however, was given to the island region between the major airfields. For more than a year, Japanese activity was restricted largely to the Buin area in the south and the Buka-Bonis pas-

---

[4] *IMAC AR–I*, C–2 Est of Sit, p. 26.

[5] Rentz, *Bougainville and the Northern Solomons*, Appendix V, contains an excellent description of the island of Bougainville and its people, terrain, and geographic features.

sage in the north.[6] There was little overland travel. Barges moving along the coastline served most of the transportation needs. As a result, few roads were improved and in the later defense of the island, this proved an important oversight.

Australian coastwatchers and a few friendly natives maintained observation posts on the southern part of the island until the summer months of 1943. Then aggressive Japanese patrols, assisted by unfriendly native guides, forced the Allied scouts to abandon their radio equipment and withdraw to the interior.[7] As a result, military information about the island and its defenders was cut off abruptly just when it became most needed.

Intelligence estimates on the number of Japanese soldiers and sailors in the area varied widely. Interceptions of radio messages provided most of the information on troop dispositions, and this intelligence was augmented and checked by enemy documents captured in the Central Solomons and by prisoner of war interrogations. Allied guesses placed the total number of defenders in the vicinity at 98,000—2,000 at Choiseul, 35,000 or more at Bougainville and the Shortlands, 5,000 at New Ireland, and the remaining 56,000 at Rabaul. The estimates on the strength of the Bougainville forces, based on order of battle information from prisoners, ranged between 35,000 and 44,000. The biggest concentration of defenders was in the southern part of the island, where an estimated 17,000 soldiers of the *Seventeenth Army* were headquartered. Another 5,000 troops were believed to be in the Buka-Bonis area, with a similar number at Kieta. The only known enemy concentration on the west coast was at Mosigetta, where about 1,000 Japanese—believed to be laborers—were engaged in cultivating the extensive rice fields of that coast. Less than 300 troops were estimated to be in the Cape Torokina vicinity of Empress Augusta Bay. The Shortlands defense force was estimated at 3,000 to 6,000, most of these naval personnel from *Eighth Fleet* headquarters and SNLF units.

Postwar records of the Japanese indicate that the Allied estimates were close. The Buin area actually had about 15,000 troops of the *Seventeenth Army* and 6,800 of the *Eighth Fleet* headquarters and base force personnel, primarily for defense of Kahili and Kara airfields. About 5,000 men were deployed in the Shortlands. The airfield on Ballale was defended entirely by naval personnel with seacoast artillery.

The troop dispositions were in line with the enemy conception of the plans of the Allies. After Guadalcanal was evacuated, and the Japanese became aware of the responsibility of defending what they had so easily grabbed, the Shortland Bay area was decided upon as the strategic key in the defense of the Northern Solomons. Accordingly, the southern part of Bougainville and the Shortland Islands received first priority in troop allotments. At that early date, the enemy believed that any Allied offensive would be directed against the airfields in the southern portion of the island with a possible subsidiary action in the Buka area. Troop strength elsewhere was proportionate to

[6] USAFISPA Obj Folder, Bougainville Island, dtd 1Aug43 (Bougainville AreaOpsFile, HistBr, HQMC).

[7] *ComSoPac WarD*, 22Jun43. An interesting sequel to this instance is the 13 December entry: "The native responsible for the capture of Allied personnel in Bougainville in June was seized and shot." *ComSoPac Dec43 WarD.*

the Japanese estimate of the Allies' ability to hit each area. Bougainville's defense was based on the premise that a landing anywhere on the island could be met by a transfer of ground troops and a counter-landing by an amphibious group.

Responsible for the defense of Bougainville and the adjacent islands was an old adversary, Lieutenant General Haruyoshi Hyakutake,[8] who had commanded the Japanese forces on Guadalcanal. The general had apparently lost little prestige with the *Imperial Staff* through the defeat, for he was still in command of the *Seventeenth Army*. His forces, however, had not shared his fortune. The *2d Division* was almost wiped out at Guadalcanal, and the *38th Division* had lost heavily at Guadalcanal and New Georgia. His sole remaining division, the *6th*, commanded by Lieutenant General Masatane Kanda, was still in fighting shape. Two regiments, the *23d* and *45th*, were near top-strength, but the third regiment of the division, the *13th Infantry*, had been badly mauled in the Central Solomons. To this division, Hyakutake could add detachments of SNLF units, plus the scattered remnants of other infantry regiments which were trickling into Bougainville after the withdrawal from the Central Solomons.

The Allied estimate of the Japanese ships in the immediate vicinity (Buka, Bougainville) was 2 cruisers, 8 to 10 destroyers, 21 personnel transports, and 12 submarines, plus a variety of smaller craft. The Imperial Navy also had a healthy reserve of warships at Truk and Rabaul. Air support in the Northern Solomons was believed to be about 160 fighters, 120 dive bombers, 120 medium bombers, and 39 float planes.[9]

There were definite signs in September that the Japanese expected an assault on Bougainville. Despite increasing air attrition, plane strength at Buka, Kara, Kahili, and Ballale remained fairly steady as the Japanese replaced their losses. Airfields were improved and expanded despite steady pounding by Allied bombers, and supply routes to the island were maintained in spite of losses incurred through harassment by Allied patrol bombers and torpedo boats. In late October, as the Allies completed their plans for attack, even the long-neglected west coast of Bougainville was given some attention by the Japanese. Intelligence photos for the first time revealed evidence of military activity near Empress Augusta Bay. Some minor construction with a few scattered defensive installations were discovered behind Cape Torokina, but since the improvements were limited and no additional troops appeared to have been moved into the area, the Allies remained convinced that the Japanese had not altered their basic defensive plans and that they had not awakened to the dangers inherent in an undefended coastline.

HALSEY'S DECISION[10]

Before a final decision was made on the direction of the SoPac attack, reconnais-

[8] Miller, *Reduction of Rabaul*, p. 235n. Early intelligence reports translated Hyakutake's given name variously as Siekichi or Harukichi.

[9] ThirdFlt OPlan 14–43, Annex A, dtd 15Oct43 (COA, NHD).

[10] Unless otherwise noted, the material in this section is derived from: *ComSoPac Sep43 WarD; ThirdFlt NarrRept; IIIPhibFor AR; IMAC AR–I; HistDiv Acct;* Rentz, *Bougainville and the Northern Solomons;* Morison, *Breaking the Bismarcks Barrier;* Miller, *Reduction of Rabaul;* Halsey and Bryan, *Halsey's Story.*

sance efforts to obtain every possible scrap of information and intelligence about the various projected landing sites continued. As at Guadalcanal and New Georgia, prior knowledge of the Bougainville area was limited to sketchy reports from former residents, planters, medical officers, schooner masters, and missionaries. Although these reports were valuable in regard to general conditions and physical improvements in certain areas, few facts of military significance were obtained. As before, personal reconnaissance by trained observers was required to accumulate the necessary detailed geographic and hydrographic information upon which to base a decision. The first intelligence efforts covered the entire Northern Solomons—Santa Isabel, Choiseul, the Treasurys, Shortlands, and Bougainville. Later, as some islands were eliminated and the choice of targets narrowed to either the east or west coast of Bougainville, reconnaissance activities were concentrated on the Kieta and Empress Augusta Bay areas.

Initial combat intelligence was gathered from air reconnaissance and submarine patrols. Aerial photography was limited because of unfavorable weather, enemy air interference, and lack of fighter plane escorts. Submarines moved in close to the island to shoot pictures through raised periscopes, but this practice was hazardous since the outdated hydrographic charts then available failed to show the exact location of dangerous coral outcroppings and reefs known to exist. Besides, the pictures did not reveal much except a good profile shot of Bougainville's rugged peaks. In time, patrols from submarines, torpedo boats, and seaplanes slipped ashore to scout various areas, and the information gained from physical reconnaissance and personal observation was added to ComSoPac's growing file.

Two patrols, dispatched to Bougainville in September after Halsey announced his two-part alert, helped the SoPac commander decide on the final choice of objectives. One Marine-Navy team, with an Australian officer and four natives as guides, remained four days in the Kieta vicinity, prowling the northeast coast of the island during the night and spending the daylight hours underwater in the submarine USS *Gato*. Considerable Japanese troop activity was observed; and despite several close scrapes from patrolling enemy barges, the group measured beach distances, took depth soundings, and scouted the area inland. On 28 September, the patrol boarded the submarine for the last time and returned to Guadalcanal. The patrol's report was generally unfavorable,[11] indicating that the harbor had many reefs and coral outcroppings, and that the area inland was not suitable for airfields since the Japanese had apparently given up on Kieta.

Another patrol similarly organized landed from the submarine USS *Guardfish* near the Laruma River in north Empress Augusta Bay. Here the terrain was found to be fairly solid with thick bush and a dense rain forest inland. Reluctant to arouse the Japanese to any Allied intentions, the patrol studied Cape Torokina through binoculars and took photographs through telescopic lens. The long-range examination revealed a narrow beach strip some 10,000 yards long with the expected coastal swamps inland. Tidal range in the bay was moderate, about 3½ feet. A coco-

[11] IIIPhibFor Rept of Reconnaissance of the Northeast Bougainville Coast, 23–27Sep43, n.d. (Bougainville AreaOpFile, HistBr, HQMC).

nut grove on Cape Torokina looked like a favorable spot for an airfield, the patrol decided, since the area appeared dry enough and long enough to support a fighter plane strip. Unable to obtain a soil sample of the area, the patrol did the next best thing and brought back soil from a similar area. The scouts then turned back to the Laruma river and headed into the bush in a wide circle that ended four days later in a rendezvous with the *Guardfish* in Atsinima Bay, some distance to the north. The only enemy sighted were a lone sentry on post near the Laruma river, and a number of Japanese reconnaissance planes flying patrol duty over Empress Augusta Bay beaches.[12]

The endeavor had one big dividend. While waiting for the patrol, the submarine commander checked his position and discovered that the navigation chart then in use was about seven miles in error in its location of Cape Torokina. Undetected, this one factor might well have jeopardized any future operations. Too, the soil sample returned by the patrol was declared favorable for construction of an airfield.

The prospect of landing in a lightly defended area close to an acceptable airfield site appealed to Halsey and his SoPac planners. To confirm the patrol's recommendation, a number of low-level aerial reconnaissance flights were made by IMAC staff members. Their quick glimpses of the Cape Torokina area convinced them that the operation could be a success. On 1 October, Halsey notified MacArthur that Cape Torokina was the main objective and that D-Day would be 1 November. The SoWesPac commander expressed his complete agreement with this decision.[13]

After 12 days of planning, Halsey's ComSoPac headquarters issued the orders which outlined the missions of the sea, air, and ground forces under the admiral's command. Specifically, ComSoPac directed the III Amphibious Force to seize and hold (on D-minus five days) the Treasury Islands as a staging area and advanced naval base, prior to establishing a beachhead on D-Day in the northern Empress Augusta Bay area for construction of airfields and another advanced naval base. The ultimate aim was strangulation of enemy operations in south Bougainville and preparations for further offensives against Rabaul.

Code names selected by ComSoPac for the planning phases were DIPPER, denoting the entire Northern Solomons operation, and CHERRYBLOSSOM, the Empress Augusta Bay area. The Treasury Islands phase of the operation was labeled GOODTIME. Later the code name DIPPER was applied to the Bougainville operation and the island, while the Treasury Islands landing retained the GOODTIME designation.

Selection of Cape Torokina as a landing site despite its disadvantages was tactically and logistically sound. The location fitted well into the plans for neutralization of Rabaul by air, and a beachhead on the western side of Bougainville made logistic support much easier. Moreover, the attack was aimed at a weak point in the Japanese defenses, thus avoiding a direct assault on main enemy defenses in the southern and eastern coasts of the island. Success of the venture depended upon the

[12] IMAC Reconnaissance Rept of Empress Augusta Bay area, 23–26 Sept43, dtd 1Oct43 (Bougainville AreaOpsFile, HistBr, HQMC).

[13] *ComSoPac Oct43 WarD.*

ability of the Allies to beat back any determined sea and air offensive by the Japanese during the critical stages of landing and establishing a beachhead. ComSoPac planners admitted that a strong enemy reaction was highly probable, but this threat was accepted as a calculated risk. The projected operation was "no better demonstration of the firmly held, but at times sorely tried, belief in the Allied superiority over the enemy in the South Pacific." [14]

Less obvious in the choice of Empress Augusta Bay was the fact that the Cape Torokina plain, bordered by the natural obstacles of the Laruma River to the northwest, the mountains inland, and the Torokina river to the southeast, fell into an ideal defensive area about six miles deep and eight miles long which could be defended by the Allied forces then available. The location, too, was believed so isolated from known Japanese dispositions by the nature of the island's terrain that at least three months would be required before a strong force moving overland could seriously threaten the beachhead. In short, large bodies of reinforcing enemy troops could come to the area only from the sea, and the Third Fleet felt confident that it could handle that threat. The Allies had no desire to capture the entire island—the size of Bougainville and the rough terrain precluded such ambitions — but two infantry divisions could hold the Cape Torokina area against any enemy forces in the immediate area or likely reinforcements.

These facts did little to increase enthusiasm for such bold plans. Sentiment was mixed — some optimism, some hesitation, some reluctance — but with D-Day less than a month away, all hands bent to the task of preparing for the assault on the Treasurys and Bougainville.

AMPHIBIOUS PLANNING [15]

The planning team that directed the preparations for the Bougainville operation was essentially the same command lineup that outlined the New Georgia attack—with one important change. As before, Admiral Halsey retained personal control of the proceedings, dividing his attention between completion of the Central Solomons campaign and the development of a new operation. From his headquarters in Noumea, Halsey coordinated the planning activities of Admiral Fitch on Espiritu Santo and Admiral Wilkinson and General Harmon on Guadalcanal. A new member of the planning staff was Lieutenant General Alexander A. Vandegrift, who had replaced General Vogel as commanding general of the I Marine Amphibious Corps, the counterpart to the Army's ground force command under Harmon. As such, Vandegrift held administrative responsibility over practically all Marine Corps personnel in the South Pacific, the exceptions being land-based air units under General Mitchell's command (Marine Air, South Pacific) and ships' detachments. The IMAC staff had not participated in the New Georgia planning since the bulk of the troops were to be furnished

[14] *IMAC AR–I*, p. 2.

[15] Unless otherwise noted, the material in this section is derived from: *ComSoPac Sep–Oct43 WarDs; IIIPhibFor AR;* CTF 31 WarDs, Sep–Oct43, n.d. (COA, NHD), hereafter *CTF 31 WarD* with month; *IMAC AR–I; ThirdFlt NarrRept; HistDiv Acct;* Rentz, *Bougainville and the Northern Solomons;* Isely and Crowl, *Marines and Amphibious War;* Morison, *Breaking the Bismarcks Barrier;* Miller, *Reduction of Rabaul.*

by Griswold's XIV Corps, but Halsey's first planning directive designated the Northern Solomons as an IMAC assignment. As a result, the Marine command became tactical as well as administrative.

Vandegrift, who had won a Medal of Honor for his leadership of the 1st Marine Division at Guadalcanal, was the first of three IMAC commanders to participate in the Bougainville operation. After completing preliminary plans, Vandegrift was relieved by Major General Charles D. Barrett on 15 September and was en route to the United States to become the 18th Commandant of the Marine Corps when his return was abruptly sidetracked.[16] General Barrett had suffered a cerebral hemorrhage in an accidental fall, and his untimely death left IMAC planning in midair. The operational order for IMAC's part in the seizure of the Treasurys and Bougainville had been drafted before Barrett's death, and Vandegrift resumed command on 18 October in time to sign the completed order[17]—noting, meanwhile, that the entire concept of the operation had changed within the space of one month.[18] Vandegrift completed the planning and witnessed the start of the operation before relinquishing command of IMAC to Major General Roy S. Geiger on 9 November, well after the success of the beachhead was assured.

Even as the Northern Solomons planning shifted through a number of changes before the final draft emerged, so did the task organization assigned to the IMAC landing force. Of the three Marine divisions in the Pacific in 1943, only the 3d was available for the Bougainville operation. The 1st Marine Division was scheduled by MacArthur to spearhead the Cape Gloucester beachhead on New Britain, and the 2d Marine Division had been shifted to the Central Pacific for the Gilbert Islands assault. To augment the 3d Marine Division and IMAC troops, the Army's 25th Division and the 1st Marine Raider Regiment were tentatively assigned to the Bougainville venture; but, as the Central Solomons campaign wore on, both organizations were committed to action far beyond original plans, with the result that neither was available for the Northern Solomons.

A number of provisional units were formed from scattered battalions, and these—with the later addition of the 37th Infantry Division—were assigned to IMAC. Vandegrift's command eventually included his own headquarters and corps troops, the 3d Marine Division, the Army's 37th Division, the 8th Brigade Group of the 3d New Zealand Division, the 3d Defense Battalion, the 2d Marine Raider Regiment (Provisional), the 1st Marine Parachute Regiment, the Army's 190th Coast Artillery Regiment (Antiaircraft), and varied naval small craft, construction, and communication units. The Marine parachute regiment, then on Vella Lavella, was designated the corps' reserve force. In area reserve were several coast

---

[16] Vandegrift was informed in January that his nomination to succeed General Holcomb as Commandant had been approved by Admiral King and the Secretary of the Navy. The general, accompanied by Colonel Gerald B. Thomas, had reached Pearl Harbor when his retention in the South Pacific was requested by Halsey to King through Nimitz. Gen Alexander A. Vandegrift interview by HistBr, G–3, HQMC, dtd 5Aug59 (WWII OpHistFile, HistBr, HQMC), hereafter *Vandegrift Interview*.

[17] BGen James Snedeker ltr to CMC, dtd 14May48 (Bougainville Monograph Comment File, HistBr, HQMC).

[18] *Vandegrift Interview*.

artillery battalions on Guadalcanal and the Army's Americal Division, then in the Fiji Islands. This latter division, however, could be committed only on Halsey's approval.

After the target had been defined in late September, Halsey established his operational chain of command with his amphibious force commander, Admiral Wilkinson, in charge of the entire Bougainville-Treasurys expedition. Wilkinson maintained control of Task Force 31 (III Amphibious Force) with the ground force and transport groups as subordinate commands. As directed by Halsey, Wilkinson would continue to command all forces afloat and ashore until the landing force commander was ashore and had indicated that he was able to take command of the ground forces.[19] The IMAC commander, at first Barrett and then Vandegrift, was responsible for the scheme of maneuver ashore at both Cape Torokina and the Treasurys, and was to exercise command over all units ashore, whether Allied forces, Marine Corps, Army, Navy, or ground echelons of air units. As at New Georgia, the actual employment of aerial support was to be under the operational control of air officers. This air echelon, Commander Aircraft Northern Solomons (ComAirNorSols) was a subordinate command of ComAirSols and was under the direction of Marine Brigadier General Field Harris. His tasks included the responsibility for active air defense of the Bougainville region (Torokina and the Treasurys) as well as operational control of all supporting aircraft entering this region. Harris' responsibilities also included establishment of an air warning system for both Torokina and the Treasurys and organization of air support control procedures for both areas.

For the Treasury Islands phase of the operation, Wilkinson added another echelon to his command, dividing his task force into a Northern Force for the Empress Augusta Bay landings and a Southern Force (Task Force 31.1) for the diversionary landings. Wilkinson retained command of the Northern Force (actually, a name designating the main units of Task Force 31) and placed Admiral Fort in command of the Southern Force. The Treasurys landing force, comprising mostly troops from the 8th Brigade Group of the 3d New Zealand Division, was commanded by Brigadier R. A. Row, under the general direction of the IMAC commander.[20]

By mid-October, all subordinate echelons of Task Force 31 and IMAC had issued operational orders, and the diverse sea and air elements under Third Fleet command had been assigned general missions in support of Wilkinson's task force in the Bougainville-Treasurys venture. Land-based air units of ComAirSols

[19] This concept of command evolved during the Guadalcanal campaign after disagreement between Vandegrift and the commander of the Amphibious Force, South Pacific, Rear Admiral Turner. Many naval officers considered the forces ashore as an extension of the forces afloat, administratively and militarily subordinate to the amphibious commander at all times. Vandegrift successfully contended that the landing force commander was more experienced and better qualified in ground operations and should have undivided responsibility once the troops were ashore. LtCol Frank O. Hough, Maj Verle E. Ludwig, and Henry I. Shaw, Jr., *Pearl Harbor to Guadalcanal—History of U.S. Marine Corps Operations in World War II*, v. 1, (Washington: HistBr, G-3, HQMC, 1958), pp. 240–241.

[20] The planning and conduct of the Treasurys operation is related in the following chapter.

(Task Force 33) were ordered to continue the general missions of reconnaissance and destruction of enemy ships and aircraft with the added duties of providing air cover and support for the land and sea forces involved in the Northern Solomons assault. Rear Admiral Frederick C. Sherman's carriers (Task Force 38) were directed to support the expedition by air strikes at Buka and Bonis airfields which were beyond the effective range of fighter planes from the now-completed airfields in the New Georgia area.

Cruisers and destroyers of Rear Admiral Merrill's Task Force 39 were to furnish protection for the amphibious force as well as bombard enemy installations in the Buka-Bonis and Shortlands area before the operation. Task Force 72, the submarines under the command of Captain James Fife, Jr., was to carry out offensive reconnaissance missions north of the Bougainville area.

Final planning for the actual assaults at Cape Torokina and the Treasurys was facilitated by the move of IMAC headquarters from Noumea to Guadalcanal, where the amphibious forces were in training. Vandegrift's command post was established in a coconut grove near Tetere, just a short distance from the headquarters of the IIIPhibFor and the 3d Marine Division bivouac area. The close proximity of the three major heaquarters responsible for coordinating the efforts of the forces involved eased considerably the problems that arose during the last weeks of preparation. And problems arose—many of them. In effect, the Northern Solomons operations had become two operations—each one with resultant problems of transportation, reinforcement, and resupply.

Tactical limitations in launching the Bougainville operation became apparent early.[21] In fact, the entire venture bore the title of "Shoestring No. 2," a reference to the general paucity of means which characterized the Guadalcanal operation in 1942.[22] Early in August, Halsey had reviewed the shipping situation in the South Pacific and estimated that he would need six additional troopships to complete preparations for the next operation, declaring that at least 14 APAs and 6 AKAs would be required from D minus 15 to D plus 5 for the assault on Bougainville.[23] Other commitments, particularly the Central Pacific campaign, had claimed all but eight APAs and four AKAs from the SoPac area; and this limited amount of amphibious shipping was the specter which haunted the Bougainville preparations. The acute shortage seriously restricted the number of troops and supplies that could be transported in the initial assault and prohibited a rapid buildup at the objective. Further, any loss of ships could not be replaced immediately; and, since these 12 ships were a bare minimum for transporting an assault division with the necessary artillery, air service personnel, engineers, and heavy equipment to construct and maintain an airfield, the sinking of any of these ships

[21] When Vandegrift began initial planning, he emphasized that his submitted tentative plans for operations in the Northern Solomons envisioned conduct under optimum conditions, and added that it was "readily foreseeable that the means required may exceed the resources available in this area, particularly in respect to transports and landing craft." CG, IMAC ltr to ComSoPac, dtd 4Aug43 (Bougainville AreaOpsFile, HistBr, HQMC).

[22] Morison, *Breaking the Bismarcks Barrier*, p. 288.

[23] *ComSoPac Aug43 WarD.*

during the beachhead phase would seriously threaten the entire operation. The definite possibility of quick and effective air retaliation by the enemy prompted a decision by Wilkinson that the slow-moving LSTs under his command would not be risked during the early phases of the operation.

The glaring contrast between the conduct of the Guadalcanal landing and the New Georgia operation was emphasized in the concept of the Bougainville assault. Vandegrift had the experience and tactical foresight required for such an operation. In Wilkinson he found a partner with a good grasp of amphibious support of operations ashore. Together the two commanders were a good team, and the plans they evolved were practical applications of the available means to the situation. Guadalcanal had been an assault followed by the establishment of a protective perimeter thrown about a captured airfield. The Japanese had dashed themselves against this perimeter, suffering ruinous casualties in the process. New Georgia, however, had been the reverse—a landing, then a protracted overland attack that had been a tiring, bloody smash against a Japanese perimeter. Given these examples, Vandegrift and Wilkinson were determined that the tactical errors of New Georgia would not be repeated. At Bougainville, the Marines would land first, establish and expand a beachhead, and—when the fighting widened into extended land operations—the Army would take over. Airfield construction would commence upon landing, with completion of air strips expected in time to help defend against any determined assault by the Japanese.

In order to take advantage of the extensive stretch of beach north of Cape Torokina, a simultaneous landing of all troops and supplies over multiple beaches was scheduled. This would allow the vitally important ships to drop anchor, unload, and depart without undue delay in the objective area. To further reduce unloading time, all cargo ships would be restricted to short loads, and assault troops would be diverted to hasten the unloading activities. Twelve beaches were to be employed in the assault, eleven extending northwest from Cape Torokina toward the Laruma River and the twelfth located on the northern (inner) shoreline of Puruata Island, the larger of two islets lying off Cape Torokina.

Three assault units of four landing teams each were to land abreast over these 12 beaches. One task unit was to land in the right sector of the beach area with a second task unit landing in the left sector. The third task unit was to be divided, two landing teams landing in each sector. Each infantry unit was to overrun any enemy defenses and seize a broad but shallow beachhead. Reconnaissance was to be started immediately to the front and flanks, while unloading was completed. Beach and antiaircraft defenses were to be prepared immediately to insure protection from a possible enemy counterlanding or the expected enemy air attacks.

Barring unforeseen difficulties, at least 13,900 men would go ashore in the D-Day landings at Cape Torokina. Unloading time for the troops and the 6,200 tons of embarked supplies and equipment was set at no more than six hours. No floating reserve for the Bougainville assault was planned, since the enemy situation ashore did not seem to warrant this risk of embarked troops and supplies. Initial plans called for the beachhead to be reinforced and resupplied by five LST echelons five days apart, each echelon bringing in another 3,000 troops and 5,500 tons of cargo. Vandegrift, upon resumption of the IMAC command, objected to this slow buildup and insisted that reinforcement of the beachhead proceed at an accelerated rate. He proposed that the 37th Division follow the 3d Division ashore no later than D plus 7.[24] This change in plans was agreed upon. The IMAC general also assumed responsibility for beachhead logistics, and, in planning the rapid unloading sequence, Vandegrift assigned about one-half of his combat Marines the temporary task of getting supplies ashore. During the initial stages, these Marines would work in the holds of the ships and on the beach as a shore party before returning to their units for combat operations.

Such preparations were not idle gestures. Once ashore, the IMAC troops faced the prospect of having an extremely tenuous supply chain cut behind them by enemy counteractions. To forestall any emergency later, Wilkinson decided to use all available ships in supply operations before the landing. Accordingly, an attempt was made to bring a 30-day level of all classes of supply forward to Vella Lavella, considerably closer to the objective area than the staging-storage-bivouac areas on Guadalcanal or in the Russells. Thus, a steady flow of supplies to the beachhead could be moved by the means available—LSTs, LCTs, APDs, or even DDs. The area picked was Ruravai on the east coast of Vella Lavella, then still under attack by 25th Division and New Zealand troops. The landing at Ruravai, however, was bombed with effectiveness by the Japanese, and the inability to divert a sufficient number of LSTs from the New Georgia supply chain ultimately resulted in a stockpile considerably smaller than that proposed—about 10-days' level, in fact. Cape Torokina, IMAC later discovered, was outside the range of small craft from Vella Lavella, a fact which further reduced the value of the IMAC supply station at Ruravai. The end result was that, despite seemingly adequate preparations, the supply problem remained a major obstacle throughout the operational planning.

Equally vital to the operation's success was the speed with which airfields could be constructed. Although the earlier reconnaissance patrols had located a possible airfield site, the odds were still high that this area might prove too swampy for quick completion of a strip. At least three bomber fields and two fighter strips were deemed essential to threaten Rabaul from Bougainville, and the problem of locating suitable areas as well as having a sufficient number of naval construction battalions assigned to the task were monumental worries. The airfield annex to the IMAC operation order called for two strips to be established as soon as possible[25] with the remaining fields to get underway as the beachhead widened. To assist Vandegrift, an experienced engineer staff was organized within IMAC, this group comprising Marine and Navy officers who had directed airfield construction under combat conditions at Guadalcanal and New

[24] *Vandegrift Interview.*

[25] IMAC OpO No. 1, Anx B, IMAC Airfield Plan, dtd 15Oct43 (Bougainville AreaOpsFile, HistBr, HQMC), hereafter *IMAC OpO No. 1.*

Georgia[26]—a situation likely to be repeated on Bougainville. Since the Cape Torokina area was still relatively unscouted, and the existence of coral a debatable question, plans were made to use pierced planking for all fields. This added a fresh burden to the limited shipping, but provided insurance that runways could be fashioned for at least local air support.[27] The proposed fighter runway, located in the plantation area at Cape Torokina, was to be about 4,500 feet long and about 600 feet wide. A bomber strip inland about 2,000 to 7,500 yards would require an extensive amount of clearing. Even though swamps were prevalent, drainage was believed possible.[28] The remaining fields would have to wait for the widening of the beachhead.

Another facet of the landing—naval gunfire support—also received close attention during the last weeks of preparation. Coordination between the amphibious force and the landing force had improved markedly since the first days of Guadalcanal, but many imperfections unfortunately still existed in the fire support of ground forces by ships' guns. The use of naval gunfire by Allied ships at Guadalcanal had little effect on the progress of the initial landings, and bombardments to cover a landing force in the Central Solomons had been impromptu affairs; *i.e.*, returning the fire of the Munda batteries during the Rendova landing, and the shelling of the Enogai garrison during the Marine raider operations at Rice Anchorage.

New enthusiasm, however, had been generated for naval shore bombardments after Kolombangara's capture. The island's fortified defenses were still intact after the Japanese withdrawal, and IIIPhibFor was granted permission to test new gunfire procedures on these bunkers and pillboxes. One of Wilkinson's destroyer squadrons did the shooting, with all available IMAC and 3d Marine Division gunfire liaison officers and spotters aboard to witness the demonstration.[29] The destroyers pounded the beach defenses in a simulated pre-H-Hour strike before one destroyer moved inshore to act as a spotting ship for on-call missions. At the conclusion of the firing, ships' officers and the observers went ashore. The preparation fires, they found, had blasted the beach areas, and the adjusted firing missions had knocked out other bunkers and fortifications.[30] As a result of this convincing display, IMAC officers were eager to give naval gunfire support a prominent part in the assault plans for Bougainville.

To retain the element of surprise, no pre-D-Day shelling of Cape Torokina was scheduled. This further increased the necessity for effective gunfire which would knock out beach defenses before the Marines went ashore and provide as well for quick fire support to reduce any undiscovered Japanese defenses which might hold up the assault waves long enough for the expected enemy counteraction to threaten the landing force. The IMAC gunfire officers initially made plans for a support group of about 4 heavy or light

[26] Col Francis M. McAlister ltr to CMC, dtd 29May48 (Bougainville Monograph Comments, HistBr, HQMC).

[27] *Ibid.*

[28] *IMAC OpO No. 1.*

[29] Col Frederick P. Henderson, "Naval Gunfire Support in the Solomon Islands Campaign," MS, 1954 (Bougainville AreaOpsFile, HistBr, HQMC), pp. 51–52, hereafter Henderson, "Naval Gunfire Support."

[30] *Ibid.*, p. 52.

cruisers and at least 10 destroyers;[31] but in light of the multiple commitments of the Third Fleet, this was a prohibitive request. Eventually, four destroyers, none of them practised in shore bombardment, were scheduled for pre-H-Hour and post-H-Hour gunfire support. In all truthfulness, IMAC liaison officers were unhappy with the amount of assistance offered by the IIIPhibFor, but accepted the proferred support with the realization that Wilkinson's force had other missions, too.

The final gunfire plan [32] positioned three destroyers (*Sigourney*, *Anthony*, *Wadsworth*) on the southeast flank of the landing area, with the fourth ship (*Terry*) stationed on the opposite or extreme northwest flank. These ships were to open fire at 10,000 yards before closing to 3,000 yards for close support and on-call missions. Target designation was to be taken from a photo mosaic of the Cape Torokina coastline, photographed from about 1,000 yards off-shore by a low-flying plane. After one final rehearsal in the New Hebrides area where the four ships tried to approximate area, range, bearings, and maneuvers as close as possible to those expected on D-Day,[33] the destroyer squadron joined the assault echelons for the move to the objective.

FINAL PREPARATIONS [34]

Major General Allen H. Turnage's 3d Marine Division was a well-trained, albeit inexperienced, outfit, a fusion of a number of combat teams and supporting troops. Three infantry regiments—the 3d, 9th, and 21st Marines—together with the 12th Marines (artillery) and the 19th Marines (engineers and Seabees) formed the nucleus for the division which was first assembled as a unit in May 1943. To these major commands was added a number of service and support groups—tanks, special weapons, amphibian tractors, motor transport, signal, and medical battalions. With a background of extensive jungle warfare training in Samoa and Guadalcanal, the 3d Marine Division was fully expected to be capable of meeting the rigors of the Bougainville operation.

The missions assigned the 3d Marine Division were an extension of IMAC tasks: capture or destroy enemy forces in the area; establish a beachhead in the Cape Torokina area about 2,250 yards inland and about 7,350 yards wide to include the two small islands offshore; commence selection of airfield sites and construction of air strips; establish long-range radar points and an advance naval base to include operating facilities for torpedo boats; expand the beachhead on corps order.

After Turnage had been alerted to the major role of the division in the forthcoming landings,[35] the task organization as-

[31] *Ibid.*, p. 58.

[32] CTF OpO No. A15-43, dtd 18Oct43, Anx A (Bougainville AreaOpsFile, HistBr, HQMC); Henderson, "Naval Gunfire Support," pp. 61–62.

[33] ComDesDiv 90 AR of 1Nov43, dtd 14Jan44 (COA, NHD).

[34] Unless otherwise noted, the material in this section is derived from: *IIIPhibFor AR; IMAC AR–I;* 3d MarDiv AR, 1–11 Nov43, dtd 18Dec43, hereafter *3d MarDiv AR; HistDiv Acct;* Rentz, *Bougainville and the Northern Solomons;* Isely and Crowl, *Marines and Amphibious War;* Miller, *Reduction of Rabaul;* Morison, *Breaking the Bismarcks Barrier;* Aurthur and Cohlmia, *3d MarDivHist.*

[35] CG, IMAC ltr of Instrn to CG, 3d MarDiv, dtd 27Sep43.

signed to the division grew rapidly with the addition of a number of corps troops and provisional battalions, some of them still in the process of forming. The major attachments were the 2d and 3d Raider Battalions, now joined in a provisional regiment, and the 3d Defense Battalion. As expected, the combat experience of these corps and division forces varied. Elements of the 2d Raider Battalion had raided Makin Island in 1942 before going to Guadalcanal, and the 3d Defense Battalion was battle-tested at Tulagi and Guadalcanal. But the rest of the IMAC force—with the exception of the 37th Division—were without prior experience. The Army division, however, had been blooded in the Munda campaign.

Early in October, the diverse elements of IMAC and the 3d Marine Division were assembled at Guadalcanal where the task of welding them into landing teams began. The three task units were based on the reinforced 3d and 9th Marines and the 3d Defense Battalion. Task Unit A–1, four landing teams comprising the 3d Marines and the 2d Raider Battalion, was assigned six beaches in the vicinity of Cape Torokina. Task Unit A–2 (9th Marines) was scheduled to land over five beaches on the left (northwest) flank, with the Marine 3d Raider Battalion, attached to the 9th Marines, to go ashore on Puruata Island. Task Unit A–3, the antiaircraft batteries of the 3d Defense Battalion and supporting troops, was to land behind the assault troops in each sector. Each task unit included artillery, air liaison and signal personnel, engineers, and naval base construction troops. As directed, every landing team was self-sustaining and self-supporting until the division as a whole could be consolidated behind the contemplated force beachhead line.[36]

The 21st Marines plus the remaining artillery units and supporting troops were formed into task units which would be brought to the objective area after D plus 3 days. By 7 November, all elements of the 3d Marine Division would be ashore with the last increments of the IMAC headquarters troops slated for arrival by 15 November. The first unit of the 37th Infantry Division—the 148th Combat Team—was to be unloaded starting 7 November with the remaining combat teams—the 129th and the 145th—scheduled to be on Bougainville by 22 November. In all, an additional 13,000 troops and another 26,672 tons of cargo were to be brought to the Torokina beachhead to reinforce and resupply the assault elements.[37] Originally these echelons were to be transported by LSTs and APDs, but later Allied sea and air victories permitted the continued use of the larger APAs and AKAs in reinforcing the Bougainville beachhead.

With only limited shipping space available to the combat troops, the assault echelons carried only enough supplies—three units of fire, B-rations, and fuel—to continue operations ashore for 10 days. The rest of the allotted cargo space contained additional ammunition, rations, fuel, engineering tools, and equipment which could be unloaded quickly if the situation permitted. The heavier equipment and facilities materiel would be

---

[36] CofS, 3d MarDiv ltr to CO, 3d Mar, dtd 16Oct43.

[37] *IMAC OpO No. 1*; IMAC, Movement of Task Organization, dtd 29Oct43 (BougainvilleArea-OpsFile, HistBr, HQMC); NorFor LoadingOs Nos. 501–43 through 509–43, dtd 12Oct–15Nov43 in *IIIPhibFor Oct-Nov43 WarDs*.

brought ashore in later echelons. The IMAC troops making the initial landings would carry a haversack only; knapsacks and blanket rolls were to be carried ashore with organizational gear. Officers of IMAC and the 3d Marine Division ordered all seabags, cots, and mattresses to be stored at Guadalcanal; and these were never embarked.

The three assault task units and equipment were embarked at Guadalcanal. Task Unit A–1 went on board four transports on 13 October for rehearsals at Efate, and then stood by at Espiritu Santo to await the assembly date. Task Unit A–2, after embarking 18 October on the remaining four transports, also rehearsed the landing operation for four days at Efate before heading back to the Guadalcanal area for the rendezvous. The third task unit continued training and rehearsals at Guadalcanal until 26 October, at which time the troops and equipment were taken on board the four AKAs.[38] The rehearsals indicated that, with a 2,500-yard run to the beach and each ship restricted to about 500 tons, uninterrupted unloading could be accomplished in about 4½ hours.[39] With allowance for time losses by air alerts while underway and during the assault phase, the amphibious force commander was sure that troops and supplies could be ashore before the end of D-Day. On 30 October, the various elements of the Northern Landing Force—transports, cargo ships, mine sweepers, and destroyers—began steaming toward the rendezvous point off Guadalcanal for the final run toward Bougainville.

[38] CG, 3d MarDiv Ltrs of Instrn, dtd 7Oct43 and 14Oct43 (BougainvilleAreaOpsFile, HistBr, HQMC).

[39] ComTransGru, IIIPhibFor, Rept of LandingOps, Empress Augusta Bay area, Bougainville Island, 1–2Nov43, dtd 22Dec43 (Bougainville AreaOpsFile, HistBr, HQMC).

ISOLATING THE TARGET [40]

As Wilkinson's transports and screening elements formed into the main convoy to open the second phase of Operation DIPPER, the final naval bombardment and air strikes aimed at the complete neutralization of Bougainville airfields began. For nearly a month the island's defenders had been on the receiving end of frequent bombing and strafing attacks as a preliminary to the actual landings. The final strikes were calculated to insure negligible air interference by the Japanese during the amphibious assault the next day.

Although the Allied air power was recognized as greater than that which could be mustered by the Japanese in the area, the fact that the Allies were moving closer to the main enemy strength was an equalizing factor. Late September estimates placed the Japanese air strength in the Northern Solomons at about 154 planes, or less than ⅓ of the estimated 476 in the Rabaul-Bougainville area. Against this number, the Allied air command—ComAirSols—could ante 728 planes of all

[40] Unless otherwise noted, the material in this section is derived from: MarCorpsBd Study and Evaluation of AirOps Affecting the USMC during the War With Japan, dtd 31Dec45 (HistBr, HQMC); *ComSoPac Oct43 WarD; ThirdFlt NarrRept;* ComAirSols StrikeComd WarD, 26Jul–19Nov43, dtd 17Jan44 (COA, NHD), hereafter *StrikeComd WarD;* ComAirSols IntelSec Repts for Oct43, n.d. (COA, NHD); *IIIPhibFor AR; IMAC AR–I; HistDiv Acct; SE Area NavOps—III;* Miller, *Reduction of Rabaul;* Morison, *Breaking the Bismarcks Barrier;* Rentz, *Bougainville and the Northern Solomons;* Sherrod, *MarAirHist.*

types. This composite group included 14 Marine Corps squadrons with 181 planes, 19 Navy squadrons with 274 planes, 16 Army squadrons with 216 planes, and 3 New Zealand squadrons with 57 planes. By mid-October, at least 314 fighters and 317 bombers were being directed against the enemy bases. The recent acquisition of airfields on New Georgia and Vella Lavella was a decided asset to ComAirSols operations; for, although most of the bombers were still based on Guadalcanal and the Russells, nearly all of the fighter strength had been moved forward within striking distance of Bougainville. This included 31 fighters and 148 bombers at Munda, 103 fighters at nearby Ondonga, 48 at Segi, and 60 at Barakoma.

The air activity over Bougainville had steadily increased as D-Day moved closer. During the month of October, ComAirSols planes ranged northward on 21 of the 31 days in raids that ranged from tree-level strafing runs by torpedo and scout bombers and fighter planes to high-level bombing attacks by B-24s and B-25s. As the month neared its end, intermittent attacks became almost daily occurrences to the beleaguered Japanese defenders. Kahili and Kara were hit most often, 23 times and 17 times respectively, but Buka, Ballale, Kieta, and the Treasurys also were raked and cratered by Allied planes. Choiseul, too, was rocked occasionally by ComAirSols craft with Choiseul Bay and the Kakasa submarine base the favorite targets.

By 18 October, jubilant aviators had reported that Ballale's airfield was "pulverized."[41] Photographs verified their claims by showing 98 bomb craters on the runway, 23 of them in the center of the strip. By the 22d, Kahili was likewise inoperable. A week later, Ballale was hit again, and the enemy's repair work rendered ineffective. Postwar enemy records also attest to the attention given the enemy airfields by ComAirSols. The Japanese admitted that, just prior to the invasion, the airfields on Bougainville were useless.

Rabaul, too, was feeling the sting of Allied bombings. General Kenney's Allied Air Forces, committed by General MacArthur to lend all assistance to the neutralization of Rabaul by air, roared out of New Guinea bases on 12 October to slug Simpson Harbor and the Rabaul airfields with a 349-plane raid. This blast was encored by an equally large attempt six days later, but this time bad weather closed in over New Britain and only 54 bombers completed their mission. Daylight attacks by fighter-bomber groups were repeated on the 20th, 24th, and 25th of October. Kenney's fliers insisted that the bombings had crippled Rabaul, and optimistic reports of damage inflicted and enemy aircraft destroyed were relayed to the invasion forces. Later assessments, however, discounted these evaluations. Far from being a smoking ruin, the Japanese stronghold with its five airfields was still very much operational and still a factor to be reckoned with in the Bougainville seizure.

As a matter of record, however, the SoWesPac bombers did considerable damage to enemy installations and considerably reduced the Japanese ability to strike at the Bougainville assault forces. Further, this damage was accomplished on restricted targets in the face of hot receptions by enemy fighter craft and accurate antiaircraft defenses. Unfortunately, foul weather near D-Day prevented the

[41] *StrikeComd WarD*, 18Oct43.

Allied Air Forces from continuing their attacks which would have further diverted the Japanese attention from the Bougainville landings. As it turned out, the SoWesPac fliers were able to strike another solid blow against Rabaul only after the IMAC troops were already ashore at Cape Torokina.

As part of ComSoPac's program to stun the Bougainville defenses during the initial stages of the operation, the supporting task forces of Admirals Merrill and Sherman took up the cudgel for a whack at the Japanese airfields prior to D-Day. Merrill, with four light cruisers and eight destroyers, steamed close inshore to the Buka Passage shortly after midnight, 31 October–1 November, to rake Buka and Bonis airfields with a 30-minute bombardment. The gunfire was given added accuracy by the spotting reports from two ComAirSols planes overhead. The task force then retired, harassed but unscathed by enemy planes. Two hours and 60 miles later, the Buka flames were still visible to the task force.

Merrill was barely out of the neighborhood before Sherman arrived in the area with the carriers *Saratoga* and *Princeton*, escorted by 2 light cruisers and 10 destroyers. Undetected by several enemy flights, the carrier planes took off and dropped an additional 20 tons of bombs on the two airfields before the carrier task force retired unopposed and unchallenged. Admiral Merrill, meanwhile, was completing the second part of his mission. The already-riddled airstrip at Ballale was shelled by his task force, as was Faisi and several of the smaller islands. Still unopposed, Merrill's force headed for Vella Lavella to refuel and rearm, satisfied that the supporting bombardments had successfully launched the assault on Bougainville.

Diversionary Assaults

TREASURY ISLAND LANDINGS[42]

If the initial plans for the direct assault on the Buin area or the Shortlands had been carried out, the two small islands of the Treasury Group would have been bypassed and left in the backwash of the campaign. Instead, with the change in plans to strike directly at Empress Augusta Bay, the islands of Mono and Stirling became important as long-range radar sites and torpedo boat anchorages. Moreover, in an attempt to deceive the enemy as to the direction of the attack on Bougainville and convince him that the ultimate Allied aim might be the Buin area or the Shortlands, the seizure of the Treasurys was given added emphasis by being set as a preliminary to the Torokina landings. To help this deception succeed, reconnaissance patrols to the Shortlands and diversionary operations on the island of Choiseul—plus low-flying photo missions over the Shortlands—were scheduled by IMAC to increase the enemy's conviction that the follow-up objective was the Shortlands.

This could have been a natural assumption by the enemy. The Treasurys are about 60 miles northwest of Vella Lavella and only 18 miles south of the Shortlands. While the size of the Treasurys limited consideration as a major target, Mono and Stirling were close enough to Shortland Island to cause the Japanese some concern that they might be used as handy stepping stones by SoPac forces. But then again, the Treasurys are only 75 miles from Cape Torokina—a fact which the Allies hoped might be lost on Bougainville's defenders.

The Treasury Islands are typical of other small islands jutting out of the sea in the Solomons chain. Mono is a thickly forested prominence of volcanic origin, with abrupt peaks and hill masses more than 1,000 feet high in the southern part. These heights slope gradually in an ever-widening fan to the west, north, and east coasts. The shores are firm, with few swamps, and rain waters drain rapidly through deep gorges. The island is small, about four miles north to south and less than seven miles lengthwise.

Stirling Island to the south is smaller, more misshapen. Fairly level, this island is about four miles long and varies from 300 yards to nearly a mile in width. There

[42] Unless otherwise noted, the material in this section is derived from: *ComSoPac Oct–Nov43 WarDs; ThirdFlt NarrRept; IMAC AR–I;* IMAC C–2 Repts, 27Oct–13Dec43 (Bougainville Area OpFile, HistBr, HQMC), hereafter *IMAC C–2 Repts;* IMAC C–2 Jnl, 27Oct–27Nov43 (Bougainville AreaOpFile, HistBr, HQMC), hereafter *IMAC C–2 Jnl;* IMAC D–2 MiscRepts (Treasury Is), 27Oct43, dtd 10Nov43 (TreasuryIs AreaOpFile, HistBr, HQMC); 8(NZ) BrigGru Rept on Ops, TreasuryIs (Op GOODTIME), dtd 30Nov43 (TreasuryIs AreaOpFile, HistBr, HQMC); 8(NZ) BrigGru OpO No. 1, Op GOODTIME, dtd 21Oct43 (TreasuryIs AreaOpFile, HistBr, HQMC); ONI, *Combat Narrative XII;* Henderson, "Naval Gunfire Support;" Rentz, *Bougainville and the Northern Solomons;* Miller, *Reduction of Rabaul;* Morison, *Breaking the Bismarcks Barrier.*

are several small, brackish lakes inland, but the island is easily traversed and, once cleared of its covering forest, would be an excellent site for an airfield. Between these two islands is a mile or more of deep, sheltered water—one of the many anchorages in the Solomon Islands to bear the name Blanche Harbor. The combination of these features—airfield site, radar points, good anchorage—was the factor which resulted in the seizure of the Treasurys as part of the Bougainville operation.

Early information about the islands was obtained by an IMAC patrol which spent six days in the Treasurys in August, scouting the area, observing the movement of the Japanese defenders, and interrogating the natives. In this latter instance, the loyal and friendly people of the Treasurys were a remarkable contrast to the suspicious and hostile Bougainville inhabitants. Additional details were received from rescued aviators who found Mono Island a safe hiding place after their planes had been forced down by damage incurred in raids over Buin and the Shortlands. This first-hand intelligence was augmented by aerial photographs. The reports and photos indicated that the best landing beaches were inside Blanche Harbor, on opposite shores of Mono and Stirling. The only beaches suitable for LSTs, however, were on Mono between the Saveke River and a small promontory, Falamai Point.

As limited as this information was, the amount of intelligence on the enemy dispositions on the two islands was even more meager. The Japanese strength was estimated at 135 men, lightly armed. These were bivouacked near Falamai but maintained a radio station and observation posts in various areas. Natives reported that much of the time the Japanese moved about Mono armed only with swords or hand guns. Stirling Island was apparently undefended.

The 8th New Zealand Brigade Group, attached to I Marine Amphibious Corps for the seizure and occupation of these islands, arrived at Guadalcanal from New Caledonia in mid-September. Although the New Zealanders would form the bulk of the assault troops, the GOODTIME operation was IMAC-directed and IMAC-supported. The landing force comprised about 7,700 officers and men, of whom about 1,900 were from I Marine Amphibious Corps support troops—antiaircraft artillery, construction battalions, signal, and boat pool personnel. Marines attached to the brigade task organization included a detachment from the IMAC Signal Battalion and an air-ground liaison team from General Harris' ComAirNorSols headquarters.

On 28 September, Brigadier Row, the landing force commander, was informed of the general nature of the GOODTIME operation, and planning in conjunction with Admiral Fort began immediately, although there was only enough information available to the commanders of the task group and the landing force to formulate a plan in broad outline. The task was far from easy, for the Southern Force was confronted with the same logistical and transportation problems that faced the Empress Augusta Bay operation.

Fort and Row decided that the main assault would be made in the area of Falamai, where beaches were suitable for LSTs. Stirling Island would be taken concurrently for artillery positions. No other landings were planned; but after Row was informed that the long-range radar would have to be positioned on the northern side of Mono to be of benefit to

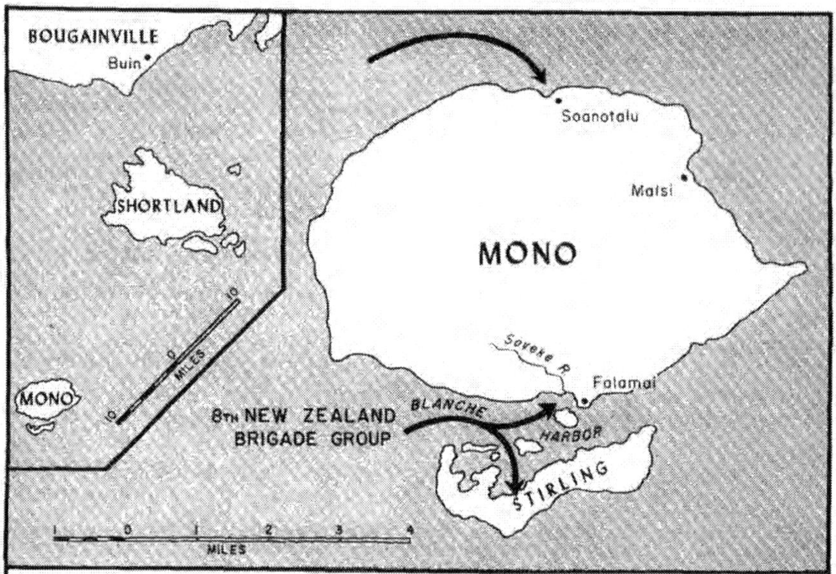

TREASURY ISLANDS LANDINGS
I MARINE AMPHIBIOUS CORPS
27 OCTOBER 1943

MAP 12

the Bougainville operation, another landing at Soanotalu on the north coast was written into the plans.

Final shipping allocation to Fort's Southern Force included 31 ships of six different types—8 APDs, 8 LCIs, 2 LSTs, and 3 LCTs for landing troops and supplies, 8 LCMs and 2 APCs for heavy equipment and cargo. The limited troop and cargo capacity of this collection of ships and landing craft restricted the Southern Force's ability to put more than a minimum of troops and supplies ashore initially, but this problem was solved by reducing the strength of the brigade's battalions and limiting the number of artillery weapons, motor transport, and engineering equipment in the first echelon. The brigade's assault units included 3,795 troops with 1,785 tons of supplies and equipment. Succeeding echelons were scheduled to sail forward at intervals of five days.

The final plans, issued by Row's headquarters on 21 October, directed the 29th and 36th Battalions to land nearly abreast near Falami Point, with the 34th Battalion landing on Stirling Island. Simultaneously, a reinforced infantry company accompanied by radar personnel and Seabees would go ashore at Soanotalu in the north. The two battalions on Mono would then drive across the island to link up with the Soanotalu landing force while naval base construction got underway at Stirling.

The initial landings in Blanche Harbor were to be covered by a naval gunfire support group of two destroyers, the *Pringle* and *Philip*. Liaison officers of IMAC planned the gunfire support, as the New Zealand officers had no experience in this phase of operations. While the brigade group expected to have no trouble in seizing the islands, the naval support was scheduled to cover any unforeseen difficulties. The gunfire plan called for the two destroyers to fire preparation salvos from the entrance to Blanche Harbor before moving in toward the beaches with the landing waves to take targets under direct fire. The IIIPhibFor, however, took a dim view of risking destroyers in such restricted waters. The desired close-in support mission was then assigned to the newly devised LCI(G)—gunboats armed with three 40mm, two 20mm, and five .50 caliber machine guns—which were making their first appearance in combat. Two of these deadly landing craft were to accompany the assault waves to the beaches.

After one final practice landing on Florida Island, the brigade group began loading supplies and embarking troops for the run to the target area. Admiral Fort's Southern Force was divided into five transport groups under separate commanders, and these groups departed independently when loaded. The slower LSTs and LCMs left first, on the 23d and 24th of October, and were followed the next day by the LCIs. The APDs sailed on 26 October

The Southern Force departed with a message which delighted the New Zealanders as typical of the remarks to which Americans at war seemed addicted: "Shoot calmly, shoot fast, and shoot straight." [2]

At 0540 on the 27th, the seven APDs of the first transport group lay to just outside the entrance to Blanche Harbor and began putting troops over the side into landing craft. Heavy rain and overcast weather obscured the beaches, but the pre-assault bombardment by the *Pringle* and *Philip* began on schedule. The USS *Eaton* moved to the harbor's mouth and took up station as fighter-director ship as the destroyers registered on Mono Island. The firing was accomplished without assistance of an air spotter, who later reported radio failure at the critical moment. This probably accounts for the disappointing results of the preparatory bombardments, which proved to be of little value except to boost the morale of the assault troops. The *Pringle's* fire was later declared to be too far back of the beach area to be helpful, and the bombardment by the *Philip* left a great deal to be desired in accuracy, timing, and quantity.

A fighter cover of 32 planes arrived promptly on station over the Treasurys at 0600, and, under this protective screen, the assault waves formed into two columns for the dash through Blanche Harbor to the beaches. Unexpectedly, enemy machine gun fire from Falamai and Stirling greeted the assault boats as they ploughed through the channel. At 0623, just three minutes before the landing craft nosed into the beaches on opposite sides of the harbor, the pre-assault cannonading ceased; and the two LCI gunboats—one on each

---

[2] Quoted in Gillespie, *New Zealand History*, p. 148.

flank of the assault wave—took over the task of close support for the landing forces. At least one 40mm twin-mount gun, several machine guns, and several enemy bunkers were knocked out by the accurate fire of these two ships. Promptly at 0626, the announced H-Hour, New Zealand troops went ashore on Mono and Stirling.

At Falamai, the 29th and 36th Battalions moved inland quickly against light rifle and machine gun fire, mostly from the high ground near the Saveke River. Casualties in the first wave were light—one New Zealand officer and five sailors wounded—and the second wave had no casualties.

The New Zealanders began to widen the perimeter as more troops were unloaded. At 0735, enemy mortar and medium artillery fire registered on the beach area, causing a number of casualties and disrupting unloading operations. Both LSTs were hit, with one ship reporting 2 dead and 18 wounded among the sailors and soldiers aboard. The other ship reported 12 wounded. Source of the enemy fire could not be determined. The *Eaton*, with Admiral Fort on board, ignored a previous decision not to enter Blanche Harbor and resolutely steamed between the two islands. This venture ended, however, when enemy planes were reported on the way, and the *Eaton* reversed course to head for more maneuvering room outside the harbor. Assured that the air raid was a false alarm, the destroyer returned to Blanche Harbor and added its salvos to those of the LCI gunboats. This fire, directed at likely targets, abruptly ended the Japanese exchange.

By 1800, the two battalions had established a perimeter on Mono Island and were dug in, trying to find some comfort in a dismal rain which had begun again after a clear afternoon. Evacuation of casualties began with the departure of the LSTs. With the exception of one LST, which still had 34 tons of supplies aboard when it retracted, all ships and landing craft had been unloaded and were on their way back to Guadalcanal by the end of D-Day. The casualties were 21 New Zealanders killed and 70 wounded, 9 Americans killed and another 15 wounded.

The landings at Stirling and Soanotalu were uneventful and without opposition. There were no casualties at either beachhead. At Stirling, the 34th Battalion immediately began active patrolling as soon as the command was established ashore. The Soanotalu landings proceeded in a similarly unhindered manner. A perimeter was established quickly, and bulldozers immediately went to work constructing a position for the radar equipment which was to arrive the next day.

The fighter cover throughout the day had shielded the troops ashore from enemy air attacks. The escorting destroyers, however, were hit by an enemy force of 25 medium and dive bombers at about 1530, and the USS *Cony* took two hits. Eight crewmen were killed and 10 wounded. The fire from the destroyer screen and the fighter cover downed 12 of the enemy planes. That night the bombers returned to pound the Mono Island side of Blanche Harbor and, in two raids, killed two New Zealanders and wounded nine.

Action along the Falamai perimeter the night of 27 October was concentrated mainly on the left flank near the Saveke River, the former site of the Japanese headquarters, and several attacks were beaten back. The following day, patrols moved forward of the perimeter seeking the enemy, and one reinforced company set out cross-country to occupy the village of Malsi on the northeast coast. There was little contact. Japanese ground activity on the night of the 28th was light, and enemy air activity was limited to one low-level strafing attack and several quick bombing raids—all without damage to the brigade group.

By 31 October, the entire situation was stable. The perimeter at Falamai was secure, Malsi had been occupied without opposition, and radar equipment at Soanotalu had been installed and was in operation. With the arrival of the second echelon on 1 November, the New Zealanders began an extensive sweep of the island to search out all remaining enemy troops on the island. The going was rough in the high, rugged mountain areas, but, by 5 November, enemy stragglers in groups of 10 to 12 had been tracked down and killed. The New Zealanders lost one killed and four wounded in these mop-up operations.

Undisturbed for some time, the perimeter at Soanotalu was later subjected to a number of sharp attacks, each one growing in intensity. The Soanotalu force was struck first on 29 October by small groups of Japanese who were trying to reach the beach after traveling across the island from Falamai. These attacks continued throughout the afternoon until a final charge by about 20 Japanese was hurled back. Construction of the radar station continued throughout the fighting. Enemy contact on the next two days was light, and the first radar station was completed and a second one begun.

On the night of 1 November, a strong force of about 80 to 90 Japanese suddenly struck the perimeter in an organized attack, apparently determined to break through the New Zealand defense to seize a landing craft and escape the island. The fight, punctuated by grenade bursts and mortar fire, raged for nearly five hours in the darkness. One small group of enemy penetrated the defenses as far as the beach before being destroyed by a command group. About 40 Japanese were killed in the attack. The Soanotalu defenders lost one killed and nine wounded. The following night, 2 November, another attempt by a smaller Japanese force was made and this attack was also beaten back. This was the last organized assault on the Soanotalu force, and the remainder of the Japanese on the island were searched out and killed by the New Zealand patrols striking overland.

By 12 November, the New Zealanders had occupied the island. Japanese dead counted in the various actions totaled 205; the New Zealanders took 8 prisoners. It is doubtful that any Japanese escaped the island by native canoe or swimming. In addition, all enemy weapons, equipment, and rations on the island were captured. The Allied casualties in this preliminary to the Bougainville operation were 40 New Zealanders killed and 145 wounded. Twelve Americans were killed and 29 wounded.

During the period of fighting on Mono Island, activity on Stirling was directed toward the establishment of supply dumps, the building of roads, and the construction of advance naval base and boat pool facilities. Although several minor enemy air raids damaged installations in the early phases of the operation, the landing at Empress Augusta Bay diverted the attention of the enemy to that area and ended all Japanese attempts to destroy the force in the Treasurys.

RAID ON CHOISEUL ISLAND[3]

If the Japanese had opportunity to speculate on the significance of the Treasurys invasion, the problem may have been complicated a few hours later by a landing of an Allied force on the northwest coast of Choiseul Island, just 45 miles from the southeastern coast of Bougainville. The landing was another ruse to draw Japanese attention from the Treasurys, point away from the Allies' general line of attack, and divert the enemy's interest—if not effort—toward the defense of another area. More specifically, the Choiseul diversion was calculated to convince the Japanese that the southern end of Bougainville was in imminent danger of attack from another direction. The salient facts which the Allies hoped to conceal were that the real objective was Empress Augusta Bay, and that the Choiseul landing force consisted only of a reinforced battalion of Marine parachute troops.

Actually, the raid on Choiseul was a small-scale enactment of landing plans which had been discarded earlier. Choiseul was considered as a possible objective for the main Northern Solomons attack; but when the decision was made that the Allied attack would strike directly amidships on the western coast of Bougainville, the Choiseul idea was dropped. Then, when the suggestion was advanced by Major James C. Murray, IMAC Staff Secretary, that, because of the size and location of Choiseul, a feint toward that island might further deceive the Japanese as to the Allies' intentions, the diversionary raid was added to the Northern Solomons operation.

Choiseul is one of the islands forming the eastern barrier to The Slot; and as one of the Solomon Islands, it shares the high rainfall total, the uniform high heat and humidity common to other islands of the chain. About 80 miles long and 20 miles wide at the widest point, Choiseul is joined by reefs at the southern end of two small islands (Rob Roy and Wagina) which seems to extend Choiseul's length another 20 miles. The big island is not as rugged as Bougainville and the mountain peaks are not as high, but Choiseul is fully as overgrown and choked with rank, impenetrable jungle and rain forest. The mountain ranges in the center of the island extend long spurs and ridges toward the coasts, thus effectively dividing the island into a series of large compartments. The beaches, where existent, vary from wide, sandy areas to narrow, rocky shores with heavy foliage growing almost to the water's edge. Other compartments end in high, broken cliffs, pounded by the sea.

The island was populated by nearly 5,000 natives, most of whom (before the war) were under the teachings of missionaries of various faiths. With the exception of a small minority, these natives remained militantly loyal to the Australian government and its representatives. As a result,

---

[3] Unless otherwise noted, the material in this section is derived from: *ComSoPac Oct–Nov43 WarD;* SoPacFor CIC, Study of Choiseul Island, dtd 19Sep43 (Choiseul AreaOpFile, HistBr, HQMC); *IIIPhibFor AR; IMAC AR–I*, Anx Q, BLISSFUL; *IMAC C–2 Jnl;* IMAC OpO No. 2, dtd 22Oct43 (Choiseul AreaOpFile, HistBr, HQMC); CO, 2d ParaBn, PrelimRept, Op BLISSFUL, dtd 5Nov43 (Choiseul AreaOpFile, HistBr, HQMC); 2d ParaBn UnitJnl, 27Oct–4Nov43 (Choiseul AreaOpFile, HistBr, HQMC); 2d ParaBn OpO No. 1, dtd 23Oct43 (Choiseul AreaOpFile, HistBr, HQMC); MajGen Victor A. Krulak ltr to ACofS, G–3, HQMC, dtd 17Oct60, hereafter *Krulak ltr;* Rentz, *Bougainville and the Northern Solomons;* Isely and Crowl, *Marines and Amphibious War.*

coastwatching activities on Choiseul were given valuable assistance and protection.

Combat intelligance about the island was obtained by patrols which scouted various areas. One group, landed from a PT boat on the southwest coast of Choiseul, moved northward along The Slot side of the island toward the Japanese base at Kakasa before turning inland. After crossing the island to the coastwatcher station at Kanaga, the patrol was evacuated by a Navy patrol bomber on 12 September after six days on the island.

Two other patrols, comprising Marines, naval officers, and New Zealanders, scouted the northern end of the island and Choiseul Bay for eight days (22–30 September) before being withdrawn. Their reports indicated that the main enemy strength was at Kakasa where nearly 1,000 Japanese were stationed and Choiseul Bay where another 300 troops maintained a barge anchorage. Several fair airfield sites were observed near Choiseul Harbor, and a number of good beaches suitable for landing purposes were marked. Japanese activity, the patrols noted, was generally restricted to Kakasa and Choiseul Bay.[4]

During the enemy evacuation of the Central Solomons, Choiseul bridged the gap between the New Georgia Group and Bougainville. The retreating Japanese, deposited by barges on the southern end of Choiseul, moved overland along the coast to Choiseul Bay where the second half of the barge relay to Bougainville was completed. This traffic was checked and reported upon by two active coastwatchers, Charles J. Waddell and Sub-Lieutenant C. W. Seton, Royal Australian Navy, who maintained radio contact with Guadalcanal.

Seton, on 13 October, reported the southern end of Choiseul free of Japanese, but added that at least 3,000 to 4,000 enemy had passed Bambatana Mission about 35 miles south of Choiseul Bay. On 19 October, the coastwatcher reported that the enemy camps in the vicinity of Choiseul Bay and Sangigai (about 10 miles north of Bambatana Mission) held about 3,000 Japanese who were apparently waiting for barge transportation to Bougainville. Seton indicated that the Japanese were disorganized, living in dispersed huts, and were short of rations. They had looted native gardens and searched the jungle for food. Further, the Japanese were edgy. All trails had been blocked, security had been tightened, and sentries fired into the jungle at random sounds.[5]

After this information was received at IMAC headquarters, Vandegrift and Wilkinson decided that a diversionary raid on Choiseul would be staged. On 20 October, Lieutenant Colonel Robert H. Williams, commanding the 1st Marine Parachute Regiment, and the commanding officer of his 2d Battalion, Lieutenant Colonel Victor H. Krulak, were summoned from Vella Lavella to Guadalcanal. At IMAC headquarters, Williams and Krulak conferred with Vandegrift and his staff. The orders to Krulak were simple: Get ashore on Choiseul and make as big a demonstration as possible to convince the

[4] IMAC Patrol Rept. on Choiseul Bay, 22–30Sep43, dtd 4Oct43; 3d MarDiv Rept of Patrol to Kakasa, 6–13Sep43, dtd 16Sep43 (Bougainville AreaOpFile, HistBr, HQMC).

[5] The Japanese uneasiness had an excellent basis. Sub-Lieutenant Seton "had organized 25 natives into a quasi-military force, armed them (Japanese weapons) and, on 2 October, ambushed an [enemy] group in a landing craft, killing seven." *Krulak ltr.*

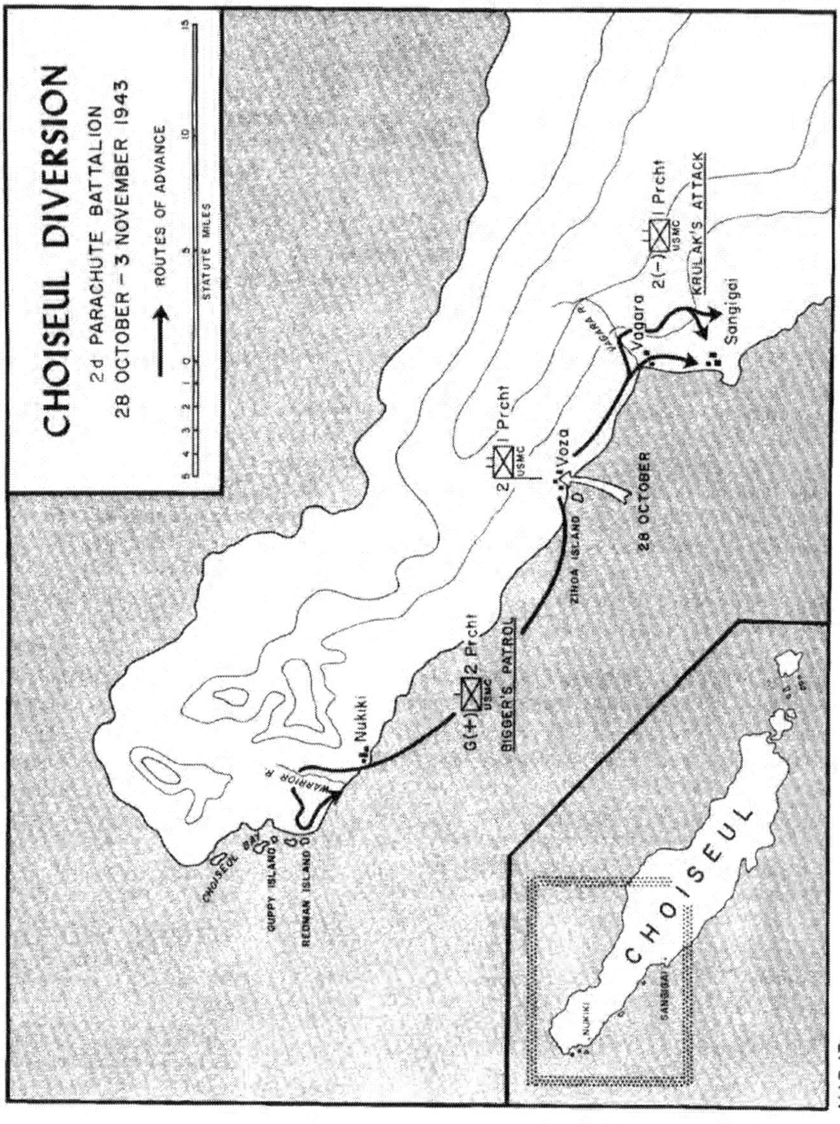
CHOISEUL DIVERSION
2d PARACHUTE BATTALION
28 OCTOBER – 3 NOVEMBER 1943
ROUTES OF ADVANCE
STATUTE MILES
5 4 3 2 1 0 5 10 15
CHOISEUL BAY
GUPPY ISLAND
REDMAN ISLAND
WARRIOR R.
Nukiki
G(+) USMC 2 Prcht
BIGGER'S PATROL
ZINOA ISLAND
2 USMC 1 Prcht
Voza
D
28 OCTOBER
VAGARA R.
Vagara
2(–) USMC 1 Prcht
KRULAK'S ATTACK
Sangigai
CHOISEUL
NUKIKI
SANGIGAI
MAP 13

Japanese that a major landing was in progress. In addition, reconnaissance would be conducted to determine possible sites for a torpedo boat patrol base.

The IMAC operation order, giving the code name BLISSFUL to the Choiseul diversion, was issued on 22 October. Based on information and recommendations from Seton, the Marines' landing was set for the beaches in the vicinity of Voza village, about midway between Choiseul Bay and Bambatana Mission. There the beaches were good, friendly natives would help the invading forces, and there reportedly were no enemy troops. Moreover, it was firmly astride the main route of evacuation of the Japanese stragglers from Kolombangara and points south. After receiving the order, Krulak went to the airstrip on Guadalcanal, and, while waiting for a plane to take him back to his command, wrote out the operation order for his battalion's landing.

This was to be the first combat operation of the 2d Battalion as well as its first amphibious venture. Although equipped and trained for special assignments behind enemy lines, these Marines—known as Paramarines to their comrades—never chuted into action because suitable objectives were usually beyond the range of airborne troops and the necessary transport planes were in chronically short supply. The 1st Parachute Battalion, however, had taken Gavutu and Tanambogo Islands before going to Guadalcanal to take part in the defense of the airfield there in 1942. This battalion had then formed the nucleus for the present 1st Parachute Regiment, now consisting of three battalions in IMAC reserve at Vella Lavella. Each battalion, of three rifle companies each, was armed with a preponderance of light automatic and semi-automatic weapons. The nine-man squads in Lieutenant Colonel Krulak's rifle platoons carried three Johnson light machine guns[6] and six Johnson semi-automatic rifles; each company had, in addition, six 60mm mortars.

Krulak's return to his command set off a flurry of near-frenzied activity, since the battalion had a minimum of time for preparation. For the next four days, officers and men worked almost around the clock to assemble equipment, make final plans, and brief themselves on the task ahead. On the 24th, Coastwatcher Seton and two of his native guides arrived at Vella Lavella to meet Krulak's officers and men and give them last-minute information. After being briefed by Seton, Krulak requested and was given authority by IMAC to operate in any direction on Choiseul, if consistent with his mission.

Equipment and supplies for the operation were pre-sorted into four stacks; and late on the afternoon of the 27th of October the parachute battalion and its gear was embarked on board eight LCMs borrowed from the Vella Lavella boat pool. Krulak's three companies were reinforced by a communications platoon, a regimental weapons company with mortars and light machine guns, and a detachment from an experimental rocket platoon (bazookas and rockets) from IMAC. Total battalion strength was 30 officers and 626 men. In addition, one naval officer accompanied the battalion for reconnaissance purposes related to the possible establishment of a torpedo boat base.

At dusk, when four APDs which had just completed the Treasury landings ar-

---

[6] "The Johnson light machine gun was more an auto-rifle than a machine gun; more a machine gun than the BAR." *Ibid.*

rived off Vella Lavella, the troops and equipment were transferred from the LCMs to the *McKean*, *Crosby*, *Kilty*, and *Ward* in a quick operation that was completed in less than 45 minutes. The destroyer division, with the USS *Conway* acting as escort, sailed from Vella Lavella at 1921. The *Conway*'s radar would locate the landing point in the dark.

Moving in column through the night, the convoy was sighted shortly after 2300 by an enemy snooper plane which dropped one bomb, scoring a near miss on the last APD in line. Shortly before midnight, at a point some 2,000 yards off the northwest coast of Choiseul, the convoy stopped, and a reconnaissance party in a rubber boat headed toward shore to scout the landing area. A signal light was to be shown if no enemy defenders were encountered. While waiting for the signal, Krulak ordered Companies F and G into the landing boats.

After waiting until 0019 (28 October), Company F headed toward the beach with Company G close behind. The operation order had directed Company G to make the initial assault, but the APDs had drifted apart and the *Kilty* with Company F embarked was closer to shore. Since no light on shore was yet discernible, the Marines expected opposition. The landing, however, was uneventful, and the patrol was waiting on shore. Observers on ship reported later that the light was visible at 0023, just four minutes after the parachute companies began the run for the beach. After setting the troops ashore, the landing craft immediately returned to the transports to bring in a load of supplies.[7]

A lone enemy plane detected the *Conway* standing patrol duty seaward, and dropped two bombs near the ship. The *Conway*, reluctant to draw attention to the landing, did not return the fire, and the enemy plane droned away. An Allied escort plane, assigned to protect the convoy against such attacks, drew considerable criticism, however, for not remaining low enough to spot such bombing runs.

Two hours after arrival in the area, the convoy reversed course and steamed back to Vella Lavella, leaving behind four landing craft (LCP(R)) with their crews for the battalion's use. These craft were dispersed under the cover of overhanging mangroves near the offshore island of Zinoa, and the Marines turned to moving supplies off the beach. Seton, who landed on Choiseul with the battalion, disappeared into the bush and returned almost immediately with a group of native bearers. With their help, the Marines moved into the jungle. The transfer was none too soon; enemy reconnaissance planes appeared at dawn to bomb the area but without success.

Early on the morning of the 28th, a base of operations was established on a high jungle plateau about a mile to the northwest of Voza and outposts were set up on the beach north and south of the village. Security was established and wire communications installed. The plateau, behind natural barriers of rivers and high cliffs, was an ideal defensive spot and a necessary base camp for the heavy radio gear with which IMAC had equipped the parachute battalion.

---

[7] DesDiv 44 AR for night of 27–28Oct, Initial Landing of Marine Paratroopers on Choiseul Island, dtd Nov43; USS *Conway* AR, 27–28Oct43, dtd 25Nov43 (Choiseul AreaOpFile, HistBr, HQMC).

During the day of 28 October, while the Marines established their camp, another enemy flight appeared and raked the beachhead with a strafing and bombing attack. The effect was wasted. The Marines had dispersed; their equipment had been moved; and good camouflage discipline had been observed. Too, the natives had obliterated every sign of a landing at Voza and established a dummy beachhead several miles to the north for the special benefit of Japanese planes seeking a target.

Informed by Seton's guides that the Marine battalion was bivouacked between a barge staging-replenishment base at Sangigai about eight miles to the south and an enemy outpost at the Warrior River about 17 miles to the north, Krulak on the morning of the 29th sent out reconnaissance patrols to the north and south. These groups were to locate trails, scout any enemy dispositions, and become familiar with the area.

Krulak personally led one patrol toward Sangigai, going overland toward the Vagara River which was about halfway between the Marine camp and the enemy base. While part of the patrol headed inland toward the high ground to the rear of Sangigai, to sketch the approaches to the village, the Marine comander led the rest of the patrol to the river. There the hidden Marines silently watched a group of about 10 Japanese unloading a barge; and since this appeared to be an excellent opportunity to announce the aggressive intentions of the Krulak force, the Americans opened fire. Seven of the Japanese were killed, and the barge sunk. Krulak's section then returned to the base, followed shortly by the other half of the patrol. After the attack order was issued, a squad was sent back over the trail to the Vagara to hold a landing point for Krulak's boats and to block the Japanese who might be following the Marines' track. The patrol ran into a platoon of the enemy about three-quarters of a mile from the original Marine landing point and drove the Japanese off.[8]

At 0400 the following morning, 30 October, Krulak led Companies E and F, plus the rocket detachment, toward Voza for an attack on Sangigai. The barge base had been marked as a target since 22 October. To help him in his assault, and give the impression of a larger attacking force, Krulak requested a preparatory air strike on reported Japanese positions northwest of the base. Estimated enemy strength was about 150 defenders, although Seton warned that Sangigai could have been reinforced easily from the southwest since the Marines' landing.

Krulak's attack plans were changed at Voza, however, since one of the four boats had been damaged a few minutes earlier in an attack by Allied planes. The strafing ended when the fighter pilots discovered their error and apologized to the boat crews with a final pass and a clearly visible "thumbs-up" signal. The requested air strike at Sangigai hit at 0610 with better results. While 26 fighters flew escort, 12 TBFs dropped a total of more than two tons of bombs on enemy dispositions.

Unable to use the boats for passage to the Vagara, Krulak ordered his troops to begin a route march overland from Voza. Seton and his native guides led the way, followed by Company F (Captain Spencer H. Pratt) with a section of machine guns

[8] This encounter left Krulak "in no doubt that we needed to go at them quickly, because they were obviously aggressive." *Krulak ltr.*

and the rocket detachment. Company E (Captain Robert E. Manchester) and attached units followed. At about 1100, Japanese outposts on the Vagara opened fire on the Marine column. Brisk return fire from the parachutists forced the enemy pickets to withdraw towards Sangigai.

Following the envelopment plan he had formulated on the 29th, Krulak sent Company E along the coastline to launch an attack on Sangigai from that direction while the remainder of the force, under his command, moved inland to attack from the high ground to the rear and east of Sangigai. The assault was set for 1400, but as that hour drew near, the group in the interior found that it was still a considerable distance from the village. The mountainous terrain, tangled closely by jungle creepers and cut by rushing streams, slowed Krulak's force, and, by H-Hour, the column was still not in position to make its attack effort. When the sound of firing came from the direction of Sangigai village, the second force was still moving towards its designated jump-off point. Seton's natives, however, indicated that the enemy were just ahead.

Company E, moving along the beach, reached its attack position without trouble. Although the assault was delayed a few minutes, the company opened with an effective rocket spread and mortar fire. As the Marines moved forward, the Japanese defenders hastily withdrew, abandoning the base and the village to flee to the high ground inland. The Marine company entered the village without opposition.[9]

The enemy's withdrawal to prepared positions inland fitted perfectly into Krulak's scheme of maneuver. The Japanese moved from the village straight into the fire of Company F, and a pitched battle that lasted for nearly an hour ensued. An enveloping movement by the Marines behind the effective fire of light machine guns forced the Japanese into several uncoordinated *banzai* charges which resulted in further enemy casualties. As the Marines moved once more to turn the enemy's right flank, the Japanese disengaged and about 40 survivors escaped into the jungle. A final count showed 72 enemy bodies in the area. Krulak's force lost four killed. Twelve others, including Krulak and Pratt, were wounded.

Company E, possessors of Sangigai, had been busy in the interim. Manchester's company, using demolitions, destroyed the village, the Japanese base and all enemy supplies, scuttled a new barge, and captured a number of documents, including a chart of enemy mine fields off southern Bougainville. The Marines then withdrew to the Vagara to board the four landing craft (the disabled boat had been repaired) for the return to Voza. Krulak's force, after burying its dead, retraced its path to the Vagara and spent the night in a tight defensive perimeter.[10] Early the

[9] The Marines with Krulak saw the first enemy position "within a few minutes of E Company's opening fire. In this sense, the timing was extremely lucky. Had the enveloping column been 30 minutes slower, the Japs would have gotten away from E Co into the bush. As it was, the sentence in the operation order 'Prevent enemy withdrawal into the mountains' (War Diary—1600 29 Dec) worked out well." *Ibid.*

[10] The original plan was for the boats to make two trips on the 30th, but by the time Company E got back to base it was getting dark. The battalion executive officer cancelled the return trip in view of the dangers of running the boats along the reef shelf at night. Krulak's radio had broken down and so he had no way of learning of this decision, although he guessed that this was the case. Still, it was an anxious night.

next morning, 31 October, the landing craft returned to carry the parachutists to Voza and the base camp.

With the battalion reassembled once more, the Marines prepared ambushes to forestall any Japanese retaliatory attacks, and aggressive patrols were pushed out along the coast to determine if the Japanese were threatening and to keep the enemy off balance and uncertain about Marine strength. A Navy PBY landed near Voza the following day to evacuate the wounded Marines and the captured documents; and, on the same day, in answer to an urgent request by Krulak, 1,000 pounds of rice for the natives and 250 hand grenades and 500 pounds of TNT were air dropped near Voza. Several brisk engagements between opposing patrols were reported on this day, 1 November, but the base camp was not attacked. Seton's natives, however, reported that Sangigai had again been occupied by the Japanese.

After Krulak returned to the base camp on 31 October, his executive officer, Major Warner T. Bigger, led a patrol to Nukiki Village, about 10 miles to the north. No opposition was encountered. On the following day, 1 November, Bigger led 87 Marines from Company G (Captain William H. Day) toward Nukiki again to investigate prior reports of a large enemy installation on the Warrior River. Bigger's instructions were to move from Nukiki across the Warrior River, destroying any enemy or bases encountered, and then move as far north as possible to bring the main Japanese base at Choiseul Bay under 60mm mortar fire. Enemy installations on Guppy Island in Choiseul Bay were an alternate target.

The patrol moved past Nukiki without opposition, although the landing craft carrying the Marines beached continually in the shallow mouth of the Warrior River. Since the sound of the coxswains gunning the boats' motors to clear obstructions was undoubtedly heard by any enemy in the area, Bigger ordered the Marines to disembark. The boats were then sent downriver to be hidden in a cove near Nukiki. Bigger's force, meanwhile, left four men and a radio on the east bank of the river, and all excess gear including demolitions was cached. Mortar ammunition was distributed among all the Marines. The patrol then headed upriver along the east bank, and the Warrior was crossed later at a point considerably inland from the coast.

By midafternoon, the natives leading the patrol confessed to Bigger that they were lost. Although in the midst of a swamp, the Marine commander decided to bivouac in that spot while a smaller patrol retraced the route back to the Warrior River to report to Krulak by radio and to order the boats at Nukiki to return to Voza. In response to Bigger's message, Krulak asked Seton if he had any natives more familiar with the country north of the Warrior River; the only man who had visited the region was sent to guide the lost Marines.

The smaller patrol bivouacked at the radio site on the night of 1–2 November and awoke the next morning to the realization that a Japanese force of about 30 men had slipped between the two Marine groups and that their small camp was virtually surrounded. Stealthily slipping past enemy outposts, the patrol members moved to Nukiki, boarded the boats, and returned to Voza. After hearing the patrol's report, Krulak then radioed IMAC for fighter cover and PT boat support to withdraw the group from the Choiseul Bay area.

Bigger was unaware of the activity behind him. Intent upon his mission, he decided to continue toward Choiseul Bay. After determining his position, Bigger ordered another small patrol to make its way to the river base camp and radio a request that the boats pick up his force that afternoon, 2 November. This second patrol soon discovered the presence of an enemy force to Bigger's rear, and was forced to fight its way towards Nukiki. This patrol was waiting there when the landing craft returned to Nukiki.

The main force, meanwhile, followed the new guide to the coast and then turned north along the beach toward Choiseul Bay. Opposite Redman Island, a small offshore islet, a four-man Japanese outpost suddenly opened fire. The Marines quickly knocked out this opposition, but one Japanese escaped—undoubtedly to give the alarm.

Because any element of surprise was lost and thinning jungle towards Choiseul Bay provided less protection and cover for an attacking force, Bigger decided to execute his alternate mission of shelling Guppy Island. Jungle vegetation growing down to the edge of the water masked the fire of the 60mm mortars, so Bigger ordered the weapons moved offshore. The shelling of Guppy was then accomplished with the mortars emplaced on the beach with part of the baseplates under water. The enemy supply center and fuel base was hit with 143 rounds of high explosives. Two large fires were observed, one of them obviously a fuel dump. Bigger's force, under return enemy fire, turned around and headed back toward the Warrior River.

The Japanese, attempting to cut off Bigger's retirement, landed troops from barges along the coastline; and the Marine force was under attack four separate times before it successfully reached the Warrior River. There the patrol set up a perimeter on the west bank and waited for the expected boats.

Several men were in the river washing the slime and muck of the jungle march from their clothing when a fusillade of shots from the opposite bank hit the Marine force. The patrol at first thought it was being fired upon by its own base camp, but when display of a small American flag drew increased fire, the Marines dove for cover. Heavy return fire from the Marine side of the river forced the enemy to withdraw. Seizing this opportunity, Bigger directed three Marines to swim across the Warrior to contact the expected boats and warn the rescuers of the ambush. Before the trio could reach the opposite shore, though, the Japanese returned to the fight, and only one survivor managed to return to the Marine perimeter.

Even as the fierce exchange continued, the Marines sighted the four boats making for the Warrior River from the sea. An approaching storm, kicking up a heavy surf, added to the difficulty of rescue. Under cover of the Marines' fire, the landing craft finally beached on the west shore, and the Bigger patrol clambered aboard.

One boat, its motor swamped by surf, drifted toward the enemy shore but was stopped by a coral head. The rescue was completed, though, by the timely arrival of two PT boats—which came on the scene with all guns blazing.[11] While the

---

[11] One of these boats was commanded by Lieutenant John F. Kennedy, USNR, later the 35th President of the United States. *Krulak ltr.*

patrol boats raked the jungle opposite with 20mm and .50 caliber fire, the Marines transferred from the stalled craft to the rescue ships and all craft then withdrew. A timely rain squall helped shield the retirement. Aircraft from Munda and PT boats provided cover for the return to Voza.

The time for withdrawal of the battalion from Choiseul was near, however, despite the fact that Krulak's force had planned to stay 8–10 days on the island. On 1 November, another strong patrol, one of a series sent out from the Voza camp to keep the enemy from closing in, returned to the Vagara to drive a strong Japanese force back towards Sangigai. From all indications, the Japanese defenders now had a good idea of the size of the Krulak force, and aggressive enemy patrols were slowly closing in on the Marines. Seton's natives on 3 November reported that 800 to 1,000 Japanese were at Sangigai and that another strong force was at Moli Point north of Voza.

After the recovery of the Bigger patrol from Nukiki, IMAC asked Krulak to make a frank suggestion as to whether the original plan should be completed or whether the Marine battalion should be removed. The Cape Torokina operation was well underway by this time, and IMAC added in its message to Krulak that Vandegrift's headquarters considered that the mission of the parachute battalion had been accomplished. Krulak, on 3 November, radioed that the Japanese aggressiveness was forced by their urgent need of the coastal route for evacuation, and that large forces on either side of the battalion indicated that the Japanese were aware of the size of his force and that a strong attack, probably within 48 hours, was likely. The Marine commander stated that he had food for seven days, adequate ammunition, and a strong position; but that if IMAC considered his mission accomplished, he recommended withdrawal.

Commenting later on his situation at this time, Krulak remarked:

> As a matter of fact, I felt we'd not possibly be withdrawn before the Japs cut the beach route. However, we were so much better off than the Japs that it was not too worrisome (I say now!) The natives were on our side—we could move across the island far faster than the Japs could follow, and I felt if we were not picked up on the Voza side, we could make it on the other side. Seton agreed, and we had already planned such a move. Besides that we felt confident that our position was strong enough to hold in place if necessary.[13]

On the night of 3 November, three LCIs appeared offshore at a designated spot north of Voza to embark the withdrawing Marines. In order to delay an expected enemy attack, the Marines rigged mine fields and booby traps. During the embarkation, the sounds of exploding mines were clearly audible. Much to the parachutists' amusement, the LCI crews nervously tried to hurry embarkation, expecting enemy fire momentarily. Krulak's battalion, however, loaded all supplies and equipment except rations, which were given to the coastwatchers and the natives. Embarkation was completed in less than 15 minutes, and, shortly after dawn on the 4th of October, the Marine parachute battalion was back on Vella Lavella.

Krulak's estimate of the Japanese intentions was correct. Within hours of the Marines' departure, strong Japanese forces closed in on the area where the parachute battalion had been camped. The enemy had been surprised by the landing

---

[13] *Ibid.*

and undoubtedly had been duped regarding the size of the landing force by the swift activity of the battalion over a 25-mile front. Then, after the operation at Empress Augusta Bay got underway, the Choiseul ruse became apparent to the Japanese, who began prompt and aggressive action to wipe out the Marine force. The continued presence of the Allied group on Choiseul complicated the evacuation program of the Japanese, and, once aware of the size of the Krulak force, the enemy lost no time in moving to erase that complication.

Before the battalion withdrew, though, it had killed at least 143 Japanese in the engagement at Sangigai and the Warrior River, sunk two barges, destroyed more than 180 tons of stores and equipment, and demolished the base at Sangigai. Unknown amounts of supplies and fuel had been blasted and burned at Guppy Island. Mine field coordinates shown on the charts captured at Sangigai were radioed to the task force en route to Cape Torokina, vastly easing the thoughts of naval commanders who had learned of the existence of the mines but not their location. Later, the charts were used to mine channels in southern Bougainville waters that the Japanese believed to be free of danger.

The destruction of enemy troops and equipment on Choiseul was accomplished at the loss of 9 Marines killed, 15 wounded, and 2 missing in action. The latter two Marines were declared killed in action at the end of the war.[13]

The effect of the diversionary attack upon the success of the Cape Torokina operation was slight. The Japanese expected an attack on Choiseul; the raid merely confirmed their confidence in their ability to outguess the Allies. In this respect, the Japanese were guilty of basing their planning on their opponents' intentions, not the capabilities. There is little indication that enemy forces in Bougainville were drawn off balance by the Choiseul episode, and enemy records of that period attach little significance to the Choiseul attack.

This may be explained by the fact that the main landing at Cape Torokina took place close on the heels of Krulak's venture and the ruse toward Choiseul became apparent before the Japanese reacted sufficiently to prepare a counterstroke to it. Certainly, the size and scope of the landing operations at Empress Augusta Bay were evidence enough that Choiseul was only a diversionary effort.

THE JAPANESE [14]

Enemy reaction to the Allied moves was a bit slow. The Japanese knew that an offensive against them was brewing; what they could not decide was where or when. The *Seventeenth Army* was cautioned again to keep a watchful eye on Kieta and

[13] This is the casualty figure given by Rentz, *Bougainville and the Northern Solomons*, p. 114. Few accounts of the Choiseul attack are in accord on Marine casualties. Muster rolls of the battalion indicate 9 KIA, 12 WIA, and 5 MIA. Of those missing, four were later declared dead and one believed a prisoner of war. IMAC C–2 Jnl, 4Nov43, and the report of the diversion attack, Operation BLISSFUL, p. 4, indicate that 9 KIA and 16 WIA is correct. *IIIPhibFor AR*, pp. 3–4, states that 8 KIA, 14 WIA, 1 MIA, and 1 captured is correct. *IMAC AR–I*, p. 11, gives the casualties as 8 KIA, 14 WIA, and 1 MIA. ONI, *Combat Narrative XII*, p. 24, gives the losses as 9 KIA, 15 WIA, and 2 MIA. The figure given by Rentz undoubtedly takes into cognizance a 13Dec43 message from Coastwatcher Seton to the effect that the bodies of two Marines, one of them bound as a prisoner, had been found near the Warrior River. *ComSoPac Dec43 WarD*.

Buka, and General Hyakutake in turn directed the *6th Division* to maintain a firm hold on Choiseul as well as strong positions in the Shortland Islands. Then, the Japanese defenders on Bougainville waited for the next developments.

After the Allied landings in the Treasurys, the Japanese thinking crystallized: Munda was operational; Vella Lavella was not. Therefore, the only targets within range of New Georgia were the Shortlands or Choiseul. And based upon this reasoning, the Allies scarcely would attempt a landing on Bougainville before staging bases on Mono or Choiseul were completed. Reassured by this assumption, the Japanese relaxed, confident that the next Allied move would come during the dark quarters of the moon—probably late in November.

With the Allied move toward Choiseul, the Japanese were more convinced that the Allied pattern was predictable. With a firm foothold on Mono and Choiseul, the Allies would now move to cut Japanese lines and then land on the southern part of Bougainville in an attempt to seize the island's airfields. Basing their estimates on the increased number of Allied air strikes on Buka and the Shortlands, the Bougainville defenders decided that these were the threatened areas. All signs pointed to a big offensive soon—probably, the Japanese agreed—on 8 December, the second anniversary of the declaration of war.

[14] Unless otherwise noted, the material in this section is derived from: *SE Area NavOps—III; Seventeenth ArmyOps—II;* Rentz, *Bougainville and the Northern Solomons;* Miller, *Reduction of Rabaul;* Morison, *Breaking the Bismarcks Barrier.*

The enemy had no hint that such an unlikely area as Empress Augusta Bay would be attacked. The defense installations were concessions to orders directing that the western coast be defended, and the troops at Mosigetta—the only force capable of immediate reinforcement to the Cape Torokina area—were alerted only to the possibility that they might be diverted on short notice to the southern area to defend against an assault there.

Japanese sea and air strength was likewise out of position to defend against the Bougainville thrust. Admiral Mineichi Koga, commander in chief of the *Combined Fleet* at Truk, had decided earlier to reinforce Vice Admiral Jinichi Kusaka's *Southeast Area Fleet* and the land-based planes of the *Eleventh Air Fleet* at Rabaul so that a new air campaign could be aimed at the Allies in the South Pacific. This operation, *Ro,* to start in mid-October, was to short-circuit Allied intentions by cutting supply lines and crushing any preparations for an offensive. To Kusaka's dwindling array of fighters, bombers, and attack planes, Koga added the planes from the carriers *Zuikaku, Shokaku,* and *Zuiho*—82 fighters, 45 dive bombers, 40 torpedo bombers, and 6 reconnaissance planes.

Koga's campaign, though, was delayed. Allied radio traffic indicated that either Wake or the Marshall Islands would be hit next, and to counter this threat in the Central Pacific, Koga sent his fleet and carrier groups toward Eniwetok to set an ambush for the Pacific Fleet. After a week of fruitless steaming back and forth, the Japanese force returned to Truk, and the carrier groups moved on to Rabaul. The Japanese admiral had at first decided to deliver his main attack against New Guinea, but the Treasurys landings caused him to swerve towards the Solomons. Then, when Allied activities between 27 October and 1 November dwindled, the fleet again turned toward New Guinea to take up the long-delayed *Ro* operation. The elements of the Japanese fleet reached the area north of the Bismarcks on 1 November, just in time to head back towards the Solomons to try to interrupt the landings at Cape Torokina.

Assault of Cape Torokina

"... *THE TROOPS ARE MAGNIFICENT.*"[1]

The Northern Landing Force arrived off Empress Augusta Bay for the assault of Cape Torokina shortly after a bright dawn on 1 November, D-Day. The approach to the objective area had been uneventful. After rendezvousing near Guadalcanal, the transports steamed around the southern and western coasts of Rendova and Vella Lavella toward the Shortland Islands. ComAirSols fighter planes provided protection overhead and destroyer squadrons screened the flanks. Submarines ranged ahead of the convoy to warn of any interception attempt by the enemy.

When darkness fell on 31 October, the convoy abruptly changed course and, picking up speed, started the final sprint toward Empress Augusta Bay. Mine sweepers probed ahead for mine fields and uncharted shoals, while Navy patrol bombers and night fighters took station over the long line of transports. Eight air alerts were sounded during the night. Each time the night fighters, directed by the destroyers, intercepted and chased the enemy snooper planes away from the convoy. The amphibious force, moving direct as an arrow toward the coast of Bougainville, was never attacked.

Nearing Empress Augusta Bay, the convoy slowed so that the final movement into the objective area could be made in daylight. General quarters was sounded at 0500, and, after the sun came up, the assault troops on the transports could see the dark shoreline and rugged peaks of Bougainville directly ahead. Only a thin cloud mist hung over the island, scant concealment for enemy planes which could have been waiting to ambush the amphibious force. The element of surprise, which had been zealously guarded during all preparations for the offensive, apparently had been retained. The conflicting re-

[1] Unless otherwise noted, the material in this section is derived from: *ComSoPac Nov43 WarD; ThirdFlt NarrRept; IIIPhibFor AR; IIIPhibFor Nov43 WarD;* ComTransGru, IIIPhibFor, Rept of LandingOps, Empress Augusta Bay, Bougainville Island, 1–2Nov43, dtd 22Dec43 (COA, NHD); *IMAC AR–I; IMAC C–2 Repts; IMAC C–2 Jnl;* 3d MarDiv, Combat Rept of the 3d MarDiv in the BougainvilleOps, 1Nov–28Dec43, dtd 21Mar44 (Bougainville AreaOpFile, HistBr, HQMC), hereafter *3d MarDiv CombatRept; 3d MarDiv AR;* 3d MarDiv D–1 Jnl, 1Oct–14Nov43 (Bougainville AreaOpFile, HistBr, HQMC); 3d MarDiv D–2 SAR, Empress Augusta Bay Ops dtd 1Feb44 (Bougainville AreaOpFile, HistBr, HQMC), hereafter *3d MarDiv D–2 SAR;* 3d MarDiv D–2 Jnl, 28Oct–28Dec43 (Bougainville AreaOpFile, HistBr, HQMC), hereafter *3d MarDiv D–2 Jnl;* 3d MarDiv D–3 Jnl, 31Oct–28Dec43 (Bougainville AreaOpFile, HistBr, HQMC), hereafter *3d MarDiv D–3 Jnl;* 3d MarDiv D–3 PeriodicRepts, 6Jul–27Dec43 (Bougainville AreaOpFile, HistBr, HQMC), hereafter *3d MarDiv D–3 Repts; HistDiv Acct;* ONI, *Combat Narrative XII;* Henderson, "Naval Gunfire Support;" Aurthur and Cohlmia, *3d MarDivHist;* Rentz, *Bougainville and the Northern Solomons;* Isely and Crowl, *Marines and Amphibious War;* Miller, *Reduction of Rabaul.*

ports by Japanese snooper planes of task forces observed at various points from Buka to the Shortlands and Vella Lavella had the general effect of confusing the Japanese.[2]

At 0545, mine sweepers and the destroyer *Wadsworth* opened fire on the beaches north of Cape Torokina to cover their own mine-sweeping operations. As the *Wadsworth* slowly closed to within 3,000 yards to fire directly into enemy installations, the busy mine sweepers scouted the bay. Thirty minutes later, advised that no mines had been found, the transports moved into the area. Off Cape Torokina, each APA shelled the promontory with ranging 3-inch fire before turning hard to port to take Puruata Island under fire with 20mm guns. At 0645, the eight troop transports were on line about 3,000 yards from the beach and parallel to the shoreline. Behind them, in a similar line, were the four cargo transports with the destroyer squadrons as a protective screen seaward.

On board the transports, observers peered anxiously toward the beaches near the Laruma River. A two-man patrol had been landed on Bougainville on D minus 4 days (27 October) with the mission of radioing information or lighting a signal fire near the Laruma if the Cape Torokina area was defended by less than 300 Japanese. Concern mounted as H-Hour approached without the expected message or signal. The alternatives were that the patrol had been captured or that the cape area was unexpectedly reinforced by the enemy. Because the landing waves had been organized to handle cargo and supplies at the expense of initial combat strength, any change in the enemy situation at this late date was cause for worry. H-Hour, set for 0715, was postponed for 15 minutes on signal from Admiral Wilkinson, but the landing was ordered as planned. (The patrol later reported unharmed, citing radio failure and terrain difficulties for the lack of messages.)

Preparatory fires by the main support group began as soon as visibility permitted identification of targets. From their firing positions south of the transport area, the *Anthony* and *Sigourney*—and later the *Wadsworth* — poured 5-inch shells into Puruata Island and the beaches north of Cape Torokina. The *Terry*, on the left flank of the transport area, fired into known enemy installations on the north shoulder of the cape. The effect was a crossfire, centered on the beaches north of Cape Torokina. This indirect fire on area targets was controlled by spotter aircraft.

Debarkation of troops began after the transports anchored in position and while the pre-assault bombardment crashed along the shoreline. The order to land the landing force was given at 0645, and within minutes assault craft were lowered into the sea, and embarkation nets tossed over the side of the transports. Marines clambered down the nets into the boats, and, as each LCVP was loaded, it joined the circling parade of landing craft in the rendezvous circles, waiting for the signal to form into waves for the final run to the beach. Nearly 7,500 Marines, more than half of the assault force, were boated for the simultaneous landing over the 12 beaches.

At 0710, the gunfire bombardment shifted to prearranged targets, and five minutes later the first boats from the

[2] *SE Area NavOps—III*, p. 12.

LANDING CRAFT *is lowered over the side of the APA* George Clymer *on D-Day at Bougainville while Marines watch in the foreground.* (*USN 80-G-55810*)

APAs on the south flank of the transport area started for shore. The support ships continued to shoot at beach targets until 0721, when the shelling was lifted to cover targets to the rear of the immediate shoreline. As the fire lifted, 31 torpedo and scout bombers from Munda streaked over the beaches, bombing and strafing the shoreline just ahead of the assault boats. The planes, from VMTB-143, -232, and -233, and VMSB-144, were covered by VMF-215 and -221 and a Navy fighter squadron, VF-17.[3] The air strike lasted until 0726, cut short four minutes by the early arrival of the first landing craft at the beaches.

The 9th Marines (Colonel Edward A. Craig) landed unopposed over the five northernmost beaches—Red 3, Red 2, Yellow 4, Red 1, and Yellow 3. Although no enemy fire greeted the approach of the boats, the landing was unexpectedly hazardous. Rolling surf, higher and rougher than anticipated, tossed the landing craft at the beaches. The LCVPs and the LCMs, caught in the pounding breakers, broached to and were smashed against shoals, the beach, and other landing craft. The narrow shoreline, backed by a steep 12-foot embankment, prevented the landing craft from grounding properly, and this further complicated the landing.

Some boats, unable to get near the shore because of rough surf and wrecked boats, unloaded the Marines in chest-deep water. Other Marines, in LCVPs with collision-damaged ramps, jumped over the sides of the boats and made their way to shore. In spite of these difficulties, the battalion landing teams managed to get ashore quickly, and, by 0750, several white parachute flares fired by the assault troops indicated to observers on board ship that the landing was successful.

Once ashore, combat units of the 9th Marines completed initial reorganization and moved inland to set up a perimeter around the five beaches. Active patrolling was started immediately, and a strong outpost was set up on the west bank of the Laruma River. Other Marines remained on the beach to help unload the tank lighters and personnel boats which continued to arrive despite the obvious inadequacy of the beaches and the difficult surf. At least 32 boats were wrecked in the initial assault and lay smashed and awash along the beach. By mid-morning, hulks of 64 LCVPs and 22 LCMs—many of them beyond repair—littered the five beaches.[4]

The landings on the six southern beaches (Yellow 2, Blue 3, Blue 2, Green 2, Yellow 1, and Blue 1) and the single beach on Puruata Island (Green 1) were in stark contrast to the northern zone. Enemy resistance in this area was evident almost as soon as the boat groups from the right-flank transports came within range. The 2d and 3d Battalions of Colonel George W. McHenry's 3d Marines landed on the three beaches south of the Koromokina River against small-arms fire. Surf was high but not difficult, and no boats were lost. The Marines, disembarking without delay, sprinted across the narrow beach to take cover in the jungle. Reorganization was completed quickly, and the battalions started to dig out the small number of Japanese defenders attempting to hold back the assault from hastily pre-

[3] Sherrod, *MarAirHist*, p. 181.

[4] "Many automatic weapons were mounted on these landing boats. These weapons were salvaged by the Marines and used later to reinforce their normal arms when the final defensive line was established." LtGen Edward A. Craig ltr to CMC, dtd 24Oct60, hereafter *Craig ltr*.

pared positions. In a few minutes, the scattered enemy in the area had been killed, and sniper patrols began moving inland. Contact was established with the 9th Marines on the left, but a wide swamp prevented linkup with the 2d Raider Battalion on the right.

The raiders, led by Major Richard T. Washburn, went ashore in the face of heavy machine gun and rifle fire from two enemy bunkers and a number of supporting trenches about 30 yards inland. Japanese defenders were estimated at about a reinforced platoon. After the first savage resistance, the enemy fire slackened as the raiders blasted the bunkers apart to kill the occupants. Other enemy soldiers retreated into the jungle. Only after the beach area was secured did the raiders discover that the regimental executive officer, Lieutenant Colonel Joseph W. McCaffery, had been fatally wounded while coordinating the assault of combat units against the enemy dispositions.[5]

Extensive lagoons and swampland backing the narrow beach limited reconnaissance efforts, and reorganization of the assault platoons and companies was hindered by constant sniper fire. Despite these handicaps, the raiders pushed slowly into the jungle and, by 1100, had wiped out all remaining enemy resistance. Raider Company M, attached to the 2d Raider Battalion for the job of setting up a trail block farther inland to stall any enemy attempt to reinforce the beachhead, moved out along the well-marked Mission Trail and was soon far out ahead of the raider perimeter.

The 1st Battalion of the 3d Marines hit the hot spot of the enemy defenses. As the waves of boat groups rounded the western tip of Puruata Island, they were caught in a vicious criss-cross of machine gun and artillery fire from Cape Torokina and Puruata and Torokina Islands. Heading toward the extreme right of the landing area over beaches which included Cape Torokina, the 1st Battalion ploughed ashore straight through this deadly crossfire. An enemy 75mm artillery piece, which had tried earlier to hit one of the transports, remained under cover during the aerial bombing and opened fire again only after the assault boats reached a point some 500 yards offshore. Its location was such that all boats heading toward the beach had to cross the firing lane of this gun.[6]

One of the first casualties in the assault waves was the LCP carrying the boat group commander. The command boat, blasted by a direct hit, sank immediately. The explosion resulted in dispersion, disorganization, and confusion among the boat group. In a split second, the approach formation was broken by landing craft taking evasive action to avoid the antiboat fire.

The result was a complete mixup of assault waves. A total of six boats were hit within a few minutes; only four of them managed to make the beach. As the first waves of boats grounded on the beaches, the Japanese opened up with machine gun and rifle fire, and mortar bursts began to range along the shoreline. A withering fire poured from a concealed complex of log and sand bunkers connected by a series of rifle pits and trenches.

---

[5] MajGen Alan Shapley ltr to ACofS, G-3, HQMC, dtd 13Oct60, hereafter *Shapley ltr.*

[6] Col George O. Van Orden ltr to CMC, dtd 27May48 (Bougainville Monograph Comment File, HistBr, HQMC) places the location of this gun well within the limits of Blue Beach 1, the landing area of the 1st Battalion, 3d Marines.

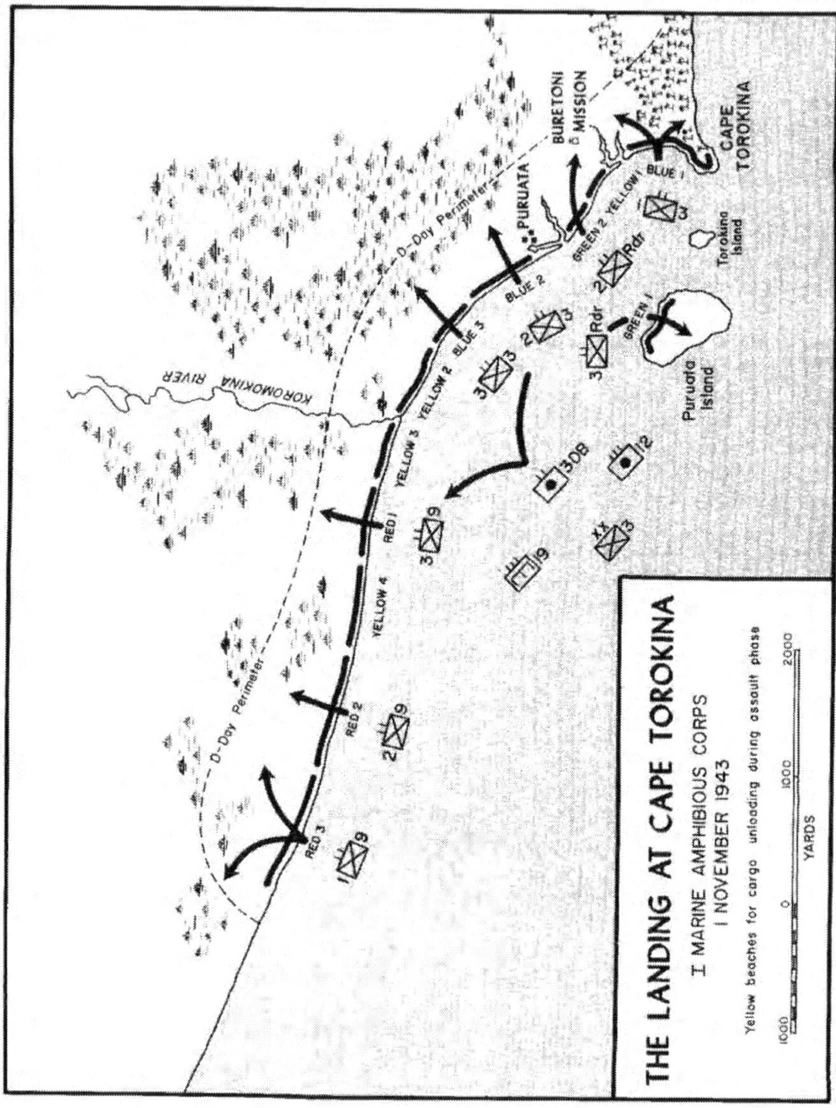
THE LANDING AT CAPE TOROKINA
I MARINE AMPHIBIOUS CORPS
1 NOVEMBER 1943
Yellow beaches for cargo unloading during assault phase
1000 0 1000 2000
YARDS
KOROMOKINA RIVER
D-Day Perimeter
D-Day Perimeter
RED 3
RED 2
RED 1
YELLOW 4
YELLOW 3
YELLOW 2
BLUE 3
BLUE 2
GREEN 2
YELLOW 1
BLUE 1
GREEN 1
PURUATA
BURETONI MISSION
CAPE TOROKINA
Torokina Island
Puruata Island

MAP 14

The enemy emplacements, barely above ground level and hidden beneath the tangled underbrush along the shoreline, were sited to cover the beaches and bay with interlocking bands of fire. The pre-assault bombardment by gunfire ships and planes had not knocked out the enemy fortifications; in most cases it had not even hit them.

The Marines, with all tactical integrity and coordination lost, plunged across the thin strip of beach to take cover in the jungle. An orderly landing against such concentrated fire had been impossible. After the scrambling of the assault waves, units from the battalion landing team had gone ashore where possible and practically every unit was out of position. Contributing to this confusion was the fact that the majority of the boats hit were LCPs carrying boat group commanders. The Japanese, correctly surmising that the more distinctive LCPs were command craft, directed most of their fire on these boats.

The initial reorganization of the elements of the battalion landing team was handicapped further by the wounding and later evacuation of the battalion commander, Lieutenant Colonel Leonard M. Mason. Control of boat teams was difficult under the pounding of 90mm and "knee mortar"[7] bursts mixed with the raking fire of machine guns and rifles. Platoons and squads from all companies were mixed along the beach. The original plans directed Company A to land on Cape Torokina, but after the assault waves were dispersed and tangled by the effective fire of the Japanese 75mm artillery piece, elements of Company C landed on the promontory. Several squads from Company F of the 2d Battalion also landed in this area and were forced to fight their way along the beach to reach their parent unit. Only Company B of the 1st Battalion landed on its assigned beach. Casualties were fairly light, though, despite the intense fire of the enemy. In addition to at least 14 men lost in the landing craft which had been sunk, fewer than a dozen Marines had been killed on the beach.

The 1st Battalion hesitated only a short time; then the extensive schooling of the past asserted itself. Training in small unit tactics against a fortified position now paid big dividends. Rifle groups began to form under ranking men, and the fight along the shoreline became a number of small battles as the Marines fought to widen their beachhead against the enemy fire. As the Marines became oriented to their location and some semblance of tactical integrity was restored, the pace of the assault quickened.

Before the operation, all units had been thoroughly briefed on the mission of each assaulting element, and each squad, platoon, and company was acquainted with the missions of other units in the area. In addition, each Marine was given a sketch map of the Cape Torokina shoreline. Small groups formed under the leadership and initiative of junior officers and staff noncommissioned officers, and these groups, in turn, were consolidated under one command by other officers. Bunker after bunker began to fall to the coordinated and well-executed attacks of these groups. As the Japanese defensive complex slowly cracked, the 1st Battalion command was established under the battalion executive officer, and the hastily re-formed companies took over the mission of the area in which they found themselves.

[7] The common term for the Japanese 50mm grenade discharger.

The efficient reduction of the enemy's defensive position added another testimonial to prior training and planning. Officers of the 3d Marine Division had studied the Japanese system of mutually protecting bunkers on New Georgia and decided that in such a defensive complex the reduction of one bunker would lead to the elimination of another. In effect, one bunker unlocked the entire position. The quickest way to knock out such pillboxes with the fewest casualties to the attacking force was for automatic riflemen to place fire on the embrasures of the bunker while other Marines raced to its blind side to drop grenades down the ventilators or pour automatic rifle fire into the rear entrance.

By midmorning, through such coordinated attacks, most of the Japanese bunkers on Cape Torokina had been knocked out. The position containing the murderous 75mm gun was eliminated by one Marine who, directing the assault of a rifle group, crept up to the bunker and killed the gun crew and bunker occupants before falling dead of his own wounds. After the last emplacement was silenced late that afternoon, Marines counted 153 dead Japanese in the Cape Torokina area.

For a while, the situation on the right flank had been touch and go. One hour after the landing, a variation of the time-honored Marine Corps phrase was flashed from the Cape Torokina beach. "The situation appears to be in hand," was the first message, but a few minutes later a more history-conscious officer flashed an amended signal: "Old Glory flies on Torokina cape. Situation well in hand." The most expressive message, however, to observers on board the transports was the report from a young officer to Colonel McHenry: " . . . the troops are magnificent." [8] The Marine officers who had directed the assault on the fortified positions added sincere endorsements to this expression of admiration.

On Puruata Island, the 3d Raider Battalion (Lieutenant Colonel Fred D. Beans) landed with one reinforced company in the assault and the remainder of the battalion as reserve and shore party. Only sporadic fire hit the boats as they neared the island. By 0930, the raiders had established a perimeter about 125 yards inland against hidden snipers and accurate machine gun and mortar fire. The Japanese, obviously outnumbered, gave little indication of yielding, and, by 1330, the reserve platoons of the battalion were committed to the attack. The raiders, with the added support of several self-propelled 75mm guns attached from the 9th Marines, then moved about halfway across the island.

Puruata was not declared secured until midafternoon of the following day. A two-pronged attack, launched by the raiders early on the morning of 2 November, swept over the island against only sporadic rifle fire, and, by 1530, all Japanese resistance on the island had been erased. Only 29 dead enemy were found, although at least 70 were estimated to have been on the island. The remainder had apparently escaped to the mainland. The raiders lost 5 men killed and 32 wounded in the attack.[9]

---

[8] 3d MarRegt Jnl, 1Nov43 (Bougainville Area-OpFile, HistBr, HQMC).

[9] This is the number given in *3d MarDiv Combat Rept*, p. 343. The same report, on p. 153, also gives 6 KIA and 18 WIA for this part of the operation. In cases of discrepancies such as this, the report of the unit engaged is given in this account.

MARINES WADING ASHORE *on D-Day at Bougainville, as seen from a beached LCVP. (USN 80-G-54348)*

PURUATA ISLAND *in the foreground, Torokina airfield in the background, appear in an aerial photograph taken on 13 December 1943. (USMC 68047)*

691-360 O—63——15

The Marines' fight uncovered extensive enemy defenses which were not disclosed in aerial photographs taken before the operation. The entire headland was ringed by 15 bunkers, 9 of them facing to the west and 6 of them overlooking the beaches on the east side of the cape. Behind this protective line and farther inland was another defensive line of eight bunkers which covered the first line of fortifications. Two other bunkers, about 750 yards inland, provided additional cover to the first two lines.

Constructed of ironwood and coconut logs two feet thick, the bunkers were bulwarked by sandbags and set low into the ground. Camouflaged by sand and tangled underbrush, the bunkers were hard to detect and difficult to knock out without flamethrowers or demolitions. Despite this, the 3d Marines suffered few casualties in destroying this defensive installation. Twenty of the bunkers had been eliminated by the coordinated fire and maneuver of individual Marines; the remaining five were blasted apart by self-propelled tank destroyers firing 75mm armor-piercing shells directly into the embrasures.

The enemy 75mm artillery piece sited as a boat gun hit 14 boats during the initial landings before it was put out of action. Only four of the boats sank. Despite the high velocity of the shells and the slow speed of the landing craft, the 50 or more rounds fired by the enemy scored remarkably few hits. This was attributed to two factors: the poor accuracy of the Japanese gunners and the limited traverse of the gun. Marines found, after knocking out the bunker, that the aperture in the pillbox permitted the muzzle of the gun to be moved only three degrees either way from center. This prohibited the gun from bringing enfilade fire to bear on the beaches. Had this been possible, the large number of boats along the shoreline would have been sitting targets which even poor gunners could not miss, and the casualties to the landing force would have been correspondingly greater.

The unexpected resistance on Cape Torokina and Puruata island after the naval gunfire bombardment and bombing was a sharp disappointment to IMAC officers who had requested much more extensive preparatory fires. The gunfire plan, which was intended to knock out or stun enemy defenses that might delay the landing, had accomplished nothing. The *Anthony*, firing on Puruata Island, reported that its target had been well covered; but the raider battalion, which had to dig the defenders out of the emplacements on the island, reported that few enemy installations had been damaged.

The *Wadsworth* and *Sigourney*, firing at ranges opening at 11,000 to 13,000 yards, had difficulty hitting the area and many shots fell short of the intended targets. The *Terry*, closest to the shore but firing at an angle into the northwestern face of Cape Torokina, was poorly positioned for effective work. None of the 25 bunkers facing the landing teams on the right had been knocked out by gunfire, and only a few of the Japanese huts and buildings inland were blasted by the ships' fire. The gunnery performance of the destroyers left much to be desired, IIIPhibFor admitted later. Particularly criticized was the fact that some ships fired short for almost five minutes with all salvos hitting the water. After two or three rounds, the range should have been adjusted, but apparently the practice bombardment at Efate had not been sufficient.

The long-range sniping at Cape Torokina with inconclusive results was vindication for the IMAC requests prior to the operation that the destroyers move as close to the shoreline as possible for direct fire.

> The sad thing about the whole show, to the corps and division gunfire planners, was that the means actually were available to give us just what we wanted, but were dissipated elsewhere in what we felt was fruitless cannonading.[19]

Valuable lessons in gunfire support were learned at Bougainville that D-Day. For one thing, the line of flat trajectory fire in some places passed through a fringe of tall palm trees which exploded the shells prematurely and denied direct observation of the target area. Further, the ships had trouble seeing the shoreline through the combination of early morning haze and the smoke and dust of exploding shells and bombs, rising against a mountainous background.

Although the enemy airfields in the Bougainville area were knocked out by Admiral Merrill's final bombardment and the prior action of ComAirSols bombing strikes, the Japanese reaction to the landing came swiftly. At 0718, less than two hours after the transports appeared off Cape Torokina and about eight minutes before the first assault boats hit the beach, a large flight of Japanese planes was detected winging toward Empress Augusta Bay. The transports, most of them trailing embarkation nets, immediately pulled out of the bay toward the sea to take evasive action again.

The first enemy flight of about 30 planes, evidently fighters from the naval carrier groups land-based in New Britain, was intercepted at about 0800 by a New Zealand fighter squadron flying cover over the beachhead. Seven of the Japanese planes were knocked down, but not before a few enemy raiders strafed the beaches and dive-bombed the frantically maneuvering APAs and AKAs. Ten minutes later, another flight of enemy fighters and bombers struck the area in a determined attack, but were turned away by the fierce interference of other ComAirSols planes, including Marine fighters from VMF-215 and VMF-221. Radical evasive tactics by the transports—aided by excellent antiaircraft gunnery by the destroyer screen and savage pursuit by the fighter cover—prevented the loss of any ships, although the *Wadsworth* took some casualties from a near miss. The fighter cover downed eight planes, and the destroyer screen claimed another four raiders.

Two hours after the attack began, the APAs and AKAs returned to resume operations. Valuable time, however, had been lost. Intruding enemy planes continued to harass the transports, but unloading operations kept up until about 1300, when the arrival of another large formation of about 70 enemy planes put the ships into action again.

One APA, the *American Legion*, grounded on a shoal and remained there during the attack despite the persistent efforts of two tugs which attempted to free it. A destroyer resolutely stood guard, pumping antiaircraft fire into attacking planes. The ship was pulled free before the air attack was driven off. As before, the aggressive fighter cover and heavy fire from the destroyer screen prevented damage to the amphibious force, and the ships turned back to the task of unloading. During the attacks, the Allies claimed 26 enemy planes as shot down—four more

[19] Henderson, "Naval Gunfire Support," pp. 61-62.

than the Japanese records indicate—with the loss of four planes and one pilot. For the first day, at least, the threat of enemy air retaliation had been turned back.

ESTABLISHING A BEACHHEAD[11]

Ashore, the defensive perimeter now stretched a long, irregular semi-circle over the area from the Laruma River past Cape Torokina, a distance of about four miles. Only the northern beaches were quiet; the area around the cape was still being contested by snipers within this perimeter and by small groups of enemy in the jungle outside the line. Within this area, the logistics situation was beginning to be cause for concern.

Confusion began after wrecked tank lighters and personnel boats were broken on the northern beaches, closing those areas to further traffic. When unloading operations began once more after the first air raid, the northern beaches were ordered abandoned and all cargo destined for the 9th Marines sector was diverted to beaches south of the Koromokina River. This change, the only move possible in view of the difficult surf conditions, led to further complications because the beaches in the 3d Marines' sector were already crowded, and the coxswains on the landing craft had no instructions regarding where the supplies should be dumped.

The landing beaches in the 3d Marines' sector were hardly an improvement. Few had any depth, and from the outset it was apparent that the swampy jungle would stall operations past the beaches. The only means of movement was laterally along the thin beach, and the gear already stacked along the shoreline was causing congestion along this route. The difficult terrain inland made the formation of dumps impractical, so all cargo was placed above the high water mark and some degree of orderliness attempted. Despite this, the 9th Marines lost much organizational property and supplies, most of which was never recovered.

As cargo and supplies mounted on the beaches assigned to the 2d and 3d Battalions of the 3d Marines, the bulk cargo was diverted further to Puruata Island. The landing craft were unloaded within a few hundred yards of the battle between the raiders and the small but determined group of defenders emplaced there. Almost 30 percent of the total cargo carried by the 12 transports was unloaded by 1130 of D-Day, and this figure was extended to almost 50 percent completion by the time that the APAs and AKAs had to depart the area for the second time. The cargo remaining on board was varied; some ships had unloaded all rations but little ammunition. Other transports had unloaded ammunition first and were just starting to move the other supplies.

While the combat troops ashore prepared to defend the newly won beachhead,

[11] Unless otherwise noted, the material in this section is derived from: *CincPac–CincPOA Nov43 WarD; ComSoPac Nov43 WarD; ThirdFlt NarrRept; IIIPhibFor AR; IIIPhibFor Nov43 WarD; IMAC AR–I; 3d MarDiv Combat Rept; 3d MarDiv AR:* 3d MarDiv D–4 Jnl 1–16 Nov43 (Bougainville AreaOpFile, HistBr, HQMC), hereafter *3d MarDiv D–4 Jnl;* 3d MarDiv ServTrps Rept of Ops, Nov–Dec43, dtd 27Jan44 (Bougainville AreaOpFile, HistBr, HQMC), hereafter *3d MarDiv ServTrps Rept;* 3d MarDiv Supply and Evac Rept, DIPPER Operation, dtd 29Jan44 (Bougainville AreaOpFile, HistBr, HQMC); *HistDiv Acct;* ONI, *Combat Narrative XII;* Aurthur and Cohlmia, *3d MarDivHist;* Rentz, *Bougainville and the Northern Solomons;* Morison, *Breaking the Bismarcks Barrier.*

the transport groups proceeded to unload as rapidly as the air attacks, loss of boats, and elimination of a number of beaches would permit. By 1600 on D-Day, only the four northernmost transports—the ones most affected by the boat mishaps and the unsuitable beaches—still had cargo on board.

The quick unloading of the other five APAs and three AKAs, despite the interruptions, was a reflection of the measures taken by Admiral Wilkinson and General Vandegrift to insure rapid movement of supplies ashore. Embarked troops on each APA and AKA had been required to furnish a complete shore party of about 500 men. During the unloading, 6 officers and 120 men remained on board ship to act as cargo handlers while a further 60 were boat riders to direct the supplies to the proper beaches. Another 200 Marines stayed on each beach to help unload the landing craft. The remaining personnel were used as beach guides, vehicle drivers, cargo handlers, and supervisors.[12] The 3d Service Battalion, augmented by supporting troops—artillerymen, engineers, military police, signal men, tank men, communicators, and Seabees—formed the bulk of these working parties. In some instances, these supporting troops were not released to their units until several days after the beachhead had been established. In all, about 40 percent of the entire landing force was engaged initially in shore party activities.

By late afternoon, each landing team reported its mission accomplished. In the absence of any identifiable terrain features in the interior, the landing teams had been directed to extend the beachhead certain distances, and, by the end of D-Day, each of the battalions was established in a rough perimeter along the first of these designated Inland Defense Lines. The division front lines extended into the jungle about 600 yards near the Laruma River and about 1,000 yards in front of Cape Torokina. Although the 1st Battalion, 3d Marines, in the area of the cape plantation, and the 3d Raiders on Puruata Island were still receiving occasional sniper fire, the remainder of the perimeter was quiet and defense was not a special problem.

There was, however, still congestion on the shoreline. In order to bring some order out of the near chaos on the beaches and to reduce the paralyzing effect of the mountains of supply piled helter-skelter, additional Marines from the combat forces were detailed as labor gangs to sort the supplies and haul them to the front-line units. This placed a double-burden on some units who were already near half-strength by the assignment of troops to the shore party work.

An additional problem, late on D-Day, was the correlation and coordination of the defensive positions and missions of the many assorted and unrelated supporting units which had landed during the day. These included echelons of artillery, antiaircraft artillery, and seacoast defense units. The 12th Marines, decentralized with a battery attached to each landing team, was in varying stages of readiness for defense of the beachhead. Battery B, in the 9th Marines area, was in position by early afternoon but was so engaged in cargo hauling that the first requests for a firing mission could not be completed. Other batteries were also in position by the end of D-Day, and several had fired registration shots and were available for intermittent fires during the first night.

[12] IMAC AdminO No. 1, dtd 15Oct43 in *IMAC AR-I*.

The remaining batteries were ready for support missions the following day.

Selection of positions in most areas was difficult. The battery supporting the 2d Raider Battalion was forced to move inland about 100 yards through a lagoon before a position could be located. Two amphibian tractors ferried the guns and most of the ammunition across the water, and the artillerymen transferred the remaining ammunition from the beach to the gun position by rubber boats. This battery registered on Piva Village by air spot, and the next day fired 124 rounds on suspected enemy positions in the vicinity of that village.

Antiaircraft batteries (90mm) and the Special Weapons Group of the 3d Defense Battalion landed right behind the assault units. Advance details of the seacoast defense battery also moved ashore early and immediately began seeking suitable positions to mount the big guns. After the first air raid on the morning of D-Day, the remaining antiaircraft guns of the Marine defense battalion were hurried ashore so that protection of the beachhead could be increased as soon as possible. By nightfall of D-Day, 20 40mm guns, 8 20mm guns, and the .50 and .30 caliber machine guns of the battalion were integrated into the defense of the perimeter and were ready for action.

As nightfall approached, the frontline units sited all weapons along fixed lines to coordinate their fire with adjacent units, and all companies set up an all-around defense. Supporting units on the beach also established small perimeters within this defensive line. There was to be no unnecessary firing and no movement. Marines were to resort to bayonets and knives when needed, and any Japanese infiltrators were to be left unchallenged and then eliminated at daybreak. An open-wire telephone watch was kept by all units, and radios were set to receive messages but no generators were started for transmissions.

The night passed as expected—Marines huddling three to a foxhole with one man awake at all times. A dispiriting drizzle, which began late on D-Day afternoon, continued through the night. Japanese infiltrators were busy, and several brief skirmishes occurred. An attack on a casualty clearing station was repulsed by gunfire from corpsmen and wounded Marines; and one battalion command post, directly behind the front lines, was hit by an enemy patrol. The attackers were turned back by the battalion commander, executive officer, and the battalion surgeon who wielded knives to defend their foxhole.

While the Marines ashore had busied themselves getting ready for the first night of defense of the beachhead, the transport groups proceeded with the unloading details. At 1645, the transports were advised to debark all weapons, boat pool personnel, and cargo handlers and leave the area at 1700. The four transports still with supplies aboard (the *Alchiba*, *American Legion*, *Hunter Liggett*, and the *Crescent City*) were to keep working until the final moment and then leave with the rest of the transports despite any Marine working parties still on board.

Admiral Wilkinson, aware that the situation ashore was well under control, had decided that all ships would retire for the night and return the next day. In event of a night attack, the transports in Empress Augusta Bay would be sitting ducks. The admiral felt that his ships could not maneuver in uncharted waters at night, and that night unloading operations were not feasible. The admiral had another reason, too. An enemy task force of four cruisers and six destroyers was reported heading toward Rabaul from Truk, and these ships, after one refueling stop, could be expected near Bougainville later that evening or early the next morning. The amphibious force, as directed, moved out to sea for more protection.

At 2300 that night, 1 November, the four transports which were still to be unloaded were ordered to reverse course and head back toward Empress Augusta Bay while the rest of the transports continued toward Guadalcanal. The four transports, screened by destroyers, regulated their speed and direction so as to reach the Cape Torokina area after daybreak. A short time later, alerted to the fact that a large enemy fleet was in the area, the transports headed back toward Guadalcanal again.

Admiral Merrill's Task Force 39, after the successful bombardment of Buka and the Shortlands which opened the Bougainville operation, had moved north of Vella Lavella to cover the retirement of the transport group. At this particular time, Merrill's concern was the condition of his force which had been underway for 29 hours, steaming about 766 miles at near-maximum speed. Although the cruisers were still able to fight, the fuel oil supply in the destroyers was below the level required for anything but small engagements at moderate speeds.

So, while Merrill's cruisers waited, one of the two destroyer divisions in the task force turned and headed for New Georgia to refuel. That afternoon, 1 November, while an oil barge was pumping oil into the destroyers at maximum rate, the report of the Japanese fleet bearing down on Bougainville was received. The destroyers, impatient to get going, hurried through the refueling.

At 1800, all destroyers raced out of Kula Gulf to rejoin Merrill. The 108-mile trip was made at 32 knots, although the engines of two of the destroyers were on the verge of breakdown. By 2330, the ships joined Merrill's cruisers south of the Treasurys, and the entire task force headed toward Bougainville where it interposed itself between the departing transports and the oncoming enemy fleet. Allied patrol planes had kept the attack force under surveillance all day, and, by nightfall, the direction of the Japanese ships was well established. If the Allied thinking was correct, another trap had been baited for the Japanese. The enemy, guessing that the same task force that hit Buka had provided the shore bombardment for the Cape Torokina landing, might be lured into assuming that the fighting ships were now low on fuel and ammunition and had retired with the transports. If that was the enemy assumption, then Merrill was in position for a successful ambush.

Moving slowly to leave scant wake for enemy snooper planes to detect, Merrill's force was off Bougainville by 0100, 2 November, and beginning to maneuver into position to intercept the enemy fleet. At that time, the enemy was about 83 miles distant. Merrill's basic plan was to stop the enemy at all costs, striking the Japanese ships from the east so that the sea engagement would be deflected toward the west, away from Bougainville. This would give his ships more room to maneuver as well as allow any damaged ships to retire to the east on the disengaged side. Further, Merrill respected the Japanese torpedoes and felt that his best chance to divert the enemy force and turn it back—possibly without loss to his own force—was by long-range, radar-directed gunfire.

The naval battle of Empress Augusta Bay began just 45 miles offshore from the beachhead whose safety depended upon Task Force 39. Merrill's cruisers opened fire at 0250 at ranges of 16,000 to 20,000 yards. The enemy fleet, spread out over a distance of eight miles, appeared to be in three columns with a light cruiser and destroyers in each of the northern and southern groups and two heavy cruisers and two destroyers in the center. Detection was difficult because, with the enemy so spread out, the radar on Merrill's ships could not cover the entire force at one time.

The enemy's northern force was hit first, the van destroyers of Task Force 39 engaging this section while the rest of the American ships turned toward the center and southern groups. As planned, the attack struck from the east. Task Force 39 scored hits immediately, drawing short and inaccurate salvos in return. The Japanese, relying on optical control of gunfire, lighted the skies with starshells and airplane flares; but this also helped Task Force 39, since the enemy's flashless powder made visual detection of the Japanese ships almost impossible without light.

The two forces groped for each other with torpedoes and gunfire. In the dark night, coordination of units was difficult and identification of ships impossible. The maneuvering of Merrill's task units for firing positions, as well as the frantic scattering of the enemy force, spread the battle over a wide area, which further increased problems of control and identification. On at least one occasion, Task Force 39 ships opened fire on each other before discovering their error.

In such confused circumstances, estimation of damage to either force was almost impossible, although some of the American destroyers believed that their torpedoes had found Japanese targets, and other enemy ships were believed to have been hit by gunfire. In the scramble for positions to take new targets under fire, two destroyers of Merrill's force scraped past each other with some damage, and several other close collisions between other destroyers were narrowly averted. One American destroyer, the *Foote*, reported itself disabled by an enemy torpedo and two other destroyers were hit by gunfire but remained in action. The only cruiser damaged was the *Denver*, which took three 8-inch shells and was forced to disengage for a short time before returning to the fight.

By 0332, Task Force 39 was plainly in possession of the field. The enemy force, routed in all directions, had ceased firing and was retiring at high speed. Merrill's cruiser division ceased firing at 0349 on one last target at ranges over 23,000 yards. This ended the main battle, although the TF 39 destroyers continued to scout the area for additional targets and disabled enemy ships. At daybreak, TF 39 was reassembled and a flight of friendly aircraft appeared to provide escort for its retirement. The *Foote* was taken under tow and the return to Guadalcanal started. The Merrill force believed that it had sunk at least one enemy light cruiser and one destroyer and inflicted damage on a number of other ships. This estimate was later found correct.[13] In addition, the

[13] *SE Area NavOps—III*, p. 14.

Japanese also had several ships damaged in collisions.

Task Force 39 was struck a few hours later by a furious air attack from more than 70 enemy planes, but the Japanese made a mistake in heading for the cruisers instead of the destroyers guarding the disabled *Foote*. The heavy antiaircraft fire and the aggressive protection of the ComAirSols fighter cover forced the enemy planes away. The air cover shot down 10 planes, and the ships reported 7 enemy aircraft downed. Only one American cruiser, the *Montpelier*, was hit by bombs but it was able to continue. While the air battle raged, the amphibious force's transports reversed course once more and returned to Cape Torokina without interference and completed the unloading. The sea and air offensive by the Japanese had been stopped cold by the combined action of ComSoPac's air and sea forces.

THE JAPANESE[14]

To the Japanese defenders, the sudden appearance of a number of transports off Cape Torokina on the morning of 1 November came as something of a shock. All Japanese plans for the island had discounted a landing north of Cape Torokina because of the nature of the beaches and the terrain. If the Allies attacked the western coast of Bougainville, the enemy thought the logical place would be southern Empress Augusta Bay around Cape Mutupena. Japanese defensive installations, of a limited nature, were positioned to repel an Allied landing in this area.

But the small garrison in the vicinity of Cape Torokina, about 270 men from the *2d Company, 23d Regiment*, with a regimental weapons platoon attached, was well trained. From the time that the alarm was sounded shortly after dawn on 1 November, the Japanese soldiers took up their defensive positions around the cape and prepared to make the invading forces pay as dearly as possible for a beachhead.

The invading Marines found the island's defenders dressed in spotless, well-pressed uniforms with rank marks and service ribbons, an indication that the Cape Torokina garrison was a disciplined, trained force with high morale, willing to fight to the death to defend its area. But after the first day, when the Japanese were knocked out of the concentrated defenses on Cape Torokina, the enemy resistance was almost negligible. A wounded Japanese sergeant major, captured by Marines the second day, reported that the understrength garrison had been wiped almost out of existence. The prisoner confirmed that the Japanese had expected an attack on Bougainville for about three days—but not at Cape Torokina.

With the notice of the Allied operations against Bougainville, all available Japanese air power was rushed toward Rabaul, and Admiral Kusaka ordered the interception operations of the *Southeast Area Fleet* (the *Ro* operation) shifted from New Guinea to the Solomons. Because all planes of the *1st Air Squadron* and additional ships were already en route to Rabaul, this action placed the entire mobile surface and air strength of the *Combined Fleet* under the direction of the commander of the *Southeast Area Fleet*.

[14] Unless otherwise noted, the material in this section is derived from: *IMAC AR–I*; *3d MarDiv CombatRept*; *IMAC C–2 Jnl*; *3d MarDiv AR*; *SE Area NavOps—III*; *Seventeenth Army Ops—II*; USSBS, *Campaigns*; USSBS, *Interrogations*; *HistDiv Acct*.

The protests of some commanders against the use of surface vessels in the area south of New Britain—which was well within the range of the area dominated by the planes of ComAirSols—were brushed aside. *Combined Fleet Headquarters*, convinced that this was the last opportunity to take advantage of the strategic situation in the southeast, was determined to strike a decisive blow at the Allied surface strength in the Solomons and directed Kusaka to continue the operation.

After the battle of Empress Augusta Bay, however, the defeated Japanese retired from the area with the realization that combined sea and air operations were difficult with limited air resources, especially "in a region where friendly and enemy aerial supremacy spheres overlapped broadly." [15]

The *Seventeenth Army*, charged with the actual defense of Bougainville, took the news of the Allied invasion a bit more blandly:

> In formulating its operation plan, the Seventeenth Army planned to employ its main force only on the occasion of an army invasion in the southern or northern region, or the Kieta sector. Therefore, at the outset of the enemy landing in the vicinity of Torokina Point, the Seventeenth Army was lacking in determination to destroy the enemy. The army's intention at that point was only to obstruct the enemy landing. [16]

There were many avenues of obstruction open to the Japanese, despite the fact that the Allied sea and air activity probably discouraged the enemy from many aggressive overtures. Deceived originally as to the intentions of the Allies, the Japanese apparently remained in doubt for some time as to the strategical and tactical importance of the operations at Cape Torokina. The enemy could have counterlanded or prevented extension of the defensive positions and occupation of the projected airfield sites by shelling or air bombardments. But none of these courses of action were initiated immediately or carried out with sufficient determination to jeopardize the beachhead seized by the IMAC forces.

The chief threat to the Cape Torokina perimeter seemed to be from the right flank. Operation orders, taken from the bodies of dead Japanese at Cape Torokina, indicated that forces in the area southeast of the Cape could strike from that direction, and it was to this side that the IMAC forces ashore pointed most of their combat strength.

[15] *SE Area NavOps—II*, p. 18.

[16] *Seventeenth Army Ops—III*, p. 103.

CHAPTER 4

Holding the Beachhead

EXPANSION OF THE PERIMETER[1]

The landings on Bougainville had proceeded as planned, except that the 3d Marines encountered unexpectedly stiff resistance initially and the 9th Marines landed over surf and beaches which were later described as being as rough as any encountered in the South Pacific. The success of the operation, though, was obvious, and General Vandegrift, leaving Turnage in tactical control of all IMAC troops ashore, confidently returned to Guadalcanal with Wilkinson.[2]

At daybreak the second day, the Marines began expansion of the beachhead. Flank patrols along the entire perimeter established a cohesive defensive front, and from this position a number of reconnaissance patrols were pushed forward. There was no enemy activity except occasional sniper fire in the vicinity of Torokina plantation and on Puruata Island, where the raiders were still engaged.

Patrols had to thread their way through a maze of swamps that stood just back of the beaches, particularly in the 3d Marines' zone of action. Brackish water, much of it knee to waist deep, flooded inland for hundreds of yards. The bottom of these swamps was a fine, volcanic ash which had a quicksand substance to it. Marines had trouble wading these areas, and bulldozers and half-tracks all but disappeared under the water.[3]

Only two paths led out of this morass, both of them in the sector of the 3d Marines. These two trails, in some places only inches above the swampland, extended inland about 750 yards before joining firmer ground. One of the pathways, the Mission Trail toward the Piva River and Piva Village, was believed to be the main route of travel by the Japanese forces, and this trail was blocked on D-Day by the quick action of raider Company M. More detailed information on the extent of swamps and the location of higher ground would have been invaluable at this stage of beachhead expansion, but the aerial photographs and hasty terrain maps available furnished few clues

[1] Unless otherwise noted, the material in this section is derived from: *ComSoPac Nov43 WarD; ThirdFlt NarrRept; IIIPhibFor AR; IIIPhibFor Nov43 WarD; IMAC AR-I;* IMAC Rept on Bougainville Operation, Phase II, dtd 21Mar43 (Bougainville AreaOpFile, HistBr, HQMC), hereafter *IMAC AR-II; IMAC C-2 Repts; IMAC C-2 Jnl; 3d MarDiv CombatRept; 3dMarDiv AR; 3d MarDiv D-2 SAR; 3d MarDiv D-2 Jnl; 3d MarDiv D-3 Repts; HistDiv Acct;* Rentz, *Bougainville and the Northern Solomons;* Isely and Crowl, *Marines and Amphibious War;* Aurthur and Cohlmia, *3d MarDivHist.*

[2] *Vandegrift interview.*

[3] Col Francis M. McAlister ltr to CMC, dtd 29Mar48, hereafter *McAlister ltr.* The fact that the 9th Marines was able to dig in deeply and well shortly after it landed "indicated that there was a good deal of firm ground back of the beaches," much more than is generally recognized or remembered by men who got soaked wading the swamps. *Craig ltr.*

as to suitable locations for defensive positions or supply dumps. The result was complete dependence upon reports of the reconnaissance patrols.[4]

Because the routes inland were passable only to tracked vehicles, road building became the priority activity of the operation. Wheeled vehicles and half-tracks could be used only along the beaches, and in places where the shoreline was only a few yards wide, the supplies piled around had a paralyzing effect on traffic. The 53d and 71st Seabees, which landed in various echelons with the assault troops, began construction of a lateral road along the beach while waiting for the beachhead to expand.

Construction of a similar road along the Mission Trail was handicapped by the swamps and the lack of bridging material. Progress was slow. Airfield reconnaissance on Cape Torokina was begun at daybreak despite the sporadic sniper fire, but plans for immediate patrols to seek airfield sites further inland was postponed by the need for roads and the limited beachhead.

Early on the morning of 2 November, a shift in the tactical lineup along the perimeter was ordered in an attempt to pull in the flanks of the beachhead and constitute a reserve force for IMAC. Prior to the operation, General Turnage had been concerned about the possibility of his lengthy but narrow beachhead being rolled up like a rug by enemy action. Without a reserve and unable to organize a defense in depth to either flank, the division commander had planned to move individual battalions laterally to meet enemy threats as they developed. Now, with the beaches in the 9th Marines' sector unsuitable for continued use, and the 3d Marines needing some relief after the tough battle to take Cape Torokina, redisposition of certain units was directed.[5]

At 0830, the 2d Battalion, 9th Marines started drawing in its left flank toward the beach while 1/9, the extreme left flank landing team, began withdrawal to the vicinity of Cape Torokina. Some units moved along the beach; others were lifted by amphibian tractors. By nightfall, 1/9 was under the operational control of the 3d Marines and was in reserve positions behind 1/3 on the right flank. Two artillery batteries of the 12th Marines registered fire on the Laruma River to provide support for 2/9 on the left flank.

The following day, 3 November, 2/9 made the same withdrawal and moved into positions to the right of the 2d Raider Battalion. During the day, 1/9 relieved 1/3 on the right flank of the perimeter, and at 1800 operational control of Cape Torokina passed to the 9th Marines. At this time, 3/9 with its left flank anchored to the beach north of the Koromokina River, was placed under the control of the 3d Marines; 1/3, withdrawn from action, was designated the reserve unit of the

[4] General Craig recalls: "It was almost impossible to spot our troop dispositions on operations maps at times. The maps were very poor and there were few identifying marks on the terrain until we got to the high ground. At one time I had each company on the line put up weather balloons (small ones) above the treetops in the jungle and then had a plane photograph the area. The small white dots made by the balloons gave a true picture finally of just how my defensive lines ran in a particularly thick part of the jungle. It was the only time during the early part of the campaign that I got a really good idea as to exactly how my lines ran." *Craig ltr.*

[5] LtCol Alpha L. Bowser, Jr., ltr to CMC, dtd 10May48 (Bougainville Monograph Comment File, HistBr, HQMC), hereafter *Bowser ltr.*

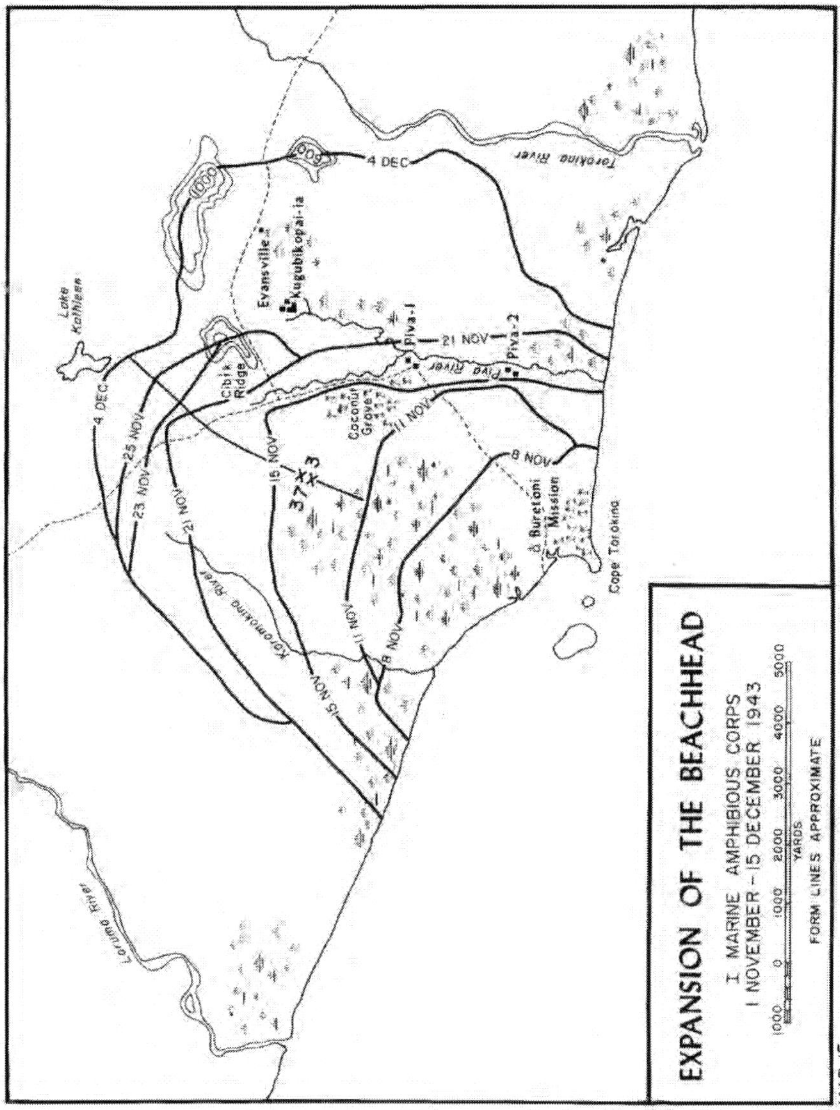
EXPANSION OF THE BEACHHEAD
I MARINE AMPHIBIOUS CORPS
1 NOVEMBER - 15 DECEMBER 1943
1000 0 1000 2000 3000 4000 5000
YARDS
FORM LINES APPROXIMATE
Lake Kathleen
Evansville
Kugubikopai-ia
Cibik Ridge
Coconut Grove
Piva-1
Piva-2
Piva River
Torokina River
Koromokina River
Laruma River
Buretoni Mission
Cape Torokina
37
3
4 DEC
25 NOV
23 NOV
21 NOV
15 NOV
11 NOV
8 NOV

MAP 15

9th Marines. In effect, after three days, the two assault regiments had traded positions on the perimeter and had exchanged one battalion.

The last Japanese resistance within the perimeter had been eliminated, also. On 3 November, after several instances of sporadic rifle fire from Torokina Island, the artillery pieces of a battery from the 12th Marines, as well as 40mm and 20mm guns of the 3d Defense Battalion, were turned on Torokina Island for a 15-minute bombardment. A small detachment of the 3d Raiders landed behind this shelling found no live Japanese but 8–10 freshly dug graves. The enemy had apparently been forced to abandon the island.

Extension of the perimeter continued despite the shuffling of front-line units. On the left, 3/9 and the two battalions of the 3d Marines continued to press inland in a course of advance generally north by northeast. Contact was maintained with the 2d Raider Battalion on the right of the 3d Marines, and patrols scouted ahead with the mission of locating the route for a lateral road from the left flank to the right flank. The 2d Raiders, moving along the Mission Trail in front of the 9th Marines, extended the beachhead almost 1,500 yards. In this respect, the addition of a war dog platoon to the front-line units was invaluable. Not only did the alert Doberman Pinschers and German Shepherds smell out hidden Japanese, but their presence with the patrols gave confidence to the Marines.

On 4 November, extensive patrolling beyond the perimeter north to the Laruma River and south to the Torokina River was ordered, but enemy contact was light and only occasional sniper fire was encountered. The 1/9 patrol to the Torokina River killed one sniper near the Piva, and some enemy activity was reported in front of 2/9; other than that, enemy resistance had vanished.

By the end of the following day, 5 November, the IMAC beachhead extended about 10,000 yards along the beach around Cape Torokina and about 5,000 yards inland. Defending along the perimeter were five battalions (3/9, 3/3, 2/3, 2/9, and 1/9) with 1/3 in reserve. The 2d Raiders, leaving one company blocking Mission Trail, and the 3d Raiders were assembled under control of the 2d Raider Regiment commander, Lieutenant Colonel Alan Shapley, in IMAC reserve on Puruata Island and Cape Torokina.

All other IMAC units which had been engaged in shore party operations also reverted to parent control. By this time, two battalions of the 12th Marines were in position to provide artillery support to the beachhead and the antiaircraft batteries of the 3d Defense Battalion were operating with radar direction. With all units now able to make a muster of personnel, the IMAC casualties for the initial landings and widening of the beachhead were set at 39 killed and 104 wounded. Another 39 Marines were reported missing. The Japanese dead totaled 202.

Nearly 6,200 tons of supplies and equipment had been carried to Empress Augusta Bay by the IIIPhibFor transports. By the end of D-Day, more than 90 percent of this cargo was stacked along the beaches in varying stages of organization and orderliness. The problem was complicated further by the final unloading of the four transports on D plus 1. Practically every foot of dry area not already occupied by troop bivouacs or gun emplacements was piled high with cargo, and the troops still serving as the shore party were hard-pressed to find additional storage areas within the narrow perimeter. Ammunition and fuel dumps had to be fitted in temporarily with other supply dumps, and these in turn were situated where terrain permitted. The result was a series of dumps with explosives and fuels dangerously close to each other and to troop areas and beach defense installations.

Main source of trouble was the lack of beach exits. Some use of the two trails had been attempted, but these had broken down quickly under the continual drizzle of rain and the churning of tracked vehicles. The lateral road along the beach was practically impassable at high tide, and at all times trucks were forced to operate in sea water several inches to several feet deep. Seabees and engineers, attempting to corduroy some of the worst stretches, had their efforts washed out.

The bordering swamplands, which restricted the use of wheeled vehicles and half-tracks in most instances, forced the discovery of the amphibian tractor as the most versatile and valuable addition to the landing force. Already these lumbering land-sea vehicles had proven their worth in carrying cargo, ferrying guns, and evacuating wounded men through the marsh lands and the lagoons, and the variations of their capabilities under such extreme circumstances were just beginning to be realized and appreciated. The arrival of the LST echelons later brought more of these welcome machines to the beachhead.

IMAC and 3d Marine Division engineers landed in the first echelons on D-Day, but reconnaissance to seek supply routes into the interior and supply dump locations was handicapped by the limited expansion of the beachhead. Many survey missions bumped into the combat battalions and were discouraged from patrolling in front of the defensive positions by the Marines who preferred to have no one in advance of the lines except Japanese. This problem was solved by the frontline units furnishing combat patrols for engineer survey parties who moved ahead and then worked a survey back to the old positions.

The swamplands were successfully attacked by a series of drainage ditches into the sea. As the ground dried, the volcanic ash was spread back as fill dirt. The airfield work was slowed by the many supply dumps and gun emplacements which had been placed in the vicinity of the plantation, one of the few dry spaces around. Many division dumps and artillery positions had to be moved from the area, including one battery of 90mm guns of the 3d Defense Battalion which occupied a position in the middle of the projected runway.

By the time the first supporting echelon of troops and cargo arrived on 6 November, the Torokina beachhead was still handicapped by the lack of good beach facilities and roads. Airfield construction had slowed the development of roads, and vice versa, and the loose sand and heavy surf action along the Cape Torokina beaches resisted efforts to construct proper docking facilities for the expected arrival of the supporting echelons.[8] Coconut log ramps, lashed together by cables, were extended about 30 feet from the shoreline, but these required constant rebuilding. Later, sections of bridges were used and these proved adaptable to beach use.

The reinforcements arrived at Torokina early on the morning of 6 November. The

[8] LtCol Harold B. West ltr to CMC, dtd 28May48 (Bougainville Monograph Comment File, HistBr, HQMC).

3,548 men and 6,080 tons of cargo were embarked at Guadalcanal on the 4th on eight LSTs and eight APDs, which were escorted to Cape Torokina by six destroyers. The APDs unloaded a battalion landing team from the 21st Marines and other division elements quickly and then headed back to Guadalcanal for another echelon. But the unloading of the LSTs was slowed by the crowded conditions of the main beaches and the lack of beach facilities. Most of the cargo was unloaded at Puruata Island where the tank landing ships could beach adequately, and the cargo was then transshipped to the mainland.[7] This created something of a problem, too, for supplies were poured onto Puruata without a shore party to organize the cargo; and this condition was barely cleared up before another echelon of troops and supplies arrived.

COUNTERLANDING AT KOROMOKINA[8]

After the first desperate defense of Cape Torokina, the enemy had offered no resistance. Then, on 7 November, the Japanese suddenly launched a counterlanding against the left flank of the beachhead. The move caught the 3d Marine Division in the midst of reorganization of the perimeter to meet the expected threat on the right flank.

The counterlanding fulfilled an ambition long cherished by the Japanese. At Rendova, an attempted landing against the Allied invasion forces fizzled out during a downpour that prevented the rendezvous of the Japanese assault force; and at New Georgia, the enemy considered—then rejected—an idea to land behind the 43d Division on Zanana Beach. The Cape Torokina operation, however, gave the Japanese a chance to try this favored counterstroke. Such a maneuver was in line with the basic policy of defense of Bougainville by mobile striking forces, and, in fact, a provisional battalion was in readiness at Rabaul for just such a counterlanding attempt.

This raiding unit was actually a miscellany of troops from several regiments of the *17th Division*. Specially trained, the battalion included the *5th Company, 54th Infantry;* the *6th Company, 53d Infantry;* a platoon from the *7th Company, 54th Infantry;* and a machine gun company from the *54th Infantry*, plus some service troops. Japanese records place the strength of the battalion at about 850 men.[9]

The attack force started for Cape Torokina on the night of 1 November, but the reported presence of an Allied surface fleet of battleships and cruisers in the area, and the threat of being discovered by Allied planes, convinced the Japanese that a counterlanding at this time would be difficult. Accordingly, the attempt was postponed, and the troops returned to Rabaul while Admiral Kusaka's *Southeast Area Fleet* concentrated on destroying the Allied interference before another try was made. The landing party finally departed Rabaul on 6 November, the four troop

[7] *Ibid.*

[8] Unless otherwise noted, the material in this section is derived from: *ComSoPac Nov43 WarD; ThirdFlt NarrRept; IMAC AR-II; IMAC C-2 Repts; IMAC C-2 Jnl; 3d MarDiv CombatRepts; 3d MarDiv AR; 3d MarDiv D-2 SAR; 3d MarDiv D-2 Jnl; 3d MarDiv D-3 Repts; HistDiv Acct; SE Area NavOps—III; Seventeenth Army Ops—II;* Rentz, *Bougainville and the Northern Solomons;* Aurthur and Cohlmia, *3d MarDivHist.*

[9] *IMAC C-2 Jnl; SE Area NavOps—III*, p. 20.

CALF-DEEP MUD *clings to a column of Marine ammunition carriers as they move toward the front lines on Bougainville. (USMC 68247)*

ADMIRAL HALSEY AND GENERAL GEIGER *watch Army reinforcements file along the shore at Bougainville. (USMC 65494)*

691-360 O—63——16

destroyers screened by a cruiser and eight escort destroyers.

Shortly after midnight, the transport group entered the objective area, but the first landing attempt was hurriedly abandoned when Allied ships were discovered blocking the way. The destroyers headed north again, then back-tracked closer to the shoreline for a second try. This time the troop destroyers managed to unload the troops about two miles from the beaches. The landing force demanded protective gunfire from the destroyers, but the Japanese skippers, considering the Allied fleet nearby, paid little heed. The troops, loaded in 21 ramp boats, cutters, and motor boats,[10] were landed at dawn near the Laruma River, just outside the left limits of the IMAC perimeter.

First indication that a Japanese counterlanding was in progress came from one of the ships at anchor which reported sighting what appeared to be a Japanese barge about four miles north of Cape Torokina.[11] Before a PT boat could race out to check this report, 3d Marine Division troops on that flank of the beachhead confirmed the fact that enemy barges were landing troops at scattered points along the shoreline and that the Marines were engaging them.

The first landings were made without opposition. A Marine antitank platoon, sited in defensive positions along the beach, did not open fire immediately because of confusion as to the identity of the landing craft. The Marines who witnessed the landings said that the Japanese ramp boats looked exactly similar to American boats, including numbers in white paint on the bow.[12] In the early dawn mist, such resemblance in silhouette was enough to allay the suspicions of the sentries. Once the alarm was sounded, artillery pieces of the 12th Marines and coast defense guns—including the 90mm antiaircraft batteries—of the 3d Defense Battalion were turned on the enemy barges and landing beaches.[13]

Instead of landing as a cohesive unit, however, the Japanese raiding force found itself scattered over a wide area, a victim of the darkness and the same surf troubles that earlier had plagued the Marines. Troops were distributed on either side of the Laruma and were unable to reassemble quickly. The Japanese were faced with the problem of attacking with the forces on hand or waiting to reorganize into tactical units. Under fire already, deciding that further delay would be useless, the enemy began the counterattack almost at once. Less than 100 enemy soldiers made the first assault.

The 3d Battalion, 9th Marines (Lieutenant Colonel Walter Asmuth, Jr.)—occupying the left-flank positions while waiting to be moved to the Cape Torokina area—drew the assignment of stopping the enemy counterthrust. Artillery support fire was placed in front of the perimeter and along the beach. At 0820, Company K, 3/9, with a platoon from regimental weapons company attached, moved forward to blunt the Japanese counterattack. About 150 yards from the main line of resistance (MLR), the advancing Marines hit the front of the enemy force. The Japanese, seeking cover from the artillery fire, had dug in rapidly and, by taking ad-

[10] *3d MarDiv D-3 Jnl*, 7Nov43.

[11] *Ibid.*

[12] *Ibid.*

[13] LtCol Jack Tabor ltr to CMC, dtd 7Jun48 (Bougainville Monograph Comment File, HistBr, HQMC).

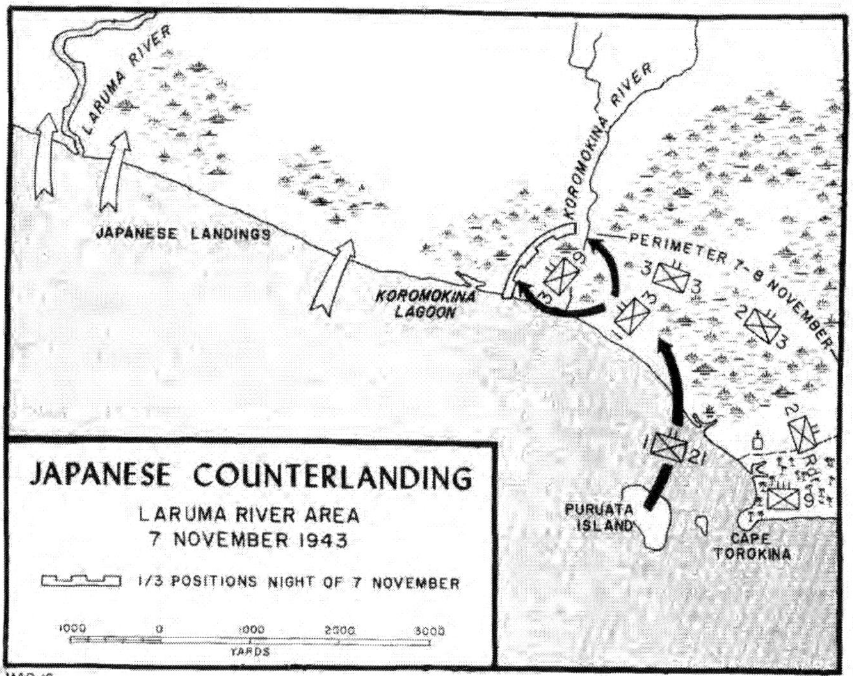

MAP 16

R.F. STIBIL

vantage of abandoned foxholes and emplacements of the departed 1/9 and 2/9, had established a hasty but effective defensive position.

Heavy fighting broke out immediately, the Japanese firing light machine guns from well-concealed fortifications covered by automatic rifle fire from tree snipers. Against this blaze of fire, the Marine company's attack stalled. The left platoon was pinned down almost at once, and, when the right and center platoons tried to envelop the defensive positions, the dense jungle and enemy fire stopped their advance. The Japanese resistance increased as reinforcements from the remainder of the counterlanding force began to arrive. At 1315, the 1st Battalion of the 3d Marines in reserve positions in the left sector was ordered into the fight.

While Company K held the Japanese engaged, Company B of 1/3 moved across the MLR on the left flank and passed through Company K to take up the fight. At the same time, Company C of 1/3 moved forward on the right. The 9th Marines' company withdrew to the MLR leaving the battle to 1/3, now commanded by Major John P. Brody. In the five hours that Company K resisted the Japanese counterlanding, it lost 5 killed and 13 wounded, 2 of whom later died.

The two companies of 1/3 found the going no easier. The Japanese were well-hidden, with a high proportion of machine guns and automatic weapons, and the Marine attack was met shot for shot and grenade for grenade. In some instances, Marines knocked out machine gun emplacements that were almost invisible in the thick jungle at distances greater than five yards. Tanks moved up to help with the assault, and the Marine advance inched along as the 37mm canister shells stripped foliage from the enemy positions. High explosive shells, fired nearly point blank, erased many of the enemy emplacements, and in some cases the HE shells—striking ironwood trees—knocked enemy snipers out of the branches.

Late in the afternoon, the advance was halted and a heavy artillery concentration, in preparation for a full-scale attack by the 1st Battalion, 21st Marines, was placed on enemy defenses in front of the Marines. The artillery fire raged through the enemy positions; and to keep the Japanese from seeking cover and safety in the area between the artillery fire and the Marine lines, Companies B and C placed mortar fire almost on top of their own positions.

The attack by 1/21 (Lieutenant Colonel Ernest W. Fry, Jr.) was set for 1700 on the 7th, but the effective artillery-mortar fire and the approaching darkness postponed the attack until the following morning. Fry's battalion, which landed on Puruata the previous day, was moved to the mainland to be available for such reserve work after the Japanese struck. The battalion spent the night behind the 1/3 perimeter, which, by the end of 7 November, was several hundred yards past the original perimeter position of 3/9 that morning.

The enemy's action in landing at scattered points along the shoreline resulted in several Marine units being cut off from the main forces during the day. One platoon from Company K, 3/9, scouting the upper Laruma River region, ambushed a pursuing Japanese patrol several times before escaping into the interior. This platoon returned to the main lines about 30 hours later with one man wounded and one man missing after inflicting a number of casualties on the enemy landing force. Another outpost patrol from Company M, 3/9, was cut off on the beach between two enemy forces. Unfortunately, the radio of the artillery officer with the patrol did not function, and so support could not be summoned.

The artillery officer found his way back to the main lines where he directed an artillery mission that landed perfectly on the Japanese position to the left. The patrol then moved toward the division main lines, only to find the beach blocked by enemy forces opposing Company K. A message scratched on the beach [14] called an air spotter's attention to the patrol's plight, and, late that afternoon, two tank lighters dashed in to the beach to pick up the patrol. Sixty men were evacuated successfully after killing an estimated 35 Japanese. Only two of the Marines had been wounded.

Two other Marine groups became isolated in the fighting along the perimeter. One platoon from 1/3, scouting the enemy's flank position, slipped through the jungle and passed by the enemy force without being observed. Choosing to head for the beach instead of the interior, the platoon struggled to the coast. There the patrol cleaned its weapons with gasoline from a wrecked barge, and spent the night in the jungle. The next morning, the attention of an Allied plane was attracted

[14] *Ibid.*

and within an hour the platoon was picked up by a tank lighter and returned to the main lines.[15] The other isolated unit, a patrol from Company B, was cut off from the rest of the battalion during the fighting and spent the night of 7–8 November behind the enemy's lines without detection.

On the morning of 8 November, after a 20-minute preparation by five batteries of artillery augmented by machine guns, mortars, and antitank guns, 1/21 passed through the lines of the 1/3 companies and began the attack. Light tanks, protected by the infantrymen, spearheaded the front. Only a few dazed survivors of two concentrated artillery preparations contested the advance, and these were killed or captured. More than 250 dead Japanese, some of them killed the previous day, were found in the area.[16] The battalion from the 21st Marines moved about 1,500 yards through the jungle paralleling the shoreline. No opposition was encountered. That afternoon, 1/21 established a defensive line behind an extensive lagoon[17] and sent out strong patrols on mop-up duties. There was no enemy contact.

The following morning, 9 November, the area between the Marine positions and the Laruma River was bombed and strafed by dive bombers from Munda. The air strike completed the annihilation of the Japanese landing force. Patrols from 1/21 later found the bodies of many Japanese in the area, apparently survivors of the attacks of 7 and 8 November who had taken refuge in the Laruma River area. There was no further enemy activity on the left flank of the perimeter, and, at noon of that day, control of the sector passed to the 148th Infantry Regiment of the 37th Division, which had arrived the preceding day. The battalion from the 9th Marines moved to the right flank, and 1/3 returned to regimental reserve in the 3d Marines area. Fry's battalion, holding down the left-flank position, remained under operational control of the 148th Regiment until other units of the 37th Division arrived.

The Japanese attempt to destroy the IMAC forces by counterlanding had ended in abject failure. The landing force, woefully small to tackle a bristling defensive position, had only limited chances for success, and these were crushed by the prompt action of Company K, 3/9, and the rapid employment of the available reserve forces, 1/3 and 1/21.

Estimates differ as to the size of the raiding unit which the Japanese sent against a force they believed numbered no more than 5,000 men. Japanese records indicate that 850 men were landed, but IMAC intelligence officers believed that no more than 475 Japanese soldiers were thrown against the defensive perimeter. Most of these were killed in the artillery barrages and the air strike on 7–9 November. The landing site was an unfortunate choice, also. The Japanese had no idea of the exact location of the Allied beachhead and believed it to be farther east around Cape Torokina. The landing was not planned for an area so close to the beachhead. With all tactical integrity lost, forced to attack before they were ready and reorganized, the Japanese were handicapped from the first.

[15] Maj Robert D. Kennedy ltr to CMC, dtd 21May48 (Bougainville Monograph Comment File, HistBr, HQMC).

[16] Two different counts are given for enemy casualties. *3d MarDiv AR*, p. 7, gives 254 enemy dead; *IMAC AR-II*, p. 9, gives 277 enemy dead.

[17] This engagement is called the Battle of Koromokina Lagoon in some accounts.

Another factor in the defeat was the Japanese inability to coordinate this counterattack with a full-scale attack on the opposite side of the perimeter, although this was the original intention of the counterlanding. The enemy's error of carrying situation maps and operation orders into combat was repeated in the Laruma River landing. Within hours of the attack by 1/21 on 8 November, IMAC intelligence officers had the Japanese plan of maneuver against the entire beachhead and were able to recommend action to thwart the enemy strategy.

PIVA TRAIL BATTLE [18]

The enemy pressure on the right flank of the perimeter began as a series of small probing attacks along the Piva Trail leading into the beachhead fronting Cape Torokina. Japanese activity on this flank, in contrast to the counterlanding effort, was entirely expected. Since D-Day, the 2d Raider Battalion with Company M of the 3d Raider Battalion attached had slowly but steadily pressed inland astride the trail leading from the Buretoni Mission towards the Piva River. This trail, hardly more than a discernible pathway through the jungle, was the main link between the Cape Torokina area and the Numa Numa trail; and if the Japanese mounted a serious counterstroke, it would probably be aimed along this route.

Advance defensive positions were pushed progressively deeper along this path by raider companies, and, by 5 November, the Marines had established a strong trail block about 300 yards west of the junction of the Piva-Numa Numa trails. Although the responsibility for the defense of this sector now belonged to the 9th Marines, the 2d Raider Battalion still maintained the trail block. Until the night of 5–6 November, there had been no interference from the enemy except occasional sniper fire. That night, with Company E of the 2d Raiders manning the defensive position, the Japanese struck twice in sharp attacks. Company E managed to repulse both attacks, killing 10 Japanese, but during the fight an undetermined number of enemy soldiers managed to evade the trail block and infiltrate to the rear of the raiders.

The following day was quiet, but anticipating further attempts by the Japanese to steamroller past the road block, the 2d and 3d Raider Battalions, under regimental control of the 2d Raider Regiment, were moved into position to give ready support of the road block. The raiders remained attached to the 9th Marines, and Colonel Craig continued to control operations of both regiments.[19]

The first enemy thrust came during the early part of the afternoon of 7 November, shortly after Company H of the 2d Raiders had moved up to the trail block to relieve Company F which had been in position the night before. A force of about one company struck the defensive block first; but Company H, aided by quick and effective 81mm mortar fire from 2/9 in the defensive perimeter to the rear of the trail block, turned the enemy's assault. One platoon from Company F, 2d Raiders, then rushed to the trail block to reinforce Com-

[18] Unless otherwise noted, the material in this section is derived from: *IMAC AR–II; IMAC C–2 Repts; IMAC C–2 Jnl; 3d MarDiv CombatRept; 3d MarDiv AR; 3d MarDiv D–2 SAR; 3d MarDiv D–3 Repts; HistDiv Acct;* Rentz, *Bougainville and the Northern Solomons;* Aurthur and Cohlmia, *3d MarDivHist.*

[19] *Craig ltr.*

pany H until another raider unit, Company G, was in position to help defend the trail. The enemy, unable to penetrate the Marine position after several furious attacks, withdrew about 1530, and was observed digging in around Piva Village, some 1,000 yards east. The Japanese force was estimated at about battalion strength.

Several small-scale attacks were started later that afternoon by the Japanese, but each time the two raider companies called for mortar concentrations from 2/9 and the assaults were beaten back. One determined attempt by the Japanese to cut the trail between the road block and the IMAC perimeter was repulsed by Company G. During the night, the enemy rained 90mm mortar fire on the trail block and sent infiltrating groups into the Marine lines, but the two raider companies, sticking to their foxholes, inflicted heavy casualties by withholding return fire until the enemy was at point-blank range. One Marine was killed.

Early the next morning, 8 November, Company M of the 3d Raiders hurried forward to relieve Company H while Company G took over the responsibility for the trail block. Company M took up positions behind the trail block and deployed with two platoons on the left side of the trail and one platoon on the right. Before Company H could leave the area, however, enemy activity in front of Company G increased and returning patrols reported that a large-scale attack could be expected at any time. Reluctant to leave a fight, Company H remained at the trail block. The Japanese assault was not long in coming. Elements of two battalions, later identified as the *1st* and *3d Battalions* of the *23d Infantry* from the Buin area, began pressing forward behind a heavy mortar barrage and machine gun fire. By 1100, the trail block was enveloped on all sides by a blaze of gunfire as the Marine units sought to push the attackers back. Company G, solidly astride the trail, bore the brunt of the enemy's assault.

Shortly after 1100, Company E moved from a reserve area into the trail block and took up positions on the right of Company G. The platoons of Company E were then extended to the right rear to refuse that flank. At this time, the combined fronts of G and E Companies astride the Piva Trail measured about 400 yards. An hour later, at noon, Company L of the 3d Raiders also advanced from a reserve area and stationed itself on the left flank of Company G. The Marine position now resembled a rough horseshoe, with Companies E, G, and L holding the front and flanks and Companies H and M connecting the trail block to the main IMAC perimeter.

Flanking movements by either the attackers or the defending Marines were impossible because of the swampy ground on either side of the trail, and two attempts by enemy groups to envelop the flanks of the Marine position ended as near-frontal attacks with heavy casualties to the attacking troops. In each instance, the Japanese were exposed to the direct fire of a Marine company in defensive positions. Both attacks were beaten back.

At 1300, with the enemy assault perceptibly stalled, the 2d Raiders attempted a counterattack. Company F, returned to the trail block from a reserve area, together with Company E began a flanking maneuver from the right. After struggling through the swamps for only 50 yards, the two raider companies struck a large force of Japanese, and the fight for possession of the trail began once more.

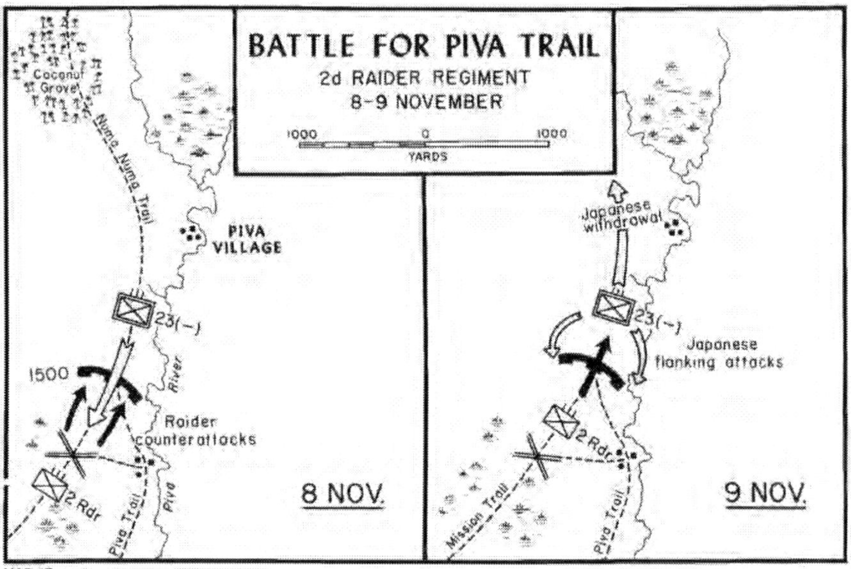

MAP 17

The enemy soldiers, attempting another counterattack, ran full into the fire of Company G's machine guns and once again took heavy casualties. Half-tracks of the 9th Marines Weapons Company, with two supporting tanks, moved forward to help the Marine attack gain impetus, but the thick jungle and the muddy swamps defeated the attempts of the machines to reinforce the front lines. Unable to help, the machines began evacuating wounded. By 1600, the fight at the trail block was a stalemate. The Marines were unable to move forward, and the enemy force had been effectively stalled. Another Japanese counterattack, noticeably less fierce than the first, was turned back with additional casualties to the enemy.

With darkness approaching, the raider companies were ordered to return to their prepared lines, and the Marines began to withdraw through the trail block. Company F covered the disengagement and beat back one final enemy attempt before the withdrawal was completed. The raider casualties were 8 killed and 27 wounded. The Marines estimated that at least 125 Japanese had been killed in the day's fighting.

That night, General Turnage directed Colonel Craig to clear the enemy from the area in front of the 9th Marines and the trail block so that the perimeter could be advanced. Craig, planning an attack with an extensive artillery preparation, decided to use Shapley's 2d Raider Regiment again because the raiders were already familiar with the terrain. The attack was to be supported by 2/9 with a section of tanks and half-tracks attached.

At 0620 the following morning, 9 November, the raider units returned to the trail block area which had been held overnight by Company M and a fresh unit, Company I. The two assault companies deployed behind Company I with Company L taking positions on the left of the trail and Company F on the right side of the trail. At 0730, the artillery preparation by 1/12 began to pound into the Japanese positions ahead of the trail block. More than 800 rounds were fired as close as 250 yards from the Marine lines to prepare the way for the attack by the two raider companies.

The Japanese, though, had not waited to be attacked. At first light, the enemy started strong action to overrun the trail block and moved to within 100 yards of the Marine position. There they had established a similar trail block with both flanks resting on an impassable swamp. Other enemy soldiers, who had crept up to within 25 yards of the front lines during the night, remained hidden until the artillery fires ceased and the raider companies began the attack. Then the Japanese opened up with short-range machine gun fire and automatic rifle fire.

The enemy's action delayed part of Company F, with the result that, when Company L began the attack at 0800, only half of Company F moved forward. Coordination between the two attacking units was not regained, and, by 0930, the raider attack had covered only a few yards. The two companies were forced to move along a narrow front between the swamps, and the enemy fire from a large number of machine guns and "knee mortars" stalled the Marine attack.

Neither the tanks nor the half-tracks could negotiate the muddy corridor to reinforce the Marine attack. Unable to flank the enemy position, the raiders could move forward only on the strength of a concentrated frontal attack. The fight along the corridor became a toe-to-toe slugging match, the Marines and Japanese screaming at each other in the midst of continual mortar bursts and gunfire. Slowly at first, then with increasing speed, the Marine firepower overcame that of the Japanese. The raider attack, stalled at first, began to move.

Threatened by a desperate enemy counteraction on the right flank, Colonel Craig—personally directing the attack of the raiders—moved Company K into the gap between Companies L and F and deployed the Weapons Company of the 9th Marines on the right rear of the trail block for additional support. These moves stopped the Japanese counterattack on that flank. Later, another platoon from Company M moved into the front lines to lend its firepower to the raider advance.

Suddenly, at 1230, the Japanese resistance crumbled and the raider companies pressed forward against only scattered snipers and stragglers. By 1500, the junction of the Piva-Numa Numa Trail was reached, and, since no enemy had been seen for more than an hour, the assault units halted. Defensive lines were dug, and patrols began moving through the jungle and along the Numa Numa Trail. There was no contact, and a large enemy bivouac area along the Numa Numa Trail was discovered abandoned. More than 100 dead Japanese were found after the attack. The Marines lost 12 killed and 30 wounded in the operation.

An air strike set for early the next morning, 10 November, was delayed for a short time by the late return of a patrol from Company K, 3d Raiders, which had been on an all-night scouting mission to Piva Village. The patrol reported no contacts. Twelve torpedo bombers from Marine squadrons VMTB–143 and –233 based at Munda then bombed and strafed the area from the Marine position to Piva Village. The front lines were marked by white smoke grenades and a Marine air liaison party guided the pilots in their strike. The first bomb fell within 150 yards of the markers. A 50-yard strip on both sides of the Numa Numa Trail was worked over by the planes, and, at 1015, the infantry began moving toward Piva Village.[20]

Lieutenant Colonel Robert E. Cushman's 2/9, followed closely by 1/9 commanded by Lieutenant Colonel Jaime Sabater, passed through the raider companies and moved along the trail. The advance was unopposed, although scattered enemy equipment, ammunition, and weapons—including a 75mm gun and a 37mm gun as well as rifles and machine guns—were found. Another 30–40 dead Japanese were also found in the area, apparent victims of the extensive air and artillery support of the Marines.

By 1300, the two battalions of the 9th Marines had moved through Piva Village and into defensive positions along the Numa Numa Trail. Aggressive patrols began fanning out toward the Piva River and along the trail, seeking the enemy. The IMAC beachhead, by the end of the day, was extended another 800 yards inland and contact had been established with the 3d Marines to the left.

The 2d Raider Regiment, which had taken the full force of the enemy's attack on the right flank, returned to bivouac positions within the perimeter as the division reserve force. In the space of three days, the threat to the beachhead from either flank had been wiped out by the immediate offensive reactions of the 3d Marine Division. The attempted mouse-trap play by the Japanese to draw the Marine forces off balance towards the Koromokina flank, to set the stage for a strike from the Piva River area, had been erased by well conducted and aggressive attacks supported by artillery and air. The landing force of nearly 475 Japanese on the left flank had been almost annihilated, and at least 411 Japanese died in the attacks on the right flank.

Another factor in the success of the beachhead was the continued arrival of reinforcements, a testimonial to the foresight of General Vandegrift who had insisted that the buildup of the forces ashore not wait the 30-day interval which had been planned. The 148th Regimental Combat Team of the 37th Division began arriving on 8 November, in time to take over responsibility for the left sector of the perimeter, allowing Marine units in that area to revert to their parent units and bolster the right flank defense. In addition, the arrival of these troops and additional equipment and supplies allowed the perimeter to expand to include a center sector.

[20] Because of the swampy nature of the ground over which the advance was made, an amphibian tractor company was attached to the 9th Marines. Colonel Craig used the LVTs to carry two days rations and supplies for the regiment and to transport radio jeeps for the air liaison party and his own and the battalions' headquarters. *Craig ltr.*

THE COCONUT GROVE [21]

The second major battle in the vicinity of the Numa Numa Trail began after a two-day lull following the seizure of Piva Village. During that interval, only minor skirmishes occurred, most of them inadvertent brushes between Marine scouting patrols and Japanese stragglers. Although contact with the main force of the enemy had been lost, there was little doubt that the enemy was still present in large numbers north of the Piva River. The 9th Marines, holding the area around Piva Village, concentrated on improving the supply routes into its position. Defensive installations and barriers were also extended and strengthened.

As the beachhead slowly widened behind the 3d and 9th Marines, airfield reconnaissance efforts were extended, and, during the time that the trail block fighting was underway, a group of Navy and Marine engineers with construction battalion personnel were busy making a personal ground reconnaissance of an area which had earlier been selected as a possible airfield site. This location, about midway between the Koromokina and Piva Rivers, was about 5,500 yards inland or about 1,500 yards in advance of the 3d Marine Division positions.

The engineers, accompanied by a strong combat patrol, managed to cut two 5,000-foot survey lanes nearly east to west across the front of the IMAC perimeter. The patrol then returned to report that at least one bomber strip and one fighter strip could be constructed in the area scouted. The survey party was unchallenged by the enemy, although a combat patrol the following day clashed with a Japanese patrol near the same area.

Because the numerous swamps and difficult jungle terrain prohibited the possibility of extending the beachhead immediately to cover the proposed airfield site, General Turnage decided that a combat outpost, capable of sustaining itself and defending the selected area until the front lines could be lengthened to include it, should be established at the junction of the Numa Numa and East-West Trails. On 12 November, the division commander directed the 21st Marines (Colonel Evans O. Ames) to send a company-sized patrol up the Numa Numa Trail the following morning. This group was to move to the junction of the two trails and reconnoiter each trail for a distance of 1,000 yards. This would delay any Japanese attempts to occupy the area, and would prevent having to fight an extended battle later for its possession.

At this time, the 21st Marines had two battalions ashore and a third due to land within the next few days. Fry's battalion was still in support of the 37th Division on the left, and 2/21 (Lieutenant Colonel Eustace R. Smoak) was then in bivouac near Cape Torokina. Smoak's battalion, with the regimental command post group, had arrived on 11 November. Alerted for action, 2/21 moved to a new bivouac area about 400 yards behind the 9th Marines and waited for orders.

On the night of 12 November, the division chief of staff (Colonel Robert E. Blake) directed that the size of the patrol be increased to two companies with a suitable command group and artillery observers to establish a strong outpost at the trail

---

[21] Unless otherwise noted, the material in this section is derived from: *IMAC AR–II*, *IMAC C–2 Repts*; *IMAC C–2 Jnl*; *3d MarDiv CombatRept*; *3d MarDiv AR*; *3d MarDiv D–3 Repts*; *HistDiv Acct*; Rentz, *Bougainville and the Northern Solomons*; Aurthur and Cohlmia, *3d MarDivHist*.

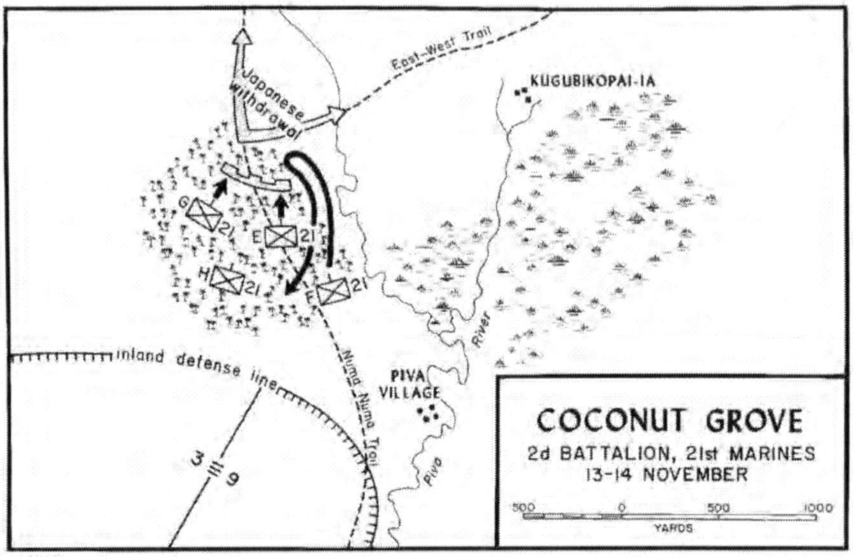

MAP 16

junction. Aware of the importance of the mission, Smoak requested and received permission to use his entire battalion in the assignment.

At 0630 the following morning, 13 November, Company E as the advance unit of 2/21 moved to the assembly area behind the 9th Marines but was ordered to hold up at this point. An hour later, with the remainder of the battalion still engaged in drawing ammunition, water, and rations, Company E was directed to begin the advance. The remainder of the battalion would follow as soon as possible. An artillery observer party, attached to the battalion, failed to arrive until after Company E had departed.

The rifle company cleared the 9th Marines perimeter at 0800, and three hours later was ambushed by a sizeable enemy force located in an overgrown coconut palm grove about 200 yards south of the trail junction which was the objective. Company E deployed to return the fire, but mortar shells and machine gun fire restricted movement and casualties began to mount.

The enemy had won the race for the trail junction.

Although it is possible that the Japanese had been in an organized position in the coconut grove for some time, it is unlikely that the airfield reconnaissance patrol would have been allowed to operate without attack if such were the case. A better possibility is that the Japanese moved into the position coincidental with the decision by Turnage to establish an outpost there.

Smoak's battalion, at that time some 1,200 yards to the rear of Company E, received word of the engagement at 1200. The battalion, pulling in its slower moving flank security patrols, hurried up the trail toward the fight. By 1245, 2/21 was about 200 yards behind Company E and a number of disturbing and conflicting reports were being received. The battalion commander was told that Company E was pinned down by heavy fire and slowly being annihilated. A personal reconnaissance by an officer indicated that the company had taken severe casualties and needed help immediately. While artillery assistance was ordered, Smoak sent Company G forward to help the beleaguered Company E, and Company H was ordered to set up 81mm mortars for additional support.

In the meantime, more conflicting reports were received as to the enemy's location and the plight of Company E. As might be expected, several of the messages bordered on panic. Smoak then moved his own command group nearer to the fire fight, and sent Company F forward so that Company E could disengage and withdraw to protect the battalion's right flank. Company G was directed to maintain its position on the left.

In a matter of moments, the combat situation deteriorated from serious to critical. Company F failed to make contact with Company E, the battalion executive officer became a casualty, and a gaping hole widened in the Marines' front lines. Company E, not as badly hurt as had been first reported, was rushed back into the lines and established contact with Company G. There was no sign of Company F. At 1630, with communication to the regimental command post and the artillery battalions knocked out, Smoak ordered his companies to disengage and withdraw from the coconut grove. A defensive line was established several hundred yards from the enemy position.

Shortly after the Marines began to dig in along the trail, a runner from Company F returned to the lines to report that Company F—failing to make contact with Company E—had continued into the Japanese position and had penetrated the enemy lines. The company had taken heavy casualties, was disorganized, and seeking to return to the 2/21 lines. Smoak ordered the runner to guide the company around the right flank of the Marine position into the rear of the lines. The missing company returned, as directed, about 1745. At 1830, communication with the regimental CP and the artillery battalions was restored, and artillery support requested. Concentrations from 2/12 were placed on the north, east, and west sides of the battalion's lines; and the 2d Raider Battalion, now attached to the 21st Marines, was rushed forward to protect the communication and supply lines between 2/21 and the regiment. There were no enemy attacks and only sporadic firing during the night.

The following morning, despite sniper fire, all companies established outposts and sent out patrols in preparation for a coordinated attack with tank support. A scheduled air strike was delayed until the last of these patrols were recalled to positions within the Marine lines. At 0905, the 18 Navy torpedo bombers then on station began bombing the coconut grove and the area between the enemy position and the Marine perimeter. A Marine air-ground liaison team directed the strike. Artillery smoke-shells marked the position for the aviators, who reported that 95 percent of the bombs fell within the target area. Bombs were dropped as near as 100 yards from the forward Marine foxholes.

Unfortunately, the ground attack was delayed until 1100 by the need to get water to the troops, so that the effect of the air strike was lost. A break in communications further delayed the attack, and new plans were made for an attack at 1155. A 20-minute artillery preparation followed by a rolling barrage preceded the assault. At 1155, 2/21 began moving forward, Company E on the left and Company G on the right with Companies F and H in reserve. Five tanks from Company B, 3d Tank Battalion, were spaced on line with the two assault companies.

In a short time, the attack had stalled. The Japanese soldiers had reoccupied their positions; and the enemy fire, plus the noise of the tanks and the rolling barrage, resulted in momentary loss of attack control. The tanks, depending upon the Marine infantry for vision, lost direction and at one point were directing fire at Marines on the flank. One tank was knocked out of commission by an enemy mine, and another was stalled by a hit from a large caliber shell. The battalion commander, seeing the confusion, ordered the attack to cease and the companies to halt in place. This act restored control, and after the three remaining tanks were returned to a reserve position, the attack was continued behind a coordinated front. The enemy positions were overrun, and the defenders killed. Mop-up operations were completed by 1530, and a perimeter around the position was established. Only about 40 dead Japanese were found, although the extent of the defensive position indicated that the enemy strength had been greater. The Marines lost 20 killed (including 5 officers) and 39 wounded in the two days of fighting.

The 2d Battalion emerged from this battle as a combat-wise unit. A series of events, unimportant on the surface, had resulted in serious consequences. The attack on 13 November with companies committed to action successively without prior reconnaissance or adequate knowledge of the situation was not tactically sound. Company E was beyond close supporting distance when attacked, and the conflicting reports on the number of casualties forced the battalion commander to push his remaining strength forward as quickly as possible. These units were engaged prematurely and without plan. The orderly withdrawal on 13th of November, and the prompt cessation of the attack on 14 November when control was nearly lost, was convincing evidence that 2/21 was rapidly gaining combat stability. The last well-coordinated attack was final proof.

In view of the bitter fighting later, the lack of preparatory artillery fires before Company E began its advance on 12 November has been pointed out as a costly omission. Actually, had the presence of the extensive and well-organized Japanese position been determined by prior reconnaissance, the support of this valued arm would have been used. Marine commanders were well aware that infantry attacking prepared defenses would sustain heavy casualties unless the assaults were preceded by an effective combination of the supporting arms—air, artillery, or mortars.

The seizure of the coconut grove area allowed the entire beachhead to leap forward another 1,000 to 1,500 yards. By 15 November, the IMAC perimeter extended to the phase line previously established as Inland Defense Line D.

DEFENSE OF THE CAPE TOROKINA AREA [22]

In the first two weeks of operations on Bougainville, the Marine-Army perimeter had progressed to the point where nothing less than an all-out effort by major Japanese forces could endanger its continued success. From the long and shallow toehold along Empress Augusta Bay on D-Day, the IMAC perimeter gradually crept inland until, on 15 November, it covered an area about 5,000 yards deep with a 7,000-yard base along the beach. Included within this defensive area were the projected sites of a fighter strip at Torokina and fighter and bomber strips near the coconut grove.

The expansion of the beachhead and the arrival of the first echelons of the 37th Division marked a change in the command of the troops ashore. General Turnage had been in command of the 3d Marine Division and all IMAC troops on the beachhead since D-Day; [23] but after the arrival of the Army troops, IMAC once more took up the command of all forces ashore. On 9 November, Vandegrift relinquished command of the Marine amphibious corps to Major General Roy S. Geiger, another Guadalcanal veteran, and returned to the United States. With the arrival of the second echelon of the 37th Division on 13 November, its commander, Major General Robert S. Beightler, assumed command of the Army sector of the perimeter.

The enemy's attempts to bomb the beachhead after D-Day were sporadic and uncoordinated. The fighter cover of ComAirSols, which included Marine Fighter Squadrons -211, -212, -215, and -221, permitted few interlopers to penetrate the tight screen; and the Japanese—after the losses taken in the strikes of 1 and 2 November—could not mount an air attack of sufficient size and numbers to affect the beachhead defenders. The enemy air interference over Cape Torokina was limited to a few night raids, and these were intercepted by Marine planes from VMF(N)-531.

During the first 15 days of the beachhead, there were 52 enemy alerts, 11 bombings, and 2 strafing attacks. The only significant damage was done in a daylight raid of 8 November during the unloading of a follow-up echelon of troops and supplies. More than 100 Japanese fighters and carrier bombers jumped the 28 badly outnumbered AirSols planes, and, during the air melee over the beachhead, the transport *Fuller* was bombed. Five men were killed and 20 wounded. A total of 26 Japanese planes were claimed by the Allied fighters. Eight AirSols planes, including one from VMF-212, were lost.

In the first days of the beachhead, the responsibility for turning back any coordinated sea and air operations by the Japanese rested with the overworked cruiser-destroyer forces of Admiral Merrill and the planes of ComAirSols. Admiral Halsey, weighing the risk of carriers in enemy waters against the need to cripple further the enemy's strength at Rabaul, on 5 No-

---

[22] Unless otherwise noted, the material in this section is derived from: *ComSoPac Nov43 WarD; ThirdFlt NarrRept; IMAC AR-II; 3d MarDiv CombatRept; HistDiv Acct;* Rentz, *Bougainville and the Northern Solomons;* Aurthur and Cohlmia, *3d MarDivHist;* Isley and Crowl, *Marines and Amphibious War;* Morison, *Breaking the Bismarcks Barrier;* Miller, *Reduction of Rabaul.*

[23] *Vandegrift interview.*

vember sent Admiral Frederick C. Sherman on a dawn raid against New Britain with the carriers *Saratoga* and *Princeton*. Despite foul weather, the carrier planes—97 in all—found a hole in the clouds and poured through to strike the enemy fleet at anchor in Simpson Harbor. The planes reported damage to four heavy and two light cruisers and two destroyers.

Six days later, three carriers (*Essex*, *Bunker Hill*, and *Independence*) on temporary loan from Nimitz' Central Pacific fleet struck from the east while Sherman's force hit from the south. The 11 November strike found few targets. The enemy fleet was absent from Rabaul; but the carrier planes knocked 50 Japanese interceptors out of the air and worked over the few ships in the harbor. The two raids ended the Japanese attempts to destroy the Bougainville beachhead by concerted air and sea action.

While the perimeter had been slowly pushed inland, the arrival of additional troops and supplies strengthened the IMAC position. By the time of the arrival of the third echelon on 11 November, beach conditions were more favorable and facilities to allow quick unloading were developed. The third and fourth echelons were unloaded and the ships headed back towards Guadalcanal within the space of a day. During the period 1–13 November, the following troops, equipment, and supplies were delivered to the beachhead:[24]

Date	Echelon	Ships	Troops	Cargo tons
1 Nov	1	8 APA, 4 AKA	14, 321	6, 177
6 Nov	2	8 APD, 8 LST	3, 548	5, 080
8 Nov	2A	4 APA, 2 AKA	5, 715	3, 160
11 Nov	3	8 APD, 8 LST	3, 599	5, 785
13 Nov	4	4 APA, 2 AKA	6, 678	2, 935
Total			33, 861	23, 137

[24] *III PhibFor AR*, pp. 11–12.

Advance to Piva Forks

THE JAPANESE VIEWPOINT [1]

Throughout the first weeks of operations on Bougainville, there was no indication that the Japanese were aware of the true intentions of the I Marine Amphibious Corps and its activities at Cape Torokina. Had the enemy guessed that the Allied purpose was limited only to the construction and defense of several airfields and a naval base in preparation for further operations, the Japanese might have objected more strenuously to the presence of uninvited co-tenants. But the *Seventeenth Army*, hesitating to commit the forces available at Buin before being more certain of Allied plans, held back.

The lack of immediate and continued aggressive action against the IMAC beachhead was a sore point between the Japanese sea command in the Southeast Area and the *Seventeenth Army*, which still chose to take a lighter, more optimistic view of the situation than the Navy. Admiral Kusaka's *Southeast Area Fleet* contended that if the Allies constructed an airfield at Torokina, further Japanese operations on Bougainville would be impractical and sea movements impossible. General Hyakutake, though, argued that the Allies would occupy a base of operations and then at the first opportunity attempt to occupy the Buin sector with the main force while striking the Buka sector with other elements. In such a case, the *Seventeenth Army* explained, it was better to intercept such movements from prepared positions in the Buin and Buka sectors than to abandon these established positions to counterattack at Torokina. [2]

This may have been wishful thinking. Hyakutake was well aware of his own situation—there were no good roads leading into the Allied position over which the *Seventeenth Army* could mount a counteroffensive, and barges were in short supply. Two attempts to wipe out the beachhead had resulted in crushing defeats, and the Navy's ill-timed *Ro* offensive had likewise ended with heavy losses. Reluctantly, the Japanese finally admitted what Allied planners had gambled on some time before—that a decisive counterstroke against the beachhead could not be undertaken for some time.

Despite this estimate, the Allies kept a wary eye on the enemy dispositions in the Bougainville area. Aerial reconnaissance to the north disclosed that the Japanese were constructing extensive defenses in the Buka area to keep their one remaining airfield in operation. The Allies reasoned that if the enemy was committed to a defense of Buka, then he was not likely to

---

[1] Unless otherwise noted, the material in this section is derived from: *ThirdFlt NarrRept; IIIPhibFor Nov43 WarD; IMAC AR–II; IMAC C–2 Repts; IMAC C–2 Jnl; HistDiv Acct; SE Area NavOps—III; Seventeenth Army Ops—II;* Rentz, *Bougainville and the Northern Solomons;* Morison, *Breaking the Bismarks Barrier.*

[2] *SE Area NavOps—III*, pp. 30–31.

691-360 O—63——17

draw troops from there for an offensive in the Empress Augusta Bay area. This removed one threat to the beachhead.

The main danger to the Allied position, however, was from the south where the bulk of the *6th Division* and, therefore, most of the *Seventeenth Army* was located. The Japanese, moving by barge from Buin to Mawareka could strike overland from that point. The meager trail net from Mosigetta and Mawareka was the logical route of approach to Cape Torokina, and reliable intelligence reports indicated that these paths could be traveled by pack animals as well as by troops. This gave the Japanese the added capability of packing artillery into the area to support an attack. The overhanging jungle foliage would screen any movements of troops and make the task of detection more difficult.

A coastwatcher patrol kept the trails to Mosigetta and Mowareka under close surveillance. Daily air searches and photographs were made of the beaches in southern Empress Augusta Bay to detect evidence of enemy landings during the night. In addition, captured enemy letters, diaries, notebooks, and plans were processed and interpreted by intelligence officers for further information. These documents and interrogations of a few prisoners gave a comprehensive order of battle for the immediate area and some approximation of forces. The Japanese apparently had no immediate plans for a counterstroke. The constant and alert protection of the combat air patrol over Bougainville and the expanded and increased activity of Allied ships in southern Bougainville waters undoubtedly played a major role in discouraging the enemy from exercising this capability.

SUPPLY PROBLEMS [3]

A number of changes in the disposition of IMAC units within the perimeter had been made during the widening of the beachhead to the 15 November line. After General Geiger took command of the Marine amphibious corps, all units temporarily attached to the 3d Marine Division for the landing reverted to IMAC control once more, and defensive installations within the beachhead were improved and strengthened.

The Marine 3d Defense Battalion, supported by long-range radar installations, continued to provide antiaircraft and seacoast artillery protection for the beachhead and offshore islands. All field artillery units—both Marine Corps and Army—were placed under central command as an IMAC artillery group to be available as massed fires for interdiction, neutralization, counterbattery, beach defense, or attack support. Brigadier General Leo M. Kreber of the 37th Division was designated commander of the artillery group. A corps reserve was established by withdrawing most of Lieutenant Colonel Shapley's 2d Raider Regiment from the front lines. This reserve was then held in readiness for counterattacks in any sector of the perimeter or for quick reinforcement of the front line defenses.

Following the battle of the Coconut Grove, contact with the main forces of the enemy was lost once more and the period was one of relative inactivity by

[3] Unless otherwise noted, the material in this section is derived from: *IMAC AR–II; 3d MarDiv CombatRept; 3d MarDiv D–3 Jnl; 3d MarDiv D–3 Repts; 3d MarDiv D–4 Repts; 3d MarDiv ServTrps Rept;* Rentz, *Bougainville and the Northern Solomons;* Aurthur and Cohlmia, *3d MarDivHist.*

the Japanese. Perimeter units of the 3d Marine Division and the 37th Division continued active combat patrolling, but there were few enemy contacts. The Japanese had apparently withdrawn. The activity by the IMAC forces was mainly to fix the location of enemy troops and to obtain information on the terrain ahead in preparation for the continuing expansion of the beachhead. After 15 November, these offensive moves were made to improve the defensive positions of the perimeter, and the attack objectives were usually just lines drawn on a map a certain distance from an established position. These moves to new phase lines were more in the nature of an active defense.

The 37th Infantry Division, during this period, found the expansion of the perimeter in its sector much less difficult than the Marines did in their sector. There was little enemy activity in front of the 148th and 129th Infantry Regiments after the Koromokina engagement, and the Army units received only glancing blows from scattered Japanese groups. Once the beachhead was carried past the outer limits of the swampy plains toward higher ground, the infantry regiments were on fairly firm terrain and could move without too much trouble. This sector of the beachhead also took on added strength as more Army support units continued to arrive with later echelons of shipping. After the movement to Inland Defense Line D, General Geiger allowed General Beightler to expand the 37th Division sector of the beachhead, coordinating with the Marine efforts only at the central limiting point on the boundary line between divisions. The lack of aggressive enemy action in front of the two Army regiments permitted the perimeter in this sector to advance more rapidly. This situation, in regard to enemy opposition, continued throughout the campaign until March 1944.

The Marine half of the perimeter at this time, in contrast to the area held by the 37th Division, was still marked by lagoons and swamplands. In most places, the front lines could be reached only by wading through water and slimy mud which was usually knee deep, was often waist deep, and sometimes was up to the arm pits. The defensive perimeter in the Marine sector actually consisted of a number of isolated positions, small islands of men located in what was known locally as "dry swamp"—meaning that it was only shoe-top deep.[4] The frequent downpours discouraged attempts to dig foxholes or gun emplacements. Machine guns were lashed to trees, and Marines huddled in the water. In this sultry heat and jungle slime, travel along the line was extremely difficult, and resupply of the frontline units was a constant problem.[5]

Improvement of the supply lines to the perimeter positions was the greatest concern of IMAC at this time. The seemingly bottomless swamps through which supply roads had to be constructed were a dilemma whose early solution appeared at times to be beyond the capabilities of the available road-building equipment and material. The move to Line D took in the site of the projected bomber and fighter strips near Piva, and although the

[4] *Bowser ltr.*

[5] In general, the former commander of the 9th Marines feels that the terrain situation was less of a problem than it is usually found described in contemporary accounts. He recalls, "I never found it too difficult to get around to my various units on foot each day." He considers the area around Hill 1000 to have furnished the hardest travelling and remembers that many Marines got lost in the deep ravines and heavy trackless underbrush that abounded there. *Craig ltr.*

FIELD TELEPHONE LINES, *the primary means of communication in the jungle, are laid by armed Marine wiremen on Bougainville.* (*USMC 67228B*)

bomber field was already surveyed, construction was held up by the lack of access roads to the area. The diversion of equipment and resources to the construction of roads and supply trails instead of airfields handicapped the work which had begun on the fighter strip at Torokina and delayed the start of the Piva bomber strip, but the problem of supply was too pressing to be ignored.

By 16 November, the lateral road across the front of the perimeter was completed after two weeks of feverish activity. During the time of the Piva Trail and Coconut Grove engagements, the 3d Battalion of the 3d Marines had pushed the construction of this supply road as fast as the limits of men and machines would permit. The speed was dictated by the need to keep pace with the assault battalions which were seeking the main enemy positions before the Japanese could consolidate forces and prepare an established defense in depth. Engineers moved along with the 3d Battalion as the Marines moved inland. On more than one occasion, bulldozer operators had to quit the machines and take cover while Marine patrols skirmished with enemy groups in a dispute over the right of way.

The end product was a rough but passable one-lane roadway which followed the path of least resistance, skirting along the edge of the swampy area. The road began near the Koromokina beaches, then wound inland for several thousand yards before cutting to the southeast toward the coconut grove and the Piva River. Small streams were bridged with hand-hewn timbers, and muddy areas were corduroyed with the trunks of fallen trees. In many instances, trucks were used to help batter down brush and small trees, with resultant damage to vitally needed motor transport. Dispersal areas were limited, and there was much needed work to be done on access, turn-around, and loop roads. But this rutted and muddy roadway joined the two sectors of the beachhead to the dumps along the shoreline and greatly aided the supply and evacuation problems of the frontline battalions.

As the lateral road was cut in front of 2/3, this battalion advanced about 1,000 yards inland to protect the roadway and to cover the widening gap created between the two divisions by the continual progress of 3/3 toward the Piva River. The road construction force and 3/3 broke out of the jungle at the junction of the Numa Numa and Piva trails on 16 November, having connected the lateral road with the amphibian tractor trails from the Cape Torokina area. Although rains sometimes washed out the crude trailway and mired trucks often stalled an entire supply operation, the roadway was assurance that the IMAC forces could now make another offensive-defensive advance, confident that the essential supplies would reach the front lines.

The critical supply situation had been corrected by an abrupt revision of the original plans. The rapidly changing tactical circumstances and the redisposition of combat elements along the beachhead left the beaches cluttered with all classes of supplies and equipment. After some semblance of order had been restored, it was apparent that the landing teams could not handle and transport their own supplies as had been planned. The battalions, striking swiftly at the Japanese, moved inland with what they could carry. Within a short time, most of the units were miles from their original shore party dumps. These were practically abandoned and became a source of supply on a first-come, first-served basis to all units of the I Marine Amphibious Corps. Rations and ammunition were picked up by most units at the first available source.

The first corrective action by the Marine division's G-4 and the division quartermaster was to direct that all shore party dumps revert to division control. A new plan was outlined under which the division quartermaster assumed responsibility for control and issue of all supplies in the dumps and on Puruata Island.[6] A division dump or distribution point was established adjacent to the plantation area on Cape Torokina. All supplies littering the beach were recovered and returned to this area. Succeeding echelons of supplies and equipment arriving at the beachhead were also placed in this dump for issue by the division quartermaster.

Before the completion of the lateral road and control of supplies by the division quartermaster, the battalions holding the perimeter were supplied on a haphazard schedule by the versatile amphibian tractors. When the new program was effected, supplies were virtually leapfrogged forward in a relay system that involved handling of the same stocks as many as four times. This system, however, provided for an equitable distribution of ammunition and rations to all units. From the division dump at the beach, supplies were carried to regimental dumps, which in turn issued to the battalions. Trucks carried the supplies as far forward as possible, then amphibian tractors took over. As the battalions advanced, forward supply points were set up. An attempt was made to build up an emergency supply level at each of these forward points. The front lines, however, moved ahead so steadily that usually an untracked jungle stretched between the troops and their supply dumps. The LVTs, when possible, skipped these forward points to continue as close to the front lines as they could manage.

A total of 29 of these LVTs had been landed with the assault waves on D-Day and more arrived in later echelons. Their contribution to the success of the beachhead, however, was in far greater proportion than their number. Without the 3d Amphibian Tractor Battalion (Major Sylvester L. Stephen), the operation as planned could not have been carried beyond the initial beachhead stages; and it was the work of the LVT companies and the skill of the amtrac operators that made possible the rapid advance of the IMAC forces during the first two weeks. The tractors broke trails through the swamps and marshes, ploughing along with vital loads of rations, water, ammunition, weapons, medical supplies, engineer equipment, and construction materials. Even towing the big Athey trailers, the LVTs were able to move over muddy trails which defeated all wheeled vehicles; and, in fact, the broad treads of the trailers sometimes rolled out and restored rutted sections of roadways so that jeeps could follow.

As might be expected, the maintenance of these machines under such conditions of operation became a problem. Many amtracs were in use continually with virtually no repairs or new parts. As a result, numerous tractors were sidelined be-

---

[6] When the 37th Division took over its own sector of the IMAC perimeter, corps took charge of the dumps on Puruata and handled supply distribution to both frontline divisions. In getting needed supplies up to forward elements, the Army unit used essentially the same system of delivery as that described for the 3d Marine Division. *Beightler ltr.*

cause of excessive wear on channels and tracks caused by the constant operation through jungle mud. The largest number of machines available at one time was 64, but the number of tractors still in service declined rapidly after the first two weeks. Ironically, by the time that a major battle between the Marine forces and the Japanese appeared likely (24 November), the number of amtracs available for use was 29—the same number that was available on D-Day.

COMBAT LESSONS [7]

Throughout this period, the individual Marine (and his Army counterpart in the 37th Division) learned how to battle both the Japanese and the jungle. For two weeks the Marines had struggled through swamps of varying depth, matching training and skill against a tenacious and fanatic enemy. This fight for survival against enemy and hardship in the midst of a sodden, almost impenetrable jungle had molded a battlewise and resourceful soldier, one who faced the threat of death with the same fortitude with which he regarded the endless swamps and forest and the continual rain. Danger was constant, and there were few comforts even in reserve bivouac positions.

The combat Marine lived out of his marching pack with only a few necessities—socks, underwear, and shaving gear—and a veritable drug store of jungle aids such as atabrine tablets, sulpha powders, aspirin, salt tablets, iodine (for water purification as well as jungle cuts and scratches), vitamin pills, and insect repellent. Dry clothes were a luxury seldom experienced and then only when gratuitous issues of dungarees, underwear, and shoes were made. Knapsacks and blanket rolls seldom caught up with the advancing Marines, and most bivouacs were made in muddy foxholes without the aid of covering except the poncho—which served a variety of uses.

Troops received few hot meals, since food could not be carried from kitchens through the swamps and jungle to the perimeter positions. Besides, there were no facilities for heating hot water for washing mess kits if hot food could have been brought forward. Troops generally ate dry rations, augmented by canned fruit and fruit juices, and waited for cooked food until they were in reserve positions. When the combat situation and the bivouac areas in the swamps permitted, Marines sometimes combined talents and rations and prepared community stews of C-rations, bouillon powder, and tomato juice which was heated in a helmet hung over a fire. Only the canned meat or cheese, the candy bar, and the cigarettes were taken from the K-rations; the hardtack biscuits found little favor and were usually thrown away. After the beachhead became more fully established, bread was supplied by regimental bakery units and delivered to the front lines. The bread was baked daily in the form of handy rolls, instead of large loaves, which helped solve the problem of distribution to Marines in scattered positions.

Heat tabs met with varied reaction until the Marines found that at least two tabs were required to boil a canteen cup of water. Experience also taught that C-rations could be cooked twice as fast over one heat tab if the ration was divided in half. The first half-can could be heated,

[7] Unless otherwise noted, the material in this section is derived from: *3d MarDiv CombatRept*; Rentz, *Bougainville and the Northern Solomons*; Aurthur and Cohlmia, *3d MarDivHist*.

then eaten while the second half-can was heating.

During the Marine advance, there was little water brought forward, and most of the drinking water was obtained from swamp holes and streams. This was purified individually by iodine or other chemicals supplied by the Navy Corpsman with each platoon. Despite this crude sanitation and continued exposure to jungle maladies, there were few cases of dysentery. The 3d Marine Division, as a whole, maintained a healthy state of combat efficiency and high morale throughout the entire campaign.

The Marines, after defeating the Japanese in three engagements, were becoming increasingly skilled jungle fighters, taking cover quickly and quietly when attacked and using supporting weapons with full effectiveness. Targets were marked by tracer bullets, and the Marines learned that machine guns could be used to spray the branches of trees ahead during an advance. This practice knocked out many enemy snipers who had climbed trees to scout the Marine attack. Although visibility was usually restricted by the close jungle foliage, the Marines learned to take advantage of this dense underbrush to adjust supporting fires almost on top of their own positions. This close adjustment discouraged the Japanese from moving toward the Marine lines to seek cover during a mortar or artillery barrage.

In the jungle, 60mm mortars could be registered within 25 yards of the Marine positions, 81mm mortars and 75mm pack howitzers within 50 yards, and 105mm howitzers within 150 yards. The latter shell was particularly effective in jungle work, as were the canister shells used in direct fire by the 37mm antitank guns. Both stripped foliage from hidden enemy positions, exposing the emplacements to a coordinated attack. Although the 60mm and 81mm mortars were virtually ineffective against emplacements with overhead cover, both shells were valuable in stopping attacks by troops in the open and in keeping the Japanese pinned to an area being hit by artillery.

Artillery was usually adjusted by sound ranging. The artillery forward observer, estimating his position on the map by inspection, requested one round at an obvious greater range and then adjusted the fire by sound into the target. The location of the target was then determined by replot, and the observer was able then to locate his position as well as the front lines.

Mortar fire was restricted in many cases by the overhanging jungle. Because most fighting was conducted at extremely close range, the mortar rounds in support were fired almost vertically with no increments. When there was any doubt about foliage masking the trajectory, a shell without the arming pin removed was fired. If the unarmed shell cleared, live rounds followed immediately.

Movement through the jungle toward Japanese positions was usually made in a formation which the 3d Marine Division called "contact imminent." This formation, which insured a steady, controlled advance, had many variations, but the main idea was a column of units with flank guards covering the widest front possible under conditions at the time. Trails were avoided. A security patrol led the formation; and as the column moved, telephone wire was unrolled at the head of the formation and reeled in at the rear. At the instant of stopping, or contact with the enemy, company commanders and supporting weapons groups clipped hand telephones onto the line and were in immediate contact with the column commander. Direction and speed of the advance was controlled by the officer at the head of the main body of troops. A command using this formation could expect to make about 500 yards an hour through most swamps. Such a column was able to fend off small attacks without delaying forward movement, yet was flexible enough to permit rapid deployment for combat to flanks, front, or rear. This formation was usually employed in most advances extending the defensive sectors of the perimeter.

Holding the Marine front lines at this time were the 3d Marines on the left and the 9th Marines on the right. Although Colonel McHenry's 3d Marines had responsibility for the left subsector, only one battalion, the 3d, was occupying perimeter positions. The 1st Battalion was in reserve behind 3/3, and the remaining battalion, 2/3, was attached temporarily to the 129th Infantry in the Army sector. During this time, however, two battalions of the 21st Marines were attached to Colonel McHenry's command for patrol operations. Elements of 2/21 took part in numerous scouting actions along the East-West trail past Piva Village to develop the enemy situation in that area; 1/21 moved into reserve bivouac positions behind the 3d Marines.

On the 17th of November, the convoy bearing 3/21 (Lieutenant Colonel Archie V. Gerard) was attacked by Japanese aircraft off Empress Augusta Bay, and the APD *McLean* was hit and sunk. At least 38 Marines from 3/21 were lost at sea. Two days later, as 3/21 prepared to join the remainder of the regiment near the front lines, the battalion's bivouac position near the beach was bombed by the Japanese and another five Marines were killed and six wounded. Gerard's battalion joined the 3d Marines for operations the same day. Without having been in action against the enemy, 3/21 had already lost as many men as most frontline battalions.

The 9th Marines, at this time, occupied positions generally along the west bank of the Piva River. Amtracs were the only vehicles which could negotiate the swamp trails from the beaches, and the supply situation in this sector was critical. Most of the 9th Marines' units were forced to take working parties off the front lines to hand-carry supplies forward and to break supply trails into the regiment's position. Evacuation of wounded was also by hand-carry. The period after the movement to Phase Line D was spent improving the defensive position, seeking enemy activity, and gathering trail information. A number of patrols moved across the Piva River looking for enemy action, but there were few contacts in the several days following the final Coconut Grove action.

THE BATTLE OF PIVA FORKS[8]

Combat activity in the Marine sector picked up again on the 17th and 18th of November after all units had devoted several days to organization of the defensive perimeter and extension and improvement

[8] Unless otherwise noted, the material in this section is derived from: *IMAC AR–II*; *IMAC C–2 Repts*; *IMAC C–2 Jnl*; *3d MarDiv Combat-Rept*; *3d MarDiv D–2 SAR*; *3d MarDiv D–2 Jnl*; *3d MarDiv D–3 Jnl*; *3d MarDiv D–3 Repts*; *3d MarDiv D–4 Repts*; *Snedeker ltr*; *Bowser ltr*; *McAlister ltr*; BGen John S. Letcher ltr to CMC, dtd 1Jun48 (Bougainville Monograph Comment File, HistBr, HQMC), hereafter *Letcher ltr*; LtCol Jack Tabor ltr to CMC, dtd 7Jun48 (Bou-

of supply lines. The 37th Division sector remained relatively inactive, with few reports of enemy sighted. Marine units started aggressive patrolling in search of routes of advance and terrain information as far out ahead as the next phase line to be occupied by the 21st of November (Inland Defense Line E). There were minor skirmishes with enemy outposts as the Marines scouted the jungle, but the flareups were brief and there were few casualties to either side. In the 9th Marines subsector, both 1/9 and 2/9 reported that enemy activity had increased, and the 3d Marines reported that all units along the line had been in contact with small parties of Japanese. A patrol from 3/3 successfully ambushed a Japanese group, killing eight enemy soldiers and one officer who had in his possession a sketch of Japanese dispositions to the immediate front. The drawing, and other captured documents, indicated that the enemy was preparing extensive defenses along both the Numa Numa and the East-West trail.

Another patrol from 3/3, moving down the Numa Numa trail on 18 November, discovered an enemy road block about 1,000 yards to the front. A patrol from 1/21, probing along the East-West trail, encountered a similar enemy position about halfway between the two branches of the Piva River. This was further evidence of Japanese intentions for a determined defense of this area, and plans were made for an immediate attack. The 3d Raider Battalion was attached to the 3d Marines to release 3/3 (Lieutenant Colonel Ralph M. King) for the reduction of the Numa Numa trail position the following day.

King's battalion, accompanied by light tanks, cut through the jungle to the left in front of the 129th Infantry subsector. After an artillery preparation, the battalion struck the enemy position in a flanking attack that completely routed the Japanese. A total of 16 dead enemy were found, although more than 100 foxholes indicated that at least a reinforced company had occupied the position. King's battalion immediately took possession of the trail block and established a perimeter defense at the junction of the Numa Numa trail and the Piva River. Meanwhile, 1/3 and 1/21 had advanced without difficulty, opposed only by a few bypassed survivors from King's attack. The 3d Raiders then moved forward to be available for support, and 2/3—released from operational control by the 129th Infantry—also started east behind the Numa Numa trail toward an assembly area. The march was made under fire; the Japanese sporadically shelled the advancing battalion with 90mm mortars.

The following morning, 20 November, the same Japanese company that had been forced to withdraw the previous day came bouncing back, full of fight. The enemy attempted to outflank the Marine positions along the trail, but King's battalion drove the enemy back again. The Japanese then undertook to harass the Marines by sniper fire and mortar concentrations, and the resistance grew more determined when King's force started a counterattack.

---

gainville Monograph Comment File, HistBr, HQMC), hereafter *Tabor ltr;* Capt Richard C. Peck ltr to HistDiv, HQMC, dtd 3Jun48 (Bougainville Monograph Comment File, HistBr, HQMC), hereafter *Peck ltr; HistDiv Acct;* Rentz, *Bougainville and the Northern Solomons;* Aurthur and Cohlmia, *3d MarDivHist;* Maj Harry W. Edwards, "Cibik Ridge—Prelude to Victory," *Marine Corps Gazette*, v. 35, no. 3 (Mar51); Maj Donald M. Schmuck, "The Battle of Piva Forks," *Marine Corps Gazette*, v. 28, no. 6 (Jun44).

Two of the light Marine tanks were disabled in the close fighting along the trail before the Marine battalion could advance. The general course of attack by 3/3 was east along the Numa Numa trail toward the two forks of the Piva River.

A number of changes in the front line dispositions were ordered as 3/3 advanced. The 3d Raider Battalion moved out of reserve positions to cover the slowly widening gap between the 129th Infantry and the 3d Marines. At the same time, Lieutenant Colonel Hector de Zayas' 2/3 on the right of 3/3 passed through the front lines of 2/21 to advance across the west fork of the Piva River. The objective of 2/3 was the enemy position reported earlier between the two forks of the Piva River. The Piva crossing was made over a hasty bridge of mahogany timbers thrown across the stream by engineers. The enemy outpost was then discovered abandoned but clumsily booby-trapped. The only opposition to the attack by 2/3 was scattered snipers and several machine gun nests. By late afternoon of the 20th, de Zayas' battalion was firmly astride the East-West trail between the two forks of the Piva River. Elements of the 21st Marines, now in reserve positions behind the two battalions of the 3d Marines, moved forward to take up blocking positions behind Colonel McHenry's regiment.

As the Marine forces prepared to continue the attack, the opportune discovery of a small forward ridge was a stroke of good fortune that ultimately assured the success of the Marine advance past the Piva River. This small terrain feature, which was later named Cibik Ridge in honor of the platoon leader whose patrol held the ridge against repeated Japanese assaults, was reported late on the afternoon of the 20th. The area fronting the 3d Marine positions had been scouted earlier, but this jungle-shrouded elevation had escaped detection. Although the height of this ridge was only 400 feet or so, the retention of this position had important aspects, since it was the first high ground discovered near the Marine front lines and eventually provided the first ground observation posts for artillery during the Bougainville campaign. There is no doubt that the enemy's desperate attempts to regain this ground were due to the fact that the ridge permitted observation of the entire Empress Augusta Bay area and dominated the East-West trail and the Piva Forks area.

All this, however, was unknown when First Lieutenant Steve J. Cibik was directed to occupy this newly discovered ridge. His platoon, quickly augmented by communicators and a section of heavy machine guns, began the struggle up the steep ridge late in the afternoon of the 20th. Telephone wire was reeled out as the platoon climbed. Just before sunset, the Marines reached the crest for the first look at the terrain in 20 days of fighting. Daylight was waning and the Marines did not waste time in sightseeing. The remaining light was used to establish a hasty defense, with machine guns sited along the likely avenues of approach. Then the Marines spent a wary night listening for sounds of enemy.

The next morning Cibik's men discovered that the crest of the ridge was actually a Japanese outpost position, used during the day as an observation post and abandoned at night. This was confirmed when Japanese soldiers straggled up the opposite slope of the ridge shortly after daybreak. The enemy, surprised by the unexpected blaze of fire from their own outpost, turned and fled down the hill.

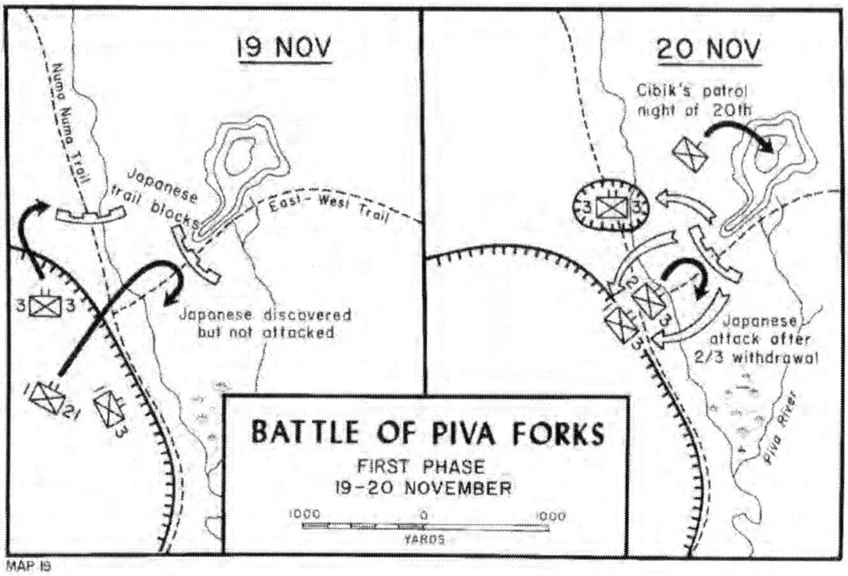

MAP 19

After that opening move, however, the enemy attacks were organized and in considerable strength. Cibik's platoon, hastily reinforced by more machine guns and mortars, held the crest despite fanatical attempts by the Japanese to reoccupy the position. The Marines, grimly hanging to their perch above the enemy positions, hurled back three attacks during the day.

The expansion of the beachhead to Inland Defense Line E jumped off at 0730 on the morning of 21 November. The general plan called for a gradual widening of the perimeter to allow the 21st Marines to wedge a defensive sector between the 3d and 9th Marines. This action would then put all three Marine infantry regiments on the front lines. Colonel Ames' 21st Marines passed through the junction of the 3d and 9th Marines and crossed the Piva River without difficulty. By early afternoon, the two assault battalions (1/21 and 3/21) had reached the designated line, and the attack was held up to await further orders. The approach march had been made without enemy interference, except on the extreme left flank where a reinforced platoon, acting as the contact between the 21st Marines and the 3d Marines, was hit by a strong Japanese patrol. The Marine platoon managed to repulse this attack with heavy losses to the enemy. Important documents, outlining the Japanese defenses ahead, were obtained from the body of a dead Japanese officer.

By 1425, the 21st Marines had established a new defensive sector, and contact between 3/21 and the 9th Marines had been established. There was, however, no contact between 1/21 and 3/21 along the front lines. The remaining battalion, 2/21, was then released from operational control by the 3d Marines, and this unit moved into reserve positions behind 3/21 and 1/21 to block the gap between the battalions.

The enemy resistance in the 3d Marines' sector, however, was unexpectedly strong. All three battalions were engaged with the Japanese during the course of the advance. The left battalion, 3/3, crossed the Piva River without trouble and advanced toward a slight rise. As the 3/3 scouts came over the top of this ridge, the Japanese opened fire from reverse slope positions. The scouts were pinned down by this sudden outburst, but after the rest of the battalion moved forward a strong charge over the ridge cleared the area of all Japanese. Before the battalion could consolidate the position, though, enemy 90mm mortars registered on the slope, and the Marines were forced to seek shelter in the 200 or more foxholes which dotted the area. These enemy emplacements and the steep slope prevented many casualties. The 3d Battalion decided to halt in this position and a defensive perimeter was set up for the night.

The 2d Battalion, making a reconnaissance in force in front of the 1/3 positions, bumped into a strong enemy position astride the East-West trail near the east fork of the Piva. About 18 to 20 pillboxes were counted, each of them spitting rifle and machine gun fire. De Zayas' battalion managed to crack the first line of bunkers after some fighting at close range, but could make no further headway. Company E, attempting to flank the enemy positions to relieve the intense fire directed at Company G, was knocked back by the Japanese defenders. Aware now that the enemy was organized in considerable depth, the battalion commander ordered a withdrawal to allow artillery to soften up the enemy positions.

The retrograde movement was difficult since there were many wounded Marines and the terrain was rugged, but the withdrawal was managed despite the determined efforts by the Japanese to prevent such a disengagement. After de Zayas' battalion had reentered the lines of 1/3, the Japanese attempted a double envelopment of the position held by the 1st Battalion (now commanded by Major Charles J. Bailey, Jr.). This was a mistake. The enemy followed the obvious routes of approach down the East-West Trail, and his effort perished in front of the machine guns sited along this route by 1/3. Bailey's battalion then extended to the left toward Cibik Ridge.

The 9th Marines, meanwhile, had crossed the Piva River in the right sector and were now occupying a new line of defense about 1,000 yards east of the river. The new positions extended from the beach to the 21st Marines in the center sector. The 129th Infantry, completing the general advance of the Marine-Army perimeter to Inland Defense Line E, also moved forward another 1,000 yards. The 37th Division unit was also unopposed.

Action along the entire beachhead dwindled on the 22d of November. The 21st Marines bridged the gap between the two front line battalions by shifting 3/21 about 400 yards to the right to make contact with 1/21. A considerable gap still existed between the 21st Marines and the 3d Marines. This break in the defensive lines was caused by the fact that the frontage of the Marine positions was greater than anticipated because of map inaccuracies.

The expansion of the perimeter was halted on these lines while a concerted at-

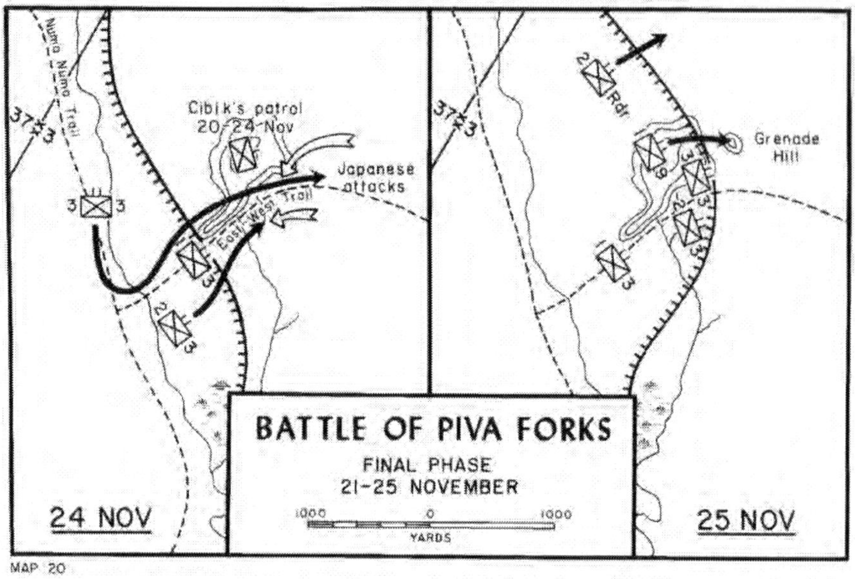

MAP 20

tack was planned to push the Japanese out of the strongly entrenched positions ahead. The enemy fortifications, which faced nearly south because of the twists of the trail, would be assaulted from the west to east in a flanking attack. To insure a coordinated advance, the attack was set for 24 November, with the East-West trail as the boundary between the assaulting regiments.

It was now apparent that the main Japanese dispositions had been reached, and intensive preparations for the full-scale assault on the enemy forces were rushed. All available tanks and supporting weapons were moved forward into positions behind the 3d Marines as fast as the inadequate trail net would permit. Engineers and Seabees worked to extend the road as close to the Piva River forks as possible, erecting hasty bridges across the Piva River despite intense sniper fire and harassment by enemy mortars. Supply sections moved huge quantities of ammunition, rations, and medical supplies forward in a relay system that began with trucks and amtracs and ended with Marines hand-carrying the supplies to the front lines. A medical station was established near the terminus of the road to facilitate evacuation of the wounded. All signs indicated that the 3d Marines, scheduled to advance on the 24th, would be meeting a strong enemy force.

By the evening of the 22d, several changes had been made in the sector of the 3d Marines. The 2d Raider Battalion, now attached to the 3d Marines, was ordered to relieve King's battalion on the small hill which had been taken the day before, and 3/3 then moved to a reserve bivouac area behind 1/3 and nearly abreast of 2/3. The dispositions of the 3d Marines at this time resembled a triangle with the apex pointed along the East-West trail toward the Japanese positions. The 1st Battalion was in front, with 3/3 on the left of the trail and 2/3 on the right. Cibik's force, holding a position in front of the perimeter, was reinforced with a company of raiders and a platoon from the 3d Marines Weapons Company. By this time, the observation post was defended by more than 200 Marines and bristled with supporting weapons. The Japanese, to reclaim this position, would have to pay a terrible price.

On 23 November, artillery observers moved to the crest of Cibik Ridge to adjust fires in preparation for the attack the next day. The Marines holding the front lines marked their positions with colored smoke grenades, and both artillery and mortars were then registered in the area ahead. The sighting rounds caused some confusion when several explosions occurred within the Marine positions. It was then realized that the Japanese were firing in return and using the same smoke signals for registration on the Marine lines.

Shells from long-range enemy guns were also falling on Torokina strip and an echelon of LSTs unloading near the cape. The observers on Cibik Ridge shifted registration fires toward several likely artillery positions and the enemy fire ceased. The news that the enemy had artillery support for the defense of his positions was disturbing, though. Scouts had estimated that the enemy force, located in the area around the village of Kogubikopai-ai, numbered about 1,200 to 1,500. The addition of artillery support would make the job of reducing this strong position even more difficult.

The attack order for 24 November directed the two battalions, 2/3 and 3/3, to advance abreast along the East-West trail and attack for about 800 yards beyond the east fork of the Piva River. Seven battalions of artillery—four Marine and three Army—would provide support for the attack after an opening concentration of 20 minutes fire on an area about 800 yards square. During the day of 23 November, while the artillery group registered on all probable enemy positions, Bailey's 1/3 moved every available weapon, including captured Japanese guns, into the front lines. By nightfall, 1/3 had emplaced 44 machine guns across the trail and had registered the concentrations of a dozen 81mm mortars and 9 60mm mortars along the zone of action of the attacking battalions.

Early the next morning, the two battalions began moving out of bivouac and up the trail toward the front lines held by 1/3. It was Thanksgiving Day back in the States; but on Bougainville this was just another day of possible death, another day of attack against a hidden, determined enemy. Few Marines gave the holiday any thought—the preparations for this advance during the last two days had built up too much tension for anything but the job ahead. Behind the Marines, in the early dawn mist, trucks and amtracs churned along muddy trails, bringing forward last loads of rations, ammunition, and medical supplies. Tanks, assigned to a secondary role in this attack, clanked toward the front lines to move into support positions.

At 0835, just 25 minutes before the attack hour, the seven battalions of the artillery group opened fire on the Japanese positions in front of the 3d Marines. From the opening salvo, the roar of the cannon fire and the sharp blasts of the explosions in the jungle ahead merged into a near-deafening thunder. In the next 20 minutes or so, more than 5,600 shells from 75mm and 105mm howitzers hammered into the Japanese positions. The enemy area was jarred and shattered by more than 60 tons of explosives in that short time. At the same time, smoke shells hitting along the hills east of the Torokina River cut down enemy observation into the Marine positions.

As H-Hour approached, Bailey's battalion, from the base of fire position astride the East-West trail, opened up with close-in mortar concentrations and sustained machine gun fire which shredded the jungle ahead, preventing the Japanese from seeking protection next to the Marine lines. But just before the attack was to jump off, Japanese artillery began a counterbarrage which blasted the Marine lines, pounding the 1/3 positions and the assembly areas of the assault battalions. The extremely accurate fire threatened to force a halt to the attack plans. At this point, the value of Cibik Ridge was brought into full prominence. The forward observer team on the ridge discovered the location of a Japanese firing battery and requested counterbattery fire. There were several anxious moments when communications abruptly failed, but the line break was found and repaired in time.

The enemy guns were located on the forward slope of a small coconut grove area some several thousand yards from the Piva River. As the two Marine officers watched, the return fire from the 155mm howitzer battery of the 37th Division began to explode near this grove. Fire was adjusted quickly by direct observation, and in a matter of moments the enemy battery had been knocked out of action.[9]

Even as the last Japanese artillery shells were exploding along the Marine lines, the two assault battalions began forming into attack formation behind the line of departure. At 0900, as the continuous hammering of machine guns, mortars, and artillery slowly dwindled to a stop, the two battalions moved through the 1/3's lines and advanced.

After the continuous roar of firing and explosions for more than 20 minutes, a strange stillness took over. The only sounds were those of Marines moving through the jungle. The neutralization of the enemy positions within the beaten zone of the artillery preparation had been complete. The first few hundred yards of the enemy positions were carried without difficulty in the incredible stillness, the Marines picking their way cautiously through the shattered and cratered jungle. Blasted and torn bodies of dead Japanese gave mute evidence of the impact of massed artillery fires. Enemy snipers, lashed into positions in tree tops, draped from shattered branches.

This lull in the battle noise was only temporary. Gradually, as the stunned survivors of the concentrated bombard-

[9] Although this account of the silencing of the enemy battery agrees with contemporary records, the former executive officer of the 12th Marines questioned its accuracy, recalling: "The [line] break was not found or repaired for more than one hour and the Japanese battery firing from a position in full view of the observer on the nose of Cibik Hill was able to withdraw behind the crest of what was later to be known as Hill 1000 before counterbattery fire could be brought to bear on the position." BGen John S. Letcher ltr to Head, HistBr, G-3, HQMC, dtd 6Oct60.

ment began to fight back, the enemy resistance swelled from a few scattered rifle shots into a furious, fanatical defense. Japanese troops from reserve bivouac areas outside the beaten zone were rushed into position and opened fire on the advancing Marines. Enemy artillery bursts blasted along the line, traversing the front of the advancing Marines. Extremely accurate 90mm mortar fire rocked the attacking companies.

Enemy fire was particularly heavy in the zone of 2/3 on the right of the East-West trail, and de Zaya's battalion, after moving about 250 yards, reported 70 casualties. A small stream near the trail meandered back and forth across the zone of advance, and the attacking Marines were forced to cross the stream eight times during the morning's movement. At least three enemy pillboxes were located in triangular formation in each bend of the stream, and each of these emplacements had to be isolated and destroyed. A number of engineers equipped with flame throwers moved along with the assault companies, and these weapons were used effectively on most bunkers. The Japanese, fully aware of the death-dealing capabilities of the flame throwers, directed most of their fire toward these weapons. Many engineers were killed trying to get close enough to enemy emplacements to direct the flame into the embrasures.

The attack by King's battalion (3/3) on the left of the East-West trail encountered less resistance, and the battalion was able to continue its advance without pause. Many dazed and shocked survivors of the bombardment were killed by the attacking Marines before the Japanese could recover from the effects of the artillery fire. But by the time the battalion had moved nearly 500 yards from the line of departure, the enemy forces had managed to reorganize and launch a desperate counterattack which King's men met in full stride. Without stopping, 3/3 drove straight through the enemy flanking attempt in a violent hand-to-hand and tree-to-tree struggle that completely destroyed the Japanese force.

By 1200, after three full hours of furious fighting, the two battalions had reached the initial objectives, and the attack was held up for a brief time for reorganization and to reestablish contact between units. Following a short rest, the Marines started forward again toward the final objective some 350–400 yards farther on. Meanwhile, artillery again pounded ahead of the Marine forces and mortars were moved forward. The final attack was supported by 81mm mortars; but as the advance began again, enemy counter-mortar fire rained on the Marines. The attack was continued under this exchange of supporting and defensive fires.

King's battalion was hit hard once more, but managed to keep going. Enemy machine gun and rifle fire from positions on high ground bordering a swampy area raked through the attacking Marines, forcing them to seek cover in the knee-deep mud and slime. Company L, on the extreme left, was hit hardest. Reinforced quickly with a platoon from the reserve unit, Company K, the company managed to fight its way through heavy enemy fire to the foot of a small knoll. Company I, with the battalion command group attached, came up to help. Together, the two companies staged a final rush and captured the rising ground. After clearing this small elevation of all enemy, the bat-

691-360 O—63——18

talion organized a perimeter defense and waited for 2/3 to come up alongside.

The 2d Battalion, moving toward the final objective, was slowed by strong enemy reinforcements as it neared its goal. Quickly requesting 60mm and 81mm mortar concentrations, the 2d Battalion overcame the enemy opposition and lunged forward. The final stand by the Japanese on the objective was desperate and determined, but as the Marines struggled ahead, the resistance dwindled and died. The 2d Battalion then mopped up and went into a perimeter defense to wait out the night. Behind the two front battalions, however, the battle continued well into the night as isolated enemy riflemen and machine gunners that had been overrun attempted to ambush and kill ammunition carriers and stretcher bearers.

During the day, the corps artillery group, providing support for the Marine attack, fired a total of 52 general support missions in addition to the opening bombardment. Nine other close-in missions were fired as the 37th Division also moved its perimeter forward. In all, during the attack on 24 November, the artillery group fired 4,131 rounds of 75mm, 2,534 rounds of 105mm, and 688 rounds of 155mm ammunition.

The casualties during this attack also reflected the intensity of the combat. After the conclusion of the advance by the Marines, at least 1,071 dead bodies of Japanese were counted. The Marine casualties were 115 dead and wounded.

For some Marines, the day was Thanksgiving Day after all. A large shipment of turkeys was received at the beachhead, and, unable to store the birds, the division cooks roasted the entire shipment and packed the turkeys for distribution to front line units.[10] Not every isolated defensive position was reached, but most of the Marines had a piece of turkey to remind them of the day.

GRENADE HILL[11]

The following morning, 25 November, the 3d Marines remained on the newly-taken ground, while a number of changes were made in the lineup along the perimeter. Two days earlier, General Turnage had directed that the 3d Marines and the 9th Marines exchange sectors as soon as possible. This move would allow Colonel McHenry's 3d Marines, by now badly depleted by battle casualties, sickness, and exhaustion, to take over a relatively quiet sector while the 9th Marines returned to action.

The changes had been started on the 24th while the 3d Marines were engaged in the battle for Piva Forks. The 1st Battalion of the 9th Marines, now commanded by Lieutenant Colonel Carey A. Randall, was in regimental reserve positions behind 2/9 and 3/9 on the right flank when alerted by a warning order that the battalion would move on 30-minute's notice.

[10] In the 9th Marines' sector, company cooks did the honors for the holiday birds using 50-gallon drums in which the turkeys were boiled together with rice. "The turkeys together with any other Thanksgiving extras were then delivered boiled to units in position on the lines." *Craig ltr.*

[11] Unless otherwise noted, the material in this section is derived from: *IMAC AR–II; 3d MarDiv CombatRept; 3d MarDiv D–2 SAR; 3d MarDiv D–2 Jnl; 3d MarDiv D–3 Jnl; 3d MarDiv D–3 Repts; Letcher ltr; Snedeker ltr;* LtCol Harold C. Boehm ltr to CMC, dtd 9Aug48 (Bougainville Monograph Comment File, HistBr, HQMC); *Peck ltr; HistDiv Acct;* Rentz, *Bougainville and the Northern Solomons;* Aurthur and Cohlmia, *3d MarDivHist.*

Late in the afternoon, Randall's battalion was ordered to move north along the Piva River to report to Colonel McHenry's regiment for operational control. Shortly before dark, Randall reported to McHenry and was directed to an assembly area. The battalion was to be prepared for relief of the front lines as soon after daybreak as possible. The 2d Battalion of the raider regiment, with two companies of the 3d Raiders attached, was also ordered to move up behind the 3d Marines for commitment to action.

On the morning of the 25th, as the 2d Raiders and 1/9 moved toward the front lines to extend the perimeter, a number of other changes were directed. De Zayas' 2/3, south of the East-West trail, extended its right flank to the southeast to make contact with the 21st Marines. King's battalion, 3/3, organized defensive positions on the left of the East-West trail to make contact with Cibik Ridge where a reinforced company was holding. As the front lines were straightened, 1/3 was withdrawn from action and 3/9 moved into reserve positions behind the 3d Marines. To substitute for the loss of 1/9 and 3/9 to the left sector, a newly arrived battalion of the 37th Division was attached to the 9th Marines. This unit, the 1st Battalion of the 145th Infantry, was placed in reserve positions in the extreme right sector near Piva Village.

Meanwhile, as these changes were made, the 2d Raiders and 1/9 moved east along the East-West trail to begin the day's attack. Randall's 1/9 was to move up Cibik Ridge and then attack almost directly east on a front of 400 yards to extend the left flank of 3/3. The 2d Raiders were to attack on the left flank of 1/9 on a front of about 800 yards. The objective, an area of high ground north of the East-West trail, was about 800 yards distant.

Randall's battalion, guided by a patrol from Cibik Ridge, proceeded single file to the crest. There the Marines could see the attack objective ahead. At 1000, after another crushing 10-minute artillery preparation, the attack was started straight down the opposite side of Cibik Ridge. The assault companies lined up with A on the left and C on the right. Company B was to follow on order. Attached machine guns supported each company, and a heavy mortar barrage from Cibik Ridge pounded ahead as the Marines attacked. At the foot of the ridge, both attacking companies were held up by extremely heavy machine gun fire from concealed positions on a small knoll just ahead. The fight for this knob of ground continued the rest of the day.

The 2d Raiders, meanwhile, had advanced against sporadic resistance. The attack was held up several times by enemy groups; but, as the raiders prepared to assault the defenses, the enemy suddenly gave ground to retire to new positions. By afternoon, Major Washburn's battalion had occupied the hill mass dominating the East-West trail and established a strong perimeter defense on the objective to wait for the battalion from the 9th Marines to come up on line.

Randall's battalion, however, had its hands full. Both attack companies had committed their reserve platoons to the assault of the small knoll facing them without making headway. The enemy was well dug-in with a complete, all-around defense. Marines estimated that the small hill was held by at least 70 Japanese with 4 heavy (13mm) machine guns and about 12 Nambu (6.5mm) machine guns. In addition, the Japanese apparently had plenty of grenades, since a continual rain of explosives was hurled from the enemy positions. The Marines, unable to advance against this formidable strongpoint, dubbed the knoll "Grenade Hill."

Many attempts to envelop this position were repulsed. At some points, the Marine attackers were only five yards from the enemy emplacements, engaged in a hot exchange of small-arms fire and grenades, but unable to carry the last few yards. The fight was conducted at such close quarters that the mortars on Cibik Ridge could not be registered on the enemy position. Many of the dugouts along the side of the hill were destroyed by the Marine attacks, but the crown of the hill was never carried. One platoon from Company A, circling the knoll to the left, managed to fight up a small trail into the position. Fierce enemy fire forced the Marines back before the crest of the ridge was reached. Fourteen Japanese were killed by the Marine platoon in this attempt to take the hill.

By midafternoon, Company B was ordered from Cibik Ridge to plug the gap between the 1/9 attack and the positions of 3/3 on the right. Company B moved down the slope of Cibik Ridge toward the East-West trail south of Grenade Hill and continued east on the trail for several hundred yards in an attempt to locate the left flank of the 3d Marines. Shortly before dusk, the company abruptly ran into a Japanese force. After an intense but short fire fight, the Marine company decided that the Japanese position on higher ground was too strong to overrun and broke off contact. The company then withdrew to a defensible position closer to Cibik Ridge. The Japanese made a similar decision and also withdrew. Company B, out of touch with 1/9 and unable to locate the left flank of 3/3 before dark, organized a defense position across the trail and settled down to wait for morning.

Meanwhile, the fight for Grenade Hill had dwindled and stopped. The two companies of 1/9, unable to capture the hill, dug in around the base of the knob to wait for another day. The 2d Raiders, on the objective, remained in front of the lines in a tight defensive ring.

The next morning, 26 November, scouts from 1/9 reported that the Japanese had quietly withdrawn from Grenade Hill during the night and the small knoll was abandoned. The two assault companies rushed for the hill at once, taking over the enemy positions along the crown of the knoll. The small knob of ground, about 60 feet across at the base and hardly more than 20 feet high, was dotted with a number of well-constructed and concealed rifle pits and bunkers. Each bunker was large enough for at least three enemy soldiers. Only 32 dead Japanese were found on the hill. At 1015, the attack was pushed forward again and contact was made with Company B. This company, during the morning, had linked up with the left flank of the 3d Marines, thereby establishing contact along the line. Company B then joined with 1/9 to complete the move to the final objective. By nightfall, the ridgeline blocking the East-West trail was in Marine hands.

During the attacks of 25–26 November, the Marines lost 5 killed and 42 wounded. At least 32 Japanese had been killed in the assaults on Grenade Hill, and there had undoubtedly been additional casualties in the attacks in other areas. The number of enemy killed during the period 18–26 November in the 3d Division sector was at least 1,196, although the total number of casualties must have been considerably higher than that figure.[12]

The fight for expansion of the beachhead in this area was recorded as the Battle of Piva Forks, and marked the temporary end of serious enemy opposition to the occupation and development of the Cape Torokina area. The only high ground from which the enemy could threaten or harass the beachhead was now held by IMAC forces, and possession of the commanding terrain facing the Piva River gave the Marine regiments an advantage in defending that sector.

After the objective of 26 November had been secured, the remainder of the directed reliefs were completed. The 3d Battalion of the 9th Marines relieved 3/3 in the front lines, and this battalion then became the reserve unit behind 1/9 and 3/9. Control of 1/3 was then shifted to the 9th Marines, and at 1600 on the afternoon of the 26th, the 3d Marines and 9th Marines exchanged commands in the left and right subsectors. The 21st Marines, maintaining positions in the center subsector, moved forward about 500 yards after the attack on Grenade Hill. The IMAC dispositions at the conclusion of the fighting on 26 November had the 148th and 129th Infantry Regiments on line in the 37th Division sector, and the 9th Marines, the 21st Marines, and the 3d Marines on line, left to right, in the Marine sector.

For the 3d Marines, this last move to the right sector completed a full cycle of the beachhead which was begun on D-Day. Following the landing, the regiment moved toward the Koromokina River, then traveled inland through the jungle and swamps to the Piva Forks area. Finally, after 27 days of jungle fighting, the regiment was back near Cape Torokina within the limits of patrols of the first two days ashore.

In the right sector, the exhausted infantry battalions of the 3d Marines were given a respite by the formation of a composite battalion from among the Regimental Weapons Company, the Scout Company, several headquarters companies, and supporting service troops. This makeshift battalion took over a position along the Marine lines on the 28th of November and maintained the defense until early in December so that Colonel McHenry's 3d Marines could reorganize and rest. The Army battalion, 1/145, assigned to this sector was also placed in frontline defenses under the operational control of the 3d Marines and aggressively patrolled past the Torokina River seeking the enemy.

The Japanese, however, evidently intended to do no more than keep this area under observation. Other than a few brushes with enemy outposts, the 3d Marines were out of contact with the Japanese for the remainder of the campaign. The problem, as before, was mainly one of maintaining and supplying the Marine fighting units in the midst of swamplands.

On the 28th of November, General Geiger ordered that the IMAC perimeter in the center and left subsectors be moved forward to Inland Defense Line F, and, after this line was occupied, artillery was displaced forward to defend the area seized and to support the last push to the final beachhead line.[13]

The area fronting the Marine perimeter at this time varied from mountainous terrain to deep swamps and dense jungle.

---

[12] *IMAC AR-II*, p. 14.

[13] The 148th Infantry and most of the 129th had already reached the final beachhead line by 25 November. *Beightler ltr.*

NUMA NUMA TRAIL POSITION *in the swamp below Grenade Hill held by Marines of Company E, 2/21.* (*USMC 69394*)

MARINE WOUNDED *are carried down a steaming jungle trail from Hill 1000 during the fighting in early December.* (*USMC 71380*)

The 9th Marines, on the right flank of the 37th Division, reported rugged terrain in this subsector with many ridges and deep gorges cut by water falls and streams. The 21st and 3d Marines, however, still faced the task of holding areas in the midst of swamps. These, the Marines reported, were not impassable, but it was certain that large forces of enemy could not advance through these swamps without detection by one of the many patrols which roved back and forth between units during the day. At night, small groups moved into the swamps as listening posts.

Every possible battle position, however, was wired and mined. Gaps in the front-line defense were covered by automatic weapons. The Japanese, however, never attempted an infiltration during this time, and only scattered groups of enemy were encountered. These were evidently only scouts who were trying to keep the Marines' progress under surveillance.

The Marines continued to gain combat experience. Two rules of jungle warfare were found invaluable during this period. The first rule was to avoid using the same trail twice in a row—because the second time the trail would be ambushed. The second rule was to avoid heckling a Japanese outpost twice unless prepared to fight a full-scale battle on the second go-around. Inevitably, in these brushes with the Japanese, the second fight was more vicious and determined than the first. While the enemy did not actively seek out the Marine units for battle, the small outpost engagements convinced the Marines that the enemy was still in the area in force and prepared to fight any further expansion of the beachhead.

End of a Mission

THE KOIARI RAID [1]

Throughout the expansion of the beachhead past the Piva River forks, the possibility of a major counterattack by the Japanese along the right flank was a constant threat. To short-circuit possible enemy plans to carry out a full-scale reinforcement effort along this route, a raid on Japanese lines in the southern part of Empress Augusta Bay was proposed. This would disrupt enemy communications, destroy installations and supplies, and obtain information about any troop movements towards the beachhead. The foray was aimed at reported Japanese installations near Koiari, about 10 miles down the coast from Cape Torokina.

The unit selected for this operation was the 1st Parachute Battalion (Major Richard Fagan), which had arrived on Bougainville from Vella Lavella on the 23d of November. Fagan's battalion was to operate much in the manner of Krulak's group on Choiseul. The parachute battalion was to harass enemy units as far inland as the East-West trail but was to avoid a decisive engagement with major Japanese forces. A boat would rendezvous each night with the raiding unit if communications failed. The orders for withdrawal would be given by IMAC headquarters.[2]

The raid was originally scheduled for the morning of 28 November so that escorting destroyers of a shipping echelon could provide naval gunfire support if needed. A trial landing on the selected beach, about 3,000 yards north of Koiari, was made after dark on the 27th by one boat, whose crew then returned to report that there was no evidence of enemy in the area. Because of delay in the actual embarkation of the parachute battalion, however, the entire operation was postponed until the 29th.

Destroyer support would no longer be available, but this lack was not disturbing. One 155mm howitzer battery from the 37th Division was in position near Cape Torokina to support the parachute battalion with long-range fire, and air support could be expected during the day. Two LCI gunboats, which had proved successful during the Treasurys operation, were also available. General Geiger, taking account of the fire support at hand, decided that one day's postponement would not jeopardize the operation and another reconnaissance boat was sent to the selected beach late on the evening of 28 November. The second report was similar to the first: no enemy sighted. In view of later devel-

[1] Unless otherwise noted, the material in this section is derived from: *IMAC AR-II*; *3d MarDiv CombatRept*; *3d Mar Div D-3 Jnl*; *3d MarDiv D-3 Repts*; 1st MarParaBn Unit Rept, Koiari Raid, dtd 30Nov43 (Bougainville AreaOpFile, HistBr, HQMC); *Snedeker ltr*; Henderson, "Naval Gunfire Support;" Rentz, *Bougainville and the Northern Solomons*.

[2] IMAC OpO No. 5, dtd 27Nov43 (Bougainville AreaOpFile, HistBr, HQMC).

opments, it is doubtful that either of the reconnaissance boats landed at the designated beach.

Fagan's battalion, with Company M of the 3d Raider Battalion and a forward observer team from the 12th Marines attached, embarked on board LCMs and LCVPs at Cape Torokina early on the morning of 29 November. One hour later, at 0400, the boats moved in toward the Koiari beach and the Marines were landed virtually in the middle of a Japanese supply dump. The surprise was mutual. A Japanese officer, armed only with a sword, and apparently expecting Japanese boats, greeted the first Marines ashore. His demise and the realization of his mistake were almost simultaneous. The Marines, now committed to establishing a beachhead in the midst of an enemy camp, dug in as quickly as possible to develop the situation.

Before the Japanese recovered from the sudden shock of American forces landing in the middle of the supply dump, the parachute battalion had pushed out a perimeter which extended roughly about 350 yards along the beach and about 180 yards inland. The force was split—the raider company and the headquarters company of the parachute battalion had landed about 1000 yards below the main force and outside the dump area. The Japanese reaction, when it came, was a furious hail of 90mm mortar shells and grenades from "knee mortars." The entire beachhead was raked by continual machine gun and rifle fire. Periodically the Japanese mounted a determined rush against one flank or the other. The Marines, taking cover in hastily dug slit-trenches in the sand, returned the fire as best they could. Casualties mounted alarmingly.

The battalion commander, meanwhile, radioed IMAC headquarters of the parachutists' plight and requested that the unit be withdrawn since accomplishment of the mission was obviously impossible and the slow annihilation of the battalion extremely likely. General Geiger immediately took steps to set up a rescue attempt, but unfortunately the return radio transmission was never received by Fagan. By midmorning, Fagan's radio failed and although it would transmit, it could not receive incoming messages. Contact with IMAC was later made over the artillery net to the forward observer team. At 0930, the beachhead was strengthened by the arrival of the two companies which had been separated. These Marines had fought their way along the shoreline to reach the main party, losing 13 men during the march.

As time passed, Fagan became more convinced that his battalion was in a tighter spot than IMAC headquarters realized. The battalion commander estimated that the Japanese force numbered about 1,200 men, with better positions than the Marines for continued fighting. When the first rescue attempt by the landing craft was beaten back by an intense artillery concentration along the shoreline, the situation looked even more grim. When a second rescue attempt was also repulsed by the Japanese artillery, the Marines resigned themselves to a long fight.

Late in the afternoon, when enemy trucks were heard approaching the perimeter from the south, the parachutists guessed that an all-out attack that night or early the next morning would attempt to wipe out the Marine beachhead. Taking stock of the situation—their backs to the sea, nearly out of ammunition, without close support weapons—the Marine parachutists reluctantly but realistically concluded that the enemy's chances for success were quite good.

During the day, 155mm guns of the 3d Defense Battalion at Cape Torokina registered along the forward edge of the parachutists' perimeter, keeping the enemy from making a sustained effort from that direction. IMAC, in the meantime, sent an emergency message to the task force escorting transports back to Guadalcanal, and three of the screening destroyers immediately reversed course and steamed at flank speed for Bougainville. The three support ships—the *Fullam*, *Landsdowne*, and *Lardner* and one of the LCI gunboats arrived shortly before 1800. The *Fullam*, first to arrive, opened fire immediately under the direction of two gunfire officers from IMAC who had raced for the beleaguered beachhead in a PT boat.

There was little daylight left for shooting on point targets by the time the destroyers arrived. Unable to see the beach, they stationed themselves by radar navigation and opened up with unobserved fires which the gunfire officers adjusted by sound. The ships fired directly to the flanks of the Marine beachhead, while the 155mm guns at Cape Torokina fired parallel to the beach. The effect was a three-sided box which threw a protective wall of fire around the Marine position. Behind this shield, the rescue boats made a dash for the beach. The Marine battalion, which had been alerted for the evacuation through the artillery radio net, was waiting patiently despite the fact that its last three radio messages had indicated that it was out of ammunition.

For some reason—probably because the Japanese were busy seeking cover from naval gunfire—there was no return enemy fire. As the rescue boats beached, the Marines slowly retired toward the shore. There was no stampede, no panic. The withdrawal was orderly and deliberate. After waiting to insure that all Marines were off the perimeter, the battalion commander gave the signal to clear the beach and at 2040 the last boat pulled away without drawing a single enemy shot. The artillery battery and the gunfire support ships then worked over the entire beach, hoping to destroy the Japanese force by random fires.

The attempt to raid the Japanese system of communications and supply along the Bougainville coast ended in a dismal failure. Although the Marines had landed in an area where great destruction could have been accomplished, they were never able to do more than hug the shoreline and attempt to defend their meager toehold with dwindling ammunition until rescued. In the pitch darkness at the time of evacuation, much of the Marine equipment was lost. Although the withdrawal was orderly, some crew-served weapons, rifles, and packs were left behind. Enemy supplies destroyed would have to be credited to the bombardment by Allied artillery and destroyers after the evacuation. The Marines estimated that the Japanese had lost about 291 men, about one-half of whom were probably killed and the others wounded. The Marine parachute battalion, which landed a total of 24 officers and 505 enlisted men plus 4 officers and 81 enlisted men from the 3d Raiders, listed casualties as 15 killed or died of wounds, 99 wounded, and 7 missing.[3]

[3] 1st MarParaBn UnitRept; *IMAC AR–II*, p. 15, gives 15 killed and 71 wounded for this operation.

HELLZAPOPPIN RIDGE AND HILL 600A[4]

After the move to Inland Defense Line F, corps headquarters kept its eyes on a hill mass some 2,000 yards to the front which dominated the area between the Piva and Torokina Rivers. These hills, if held by the Japanese, would give them observation of the entire Cape Torokina area and a favorable position from which to launch an attack against the IMAC beachhead. Geiger's headquarters felt these hills were a continual threat to the perimeter.

On the other hand, Marine occupation of the hills would provide a strong natural defensive position blocking the East-West trail at its Torokina River crossing and would greatly strengthen the final Inland Defense Line. That line, which included the hill mass, was to have been occupied by 30 November, but the supply problems through the swamps and the enemy action had caused an unavoidable delay. Despite the added logistical difficulties which occupation of this hill mass would involve, corps headquarters directed the 21st Marines to maintain a combat outpost on one of the hills until the perimeter could be extended to include this area.

On 27 November, a patrol of 1 officer and 21 men from Colonel Ames' regiment moved to Hill 600 where observation could be maintained over the other two hills and the Torokina River. Hill 600, just south of the East-West trail, was bordered on the left by a higher, longer ridge about 1,000 feet high. A smaller hill, about 500 feet high, was farther south.

As a preliminary to eventual occupation of the hills by IMAC forces, the 1st Marine Parachute Regiment, commanded by Lieutenant Colonel Robert H. Williams, was moved from reserve positions on Vella Lavella on 3 December. Two days later, this unit—actually just the regimental headquarters, the weapons company, and the 3d Parachute Battalion—moved toward the largest of the three hills, Hill 1000. The parachute regiment made a forced march with a half-unit of fire and only three days' rations. This shortage of food was later partially solved by an airdrop on the regiment's position. Williams' regiment pushed forward along the East-West trail almost to the Torokina River before turning to the north to start the climb of the ridgeline toward Hill 1000.

By the end of 5 December, the 1st Parachute Regiment with units of the 3d, 9th, and 21st Marines in support, had established a general outpost line stretching from Hill 1000 to the junction of the East-West trail at the Torokina River. A Provisional Parachute Battalion, under Major George R. Stallings, was formed by Williams from regimental headquarters and Company I to occupy the left sector of his defenses.[5] Williams' command,

[4] Unless otherwise noted, the material in this section is derived from: IMAC Rept. on Bougainville Operation, Phase III, 1–15Dec43, dtd 21Mar44 (Bougainville AreaOpFile, HistBr, HQMC), hereafter *IMAC AR-III; IMAC C-2 Repts; IMAC C-2 Jnl; 3d MarDiv CombatRept; 3d MarDiv D-2 SAR; 3d MarDiv D-2 Jnl; 3d MarDiv D-3 Repts; 3d MarDiv D-4 Repts;* 1st MarParaRegt UnitRept, 6–13Dec43, n.d. (Bougainville AreaOpFile, HistBr, HQMC), hereafter *ParaRegt Rept;* LtCol Robert T. Vance ltr to CMC, dtd 27May48 (Bougainville Monograph Comment File, HistBr, HQMC); *Snedeker ltr; McAlister ltr; Bowser ltr; Letcher ltr;* Rentz, *Bougainville and the Northern Solomons;* Aurthur and Cohlmia, *3d MarDivHist.*

[5] Col Robert T. Vance ltr to CMC, dtd 6Oct60, hereafter *Vance ltr.*

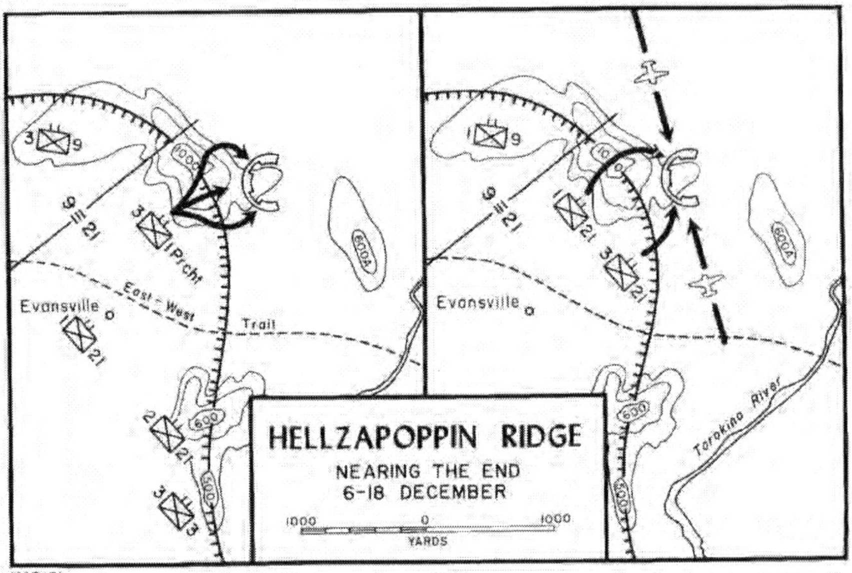

with about 900 men, was strung over a thinly held line about 3,000 yards in length. Meanwhile, the 21st Marines outpost on Hill 600 was reinforced. A full rifle company with a machine gun platoon and the IMAC experimental rocket platoon attached was moved forward to bolster the Marine defense on the two hills blocking the East-West trail.

The move to this hill mass overlooking the Torokina River was made at about the time that the Japanse evidenced a strong interest in these positions. During the period 27 November to 4 December, the enemy activity was confined to minor contacts along the perimeter with the Japanese using mostly long-range artillery fire to make their continued presence known. On 29 November, the Cape Torokina airfield site was hit by a number of 15cm howitzer shells from extreme range. On 3 December, the Japanese attempted to shell the beachhead once more. This time the artillery was emplaced on a forward slope, and a fierce counterbattery fire from three battalions of 75mm howitzers, one battalion of 105mm howitzers, and one battery of 155mm guns quickly smashed the Japanese position. The artillery spotting problem was further aided on 4 December by the delivery of two light planes. These slow-moving aircraft increased the effectiveness of artillery fire by allowing observers to spot targets of opportunity and request and adjust fires. The scout bombers which had been used as spotting planes flew too fast and had too many blind spots to be good observation planes.

The first enemy contact of any consequence since Grenade Hill came in the sector of the 9th Marines on 5 December. Colonel Craig's regiment had been ordered to expand the perimeter to make contact with the parachute regiment on Hill 1000, and, as a small patrol from 2/9 moved out, it was ambushed by about 10 Japanese. The Marines lost two killed and two wounded in the first exchange. The Japanese then withdrew. The following day a 40-man patrol from the 9th Marines aggressively scouted ahead of the regiment but did not encounter any enemy forces. This day, the entire beachhead took another jump forward as advance units of the three regiments pushed inland. On the right flank, 3/3 advanced from positions which had been occupied since 21 November and put a patrol on Hill 500 south of the strong 21st Marines outpost on Hill 600. This position was then strengthened by the extension of an amphibian tractor trail from the swamps to Hill 600, assuring an adequate supply route. The 9th Marines, on the far left, moved up to make contact with the Marines holding Hill 1000.

A general line of defensive positions now stretched from the area north of Hill 1000 along this ridge to the East-West trail and then to the two smaller hills, 600 and 500, south of the trail. With the extension of the supply lines, a growing supply dump called Evansville was established to the rear of Hill 600 near the East-West trail to insure supply of the final defensive line. The advance to Inland Defense Line H came on a day when the entire island was rocked by a violent earthquake. Earthworks and trenches were crumbled and gigantic trees swayed as the ground trembled and rolled. Persons standing were thrown to the ground by the force of the quake. Other earth tremors were recorded later, but none achieved the force of the quake of 6 December.

Movement to the final defense line came as minor patrol clashes were reported along the entire perimeter. On 7 December, the Provisional Parachute Battalion discovered abandoned positions on a 650-foot ridge which extended east from Hill 1000 toward the Torokina River, much in the manner that Grenade Hill was an offshoot of Cibik Ridge.[6] Patrols returned to report that a number of well dug-in and concealed emplacements had been found on the ridge, and Williams made plans to occupy this area with part of his force the next morning. A patrol from the Provisional Parachute Battalion started down the spur on the 8th of December but was driven back by unexpected enemy fire. The Japanese, repeating a favorite maneuver, had returned to occupy the positions during the night.

The patrol reorganized and made a second attempt to seize the enemy position. No headway was made during a sharp exchange of fire. After eight Marines had been wounded, the patrol returned to the front lines. On the morning of 9 December, three patrols of the 3d Parachute Battalion converged on the spur; each encountered light resistance and reported that the enemy broke contact and withdrew. Major Vance ordered Company K to move forward and secure the area. The advancing Marines soon discovered that, "far from withdrawing, the Japs were still there and in considerable strength."[7]

Company K managed to penetrate the Japanese positions, but continued heavy casualties forced a withdrawal before the entire ridge could be captured. During this attack, the enemy fire showed few signs of slackening. Vance then ordered

---

[6] *Ibid.*

[7] *Ibid.*

Company L to outflank the Japanese lines, requesting the Provisional Battalion to support with Company I on the left flank. "Neither of these two units could make headway against the almost sheer sides of the ridge and the heavy fire of the enemy."[8] Several patrols from Company I were ambushed by Japanese in strong emplacements, and, during the confused close-range battle, one light machinegun squad became separated from the rest of the patrol. This squad did not return with the rest of the company, and an immediate search by a strong patrol failed to locate the missing men. Three of the Marines later returned to Hill 1000, but the rest of the squad were carried as casualties.

Despite a number of sharp attacks by the parachute regiment, the enemy position remained as strong as before. The Marines, eyeing the Japanese bastion, called for reinforcements to plug the weak spots caused by the casualties in the overextended lines, and two rifle companies from adjacent units were rushed to Hill 1000. Company C, 1/21, at Evansville, was attached to Vance's unit and moved into the 3d Parachute Battalion's positions bringing much needed ammunition and supplies. At the same time, Company B, 1/9 advanced to cover the gap on the left between 3/9 and the lines on Hill 1000.

That afternoon, the parachute regiment was hit suddenly by a strong Japanese counterattack aimed at the center of the Marine positions on Hill 1000. An estimated reinforced company made the charge, but an artillery concentration centered on the saddle between Hill 1000 and the enemy-held spur broke the back of the Japanese rush. The quick support by 105mm and 75mm howitzers scattered the Japanese soldiers and ended the attack. The Marines, though, had 12 men killed and 26 wounded in the brief struggle.

Evacuation of wounded from this battle area was particularly difficult. Only two trails led to the top of the hill—one a hazardous crawl up a steep slope, and the other a wider jeep trail leading toward the Torokina River. Torrential rains made both virtually impassable. At least 12 men were required to manhandle each stretcher case to an aid station set up about half-way down the rear slope. There, blood plasma and emergency care were given the wounded before they were carried to an amtrac trail at the foot of the hill. The wounded men were then moved across the swamps to roads where jeep ambulances were waiting.[9]

On the morning of 10 December, all units completed the advance to the final defense line along the general line of Hill 1000 to Hill 500, and the 1st Battalion of the 21st Marines moved forward under enemy mortar fire to take over responsibility for the defense of Hill 1000. There was no enemy action in the sectors of the 9th and 3d Marines on either flank. At 0900, Lieutenant Colonel Williams met the commanders of 1/9 and 1/21 at his command post on Hill 1000, and the details of the relief of the 1st Parachute Regiment by 1/21 and of contact with 1/9 were worked out. There was still action by the enemy, though. A small Marine patrol scouting ahead of the lines to pick up the bodies of dead Marines was fired upon by an enemy force. The Marines drew back and called for an artillery concentration on the area before continuing the patrol. This time

[8] *Ibid.*

[9] Sgt Charles E. McKenna, "Saving Lives on Bougainville," *Marine Corps Gazette*, v. 28, no. 3 (Mar44).

there was only sporadic sniper fire. That afternoon the scattered elements of the parachute regiment moved off Hill 1000, leaving the unfinished business of the enemy stronghold and the difficulties of supply and evacuation to 1/21 and 1/9. The parachute troops, obviously ill-equipped and understrength for such sustained combat, moved into reserve positions behind the IMAC perimeter.

The strong Japanese position on the spur extending east from the heights of Hill 1000 resisted all efforts by the Marines for the next six days. The bitter fighting on this small ridge, which did not even show on the terrain maps provided by IMAC, earned for this stronghold the name of "Hellzapoppin Ridge." Although Marines sometimes called the enemy position "Fry's Nose" for the commander of 1/21, Lieutenant Colonel Fry, or "Snuffy's Nose," for Colonel Ames, the 21st Marines' commander, the name of Hellzapoppin Ridge is more indicative of the fierce combat that marked the attempts to capture this spur.

This position, abandoned at the time, was first discovered on 7 December when an enemy operations map was picked up and interpreted. This document indicated that the ridge positions were those of a reinforced company from the Japanese *23d Infantry*. Enemy strength was estimated at about 235 men, but, beyond this, the Marines had no further information on the area which the enemy so stoutly defended. The ridge was about 300 yards long with steep slopes leading to a narrow crest some 40 yards wide. Combat patrols sought to uncover more information about this natural fortress, but the enemy resisted every attempt. All companies of 1/21 launched attacks against this position, but the enemy's fierce fire drove them back. The Japanese appeared to be well dug in, with overhead cover, in a carefully prepared all-around defense.

Unable to define the limits of the Japanese position, the Marines were unable to bring any supporting weapons except 60mm and 81mm mortars to bear on the emplacements. The 60mm mortars proved too light to open holes for the Marine attacks, and the heavier mortars also appeared to have little effect on the enemy positions. During the first stages of the repeated attacks on Hellzapoppin Ridge, and until the final assault on 18 December, artillery fire was also ineffective. Because the ridge was on the reverse slope of Hill 1000, the artillery batteries had trouble adjusting the angle of fire to hit the spur. The huge trees lining the ridge caused tree bursts, with no effect on the enemy bunkers. The enemy fortress, effectively defiladed by most of Hill 1000, was relatively immune to shelling.

At this time the Marine artillery firing positions were near the bomber airstrip in the vicinity of the Coconut Grove. The direction of fire was almost due east. This location, however, resulted in many "overs" on the enemy position or tree bursts in the jungle along the ridge of Hill 1000 where the front lines of 1/21 were located. Some casualties to Marine personnel were taken during attempts to bring the fire to bear on the enemy fortifications.

On the 13th of December, after repeated attempts to knock the Japanese off Hellzapoppin Ridge by artillery fire had failed, a request for direct air support was made to ComAirNorSols by General Geiger. The IMAC commander requested that the three scout bombers and three torpedo bombers which had just landed at the newly completed airstrip at Torokina be used that afternoon. The six Marine planes, loaded with 100-pound bombs, made a run on the enemy position just at dusk after the target had been marked by smoke shells from 81mm mortars. Four planes managed to hit the target area; but the fifth plane dropped its bombs well behind the Marine front line about 600 yards north of the enemy ridge. The explosions killed two Marines and wounded five. The sixth plane returned to Torokina without completing the mission. The Marines, somewhat dubiously, requested another strike for the next day.

For this mission, 17 torpedo bombers from VMTB–232 landed at Torokina airfield for a pre-strike briefing. The locations of the Marine lines and the target area were described for the pilots before the planes took off again. This time the Marine lines were marked with colored smoke grenades and the target area with white smoke. The results were considerably better. The planes made the strike in column formation parallel to the front lines. About 90 percent of the bombs landed in the target area.

Another strike the following day, 15 December, was conducted in the same manner. Pilots of 18 TBFs (VMTB–134) landed at Torokina for an extensive briefing by the strike operations officer and a ground troop commander who then led the strike in an SBD. Smoke shells again marked the front lines and the target area, and the torpedo bombers hit the ridge with another successful bombing run. On the 18th of December, two final strikes completed the softening process on the enemy-held positions. Six planes from VMTB–134, each with a dozen 100-pound bombs, landed at Torokina once more for briefing before heading for the target. Each mission was led by the AirSols operations officer, Lieutenant Colonel William K. Pottinger, with the 21st Marines Executive Officer, Lieutenant Colonel Arthur H. Butler, as observer. The first strike was completed with all bombs reported landing in the target area. Later, five more TBFs made a second strike, this time dropping 36 bombs. The second strike was also guided to the target by the lead plane. At the end of this mission, the five planes continued to make strafing runs on the target and executed dummy bombing runs to hold the enemy defenders in place while the 21st Marines attacked and seized the ridge.

Prior to the two air attacks on the 18th, the I Marine Amphibious Corps and the 21st Marines made extensive preparations for the final assault on Hellzapoppin Ridge. A battery of 155mm howitzers from the 37th Division was moved by landing craft (LCMs) to new firing positions near the mouth of the Torokina River. From this position the battery could fire north along the river valley and put shells on the south side and crest of the ridge without endangering the Marine lines. The 155mm howitzers, which were moved into these temporary firing positions early on the morning of the 18th, opened fire at about 1000. Smoke shells were used for registration shots before the final concentrations were fired. The battery fired about 190 rounds in the next hour, hitting the ridge repeatedly although there was some initial difficulty in adjusting the fire to get hits along the crest of the knife-like ridge. The artillery shells cleared much of the brush and smaller jungle growth off the ridge, exposing the target to the two air strikes which followed the artillery fire.

Shortly after the final air attack,[10] units of 1/21 and 3/21 (committed to action from the regimental reserve) moved off Hill 1000 and over Hellzapoppin Ridge in a coordinated double envelopment. The air attacks and the direct artillery fire had done the job. The blasted and shredded area revealed many dead Japanese. The stunned survivors who made a token resistance were quickly eliminated. After more than six days of repeated attacks on this defensive stronghold, Hellzapoppin Ridge was captured, and the enemy force cleared from the many concealed and emplaced bunkers. The victory cost the 21st Marines 12 killed and 23 wounded. More than 50 Japanese bodies were found in the area. The remainder of the enemy defenders had apparently fled the area.

The next three days were devoted to the extension of the perimeter to include this natural fortress and to improvement of the final defensive line. There was no enemy contact until the morning of 21 December when a reconnaissance patrol reported that about 14–18 enemy soldiers had been discovered on a hill near the Torokina River. A combat patrol from the 21st Marines immediately moved out to drive the Japanese from this hill, 600–A. The Marines lost one killed and one wounded in a short action that ended in a repulse for the attackers. Directed to put an outpost on Hill 600–A, a platoon from 3/21 (Lieutenant Colonel Archie V. Gerard) started for the hill early on the 22d of December, but once more the Japanese had occupied a position in strength during the night. As the Marines reached the crown of Hill 600–A, a blast of small-arms fire from entrenched enemy forced them back down the hill. Company I of Gerard's battalion then started forward to reinforce the platoon.

A double envelopment was attempted by Company I, but Japanese defenses on the reverse slope of Hill 600–A stopped that maneuver and pinned the attackers into an area between the enemy lines and the Marine base of fire. Company I, under heavy rifle and machine gun fire, wriggled out of this predicament and withdrew to request artillery support.

Another attack the next day, 23 December, by Company K of Gerard's battalion ended up much in the same manner. The company, reinforced by a heavy machine gun platoon, attempted to break the Japanese hold on Hill 600–A by a direct attack, but the advance platoon took so much fire that the attack could not move forward. Company K withdrew and artillery support was requested. A 30-minute concentration pounded the forward slopes of the hill with the usual tree-bursts reducing the effectiveness of the fire. Then Company K started forward again. The attack was repulsed. A third attempt, after another mortar and artillery preparation, also failed. Company K then withdrew to the front lines.

The next morning, in preparation for another attack, scouts moved forward toward the enemy position. Inexplicably, after putting up a stiff fight for two days, the enemy had retreated during the night. Only one Japanese body was found in the 25 covered emplacements on the hill. Artillery fire had damaged only a few of the bunkers. The Marines, in attacking for two days, lost four killed and eight wounded.

The next several days were quiet, the Marines resting and preparing for a gen-

[10] BGen John S. Letcher ltr to Head, HistBr, G-3, HQMC, dtd 6Oct60.

eral relief along the perimeter by the units of the Army's Americal Division. There was no further action of any consequence before the Marines departed Bougainville. After the capture of Hill 600–A, the Japanese resistance consisted mostly of periodic shelling of the area around Evansville with 75mm howitzers and 90mm mortars. This firing, however, was sporadic and ineffectual. The Japanese quickly retreated past the Torokina River when combat patrols went out to eliminate the enemy fire.

Three final air attacks, two on the 25th and one on the 26th of December, apparently discouraged the Japanese from staging another attack on the perimeter. The object of the attacks was to clear out an area north of Hill 600–A where Japanese activity was reported by patrols. Eighteen torpedo bombers armed with 500-pound and 100-pound bombs blasted the target, and after the attacks patrols found the area abandoned. Trenches and installations indicated that about 800 Japanese had been in the area. A number of patrols across the Torokina River, however, failed to make contact with the enemy.

The relief of the Marine division from Bougainville had been expected since 15 December after the consolidation of the final defense line along Hill 1000 and Hill 600. As additional Army troops continued to arrive at the beachhead, Admiral Halsey directed the Army's XIV Corps to assume control of the Bougainville operation, and, on 15 December, General Geiger turned over control of the beachhead to the commanding general of the XIV Corps, Major General Oscar W. Griswold. The relief of the 3d Marine Division by the Americal Division began on the morning of 27 December when all units of the 9th Marines were relieved on frontline positions by the 164th Infantry and moved into bivouac areas in preparation for the return to Guadalcanal. The last units of the 3d Marines manning perimeter positions were relieved by the 132d Infantry on the afternoon of the 28th. With two regiments on line, command of the right sector of the beachhead was assumed by Major General John R. Hodge of the Americal Division. The 21st Marines was relieved by the 182d Infantry on the 1st and 2d of January 1944. By 16 January, the entire 3d Marine Division had returned to Guadalcanal.

The raider and parachute regiments remained on Bougainville for two weeks after the 3d Division departed as part of a composite command led by Lieutenant Colonel Shapley with Lieutenant Colonel Williams as executive officer. The provisional force manned the right flank of the perimeter along the Torokina and spent its time improving defenses and patrolling deep into enemy territory.[31] By the end of January, the raiders and parachutists had turned over their positions to the Army and joined the general exodus of IMAC troops from the island. The 3d Defense Battalion, which stayed on until 21 June, was the sole remnant of the Marine units that had taken part in the initial assault on Bougainville.

The damage to the Japanese forces in the nearly two months of Marine attacks was not overwhelming. The enemy committed his units piecemeal, and, although most of these were completely wiped out, the total loss was not staggering. The Marines estimated that 2,458 enemy soldiers lost their lives in the defense of the Cape Torokina area, the Koromokina counter-

[31] *Shapley ltr*; BGen Robert H. Williams ltr to Head, HistBr, G–3, HQMC, dtd 4Nov60.

landing, and the counterattacks in the Marine Corps sector. Prisoners numbered 25. Japanese materiel captured was also negligible and consisted of a few field pieces and infantry weapons.

The postwar compilation of Marine Corps casualties in the Bougainville operation totaled 423 killed and 1,418 wounded. The breakdown by major units was: Corps troops—6 killed and 31 wounded; 1st Parachute Regiment (less 2d Battalion)—45 killed and 121 wounded; 2d Raider Regiment (provisional)—64 killed and 204 wounded; 3d Defense Battalion—13 killed and 40 wounded; and 3d Marine Division—295 killed and 1,022 wounded.[12]

COMPLETION OF THE AIRFIELDS[13]

For a period of about 10 days after the landings on Bougainville, the Allies had almost complete air superiority in the Solomons. The Japanese bases in the Bougainville area had been worked over so well and so many times by Allied air strikes and naval bombardments that the enemy experienced extreme difficulty putting them into operation again. As a result, the Japanese were forced to contest the Cape Torokina landings from bases at Rabaul and fields on New Ireland. By the time that the beachhead had expanded to include the Piva fields, the Japanese capability to threaten the beachhead from air bases in the Northern Solomons had been partially restored, and only repeated strikes against the fields at Buka, Kahili, and Ballale kept the Japanese air threat below the dangerous point. On 20 November, ComAirNorSols estimated that at least 15 known enemy airfields within 250 miles of Empress Augusta Bay were either under construction or were repaired and operational once more. Completion of airstrips within the Cape Torokina perimeter was then rushed to meet this growing enemy threat.

During the early stages of the beachhead, the construction of the airfields had been weighed against the immediate need for a road net to insure an adequate system of supply to the front lines. The road net had been given priority, and most of the efforts of the 19th Marines had been directed to this project. After the perimeter road was completed in time to support the fight for the Piva Forks, attention was again turned toward airfield construction. The road network was still far from finished, however. When the various artillery units and support outfits occupying the projected airfield sites were asked to move out of the way of construction work, the answer was usually a succinct, "Over which roads?"[14]

The Japanese, ironically, gave the construction gangs a big assist. The enemy emplaced several 15cm howitzers in the high ground east of the Torokina River, and the construction work in the vicinity of the Coconut Grove appealed to his curiosity. As a result, whenever there was no combat air patrol over the beachhead, the Japanese were quite apt to drop shells into the airfield area. The Seabees and the

[12] Rentz, *Bougainville and the Northern Solomons*, App. II, p. 140. See Appendix H, Marine Casualties, for the final official totals.

[13] Unless otherwise noted, the material in this section is derived from: *ThirdFlt NarrRept; IMAC AR–III; 3d MarDiv Combat Rept;* Rentz, *Bougainville and the Northern Solomons;* Isely and Crowl, *Marines and Amphibious War;* Morison, *Breaking the Bismarcks Barrier;* Miller, *Reduction of Rabaul.*

[14] *McAlister ltr.*

PIVA AIRFIELDS, *the key bomber and fighter strips in the aerial offensive against Rabaul, as they appeared on 15 February 1944. (USN 80-G-250368)*

FIELD ARTILLERY *missions are fired against Japanese attacking the Torokina perimeter by 155mm seacoast defense guns of the Marine 3d Defense Battalion. (SC 190032)*

Marine engineers moved to the end of the field which was not being hit and continued to work. The other tenants, though, anxious to avoid repeated artillery shelling, vacated the area.

By the time that Inland Defense Line F had been occupied on 28 November, clearing of the projected bomber and fighter strips was well underway. The work was speeded by the arrival on the 26th of November of the 36th Naval Construction Battalion which brought all its equipment and went to work almost full time on the Piva fields. Meanwhile, the Torokina strip had been unexpectedly put into operation. About noon on the 24th, Seabees and engineers working on the airstrip were amazed to see a Marine scout-bomber preparing to land on the rough runway. Quickly clearing the strip of all heavy engineering equipment, the construction workers stood by while the Marine pilot brought his plane into a bumpy but successful landing. The emergency landing of the plane, damaged in a raid over Buka, initiated operations on the new strip.

Admiral Halsey, whose cheerful messages delighted the South Pacific Forces all through the long climb up the Solomons ladder, provided a fitting note of congratulations on the completion of the fighter strip:

> In smashing through swamp, jungle, and Japs to build that air strip, your men have proven there is neither bull nor dozing at Torokina. . . . A well done to them all. Halsey.[15]

The advance naval base and boat pool, which had been underway since the first landings, was also rushed toward completion during the month of November. This gave the III Amphibious Force torpedo boats a wider range, and these ubiquitous craft then prowled along the Bougainville coast as far north as Buka and as far south as the Shortlands in protective patrols.

The Torokina strip was finally declared operational on 10 December, just one month after the initial construction was started. At dawn on the 10th, Marine Fighter Squadron 216, with 17 fighter planes, 6 scout bombers, and 1 cargo plane, landed as the first echelon. The following day, 11 December, three Marine torpedo bombers landed and these were joined six days later by four aircraft from an Army Air Forces squadron. The first direct air support mission was flown on 13 December with Hellzapoppin Ridge as the target, while combat air patrols began flying from the former plantation site on 10 December, the day that the planes arrived. Later in the month, additional flights including night fighter patrols, began operations from the Torokina strip.

After the completion of the first field, work was rushed on the Piva bomber field, and early in December another full-strength naval construction battalion, the 77th, arrived to help with the job. As the network of roads throughout the perimeter was completed, the 71st and 53d Seabees also went to work on the airfields. The Piva bomber field received its first planes on 19 December, and was completely operational on 30 December. The Piva fighter field, delayed by lack of matting, was operational on 9 January, after the main units of the 3d Marine Division were withdrawn from the island.

With the opening of three Allied airfields on Bougainville, the Japanese capability to threaten the Allied position on

[15] ComSoPac msg to CG, IMAC, in *3d MarDiv Jnl*, 13Dec43.

Cape Torokina from the air was virtually eliminated. The tactical importance of these airstrips was demonstrated in the quick support given the ground troops in the attack on Hellzapoppin Ridge, but the advantage was strategic as well as tactical. The Torokina airfields brought Allied air power to within 220 miles of Rabaul and allowed fighter aircraft to escort bombers on air strikes against Japanese bases on New Britain and New Ireland. The completion of the three fields on Bougainville, and a fourth field at Stirling Island in the Treasurys, marked the successful achievement of the primary aim of the Bougainville operation.

The constant pressure which the expanding air strength of the Allies exerted upon the Japanese installations was reflected in the attempts which the enemy made to delay the Torokina airstrips. During the first 26 days of operation at Torokina, the beachhead had 90 enemy alerts. The vigilance of the ComAirNorSols fighter cover and the accuracy of the Marine 3d Defense Battalion's antiaircraft fire was indicated by the fact that bombs were dropped during only 22 of these alerts. Casualties from enemy air action up to 26 November were 24 killed and 96 wounded. Damage was restricted mainly to the boat pool and supply stocks on Puruata Island.

During the time that the 3d Marine Division remained on the island, the Japanese managed to bomb the perimeter only five more times, and the interval between alerts grew increasingly lengthy. For the entire period 1 November to 28 December, there were 136 enemy air alerts with bombs dropped during only 27 of these alerts. The total casualties in enemy air raids were 28 killed, 136 wounded, and 10 missing.

JAPANESE COUNTERATTACKS, MARCH 1944 [16]

Several weeks after the 3d Marine Division departed Bougainville, the XIV Corps became aware of gradually increasing enemy activity around the area of the Torokina River and the right sector of the beachhead now being held by the Americal Division. Immediate offensive efforts were directed toward the Torokina sector, and aggressive patrols by the Americal Division erased the threat to the perimeter by driving the Japanese out of prepared positions along the coast near the Torokina River. The bunkers and pillboxes encountered were destroyed. The perimeter, however, was not extended to cover this area.

The following month, Japanese patrol action was more aggressive, and throughout February the entire perimeter was subjected to a number of sharp probing attacks by small groups of enemy soldiers. The two frontline divisions, keenly aware that the perimeter might soon be tested by another determined Japanese attack, prepared extensive defenses to meet it. Fortifications in depth were constructed, and the entire front was mined and wired where possible.

---

[16] Unless otherwise noted, the material in this section is derived from: *ThirdFlt NarrRept;* XIV Corps, History of TA Operation Mar44, dtd 21Apr44 (Bougainville AreaOpFile, HistBr, HQMC); 37th InfDiv G-2 Periodic Repts, n.d. (Bougainville AreaOpFile, HistBr, HQMC); 37th InfDiv OpRept—Bougainville, 6Nov43–30Apr44, n.d. (WW II RecsDiv, FRC Alex); *3d MarDefBn SAR;* LtCol Edward H. Forney ltr to CMC, dtd 7Jun48; LtCol John G. Bouker ltr to CMC, dtd 24Jun48; Rentz, *Bougainville and the Northern Solomons;* Miller, *Reduction of Rabaul.*

The Japanese, after being forced to withdraw from the heights around the Torokina River late in December, began planning the counterstroke about the middle of January. Determined to end the Allied possession of the Cape Torokina area, the enemy readied the entire *6th Division*, plus a number of special battalions from the *Seventeenth Army*. This force began training for the operation while support troops started building roads toward the Cape Torokina area from Mawareka. This offensive was to be a joint Army-Navy effort, but when Truk Island in the Central Pacific was hit by a large Allied carrier force on 16 February 1944, the remainder of the naval air strength at Rabaul was dispatched to Truk. The *Seventeenth Army* then carried on the operation alone.

As expected by Allied intelligence officers, the trail net from Mosigetta-Mawareka was the main route of travel, and a rough road was completed during the early part of the year. This route had its share of troubles, though. Sections of the road were washed out by rains, and swollen rivers carried away the hasty bridges thrown across them by the Japanese engineers. Additionally, the Japanese activity along this road was a prime target for repeated aerial attacks from patrolling Allied planes. Japanese barge movements along the coast were so harassed by Allied torpedo boats, LCI gunboats, and patrol planes that only a few of the barges remained afloat by March. In the end, the enemy force was required to move overland through the jungle, tugging and pulling artillery and supplies behind.

The Japanese offensive, known as the *Ta* operation, included elements from five infantry regiments—the *13th*, *23d*, *45th*, *53d*, and the *81st*—and two artillery regiments. The attack force numbered about 11,700 troops out of a total force of some 15,400. The general plan of the *Ta* operation was a simultaneous attack on the Allied perimeter from the northwest, north, and east, while the artillery units pounded the objective from positions east of the Torokina River. The attack opened on 8 March with a simultaneous shelling of the Torokina and Piva airstrips and a sudden thrust at the 37th Division lines near Lake Kathleen in the center of the perimeter.

The Army division held its positions against light probing attacks and waited for the all-out assault. The following day three battalions from the *13th* and *23d Infantry* slashed at 37th Division positions in an effort to penetrate the Army lines and obtain high ground where they could emplace artillery weapons to threaten the airfields. The 145th Infantry bore the brunt of these attacks, and by the 12th of March had made three counterattacks to dislodge the enemy forces and restore the lines. During these bitter struggles, the entire enemy force was virtually annihilated; 37th Infantry Division troops counted 1,173 dead Japanese in the area after the attacks.

While this fight raged, another strong Japanese force suddenly assaulted the Americal Division in the area of the upper Torokina River. The action here was considerably less violent but more protracted than in the 37th Division area, and the Japanese forces were not driven back until the 29th of March. The enemy in this area suffered 541 casualties.

In sharp contrast to the weeks of November and December, the main enemy thrusts during the March counterattack continued in the perimeter sector held by the 37th Division. After the staggering repulse near the center of the perimeter, the Japanese forces moved toward the Laruma area, where four more determined attacks were launched against the Army positions. A coordinated attack in two places on the northwest side of the perimeter on the 13th and 14th was repulsed by the 129th Infantry, and more than 300 Japanese were killed. On the 17th of March, the Japanese struck another blow at the 37th Division positions and were met by a tank-infantry counterattack that killed another 195 enemy.

The sixth and final attack against the XIV Corps perimeter was launched on the 24th of March. Japanese fanaticism carried the attack as far as one of the battalion command posts of the 129th Infantry before the penetration was sealed off. Another 200 Japanese perished in this last attack. The following day, XIV Corps artillery shelled the retreating Japanese, and the last bid by the enemy to retake the Cape Torokina area was over. Admiral Halsey, describing the "Ta" operation later, reported:

> The attack came against positions that had been carefully prepared in depth, with well-prepared fields of fire and manned by well-disciplined, healthy, and ready-to-go troops of the 37th and Americal Division. Some damage was done by enemy artillery but the damage did not prevent aerial operations. The Japanese infantry attacks were savage, suicidal, and somewhat stupid. They were mowed down without mercy and the attack was actually broken up by the killing of a sufficient number of Japanese to render them ineffective. At this writing, over 10,000 Japanese have been buried by our forces on Bougainville. The remainder can probably manage to keep alive but their potential effectiveness and heavy weapons have been destroyed.[17]

The XIV Corps estimated that more than 6,843 Japanese had died in the futile charges against the strong Allied positions from 8 to 25 March. These casualty figures compare favorably with Japanese records which indicate that the *Ta* attack force lost 2,389 killed and 3,060 wounded. In addition, various supporting units under direct control of the *Seventeenth Army* suffered casualties of about 3,000 killed and 4,000 wounded.[18]

At the time of the attack against the XIV Corps perimeter, the Antiaircraft and Special Weapons Groups of the Marine 3d Defense Battalion were operating under the tactical control of Antiaircraft Command, Torokina, an XIV Corps grouping of all air defense weapons. For some time, enemy activity over the beachhead had been restricted because of the almost complete dominance of Allied aircraft. As a result, the Marine weapons were seldom fired. When the Japanese attack was launched, the 90mm batteries were employed as field artillery.

The Marine batteries were usually registered by an Army fire direction center on various targets beyond the range of light artillery units and adjusted by aerial observers. Most of the firing missions were targets of opportunity and night harassing registrations on enemy bivouac and supply areas. The Marine weapons fired a total of 4,951 rounds of ammunition in 61 artillery missions.[19]

---

[17] *ThirdFlt NarrRept*, p. 12.

[18] *Seventeenth Army Ops—II*, p. 111.

[19] CO, 3d DefBn SAR Covering Employment of Guns in Direct Ground Fire Missions, 14Mar–15Apr44, dtd 15Apr44 (Bougainville AreaOps-File, HistBr, HQMC).

On the 13th of March, two 90mm guns from the Marine unit were moved toward the northwest side of the perimeter with the mission of direct fire on a number of enemy field pieces and other installations located on a ridge of hills extending in front of the 129th Infantry. Emplacement of the guns in suitable positions for such restricted fire was difficult, but both guns were eventually employed against limited targets of opportunity. The guns were later used to greater advantage to support the soldiers in local counterattacks against the Japanese forces.

The 155mm seacoast defense guns of the Marine battalion were also used as field artillery in defense of the perimeter. The Marine battery was called upon for 129 firing missions in general support of the Army defenses and was usually employed against long-range area targets. The big guns fired 6,860 rounds, or 515 tons of explosives, against enemy positions.

In addition, two of the 40mm guns from the Special Weapons Group were moved into the front lines in the Americal Division sector for close support. There were few suitable targets for the guns in the slow action in this area, and only two or three bunkers were hit by direct fire. One weapon, however, fired over 1,000 rounds in countermortar fire and to strip obstructing foliage from enemy positions. Targets being scarce, the guns were sometimes used as sniping weapons against individual Japanese.[20]

The defense battalion was the last Marine Corps ground unit withdrawn from Bougainville. It departed Cape Torokina on 21 June, nearly eight months after the initial landings of the battalion on 1 November with the assault waves. The withdrawal came one week after Admiral Halsey's South Pacific Command was turned over to Vice Admiral John H. Newton and all the Solomon Islands were annexed as part of General MacArthur's Southwest Pacific Command. The South Pacific campaign against the Japanese was virtually ended. Ultimately, the Cape Torokina area was occupied by Australian forces which, by the end of the war, were closing slowly on Numa Numa and the last remnants of the Bougainville defenders.

[20] 3d DefBn Memo No. 6-44, dtd 1Apr44 (Bougainville Area OpsFile, HistBr, HQMC).

CONCLUSIONS

In an analysis of the operations against the Japanese in the Northern Solomons, three points of strategic importance are apparent. The first of these is that despite the risk inherent in attacking deep into enemy-held territory, the Allied forces successfully executed such a venture. Diversionary and subsidiary actions were so timed and executed that the Japanese were deceived as to the intentions and objective of the final attack. The South Pacific forces also gambled and won on the premise that the ever-increasing superiority of Allied air and sea power would counterbalance the vulnerability of extended supply lines attendant upon an operation so close to Rabaul.

Another point is that, although the operations in Bougainville and the Treasurys were planned, directed, and executed by South Pacific Forces, the campaign was partially a maneuver to provide flank security for the advance of the Southwest Pacific Forces along the northern coast of New Guinea. The need to strike at a point which would insure this flank security was the factor which resulted in the selection of Cape Torokina as the I Marine Amphibious Corps objective.

The third point is that the Bougainville campaign was wholly successful, with a minimum of lives lost and materiel expended. Seizure of a shallow but broad beachhead by Marine units and the expansion of this perimeter with the subsequent and continued arrival of Army reinforcements was an economical employment of both amphibious troops and infantry.

The operations in the Northern Solomons knocked the Japanese off-balance. The only enemy activity was a day-to-day reaction to Allied moves. After 15 December, Allied air superiority south of Rabaul was unchallenged, and the Japanese cancelled further naval operations in the latitude of the Bougainville operation. There were other considerations, however. The enemy was pressed by additional Allied moves in the Central Pacific, the loss of the Gilbert Islands, and the continued advance of General MacArthur's forces in the Southwest Pacific. The success at Bougainville, however, led to the eventual collapse of the enemy's defensive positions in the Bismarcks.

> The Bougainville campaign was intended to accomplish the destruction of enemy air strength in the Bismarcks; not only was this accomplished but the by-products of the campaign were so extensive that the subsequent operations at Green Island and Emirau were accomplished virtually without enemy opposition, and the entire enemy offensive potential in the Bismarcks area was destroyed. In the matter of ultimate achievement and importance in the Pacific War, the Bougainville operation was successful beyond our greatest hopes.[2]

By tactical considerations, the planning of the Northern Solomons campaign was daring, yet sound. The diversionary landing on Choiseul and the operations in the Treasurys were conceived to mislead the Japanese, and these maneuvers served as the screen behind which the I Marine Amphibious Corps moved toward the actual point of attack. The surprise achieved by the landings at Empress Augusta Bay is evidence of astute, farsighted thinking behind these operations.

The tactical stroke which decided the success of the Bougainville operation was the selection of the Cape Torokina beaches as the landing site. To pick a landing area lashed by turbulent surf, with tangled jungle and dismal swamps immediately inland, was a tactical decision which the Japanese did not believe the Allied forces would make. Amphibious patrols had failed to scout the exact beaches chosen and general knowledge of the area was unfavorable. Despite these disadvantages, the shoreline at Cape Torokina was picked because the Japanese had decided not to defend such an unlikely area in strength.

When the I Marine Amphibious Corps was relieved by XIV Corps on 15 December, General Geiger's Marine forces could leave the island confident that the beachhead was firmly anchored along the prescribed lines and that the mission was complete. Advance naval base facilities were installed and functioning; one fighter strip was in operation and another underway, and a bomber field was nearly complete. Equally as important, a road net capable of carrying all anticipated traffic was constructed or nearly finished throughout the beachhead. This last achievement, as much as the rapid completion of airfields, insured the continued existence of the Cape Torokina perimeter against Japanese attacks.

[2] *ThirdFlt NarrRept*, p. 12.

That the operations ashore were successful though faced by three formidable obstacles—Japanese forces, deep swamps, and dense jungle—is a tribute to the cooperation of the Marine Corps, Army, and Navy units assigned to the I Marine Amphibious Corps.

For the 3d Marine Division, which formed the bulk of the IMAC forces initially, the months of hard fighting established skills and practices which lasted throughout the Pacific campaign. The Marines of General Turnage's division first experienced the difficulties of maintaining supply and evacuation while under enemy fire at Cape Torokina. If there had been any misgivings about committing an untested division to such a task, these were dispelled by the manner in which the Marines conducted the operation. After the campaign, General Turnage wrote: "From its very inception it was a bold and hazardous operation. Its success was due to the planning of all echelons and the indomitable will, courage, and devotion to duty of all members of all organizations participating."[22]

By the time of the Piva Forks battle, the 3d Division Marine was a combat-wise and skilled jungle fighter capable of swift movement through swamps and effective employment of his weapons. By the end of the campaign, the Marine was a veteran soldier, capable of offering a critical appraisal of his own weapons and tactics as well as those of his opponent.

This increased fighting skill was reflected by mounting coordination between all combat units, with lessening casualties. In this respect, the 3d Medical Battalion achieved a remarkable record of rescue operations which, despite the complexities of evacuation, resulted in less than one percent of the battle casualties dying of wounds. Aid stations were located as close to the front lines as possible. Wounded Marines were given emergency treatment minutes after being hit. The casualties were removed by amtracs to hospitals in the rear, or, in the case of seriously wounded Marines, to more extensive facilities at Vella Lavella. The hospitals on the beach were subjected to air raids and shelling, but the treatment of wounded continued despite these handicaps.

Disease incidence was low, except for malaria which had already been contracted elsewhere. The construction work on the airfields and roads resulted in the draining of many adjacent swamps which aided the malaria control. There were no cases of malaria which could be traced to local infection, and dysentery and diarrhea were practically nonexistent—a rare testimonial to the sanitary regulations observed from the start of the operation. Before the campaign, many preventive measures were taken. Lectures and demonstrations which stressed the value of clothing, repellents, bug spray, and head and mosquito nets were part of the pre-operation training. Their effect and the discipline of the Marines is reflected in the fact that few men were evacuated because of illness.

The Marines learned that, with few exceptions, jungle tactics were a common sense application of standard tactical principles and methods to this type of terrain. Control of troops was difficult but could be accomplished by the "contact imminent" formation described earlier. During most of the fighting, command posts were situated as close to the front lines as

[22] *3d MarDiv CombatRept*, p. 12.

possible—sometimes as near as 75 yards. Although this close proximity sometimes forced the command post to defend itself against enemy attacks, Marine officers felt that such a location provided more security. Additionally, this close contact allowed rapid relay of information with resultant quick action. The command post was usually in two echelons—the battle or forward CP with the commanding officer and the operations and intelligence officers, and the rear command post manned by the executive officer and the supply and communications officers.

Defense in the jungle usually took the form of a thinly-stretched perimeter with a reserve in the center. Organization of the positions was seldom complete. Sometimes a double-apron fence of barbed wire blocked some avenue of approach, but usually a few strands of booby-trapped wire sufficed. The Marines took a lesson from the Japanese and cleared fields of fire by removing only a few obstructing branches from trees and underbrush.

As the fighting moved through the jungles, the Marines found that the automatic rifle (BAR) was the most effective weapon for close combat. Light and capable of being fired instantly, the BAR was considered superior to the light machine gun for most occasions. The machine gun, however, had its adherents because of its mobility and low silhouette. These rapid-fire weapons and the highly regarded M–1 rifle were used most of the time in jungle attacks. The .30 caliber carbine most Marines dismissed as too light in hitting power, too rust-prone, and too similar in sound to a Japanese weapon. The heavy machine gun was too heavy and too high for jungle work except for sustained firing in a defensive position.

Supporting weapons such as the 37mm antitank gun and the 75mm pack howitzer were effective in the dense foliage, but these guns were difficult to manhandle into position and could not keep up with the advance in a rapidly changing tactical situation. The 81mm mortars, augmented by the lighter 60mm tubes, provided most of the close support fires. These weapons achieved good results on troops in the open, but were too light for emplaced bunkers. The Marines were supported on two occasions by an Army unit with chemical mortars (4.2 inch), and these were found to be extremely effective against pillboxes and covered positions.

Several attempts were made to use flame throwers in stubborn areas, and these weapons had a demoralizing effect on the Japanese. Many bunkers were evacuated by the enemy before the weapons could be fired against the position. The flame throwers quickly snuffed out the lives of the Japanese remaining in their emplacements. Ignition of the fuel was difficult in the jungle, but Marines solved this by tossing incendiary grenades against a bunker and then spraying the position with fuel. The 2.36-inch antitank bazooka was used on enemy emplacements on Hellzapoppin Ridge, but the crews were unable to get close enough for effective work. The experimental rockets, which were used in the latter stages of the campaign, were highly inaccurate against small area targets and revealed the positions of the Marines.

The Japanese proved to be as formidable an enemy as the Marines expected. The Japanese defenses were well-placed and skillfully concealed and camouflaged. Expenditure of ammunition was small and fire from the bunkers was deadly. Foxholes were cleverly camouflaged and only narrow, inconspicuous lanes of fire were cleared. Fire was limited to short range. Most Marines were killed within 10 yards of enemy positions. In such actions, the rate of the wounded to the dead was high, and the majority of bullet wounds were in the lower extremities. This reflected the Japanese tendency to keep firing lanes low to the ground.

Grenades were used extensively, both by hand and from launchers. The concussion of these enemy grenades was great but fragmentation was poor. The Japanese weapon most feared by the Marines was the 90mm mortar which the enemy employed with great skill. The enemy shell contained an explosive with a high velocity of reaction which resulted in a concussion of tremendous force. The sound of this blast was almost as terrifying as the actual burst. One Marine regiment, assessing the effect of this weapon against the Marines, estimated that at least one-fifth of all battle casualties were inflicted either by the blast concussion or fragments from this enemy shell.

The Japanese, however, displayed what the Marines believed was amazing ineptness in the tactical use of artillery. The gunnery was excellent, and fire from the light and heavy artillery pieces was accurately placed on road junctions, observations points, front lines, and supply dumps. But the placement of firing batteries was so poor that, almost without exception, the Japanese positions were detected and destroyed within hours after firing was started. The enemy fired concentrations of short duration, but, despite this, the muzzle blasts of the weapons were detected and the positions shelled by counterbattery fire.

The enemy may have lacked shells because of the difficulty of supply through the jungle, but "there is no reasonable explanation for the Japanese repeatedly placing their guns on the forward slopes of hills under our observation and then firing them even at twilight or night when the muzzle flashes of the guns fixed their positions as surely as if they had turned a spotlight on them."[23]

As for the artillery support offered by the 12th Marines (and later the 37th Division units), the 3d Marine Division stinted no praise in reporting that the accurate artillery fire was the dominant factor in driving the Japanese forces out of the Torokina area. At least half of the enemy casualties were the direct result of artillery shelling. The preattack bombardment fired before the Battle of Piva Forks was devastating evidence of the force of sustained shelling. The Marines, aware that the all-around defensive system of the Japanese discouraged flank attacks, depended upon the fires of the supporting artillery to pave the way for the infantry to close with the enemy. This infantry-artillery cooperation was one of the highlights of the Bougainville campaign. A total of 72,643 rounds were fired by Marine artillery units in support of the attacks by the IMAC forces.

The greatest advance in supporting arms techniques was in aviation. Prior to the operation, the Marines had a tendency to regard close air support as a risk. The infantrymen wanted and needed direct support from the air, but their faith as well as their persons had been shaken on occasion by air support being too close. Ground troops felt insecure with only smoke to mark their front lines and targets, and airmen were hesitant to bomb after it became known that the Japanese

---

[23] *Ibid*, p. 145.

shot white smoke into the Marine lines to deceive the airmen into bombing their own people. Since it was known from the New Georgia campaign that the enemy troops tended to close in against the front lines to wait out a bombing attack, preparations for close air support at Bougainville started with the idea of developing techniques which would result in maximum accuracy at minimum distance from Marine lines.

Prior to the operation, three air liaison parties were organized and trained. In addition, each battalion and regiment sent a representative to the training school so that each infantry command post would have a man available to direct the close support missions. The pre-strike briefings by both the strike operations officer and a ground officer familiar with the terrain was another innovation. How well this system succeeded is demonstrated by the fact that in 10 missions requested by forces on Bougainville, the only casualties to ground troops were from bombs inexplicably dropped 600 yards from a well-marked target. The other nine strikes were highly successful and in two instances were made within 75 yards of the front lines without harm to the infantry.

Before the Cape Torokina operation, air support was employed mostly against targets beyond the reach of artillery. The Bougainville fighting showed that air could be employed as close to friendly troops and as accurately as artillery, and that it was an additional weapon which could be used to surprise and overwhelm the enemy. Much credit for developing this technique should go to Lieutenant Colonel John T. L. D. Gabbert, the 3d Division Air Officer, and the ground and flying officers who worked with him.

The work of the amphibian tractor companies is another highlight of the operation. Throughout the campaign, these machines proved invaluable. Had it not been for the amtracs, the supply problem would have been insurmountable. The 3d Amphibian Tractor Battalion transported an estimated 22,992 tons of rations, ammunition, weapons, organizational gear, medical supplies, packs, gasoline, and vehicles, as well as reinforcements and casualties. A total of 124 amtracs were landed, but the demand was so great and jungle treks so difficult that only a few were in operating order at any one time. These, however, did yeoman work, and the list of the duties and jobs performed by these versatile machines varied from rescuing downed aviators from the sea to conducting scouting trips along the front lines. The appreciation and affection felt by the Marines for these lumbering lifesavers is best expressed by this comment:

> Not once but all through the campaign the amphibian tractor bridged the vital gap between life and death, available rations and gnawing hunger, victory and defeat. They roamed their triumphant way over the beachhead. They ruined roads, tore down communication lines, revealed our combat positions to the enemy—but everywhere they were welcome.[24]

Not to be overlooked in an analysis of the campaign is the contribution to the success of the Cape Torokina operation by the 19th Marines. The Marine pioneers and engineers, with a Seabee unit attached, landed with the assault waves on D-Day and for some time functioned as shore party personnel before being released to

---

[24] John Monks, Jr., and John Falter, *A Ribbon and a Star, The Third Marines at Bougainville* (New York: Henry Holt and Company, 1945), p. 65.

perform the construction missions assigned. Only one trail to the interior existed at the time of landing; in slightly more than two weeks a rough but passable road connected all units along the perimeter. This provided a vitally necessary supply and reinforcement route for further advances.

One of the reasons the Marines gained and held the perimeter was this ability to construct quickly the necessary roads, airfields, and supply facilities. Engineering units and Seabees of the I Marine Amphibious Corps built and maintained 25 miles of high-speed roads as well as a network of lesser roads within the space of two months. This work was in addition to that on two fighter strips and a bomber field. The Japanese had not developed the Cape Torokina area and therefore could not defend it when the Marines landed. After the IMAC beachhead was established, the Japanese employed a few crude pieces of engineering equipment in an attempt to construct roads from the Buin area toward Empress Augusta Bay. The project was larger than the equipment available. The attacking force for the *Ta* operation had to struggle more than 50 miles through jungle, swamps, and rivers tugging artillery, ammunition, and rations over rough jungle trails. XIV Corps was able to meet this threat over IMAC-constructed roads, which permitted reinforcements to rush to the perimeter at high speed.[25]

The solution of the logistics muddle, which started on D-Day and continued for several weeks, was another accomplishment which marked the Bougainville campaign. With the exception of the unexpected accumulation of supplies and equipment which nearly smothered the beaches early in the operation, logistical planning proved to be as sound as conditions permitted. An advance base at Vella Lavella for a source of quick supply to augment the short loads carried ashore on D-Day proved to be unnecessary in the light of later events. The III Amphibious Force lost a minimum of ships during the Bougainville operation, a stroke of good fortune that accrued as a result of air domination by the Allies. This permitted direct and continued supply from Guadalcanal. More than 45,000 troops and 60,000 tons of cargo were unloaded in the period 1 November–15 December, with no operational losses. With the exception of the first two echelons, all supplies were unloaded and the ships had departed before nightfall.

The lack of an organized shore party for the sole purpose of directing and controlling the flow of supplies over the beachhead was partially solved by the assignment of combat troops for this duty. Luckily, the enemy situation during the early stages at Cape Torokina was such that nearly 40 percent of the landing force could be diverted to solving the problem of beach logistics. This led to many complaints by assault units, which—despite only nominal opposition by the Japanese during the landing stages—protested the loss of combat strength at such a time.

It is to General Vandegrift's credit that quick unloading was assured by the assignment of infantrymen to the task of cargo handling. This, however, was a temporary measure to meet an existing problem and would have been unfeasible in future operations where opposition was more determined. His action, though, was an indication of the growing awareness that

[25] Col Ion M. Bethel ltr to CMC, dtd 16Jul48 (Bougainville Monograph Comment File, HistBr, HQMC).

beach logistics was a vital command responsibility. It is interesting to note that after the Bougainville operation a number of units recommended that the shore party be augmented by additional troops or personnel not essential to operations elsewhere. Later in the war, combat leaders realized that nothing was more essential than an uninterrupted flow of supplies to the assault units, and that the complicated task of beach logistics must be handled by a trained shore party organization and not an assigned labor force of inexperienced personnel.

The Bougainville operation was no strategic campaign in the sense of the employment of thousands of men and a myriad of equipment requiring months of tactical and logistical maneuver. As compared with the huge forces employed in later operations, it was a series of skirmishes between forces rarely larger than a battalion. Yet, with all due sense of proportion, the principal engagements have the right to be called battles, from the fierceness and bravery with which they were contested and the important benefits resulting from their favorable outcome.[26]

The campaign for the Northern Solomons ended as the Bougainville operation drew to a close. The end of the Solomons chain had been reached, and new operations were already being planned for the combined forces of the Southwest and South Pacific commands. The general missions of the campaign had been accomplished. Rabaul was neutralized from airfields on Bougainville, and the victory won by Marines and soldiers at Cape Torokina allowed other Allied forces to continue the attack against the Japanese at other points.

The final accolade to the Bougainville forces is contained in the message General Vandegrift sent his successor, General Geiger, when the Marine forces returned to Guadalcanal:

> I want to congratulate you on the splendid work that you and your staff and the Corps did on Bougainville. The spectacular attack on Tarawa has kind of put Bougainville off the front page; but those of us who know the constant strain, danger, and hardship of continuous jungle warfare realize what was accomplished by your outfit during the two months you were there.[27]

---

[26] *HistDiv Acct*, p. 1.

[27] LtGen Alexander A. Vandergrift ltr to MajGen Roy S. Geiger, dtd 29Jan44 (Vandegrift Personal Correspondence File, HistBr, HQMC).

PART IV

The New Britain Campaign

691-360 O—63——20

New Britain Prelude

GHQ AND ALAMO PLANS[1]

By late November, the parallel drives of South and Southwest Pacific forces envisaged in ELKTON plans had reached the stage where General MacArthur was ready to move into New Britain, with the main target the enemy airfields at Cape Gloucester on the island's western tip. The timing of the attack depended largely upon the availability of assault and resupply shipping, and the completion of Allied airfields on Bougainville and in the Markham-Ramu River valley at the foot of the Huon Peninsula. Complete control of the Vitiaz Strait, the prize sought in the pending operation, would give the Allies a clear shot at the Japanese bases on the New Guinea coast and a secure approach route to the Philippines.

The airstrips building within the IMAC perimeter on Bougainville provided the means for a heightened bomber offensive against Rabaul—raids which could count on strong fighter protection. With the completion of Torokina Field expected in early December, and the first of the Piva Field bomber runways ready by the month's end, SoPac planes could throw up an air barrier against enemy counterattacks on landings on western New Britain. In like manner, the new Markham-Ramu valley fields increased the potential of American and Australian air to choke off raids by the Japanese *Fourth Air Army* based at Madang and Wewak and points west on the New Guinea coast.

While men of the Australian 9th Division drove the Japanese garrison of Finschhafen back along the shore of the Huon Peninsula toward Sio, other Australians of the 7th Division fought north through the Markham-Ramu uplands, keeping the pressure on the retreating enemy defenders. Behind the assault troops, engineers worked feverishly to complete airfields at Lae and Finschhafen, and at Nadzab and Gusap in the valley. Completion schedules were slowed by the seemingly endless rain of the New Guinea region, and the most forward strip, that at Finschhafen, could not be readied for its complement of fighters before 17 December. An all-weather road building from Lae to Nadzab, key to the heavy duty supply of the valley air bases, was not slated to be fully

[1] Unless otherwise noted, the material in this section is derived from: GHQ, SWPA G-3 Jnl and File, Sep43–May44, hereafter *GHQ G-3 Jnl;* ALAMO Force G-3 Jnl and File 23Jul43–10Feb44, in 20 parts, hereafter *ALAMO G-3 Jnl* with part number; ALAMO Force Rept of the DEXTERITY Operation, 15Dec43–10Feb44, n.d., hereafter *DEXTERITY Rept* (all in WW II Recs Div, FRC Alex); USSBS, MilAnalysisDiv, *Employment of Forces under the Southwest Pacific Command* (Washington, Feb47); Craven and Cate, *Guadalcanal to Saipan;* Miller, *Reduction of Rabaul.* Where location citations for documentary sources for this part are missing, the material is in the New Britain Area Operations File, New Britain Monograph and Comment File, or Unit Historical Report File of the Historical Branch, G-3, HQMC.

operational until the 15th. Prior to that date, all troops and equipment were airlifted into the valley by transports of the Allied Air Forces.

Until the new forward bases were ready to support attacks on the next SWPA objectives, all the shipping available to Admiral Barbey's VII Amphibious Force was tied up moving troops and supplies forward from depots at Townsville, Port Moresby, and Milne Bay. The landing craft and ships essential to a move across Vitiaz Strait against Cape Gloucester could not be released for rehearsal and loading until 21 November at the earliest.

These factors, coupled with the desire of planners to execute the movement to the target during the dark of the moon, combined to set D-Day back several times. The target date first projected for Cape Gloucester was 15 November; the date finally agreed upon was 26 December. In both cases, provision for a preliminary landing on the south coast of New Britain was also made, the advance in scheduling here being made from 9 November to 15 December. Altogether, the planning for the operation was characterized by change, not only in landing dates, but also in the targets selected and the forces involved.

General MacArthur chose to organize his troops for the operations on New Guinea and against Rabaul into two task forces. The headquarters of one, New Guinea Force, under Australian General Sir Thomas A. Blamey, conducted the Papuan campaign and directed the offensive operations on the Huon Peninsula. The first operation of the other, New Britain Force, led by Lieutenant General Walter Krueger, USA, was the seizure of Woodlark and Kiriwina. Krueger's command, known as ALAMO Force after July, was next charged with the execution of DEXTERITY (the seizure of Western New Britain).

Technically, Blamey, serving as Commander, Allied Land Forces, had operational control of the national contingents assigned to his command. This assignment included the U.S. Sixth Army, led by Krueger, the Australian Military Forces, also led by Blamey, and the Royal Netherlands East Indies Army. Actually, most of the troops in Sixth Army were assigned to ALAMO Force, which MacArthur kept directly under his General Headquarters (GHQ). The effect of this organizational setup was to make New Guinea Force an Australian command to which American troops were infrequently assigned and to fix ALAMO Force as an American command with very few Australian units.

In contrast to the situation on land, no separate national task forces were created at sea or in the air. At a comparable level with Blamey, directly under MacArthur were two American officers, Vice Admiral Arthur S. Carpender, who led Allied Naval Forces, and Lieutenant General George C. Kenney, who headed Allied Air Forces. Each man was also a national contingent commander in his own force; Carpender had the Seventh Fleet and Kenney the Fifth Air Force. The Dutch and Australian air and naval forces reported to the Allied commanders for orders.

At this time, amphibious operations in the Southwest Pacific, unlike those in Halsey's area where naval command doctrine prevailed, were not conducted under unified command lower than the GHQ level. Control was effected by cooperation and coordination of landing and support forces. Neither amphibious force nor landing force commander had sole charge during the crucial period of the landing itself; the one had responsibility for movement to the target, the other for operations ashore. This deficiency in the control pattern, which was recognized as a critical "weakness" by MacArthur's G–3, Major General Stephen J. Chamberlin, was not remedied until after DEXTERITY was officially declared successful and secured.[2]

Despite its complexity, the command setup in the Southwest Pacific had one indisputable virtue—it worked. And it worked with a dispatch that matched the efforts of Halsey's South Pacific headquarters. Both GHQ and ALAMO Force displayed a tendency to spell out tactical schemes to operating forces, but this was a practice more annoying to the commanders concerned than harmful. Since frequent staff conferences between cooperating forces and the several echelons of command was the rule in planning phases, the scheme of maneuver ordered was inevitably one acceptable to the men who had to make it work. In like manner, differences regarding the strength of assault and support forces were resolved before operations began. When, at various points in the evolution of DEXTERITY plans, the differences of opinion were strong, the resolution was predominantly in favor of the assault forces.[3]

[2] S.J.C. memo for CinC, dtd 12Feb44, Subj: OpsInstns for Manus-Kavieng Ops, in *GHQ G–3 Jnl*, 13Feb44.

[3] "In all these operations General MacArthur gave General Krueger responsibility for coordinating ground, air, and naval planning. This gave the ground force commander a preeminent position." Dr. John Miller, Jr., OCMH, ltr to Head, HistBr, G–3, HQMC, dtd 9May62.

The GHQ procedure in planning an operation was to sketch an outline plan, including forces required and objectives, and then to circulate it to the Allied commands concerned for study, comment, and correction. In the case of DEXTERITY, the principal work on this first plan was done by Lieutenant Colonel Donald W. Fuller, one of the three Marines who were assigned to MacArthur's headquarters as liaison officers shortly after the 1st Division left Guadalcanal. With the others, Lieutenant Colonels Robert O. Bowen and Frederick L. Wieseman, Fuller soon became a working member of the GHQ staff. He summarized the planning sequence at this stage by recalling:

> The routine was then to let all services study the outline plan for a prescribed period and then hold a conference in GHQ. The commanders concerned [including General MacArthur], the General Staff, and Technical Services GHQ then discussed the plan and, if any objections were made, they were resolved at that time. After everyone appeared happy, the plan was filed. It was never issued formally but merely handed out for comment. The next step was the issuance of orders which were called "operations instructions." This was actually the only directive issued to conduct an operation . . . never while I was there was any command ever directed to conduct an operation in accordance with General MacArthur's outline plan.[4]

On 6 May, ALAMO Force had received a warning order from GHQ which set a future task for it of occupying western New Britain by combined airborne and amphibious operations. Engrossed as it then was in preparations for the Woodlark-Kiriwina landings, General Krueger's headquarters had little time for ad-

[4] Col Donald W. Fuller ltr to HistDiv, HQMC, dtd 28Jan52.

vance planning, but, as the weeks wore on toward summer, attention focused on the target date in November. Although the assault troops were not yet formally named, there was no doubt that they would be drawn from the 1st Marine Division, and the division sent some of its staff officers to Brisbane in June to help formulate the original plans. These Marines served with Sixth Army's staff in the Australian city, the location of MacArthur's headquarters as well as those of the principal Allied commands. In the forward area, at Port Moresby in the case of GHQ, and Milne Bay for Krueger's task force, were advance headquarters closer to the scene of combat. Members of the Sixth Army staff were detailed to additional duty as the ALAMO Force staff, and the traffic between Brisbane and New Guinea was heavy. After the initial GHQ outline plan was circulated on 19 July, ALAMO planners were not long in coming up with an alternate scheme of their own.

Conference discussions tended to veer toward the ALAMO proposals which differed mainly in urging that more forward staging areas be used that would place troops nearer the targets selected and thus conserve shipping. A second outline plan circulated on 21 August was closer to the ALAMO concept and named the units which would furnish the assault elements: 1st Marine Division; 32d Infantry Division; 503d Parachute Infantry Regiment. By mid-September, preparations for the operation were far enough along so that detailed planning could be undertaken. MacArthur's operations instructions to cover DEXTERITY were published on the 22d; six days later, an ALAMO Force draft plan was sent to GHQ for approval.

After a summer of discussion, refinement, and change, General Krueger's plan called for seizure of a foothold on New Britain's south shore at Lindenhafen Plantation on 14 November and subsequent operations to neutralize the nearby Japanese base at Gasmata. Once the Gasmata effort was well underway, the main DEXTERITY landings would take place at Cape Gloucester with the immediate objective the enemy airfields there. The eventual goal of the operation was the seizure of control of Western New Britain to a general line including Talasea on the north coast's Willaumez Peninsula and Gasmata in the south.

The assault force chosen for the Gasmata operation (LAZARETTO) was the 126th Infantry,[5] reinforced as an RCT with other 32d Division units and Sixth Army troops. To carry out operations against Cape Gloucester (BACKHANDER), General Krueger designated the 1st Marine Division, reinforced by the 503d Parachute Infantry. In ALAMO Force reserve for all DEXTERITY operations was the remainder of the 32d Division. While the 32d was a unit of the Sixth Army assigned to ALAMO, both the Marine and parachute units were assigned from GHQ Reserve, which came immediately under MacArthur's control.

The basic scheme of maneuver proposed for BACKHANDER called for a landing by the 7th Marines (less one of its battal-

[5] The assault force initially selected was the 5th Marines, reinforced as an RCT, with additional supporting units from the Sixth Army, all under the command of Brigadier General Lemuel C. Shepherd, Jr., ADC of the 1st Marine Division. General Shepherd and his staff completed preliminary plans for the operation on board ship en route from Melbourne to Milne Bay, but found on arrival that the 126th Infantry had been designated the assault element instead. Gen Lemuel C. Shepherd, Jr., ltr to CMC, dtd 20Aug62, hereafter *Shepherd ltr*.

ions), organized as Combat Team C, on north shore beaches between the cape and Borgen Bay. Simultaneously, the remaining battalion of the 7th, suitably reinforced, would land near Tauali just south of the cape to block the trail leading to the airfields. Shortly after the Marines landed, the 503d was to jump into a drop zone near the airfields and join the assault on enemy defenses. The 1st Marines, organized as Combat Team B, would be in immediate reserve for the operation with Combat Team A (the 5th Marines) on call subject to ALAMO Force approval. The intent of the operation plan was to use as few combat troops as possible and still accomplish handily the mission assigned.

On 14 October, GHQ returned the ALAMO plan approved and directed that Combat Team B be staged well forward along the New Guinea coast, at Oro Bay or Finschhafen rather than Milne Bay, if it developed that Allied Air Forces could provide adequate daylight cover over loading operations. Similarly, authorization was given to increase the strength of the assault forces if late intelligence of the enemy garrison developed the need. In order to conserve operating time and get more value out of the shipping available, maximum use was ordered made of Oro Bay as a supply point for BACKHANDER.

After considering the ALAMO Force plan, Admiral Carpender agreed to use two of his overworked transports together with all the LSTs, LCIs, and smaller amphibious craft available to Seventh Fleet to support the operation. Close on the heels of this commitment, Admiral Barbey protested the exposure of his priceless transports to enemy attack and proposed instead "to restrict the assault ships to those which could discharge their troops directly to the beaches," in order "to reduce the turn-around time and thereby reduce the hazards from air attack."[6] The matter was brought to MacArthur's attention, but he refused to specify the equipment to be used; however, he did relay to General Krueger his intention that all "troops and supplies that you wished landed must be landed and in such order as you consider necessary."[7] The Seventh Fleet commander then made arrangements to borrow six APDs from Third Fleet in return for extending the loan period of four APDs then being used by Halsey in South Pacific operations. Once replacement ships were obtained, the attack transports were scratched from the task organization of assault shipping.

When the first ALAMO plan for DEXTERITY was being prepared, the probable enemy garrison in the target area was estimated at being between 3,000 and 4,000 men. Toward the end of October, the evidence assembled by coastwatchers, scouts, and other intelligence agencies pointed strongly to a sharp increase in the number of defenders, particularly in the Cape Gloucester vicinity. Krueger's order of battle officers now considered that there were as many as 6,300 troops to oppose the landings and probably no less than 4,100. In order to counter this new strength, it seemed imperative to the commanders concerned that the BACKHANDER landing force be reinforced.

---

[6] VAdm Daniel E. Barbey ltr to ACofS, G-3, HQMC, dtd 10May62, hereafter *Barbey ltr.* Admiral Barbey argued against the use of the transports after General Kenney said that he could not provide adequate air cover for the Gasmata operation.

[7] CofS, GHQSWPA ltr to CG, ALAMO For, dtd 18Oct43, in *ALAMO G-3 Jnl No. 3.*

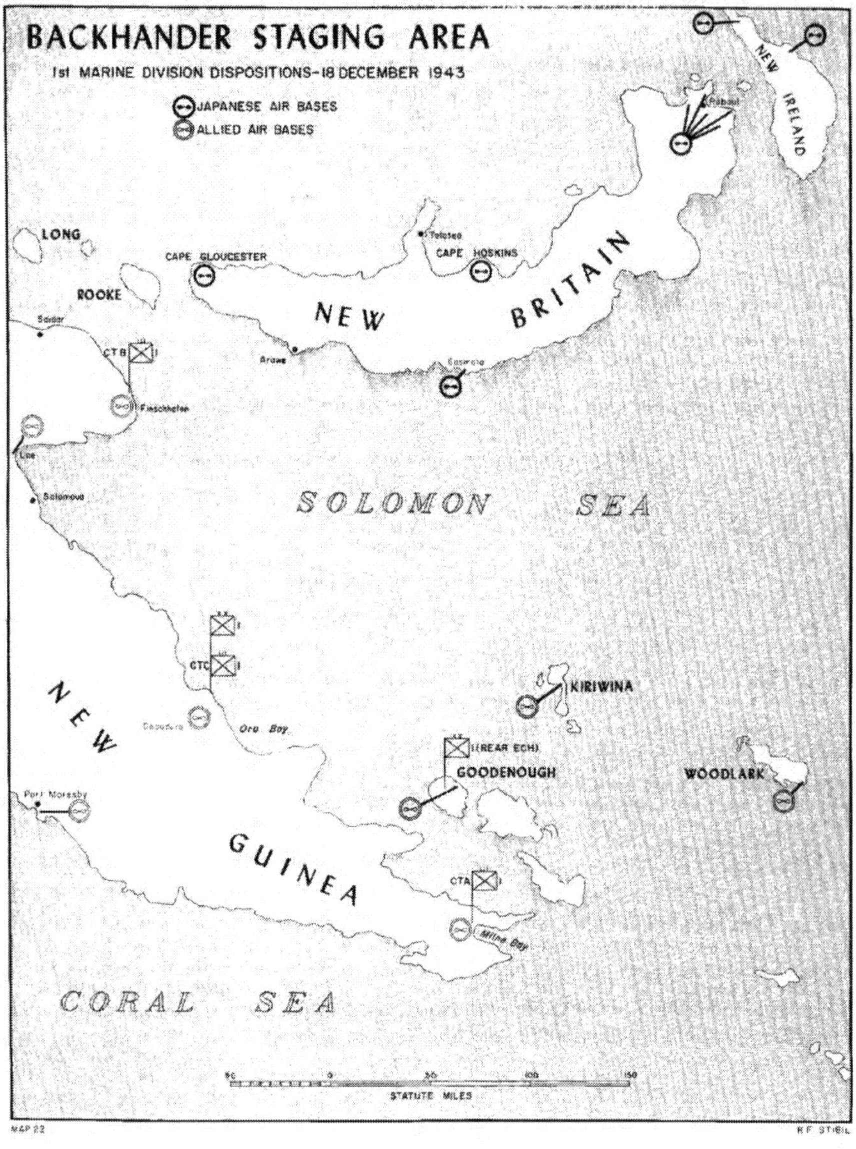
BACKHANDER STAGING AREA
1st MARINE DIVISION DISPOSITIONS-18 DECEMBER 1943
JAPANESE AIR BASES
ALLIED AIR BASES
NEW IRELAND
Rabaul
LONG
CAPE GLOUCESTER
Talasea
CAPE HOSKINS
ROOKE
NEW BRITAIN
Saidor
CTB
Arawe
Gasmata
Finschhafen
Lae
Salamaua
SOLOMON SEA
CTC
KIRIWINA
Dobodura
Oro Bay
NEW GUINEA
I(REAR ECH)
GOODENOUGH
WOODLARK
Port Moresby
CTA
Milne Bay
CORAL SEA
50 0 50 100 150
STATUTE MILES
MAP 22
R F STIBIL

On 2 November, during a conference at 1st Marine Division headquarters attended by the division commander, Major General William H. Rupertus, Admiral Barbey, and Colonel Clyde D. Eddleman, Krueger's G-3, the matter was considered at length. Under existing plans, the ratio of assault troops to defenders would be 1.8 to 1 if the 503d landed and only 1.3 to 1 if weather prevented the drop. If an additional Marine battalion landing team was committed, the ratio would rise to 2 to 1 (or 1.7 to 1 without paratroops), a balance closer to the clear superiority experience demanded for an attacking force. Eddleman recommended to Krueger that a landing team of Combat Team B be employed at Tauali and that all of Combat Team C land east of the airfield. General Rupertus seconded this finding with proposals for several alternative landing schemes, one favoring the change endorsed by Eddleman. Conclusively, the ALAMO staff came up with a new estimate of the situation on 5 November that recommended an additional battalion be employed. A query from ALAMO headquarters to GHQ at Brisbane brought a quick reply that Krueger had full authority to use the battalion if he so desired.

In addition to getting a sufficient number of assault troops ashore at BACKHANDER, Krueger's staff was gravely concerned about the possible need for reinforcement. To meet this contingency, they recommended that shipping to lift the remainder of Combat Team B to the target be available at its staging area on D-Day. The same requirement was stated for Combat Team A, then at Milne Bay and the unit of the 1st Division farthest from Cape Gloucester. Admiral Barbey was prepared to furnish the ships required, using LCIs to move Combat Team B and transports to shift Combat Team A to Oro Bay, where it could move on, should it prove necessary, in APDs and LCIs. Many of the vessels Barbey designated would have to do double duty; first at Gasmata, then at Gloucester.

With amphibious shipping heavily committed until late November to support Huon Peninsula operations, the time for rehearsal, training, and reoutfitting before DEXTERITY got underway was woefully short. The lapse of six days then figured between LAZARETTO and BACKHANDER landings gave planners little leeway, and tight scheduling for maximum shipping use made no provision for losses at Gasmata. Under the circumstances, General Krueger asked General MacArthur for permission to set back the landing date for LAZARETTO to 2 December. At the same time, in consideration of the delay in the completion of the Huon airfields and the increased demand for shipping, Krueger suggested a BACKHANDER D-Day of 26 December. MacArthur approved deferral of the landing dates for the two operations and eventually decided upon 15 and 26 December after additional changes in forces and objectives.

Even while the date for LAZARETTO was being altered, the need for undertaking the operation at all was being seriously questioned, particularly by Allied Air Forces.[8] Kenney's staff was swinging strongly to the opinion that the forward airstrip planned for the Linden-

---

[8] Major General Ennis C. Whitehead, Commander, Advanced Echelon, Fifth Air Force thought that the whole New Britain operation was unnecessary. See his letter to General Kenney, dtd 11Nov43, quoted in Craven and Cate, *Guadalcanal to Saipan*, pp. 329-330.

hafen Plantation beachhead was not necessary to future air operations. Reinforcing this conclusion was the fact that long-range fighter planes would be in short supply until February 1944. Replacement and reinforcement aircraft scheduled to arrive in the Southwest Pacific had been delayed, and, in November, Kenney had to curtail daylight strikes on Rabaul to conserve the few planes he had. The Allied air commander could promise fighter cover over the initial landings at Gasmata, but nothing thereafter. The assault troops would have to rely on antiaircraft fire to fend off Japanese attacks. Perhaps the most disquieting news was that the enemy was building up his garrison at Gasmata, apparently anticipating an Allied attack.

With General Kenney reluctant to regard LAZARETTO as an essential operation and unable to provide aerial cover after it began, the Gasmata area lost its appeal as a target. On 19 November, following a conference with air, naval, and landing force representatives, the ALAMO Force G-3 concluded that carrying through the operation would mean "that we can expect to take considerable casualties and have extensive damage to supplies and equipment as a result of enemy bombing operations subsequent to the landing."[9] An alternative objective, one closer to Allied bases, less vulnerable to enemy air attack, and more lightly defended, was sought.

Generals Krueger and Kenney and Admiral Barbey met on 21 November to decide on a new objective. They chose the Arawe Islands area off the south coast of New Britain, 90 miles closer to Cape Gloucester than Gasmata. Kenney's staff had considered placing a radar station at Arawe before the LAZARETTO project clouded and still was interested in the area as a site for early-warning radar guarding the approaches from Rabaul. The conferees agreed that Arawe would also be a good location for a motor torpedo boat base from which enemy barge traffic along the coast could be blocked. There is evidence that the Commander, Motor Torpedo Boats of the Seventh Fleet was less convinced of the need for this new base, but his objections were evidently not heeded at this time.[10] A powerful argument in favor of the choice was the denial of the site to the Japanese as a staging point through which reinforcements could be fed into the Cape Gloucester area. Then, too, the diversionary effect of the attack might draw off defenders from the main objective.[11]

General MacArthur quickly approved the findings of his field commanders and confirmed the landing date they asked for, 15 December.[12] On 22 November, amend-

---

[9] ALAMO G-3 memo to CofS, dtd 19Nov43, Subj: Rept of Conference, in *ALAMO G-3 Jnl No. 4*.

[10] Morison, *Breaking the Bismarcks Barrier*, p. 372; ALAMO ForEngr memo of conference with Cdr Mumma, dtd 5Dec43, Subj: PT Boat Requirements DIRECTOR in *ALAMO G-3 Jnl No. 5*.

[11] "The determining factor in the selection of Arawe was a statement by Kenney that he could supply air protection for the assault from his air bases on New Guinea. . . ." *Barbey ltr.*

[12] Once Arawe's landing date was set, the final decision on D-Day at Cape Gloucester could be made. At a meeting at General Krueger's headquarters on Goodenough early in December, General MacArthur suggested 21 December, but approved the 26th "so that the assault craft used in the Arawe landing could return in time for a brief training period with the 1st Marine Division before the BACKHANDER operation." RAdm Charles Adair ltr to Head, HistBr, G-3, dtd 1Jun62.

ed operations instructions for DEXTERITY were issued by GHQ cancelling LAZARETTO and substituting for it the DIRECTOR (Arawe) operation. An immediate benefit of the switch was a gain in shipping available for BACKHANDER and an increase in the number of troops ready for further operations. The suspected enemy garrison at Arawe was far weaker than that known to be at Gasmata, and the mission assigned the landing force was less demanding in men and materiel resources.

The LAZARETTO task force built around the 126th Infantry was dissolved and its elements returned to ALAMO Force reserve. A new task force, half the strength of its predecessor, was formed using troops released from garrison duties on Woodlark and Kiriwina. Named to command the DIRECTOR Force, which centered on the 112th Cavalry, was Brigadier General Julian W. Cunningham, who had commanded the regiment during the occupation of Woodlark.

The abandonment of LAZARETTO gave GHQ planners a welcome bonus of combat troops available for further DEXTERITY operations. Once the Japanese threat at Cape Gloucester was contained and both flanks of Vitiaz Strait were in Allied hands, the seizure of Saidor on the north New Guinea coast could follow swiftly. An outline plan for the Saidor operation issued on 11 December mentioned that either an RCT of the 1st Marine or the 32d Infantry Divisions might be the assault element. Within a week, changes in the composition of the BACKHANDER Force had narrowed the choice to an Army unit.

On 17 December, impressed by the evident success of DIRECTOR operations, General MacArthur ordered preparations for the capture of Saidor to get underway with a target date on or soon after 2 January. Many of the supporting ships and planes employed during the initial landings at BACKHANDER would again see service at the new objective. ALAMO Force issued its field order for Saidor on the 22d, assigning the 126th Infantry (Reinforced) the role of assault troops. Krueger saved invaluable preparation time by organizing a task force which was essentially the same as the one which had trained for LAZARETTO.

One aspect of the BACKHANDER plan of operations—the air drop of the 503d Parachute Infantry—gathered opposition from all quarters as the time of the landing grew nearer. Krueger's staff was never too enthusiastic about the inclusion of the paratroopers in the assault troops, but followed the outline put forward by GHQ. Rupertus was much less happy with the idea of having a substantial part of his force liable to be cancelled out by weathered-in fields or drop zones at the most critical stage of the operation.[13] On 8 December, Allied Air Forces added its opposition to the use of the 503d with Kenney's director of operations stating that "ComAAF does not desire to participate in the planned employment of paratroops for DEXTERITY."[14] The air commander understood that a time-consuming plane shuttle with a series of drops

[13] MajGen William H. Rupertus ltr to LtGen Alexander A. Vandegrift, dtd 7Dec43 (Vandegrift Personal Correspondence File, HQMC). Rather graphically, General Rupertus noted "if the weather is bad, then the jumper boys won't be there to help me & I'll be on the beach with a measly C. T.!"

[14] DirOps, AAF Check Sheet to G-3, GHQ, dtd 8Dec43, in *GHQ G-3 Jnl*, 8Dec43.

was planned, a maneuver which would greatly increase the chance of transports being caught by the expected Japanese aerial counterattack on D-Day. In addition to this consideration, Kenney found that the troop carriers needed to lift the 503d would crowd a heavy bomber group off the runways at Dobodura, the loading point for the air drop. The displaced bombers would then have to operate from fields at Port Moresby, where heavy weather over the Owen Stanley Mountains could keep them from supporting the operation.

The stage was set for the change which took place on 14 December at Goodenough Island where MacArthur and Krueger were present to see the DIRECTOR Force off to its target. The two generals attended a briefing on the landing plans of the 1st Marine Division where Colonel Edwin A. Pollock, the division's operations officer, forcefully stated his belief that the drop of the parachute regiment should be eliminated in favor of the D-Day landing of the rest of Combat Team B to give the division a preponderance of strength over the enemy defenders. Pollock's exposition may have swung the balance against use of the paratroops,[15] or the decision may have already been assured;[16] in either event, the order went out the next day over Krueger's signature changing the field order for BACKHANDER. The seizure of Cape Gloucester was now to be an operation conducted by a unit whose components had trained and fought together.

[15] LtCol Frank O. Hough and Maj John A. Crown, *The Campaign on New Britain* (Washington: HistBr, HQMC, 1952), p. 19, hereafter Hough and Crown, *New Britain Campaign*.

[16] MajGen Clyde D. Eddleman, USA, ltr to HistBr, G–3, HQMC, dtd 28Apr52.

TRAINING AND STAGING BACKHANDER FORCE[17]

For effective planning of DEXTERITY operations, General Krueger and the Allied Forces commanders had to strike a balance between the flexibility necessary to exploit changing situations and the exact scheduling required for effective employment of troops, ships, and supplies. The changes wrought in ALAMO Force plans were keyed, therefore, to the physical location of troops in staging and training areas, to the contents and replenishment potential of forward area depots, and to the number and type of amphibious craft available. Throughout the later stages of the planning for BACKHANDER, the variations in scheme of maneuver and number of troops employed were based upon a constant factor, the readiness of the 1st Marine Division for combat.

When the division left Guadalcanal on 9 December 1942, many of its men were walking hospital cases wracked by malarial fevers or victims of a host of other jungle diseases. All the Marines were bone tired after months under constant combat strain and a rest was called for. No more

[17] Unless otherwise noted, the material in this section is derived from: *ALAMO G–3 Jnl*; VII PhibFor ComdHist 10Jan43–23Dec45, n.d., hereafter *VII PhibFor ComdHist*; VII PhibFor Rept on Cape Gloucester Op, dtd 3Feb44, hereafter *VII PhibFor AR*; CTF 76 (VII PhibFor) WarD, Aug43, n.d. (COA, NHD), hereafter *CTF 76 Aug43 WarD* and following appropriate months; *CTF 76 Sep–Dec43 WarDs*; 1st MarDiv SAR Cape Gloucester Op, Phase I, Planning and Training, n.d., hereafter *1st MarDiv SAR* with appropriate phase or annex; 1st MarDiv WarD, Oct43, n.d., hereafter *1st MarDiv WarD* with appropriate month; *1st MarDiv Nov43 WarD*; Hough and Crown, *New Britain Campaign*.

perfect tonic could have been chosen than Australia.

Melbourne, where the division arrived on 12 January, became a second home to the men of the 1st. Fond memories of the city and of the warm reception its people gave the Marines were often called to mind in later years by those who served there. Rewarded in a hundred ways by its release from the jungles of the Solomons and a return to civilization, the 1st Division slowly worked its way back to health and battle fitness. Although at first as many as 7,500 men at a time were down with malaria or recovering from its ravages,[18] the number dwindled as climate and suppressive drugs took effect.

A training program, purposely slow-starting, was begun on 18 January with emphasis during its initial phases on the reorganization and reequipment of units and the drill and physical conditioning of individuals. By the end of March, practice in small-unit tactics was the order of the day. Through April, May, and June, all men qualified with their new basic weapon, the M–1 rifle, a semi-automatic which replaced the bolt-action Springfield M–1903 carried by American servicemen since the decade before World War I. Nostalgia for the old, reliable '03 was widespread, but the increased firepower of the M–1 could not be denied.

In April and May, battalions of the 5th and 7th Marines practiced assault landings on the beaches of Port Philip Bay near Melbourne. One of the two ships used was a converted Australian passenger liner and the other was the only American attack transport (APA) in the Southwest Pacific. Sent over from Halsey's area, the APA was assigned to Barbey's command to give him at least one of the Navy's newest transports for training purposes.[19] With the help of the experienced Marines, VII Amphibious Force officers worked out a series of standing operating procedures during the exercises which would hold for future training and combat usage.

While it was in Australia, the 1st Division had no opportunity to use the variety of specialized amphibious shipping that had come into use since its landing on Guadalcanal. Most LSTs and LCIs were sent forward to Papuan waters as soon as they arrived from the States; the LCTs, which were shipped out in sections, joined the ocean-going ships as soon as they were welded together. In the combat zone, the landing ships were urgently needed to support the operations of New Guinea and ALAMO Forces. The Marines' chance to familiarize themselves with the new equipment would come in the forward area where GHQ planned to send the division after it completed a summer of intensive field training in the broken country around Melbourne.

As a necessary preliminary to effective large unit training and operations, the division was organized into combat and landing teams on 25 May. Many of the supporting unit attachments were the same as those in regimental and battalion em-

[18] Col John E. Linch comments on draft of Hough and Crown, *New Britain Campaign*, dtd 11Apr52.

[19] Commenting on the state of this APA, the *Henry T. Allen*, Admiral Barbey noted: "It was not the newest, but if not the oldest, it was in the worst condition. It arrived in our area in need of an overhaul so extensive that Australian yards were reluctant to undertake it. It leaked oil so badly that it never could be used in the combat zone throughout the campaign. It was used as an administrative flagship in the rear areas." *Barbey ltr.*

barkation groups already in existence. In terms of the general order outlining the assignments, the combat team was described as "the normal major tactical unit of the division," and the landing team was "regarded essentially as an embarkation team rather than a tactical unit."[20] Many of the reinforcing units in the battalion embarkation groups would revert to regimental control on landing, and, similarly, the division expected to regain control very soon after landing of supporting headquarters and reserve elements included within regimental embarkation groups.

In mid-summer, while 1st Division units were either in the midst of combat team exercises, preparing to take the field, or squaring away after return, a Sixth Army inspection team made a through survey for General Krueger of the Marine organization. The Army officers came away much impressed, noting:

> [This division] is well equipped, has a high morale, a splendid esprit and approximately 75% of its personnel have had combat experience. The average age of its enlisted personnel is well below that in Army divisions. . . . At the present time, the combat efficiency of this division is considered to be excellent. In continuous operations, this condition would probably exist for two months before declining. With rest and replacements between operations, it is believed that a better than satisfactory combat rating could be maintained over a period of six months.[21]

The inspection team expressed some concern about the substantial incidence of malaria in the division's ranks, but made its finding of combat efficiency "Excellent" despite this. The very apparent high morale was attributed to the 1st's experienced leadership, as all the division staff, all regimental and battalion, company and battery commanders had served on Guadalcanal.[22] This pattern of veteran leadership was evident down through all ranks of the division with a good part of the infantry squads and artillery firing sections led by combat-wise NCOs.

The first echelons of the division to move northward to join ALAMO Force were the engineer and pioneer battalions of the 17th Marines. On 24 August, the engineers (1/17) sailed from Melbourne for Goodenough Island to begin construction of the division's major staging area. The pioneer battalion (2/17) moved by rail to Brisbane, "where it drew engineering supplies, transportation, and equipment of the regiment, and stood by to load this material on board ship if a wharf labor shortage developed."[23] It departed for Goodenough on 11 September. The 19th Naval Construction Battalion, which served as the 3d Battalion, 17th Marines, was working at the big U.S. Army Services of Supply (USASOS) base at Cairns, north of Townsville, and remained there under Army control until the end of October.

The formal movement orders were issued to the division on 31 August, setting forth the priority of movement of units and the amount and type of individual

[20] 1st MarDiv GO No. 83, dtd 25May43, in *ALAMO G-3 Jnl No. 1*.

[21] Rept of Inspection 1st MarDiv by Sixth Army Inspection Team, 9-13Aug43, dtd 28Sep43, Anx 1, p. 11. "On 22 August, General Krueger himself arrived in Melbourne for a two-day inspection of the division and upon his departure commented favorably on the observations he had made." *Shepherd ltr.*

[22] Rept of Sixth Army Inspection, dtd 28Sep43, *op. cit.*, p. 11.

[23] BGen Robert G. Ballance ltr to ACofS, G-3, HQMC, dtd 14Jun62, hereafter *Ballance ltr.*

and organizational equipment that should be taken. In general, 40-days' rations, quartermaster, and medical supplies were to be loaded, as well as a month's supply of individual and organizational equipment and 10 units of fire for all weapons. Reserve stocks of all classes of supply, and any material not available on first requisition to USASOS, were to be forwarded to the division's resupply points. The troops themselves were to take only the clothing and equipment necessary to "live and fight,"[24] storing service greens and other personal gear in sea bags and locker boxes for the better day when the pending operation would be over and a hoped-for return to Australia the reward for success.

The division's main movement began on 19 September when Combat Team C finished loading ship and sailed from Melbourne in a convoy of three Liberty ships; its destination was Cape Sudest near Oro Bay. Three more convoys carrying the remainder of the division cleared Melbourne over the next several weeks, with the last Libertys pulling out with the rear echelons of division headquarters and Combat Team B on 10 October. The 18 ships used, hastily converted from cargo carriers, were far from ideal troop transports. Galleys, showers, and heads had to be improvised on weather decks, and the holds were so crowded that many men preferred "to sleep topside, fashioning out of ponchos and shelter halves and stray pieces of line rude and flimsy canvas housing."[25] Happily, the voyage had an end before the objectionable living conditions became a health hazard.

After its move, General Rupertus' command was dispersed in three staging areas that corresponded, in nearness to the target, with the roles the combat teams were scheduled to play in BACKHANDER operations. Encamped at Cape Sudest was the main assault force, Colonel Julian N. Frisbe's 7th Marines, suitably reinforced as Combat Team C. At Milne Bay, farthest from Cape Gloucester, was Combat Team A, centered on Colonel John T. Selden's 5th Marines; Selden's troops, scheduled for a time to be the assault force at Gasmata, were now ticketed as reserves for Gloucester. On Goodenough Island, Combat Team B, under Colonel William J. Whaling, and the remainder of the division moved into camps the 17th Marines had wrested from the jungle. On 21 October, ALAMO Force Headquarters joined the division on Goodenough, moving up from Milne Bay in keeping with General Krueger's desire to keep close to the scene of combat. (See Map 22.)

Each division element, immediately after arrival in its new location, unloaded ship and turned to setting up a tent camp. Before long, combat training was again underway at Oro Bay and Goodenough with emphasis on jungle operations, a species of warfare all too familiar to Guadalcanal veterans. At Milne Bay, Combat Team A had to clear and construct its own camp area while providing 800–900-man working parties daily to help build roads and dumps in the base's supply complex. Colonel Selden rotated the major labor demand among his battalions, giving them all a chance to work as a whole to finish their living area and get a

---

[24] 1st MarDiv AdminO 4–43, dtd 7Sep43, in *1st MarDiv Oct43 WarD*.

[25] George McMillan, *The Old Breed, A History of the First Marine Division in World War II* (Washington: Infantry Journal Press, 1949), p. 162, hereafter McMillan, *The Old Breed*.

start on jungle conditioning and combat training.[26]

On 1 November, with the arrival of the Seabees of 3/17 at Goodenough, the organic units of the 1st Division were all assembled in the forward area. The Japanese attempted a lively welcome by sending bombers and scout planes over Oro Bay and Goodenough throughout the staging period; these raids, seldom made in any strength or pursued with resolution, caused no casualties or damage in Marine compounds and only slight damage to adjoining Army units. The enemy effort had nothing but nuisance value as far as stemming preparations for future operations was concerned, but it did give the Japanese a pretty fair idea of the Allied buildup along the New Guinea coast.

In order to give the troops assigned to DEXTERITY adequate opportunity to familiarize themselves with the landing craft they were to use, Admiral Barbey had to devise a system that would allow him to use a small number of craft to train a large number of men. There was no time to fit the construction of an amphibious training base on New Guinea, similar to those in Australia, into support plans for the move against New Britain. Five months was the lowest estimate of the time necessary to complete such an undertaking, and the men, materials, and ships necessary to support it could not be spared from current operations. On 16 August, Barbey recommended the establishment of a mobile training unit consisting of enough landing craft and supporting auxiliaries to perform the amphibious training mission in the forward area. Carpender and MacArthur both concurred in the recommendation, and the mobile group was organized with headquarters at Milne Bay.

Marines from all three 1st Division staging areas made their practice landings on the beaches of Taupota Bay, a jungled site on the north shore of New Guinea in the lee of the D'Entrecasteux Islands. At Taupota there was opportunity to put ashore enough troops, vehicles, and supplies to test landing and unloading techniques. On 22 October, 1/1 in two APDs and two LSTs lifted from Goodenough to the practice beaches. There the assault troops went ashore in LCVPs from the destroyer-transports and were followed by landing ships with a full deck-load of vehicles plus 40 tons of bulk stores. Since all of the division pioneers were attached to the 7th Marines at Cape Sudest, there was no experienced nucleus for the shore party. One LST took three hours to unload, the other four and a half, prompting VII Amphibious Force to observe that "unloading parties provided were quite inadequate and very little appreciation was shown for the necessity of getting the craft off the beach as quickly as possible." [27]

The critique of the faults of this landing, together with a more comprehensive application of the newly promulgated Division Shore Party SOP,[28] enabled 2/1, using the same ships and landing scheme

[26] 2/5 Record of Events, 16Sep43–9May44, entry of 20Oct44.

[27] *CTF 76 Oct43 WarD*, p. 22.

[28] When General Shepherd reported as ADC of the 1st Division, he found the Shore Party SOP to be "woefully deficient." With General Rupertus' approval, he prepared a new SOP based on that of the 3d Marine Division which he "had helped prepare under the direction of Major General Charles D. Barrett, who had made a study of the deficiencies noted during the Guadacanal landing and subsequent developments in this important phase of a landing operation." *Shepherd ltr.*

two days later, to halve unloading time. On 28 and 31 October, the 7th Marines mounting out from Oro Bay and Cape Sudest was able to land all infantry battalions and supporting elements in the pattern of assault waves coming from APDs, followed by LCIs with support troops, and LSTs with vehicles and bulk cargo. The unloading time per LST was cut to less than an hour by means of adequate troop labor details and the application of cargo handling techniques developed in training for the landings. Combat Team A was able to profit from this experience when the reinforced 5th Marines' battalions made a series of trouble-free practice landings from LSTs and LCIs between 14 and 30 November.[29]

During the amphibious training period, Marine assault troops landed as they would at Cape Gloucester in ships' boats, LCVPs and LCMs. No serious obstacles existed off the chosen beaches which could bar landing craft from nosing ashore, and the amphibian tractors organic to the 1st Division were reserved for logistical duties. That role promised to be quite important to the success of the operation. Extensive tests in the jungles that fringed the staging areas showed that the LVT could negotiate terrain, particularly swamp forest, that was an absolute barrier to other tracked vehicles, even the invaluable bulldozer. The cargo space of 1st Amphibian Tractor Battalion's standard LVT(1) Alligator could hold 4,500 pounds, and that of the few newer, larger LVT(2) Buffaloes that were received a few days before embarkation could contain 6,500 pounds.[30] The tests also revealed that the LVT was a splendid trail breaker for tractors, a fact that had immediate application in the plans for employment of artillery at Cape Gloucester.

In Combat Team C's Oro Bay training area, Alligators were used successfully to smash a path through the jungle for 4/11's prime movers and the 105mm howitzers they towed. To distribute the weapons' weight over a large area, truck wheels were mounted hub to hub with the howitzer wheels.[31] The one-ton trucks assigned to haul 1/11's 75mm pack howitzers proved unequal to the task of following in the LVTs' rugged trace, and the 11th Marines' commander, Colonel Robert H. Pepper, took immediate steps to secure light tractors from the Army as supplementary prime movers for his three pack battalions.[32] The potential of the LVTs was firmly demonstrated to the ALAMO Force artillery officer who watched them work in a swampy area of Goodenough on 4 December. He reported to the ALAMO chief of staff:

> Performance was impressive. Knocked over trees up to 8'' in diameter and broke a trail through the densest undergrowth. The branches, trunks, and brush formed a natural matting capable of supporting tractors and guns. I believe that with a limited amount of pioneer work practically any jungle country can be traversed with artillery drawn by tractors if preceded by two LVTs.[33]

As this demonstration was being conducted, the movements to final staging areas had begun. On 3 December, a detachment of Combat Team B left Good-

[29] *Ballance ltr.*

[30] ONI 226, Allied Landing Craft and Ships, dtd 7Apr44 and Supplement No. 1, *ca.* Jun45.

[31] LtCol Joe B. Russell ltr to CMC, dtd 14Mar 52.

[32] Col Louis A. Ennis comments on draft of Hough and Crown, *New Britain Campaign*, dtd 17Apr52.

[33] ALAMO ArtyO memo to CofS, dtd 4Dec43, in *ALAMO G-3 Jnl No. 5.*

691-360 O—63——21

enough for Cape Cretin near Finschhafen, followed on the 11th by the rest of the 1st Marines and its attached units. From this point nearest to the target area, the landing team assigned the role of taking and holding the trail block at Tauali, the 2d Battalion, 1st Marines, and the rest of the combat team destined to follow the 7th Marines across the main beaches, would mount in separate convoys for D-Day landings. Accompanying Combat Team B was the Assistant Division Commander, Brigadier General Lemuel C. Shepherd, Jr., and his staff.

Between 7 and 18 December, all elements of the division that would land with or under command of Combat Team C during the initial phase of BACKHANDER, moved from Goodenough to Cape Sudest. General Rupertus, the task force commander, shifted his CP to Oro Bay at this time, leaving only the division rear echelon on Goodenough. Colonel Selden's Combat Team A at Milne Bay made ready to sail for Cape Sudest on D minus one (25 December) to be in position as division reserve to answer a call for reinforcements. The team was to move from Milne in transports and transfer at the Oro Bay staging area to landing ships for further movement to Cape Cretin. ALAMO Force placed a hold order on the commitment of one 5th Marines battalion (3/5) so that it might be used to seize either Rooke or Long Islands as the site for a sentinel radar guarding the overwater approaches to the main objective from the northwest. Besides this tentative mission, another possible employment of the battalion was as part of the reserve in support of the assault units of the 1st Division.

The major part of the BACKHANDER Force assembled in the staging area for the assault phase of the operation was organic to the 1st Marine Division. One unit, in fact, was peculiarly the division's own, its air liaison detachment. Impressed by the need of a light plane squadron to handle reconnaissance and air spotting, the division commander's personal pilot, Captain Theodore A. Petras, and the division air officer, Lieutenant Colonel Kenneth H. Wier, had recommended in early summer of 1943 that such a unit be formed within the division. General Vandegrift agreed and was able to persuade General MacArthur to provide the division with 12 Army L-4 Piper Cubs. When General Rupertus took over the 1st's command, he endorsed the idea fully and directed Petras to organize the unit and run its training program.

Volunteers with flying experience were called for and 60 men applied; from this group, 12 pilots, 1 officer and 11 enlisted men, were selected. Mechanics for the Cubs were similarly chosen and the maintenance men who kept up Petras' transport served as their instructors. For two and a half months before D-Day, the makeshift air force worked intensively to reawake flying skills and to learn with the artillery suitable techniques for air spotting of targets. The most serious problem faced was the lack of adequate air-ground communication; the radios available were unreliable and a system of visual signals was developed. For movement to Cape Gloucester, the planes were dismantled and loaded on board LSTs scheduled to arrive on D-Day.[34]

In addition to the force of light planes, General Rupertus also had a pool of land-

[34] Hough and Crown, *New Britain Campaign*, App V, "The Cape Gloucester Air Force," pp. 202–203.

ing craft under his direct command, both elements which would give him greater flexibility in meeting combat emergencies or in taking advantage of sudden changes of fortune that might affect the Japanese defenders. The boat crews were not sailors, however, but Army amphibian engineers, members of a provisional boat battalion of the 592d Engineer Boat and Shore Regiment. The engineers, manning LCVPs and LCMs, were products of a special training program in the States through which the Army anticipated and met some of the problems that arose in conducting amphibious operations. Used in strength first in the Lae-Finschhafen campaign, the engineers were prepared to move supplies of troops from ship to shore or shore to shore, to act as a shore party, to man beach defenses, and, in sum, to make themselves generally useful.[35] At Cape Gloucester, since the Marines had their own shore party, the amphibian engineers provided only a part of their services, but these were of inestimable value.

A more usual attachment to BACKHANDER Force in Marine experience was the assignment of Colonel William H. Harrison's 12th Defense Battalion to the operation. The rapid-firing 40mm guns of the battalion's Special Weapons Group would augment the fire against low-level attackers put up by Battery A of the division's 1st Special Weapons Battalion. The searchlight battery and the 12th's 90mm Group would guard the beachhead from bombers making their runs above the reach of automatic weapons. A platoon of 155mm guns from the Seacoast Artillery Group was to come in with the light antiaircraft elements of the battalion,[36] while the remainder of Harrison's artillery would arrive with the garrison force.

BACKHANDER LANDING AND SUPPORT PLANS[37]

Throughout the evolution of BACKHANDER plans, the 1st Marine Division expressed its determination to preserve its tactical integrity as a unit and to place "an overwhelming force on the beach against a determined enemy."[38] The operation plan that finally governed the Cape Gloucester assault mirrored this concept of the division's most efficient employment.

The tasks set General Rupertus' force were threefold: to land in the Borgen Bay-Tauali areas, establish beachheads, and capture the Cape Gloucester airfields; to construct heavy duty landing strips at Gloucester as soon as possible and assist Commander, Allied Air Forces in establishing fighter sector, air warning, and radio navigational facilities; and to extend control over western New Britain to include the general line Itni River-Borgen

[35] BGen William F. Heavey, USA, *Down Ramp! The Story of the Army Amphibian Engineers* (Washington: Infantry Journal Press, 1947), *passim*.

[36] Col Thomas L. Randall comments on draft manuscript, included in MajGen James M. Masters, Sr., ltr to ACofS, G-3, HQMC, dtd 2Jul62.

[37] Unless otherwise noted, the material in this section is derived from: *ALAMO G-3 Jnl*; CTF 76 OPlan 3Z-43, dtd 20Dec43 (COA, NHD); *VII PhibFor ComdHist; VII PhibFor AR*; BACKHANDER For OpO 2-43, dtd 14Nov43; BACKHANDER For AdminO 2-43, dtd 14Nov43; *1st MarDiv SAR*, Phase I, Planning and Training, and Phase II, Landing and Seizure of Cape Gloucester Airfield; Hough and Crown, *New Britain Campaign*, App IV, "The Shore Party;" Craven and Cate, *Guadalcanal to Saipan*.

[38] MajGen Edwin A. Pollock comments on draft of Hough and Crown, *New Britain Campaign*, dtd 27Feb52, hereafter *Pollock comments*.

Bay, with patrols to investigate the possibility of developing an overland supply route to Gilnit village on the Itni. Within the airfield defense perimeter BACKHANDER Force was to establish, construction priority was assigned facilities to accomodate an Allied fighter-interceptor group.

Basically, the scheme of maneuver developed to capture Cape Gloucester called for simultaneous landings east and west of the airfields, each site about seven miles from the point of the cape itself. On Green Beach near Tauali, Lieutenant Colonel James M. Masters, Sr., would land his battalion, 2/1, and its attached units, seize a limited beachhead, organize it for defense, and hold it against enemy forces attempting to use the coastal trail to reach the airfields or to withdraw from them. On the opposite side of the cape, on a pair of beaches (Yellow 1 and 2) near Silimati Point, Colonel Frisbie's 1st and 3d Battalions, 7th Marines would land in assault followed by 2/7, with the mission of seizing a beachhead, organizing it for defense, and covering the landing of the rest of the assault force.

Shortly after H-Hour, 3/1 would begin landing behind 3/7 on the westernmost beach, Yellow 1. Once ashore, the battalion, resting its right on the coast, would attack west on a 500-yard front to seize the first of a series of phase lines designated to guide the attack toward the airfields. During the afternoon of D-Day, the remainder of Colonel Whaling's combat team would land, assemble just outside the right flank of the beachhead perimeter behind 3/1, and prepare to attack west on order.

After landing, Combat Teams B and C were to retain a number of the units assigned to them under direct command, but a large portion of their strength was to revert to control of the force commander. The combat teams kept their close-in supporting weapons, tanks, and antitank guns, but lost their artillery and antiaircraft guns to the force. Each team also kept its attached engineer company, its scout platoon, and its detachments from the division medical and service battalions. Since its primary mission was defense of a beachhead perimeter in tangled jungle terrain, Combat Team C could not make best use of some of its attachments. Accordingly, the force shore party was reinforced by the military police, motor transport, and amphibian tractor units once assigned to Colonel Frisbie's command. Similar attached units with Combat Team B remained under Colonel Whaling's control to support the advance up the coastal road.

In planning the disposition and employment of field artillery, Colonel Pepper's 11th Marines' staff picked out the one open area of any extent within the chosen beachhead, a patch of kunai grass on the right flank with a good all-around field of fire, as the post-landing position of 4/11. The 4th Battalion's 105mm howitzers, in direct support of Combat Team B, could reach the airfield but also could fire on counterattacking enemy anywhere along the perimeter and its approaches. With the only good artillery position inside the beachhead going to 4/11, the 75mm pack howitzers of the 1st Battalion landing in direct support of Combat Team C were left to "fend for themselves in an area that appeared fairly open with only scattered growth" near Silimati Point. [39] Both

[39] LtCol Robert B. Luckey, "Cannon, Mud, and Japs," *Marine Corps Gazette*, v. 28, no. 10 (Oct 44), p. 51, hereafter Luckey, "Cannon, Mud, and Japs."

artillery battalions were assigned a platoon of LVTs to help them reach and maintain their firing positions in what was expected to be very rugged terrain. The 75mm packs of 2/11, sailing and landing with Combat Team B, were to set up in what seemed a suitable area just off the coastal road outside the perimeter. While the field artillery landing on the Yellow Beaches would revert from combat team to 11th Marines' control, Lieutenant Colonel Masters on Green Beach was to keep command of Battery H of 3/11 as an integral part of his landing team. The 12th Defense Battalion's Seacoast Artillery Group commander was to coordinate the fire of all weapons used against seaborne targets.

The antiaircraft units assigned to BACKHANDER Force were to come under Colonel Harrison of the 12th Defense Battalion as senior antiaircraft officer. Initially, he would also control air raid warning throughout the force, but this service would become a function of the Allied Air Forces' Fighter Sector Commander, as soon as this officer landed and had his radar and communications in operation. The sector commander controlled all airborne Allied fighter units assigned to protect the Cape Gloucester area and, in addition, could order antiaircraft fire withheld, suspended, or put up in special defensive patterns through Harrison's fire direction center.

The engineer plan for BACKHANDER operations gave primary responsibility for tactical support to Colonel Harold E. Rosecrans' 17th Marines, with airfield repair and construction assigned to a base engineers group built around two Army engineer aviation battalions. In addition to the usual combat engineer tasks of construction, demolition, and repair, the lettered companies of 1/17, attached to combat teams, provided flamethrower teams to work with infantry against enemy fortifications. Colonel Rosecrans' regimental Headquarters and Service Company, 1/17 (less Companies A, B, and C), and 3/17 formed a combat engineer group which was responsible for water supply, for construction, repair, and maintenance of roads, bridges, and ship landing facilities, and for any other construction task assigned. Through contacts made by officials of the Australia-New Guinea Administrative Unit (ANGAU) accompanying the task force, it was hoped that maximum use could be made of native labor in all engineer missions.

The 17th Marines' 2d Battalion, the division's pioneers, formed the backbone of the shore party. Reinforced by two companies of replacements and the transport and traffic control units drawn from Combat Team C, the shore party under Lieutenant Colonel Robert G. Ballance was responsible for the smooth and effective unloading of all task force supplies. In order to expedite the job of getting bulk stores off LSTs, a system of overlapping dumps was planned, with each beached LST sending its cargo to its own class dumps, sharing with other ships only those on the flanks of its unloading area. Most of the supplies were mobile loaded on trucks to be run off LSTs to the dumps. In order to make mobile loading work, General Krueger's headquarters assigned 500 reconditioned 2½-ton trucks to temporary use of BACKHANDER Force, with the drivers recruited from an Army artillery battalion not actively committed to ALAMO operations. The trucks were to make a round trip circuit from LST to dump and return with the ships to the staging areas. All tractors and trucks organic to 1st Marine Division units, that were not needed for tactical purposes in the early stages of the landing, were to report to the shore party for use in moving supplies.

Assault troops headed for Cape Gloucester were to carry 20-days' replenishment supplies of all types as well as three units of fire for task force weapons. The garrison troops would bring in 30-days' supplies, three units of fire for ground weapons, and five units of fire for antiaircraft guns. In order to insure a smooth flow of supplies to Cape Gloucester, and the eventual maintenance of a 30-day level there, the 1st Division established a control system for loading and resupply. At Cape Sudest on Oro Bay, the main supply base for BACKHANDER operations, an officer from the division quartermaster's office acted as forwarding officer, his main duty to insure the movement of essential materiel to the combat zone. Supplies and personnel needed at the objective were to be funnelled through a regulating officer, who established priorities for loading and movement, and a transport quartermaster, who planned and supervised the actual loading.

Admiral Barbey's allocation of shipping to lift the BACKHANDER Force made the LST the main resupply vessel. With mobile loading of most bulk stores, a practiced shore party, and a dump plan that promised swift clearance of ships' cargo space, the amphibious force commander felt that he could risk the vulnerable LSTs in the combat area. The schedule of arrival and departure of the hulking landing ships, often dubbed Large Slow Targets by crew and passengers, was kept tight to lessen exposure to enemy air attack. Since the logistic requirements of shore-to-shore operations meant that many vessels would make repeated trips to the Yellow Beaches, the speed of unloading promised to pick up as shore party and ships' crews became more experienced.

The LST had a prominent part, too, in the medical evacuation plan for BACKHANDER. The flow of casualties during the first days of the operation would be from landing force units through the evacuation station run by the naval element of the shore party. Men hit during fighting immediately after H-Hour would be sent out by the first landing craft available to APDs riding offshore; once landing ships had beached, the wounded would be carried on board over the ramps into special areas set aside for casualty treatment. Medical officers and corpsmen from Seventh Fleet were assigned to all ships used as transports for the assault forces. To give the best possible care to the seriously wounded on the return voyage to New Guinea, Army surgical-medical teams were present on one LST of every supply echelon. At Cape Sudest, an LST equipped as an 88-bed hospital ship was ready to receive casualties from Cape Gloucester and pass them on to base hospitals ashore when their condition warranted. As soon as the combat situation permitted, casualties requiring less than a month's bed care and rest would be kept at Cape Gloucester in garrison force hospitals.

The total of shipping assigned to BACKHANDER support was not impressive in comparison to the large number of vessels needed to land and protect a division in the Central Pacific. There was no need, however, for massive invasion armadas. Because of the wider range of objectives that the large islands of the southern Pacific gave them, MacArthur and Halsey seldom had to send their assault troops against heavily fortified Japanese positions. This fortunate circumstance shaved the requirement for naval gunfire support ships, or at least gave the operations against Central Pacific fortress islands a higher priority of assignment. Similarly, the availability of land-based Allied air in significant force made the use of carrier planes wasteful when the naval pilots could be better employed against other targets.[40]

In total, Admiral Barbey as Commander, VII Amphibious Force assigned himself as Commander, Task Force 76, the BACKHANDER Attack Force, 9 APDs, 23 LSTs, 19 LCIs, 12 LCTs, and 14 LCMs to transport, land, and maintain the assault and garrison troops. The LCTs and LCMs plus five LCIs were assigned to the Western Assault Group (Green Beach) under Commander Carroll D. Reynolds; the rest of the LCIs, the APDs, and the LSTs were part of the Eastern Assault Group scheduled for the Yellow Beaches. Admiral Barbey commanded the ships at the main landing as well as the naval phases of the whole operation.

For fire support duties, Commander Reynolds had two destroyers and two rocket-equipped amphibious trucks (DUKWs) carried in LCMs. To cover the landings on the eastern side of Cape Gloucester, Barbey had 12 destroyers in addition to his flagship and two rocket-firing LCIs.[41] Escorting and supporting the attack force would be TF 74 under Vice Admiral V. A. C. Crutchley, RN, who had two Australian and two American cruisers with eight destroyers. The bombardment plan called for TF 74 to guard the approach of the main convoy against surface attack, to shell the airfield area before and immediately after the landing, and to retire westward when released by Barbey to take part in further operations.

All air units supporting the Cape Gloucester landings came under Brigadier General Frederick A. Smith, Jr.'s 1st Air Task Force with headquarters at Dobodura. From first light on D-Day, a squadron of fighters would escort the Eastern Assault Group to the target, with three squadrons successively covering the landings, and a fifth screening the planned retirement in midafternoon. Several destroyers in the attack force, including the *Conyngham*, had the special communications equipment and trained personnel to act as fighter director ships. During the time the main convoy was in the Cape Gloucester area, fighter control would remain afloat. Bomber control was a function of the Army's 1st Air Liaison Party, part of BACKHANDER Force, which was to land with General Rupertus' headquarters, establish contact with Dobodura, and direct aerial support of ground operations.

[40] Commenting on the predominence of naval support given the Central Pacific, Admiral Barbey noted that his sailors characterized the situation as "never had and won't get," but quoted Admiral Nimitz in a contemporary comment as saying: "When conflicts in timing and allocation exist, due weight should be accorded to the fact that operations in the Central Pacific promise at this time a more rapid advance toward Japan and her vital lines of communication; the earlier acquisition of bases closer to the Japanese homeland; and, of greatest importance, are more likely to precipitate a decisive engagement with the Japanese fleet." *Barbey ltr.*

[41] These LCIs were improvised support ships using Army surplus 4.5-inch rockets; techniques for their employment were worked out by practice in the vicinity of Milne Bay. VAdm Daniel E. Barbey interview by HistBr, G-3, HQMC, dtd 22May62.

In the hour immediately preceding H-Hour, high-level bombing by five squadrons of B–24s would hit defensive positions back of the Yellow Beaches while destroyers shelled the same target area. When naval gunfire was lifted, three squadrons of B–25s would streak across the coast bombing and strafing the immediate beach area, while a fourth squadron blanketed a prominent hill behind the beaches with white phosphorus bombs. At the same time, across the cape, another medium bomber squadron was slated to bomb and strafe Green Beach defenses before 2/1's landing craft touched down.

On overhead standby during the initial landings would be four squadrons of attack aircraft; if they were not called down, these A–20s would hit targets south and east of the airdrome before returning to base. Later on D-Day morning, both a heavy and a medium bomb group would attack enemy bases and routes of approach along the southern coast. Nine squadrons of B–24s and four of B–25s that took part in the morning missions were to refuel and rearm immediately after landing at their home fields on New Guinea, and strike again in the afternoon at enemy installations west of the beachhead.

As the time of the main landings at Cape Gloucester neared, the Japanese defenders of New Britain were increasingly alert to the probability of Allied attack. The enemy commanders at Rabaul, however, could only guess where and in what strength the assault would come. Cape Gloucester's airfields seemed a logical main target but so did Gasmata's, and there were a number of lesser bases on both coasts that had to be considered and defended. The terrain of the island itself may perhaps have played the largest part in determining the ALAMO Force objectives and *Eighth Area Army's* countermoves.

The Enemy: Terrain and Troops

THE OBJECTIVE [1]

Two narrow straits, Vitiaz (Dampier) and St. George's Channel, join the Bismarck and Solomon Seas. Between them lies New Britain, a 350-mile-long island that forms a crescent-shaped link between New Guinea's Huon Peninsula and New Ireland. In width New Britain varies between 20 and 60 miles, narrowing toward Cape Gloucester in the west, and in the northeast joining Gazelle Peninsula to the trunk of the island. Jutting out from the northern shore for a distance of 30 miles is the narrow Willaumez Peninsula, a natural barrier to coast-hugging small boat traffic. Japanese troop and supply craft bound to and from Rabaul frequently avoided Willaumez by making night runs across the open sea to Garove, largest of the Witu Islands group, standing some 40 miles northwest of the peninsula. (See Map III, Map Section.)

There is a fringing reef along most of New Britain's 1,000-mile coast with occasional breaks that open the shore to the sea. Beyond the reef fringing the north and southwest coasts, barrier reef formations abound. Numerous islets crop up among these reefs, many of them jungled with hilly spines, copies in miniature of the vegetation and terrain of New Britain. Prominent among these islands is the Arawe group, which clusters thickly about the Cape Merkus peninsula, a crooked finger of land lying half-way between Cape Gloucester and Gasmata.

Fourteen miles off the coast of New Britain at Grass Point, its western tip, is Rooke or Umboi Island. Rooke, like so many islands in the Southwest Pacific, is no more than the crest of a mountain range rising steeply from the sea; its bulk splits Vitiaz Strait into two parts. The reef-studded channel between Rooke and New Britain is known as Dampier Strait, a name that the Japanese applied to both Vitiaz and Dampier without distinction. In 1943, hydrographic information available to the Allies about this area was sketchy and unreliable, enough so that Barbey would not risk his larger ships in Dampier's waters. The approach route chosen to Cape Gloucester skirted Rooke on the west and passed between that island and smaller Tolokiwa to the north, offering a safer passage as well as one less likely to be discovered by the enemy.

The coastline of western New Britain is generally regular in outline, with a series of gentle capes and shallow bays marking its length. On the south shore, the most prominent land projection is tipped by

[1] Unless otherwise noted, the material in this section is derived from: MID, WD, Surv of Bismarck Archipelago (S30-675), dtd 5Oct43; Allied Geographical Sect, SWPA, Terrain Handbook No. 7—New Guinea—Cape Gloucester, dtd 24Sep43; Sixth Army G-2, Terrain Est Cape Gloucester Area, dtd 15Aug43; Hough and Crown, *New Britain Campaign*, App II, Capt Levi T. Burcham, "The Vegetation of New Britain and Its Effect on Military Operations."

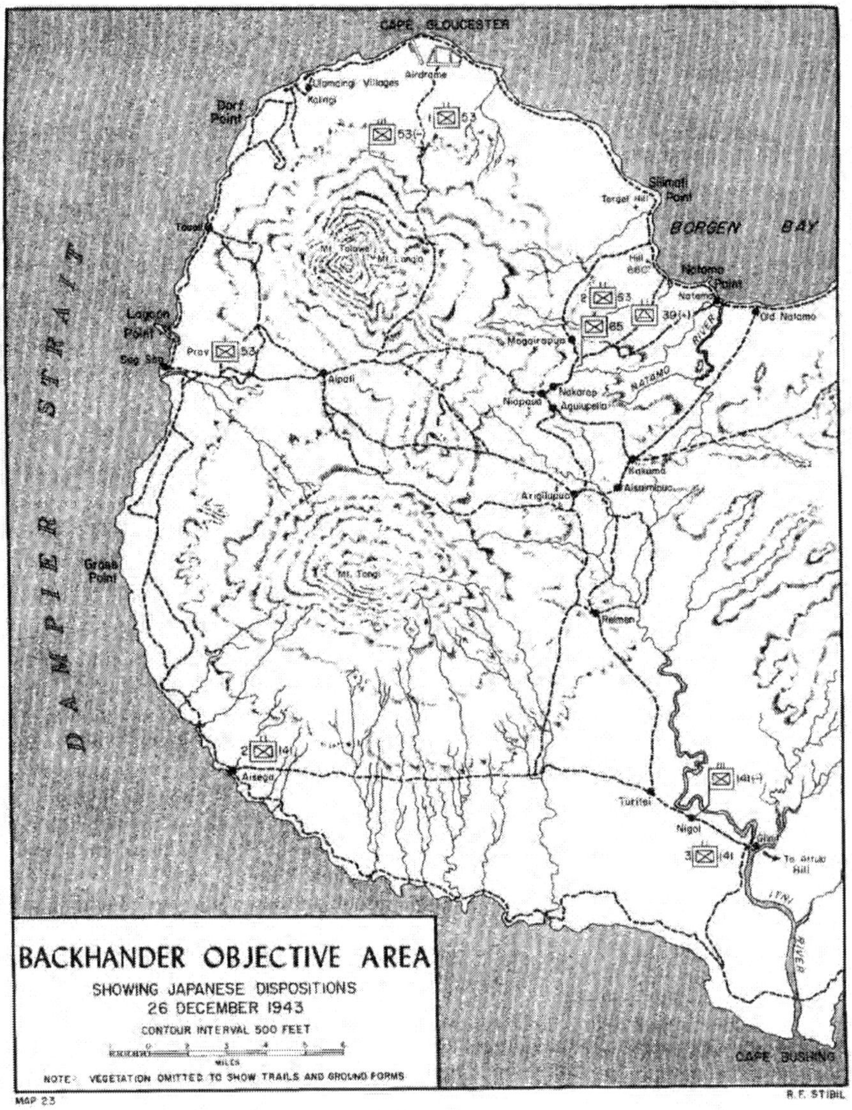

BACKHANDER OBJECTIVE AREA
SHOWING JAPANESE DISPOSITIONS
26 DECEMBER 1943
CONTOUR INTERVAL 500 FEET
MILES
NOTE: VEGETATION OMITTED TO SHOW TRAILS AND GROUND FORMS

MAP 23

R. F. STIBIL

Cape Merkus, and on the north the deepest cut forms Borgen Bay. For a mile and a half on each side of Cape Gloucester there is no fringing reef, and large ships can lie close inshore in the immediate vicinity of the cape. Throughout the area there is a firm beach of black volcanic sand strewn with large stones. Bordering the beach is a 5–10-foot red clay embankment; during normal high tides water covers the beach to the bank. In heavy weather a deep swell and high-breaking surf made the approaches dangerous and unloading operations impracticable. (See Map IV, Map Section.)

The only other areas near Cape Gloucester where a significant break occurs in the offshore reefs are at Tauali on the shore of Dampier Strait and on both sides of a small, unnamed cape a mile and a half northwest of Silimati Point. These beaches exposed by gaps in the reef were the ones chosen for the BACKHANDER assault.

Just north of the village of Tauali the reef fades away for a stretch of 400 yards, and the usual black sand beach is backed by a three-foot bank. Inland, the coastal flat is narrow and covered with secondary growth, and the ground rises sharply to bluffs overlooking the shore. The beach itself is 8–10-feet wide at high water, when the three-foot depth is located 10–15-feet offshore; at low tide, the same depth is found 40 feet from shore.

The reef around Silimati Point narrows and disappears about a half mile from the point, and for a thousand yards the beach is free of close-in obstacles. After the reef crops out again for a few hundred yards, there is another half-mile of open beach. BACKHANDER Force's Yellow Beach 1 was plotted in on this second stretch of black sand, which is quite narrow and overhung with jungle growth. Yellow 2 was planned for the sector nearest Silimati Point where the beach varies in width between 30 and 60 feet. Both beaches fall away steeply underwater, a six-foot depth being 20 feet out at low water and 25 at high. Barrier reef formations in the area are scattered and far enough offshore to give large landing ships access to the beaches.

Behind the Yellow Beaches is an area that was labeled on Marine operation maps as "damp flat" and discussed in intelligence studies as being covered at times by storm water. Forewarned, landing force commanders were prepared to encounter the area's reddish brown volcanic soil as sticky mud and even for a good deal of standing water, but no one was quite ready for the swamp that stood just back of the low embankment behind the beaches. Fortunately, the plans laid to counter known terrain problems proved adaptable to meeting the graver situation posed by the unexpected obstacle.

The reason intelligence officers did not detect the presence of the swamp was the concealment trees and undergrowth afforded against the probing aerial camera. One 1st Marine Division officer, a forester in peacetime, wrote a good description of the vegetation that covers much of the beachhead area and the ground to both flanks; he called it "the dank, steaming tropical jungle of the fiction writer," the "swamp forest" which is:

> . . . characterized by widely spaced trees of very irregular height, the tallest being upwards of 100 feet high. Quite commonly these trees have widely spreading buttress roots, which give a fluted appearance to the bottom of the trunk. Wide spacing and irregular height of the mature trees permit a

moderately dense to very dense undergrowth of varying heights. Both undergrowth and mature trees are generally thickly matted with lianas, vines, and lawyer cane. . . .

Cover and concealment are complete in swamp forest. Ground observation usually is restricted to a few yards. Movement of troops or vehicles is very difficult. Flooded areas virtually preclude movement during the wet season. Large areas of swamp forest occur in the Cape Gloucester-Borgen Bay area; the vegetation behind the landing beaches is a strip of swamp forest which extends nearly to the airdrome area on the west, as well as eastward around Borgen Bay.[2]

The coastal flat west of the beachhead widens back of Cape Gloucester to two miles of gently rolling ground covered by kunai grass and then narrows as it nears Dorf Point. One of the few major areas of New Britain not covered by jungle growth, the grassland was the site of a prewar emergency landing strip used by the planes of an island-hopping commercial line. The Japanese moved in during December 1942 and began building a new runway diagonally across the trace of the old; in April 1943, a second airstrip was started about 1,000 feet southeast of the original one. A small stream emptying into the sea east of Cape Gloucester separated the two. By October, Allied air attacks and the demand for Japanese planes to garrison existing bases had combined to arrest development of the airfields. Kunai grass had laid claim to one and was encroaching on the other in December 1943.

Southeast of the airfields the ground rises gradually then steeply toward the twin volcanic cones of Mt. Talawe (6,600 feet). Ten miles to the south, Mt. Tangi (5,600 feet) breaks the horizon, and close to Talawe, along a ridge connecting the two mountains, three smaller peaks are readily identifiable. The middle of these lesser heights, Mt. Langila (3,800 feet), is a dormant volcano from which a wispy plume of steam issues steadily; the other mountains are all extinct volcanoes. The whole complex of high ground is deeply cut by ravines, many the path of streams coursing downhill to the sea. On the southern and western slopes, ridge spurs reach out to the shore of Dampier Strait; on the north, the foothills of Mt. Talawe edge the coastal flat; and on the east, those of Mt. Tangi form one side of the broad Itni River valley.

The vegetation covering the mountains is tropical rain forest, common to all New Britain at elevations of 500 feet and above. Here the tallest trees, usually 125 to 150 feet high, form virtually a complete canopy underneath which the crowns of a second layer of smaller trees crowd. Frequently there is:

. . . an understory of brush and young trees beneath these, ranging up to twenty or thirty feet tall. Sometimes there is a fairly complete ground cover of ferns and herbs up to about two feet in height. The actual interior of such a forest is relatively open except for occasional small tangles of climbing bamboos, rattan palms, or lianas. . . .

Complete cover and concealment from air observation is afforded. Ground observation is limited—a standing man can be observed at about 50 yards; a prone man usually will be concealed at 10 yards. Foot troops can move through rain forest with little difficulty and require practically no trail cutting. Physical character of the terrain—spur ridges, deep stream channels, or the like—may make travel difficult, but the vegetation itself offers little hindrance. . . .[3]

[2] Burcham, *op. cit.*, p. 192.

[3] *Ibid.*, p. 193.

One other type of jungle vegetation, secondary growth, is frequently found in western New Britain, particularly in the vicinity of native villages. Because of the local practice of abandoning garden patches of bushy vegetables and fruits after one year's cultivation, wild plants in tangled profusion quickly reclaim the temporary clearings. Wide areas of secondary growth, all of them a formidable barrier to troop movement, occur in the regions most heavily settled by the natives.

In prewar years, about 3,000 Melanesians inhabited New Britain west of the Itni River; the only Caucasians who lived there were three missionaries, a Catholic priest who ran a mission at Kalingi near Dorf Point, and a Church of England couple who served a mission at Sag-Sag, three miles south of Tauali. The native villages were in four main groups, two near the missions and the others on the eastern slopes of the mountains, one clustered near the mouth of the Itni and the second fanned out along the upper reaches of the river.

The trail network connecting the villages is the only practical means of penetrating the interior, and the frequent ridges and streams cutting the tracks on the mountainsides limit travel to men on foot. No waterway connects the coasts, and of all the rivers and streams in western New Britain, only the Itni is navigable; the others are shallow and fordable through all their length. On the Itni, native canoes can reach Relmen, 14 air miles from the river's mouth, and small landing craft can reach a point about two-and-a half miles above Gilnit.

The location of the various trails had an important effect on the progress of the fighting at Cape Gloucester. The layout was such that the Japanese could feed reinforcements into the combat zone or withdraw their troops from contact almost at will. A coastal track, usable by motor vehicles between the airfields and Silimati Point, bordered the entire objective area. From Tauali east across the saddle between Mts. Talawe and Tangi, a cross-island trail led to Natamo on Borgen Bay. Joining this track was another, running over the shoulder of Talawe to the airfields, and a second which led southeast then south, eventually reaching Gilnit. A most important trail, its existence unknown to Allied intelligence, linked the area picked for the BACKHANDER beachhead to the Tauali-Natamo track.

The climate of western New Britain is what might be expected of a region of jungle-covered mountains and swamps. At all times during the year, the humidity is high, and the daytime temperature range hovers around 90 degrees; at night the temperature seldom drops below 72 degrees. The annual rainfall usually totals 150–200 inches and much of this, an average of 30 inches a month, comes during the period of the northwest monsoon, mid-December to mid-February. In this wettest season, rain may fall almost every day and squalls with torrential downpours are frequent. The northwest winds are strong and fairly steady, making the sea rough and the surf heavy. The dry season at Cape Gloucester occurs during the summer months when the prevailing southeast winds vent most of their force on the south slopes of the mountains. The periods between the two seasons, and the period of the southeast monsoon itself, are times of comparatively calm weather.

The campaign to seize control of western New Britain would be fought in the worst possible weather of the year. Low-lying terrain would disappear beneath a cover of standing water, and, on the higher ground, the trees, the undergrowth, and the land itself would become and remain, rainsoaked. The prospect was that attacker and defender alike, mired in combat in the dripping jungle, would curse the day they set foot on the island.

JAPANESE DEFENSES [4]

While, at times during the war, Japanese planners at *Imperial General Headquarters* seemed unreasonably optimistic in the face of repeated setbacks, that mood vanished when Tokyo appraised the strategic situation in late September 1943. The Army and Navy headquarters staffs conceded that the Allied forces were strong enough to break through the outpost cordon of defenses in the Central and Southwest Pacific and reach the inner perimeter of island bases in western New Guinea, the Marianas, Palaus, and Philippines. To stave off that event, enemy commanders of the garrisons manning the barrier positions were ordered to "do their utmost to hold out as long as possible." [5] With the time thus gained, the Japanese intended to strengthen the fortifications and reinforce the defending troops of the islands that directly barred the approaches to Japan.

In Rabaul, the decision that the *Eighth Area Army's* zone of responsibility was outside the vital inner perimeter came as no shock. General Imamura believed that the Allied advance was inevitable, but that its force could be blunted considerably. Reinforcements were dispatched to threatened bases in the northern Solomons, the Bismarcks, and eastern New Guinea, but there was a strong inclination to keep back a substantial force at Rabaul itself. In the eyes of the Japanese, Rabaul was an objective that could not be bypassed, and in estimating Allied intentions, Imamura's staff considered its major problem to be a decision whether:

> . . . the enemy will attempt to capture Rabaul immediately after the occupation of the Dampier Strait Area and Bougainville island or the enemy will invade Rabaul after isolating our forces there by severing our line of communications in the rear through the occupation of the Admiralty Islands and New Ireland, especially the Kavieng sector of the latter. However, in view of the absolute superiority of the enemy naval and air strength and the value of Rabaul from both the political and strategic viewpoint, it is more probable that the enemy will take the shortest course for capturing this valuable base, presumably in February or March of next year (1944).[6]

The Japanese shared, along with this conviction that Rabaul would be attacked, a feeling that its defenses could be held against Allied landing forces. Intelli-

[4] Unless otherwise noted, the material in this section is derived from: MilHistSec, GHQ, FEC, Japanese Monograph No. 127, Army Southeast Area OpsRec Part IV, Eighth Area Army Ops, rev. ed., dtd Jul49 and Japanese Monograph No. 128, Army Southeast Area OpsRec Part IV Supplement, 17th DivOps in Western New Britain, dtd Jul49 (OCMH), hereafter *Eighth Area Army Ops* and *17th Div Ops*; *1st MarDiv SAR*, Anx A, Intelligence: IntelSec, ADC Hq, 1st MarDiv, Documents and POW Repts Folder, n.d., hereafter *ADC IntelDocuments;* Takushiro Hattori, *Dai Toa Senso Zenshi* [*The Complete History of the Greater East Asia War*], 4 vols. (Tokyo: Masu Publishing Company, 1953—MS translation at OCMH), hereafter Hattori, *Complete War History*; Hough and Crown, *New Britain Campaign.*

[5] "Army-Navy Central Agreement on Central and South Pacific Area Operations," with IGHQ Army Dept Directive No. 1652, dtd 30Sep43, in HistDiv, MilIntelSec, GHQ, FEC, IGHQ Army Directives, v. III (OCMH).

[6] *Eighth Area Army Ops*, pp. 85–86. Although this document was compiled after the war, it is contemporary in tone and content.

gence and operations officers on both sides felt a struggle to take the base would require extensive assault forces and a bitter, costly battle. What the enemy failed to appreciate, however, was that predominant Allied air and naval strength made the seizure of Rabaul unnecessary if an encompassing ring of less important bases was captured.

Even if the Japanese had been willing to denude Rabaul's defenses to reinforce outlying positions, the problems of transferring troops and equipment were formidable. The sea was the only feasible supply route, but its use was dangerous. Allied fighters and bombers vied with torpedo boats to cut down enemy barge and destroyer traffic in the forward area, and the losses suffered by the Japanese crippled their plans to meet the expected attacks. Personnel losses were heavy, but even more damaging was the destruction of supplies and the means of transporting them. In general, the closer to the Vitiaz Strait enemy troops were stationed, the more likely they were to be scantily equipped, poorly clothed, and on short rations.

The barge traffic from Rabaul to New Guinea was largely responsible for the buildup of Japanese positions in western New Britain. Aside from the Cape Gloucester airfields and the defenses that grew up around them, enemy installations west of Gasmata and Willaumez Peninsula were mainly way stations on the coast-hugging barge route forced upon the Japanese by Allied attacks. The miscellaneous collection of motor sailers, launches, and landing craft, that were collectively known as barges, found shelter from probing aircraft wherever a gap in the fringing reef, a narrow beach, and overhanging foliage combined to give them a chance to escape detection.

On the south coast west of Cape Merkus, the principal enemy barge hideouts were located at Cape Bushing, where the Itni River emptied into the sea, and near the villages of Aisega, Sag-Sag, and Ulamaingi. Before the Allies seized control of the Huon Peninsula, a steady traffic passed through these points to Rooke Island, and then on to New Guinea. Routes along the north coast, less vulnerable to attack and therefore more heavily travelled, crossed the open sea, stopping at Garove Island, or followed the coast around Willaumez Peninsula. In succession west from Willaumez, the major barge stations were Iboki, Karai-ai, Kokopo, and Natamo. (See Map 29.)

Elements of the *1st* and *8th Shipping Engineer Regiments* manned and maintained the boats used along the coastal barge routes, mainly standard Japanese landing craft. Most of the vessels that followed the Rabaul-Garove-New Guinea course were larger, deeper-draft coastal schooners and fishing trawlers handled by crews of the *5th Sea Transport Battalion*. Although Japanese destroyers were occasionally called upon for emergency, high-speed movement of troops and priority cargoes, the burden of supply fell upon shipping manned by soldiers. Detachments of the *1st Debarkation Unit* acted as shore party at the barge bases near Cape Gloucester, while a variety of service units performed that mission at other points. The shipping engineers and other troops concerned with moving personnel and supplies were all armed and able to act as infantry.

In May 1943, General Imamura sent the *65th Brigade* from Rabaul to take over the defense of western New Britain. The *65th*, which had earned an Imperial citation for its part in the capture of Bataan, was down to a strength of one two-battalion infantry regiment and supporting signal, engineer, and medical troops when it reached New Britain. Only the *141st Infantry* came from the Philippines; the *142d Infantry*, once part of the brigade, remained there, and the *122d Infantry* was sent to the Marshalls. In addition to regimental headquarters and signal companies, the *141st* had an antitank company equipped with six 37mm guns and an artillery company armed with four 75mm mountain guns. Each infantry battalion of the regiment had a headquarters and supply train, three rifle companies, a heavy machine gun company, and a gun platoon manning two 70mm howitzers. (See Map 23.)

Throughout the summer, the *65th Brigade's* main concern was keeping the supply lines to New Guinea open. Most of the units attached to it were companies or smaller elements of service organizations concerned with the barge traffic, airfield construction and operation, or the logistical support of the garrison. One type of unit, however, antiaircraft troops, added significantly to defensive firepower, both against Allied aircraft and ground forces. Each of the *39th Field Antiaircraft Battalion's* three firing batteries had four 75mm guns that could be used against ground targets. The *28th* and *30th Machine Cannon Companies* both had eight dual-purpose 20mm guns and six heavy machine guns.

The *Eighth Area Army's* interest in the defense of western New Britain picked up considerably as the situation in the Salamaua area worsened. On 5 September, a new command, the *Matsuda Force* which took its name from its commander, Major General Iwao Matsuda, took over the *65th Brigade*, the various shipping engineer and debarkation units, and a number of troops of the *51st Division*.[7] The main body of the *51st* was fighting the Australians on New Guinea, but the division's rear echelon units as well as the survivors of transport sinkings were present on New Britain, Garove, and Rooke Islands. Of particular value to Matsuda's force as combat troops were two companies of the *115th Infantry* and a smaller detachment of the *66th Infantry*, two provisional infantry companies formed from artillery and engineer elements of the division, and about half of the *51st Reconnaissance Regiment*.[8]

General Matsuda was an infantry officer with considerable experience in the Army's shipping transportation department. He had commanded an infantry regiment in China and an infantry group in Manchuria before taking over the *4th Shipping Command* in February 1943. The command, an administrative headquarters geared to handle shipping activities, moved from Japan to Rabaul by way of eastern New Guinea, arriving in New Britain in early August. On 29 October, when Matsuda was also appointed to command the *65th Brigade*, he formed a composite headquarters for the *Matsuda Force* by merging the staffs and headquarters of the brigade and the shipping command.[9]

About a month after Matsuda established his command post near Kalingi mission at Cape Gloucester, his force came

---

[7] Chief, WarHistOff, DefAgency of Japan, ltr to Head, HistBr, G-3, HQMC, dtd 20Jun62, hereafter *Japanese Comments*.

[8] ATIS 1328, 141st InfRegt OpO A. No. 11, 2Dec43, in ATIS Current Translations No. 122, dtd 31May44 (ACSI Recs, FRC Alex).

[9] Docu No. 59881, MilServRec of Ex-LtGen Iwao Matsuda, in HistDiv, MilIntelSec, GHQ, FEC, Personal History Statements, 2 vols (OCMH); *Japanese Comments*.

under the *17th Division*. The division, newly arrived in Rabaul from China, was commanded by Lieutenant General Yasushi Sakai. Originally, the *17th* had been slated to reinforce the garrison of northern Bougainville, but General Imamura changed its mission when he was assured by the *Southeast Area Fleet* that it could keep another large body of men supplied in western New Britain. On 5 October, the *Eighth Area Army* commander added the *Matsuda Force* and the Gasmata garrison to General Sakai's command and ordered him to assume responsibility for defense of New Britain west of a line joining Commodore Bay and Vahsel Harbor. Division headquarters was established at Malalia near Cape Hoskins to the east of Willaumez Peninsula.

The first echelon of the *17th Division* left Shanghai on 24 September, arriving at Rabaul on 4–5 October. Three more convoy groups were to bring the remainder of the division, but shipping losses and the mounting danger of attack by Allied planes and submarines combined to cancel the sailing of the last convoy, which was to have lifted a number of service troops and 3,000 infantry replacements. The second and third convoys which departed from Shanghai late in October lost one ship to submarines while passing through the Ryukyu Islands, had another damaged by B–24s south of Truk, and lost a third to a mine off Kavieng. Total casualties from the three attacks were 1,173 men killed and wounded.[10]

The main strength of the *17th Division's 53d Infantry Regiment* and the regimental headquarters and *3d Battalion, 23d Field Artillery* were assigned to General Matsuda's command to reinforce the defenses in the immediate vicinity of Cape Gloucester. Also coming under *Matsuda Force* was the *1st Battalion, 81st Infantry* which was ordered to Cape Merkus. The *3d Battalion, 53d Infantry* and the *6th Company* of the *2d Battalion* were detached from the regiment to serve on Bougainville as was the rest of the *81st Infantry*, a battalion of the *54th Infantry*, and one of the *23d Field Artillery*. The remaining units of the *54th* and *23d* were distributed between Gasmata, Malalia, and Talasea.

With a battalion and a rifle company stripped from its strength, the *53d Infantry* was little stronger than the *141st*. Regimental headquarters and supporting companies in both units were the same, with the *53d* having in addition a platoon of 90mm mortars. The *1st Battalion* of the *53d* had four rifle companies, but the commitment of the *6th Company* on Bougainville pared the *2d Battalion* to three. *Eighth Area Army* kept control of two rifle companies, the 70mm howitzers, and most of the heavy machine guns of the *1st Battalion, 81st Infantry*, leaving the reinforcements for Cape Merkus at a strength of a headquarters and two rifle companies plus a machine gun platoon.

Japanese destroyers were used to transport the *17th Division* troops to western New Britain with most units landing at Karai-ai and carrying on by barge or trail to Cape Gloucester. The waters along the south coast were too close to Allied bases to risk destroyers there, so the skeleton battalion headed for the Arawe area had to march overland to its post. Since the *1st Battalion, 81st Infantry* had been on board the transport torpedoed during the move from China, it had to be reorganized

[10] *17th Div Ops*, pp. 4–5. This source does not provide a casualty breakdown.

and refitted in Rabaul. It was early December before the battalion landed at Iboki, terminus of a cross-island trail to Cape Merkus.

While the last of the *17th Division* reinforcements were moving into position in western New Britain, General Matsuda took steps to increase his combat strength by organizing a third battalion for the *141st Infantry*. The existing two battalions of the regiment furnished the men for the headquarters, one rifle company, and most of the machine gun company; the men of the *66th* and *115th Infantry* detachments formed the remaining two rifle companies and a platoon of machine guns. The new unit began organizing on 12 December and was formally joined to the regiment on the 20th.

The deployment of Matsuda's forces, as the time of the expected Allied assault approached, reflected the importance the Japanese general attached to the various objectives in his defensive area. Holding the vital airfield sector was the *1st Battalion, 53d Infantry*, reinforced by the regiment's 37mm and 75mm guns and a miscellany of service troops. Beach defenses, mainly bunkers connected by rifle trenches, were scattered along the shore on either side of Cape Gloucester. The heaviest concentration was located back of the most logical site for a large-scale landing attempt, a stretch of beach three miles southeast of the cape that led directly into the grasslands. In the foothills of Mt. Talawe, hidden by the lush vegetation, was a bunker-trench complex that commanded the airstrips.

Knowing what lay behind the beaches near Silimati Point, the Japanese paid scant attention to their defense. Just to the south, however, a strong force built around elements of the *1st Shipping Engineers* and the *39th Field Antiaircraft Battalion* held the sector where the trail net from the interior reached the coast. Matsuda's scheme of defense depended upon retaining possession of the trails to move his limited manpower to meet Allied attacks. Two small hills which dominate the area served as focal points for enemy defenses. One, a 450-foot height called Target Hill by the Allies, stands out starkly above the swamp forest, its open, grass-covered crown in sharp contrast to the surrounding sea of trees. A mile and a half south of Target Hill, Hill 660 rises out of the jungle with a cover of tangled growth that blends easily with its environs.

Across the island, in the vicinity of Cape Bushing, was the defensive sector assigned to the *141st Infantry*. The *1st Battalion* was located near the mouth of the Itni, the *2d* occupied defenses at Aisega, and the newly formed *3d* was in reserve at the regimental headquarters at Nigol, a village on the Itni about three miles above Gilnit. Farther down the coast at Cape Merkus the defending force was composed of a platoon of naval personnel manning a communication relay station [11] and two provisional infantry companies formed from elements of the *51st Division*.

The garrisons of Rooke and Garove Islands were also part of the *Matsuda Force*. On Garove, the principal unit was the *5th Sea Transport Battalion* with an antiaircraft battery and shipping service units attached. The headquarters and two companies of the *51st Reconnaissance Regiment* were still on Rooke on D-Day, but, prior to 26 December, a reinforcing 75mm gun battery and about half of the regi-

[11] *SE Area NavOps—III*, p. 39.

ment's cavalrymen were sent to New Britain to bolster the *141st Infantry's* defenses.

In November, General Matsuda moved his headquarters from Kalingi to a concealed position in the rain forest near Nakarop, a village on the main cross-island trail about seven miles inland from Borgen Bay and a thousand feet about sea level on Talawe's lower slopes. Close by the new *Matsuda Force* command post, the general stationed the *2d Battalion, 53d Infantry* as general reserve, ready to move by the mountain trails to the airfields, to Cape Bushing, or to Target Hill as the situation demanded.

By mid-December, the effective strength of Matsuda's command, including all classes of troops in the area west of an Iboki-Arawe boundary, stood close to 10,500 men. The figure represents 9,500 that Matsuda reported to General Sakai on 1 December, plus an estimate of the combined strength of the *17th Division* units—*1st Battalion, 81st Infantry; 3d Battalion, 23d Field Artillery; 2d Field Hospital*—that reported from Rabaul later in the month. Roughly half of the total force was located in positions within effective supporting range of Cape Gloucester.

The enemy leaders were well aware that the Allies had the strength to take western New Britain, but the Japanese were determined to make a bloody fight for its possession. On 12 December, in a message addressed to all the officers in his command, General Sakai warned that a landing was imminent and that in meeting the Allied assault force each man was to observe the principle of "certain death warfare to the utmost, in such a way that not even the slightest disgrace will adhere to your name."[12]

KNOWLEDGE OF THE ENEMY[13]

Aerial reconnaissance kept the Japanese informed of concentrations of Allied landing craft off the coast of New Guinea, but the traffic was so heavy that no clear picture emerged as to when a move was coming against western New Britain. False alarms were frequent as the end of 1943 approached, but all the reports of pilots of the *Eleventh Air Fleet* at Rabaul and the *Fourth Air Army* at Wewak could do was confirm the estimate that a large-scale landing was in the offing. Even more unsettling to the Japanese commanders was the knowledge that Allied scouting parties had landed repeatedly in the territory defended by the *Matsuda Force* and departed with valuable intelligence.

General Krueger, confronted by a serious lack of information about the terrain, beach conditions, and defenses of possible ALAMO Force objectives, had formed a group of scouts whose job it was to land behind enemy lines and find the answers to vital questions that plagued intelligence officers. The ALAMO Scouts, who operated directly under Krueger, were a composite force of Australians with

[12] 141st InfRegt Bul, dtd 12Dec43, in ATIS Enemy Publication No. 257, 141st InfRegt Buls and Related Papers, dtd 27Dec44. (ACSI Recs, FRC Alex).

[13] Unless otherwise noted, the material in this section is derived from: MIS, GHQ, FEC, Ops of the AlliedIntelBu, GHQ, SWPA—v. IV, Intel Series, dtd 19Aug48; *1st MarDiv SAR*, Anx A, Intelligence; 1st MarDiv AmphibRecon PtlRepts, Cape Gloucester and Talasea, 11Oct43–9Mar44, hereafter *1st MarDiv PtlRepts;* Feldt, *Coastwatchers;* Hough and Crown, *New Britain Campaign*.

experience in the islands, American intelligence personnel, and natives of proven loyalty and dependability. In preparing for BACKHANDER operations, a number of the 1st Marine Division's own scouts, led by First Lieutenant John D. Bradbeer, took part in pre-D-Day reconnaissance on New Britain with men from the ALAMO Scouts.

Usually, the scouting parties travelled to their objective by torpedo boat, landed in rubber boats, and moved inland to establish a patrol base well off the regular tracks. Fanning out from this point, the scouts contacted the local populace and tried to observe enemy dispositions. When the mission was accomplished, or, as often happened, the Japanese got wind of the presence of the scouts and started to hunt them down, the torpedo boats returned and took off the party. Occasionally, the reconnaissance was an assignment to spot-check preferred beaches and their defenses, a mission in which success depended upon the scouts' ability to obtain essential information quickly and accurately.

Three Australian officers were particularly concerned with the scouting in western New Britain—Major John V. Mather, AIF, who had served with the 1st Marine Division on Guadalcanal and had been a labor contractor in the Solomons, Sub-Lieutenant Andrew Kirkwell-Smith, RANVR, who had been a coastwatcher at Cape Gloucester before the Japanese came, and Sub-Lieutenant William G. Wiedeman, RANVR, who had been the Anglican missionary at Sag-Sag. The local knowledge of these men was invaluable to division intelligence, as was their understanding of the natives and their ability to teach others how to live and operate successfully in the jungles of New Britain.

The first party with American scouts to go ashore in the objective area was led by Bradbeer and Kirkwell-Smith. After it landed near Grass Point on 24 September, the nine-man patrol operated two weeks in the region south of Mt. Tangi looking for a trail that was supposed to wind south around the mountain and go over the saddle between Tangi and Talawe. Although the search was unsuccessful, the party brought back much useful information, particularly the welcome news that the natives were at odds with the Japanese. In July, the enemy had cleared all the inhabitants from the coastal villages and those near the airfields, forcing an exodus to the interior. Since that time, the natives had noticed that many of the Japanese were sickly and on short rations, so much so that the soldiers had begun to raid the village gardens for food. The scouts retraced their steps to the coast and left the island on 6 October after they received word that enemy patrols were looking for them.

In mid-October, a small Australian-led party, most of its personnel borrowed from the coastwatcher organization, landed on Rooke Island. In two weeks ashore, the scouts found little evidence of a large enemy garrison and pulled out undetected. A month later, a planned 24-hour patrol, led by First Lieutenant Robert B. Firm of the 5th Marines R-2 Section, went ashore below Dorf Point to check possible landing beaches in the vicinity. The enemy was too active to make any move inland feasible, but Firm was able to determine that the beach was not usable before he had to order his men back into the rubber boats.

Since Admiral Carpender did not want to risk torpedo boats north of Cape Gloucester, it proved impossible to make a reconnaissance of the main beaches near Silimati Point. An ALAMO Scout party did land at Arawe, however, on the night of 8–9 December and concluded that few Japanese were present to hinder the projected assault there. On the night of 21–22 December, Bradbeer and First Lieutenant Joseph A. Fournier of 2/1 led two parties ashore to check the beaches near Tauali. The Marine scouts confirmed the selection of Green Beach as the best that could be found for the landing; no enemy forces were encountered. Coming away from New Britain the torpedo boat transporting the scouts attacked some Japanese barges; in the exchange of fire, three American crewmen were wounded and an engine was knocked out, but the boat got back to base safely.[14]

The ALAMO Scouts were not the only group to penetrate the Japanese defenses on New Britain. As usual, the coastwatchers were there, a constant irritant to the enemy. One station set up inland from Cape Orford was in operation through the summer, giving warning of flights from Rabaul and passing on information of the Japanese forces. On 30 August, General Kenney requested that a series of air spotting stations be established across the base of Gazelle Peninsula. General MacArthur agreed and his G–2 ordered the coastwatcher directorate, Allied Intelligence Bureau, to have the spotters in position by November. Sixteen coastwatchers and 27 natives landed from a U.S. submarine on 28 September and separated into five parties to take their posts in the rain forest cover on the island's mountainous spine.

The Japanese managed to capture one of the parties, but the others kept free of searching patrols, sending their reports of enemy flights directly to fighter controls at Woodlark and Dobodura that alerted interceptors at Nadzab's fighter strips. The location of the chain of observers promised 30 to 60 minutes' advance warning of Japanese attacks on Cape Gloucester, time enough for Allied fighters to rendezvous at the most favorable altitude and meet the raiders.

The reports of scouts and coastwatchers were only a portion of the intelligence sifted to get accurate information about New Britain and its defenders. Japanese documents and prisoners taken on New Guinea and Bougainville proved a fruitful source of order of battle data, and at MacArthur's, Krueger's, and Rupertus' levels of command there was a constant re-evaluation of the strength of the defending force in western New Britain. In light of the difficulty of piecing together the various scattered segments of news, the picture of its opponents that the 1st Marine Division was able to assemble was a remarkably good one.

When the BACKHANDER Force operation plan was issued on 14 November, the intelligence estimate of enemy strength west of Arawe on New Britain and on Rooke and Garove was 7,071 men. The total moved steadily upward as elements of the *17th Division* were identified as part of the garrison. On 9 December, the various Allied headquarters published minimum-maximum enemy strength figures for western New Britain that differed considerably. While MacArthur's and Blamey's intelligence officers were content to count only the sure identifications, there was little disposition at Krueger's or Rupertus' command posts to ignore signs, however slight, of Japanese units that might dilute the superiority of attackers

[14] McMillan, *The Old Breed*, p. 168.

over the defenders. Where GHQ and Allied Land Forces found a strength range of 4,000–5,000, ALAMO Force had 5,668–9,344, and the 1st Marine Division figured 7,416–9,816 enemy were present.[15] The final division estimate before the Arawe landing placed the lowest strength at 8,400 and the highest at 12,076.[16] Although the division's educated guess proved to be high, the figures were reasonable in view of the information available.

A scattering of evidence indicated that the *65th* might have been raised from brigade to division status, providing a basis for a belief that at least one other of its regiments besides the *141st Infantry* was present. Accordingly, the *142d Infantry*, which could not be accounted for elsewhere, was added to the order of battle. Similarly, additional companies of units known to be at Cape Gloucester in some strength were counted present if they failed to turn up in other sectors. In the absence of more exact figures, Japanese tables of organization were used as the basis for strength estimates, with 20 percent deducted for losses of all types. Acting to counterbalance the inflated totals was the fact that the presence of the remnants of the *51st Division*, including the *51st Reconnaissance Regiment*, went undetected.

The location of the various elements of the *Matsuda Force* remained a worrisome problem to Allied intelligence officers. The *17th Division* reinforcements arrived so close to D-Day that their defensive assignments were unknown, and it was not clear just what combat units occupied the located defenses. Confusing the situation further was the fact that Japanese place names on western New Britain were a far cry from the names found on prewar Australian maps.

While the enemy accepted and used most of the native place names in western New Britain, the result of such use was often an unrecognizable title. Allied POW interrogators and the translators of Japanese documents had to judge how native speech had been rendered in Japanese and then, in turning the result into English, to try to come close to the original. And often this transition involved the additional problem of considering how differently native dialects would be set down by persons used to Australian and American speech. Intelligence officers, at least, became well acquainted with the Japanese name for Cape Gloucester, translated interchangeably as Tuluvu or Tsurubu. Other names could not be identified. The headquarters of General Matsuda appeared as Egaroppu when translated, a fact that prevented its recognition as Nakarop on prewar maps.[17] And there were puzzling references to villages that could not be plotted anywhere along known trails.

On the whole, however, the maps available to BACKHANDER Force were useful, if not precisely accurate. A German survey of the island, made prior to the first World War, was the base on which later maps were constructed. On these, the trails along the coast were plotted with some care, but the location of trails crossing the interior were at best approximations. As western New Britain attracted Allied attention as a possible objective, repeated aerial photographic missions were flown over it in an effort to improve knowledge

[15] ALAMO Force G–2 PeriodicRept No. 18, dtd 9Dec43, in *ALAMO G–3 Jnl No. 6*.

[16] 1st MarDiv D–2 Sect, Enemy O/B Estimate, dtd 13Dec43.

[17] In the text, the names used by BACKHANDER Force for locations on New Britain have been used.

of its terrain. When MacArthur's cartographers were ready to draft operation maps in October 1943, the aerial photographs gave them a clear picture of coastline and offshore reefs, terrain in the few open areas, and prominent heights, but the jungle trees gave up little of the secrets that lay beneath their branches. Information needed to plot inland trails, villages, and streams was often taken from old maps and charts and from the memories of islanders who had visited or lived on New Britain.

The map most often used by planners and combat troops was drawn in a scale of 1:20,000 with a thousand-yard military grid superimposed. On the back of each of the seven sheets it took to cover the coast from Borgen Bay to Cape Bach and the interior to include Mt. Talawe was the photo mosaic used in making that segment. Two other maps frequently used by higher headquarters were also printed in quantity; one covered the immediate objective area in a one-mile scale and the other showed all of New Britain at an inch to every four miles. To help troop leaders visualize the terrain where they would be fighting, the 1st Division's relief mapping section modeled the ground forms of the Yellow Beach, Green Beach, and airfield areas in several small scales. Copies of these relief maps were then moulded and distributed to division units and other elements of the landing force.

While Marine intelligence officers worked hard to assemble information on the enemy and the objective in order to brief assault troops, they devoted as much attention to the problem of procuring information after the landing. Deliberate and repeated stress was given in all division training to the need for passing along promptly any enemy papers or material that were found. Emphasis was laid, through demonstrations and review of combat experience, on the fact that the most insignificant appearing document might provide the key that would shorten the battle and save lives. The Marines were reminded of the importance of taking prisoners and of the ordinary Japanese soldier's willingness to cooperate with his captors in providing military information once he had surrendered.

To process the flow of intelligence on D-Day and after, the intelligence sections of combat teams and division were set up to give a quick evaluation to captured enemy maps, diaries, and orders. Information of immediate use to Marine commanders was to be extracted and the documents passed on to the ALAMO Force headquarters for definitive translation. In the same manner, prisoners were to be interrogated as soon as possible after capture to extract what intelligence they had of use to BACKHANDER Force, and then shipped out for further questioning in New Guinea. To help speed the search for useful information in enemy documents, ten Nisei were borrowed from Sixth Army to augment the division's own language section.

One other source of up-to-date information of the enemy was available before D-Day—the DIRECTOR operation. If the Arawe assault had its hoped-for diversionary effect, Japanese troops would be drawn off from the area of the main landings. Any contact General Cunningham's cavalrymen had with elements of the *Matsuda Force* promised intelligence of value to the Marines who were headed for Cape Gloucester. Exact answers to questions about the enemy's ability and will to fight and his dispositions to meet Allied attack could only be found ashore on New Britain.

DEXTERITY Landings

ARAWE—Z-DAY TO D-DAY [1]

The focal point of DIRECTOR operations was a boot-shaped peninsula with Cape Merkus as its heel. Lying offshore from the boot's sole are three islands, Ausak, Arawe, and Pilelo, which bound Arawe Harbor, an anchorage used by coastal shipping before the war. The beach chosen for the main Allied assault is about 1,000 yards due north of the cape on the harbor's shore. Behind the beach, designated Orange Beach by planners, the ground slopes up sharply through a break in the cliffs that line the peninsula's western and southern coasts. Most of the north shore of the peninsula and its bulbous toe are taken up by mangrove swamp, which occurs frequently in this part of New Britain; the high ground is occupied by the coconut trees of Amalut Plantation. (See Map 24.)

Between Cape Merkus and Sipul village, about 30 miles to the west, is a trackless region of swamp and jungle, a formidable barrier to movement. From the cape east, the coastal trail springs up again, forking at the Pulie River with one path leading cross island to Iboki and the other continuing along the shore until it disappears in dense rain forest. Four miles east of the neck of Arawe Peninsula (to give it the name used by Allied forces) is Lupin village, the site of a small prewar emergency landing ground. Unused by the Japanese, the airstrip was choked with kunai grass in December 1943.

The most practical way to travel along New Britain's south coast is by boat, a fact that played the most important part in the selection of Arawe as an ALAMO Force target. Once General Cunningham's landing force was ashore on Arawe Peninsula, it could put an end to enemy barge traffic and sever the supply and reinforcement route to Cape Gloucester. Since the Arawe-Iboki trail was known to be difficult and little-used—hardly more than a footpath—Allied planners thought it unlikely that the Japanese could move enough troops overland to threaten seriously Cunningham's position. The proven hunting ability of Allied planes and torpedo boats made remote the prospect of any large-scale enemy movement by sea against Arawe.

In planning DIRECTOR operations, General Cunningham was hampered by a lack of intelligence of both terrain and enemy dispositions. To meet the situation as he understood it, he planned two subsidiary landings before making the assault

[1] Unless otherwise noted, the material in this section is derived from: *ALAMO G-3 Jnl;* CTF 76 OPlan No. 3A-43, dtd 10Dec43; ComVIIPhibFor (CTF 76) Rept of Arawe Op, dtd 10Jan44; 112th CavRegt HistRept 24Nov43-10Feb44, dtd 10Feb44; *17th Div Ops; SE Area NavOps-III;* MilIntelSect, GHQ, FEC, Japanese Monograph No. 142, Outline of SE Area NavAirOps, Part V (Dec43–May44), n.d., hereafter *SE Area NavAirOps-V;* Craven and Cate, *Guadalcanal to Saipan;* Miller, *Reduction of Rabaul;* Hough and Crown, *New Britain Campaign.*

on Arawe Peninsula. The object of one landing was the seizure of a reported Japanese radio station and defensive position on Pilelo Island which commanded the best passages into Arawe Harbor. The second landing was designed to put a blocking force across the coastal trail just east of the foot of the peninsula. In both instances, General Cunningham strove for surprise, planning predawn assaults without air or ships' gunfire support. The troops were to land from rubber boats instead of powered landing craft.

For the main landing, both air strikes and naval bombardment were scheduled to cover assault waves in LVTs and following waves in LCVPs and LCMs. The amphibian tractors were to be manned by Marines of the 1st Division and the landing craft by Army engineers of the 2d Engineer Special Brigade. So little sure knowledge existed of reef conditions in and around Arawe Harbor that Admiral Barbey, who commanded the attack force, asked that ALAMO Force make LVTs available to carry the first waves ashore. The amphibian engineers and their boats were detailed to DIRECTOR Force to provide General Cunningham with a boat pool after the initial landings had been made.

On 30 November, General Krueger directed General Rupertus to assign 29 LVTs and their crews to DIRECTOR Force plus the Marines needed to operate 10 new armored Buffalo amtracs to be provided by the Army. The ALAMO Force commander pointed out that the tractors would be returned to division control before D-Day. The choice of the unit to fill the assignment logically fell on Company A, 1st Amphibian Tractor Battalion. The company was part of the reserve combat team of the division and had had more experience handling the new Buffaloes than any other battalion unit.[2]

Plans called for the DIRECTOR assault troops to be carried to the objective in fast ships, vessels that could unload quickly and promptly leave the area. Later echelons carrying reinforcements and supplies would use LCTs shuttling from Finschhafen. Admiral Barbey was reluctant to risk any of his LSTs at Arawe with the BACKHANDER Operation so close at hand. He did approve, however, the use of the *Carter Hall*, a landing ship dock (LSD) which had just arrived in the Southwest Pacific, and of the Austrialian transport *Westralia* to move the men, amphibious craft, and supplies involved in the main landing. Two APDs were assigned to transport Troops A and B, 1st Squadron, 112th Cavalry, chosen to make the two rubber boat landings.

General Cunningham issued his formal field order for the seizure of positions in the Arawe area on 4 December; the assault troops had been alerted to their mission earlier as they assembled at Goodenough from Woodlark and Kiriwina. In addition to the task of securing Arawe Peninsula and the islands forming Arawe Harbor, the cavalrymen were to outpost the trail leading to the Pulie River and patrol vigorously to guard against enemy attacks. As soon as the task force position was consolidated, Cunningham was to send an amphibious patrol to the Itni River and Gilnit to check the possibility of overland contact with the Marines at Cape Gloucester.

The torpedo boat base called for in GHQ, Seventh Fleet, and ALAMO Force plans for Arawe remained a vague affair

[2] *1st MarDiv SAR*, Anx D, Amphibian Tractors, p. 1.

at the operating level despite the definite language directing its establishment. In conferences with General Cunningham on 5 December, the fleet's motor torpedo boat commander, Lieutenant Commander Morton C. Mumma, stated that he would assign two boats based at Dreger Harbor on New Guinea to patrol east of Arawe each night after the landing. The boats would report to the task force intelligence officer for briefing on arrival and be available for special missions on request, but they would return to Dreger with dawn. The only naval installations Mumma asked for at Arawe were emergency fueling facilities. In addition to the torpedo boats hovering near Arawe, he said another pair from Dreger would hunt barges nightly between Tauali and Sag-Sag and two others from Kiriwina would scout the vicinity of Gasmata. No boats would patrol the sector from Arawe west to the Itni because of the poorly charted reefs and shallows among the offshore islands.

The *Carter Hall* loaded Marine tractors and crewmen at Milne Bay on 5 December, as the *Westralia* picked up a company-size task group from the Boat Battalion, 592d Boat and Shore Regiment. At Goodenough the two larger ships were joined by the APDs *Sands* and *Humphreys*, and the 112th's assault troops came on board to take part in two practice landings, one a full-scale rehearsal of the operation. The results of the training showed that boat wave timing was off, that some junior leaders were not sure of themselves in the unfamiliar amphibious role, and that unit commanders did not always have control of their men—all faults that could be corrected with further rehearsal. Time was too short for more practice, however, as Z-Day (15 December) was approaching fast and the ships and men had to make last-minute preparations for the assault.

When the DIRECTOR operation was decided upon, the Japanese defenders of Arawe were calculated at 100–150 men with the only identified units a naval antiaircraft platoon and a small radio detachment. On 5 December, reconnaissance planes sighted 12–14 enemy barges at Kumbum Island near Arawe and four more along the peninsula's shore. Although it was conceded that these craft might be taking part in a routine supply run, ALAMO intelligence officers felt it was "prudent to assume the enemy at Arawe now has at least 500 [men] and considerable reinforcement potential." [3] General Cunningham asked General Krueger for reinforcements to meet the added threat. On 10 December, a battalion of the 158th Infantry, a non-divisional regiment, was alerted as reserve for DIRECTOR.

Cunningham also asked that a 90mm gun battery be assigned to his force to supplement the antiaircraft fire of the two batteries of automatic weapons he already had. Although the strength of the Japanese ground garrison at Arawe and its defensive potential were perplexing questions, there was little doubt in the minds of Allied commanders that the enemy air reaction to the landing would be swift and powerful. The few heavy antiaircraft guns available were already committed to other operations, however, and Cunningham's force had to rely on machine guns and 20mm and 40mm cannon for aerial defense.

Loading for the Arawe landings got underway at Goodenough during the afternoon of 13 December. As Generals Mac-

---

[3] ALAMO For G-2 PeriodicRept No. 18, 1-8Dec43, dtd 9Dec43, in *ALAMO G-3 Jnl No. 6*.

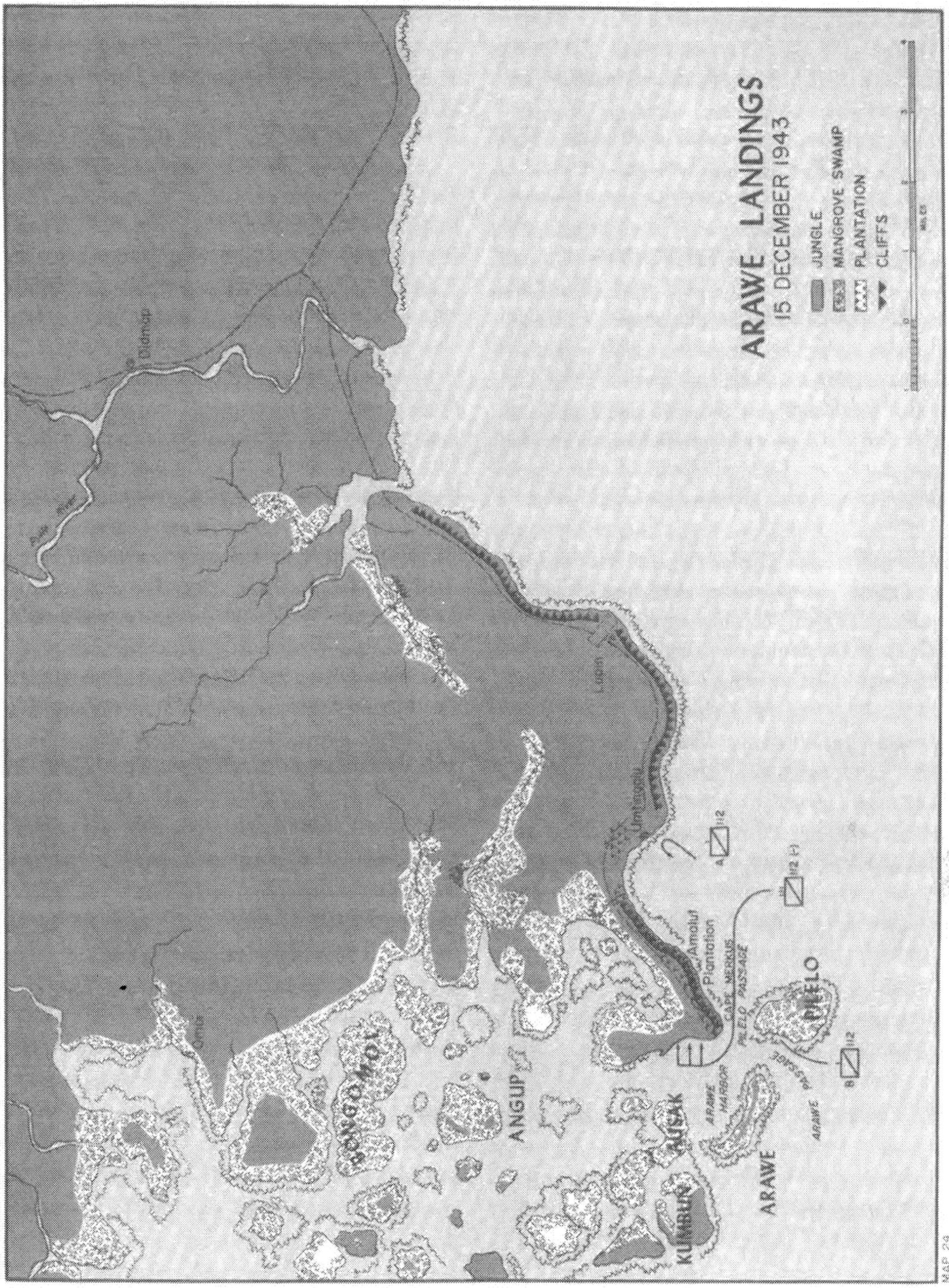

MAP 24

RF STIBIL

Arthur and Krueger looked on, the Marine LVTs and two rocket-firing DUKWs, manned by Army engineers, drove into the water and churned through the stern gate and into the flooded well deck of the *Carter Hall* standing offshore. The assault troops who would ride the tractors, the 112th's 2d Squadron and supporting units, boarded the LSD while the 1st Squadron and regimental and task force headquarters rode the *Westralia*. Replacing the regular landing craft on the Australian ship's davits were 16 LCVPs and 2 LCMs of the amphibian engineers. The troopers slated to make the rubber boat landings bedded down on the *Sands* and *Humphreys*, 150 men to each ship.

At midnight, the convoy sailed for Buna where General Cunningham boarded the destroyer *Conyngham*, Admiral Barbey's command ship for the operation. After feinting a movement toward Finschhafen, the task force turned toward its target after dusk on 14 December and headed across the Solomon Sea. Guarding the west flank were cruisers and destroyers of Admiral Crutchley's force and to the east was a cordon of Commander Mumma's motor torpedo boats. Escorting the transports were nine destroyers, five of them designated a shore bombardment unit.

About 15 minutes before the convoy arrived in the transport area, a Japanese float plane scouted the ships, dropping a bomb near one of the destroyers. There was no return fire, in accordance with Barbey's orders for dealing with night snoopers; gun flashes might reveal the presence of a ship and offer an aiming point to lurking attackers. But the Japanese pilot had reported the ships he had sighted as five destroyers and five transports, and Admiral Kusaka at Rabaul ordered planes of the *Eleventh Air Fleet* to attack. A Japanese submarine had also sighted the Allied ships late on the 14th, but its report did not reach enemy headquarters until after the landing.

The *Carter Hall* began taking on ballast before it arrived in the transport area about five miles from Orange Beach. By 0400 there was 4½ feet of water in the well deck and launching began. Within ten minutes all the vehicles had cleared the LSD. At about the same time, the *Westralia* lowered its landing craft and men began clambering down nets into the boats, while winches and rope slings handled the 40 tons of supplies and equipment figured as essential to sustain the assault troops in their first hours ashore. At 0500, the two large transports headed away in the darkness for New Guinea.

While the LVTs made ready for the run to Orange Beach and the landing craft circled waiting to follow the tractors' lead, the surprise landings got underway. The APDs carrying Troops A and B hove to about 1,000 yards offshore, unloaded the cavalrymen and their boats, and started to withdraw immediately. As they had practiced, the soldiers paddled toward their target beaches in three waves of five boats each.

Troop A, which had the mission of cutting the coastal road near Umtingalu village, never reached its objective. At 0522, enemy automatic weapons cut loose from the featureless black outline that marked the shore, raking the oncoming boats. Within a few minutes, only three of the rubber craft were still afloat. From its position 3,000 yards out from Umtingalu, the supporting destroyer *Shaw* had difficulty seeing if the soldiers struggling in the water were in its line of fire. At 0542, when the *Shaw* finally had a clear shot, two salvos of 5-inch high explosive shell were all that was needed to silence the enemy guns. The abortive landing attempt cost the 112th Cavalry 12 men killed, 17 wounded, and 4 missing in action.

When the Japanese fired on Troop A's boats, Troop B abandoned its original landing plan for Pilelo and headed for the nearest beach on the island. Moving quickly inland after they landed, the cavalrymen isolated the small enemy garrison in caves near Winguru village and wiped them out after a fight at close quarters with bazookas, flame throwers, and grenades called into use. The troop lost one man and accounted for seven defenders in seizing its objective.

While the subsidiary landings were taking place, a column of Marine amphibian tractors started on the 5,000-yard run to Orange Beach. The armored Buffaloes in the lead, traveling at six knots, soon outdistanced the Alligators which were hard put to make 4½ knots. One of the control boats for the main landing, the *SC-742*, spotted the LVTs moving into Pilelo Passage well ahead of schedule and headed them off. The Buffaloes then circled off Cape Merkus waiting for the Alligators to catch up. Meanwhile, the slower tractors had gotten off course and had to be herded back into the prescribed boat lane. The upshot of the confusion was that the landing was delayed 40 minutes.

The pre-H-Hour destroyer bombardment was extended when it became clear that the LVTs were falling behind schedule, but the cease fire was sounded on board the ships at 0641 to conserve ammunition for antiaircraft defense. When the 5-inch shells stopped falling, the standby control boat, *SC-981*, began firing a spread of rockets on Cape Merkus to cover the movement of the landing vehicles through Pilelo Passage. At 0710, the DIRECTOR Force air support officer on the *Conyngham* called down a waiting squadron of B-25s to strafe and bomb the beach. Shortly thereafter, behind a barrage of rockets fired by the engineer DUKWs and *SC-742*, the assault troops of 2/112 landed.[4]

The bombardment silenced what little opposition there was at the beach, and the squadron moved inland cautiously but easily. Two of the Marine amtracs in the first wave were able to negotiate the steep bank behind Orange Beach and accompany the soldiers to their destination, the narrow neck of the peninsula. When enemy machine guns firing from the edges of the mangrove swamp pinned down the troopers, the armored Buffaloes crushed the defenders' positions and silenced the guns. By mid-afternoon the 2d Squadron was digging in on its objective. A pocket of Japanese riflemen left behind by the assault troops was eliminated by the 112th's Headquarters Troop, and the peninsula was cleared of enemy forces by nightfall.

Survivors of the 120 Japanese soldiers and sailors who had defended Arawe Peninsula, Pilelo, and Umtingalu had withdrawn from contact with the landing force. The Army units, detachments from two temporary companies of the *51st Division*, one of infantrymen, the other of artillerymen, rejoined their parent unit in

---

[4] Contemporary reports differ considerably on the exact time of the landing, but it appears that 0728, the time reported by the LVT commander, is correct. Co A, 1st PhibTracBn Rept of PhibTrac Ops in Arawe Landing, dtd 27Dec43, in *ALAMO G-3 Jnl No. 12*.

positions near the Pulie River at Didmop.[5] The naval coastal antiaircraft platoon that had repulsed the landing attempt at Umtingalu, abandoned its guns and took off in precipitate retreat north up the overland trail. Coming south along the same trail, hurrying as best the jungle, rain, and frequent swollen streams would allow, was Major Shinjiro Komori's *1st Battalion, 81st Infantry*. Komori had arrived within earshot of the preliminary bombardment at Arawe, but was still four days' forced march from the Pulie River, illustrating the difficulties of moving a large body of men through the center of New Britain.[6]

Intensive Japanese ground opposition to the Allied landing was yet to come, but, as expected, the aerial counterattack came on the heels of the landing. At 0855, after one flight of enemy planes had engaged the covering P–38 squadron and led it in a snarling dogfight away from the target, 30–40 naval fighters and bombers struck at the beachhead. From the Japanese point of view, the time of attack was auspicious. The only heavy antiaircraft guns left in Arawe waters were those on the *Conyngham*; all the other escort, control, and bombardment vessels had departed moments before. The first follow-up echelon of five LCTs, accompanied by seven engineer-manned LCMs, had just moved into Arawe Harbor to unload. The fire of machine guns and 20mm cannon mounted on ships, LVTs, trucks, landing craft, and beach proved more than enough to beat off the attack. The total of serious damage was one LCVP blown up and a few men wounded.

Throughout the day the attacks continued, but Allied fighter squadrons were able to turn away most of the enemy planes. The vivid imaginations of the Japanese pilots supplied them with the damage that they were unable to inflict during their attacks, and Rabaul heard tales of many sunken transports and landing craft and damaged cruisers and destroyers.

Returning again and again during the next week, the Japanese planes, mostly carrier aircraft stationed at Rabaul, attempted to wipe out the Allied beachhead. Their effort was unsuccessful, although one LCT echelon arriving late on the 16th was under almost continuous air attack while it was at Arawe and had an escort coastal transport sunk, and an escort minesweeper and seven LCTs badly damaged. This was the high point of the enemy strikes; defending fighters and antiaircraft automatic weapons fire whittled down the attacker's strength at a rate of three, four, or five planes a day, losses the Japanese could ill afford. As the fourth week of December began, large-scale air raids tapered off to night bombing runs and hit-and-run daylight missions by small numbers of enemy planes.

The frequent air attacks on Z-Day and the days immediately following severely disrupted the unloading process. An inexperienced shore party and a shorthanded naval beach party, worked to exhaustion, were unable to move supplies from ships to dumps smoothly; congestion on Orange Beach was constant. The beach itself proved capable of handling only two

---

[5] ATIS Item No. 8159, Merkus Garrison Intel Rept, dtd 10Dec43, in ATIS Bul No. 614, dtd 8Jan44 (ACSI Recs, FRC Alex). Prior to D-Day, the infantry company at Arawe was designated the *10th Company, 3d Battalion, 141st Infantry*, but the association with the regiment at Cape Gloucester was strictly a paper matter.

[6] ATIS Item No. 11249, Diary of Major Komori, 16Oct43–31Mar44, in ATIS Bul No. 999, dtd 6May44 (ACSI Recs, FRC Alex), hereafter *Komori Diary*.

TROOPERS OF THE 112TH CAVALRY *wade ashore at Arawe as Marine LVTs carry in supplies on 15 December 1943. (SC 187063)*

MOVING OFF THE RAMP *of a Coast Guard-manned landing craft, Marines move ashore on D-Day at Cape Gloucester. (USN 80-G-44371)*

LCTs at a time, and other ships in each echelon had to stand by under threat of enemy air attack while waiting their turn to unload. Under the circumstances, LCT commanders were unwilling to remain at Arawe any longer than their movement orders stated, and, on occasion, cargo holds were only partially cleared before the ships headed back for Finschhafen. The problem grew steadily less acute as Japanese air raids dwindled in number and strength and the reserve supplies of the landing forces reached planned levels.

General Cunningham lost no time in making his position on Arawe Peninsula secure behind a well-dug-in defense line closing off the narrow neck. Engineer landing craft gave the general protection on his open sea flanks, and a series of combat outposts stretching up the coastal trail beyond Lupin airstrip promised adequate warning of any attempt to force his main defenses. The 112th was brought up to strength; Troop A was re-equipped by air drop on the 16th, and, two days later, APDs from Goodenough brought in the 3d Squadron. The cavalry regiment acting as infantry and a reinforcing field artillery battalion seemed quite capable of handling any Japanese units that might come against them.

Pursuing his mission of finding out all that he could about Japanese forces in western New Britain before the BACKHANDER Force landed, Cunningham dispatched an amphibious patrol toward the Itni River on the 17th. Moving in two LCVPs, the cavalry scouts reached a point near Cape Peiho, 20 miles west of Arawe, by dawn on the 18th. There, appearing suddenly from amidst the offshore islets, seven enemy barges attacked the American craft and forced them into shore. Abandoning the boats, the scouting party struggled inland through a mangrove swamp, finally reaching a native village where they were warmly received.

An Australian with the scouts was able to get hold of a canoe and report back to Arawe by the 22d to tell Cunningham that the waters between Cape Merkus and the Itni were alive with enemy barges. The natives confirmed this finding and also said that large concentrations of Japanese troops were located at the mouth of the Itni, Aisega, and Sag-Sag. Special efforts were made to insure that this intelligence reached General Rupertus, and Cunningham sent it out both by radio and an officer messenger who used a torpedo boat returning to Dreger Harbor.

The barges that the 112th's scouts ran up against on the 18th were transporting enemy troops from Cape Bushing to the scene of action at Arawe. As soon as he heard of the Allied landing, General Sakai of the *17th Division* had ordered General Matsuda to dispatch one of his battalions to Cape Merkus by sea. The unit Matsuda selected, the *1st Battalion, 141st Infantry* (less its *1st Company*), landed at the village of Omoi on the night of 18 December, bivouacked, and started overland in the morning, heading for a junction with the Japanese at Didmop. Getting lost repeatedly in the trackless jungle, pausing whenever contact with the Americans seemed imminent, the battalion took eight days to march a straight-line distance of seven miles.

Major Komori, who was designated overall commander of enemy troops in the area, reached Didmop on the 19th, having gathered in the retreating sailors from Umtingalu on his way. For a few days he paused there, organizing his forces, and waiting for *1/141* to arrive. Finally deciding to delay no longer, Komori crossed the Pulie with his main strength on Christmas Eve and arrived on the edge of the airstrip at dawn. The 112th outposts and patrols were forced back on Umtingalu and, with the troop stationed there, retired into the perimeter. A determined probing attack in company strength was made on the American main line of resistance on the night of the 25th; 12 infiltrators were killed within the cavalrymen's positions before the Japanese were repulsed.

Komori's attack emphasized one of the most successful aspects of the DIRECTOR operation. By the dawn of D-Day for the main DEXTERITY landings, two Japanese infantry battalions that might have fought at the main Allied objective were tied up defending a secondary target. What effect this shift of a thousand enemy troops, most of them combat veterans of the Philippines or China fighting, had on the Cape Gloucester operation is impossible to say with any certainty. It seems probable, however, that the casualties suffered by the 1st Marine Division would have mounted and the seizure of control of western New Britain would have been delayed had the enemy not changed his troop dispositions.

PRELANDING PREPARATIONS[7]

While action at Arawe drew the attention of the Japanese, the BACKHANDER Force made its final preparations for the landings at Cape Gloucester. Rehearsals following the pattern of the Yellow Beach landings, with all assault elements of Combat Team C embarked, took place at Cape Sudest on 20–21 December. The long months of training bore fruit, and the first waves moved from ship to shore without incident and were smoothly followed to the practice beaches by landing ships carrying supplies and reserve forces.

At Cape Gloucester, Admiral Barbey would again control elements of his amphibious force as Commander, Task Force 76 and fly his flag on the *Conyngham* as headquarters ship for the operation. From Seventh Fleet and Allied Naval Forces, now under Vice Admiral Thomas C. Kinkaid who had relieved Admiral Carpender on 26 November, would come the escort, covering, and bombardment vessels to support the landings. General Kenney's Fifth Air Force would handle all air support missions except those that could be undertaken by the 1st Division's own squadron of light planes.

The preliminary softening-up of the target was exclusively the province of Kenney's planes and pilots. Air raids were so frequent through the late fall that Japanese soldiers' diaries show little recognition of the fact that air strikes intensified after 18 December, for the condition of the men being attacked was so consistently miserable that the step-up escaped their notice. One ground crewman at the knocked-out airfields noted unhappily that "enemy airplanes flew over our area at will and it seemed as though they were carrying out bombing training."[8] Japanese interceptors and antiaircraft fire never seriously challenged the daily runs of strafers and bombers.

---

[7] Unless otherwise noted, the material in this section is derived from: *ALAMO G-3 Jnl; VII PhibFor AR; CTF 76 Dec43 WarD; 1st MarDiv SAR*, Anx C, Logistics and Supply.

[8] ATIS CurrTranslation No. 1324, Diary of a ground crewman at Rabaul and Tsurubu, 5Nov–17Dec43, in ATIS CurrTranslations No. 122, dtd 31May44 (ACSI Recs, FRC Alex).

The weight of bombs dropped on targets in the Cape Gloucester vicinity between 1 December and D-Day exceeded 3,200 tons. Over 1,500 individual sorties were flown to deliver the explosives and to search out enemy targets with cannon and machine gun fire. Up until 19 December, the bombers concentrated their attacks on the airfields and their defenses, but then target priority shifted to objectives that might impede the landings and subsequent advance of the Marines. The attacks resulted in substantial destruction of prepared positions spotted by aerial reconnaissance and in the death of scores of Japanese soldiers. The judgment of air historians viewing the campaign was that mass bombing of the invasion areas at Cape Gloucester by the Fifth Air Force virtually eliminated the combat effectiveness of the Japanese defenses.[9] Hidden by the jungle's impenetrable cover from the probing aerial camera, most of the *Matsuda Force* survived the protracted aerial assault. The damage to defenders' nerves and morale is apparent from contemporary records, however, and the widespread destruction of defensive positions undoubtedly eased the task of Marine attacking forces.

Behind the screen of air activity, the assembly of Barbey's attack force proceeded without major hitch. A complicated schedule of loadings and sailings had to be met that would enable the various echelons to reach the target on time. On the afternoon of the 24th, two LSTs with headquarters and service detachments operating directly under General Rupertus' command and reserve elements of Combat Team C embarked, left Cape Sudest for Cape Cretin near Finschhafen. At Cretin, the two ships were joined by five others loaded with troops of Combat Team B due to land on D-Day afternoon. At 0100 on Christmas morning, six LSTs of the third echelon sailed from Sudest with men of Combat Teams B and C and BACKHANDER Force supporting elements; at Cape Cretin, an LST carrying men and equipment of 3/1 joined the ships which were scheduled to nose into the Yellow Beaches as soon as the infantry-carrying LCIs of the second echelon landed their cargos.

The main convoy of 9 APDs and 11 LCIs got underway from Cape Sudest at 0600, 25 December. Accompanying the ships and their six escort destroyers was the *Conyngham* with Barbey and Rupertus on board. Not long after the assault troops departed, transports carrying reserve Combat Team A arrived in Oro Bay from Milne, and the 5th Marines and its attached units landed at Cape Sudest to await the call for employment at Cape Gloucester. The 14 LSTs which were to land on D plus one (27 December), bringing in engineer, antiaircraft, and medical units and supplies for BACKHANDER Force, loaded out on Christmas and sailed during the night in trace of the D-Day convoys. Each LST bound for Cape Gloucester was fully loaded with trucks carrying bulk cargo; those ships landing on 26 December carried an average of 150 tons of supplies apiece, and the ones due in the next day 250 tons.

To minimize the risk of sailing poorly charted waters, the five LCIs carrying 2/1's assault troops joined the main con-

[9] Craven and Cate, *Guadalcanal to Saipan*, p. 311; Dr. Robert F. Futrell, USAF HistDiv, ltr to Head, HistBr, G-3, HQMC, dtd 19Jun62.

voy off Cape Cretin to make the passage through Vitiaz Strait. A small-boat task group, 12 Navy LCTs plus 2 LCVPs and 14 LCMs crewed by amphibian engineers, cut directly through Dampier Strait toward Green Beach. Accompanying the landing craft as escorts for the 85-mile voyage from Cape Cretin to Tauali were an engineer navigation boat, a naval patrol craft, and two SCs; four torpedo boats acted as a covering force to the east flank.

While the Eastern and Western Assault Groups of Barbey's force approached their respective beaches, one possible objective of DEXTERITY operations was occupied by a reinforced boat company of the 592d Engineer Boat and Shore Regiment. On 20 December, General Krueger approved a plan to set up a long-range radar on Long Island, about 80 miles west of Cape Gloucester, after Australian coastwatchers had reported the island free of enemy troops. Acting on an operation plan published only three days before embarkation, the engineer boat group began its shore-to-shore movement from Finschhafen on Christmas afternoon. Preceded by an advance party that landed from torpedo boats, the engineers and an Australian radar station went ashore on the island on the 26th. The lodgment on Long Island removed one target from the list of those that might be hit by 3d Battalion, 5th Marines, held back from employment at Cape Gloucester.[10]

[10] General Krueger noted in his comments approving the Long Island operation that the Marine battalion slated for possible seizure of Rooke Island might be the source of a detachment to relieve the engineers on Long. ALAMO CofS memo to Gen Krueger with Krueger's comments, dtd 20Dec43, in *ALAMO G-3 Jnl No. 10.*

STONEFACE TRAIL BLOCK [11]

Although they occurred simultaneously as complementary operations within the overall BACKHANDER concept, the landings at Green and Yellow Beaches are seen clearer when examined separately. The primary mission of Lieutenant Colonel Masters' 2/1, reinforced, was a defensive one—to land, seize a trail block, and hold it against all comers. The first mission of the rest of General Rupertus' force was offensive in nature—the capture of Cape Gloucester's airfields. Masters' command, code-named the STONEFACE Group, was scheduled to rejoin the rest of the 1st Division once the airfield objective was secured. (See Map 25.)

The LCIs with the STONEFACE Group embarked broke off from the convoy headed for the Yellow Beaches at 0422 on D-Day morning. Accompanied by two escort and bombardment destroyers, the *Reid* and *Smith*, the troop-laden landing craft headed for a rendezvous point about four miles off the New Britain coast opposite Tauali village. Contact with the small-boat group that had made the voyage through Dampier Strait was made later than had been planned, but the transfer of assault troops to LCMs began

[11] Unless otherwise noted, the material in this section is derived from: *1st MarDiv SAR*, Phase II, Part II, Green Beach Landing; CTG 76.3 (ComDesDiv 10) AR—Bombardment and Landing Ops at Tauali (Cape Gloucester), New Britain 26Dec43, dtd 29Dec43 (COA, NHD); CO, Det, 2d ESB Rept to CG, 2d ESB, dtd 20Jan44, Subj: Ops CTF 76 LT 21 (Green Beach), in *ALAMO G-3 Jnl No. 19;* OCE, GHQ, AFP, *Amphibian Engineers Operations—Engineers in the Southwest Pacific 1941–45*, v. IV (Washington: GPO, 1959), hereafter OCE, GHQ, AFP, *Amphibian Engineers;* Hough and Crown, *New Britain Campaign.*

immediately, and the lost time was recovered.

Engineer coxswains, moving in successive groups of four, brought their boats alongside the three LCIs that carried the two assault companies, E and F. The Marines were over the side and headed for the line of departure within 10 minutes. Accompanying the four landing craft that formed the first wave were two other LCMs carrying rocket DUKWs of the 2d Engineer Special Brigade's Support Battery; the rockets were intended for a beach barrage to silence opposition in the few minutes just before the landing.

As the column of LCMs moved lazily toward the line of departure, throttled down to keep from getting ahead of the landing schedule, the troops could hear the sound of naval gunfire at Cape Gloucester, seven miles away. Falling into formation behind the assault craft, two LCIs with most of Companies G and H on board made up the fourth wave. Bringing up the rear with a cargo of guns and vehicles and 575 tons of bulk stores were 12 LCTs organized in waves of three. The ships carried 20-days' supplies and five units of fire for Masters' 1,500-man force; the men themselves carried one unit of fire and a day's rations.

From his headquarters on the *Reid*, the Commander, Western Assault Group, Commander Carroll D. Reynolds, ordered the prelanding supporting fires to begin on schedule. The two destroyers cruising 5,400 yards offshore fired 675 rounds of 5-inch at targets behind the beach and to its flanks as far as Dorf Point on the north and Sag-Sag to the south. After 20 minutes, the ships' fire was lifted and Commander Reynolds radioed a waiting squadron of B–25s to attack; for insurance the *Reid* fired two star shells as a visual signal. As the medium bombers began making bombing and strafing runs along the long axis of the beach, the first wave of LCMs left the line of departure. The planes were scheduled to cease their bombardment when the landing craft were 500 yards from Green Beach, an event calculated to occur at 0743. At that moment, the engineer DUKWs began firing the first of 240 rockets they arched ashore.[12] Two strafers made a last minute pass at beach targets after the DUKWs opened up but fortunately avoided the plunging rockets.[13] At 0748, the first LCMs grounded on the beach and dropped their ramps.

The Marines of the assault companies moved quickly across the volcanic sand, mounted the slight bank bordering the beach, and then advanced cautiously into the secondary growth that covered the rising ground beyond. There had been no enemy response from the beach during the bombardment and there was none now that the Americans were ashore. Well-dug trenches and gun pits commanding the seaward approaches had been abandoned, and there was no sign of active Japanese opposition. The second and third waves of LCMs landed five minutes apart, adding their men to the swelling force ashore. At

[12] Oddly enough, this barrage of rockets made little impression on some of the troops it supported: it goes unmentioned in the division action report and escaped the notice completely of at least one officer who witnessed the landing. Maj Theodore R. Galysh ltr to CMC, dtd 16Feb52.

[13] The engineer brigade commander had suggested to Admiral Barbey "that the last plane in the final strafing run drop a flare to indicate the way was clear for the rocket barrage to start. The reply was that the airmen 'preferred to work on a strictly time basis,' accordingly no signal was given." OCE, GHQ, AFP, *Amphibian Engineers*, p. 170, n. 96(2).

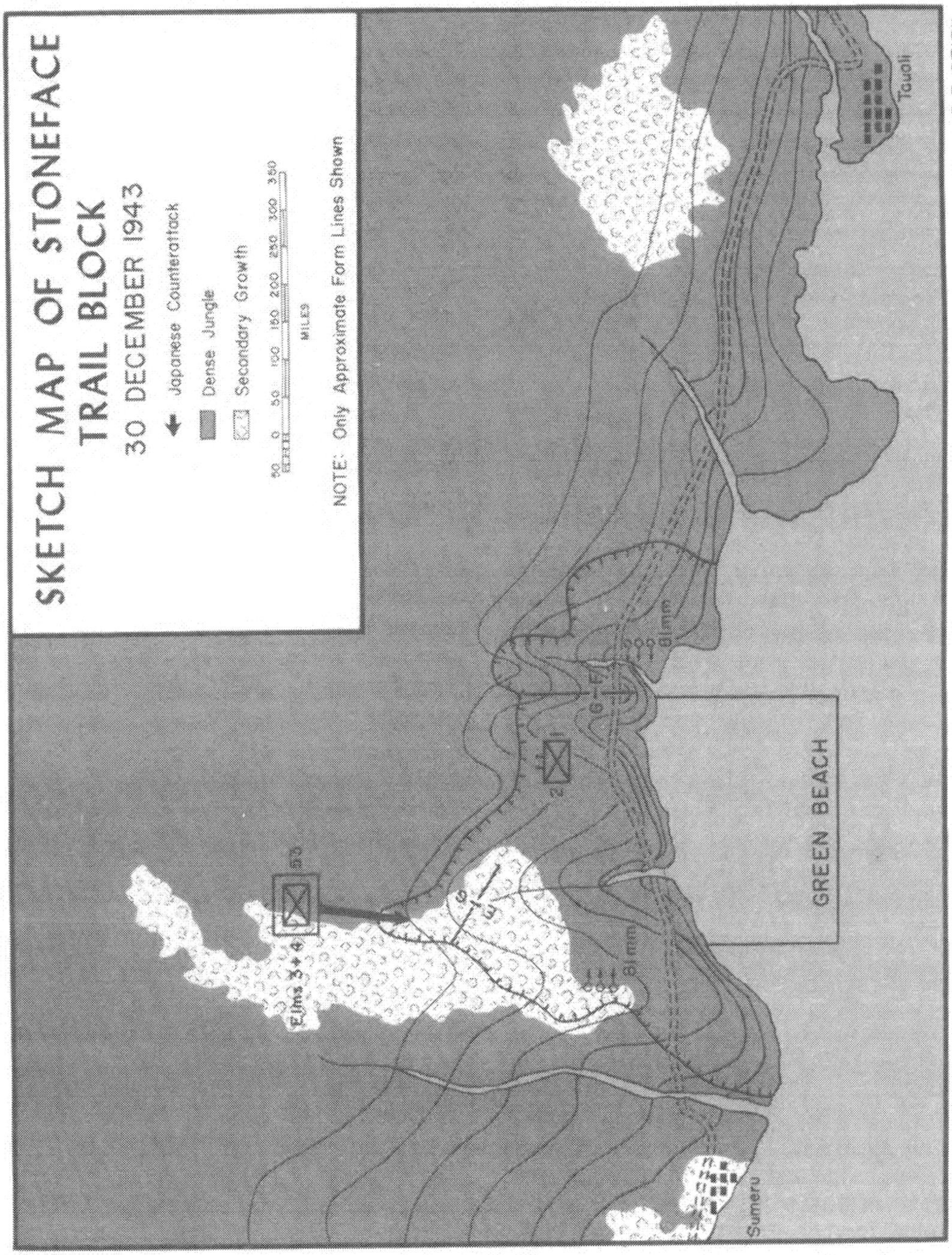

MAP 25

R.F. STIBIL

0754, an amber star cluster was fired to signal a successful landing to the waiting ships.

The engineer LCMs retracted after delivering the assault troops and headed out to a rendezvous point a mile and a half from shore. When the beach was clear, the LCIs landed, dropped their twin bow ramps, and the Marines of Companies G and H filed down into shallow water. At 0830, the first of the LCTs hit, and the others followed in rapid succession as general unloading began.

The shore party was composed of a headquarters under 2/1's operations officer, a labor platoon of 130 men drawn from all the major elements of the STONEFACE Group, and a beach party of amphibian engineers.[14] Supplies were channelled through four predesignated unloading points into as many dumps set up in the area just off the beach. In each dump, similar amounts of rations, fuel, ammunition, and organizational equipment were further segregated to cut down possible losses to enemy air action. The absence of Japanese planes on D-Day, combined with unseasonably fine weather and a light surf, did much to ease the task of the inexperienced shore party. By 1715 the last LCT had been emptied and was assembled in the rendezvous area for the voyage back to New Guinea.

The convoy departed at 1900, leaving behind two LCVPs and four LCMs for offshore defense. The engineer boats anchored inside a reef outcrop about 200 yards from the beach with their stern guns pointing seaward. Lieutenant Colonel Masters gave the job of organizing the beach defenses to the Army engineer detachment commander, Major Rex K. Shaul, and assigned a platoon of LVTs and a platoon of the 1st Marines Weapons Company to his command. Inland, the STONEFACE commander disposed his three rifle companies along a 1,200-yard-long perimeter that ran along the ridges overlooking the beach. At its deepest point, a salient where a causeway led into the Marine lines, the perimeter was 500 yards from shore. The riflemen were carefully dug in by nightfall, and Company H's heavy machine guns were emplaced amidst the foxholes where their fire could be most effective. The battalion weapons company's 81mm mortars were set up to fire in front of any part of the defensive position. In reserve behind each rifle company was a platoon of artillerymen. These provisional units had been formed on Masters' order as soon as it became apparent that Battery H of 3/11 could not find a firing position that was not masked by precipitous ridges.

By nightfall on D-Day, the STONEFACE Group had established a tight perimeter defense encompassing all the objectives assigned it. The main coastal track lay within the Marine lines, denied to enemy use. Probing patrols into the jungle and along the coast did not discover any Japanese troops until dusk, when there was a brief brush with a small enemy group near the village of Sumeru, just north of the beachhead. The Japanese faded away into the jungle when the Marine patrol opened fire.

[14] General Rupertus asked that an Army engineer shore company be assigned to support the Green Beach landing, and 2/1's operation order was amended on 23 December to include one in the shore party, but the tentative assignment was never completed. CG, BACKHANDER For msg to CG, ALAMO For, dtd 20Dec43, in *ALAMO G-3 Jnl No. 7*; Anx C, 2/1 OpO 3-43, dtd. 23Dec43, in *1st MarDiv SAR*, Phase II, Anx D.

Lieutenant Colonel Masters was unable to contact General Rupertus directly to report his situation. Mt. Talawe, looming between the two Marine beachheads, proved an impenetrable barrier to the battalion's radios. Since the set of the amphibian engineers was intermittently able to reach ALAMO Force, word of the STONEFACE Group's dispositions eventually reached the division CP on the morning of the 27th. Except for the fact that his force was located in a radio "dead spot" for overland communication,[15] Masters was in excellent shape to accomplish his mission.

ASHORE AT THE YELLOW BEACHES[16]

On Christmas morning, a Japanese coastwatcher hiding out in the hills above Cape Ward Hunt on New Guinea spotted the Eastern Assault Group en route to its target. This enemy observer's report of the ships' passage, sightings by a Japanese submarine scouting the area, or, perhaps, the last frantic message from a reconnaissance plane that was shot down shortly after noon while it was skirting the convoy, may one or all have been responsible for Rabaul's deduction that the assault group was headed for Cape Merkus. On the strength of this faulty judgment, the *Eleventh Air Fleet* and *Fourth Air Army* planned a hot reception for the ships at Arawe on the 26th.[17]

The path of the invasion convoy during daylight hours had been plotted to mislead the Japanese. Once night fell, the moonless dark that ALAMO planners had waited for, the ships shifted their course to a more direct route through Vitiaz Strait. The destroyers, transports, landing ships, and escorts steamed along at 12 knots while Admiral Crutchley's cruiser force ranged ahead to take position in gunfire support areas lying off the Cape Gloucester airfields.

Maintaining the convoy's pace proved to be too much of a strain on the engines of a harbor mine sweeper (YMS) which was scheduled to clear the waters off the Yellow Beaches. The ship broke down finally and fell out at 2120, returning to Cape Cretin; earlier, another YMS had been taken in tow by an APD to keep it in the convoy. As Admiral Barbey's ships began to move into the transport area, the two YMSs that had completed the voyage, two destroyers, and two SCs operated together to locate, clear, and buoy the channel leading to the Yellow Beaches. (See Map IV, Map Section.)

The destroyers, taking a radar fix on the wreck of a Japanese destroyer that was hung up on the reef 7,000 yards offshore, found the entrance to the channel in the darkness. Using their sound gear to locate shoals along the passage, the destroyers gave their bearing and distance to the YMSs which would then "strike off for the shoal as a dog after a bone."[18] The mine

[15] LtCol Robert Hall ltr to CMC, dtd 4Mar52.

[16] Unless otherwise noted, the material in this section is derived from: *ALAMO G-3 Jnl; VII PhibFor ComdHist; VII PhibFor AR; CTF 76 Dec43 WarD; 1st MarDiv SAR*, Phase II, Part I, Yellow Beach Landings and Plan Annexes; 1st MarDiv D-3 Jnl, 24Dec43–29Feb44, hereafter *1st MarDiv D-3 Jnl—I; 17th DivOps;* Craven and Cate, *Guadalcanal to Saipan;* Hough and Crown, *New Britain Campaign;* Morison, *Breaking the Bismarcks Barrier*.

[17] *SE Area NavOps—III*, pp. 47–48; *SE Area AirOps*, pp. 28–30.

[18] CTU 76.1.42 (CO, USS *Flusser*) AR of Cape Gloucester Op on 26Dec43, dtd 29Dec43 (COA, NHD).

sweepers and a destroyer's whaleboat buoyed the three principal reef obstacles on schedule, and one of the YMSs swept the channel which proved to be free of mines. Immediately, the destroyers moved inside the outlying reefs to deliver close-in fire on the eastern flank of Yellow 2. The SCs, which had also controlled the Arawe landing, proceeded to line of departure and standby stations to handle the waves of landing craft.

Throughout the clearance and marking of the channel, a mounting drumfire of naval bombardment marked the approach of H-Hour (0745). At 0600, the two Australian heavy cruisers of Task Force 74 opened up on targets in the vicinity of the airfields with the first of 730 rounds of 8-inch high explosive fired during the next 90 minutes. The 6-inch guns on the American light cruisers sounded next, firing 2,000 rounds between 0625 and 0727 at Target Hill and the Yellow Beach area. Escort destroyers with the cruiser force and those that had worked with the harbor control unit exploded 875 5-inch shells ashore while the larger ships were firing and, later, in the few moments immediately preceding H-Hour. The ground just inland from the beaches and the hills to the southeast drew most of the destroyer fire. Targets of opportunity were few since no evidence of enemy movement or opposition developed ashore during the Allied bombardment.

From 0700 to 0720, while naval gunfire was concentrated on the beaches and airfields, five squadrons of B–24s dropped 500-pound bombs on the Target Hill area. Then, on schedule, at the call of a command plane aloft, naval gunfire on the beaches lifted and a squadron of B–25s streaked in over Target Hill to let go eight tons of white phosphorous bombs on its naked crest. Smoke soon obscured the vision of any enemy who might still have been using the hill as an observation post, and three medium bomber squadrons began working over the beaches.

As General Rupertus had feared when he argued against its use,[19] smoke from Target Hill drifted down across the landing beaches pushed on by a gentle southeast breeze. By H-Hour the shoreline had disappeared in a heavy haze, and within another half hour the approach lanes were obscured as far as the line of departure, 3,000 yards out. Coxswains of the leading boat waves handled their craft boldly, however, and the smoke posed no severe problem of control or orientation.

The first four waves of LCVPs assembled in succession as the APDs carrying the assault troops sailed into the transport area, two and three ships at a time. Once the Marines of 1/7 and 3/7 were over the side, the APDs left for positions outside the reef to await the return of their boats. The 12 landing craft in the first wave, 6 for each beach, moved from column into line at the control boat centered on the line of departure. At about 0730, the *SC–981* dropped the wave number flag it was flying, and the assault platoons headed for the smoke-shrouded beaches. Following waves were dispatched at five-minute intervals.

Taking station on the flanks of the first wave as it moved shoreward, two rocket LCIs made ready to fire a stunning barrage onto possible beach defenses to cover

---

[19] The 1st Division commander was convinced that smoke and dust from exploding fragmentation bombs, rockets, and shells would provide enough concealment for approaching landing craft. *Pollock comments.*

SHORE PARTY MARINES *struggle to build a sandbag ramp for LSTs in the surf at Cape Gloucester's Yellow Beaches.* (*USMC 68999*)

105MM HOWITZERS *of 4/11, set up in a kunai grass clearing, fire in support of Marines attacking toward Cape Gloucester's airfields.* (*USMC 69011*)

the interval between the last B–25 strafing run and the wave's landing. When the boats were 500 yards out, the LCI rockets began dropping ashore and worked inland as the Marines approached the beach. At 0746 on Yellow 1 and 0748 on Yellow 2, the LCVPs grounded and dropped their ramps.

Charging ashore to the sound of their own shouts, the Marines splashed through knee-deep water onto narrow strands of black sand. There was no enemy response—no sign of human opposition—just a dense wall of jungle vegetation. On many stretches of Yellow 1, the overhanging brush and vines touched the water; there was only a hint of beach. Led by scouts forced to travel machete in one hand, rifle in the other, the assault platoons hacked their way through the tangled mass, won through to the coastal trail, and crossed it into the jungle again. Once they had passed over the thin strip of raised ground back of the beaches, the men encountered the area marked "damp flat" on their maps. It was, as one disgusted Marine remarked, "damp up to your neck."[20]

Under the swamp's waters was a profusion of shell and bomb craters and potholes, places where a misstep could end in painful injury. An extra obstacle to the terrain's natural difficulty were the hundreds of trees knocked down by the air and naval gunfire bombardment. The roots of many of the trees that remained standing were so weakened that it took only a slight jar to send them crashing. The swamp took its own toll of casualties, dead and injured, before the campaign ended.

[20] Hough and Crown, *New Britain Campaign*, p. 48.

Enemy firing first broke out to the west of Yellow 1 where Company I of 3/7 had landed. Confused by the smoke, the LCVPs carrying the company's assault platoons reached shore about 300 yards northwest of the beach's boundary. Once the Marines had chopped through the jungle to reach the coastal trail, they came under long-range machine gun fire issuing from a series of bunkers lying hidden in the brush. The company deployed to take on the Japanese and gauge the extent of the enemy position.

The task of reducing the bunkers fell to 3/1 which was charged with leading the advance up the trail toward the airfields. The men of the 1st Marines' battalion began landing from LCIs on Yellow 1 at 0815, and the last elements were ashore and forming for the advance west by 0847. Lieutenant Colonel Joseph F. Hankins started his men forward on a two-company front 500 yards wide, passing through 3/7, which temporarily held up its advance to the beachhead line. The swamp on his left flank soon forced Hankins to narrow his assault formation to a column of companies. At 1010, the Japanese opened up on Company K in the lead, soon after it had passed through the firing line set up by Company I of 3/7. The ensuing fight was the bitterest struggle on D-Day.

For a short while the course of the battle seemed to turn against the Marines—both the commander and executive officer of Company K were killed; bazooka rockets failed to detonate in the soft sand covering the enemy bunkers; flame throwers malfunctioned when they were brought into play; and antitank and canister shot from 37mm cannon proved ineffective against the log-reinforced dugouts. The break of fortune came when an LVT which had come up from the beachhead with ammunition tried to crash into the enemy position. The vehicle got hung up between two trees, and the well-protected Japanese broke from cover to attack the cripple, killing its two machine gunners. The driver managed to work the tractor loose and caved in the nearest bunker by running over it. The Marine riflemen took advantage of the LVT's success and cracked the defensive system, killing or scattering the defenders of the four bunkers that had held up the advance.

Medium tanks had been requested for support as soon as it became apparent that 3/1 was up against a fortified position. A platoon of Shermans arrived shortly after Company K won its fight,[21] and led Company I forward through the ruined emplacements. Only a few snipers were encountered on the way to the 1-1 Line, the first objective on the coastal route to the airfields. Lieutenant Colonel Hankins reported 1-1 secure at 1325 and received orders to hold up there and dig in for the night. The brief but bitter fight for the bunkers cost 3/1 seven Marines killed and seven wounded. The bodies of 25 Japanese lay in and around their shattered defenses.

On the opposite flank of the beachhead, the 1st Battalion, 7th Marines (Lieutenant Colonel John E. Weber), making Combat Team C's main effort, had driven to Target Hill by noon and seized the left sector of the beachhead line. A platoon of Company A flushed and finished off the only opposition, the few dazed defenders of two gun positions on Silimati Point. Company B's men, picking their way through the swamp in an area where two feet of mud underlay a like cover of water,[22] took the day's prime objective and found ample evidence that the Japanese had used Target Hill as an observation post.

The task of seizing the center of the beachhead was assigned to Lieutenant Colonel Odell M. Conoley's 2d Battalion of the 7th. The unit was ashore and assembled by 0845, though the men on two LCIs that had eased their way into the beach had to wade through neck-deep surf. The sodden condition of the unfortunate Marines that had to breast the water to shore was soon matched by that of many others when 2/7 had its bout with the swamp. The battalion passed through a Japanese supply depot south of the coastal trail and ran into scattered opposition as it advanced into the jungle again. A scout in Company G was the first Marine killed and other casualties occurred as the leading companies struggled through the 900-yard width of the swamp to rising ground. By late afternoon, Conoley could report his battalion had reached dry land and was digging in. LVTs brought up ammunition, and 2/7 got set for counterattacks that irregular but hot enemy fire promised. Both flanks hung open until they were dropped back to the swamp's edge.

The 3d Battalion, 7th Marines, led by Lieutenant Colonel William R. Williams, reached its assigned portion of the beachhead line after threading its way through the swamp barrier and crossing the only

[21] Maj Hoyt C. Duncan ltr to CMC, dtd 14Mar52. The tank company commander recalls that the tank took part in the last stages of the action against the bunkers. Col Donald J. Robinson ltr to Head, HistBr, G-3, HQMC, dtd 5Jun62.

[22] CWO Sidney J. Fishel comments on draft of Hough and Crown, *New Britain Campaign*, dtd 14Feb52.

large patch of kunai grass within the planned perimeter. Williams' unit met no serious opposition during its advance, but late in the afternoon, when division ordered a shift to the west to link up with Combat Team B, a small group of enemy troops attempted to infiltrate through the gap that opened. The battalion was recalled to its original positions and dug in along the edge of the kunai grass for night defense.

When the second echelon of LSTs arrived and began unloading on D-Day afternoon, the remaining rifle battalion assigned to the BACKHANDER assault force landed and moved into position. The 1st Battalion, 1st Marines (Lieutenant Colonel Walker A. Reaves) filed up the coastal trail past long lines of trucks waiting to move into the limited dump areas available. The unit's own equipment and combat team supplies, mobile loaded in organic vehicles, followed when traffic would allow. Accompanying the infantrymen when they moved out of the beachhead to the 1–1 Line was the 2d Battalion, 11th Marines (Major Noah P. Wood, Jr.), Combat Team B's direct support pack howitzer battalion. The artillery batteries set up along the coastal trail within the perimeter established by 1/1 and 3/1. General Shepherd and Colonel Pollock visited Colonel Whaling's CP at 1800 to pass on a division attack order for 0700, 27 December.

In contrast to the relative ease with which 2/11 reached its firing site, the two artillery battalions assigned to Combat Team C spent most of the day getting their guns into effective supporting positions. The lighter 75mm howitzers of 1/11 had been loaded in LVTs in anticipation of the difficulties that would be met in the "damp flat." This forethought paid dividends, since the amphibian tractors were able to move the guns directly from the LVTs to the battery locations which had been picked from maps and aerial photographs. The battalion's preselected positions proved to be in the midst of the swamp, but Lieutenant Colonel Lewis J. Fields directed his commanders to set up their guns around the edges wherever a rise of ground gave firm footing. The LVTs knocked down trees to clear fields of fire, hauled loads of ammunition and equipment to the guns, and helped the tractor prime movers and trucks get across the swamp. Fields' howitzers began registering at 1400,[23] five and a half hours after they landed, and 1/11 was ready to fire in direct support of the 7th Marines all along the perimeter by nightfall.

Lieutenant Colonel Thomas B. Hughes' 4th Battalion's 105mm howitzers needed every bit of assistance they could get to reach the kunai grass patch. LVTs broke a path through the swamp's tangle of trees and undergrowth for the tractors and guns and provided extra pulling power where it was needed. And the need was constant. The progress of each howitzer through the swamp was a major operation which often found the men of the firing section chest deep in water hauling on drag ropes or pushing mired wheels while tractor winches and LVTs in tandem applied full power to keep the guns moving. All supplies and ammunition had to be carried in amphibian tractors, since no truck could negotiate the tortuous trail through the swamp. Despite the incredible difficulties, the first battery was in position and ready to fire at 1330, the second was registered by 1700,

[23] ArtyLiaisonO msg to ArtyO, ALAMO For, dtd 22Jan44, Subj: Rept on Arty in BACKHANDER Op, in *ALAMO G–3 Jnl No. 17*.

and the third was pulling into the clearing as darkness fell.[24]

While the division assault troops were consolidating their hold on the beachhead, Lieutenant Colonel Ballance's shore party was coping with the confusion of problems arising from the unloading of equipment, vehicles, and bulk stores brought in by the LSTs. The carefully-thought-out plan of overlapping dumps to be used by each landing ship was adapted to meet the situation posed by the swamp. All supplies had to be crowded on the narrow strip of dry land between beach and swamp, nowhere more than a couple of hundred yards wide. The coastal track was crammed with vehicles almost as soon as the first seven LSTs beached at 0840, and the jam never eased during the rest of the day.

On the beaches themselves, the shore party constructed sandbag causeways out to the LSTs that grounded and dropped their ramps in the water. Although VII Amphibious Force reported that D-Day's "four foot surf around the ramps was of no consequence," and that "thick growth and soft ground behind the beach was the retarding factor" [25] in unloading operations, the men that wrestled sandbags in the water had a different opinion. Actually, the slowness of unloading was due to a combination of factors, not the least of which was the reluctance of the drivers of the mobile loaded trucks borrowed from ALAMO Force to chance being left behind on New Britain when the LSTs pulled out. Numerous trucks were abandoned and "stranded on the beach exits for quite some time" [26] before Marines could move them to the dumps and get them unloaded.

The first echelon of LSTs retracted at 1330 with about 100 tons of bulk stores still on board in order that the D-Day landing schedule could be kept. As the seven ships in the second echelon beached and started unloading about an hour later, the long-expected Japanese aerial counterattack materialized. As the enemy planes dove through the gunfire of American fighters covering the beachhead, a squadron of B–25s bound for a routine bombing and strafing mission in the Borgen Bay area flew low over the LSTs. In a tragic mistake of identity, the ships' gunners shot down two of the American planes and badly damaged two others. Compounding the original error, the B–25s that had weathered the LSTs' fire bombed and strafed 1/11's position on Silimati Point, killing one officer and wounding 14 enlisted men.

The enemy planes that hit the beachhead were only a small portion of an attack group of 88 naval fighters and dive bombers dispatched from Rabaul after they had returned from a morning's fruitless raid on Arawe. Most of the Japanese pilots concentrated on the ships offshore and the damage they did was severe, although they suffered heavily in the process. The radar on the fighter director ship, the destroyer *Shaw*, had picked up the enemy when they were about 60 miles away and vectored two P–38 squadrons to intercept, while the escort vessels cleared the reef-restricted waters off the coast and steamed out to sea to get maneuver room. The interceptors missed contact but wheeled quickly to get on the enemy tails, and a vicious dog fight took place all over the sky as the American and Jap-

[24] *Ibid.*; LtCol Dale H. Heely ltr to CMC, dtd 1Mar52.

[25] *CTF 76 Dec43 WarD*, p. 35.

[26] Maj George J. DeBell comment on draft of Hough and Crown, *New Britain Campaign*, ca. Feb52.

anese fighters tangled. Some of the dive bombers evaded the defending fighters to strike at the destroyers. The *Shaw* was badly crippled by near misses, two other ships were damaged, and the *Brownson*, which took two bombs directly back of its stack, sank within a few minutes. One hundred and eight officers and men went down with their ship, but the rest of the crew was rescued while the enemy planes were being driven off.

The toll of lost planes, pilots, and crews was still enough so that the Japanese never again attempted a daylight raid on Cape Gloucester in comparable strength. Just what the exact enemy loss was in the several attacks mounted by Army and Navy pilots on 26 December cannot be discovered. The total was probably more than the 18 planes recalled by the Japanese in postwar years,[27] and a good deal less than the 57 planes claimed by American pilots while the heat of battle was on them.[28] From 27 December on, Allied air strikes mounted from South Pacific bases kept Rabaul's air garrison too busy flying defensive missions to devote much effort to Cape Gloucester.

By nightfall on D-Day, it was evident that the BACKHANDER Forces' main beachhead on New Britain was secure. General Rupertus had left the *Conyngham* at 0800 and was on the beach before the advance echelon of his command post set up at 1030. The CP location, like so many positions chosen prior to the landing, proved to be too "damp" for effectiveness and was moved to dry land by noon Troops landed according to plan, and 11,000 men were ashore when the second LST echelon retracted at 1800. Although 200 tons of bulk stores returned in these ships to New Guinea, they were sent back on turn-around voyages and the temporary loss was not a vital one. All guns and vehicles on the LSTs had landed. During the day's sporadic fighting, the 1st Division had lost 21 killed and 23 wounded and counted in return 50 enemy dead and a bag of 2 prisoners.

At 1700, after it became evident that the Japanese were warming up to something more effective than harassing fire, particularly on the front of 2/7, Rupertus dispatched a request to ALAMO Force that Combat Team A be sent forward immediately to Cape Gloucester. While it seemed obvious that the troops already ashore could hold the Yellow Beaches perimeter, the task was going to require all of Colonel Frisbie's combat team. The planned employment of 3/7 as Combat Team B reserve could not be made, and Rupertus considered that the 5th Marines was needed to add strength to the airfield drive and to give him a reserve to meet any contingency.

All along the front in both perimeters, the Marines were busy tying in their positions as darkness fell, cutting fire lanes through jungle growth, and laying out trip wires to warn of infiltrators. For a good part of the men in the foxholes and machine gun emplacements, the situation was familiar: Guadalcanal all over again, Americans waiting in the jungles for a Japanese night attack. The thick overhead cover was dripping as the result of an afternoon rain that drenched the beachhead and all that were in it. The dank swamp forest stank, the night air was humid and thick, and the ever present jungle noises mingled with the actual and the imagined sounds of enemy troops readying an assault.

---

[27] *SE Area AirOps*, p. 30; *SE Area NavOps—III*, p. 48.

[28] 1st MarDiv D–2 Jnl., 26Dec43–26Jan44, entry of 1035 on 27Dec43.

Capture of the Airfields

ENEMY REACTION[1]

The D-Day landings on the Yellow Beaches came as a surprise to the Japanese.[2] To the enemy, the extensive swamp along the north coast between the airfields and Borgen Bay seemed an effective barrier to a large-scale amphibious assault. Generals Sakai and Matsuda hoped and prepared for landings aimed directly at Cape Gloucester's beaches, but actually expected the blow to come elsewhere. Instead of hitting these prepared defenses, the Japanese leaders saw the principal Allied thrust being made against Cape Bushing, to be followed by overland and shore-to-shore advance to the airfields.

As soon as he received word of the actual BACKHANDER landing scheme, the *17th Division* commander ordered General Matsuda to counterattack and annihilate the Allied assault forces "at the water's edge."[3] Colonel Kouki Sumiya of the *53d Infantry*, who commanded the 1,400-man airfield defense force,[4] was directed to concentrate his troops against the main beachhead, leaving the Tauali trailblock to the attention of minor elements of his regiment's *1st Battalion.* Matsuda's reserve, the *53d*'s *2d Battalion* under Major Shinichi Takabe, started moving down the trail from Nakarop to Target Hill shortly after General Sakai's order was received. The leading elements of Takabe's battalion reached the ridges in front of the Marine line late in the afternoon of D-Day. The shipping engineer and antiaircraft units defending Borgen Bay and Hill 660 were directed to hold their positions and support Takabe's counterattack. (See Map 23.)

The strongest combat force available to General Matsuda was Colonel Kenshiro Katayama's *141st Infantry* and its reinforcing units defending the Cape Bushing sector. After the dispatch of the major part of the regiment's *1st Battalion* to Arawe, Colonel Katayama's command amounted to some 1,700 troops,[5] the majority veteran infantrymen and artillerymen. The *17th Division* operation order of 26 December that directed Matsuda to commit his reserve also called for the *141st Infantry* to move overland from Cape Bushing to help wipe out the Allied beachhead. The elements of the *51st Reconnaissance Regiment* on Rooke Island were ordered to sail for Aisega and march

[1] Unless otherwise noted, the material in this section is derived from: *ADC IntelDocuments;* ATIS Item No. 9115, Matsuda For and 141st InfRegt OpOs, 26Dec43–8Jan44, in ATIS Bul No. 721, dtd 14Feb44 (ACSI Recs, FRC Alex); *17th Div Ops.*

[2] Docu No. 52399, Statement of ex-LtGen Yasushi Sakai in HistDiv, MilIntelSect, GHQ, FEC, Statements of Japanese Officials on WW II, V. III, p. 190 (OCMH).

[3] *17th Div Ops*, p. 15.

[4] ATIS Item No. 9657, MatsudaFor IntelRept No. 3, dtd 23Dec43, in ATIS Bul No. 788, dtd 11Mar44 (ACSI Recs, FRC Alex); *Japanese comments.*

[5] ATIS Item No. 9657, *op. cit.*

from there to the north coast to join in the fighting.

Abandoning their positions at Cape Bushing, Aisega, and Nigol, most of Katayama's units were underway for an assembly area near Nakarop by the evening of D-Day. After 27 December, only a reinforced rifle platoon remained at Aisega to man the defenses once occupied by a battalion, and a scattering of service troops was all that was left in the *141st*'s posts along the Itui. The commander of the token defense force at Aisega gloomily predicted in his diary that "we shall surely make Aisega our graveyard," but his guess proved wrong, at least in location.[6]

In a week's time, this platoon, too, was ordered into the battle to contain the advancing Marines.

In the immediate area of the Yellow Beaches on D-Day, most of the Japanese troops that tried to stem the Marine advance came from small detachments of the *1st Shipping Engineers* and *1st Debarkation Unit*. These enemy soldiers, who had operated the supply depot overrun by 2/7, fell back before Combat Team C's assault platoons until they ran into Major Takabe's battalion moving up to launch a counterattack. By late afternoon of 26 December, enough of Takabe's unit had filed down the trail from Nakarop to the area opposite the center of the Marine perimeter to man a strong firing line.

Just why the position of 2/7 was chosen as the point of attack is not known, although it is logical to assume that General Matsuda considered this segment of the 1st Division's beachhead would be lightly held. The swamp to the rear of the thin band of Marine foxholes appeared to isolate the defenders from the beach. Darkness veiled the American position before enemy scouts discovered that both flanks of 2/7 were open.

HOLDING THE PERIMETER[7]

The night of D-Day was moonless, and no trace of light penetrated the jungle canopy to reach the men of the 2d Battalion, 7th Marines or the Japanese of the *2d Battalion, 53d Infantry* who faced them. Occasionally the adversaries caught a glimpse of the flash of fire at a rifle or machine gun muzzle or the momentary flare when a mortar or artillery shell exploded, but, in the main, the battle was fought by sound. One man, one gun, one group fired and drew a response from the other side aimed at the sound of the firing; then the tempo would pick up sharply and firing would break out all along the front to die away slowly and crop out again at another point. Despite the handicaps of fighting in the pitch-black gloom, Marine fire discipline was good, and Lieutenant Colonel Conoley repeatedly cautioned his company commanders to keep a tight rein on ammunition expenditure. Experience in fighting the Japanese indicated that a headlong assault would be launched at the height of the fire fight.

---

[6] Diary of 2dLt Takashiro Sato, 1st Plat, 6th Co, 141st InfRegt in *ADC IntelDocuments*. The diary was found in front of the positions of 3/5 during the fighting around Aogiri Ridge.

[7] Unless otherwise noted, the material in this section is derived from: *ALAMO For G–3 Jnl No. 11;* Maj Harry A. Stella memo to ALAMO For G–3, dtd 5Jan44, Subj: Rept of Observer with CT C, in *ALAMO G–3 Jnl No. 14; 1st MarDiv SAR*, Phase II, Part I, Yellow Beach Landing; 1st MarDiv D–2 Daily Buls Nos. 1–6, dtd 27–31-Dec43; *1st MarDiv D–3 Jnl—I;* LtCol William J. Dickinson ltr to CMC, dtd 14Mar52; Hough and Crown, *New Britain Campaign*.

Keeping an adequate reserve of ammunition on 2/7's side of the swamp was a problem handled by the regimental executive officer, Lieutenant Colonel Lewis B. Puller, and his solution for it was carrying parties. LVTs could not be used to haul cargo until daybreak gave the drivers a chance to see the obstacles in their path. In the early evening, files of men from the Regimental Headquarters and Service Company snaked their way through the swamp to Conoley's CP with belted ammunition, bandoliers of rifle clips, and loads of 60mm and 81mm mortar shells. At about 2000, Colonel Frisbie decided to commit Battery D, 1st Special Weapons Battalion as infantry to reinforce 2/7, and Lieutenant Colonel Puller told the battery commander to leave his 37mm guns behind, have each man pick up a load of ammunition at the regimental dump, and then set out across the swamp.

Soon after an officer guide from 2/7 met the ammunition-laden column, a violent storm lashed the beachhead area and all trace of landmarks and trail signs vanished in a solid downpour of wind-driven rain. The men of the antitank battery, each holding to the belt of the Marine ahead of him, spent the night struggling to get through the morass and deliver the vital ammunition. It was 0805 before the guide was able to lead Battery D into 2/7's command post. Lieutenant Colonel Conoley immediately sent the reinforcements to his right flank to plug the gap between the 2d and 3d Battalions.

The Marines of 2/7 had fought with the Japanese all through the wild night-long storm. The drenching rain filled foxholes and emplacements and forced the men to scramble for cover on top of the ground. Rifles and machine guns fouled by mud and water refused to work, and the fire of battalion and company mortars "layed by guess and by God"[8] was invaluable in beating back repeated Japanese attacks. When the storm subsided and daylight began to filter through the tree tops, the tempo of battle increased and the weight of the enemy thrusts shifted west toward 2/7's open right flank.

As they arrived at their designated position, the special weapons men of Battery D tangled with the Japanese that had infiltrated the gap between the 7th's battalions. The Marines counterattacked, threw back the enemy troops, and built up a hasty defense line. Company F on the battery's left flank took the brunt of Major Takabe's attack and in a violent, seesaw battle, during which two of its machine guns were lost and recaptured, finally forced the Japanese to withdraw. The estimate of enemy dead in and around the Marine lines was 200–235, but the opportunity to make an accurate count was lost when the action opened anew during the afternoon of 27 December.

The *2d Battalion, 53d Infantry*, reinforced by small engineer and service units, attacked Combat Team C's perimeter repeatedly during the next few days losing more and more of its strength in every futile attempt to penetrate the Marine lines. The gaps that existed on the night of D-Day were closed by dark on the 27th. The 7th Marines' Regimental Weapons Company took over that part of 1/7's defensive sector closest to the beach to enable Lieutenant Colonel Weber to stretch his men thinly around Target Hill and make contact with the 2d Battalion. On the right flank, Company F was relieved in position by Company I of 3/7 and Battery

[8] Col Odell M. Conoley ltr to CMC, dtd 7Mar52.

691-360 O—63——24

D was attached to Lieutenant Colonel Williams' command.

The center of the perimeter continued to be the focus of enemy attacks, and Colonel Frisbie was able to concentrate what few reserves he had in the area of greatest threat. Wherever the fighting reached a peak, a regimental casual detachment of 30-odd men was committed, then pulled out and used again to meet the next emergency.[9] Pioneers from the shore party manned strong points in the rear of 2/7's lines to provide defense in depth. The Japanese did not discover how sparsely manned the perimeter was to either flank and persisted in their attacks on what became its strongest sector.

In three days of intermittent but intense fighting, the 7th Marines lost 18 men killed, 58 wounded, and 3 missing in action. Surviving records indicate that in the same period the attacking Japanese suffered at least five times as many casualties and that *2/53* was badly crippled, remaining a battalion in name only.

With Combat Team C fully committed just to hold the beachhead it had seized on D-Day, General Rupertus was unable to use 3/7 as he had intended, as a reserve for the 1st Marines advance on the airfields. Neither was Colonel Frisbie in a position to mount an offensive and drive back the Japanese troops attacking his lines. This situation reinforced the opinion that the 1st Division staff had held throughout the latter stages of BACKHANDER planning—the 5th Marines were needed on New Britain. At 2314 on D-Day, Rupertus sent a request to Krueger asking that Combat Team A be sent forward to Cape Gloucester with the advance echelon transported in nine APDs. The request was repeated several times and sent by different means to insure its receipt at ALAMO Force advance headquarters at Cape Cretin.

Krueger had agreed to release two battalions of the 5th Marines to Rupertus if the Marine general asked for them after he had landed at the objective.[10] At 0751 on 27 December, Krueger sent a liaison officer to Cape Sudest with orders for Combat Team A (less 3/5) to get underway for the Yellow Beaches. At the same time, the ALAMO Force commander sent a radio dispatch to Admiral Barbey informing him of the decision. Bad weather delayed the plane carrying the messenger and played havoc with radio reception so that Barbey received word too late in the afternoon for the APDs to load Marines of 1/5 and 2/5 and still reach the beachhead by dawn on the 28th. Accordingly, the admiral, who did not wish to expose the loaded transports to the chance of daylight air attack, delayed the whole movement 24 hours.

Colonel Selden and his entire combat team stood by all through the morning of 27 December ready to load out. The first troops were already embarked in DUKWs and headed for the APDs offshore when word of the change in orders was passed. On the 28th, while LSTs loaded rear elements of the combat team and all its supplies and equipment, the 1st and 2d Battalions boarded their fast transports and left for the target. At 2100, after a day of hard work, the landing ships followed, setting a course by Cape Cretin to drop off Lieutenant Colonel David S. McDougal's 3d Battalion.

---

[9] LtCol Charles S. Nichols, Jr., interviews with HistBr, G-3, HQMC, *ca.* Jan55.

[10] *Pollock comments.*

While Combat Team A was en route, Rupertus sent Krueger "an earnest plea" [11] that no part of the 5th Marines be held back from employment at Cape Gloucester. The Marine general stated that increasing pressure on the perimeter required a landing team reserve in that area, while the remainder of the combat team was used in the airfield attack. Krueger responded almost immediately that he "had no intention to deprive you of its use" [12] should the battalion be needed and that 3/5 would be sent forward with the remainder of Combat Team A.

On 28 December, while the reinforcements he had requested were loading at Cape Sudest, the 1st Division commander moved his CP to a position behind Combat Team B's front line where he could better direct the advance west. Control of the beachhead defenses was turned over to General Shepherd, whose ADC command group paralleled the organization of the division staff. The 3d Battalion, 5th Marines was slated to report to Shepherd for orders when Combat Team A's LST-borne echelon arrived. With the addition of a sorely-needed maneuver element, the ADC planned an attack toward Borgen Bay to eliminate the Japanese menacing the perimeter.

The steady influx of supplies planned for BACKHANDER operations went on without a major hitch while Combat Team C fought to hold the beachhead and Combat Team B drove forward toward the airfields. Through hard, demanding work, Lieutenant Colonel Ballance's shore party was able to overcome the considerable handicap to its operations posed by the limited stretch of dry land between sea and swamp. The LST unloading rate improved considerably after 27 December, when it was no longer necessary to handle two echelons daily. As many as 300 deadweight tons of stores and a full shipload of vehicles were unloaded in under six hours. Ballance's 1,400-man force was able to do its job well despite the fact that the replacement companies and pioneers that made up the bulk of its strength were often committed to hold reserve positions backing up Combat Team C's lines.

Aside from the human workhorses of the beachhead, the shore party Marines, the most important logistical elements were the amphibian tractors. The versatile LVT was the only vehicle capable of negotiating the swamp unaided, and it was used to solve every conceivable cargo and personnel-carrying problem that arose. When it appeared that the tractors used at Arawe might be a maintenance liability at Cape Gloucester, a provisional company using reserve LVTs and crews from both Company A and the tractor battalion's Headquarters and Service Company was organized and loaded out to arrive with Combat Team A.[13]

Almost as useful in their way as the LVTs were the boats of the amphibian engineers, particularly the LCMs which could stand the buffeting of heavy surf more easily than the lighter and smaller LCVPs. The 592d Engineer Boat and Shore Regiment's detachment at the Yellow Beaches, including LCVPs that were carried in on ships' davits on D-Day and LCMs that arrived in tow behind LSTs on the 27th, were used mainly to transport supplies to Combat Team B. The coastal

[11] *1st MarDiv D-3 Jnl—I*, entry No. 20 of 28Dec43.

[12] *Ibid.*, entry No. 20 of 29Dec43.

[13] *1st MarDiv SAR*, Anx D, Amphibian Tractors, pp. 1–2.

road used as the axis of advance by the 1st Marines quickly became a quagmire as the result of frequent rain and the damage caused by traffic far heavier than its bed could stand. Moving directly behind the assault troops, Marine engineers with Seabees following close in their trace attempted to keep the supply route in operation. When road maintenance efforts failed, the Army boatmen gave General Rupertus the assurance of adequate supplies he needed to keep the offensive rolling forward.

CAPTURE OF HELL'S POINT [14]

Colonel Whaling's combat team spent a quiet first night ashore, its position unchallenged by the Japanese. While the rifle companies of the 7th Marines were hotly engaged repelling counterattacks on 27 December, assault companies of the 1st advanced cautiously but steadily toward Cape Gloucester. A series of phase lines, ½ to ¾ of a mile apart, marking terrain objectives in the zone of attack, were reached and passed without opposition. (See Map IV, Map Section.)

The narrow coastal corridor forced Whaling to confine his attack formation to a column of companies with 3/1's Company I in the lead. A squad of scouts acting as point was followed by a section of three medium tanks, each machine trailed in turn by a rifle squad to give it covering fire in the event of attack. Support to this advance party was furnished by another section of tanks and a rifle platoon which preceded the rest of the 3d Battalion. Peeling off the head of 3/1's column, a succession of small combat patrols took position in the swamp to guard the left flank and then fell in to the rear when the main body had passed.

During the day's advance, two belts of enemy defensive positions were overrun and destroyed. When the scouts spotted a pillbox or bunker, they signaled to the tank platoon commander who came forward on foot, located the targets, and then directed the tanks' fire to knock them out. Despite day-long rain that slowed the forward movement, the lack of opposition enabled 3/1 to reach the 4–4 Line, 5,000 yards from the morning's line of departure, by 1350. Division ordered Combat Team B to remain at 4–4 for the night and set up a perimeter defense.

The next day's objective was the final phase line designated before the landing to control the advance on the airfield. This 0–0 Line was plotted along a low, grass-covered ridge leading northeast through the jungle to a promontory, soon dubbed Hell's Point by the Marines, which formed one arm of a crescent-shaped beach. Fifth Air Force pilots had located an extensive system of bunkers and trenches in back of this beach with the heaviest concentration on Hell's Point. Although the enemy positions were obviously sited to oppose a landing, they stretched along 300 yards of the long axis of the coastal corridor and promised to be a formidable obstacle if they were defended.

At the time the 1st Division attack plan for 28 December was laid out, the arrival of the advance echelon of the 5th Marines was expected in the early morning. H-Hour was consequently moved up to

[14] Unless otherwise noted, the material in this section is derived from: *1st MarDiv SAR*, Phase II, Part I, Yellow Beach Landing; *1st MarDiv D–3 Jnl—I;* LtCol Donald J. Robinson ltr to Dir, HistDiv, HQMC, dtd 7Aug52, enclosing OpLog of Co A, 1st TkBn; Hough and Crown, *New Britain Campaign.*

permit the reinforcements to get ashore and started into position to exploit any successes won by 1st Marines assault troops. The starting time and span of air and artillery preparations for the attack were adjusted accordingly. General Krueger sent General Rupertus a dispatch explaining the 24-hour delay of Combat Team A as soon as he knew of it, but the message was received in garbled form and could not be decoded until numerous transmissions regarding corrections had passed between the two headquarters.[18] Confirmation of the 5th Marines arrival finally came in a message received at 0040 on 29 December.

On the 28th, 2/11 culminated a night of harassing and interdiction fire with an hour's heavy shelling of suspected enemy positions immediately to the front of the 4-4 Line. Then at 0900, American A-20s began strafing and bombing targets from the 0-0 Line to the airfields, observing 0-0 as a bomb line beyond which they could attack with no danger to friendly troops. At 1000, when the last of the planes drew off, a further delay in jump-off time was authorized to get additional tanks into position to support the assault.

When the order to attack was finally issued at 1100, the 3d Battalion, 1st Marines moved out in the same formation it had used during the previous day's advance. Company I was again at the head of the battalion column, working this time with a fresh platoon of medium tanks. In an effort to broaden the regimental frontage, Colonel Whaling had 1/1 simultaneously start a company through the jungle on the left flank in an attempt to reach the open ground along the 0-0 ridge.

The importance of the Marine attack objective was evident to the Japanese, and Colonel Sumiya had no intention of letting it fall without a fight. During the night of 27-28 December, the airfield defense troops who had been held under cover of the jungle on the mountain slopes moved into positions that ran roughly along the 0-0 Line. Waiting for the advancing men of Combat Team B was the major part of the *1st Battalion, 53d Infantry* and elements of the Japanese regiment's 75mm gun company.

Company A of 1/1, which drew the rugged assignment of cutting through the swamp forest, hit Sumiya's defenses first. At about 1145, the Marine unit reached the edge of an extensive clearing in the jungle about 500 yards from the coast. As the scouts and leading squads started to move through the chest-high kunai grass, a fusillade of rifle, machine gun, and mortar fire broke out from enemy positions hidden in the dense undergrowth across the open ground. Falling back quickly to the cover of the jungle on their own side, the Marines replied in kind, and the grass was whipped by a killing crossfire. It was soon apparent that the Japanese strength was at least equal to Company A's and neither side could gain an advantage.

For four hours the fire fight dragged on. The enemy force, mainly from the *1st Company, 53d Infantry*, held its ground but could not drive the Marines back.

[18] The last correction needed to enable BACKHANDER Force to break the original message was received at 1133 on 29 December well after Combat Team A's advance echelon had landed. *1st MarDiv G-3 Jnl—I*, entry No. 17, dtd 29Dec43.

Company A easily beat off two Japanese counterattacks and an attempt to turn its flank. At 1545, the Marines, who were running low on ammunition, began to disengage and pull back to the 1st Battalion's position for the night. Fire from 2/11 covered the withdrawal and discouraged any Japanese attempt to follow. Clearly the task of overcoming the enemy defenses required an attacking force of greater strength. Next morning when this position was found abandoned by advancing Marines, it contained 41 dead enemy soldiers. The cost to Company A of this hard-fought action was 8 men killed and 16 wounded.

In a sense, the skirmish in the jungle was a grim side show to the main event, the battle for Hell's Point. This Japanese strongpoint of mutually supporting bunkers and trenches was covered by belts of barbed wire and land mines. Hastily cut gun ports enabled enemy crews to train their weapons along the coastal trail to meet the attack. At least three 75mm regimental guns, a 70mm gun, a 20mm machine cannon, and a dozen or more heavy and light machine guns and mortars were brought into play against the Marines. Steady rain and thick foliage cut visibility to 10–20 yards, and Marine riflemen all too frequently stepped into the fire lanes of enemy bunkers before they spotted the Japanese positions. In this kind of blind fighting, tanks proved invaluable and helped hold down infantry casualties that might have soared had the attack been made without the assistance of armor. The support was mutual, however, and each tank's protective rifle squad kept its sides and rear free from suicide attackers.

The fight to capture the beach defense positions was joined about noon when Japanese troops opened fire on the leading platoon of Company I and its spearhead section of tanks. The tanks crushed trenches and bunkers and blasted guns and crews alike, while the infantry shot down the Japanese who tried to flee. When the 75mm guns on two tanks malfunctioned, the advance was halted while tank machine guns poured out covering fire and fresh armor was ordered into the fight.

In its advance and systematic destruction of the enemy positions in its path, Company I had veered to the left of the coastal road. As a new tank platoon lumbered up to the front lines, Lieutenant Colonel Hankins decided to use Company K on a platoon-wide front to cover the 50-yard stretch between beach and road. One section of mediums fought with Company I and another with K. The reduction of Hell's Point was the job of Company K, whose commanding officer later recounted the fight that eliminated the defenses;

> I put one squad of the Second Platoon behind each tank and deployed the Third Platoon to set up a skirmish line behind the tanks. We encountered twelve huge bunkers with a minimum of twenty Japs in each. The tanks would fire point blank into the bunkers; if the Japs stayed in the bunkers they were annihilated, if they escaped out the back entrance (actually the front as they were built to defend the beach) the infantry would swarm over the bunker and kill them with rifle fire and grenades. By the time we had knocked out twelve bunkers the Second Platoon, which had originally been behind the tanks, were out of ammunition and had been replaced by the Third Platoon and they too were out or down to a clip of ammunition per man. I called a halt and sent for the First Platoon. By the time the First Platoon arrived and ammunition was resupplied forty-five minutes had elapsed. We

continued the attack and found two more bunkers but the enemy had escaped.[16]

The third platoon of Company A, 1st Tank Battalion, to be committed during the day's action was called up and reached Hell's Point about 1630 in time to destroy the last enemy bunker on the point. It was undefended. Apparently, Colonel Sumiya had ordered a general withdrawal of the survivors of the force which had fought the 1st Marines. Combat Team B was able to occupy the entire defensive position and dig in for the night without any harassment from the Japanese. The night's only excitement was furnished by a false alarm of the approach of enemy tanks, which was countered by positioning two platoons of Marine mediums along possible approach routes with mortars ready to fire illuminating shells to help them locate targets.

After the action of 28 December, the Japanese undoubtedly wished they had tanks and plenty of them since their guns proved no match for the American armor. The count of enemy dead reported to division, before darkness stopped the search, was 266.[17] In return, the 3d Battalion, 1st Marines suffered casualties of 9 killed and 36 wounded. With the complete reduction of Hell's Point,[18] the way was open to the BACKHANDER Force's major objective—the Cape Gloucester airdrome.

[16] Maj Hoyt C. Duncan, Jr., ltr to CMC, dtd 14Mar52.

[17] At least one participant in the day's action questioned this total, and, on the basis of a personal count, estimated the enemy loss at a much lower figure—68–88. Maj William W. Wright ltr to CMC, dtd 15Feb52.

[18] General Rupertus later renamed this spot Terzi Point after the commanding officer of Company K, 3/1, who was killed on D-Day—Captain Joseph A. Terzi.

OBJECTIVE SECURED [19]

A new landing point, Blue Beach, located about four miles northwest of the beachhead and a few hundred yards behind the 0–0 Line was opened on 28 December. General Rupertus decided to land the assault battalions of the 5th Marines there in order to have them closer to their proposed zone of attack on the 29th. The order detailing the change in landing sites reached Combat Team B while some units were disembarking from their APDs and others were en route to their original destination, Yellow 2. As a result, Colonel Selden, his headquarters, two companies of 1/5, and most of 2/5 reached Blue Beach about 0730, while the rest of the assault echelon landed within the original perimeter. (See Map IV, Map Section.)

General Rupertus was present on Yellow 2 when the first units came ashore, and he got them started toward Blue Beach to rejoin the rest of the combat team.[20] The march was made by truck where possible and by foot whenever the condition of the river of mud called a road demanded. The beachmaster reported the last elements of the 5th Marines on their way west by 0935.

While his regiment regrouped behind the 1st Marines' lines, Colonel Selden conferred with Colonel Whaling, receiving a thorough briefing on the combat situation and word that the attack on 29 December

[19] Unless otherwise noted, the material in this section is derived from: *ALAMO G–3 Jnl Nos. 13–17; 1st MarDiv SAR*, Phase II, Part I, Yellow Beach Landing; *1st MarDiv D–3 Jnl—I;* BGen John T. Selden ltr to HistDiv, HQMC, dtd 27Jan 51; Hough and Crown, *New Britain Campaign*.

[20] LtCol Charles R. Baker comments on draft of Hough and Crown, *New Britain Campaign*, *ca*. Mar52.

was to be delayed until both the 1st and 5th Marines could join in the advance. General Rupertus and Colonel Pollock reached Whaling's CP shortly after this and outlined the day's operation plan. At 1200, following an air and artillery preparation, Combat Team B, with 1/1 in assault, would drive forward on the right with the coastal road as its axis of advance and Airfield No. 2 as its objective. At the same time on the left flank, Combat Team A, with 2/5 in the lead, would attack through an area in Talawe's foothills believed to contain prepared enemy defensive positions and then strike north toward the airfields.

After the attack order was issued, Selden and Whaling went up to the front lines to establish the left flank of the 1st Marines' position along the 0–0 Line. Selden then took his battalion commanders, Major William H. Barba (1/5) and Lieutenant Colonel Lewis W. Walt (2/5), up to the boundary for a brief inspection of the terrain. Deciding to advance with his assault battalion, Selden accompanied Walt when 2/5 moved to its line of departure 1,200 yards inland. Barba's orders called for his battalion to follow the 2d in attack, keeping contact with the tail of Walt's column.

On its way to the line of departure, the 2d Battalion, 5th found itself wading through a swamp, one that the regimental commander had been assured contained only a few inches of water. Instead Selden found "the water varying in depth from a few inches to 4 and 5 feet, making it quite hard for some of the youngsters who were not much more than 5 feet in height."[21] The advancing Marines had to move through the swamp and a thick bordering belt of forest in single file, and progress was unexpectedly slow. Before the leading company was in position to move out from the 0–0 Line, division had had to put off the time of attack several times. When 2/5 finally jumped off at 1500, a good part of the battalion was still in the swamp, and Barba's men were just entering. At dusk, the rear elements of 2/5 were clearing the line of departure as the leading units of 1/5 came out of the forest.

While Barba held up his men to get the companies organized in approach march formation, the 2d Battalion pulled away from the 1st and contact was lost. Patrols sent out in the gathering darkness and unfamiliar terrain were unable to link up, and 1/5's commander, whose radio to regiment was out, decided to hold up where he was for the night. The battalion established an all-around defensive perimeter in the middle of the open grassy area and spent a quiet night without enemy challenge.[22]

The terrain forward of Combat Team A's line of departure proved to be much more rugged than it appeared on operation maps. The 2d Battalion column, with Selden and Walt moving close to its head, found the stretches of kunai grass in its path were broken by ridges and bordered by gullies and ravines choked with jungle growth. As 2/5 swung north to move down to Airfield No. 2, it passed through an area of hidden trenches and bunkers all showing signs of recent occupation. There were no enemy troops to be seen. The battalion advanced unchecked and, by 1925, had reached the center of the airstrip and linked up with Combat Team B.

[21] Selden ltr, *op. cit.*

[22] Col William H. Barba ltr to HistDiv, HQMC, dtd 24Mar51.

The only enemy opposition met on 29 December was encountered by the 1st Battalion, 1st Marines, shortly after it passed through 3/1's positions on the 0–0 Line. A few scattered rifle shots fired at long range by small groups of Japanese troops did nothing to slow the advance. When the open ground along 0–0 was crossed, Lieutenant Colonel Reaves' assault companies moved as skirmishers through 300–400 yards of jungle. Supporting medium tanks, and half tracks mounting 75mm guns from the Regimental Weapons Company, were forced to stick to the coastal road and kept pace with the infantry while filing around the right flank of the forest barrier. Once they had cleared the trees, the assault troops, tanks, and half tracks joined forces again, ready for a final push to the airfield in sight ahead.

Artillery of 2/11 and 4/11 fired on suspected Japanese positions, and two rocket DUKWs of the amphibian engineers which had driven up to the front lines added 50 rounds to the preparation.[23] As if in honor of the occasion, the rain even stopped for a few brief moments. The advance was anticlimactic, and, at 1755, the 1st Battalion, 1st Marines reached the edge of Airfield No. 2.

As rain started falling and hastened the approach of darkness, 1/1 hurriedly dug in along a perimeter that looped from the coast inland to the east of the airfield. The 3d Battalion, 1st, which had followed the attacking troops in echelon to the left rear, linked up with 1/1 and extended the perimeter farther around the center of the field. When 2/5 came down out of the hills, the battalion was directed to fill in the remaining segment of the night's defense line which included all of Airfield No. 2.

Although surviving enemy records do not detail Colonel Sumiya's orders to his troops, prisoner of war interrogations and Japanese actions on 30 December provide a reasonable picture of what transpired on the 29th when the Marines were allowed to seize their objective unopposed. The enemy commander and the surviving members of his reinforced battalion had hidden in the rain forest that blanketed the hills south of the airfield. Conceding to Marine tanks and infantry the ownership of the low, open ground of the airdrome, Sumiya planned to take advantage of the terrain, his most important defensive asset. Moving at night, the remnants of the *1st Battalion, 53d Infantry* units which had fought at Hell's Point plus at least half of the *2d Company* that had not yet been engaged [24] occupied the prepared defenses hidden in the dense vegetation on the hillsides and in the cuts that laced the slopes. The most significant feature of the defensive system was Razorback Hill, a high, narrow, north-south ridge with a grass-covered crest that overshadowed both airstrips. To aerial observers and mapmakers, Razorback appeared to be just another kunai patch among many; from the ground it was clearly the key height in the hills bordering the airdrome.

On the morning of 30 December, while two assault companies of 2/5 moved out across Airfield No. 1 to investigate the village area west of the night's positions, Company F in reserve sent out a pair of scouts to locate and guide 1/5 into the

[23] OCE, GHQ, AFPac, *Amphibian Engineers*, p. 177.

[24] ATIS Item No. 9981, Diary of Cpl Ryoichiro Takano, 2d Co, 1/53, 31Oct–29Dec43, in ATIS Bul No. 808, dtd 14Mar44 (ACSI Recs. FRC Alex).

MARINE RIFLEMEN, *joined by a stray dog, pause behind a medium tank as they reach the outskirts of the Cape Gloucester airdrome. (USMC 69043)*

MEDIUM TANK *crosses Suicide Creek to blast Japanese emplacements holding up the Marine advance. (USMC 72283)*

perimeter. On the lower portion of Razorback Hill, where the 2d Battalion had passed through abandoned defenses on the 29th, the Marines surprised a group of 12 enemy soldiers just rising from their night's bivouac. The scouts withdrew after an exchange of fire, and a platoon of Company F immediately returned to wipe out what was thought to be a group of stragglers.

As the rifle platoon neared the top of a small knob in a kunai field, a heavy outburst of rifle and machine gun fire met it, coming from positions in the edge of the forest ahead. As the Marines sought cover on the hilltop, the Japanese launched a screaming charge up its sides which was bloodily repulsed. When the enemy troops withdrew, the fires of grenade launchers, mortars, and an artillery piece were added to the outpouring from the Razorback defenses. A radio message requesting support brought prompt response, and Marine mortar fire crashed down on the enemy position while the rest of Company F moved up.

The Japanese soldiers broke from cover and came racing up the hill in a second attempt to close with the Americans, just as a reinforcing platoon maneuvered to extend the firing line. Heavy fire drove the enemy back into the jungle again and scattered more bodies over the slopes. Tanks were called up and, when the first machine arrived, Company F attacked. Using the tank's gunfire and armor like a crowbar, the Marines split open the first line of the defensive system and drove on through, mopping up the Japanese with grenades and automatic rifle fire. By 1130, the hillside position, which had contained over 30 bunkers, was silent and smoking: its defenders were dead and, in many instances, buried in the ruins of their foxholes, trenches, and emplacements. A count of more than 150 Japanese bodies was reported to division,[25] while Company F had lost 13 killed and 19 wounded.

Company F's discovery that the enemy had reoccupied his defensive positions coincided with a similar unpleasant finding by the 1st Battalion, 5th Marines, which was marching in column toward the airdrome. The point of the leading company (A) was fired upon by Japanese who had manned defenses located in the jumbled ridges and ravines east of Razorback Hill. Once it came under fire, Company A deployed to continue the advance, and then, as the battalion commander later recounted the action, it:

> . . . was pinned down by heavy small arms and machine gun fire from enemy positions to the west, and enemy mortar fire began falling within the First Battalion zone of action. B Company was committed to the right of A Company in order to bring more fire to bear on the enemy and to prevent them from pushing down through to the airfield. The enemy made one sally against A Company's left flank and was repulsed with fire. Preparations were made to assault the enemy position following a mortar barrage. . . . The assault was made on the enemy position which was found to be abandoned—the remaining enemy withdrawing to the south into the hills. The First Battalion then withdrew to the airfield evacuating its dead and wounded, which I believe totaled 18, 6 dead and 12 wounded.[26]

---

[25] The Commanding Officer, Company F later estimated the dead at a lesser figure, 60 to 70. Maj John B. Doyle, Jr., comments on draft of Hough and Crown, *New Britain Campaign*, dtd 13Mar52. The battalion commander believed that "a more accurate figure would be about one hundred." BGen Lewis W. Walt ltr to HistBr, G–3, HQMC, dtd 13Jun62, hereafter *Walt ltr*.

[26] Barba ltr, *op. cit.*

The finishing blow to the Japanese defenses was delivered by Lieutenant Colonel Hankins' battalion. Working forward with medium tanks in support, 3/1 located the remainder of the occupied defenses in the area west of the strongpoints encountered by the 5th Marines. Attacking in the early afternoon along a three-company front, Hankins' men drove the enemy out of their positions, forced them to retreat up a ravine leading to Razorback's summit, and followed so close in pursuit that the Japanese had no chance to develop a defense. At a cost of one Marine killed and four wounded, 3/1 overwhelmed the last effective resistance of Colonel Sumiya's force. Artillery forward observers with the attacking troops brought down fire on the fleeing enemy to speed them on their way.

A vastly enlarged perimeter, including Razorback Hill in the center and reaching well to the west of Airfield No. 1, marked the Marine position on the night of 30 December. The 1st Battalion, 5th Marines took over a sector of the line between 3/1 and 2/5. During the day's fighting, more supporting elements of both combat teams and of the task force reached the airfield area. On the 31st, after the LST-borne echelon of Combat Team A arrived at the Yellow Beaches, the assault battalions' heavy gear and the team's light tank company and artillery battalion (5/11) came up. The American force securing the airdrome, basically four infantry battalions, two artillery battalions, and two tank companies, was far stronger than anything General Matsuda could bring against it.

At noon on 31 December, General Rupertus raised the American flag over Cape Gloucester to mark officially the capture of the airdrome. At least one unscheduled flag-raising, by Company I, 3/1 on Razorback Hill,[27] preceded this event. Undoubtedly, there were others, for combat Marines have a penchant for hoisting the Colors over hard-won heights.

Once the airfields were within Marine lines, General Rupertus radioed General Krueger to offer him the "early New Year gift [of] the complete airdrome of Cape Gloucester."[28] The ALAMO Force commander, in turn, made the present to General MacArthur as "won by the skill and gallantry of the First Marine Division brilliantly supported by our air and naval forces."[29] Both Army generals sent Rupertus their congratulations; MacArthur's was in eloquent language ending with the phrase: "Your gallant division has maintained the immortal record of the Marine Corps and covered itself with glory."[30] Perhaps the best tribute to the men who had seized the BACKHANDER operation's prime objective was written by an Army officer observer attached to Combat Team B when he reported to the ALAMO G-3:

> The front line soldier was superb. These men were in splendid physical condition and spoiling for a fight. They were like hunters, boring in relentlessly and apparently without fear. I never heard a wounded Marine moan. The aid men, unarmed, were right up in the front lines getting the wounded. Fire discipline was excellent.[31]

---

[27] LtCol George E. Bowdoin ltr to CMC, dtd 10Mar52.

[28] Quoted in *1st MarDiv SAR*, Phase II, Part I, p. 13.

[29] Quoted in *CTF 76 Dec43 WarD*, p. 43.

[30] *1st MarDiv D-3 Jnl—I*, entry No. 9 of 31Dec43.

[31] Maj J. B. Bonham memo to Col Eddleman, dtd 4Jan44, in *ALAMO G-3 Jnl No. 15*.

TRAIL BLOCK WITHDRAWN [32]

While the 1st Division's attack up the north coast was developing, the Marines who were to close the back door to the airdrome had a relatively quiet time. Lieutenant Colonel Masters' STONEFACE Group spent the first few days after it landed consolidating its hold on the Tauali beachhead and patrolling vigorously to locate the Japanese. Many of the small enemy detachments encountered cared surprisingly little about security, and 2/1's scouts and reconnaissance patrols were often able to jump the Japanese as they were resting or marching at ease, fire a few telling shots, and escape unscathed. No intelligence that the Marines uncovered indicated that a large enemy force was in the vicinity of the Green Beach perimeter. (See Map 25.)

The direct radio link with division that consistently eluded Masters on D-Day could not be made on the 27th either. On the following day, two engineer LCMs with Marine radio jeeps on board were sent north up the coast to get around the communication barrier posed by the mountains. An enemy 75mm gun firing from the Dorf Point vicinity dropped a few rounds near the boats and prompted them to turn back. The coastal voyage proved unnecessary, however, since the battalion's radios "boomed in" [33] to division receivers once the landing craft were 200 yards offshore. Starting on the 29th, the LCM-borne radios made contact with the division on a regular daily schedule.

The first inkling that the Japanese would probably attack the trail block was discovered on the 28th. An alert Marine on outpost duty spotted two Japanese observers on the ridge opposite 2/1's lines and dropped them both with his rifle. A patrol checked the bodies and found the men, carrying field glasses and maps, were officers who had apparently been reconnoitering the American position.

Toward nightfall on 29 December, scattered shots and then a growing volume of sustained fire began to strike the Marine defenses, coming from the jungle inland. A heavy downpour fell steadily, adding to the darkness and deadening the men's sense of hearing as they strained to catch the first sign of the attack that was clearly building. Finally, at 0155, the Japanese charged forward along the one route most likely to let them break through the perimeter—a natural causeway that joined the ridgeline defended by 2/1 and the higher ground inland.

At the point the enemy chose to assault, the lines of Company G formed a narrow salient. Japanese mortars and machine guns poured their fire ahead of the attackers, but in vain. The first thrust was beaten back with the help of the battalion's 60mm and 81mm mortars, and Company G's reserve platoon of artillerymen was rushed up to bolster the line. A second Japanese assault carried one of Company H's machine gun positions, but a counterattack by a mixed force of heavy weapons crewmen and artillerymen won it back. After this repulse, the ardor of the enemy troops cooled perceptibly, and two further attacks were blunted by Marine

---

[32] Unless otherwise noted, the material in this section is derived from: *1st MarDiv SAR*, Phase II, Part II, Green Beach Landing; *1st MarDiv D-3 Jnl—I;* CO, Det, 2d ESB Rept to CG, 2d ESB, dtd 20Jan44, Subj: Ops CTF 76 LT 21 (Green Beach), in *ALAMO G-3 Jnl No. 19;* Maj Theodore R. Galysh ltr to CMC, dtd 16Feb52; OCE, GHQ, AFP, *Amphibian Engineers;* Hough and Crown, *New Britain Campaign.*

[33] LtCol Robert Hall ltr to CMC, dtd 4Mar52.

fire and stopped in front of the lines. Toward dawn, the Japanese fire slacked until, by 0700, it had died away completely. The cost of this victory to 2/1 and its attached units was 6 men killed and 17 wounded.

Mop-up patrols found 89 enemy dead sprawled amongst the Marine positions and in the forest to the front. Five prisoners, all of them dazed and some wounded, were seized. No other wounded Japanese were sighted, and it appeared that the remnants of the attacking force had fled south. The prisoners identified their units as elements of the *3d* and *4th Companies* of *1/53*, and an enemy officer candidate, who surrendered during the day, estimated the attacking strength at 116 men. From intelligence gained later in the battle for western New Britain, it appears that the few men who escaped unhurt from the attack on 2/1 were ordered into the fight to contain the Marines driving southeast toward Borgen Bay.

On New Year's morning, torpedo boats which had patrolled from Grass Point to Cape Bach during the night took 10 of the most seriously wounded Marines and the enemy prisoners back to New Guinea.[24] The ammunition supplies depleted during the fight on the 30th were replenished in an air drop by Fifth Air Force bombers on the 3d. Although almost all signs pointed to a virtual abandonment of the west coast by Japanese troops, Lieutenant Colonel Masters could not chance being caught short. Through the week following the enemy assault, there was scattered opposition to Marine patrols but nothing that indicated the presence of large units.

An enemy artillery piece located about 2,500 yards east of the Tauali beachhead began firing ranging shots on the 31st. While the aim of Japanese gunners was atrocious—all of their rounds fell in the sea—Masters wasted no time getting his own artillery in position to reply. Battery H's crews used block and tackle to manhandle their guns to the top of a precipitous bluff that overhung the beach. On the 1st, when the Marine 75s began firing, the enemy gun fell silent. The pack howitzers then furnished support to the patrols that sought the enemy on all sides of the perimeter.

In response to an order from division, Masters sent a small detachment to Dorf Point on 2 January to guide Company E of 2/5 into the STONEFACE position for an overnight stay. The 5th Marines' company found no evidence of Japanese in any strength during its march from the airdrome. The enemy troops that had retreated from the airfield defenses seized by Combat Teams A and B had avoided the coastal track and used the trail that led over the eastern shoulder of Mt. Talawe. Under the circumstances, there was no longer any need for the trail block at Tauali, and General Rupertus issued orders for 2/1 to secure and rejoin.

The movement of Masters' command began on 5 January, when four loaded LCMs, one towing an LCVP, made the trip from Green to Blue Beach. The passage was a rough one for both crews and passengers. The heaving seas of the monsoon season battered the small craft, but the amphibian engineers proved equal to the feat of seamanship required to bring the boats in safely. On succeeding days, most of them equally rain- and windswept, LCMs from the detachment at the

---

[24] ComMTBRons, SeventhFlt (CTG 70.1) WarD, Dec43, dtd 27Jan44 (COA, NHD), entry of 31Dec43.

Yellow Beaches and those assigned to the STONEFACE Group shuttled supplies, ammunition, and equipment from the west to the north coast. By the 11th, most of 2/1's bulk stores and heavy gear had been transferred, and the main body of the reinforced battalion marched north toward the airdrome. A small rear-guard working party loaded the last supplies during the early afternoon. At about 1600, when several shots were fired from the jungle south of what had been the Marine lines, the remaining troops embarked. To discourage any venturesome Japanese who might want to fire on the boats as they withdrew, the LCMs sprayed Green Beach with machine gun fire as a parting present.

The 2d Battalion, 1st Marines rejoined its regiment on the 12th, after spending a wet and thoroughly miserable night in a bivouac area near Airfield No. 1. On 13 January, Masters' battalion occupied positions near the shore within Combat Team B's sector of the airdrome perimeter. With the successful completion of its mission, the STONEFACE Group was dissolved.

The Drive to Borgen Bay

Several major counterattacks were mounted by the Japanese in the course of the battle for control of western New Britain. In one, the *2d Battalion, 53d Infantry* was pared to skeleton strength by the concentrated grenade, small-arms, mortar, and artillery fire laid down by Combat Team C. In another, the STONEFACE Group crushed elements of *1/53* assaulting the Tauali perimeter. A third attack, carefully but poorly planned by the *141st Infantry* commander, Colonel Katayama, was aimed at the most prominent objective within the Yellow Beach defenses—Target Hill. The time chosen for the attempt was the early morning hours of 3 January. (See Map 26.)

Like its predecessors, the new enemy thrust had little chance of success. Although the Japanese were powerfully outnumbered and outgunned at all times, they had the opportunity to concentrate their forces and counterattack in significant strength. Why they failed to mass their resources remains a puzzle. The Target Hill assault force continued to fit the pattern of being too small to achieve results worth the cost of the effort.

For the first few days of the BACKHANDER operation, enemy intelligence officers at Matsuda's headquarters seriously underestimated the size of the Allied landing force, a fact which may account for the limited number of troops committed against Frisbie's and Masters' positions. There appears to be little reason to believe, however, that Colonel Katayama was ignorant of the real strength of the BACKHANDER Force when he selected a reinforced rifle company as the spearhead of his Target Hill assault. Any lingering doubts that he may have held regarding Allied strength must have been dispelled on 2 January when General Shepherd launched an attack to drive the Japanese back from the Marine lines.

ATTACK AND COUNTERATTACK [1]

New Year's Day had been a period of restless waiting for General Shepherd's command as preparations were made to attack the enemy troops dug in facing the Marine beachhead. The ADC's scheme of maneuver called for 3/7 to pivot on its left, where its position joined the 2d Battalion's foxholes, and to advance southeast across 2/7's front. The newly arrived 3d Battalion, 5th Marines was to move farther inland, tie into 3/7's right flank, and extend the assault frontage to 1,000 yards, far enouth south to overlap any Japanese defenses. "With 2/7 providing a base of fire to contain the enemy

[1] Unless otherwise noted, the material in this section is derived from: *1st MarDiv SAR*, Phase III, Extension of Beachhead Perimeter and Capture of Hill 660; *1st MarDiv D-3 Jnl—I;* Col John E. Weber ltr to HistBr, G-3, HQMC, dtd 27Mar52; LtCol Marshall W. Moore ltr to HistBr, G-3, HQMC, dtd 27Apr52; Hough and Crown, *New Britain Campaign*.

to their front, it was envisioned that the enveloping force would roll up the Japanese left flank." [2]

On the morning of 2 January, the movement to jump-off positions was laborious and time-consuming. Supporting fires by the 1st and 4th Battalions of the 11th Marines were concentrated in the area facing 2/7's lines; both artillery units fired through 1/11's fire direction center to simplify control and coordination.[3] At 1000, leading elements of the two assault battalions crossed the line of departure. The experience of the average Marine rifleman in this situation was vividly described by one of the 1st Division's scout officers:

> You'd step off from your line in the morning, take say ten paces, and turn around to guide on your buddy. And—nobody there, Jap or Marine. Ah, I can tell you it was a very small war and a very lonely business.[4]

For 300 yards, the two assault companies of Lieutenant Colonel Williams' battalion hacked and dodged their way through the jungle, trying to keep contact and watching warily for the first sign of the Japanese ahead. As they approached the banks of a little stream that cut through 2/7's line and extended on a north-south axis across the zone of advance, enemy small-arms fire sprayed the front. The Marines of 3/7 dived for cover, returned the fire as best they could against unseen targets, and began to inch their way forward.

On the right of the developing battle, 3/5 had to cut its way through dense fields of kunai grass during most of the morning's advance. Patrols ranging south from the open flank found no evidence of enemy troops on the rising mountain slopes. As Lieutenant Colonel McDougal's battalion left the grassy area and moved into the jungle toward the sound of the firing, its lead platoons also encountered the Japanese defenses. By midafternoon, the Marines had formed a line along the west bank of the stream, already called Suicide Creek, which bordered the enemy position. What lay ahead was a nightmare for the attacking infantrymen.

The Japanese had dug foxholes and bunkers under the arching roots of the forest giants and amidst the thick intervening brush, camouflaging the whole position so that no trace of it was revealed. Interlocking fire lanes gave enemy gunners enough of a view of ground to their front and flanks to provide targets and yet were almost impossible to detect. The few yards of open area over the stream bed was a killing ground without any concealment for the attackers. Most of the Marines' supporting artillery and mortar fire burst in the canopy of leaves and branches far overhead and had little effect on the hidden and protected enemy below.

Suicide Creek was aptly named, and, for the night of 2–3 January, the Marines dug in at approximately the same positions they had held when the Japanese first opened fire. All attempts to rush the enemy had failed; the volume of defending fire was so heavy that most of the assault troops spent the afternoon pinned to the ground. The man who rose to advance inevitably became a casualty.

Manning the Suicide Creek defenses were the survivors of the *2d Battalion, 53d Infantry* and its attached units. Major Takabe's badly shot-up command, lying in wait for the advancing Marines, could

[2] *Shepherd ltr.*

[3] Luckey, "Cannon, Mud, and Japs," pp. 52–53.

[4] Quoted in McMillan, *The Old Breed*, p. 192.

now extract grim payment for the punishment it had received in a week of futile attacks of Combat Team C's perimeter. Takabe's part in the planned counterattack on Target Hill was to create a diversion by an assault on 2/7's lines, a tactic that failed miserably in the face of the Marine battalion's practiced defensive measures.

Equally unsuccessful was the main event. The *5th Company* of *2/141*, supported by the direct fire of 20mm cannon and machine guns firing from positions in the jungle at the base of the hill and of 75mm guns emplaced near Hill 660, was to seize the crest of the hill. The unit defending, 3d Platoon, Company A, 1/7, was ready when the assault came. Long before the Japanese actually started up the lower slopes, the Marines on the narrow nose of the hill above the point of attack could hear the enemy soldiers cutting steps into the steep base of the hill which was hidden in the jungle growth. The 1st Battalion's mortars, in position back of the height, could not bear on the Japanese as they were actually as close as 20 yards to the Marines above them. The Japanese mortars and grenade dischargers were not hampered by the same limitation, and enemy shells landed all over the hill during the night to cover the attack preparations.

Toward dawn, the enemy soldiers rose out of the trenches they had dug to protect themselves from Marine fire and attempted to storm a machine gun position on the naked nose of ground. Although a Japanese mortar shell killed two men at the gun, the sole survivor of the crew stayed on and kept firing, cutting down the Japanese as they climbed into his line of sight. The rest of the Marine platoon, with the support of men from the various observation posts located on the hill's crest, used small arms and grenades to beat back every attempt of the enemy to gain the hill's upper slopes. By daylight, although the Japanese were still firing on the hill, the counterattack proper had petered out. When it was safe to move about in the open later on during the morning of the 3d, patrols were sent out to probe the area from which the attack was launched. Forty bodies were found, many of them piled in heaps in the trenches at the hill's foot; the absence of any wounded was evidence that cost of the fruitless attack was greater. The Japanese themselves counted the casualties at Target Hill as 46 killed, 54 wounded, and 2 missing in action.[5]

The prize of the night's action was the documents taken from the body of the Japanese company commander who fell attacking the Marine machine gun position. The papers helped the ADC's Intelligence Section fill in gaps in the order of battle and gave them a pretty clear picture of the movements of the troops opposing them. A fragmentary order signed by this officer, and picked up from the body of one of his platoon leaders on 4 January, gave the Marines their first inkling of the existence of Aogiri Ridge, a formidable defensive position that guarded the trail over which most of the Japanese were reaching the battle area. An accompanying rough sketch with this order gave the approximate location of both ridge and trail, but the ADC observation post on Target Hill could not pinpoint either in the maze of jungle ahead of the Marine lines.

---

[5] AET 2026, ATIS AdvEch No. 2, dtd 18Feb44, MatsudaFor MedSect Casualty Repts, 26Dec43–20Jan44 in *ADC IntelDocuments*.

Before Aogiri Ridge became a pressing problem, however, the Japanese had to be driven back from Suicide Creek. The repulse of the counterattack on Target Hill evidently had no effect on the defenders of the enemy position east of the creek, and the volume of fire stemming from the hidden bunkers showed no sign of let-up when the American assault was renewed on 3 January. During the morning, Marines of 3/7 were able to get across the creek to the flanks of the main Japanese defenses and 3/5 closed in similar manner from the southwest, but neither battalion could exploit its advantage. The attack stumbled to a halt in the face of determined opposition. Some weapon heavier than a rifle or a machine gun was needed in the forefront of the attack, preferably tanks, if they could be gotten to the front.

The engineers of Company C, 1/17, were equal to the challenge of getting the armor forward, and, in a day of incredibly hard labor, built a corduroy road across the coastal swamp to the kunai fields. Late in the afternoon of the 3d, three medium tanks of a platoon that had been dispatched from the airdrome crashed through the brush and trees to a point opposite the center of Japanese resistance. Before the Shermans could attack, however, the engineers had to cut a passage through 12-foot high banks to enable the tanks to cross the sluggish stream. The Japanese shot two drivers out of the seat of an unarmored bulldozer that came up to dig its blade into the bank and shove the dirt down into the water. A third engineer volunteer was able to operate the machine, crouching in its shelter and moving the controls with a shovel and axe handle. By nightfall, the way was clear for a tank-led attack.

On the morning of the 4th, after artillery had fired a preparation, the first medium eased its way down the earthen ramp, churned through the shallow water, and nosed up the far bank right into the heart of the enemy position. Covering Marine riflemen cut down two Japanese who attempted to lay explosives against the tank, and the rest of the battle was almost easy. Like grim executioners, the tank-infantry teams expertly destroyed the dug-in defenses with point-blank cannon fire, the crushing action of weighty treads, and the reaper-like spray of bullets from small arms which caught the few Japanese who escaped burial in their emplacements.

After pausing to reorganize, the two assault battalions swept forward nearly a thousand yards to seize an objective line in the jungle that would serve as a line of departure for the next phase of General Shepherd's attack. Japanese opposition was negligible during the day's advance, after the reduction of the Suicide Creek defenses. The forward movement of 3/7 masked 2/7's old position, and, on order from Colonel Frisbie, the 2d Battalion advanced across the trace of the attacking Marines to reach the right flank and tie in with 3/5. As Lieutenant Colonel Conoley's unit moved through the web of defenses that had been dug in facing the beachhead perimeter and the creek, 115 enemy dead were counted.

The total of estimated Japanese casualties inflicted by all of Combat Team C's units during the fighting at Target Hill and Suicide Creek was close to 500 killed and wounded. In light of the information contained in captured reports, that figure was probably not far above the actual losses. By 5 January, the two major enemy infantry units involved in the fight-

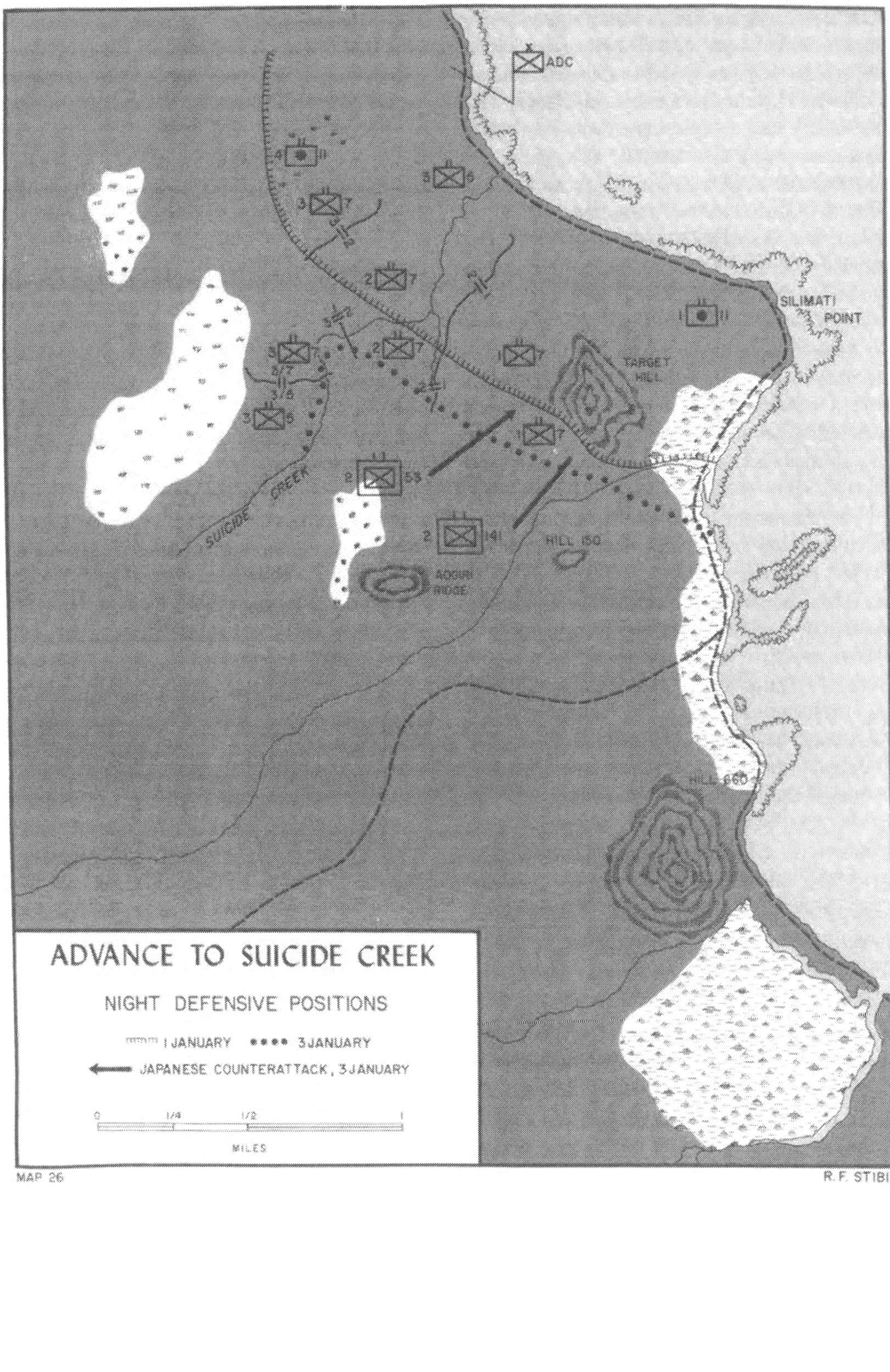

MAP 26 R. F. STIBIL

ing, *2/53* and *2/141*, were down from strengths of about 500 men each to 147 and 324, respectively.[6] The losses of attached units for this period are not known, but were logically on a comparable scale. The *53d's 2d Battalion* had been badly chewed up by 2/7 in the first few days after the landing, and Suicide Creek left its companies with an average strength of 22. Except for the losses of its *5th Company* at Target Hill, the *141st Infantry* was still pretty much intact.

The Marines lost comparatively few men on 4 January when they advanced with the support of tanks, but their casualties in the previous two days of fighting were heavy. The combined losses reported by the 7th Marines and 3/5 were 36 men killed, 218 wounded, and 5 missing in action.[7] The cost promised to be just as great in future days' actions whenever the Japanese chose to hold prepared defenses in the jungle. Characteristically, the enemy showed no lack of a will to fight, even in a hopeless cause.

Captured diaries and letters of Japanese soldiers who fought at Cape Gloucester show that most of them knew they were cut off from effective support by the remainder of New Britain's garrison. Nightly raids by a few Rabaul-based planes which occurred for the first few weeks after D-Day seem to have given little lift to enemy morale. In fact, since antiaircraft fire kept the pilots flying high and erratically, their bombs fell as often within the Japanese lines as they did inside the American. Several Marines were killed and about 30 were wounded by bomb fragments when they were caught in the random pattern of hits.

The support the *Matsuda Force* needed was not nuisance raids by a few scattered bombers, but a steady influx of men, rations, guns, and ammunition. The few Japanese barges that attempted to sneak into the eastern end of Borgen Bay to land supplies were sunk by Marine artillery. On the southern coast, Allied attack planes and torpedo boats kept the barge route permanently closed. The pittance of supplies brought forward along the north coast trail by Japanese and native carriers was not enough to sustain the troops opposing the 1st Marine Division. The average enemy soldier fought with his stomach gnawingly empty, his clothes and shoes sodden and rotting away, and his body attacked by jungle diseases. Despite the circumstances, discipline and national pride made the Japanese capable of an impassioned defense. There were no cheap victories to be had at Cape Gloucester.

AOGIRI RIDGE[8]

The Marines of General Shepherd's command spent 5 January preparing to attack, replenishing ammunition, and replacing essential items of equipment lost in the Suicide Creek fighting. During the day, the interior flanking companies of 1/7 and 3/5 closed toward each other and pinched out the 3d Battalion, 7th Marines, taking over its positions. As the 1st Bat-

---

[6] AET 2026, ATIS AdvEch No. 2, dtd 18Feb44, MatsudaDet StfTele A No. 270 to CofS, 17th Div, dtd 7Jan44, in *ADC IntelDocuments*.

[7] 7th Mar R-1 Jnl, 27Oct43-26Mar44, entries of 2-4Jan44.

[8] Unless otherwise noted, the material in this section is derived from: *1st MarDiv SAR*, Phase III, Extension of Beachhead Perimeter and Capture of Hill 660; *1st MarDiv D-3 Jnl—I; ADC IntelDocuments;* Col Lewis W. Walt comments on draft of Hough and Crown, *New Britain Campaign*, dtd 4Mar52; Hough and Crown, *New Britain Campaign*.

talion shifted to the right, Weapons Company of the 7th moved into line along the coastal road and reported to Lieutenant Colonel Weber for orders. On the opposite flank, 2/7 with Battery D, 1st Special Weapons Battalion attached, conformed to the movement of 3/5 and traced a curve through the jungle and kunai patches to face south and east. The ADC's infantry reserve was 3/7, temporarily commanded by Lieutenant Colonel Puller.[9] (See Map 27.)

The principal objectives of the 6 January attack were a small hill, Hill 150, due south of Target Hill in 1/7's zone, and the mysterious Aogiri Ridge, which might be in either 2/7's or 3/5's path. Since the jungle ahead of the Marines showed no prominent rise between Hills 150 and 660 along the coast and the mountain slopes inland, the consensus was that Aogiri would be found in a southwesterly direction toward the mountains. This guess, although logical, was inaccurate. Aogiri Ridge proved to be only a thousand yards west of Hill 150 and dead ahead of 3/5. A meandering, nameless stream which crossed the whole front of 1/7 coursed through the low ground between the two Marine objectives.

Colonel Katayama, commanding the Japanese troops directly opposing the advance, numbered his remaining combat effectives at 1,320 men. About 550 soldiers, most of them members of the *2d Battalion, 141st Infantry* or survivors of *2/53*, held a line of defenses that protected the trail to Nakarop. Dug in on Aogiri Ridge itself were the men and machine guns of one company of the *39th Field Antiaircraft Battalion*. In reserve, Katayama held another 520 men, including regimental headquarters and heavy weapons units and the weak companies of the uncommitted *3d Battalion* of the *141st*. Hill 660 was defended by antiaircraft units reinforced with rifle sections from the *6th Company* of *2/141*.

All through the 5th, Marine artillery fired on possible concentration points and suspected defensive positions in the area ahead of the American lines. In very few cases could the results of preparatory fire be observed; if shells did not burst in the tree tops, they exploded out of sight below, amidst the brush. The men manning the vantage points on Target Hill and observers in planes overhead were seldom able to do more than determine that the right area had been hit.

On 2 January, the 1st Division's own light aircraft, which had been intended for use in spotting artillery fire, began operating from a strip on Airfield No. 1. The makeshift air-ground radios with which the planes were fitted proved to be next to useless; the unit commander declared that the radios "petered out as soon as you got them more than two or three miles [from base]."[10] As a result, the airborne spotters either reported what they had seen on landing or dropped a message to the nearest artillery battalion. Fire control from the air proved to be an impossibility with the communications equipment available.

The planes of the Fifth Air Force that flew strikes in support of the attacking Marines had even greater difficulty than the artillery spotters in locating targets.

[9] The regimental executive officer took over the battalion on 4 January when Colonel Frisbie relieved its commanding officer. Puller held temporary command until 9 January when Lieutenant Colonel Henry W. Buse, Jr. took over 3/7.

[10] Maj Theodore A. Petras interview by HistDiv, HQMC, dtd 11Apr50, p. 3.

From a fast-flying plane directly overhead, the jungle in the Borgen Bay region had the appearance of an undulating sea of green with few prominent landmarks to guide strafing and bombing runs. Most air attacks were directed at objectives designated by map coordinates, but occasionally high explosive and smoke shells were used to pinpoint targets for pilots.[11] The BACKHANDER Force's requirement, that most bombing take place at least 500 yards from the nearest friendly troops,[12] usually limited air support to destruction and interdiction missions aimed at points well back of the Japanese defenses immediately facing the Marines. On 6 January, before the assault companies advanced, two squadrons of B–25s hit smoke-marked targets that included possible reserve assembly areas, routes of approach to the front, and what was hoped to be the particular stretch of jungle that hid Aogiri Ridge.

The Marines opened the next phase of their attack with a general advance that began at 1100 on the 6th, following a 15-minute preparation fired on Hill 150. On the left flank, Company A of 1/7 waded across the stream at the foot of Target Hill and advanced until it was stopped by heavy small-arms fire crackling from a road block on the coastal trail. Patrols had discovered the block on the previous day, and tanks were moving close behind the assault infantry to take care of it. When the lead tanks, fearing that they would bog down in its muddy bottom, hesitated at the banks of the stream, a Weapons Company half track splashed ahead through the water and showed the way. The Shermans followed and their 75mm guns made short work of the enemy positions.

Freed by the armor's fire, the troops on the left of 1/7 continued to advance across ground that became increasingly swampy. In the afternoon, Company A seized a trail junction at the coastal track; the path leading inland appeared on no Allied maps. The troops moving in the center of the 1st Battalion's attacking line discovered stretches of this unknown trail, too, after they had swept over Hill 150 against surprisingly light resistance. The monsoon season's heavy rains had virtually destroyed the trail and it seemed to disappear in the swamp as it led west.

Except for the brief fight at the trail block, the advance of the left and center of 1/7 was held up more by the water-logged terrain than enemy opposition. Company C on the right of the battalion's line, however, was met by a wall of fire coming from hidden positions as it attempted to push through the jungle lowland to the west of Hill 150. No effective progress could be made and the attack stalled. The 3d Battalion, 5th Marines ran up against what appeared to be a continuation of the same strongpoint. The men of 3/5 could not locate the origin of the defensive fire and were forced to dig in for protection soon after crossing the morning's line of departure. Towards nightfall, as the assault troops ceased the day's fruitless attack, elements of 3/7 were committed to cover a gap that had opened between 1/7 and 3/5.

The pattern of attack on 7 January followed closely the happenings on the 6th. The center assault companies could make no appreciable progress against concentrated enemy rifle, machine gun, and mor-

---

[11] Craven and Cate, *Guadalcanal to Saipan*, p. 343.

[12] BACKHANDER TF Cir No. 1–43, dtd 9Nov43, Subj: Air Force and Target Bomblines, in *ALAMO G–3 Jnl No. 3*.

tar fire. The units on the flanks that attempted to turn the Japanese position found terrain, densely forested swamp and gully, that heavily favored the defenders. Soon the Marine lines resembled those at Suicide Creek, but the enemy stronghold which formed a slight salient was far more extensive than the one encountered in the fighting on 2–4 January.

During the hottest part of the day's action, Lieutenant Colonel McDougal was shot while he was up with his assault platoons. General Shepherd immediately asked division to send a suitable replacement. Major Joseph S. Skoczylas, 3/5's executive officer, who took command when McDougal was wounded, was hit himself later in the afternoon. Pending the arrival of a new commanding officer, Lieutenant Colonel Puller was ordered to take charge of 3/5 as well as 3/7. The following morning, Lieutenant Colonel Lewis W. Walt, who had just been appointed executive officer of the 5th Marines,[13] reached the front lines and assumed command of the 3d Battalion, 5th.

When the Marines attacked on 8 January, there was no let-up in the vicious, blind struggle. Units on the flanks of 3/5 made little progress against their unseen, dug-in opponents. The assault companies of Walt's battalion found the jungle undergrowth to their front became, if anything, more dense and tangled, while the enemy fire grew in intensity. As the men inched forward, they could feel the ground slowly rising beneath their feet, although no hill or ridge was visible in the dank jungle ahead. Walt was convinced he had discovered Aogiri Ridge by the time he pulled his battalion back to more secure night defensive positions. The ADC reported at 1800 that the previous 24 hours of fighting had cost his command casualties of 15 killed, 161 wounded, and 5 missing in action. No one could estimate accurately how much the defenders had suffered in the return fire.

The Japanese did not dissipate their strength in counterattacks this time but waited for the Marines to come on. The terrain and weather were all in the defenders' favor. Although the Marine engineers tried desperately to build a log causeway for tanks across the swamp to the center of the front, continued rain and rising water slowed their efforts to a crawl. The heaviest direct support available in the lines opposite Aogiri Ridge was a 37mm gun which was hauled up to 3/5's position late on 9 January.[14]

The day's main attack was delayed until 1630, while the 37mm was brought forward [15] and a heavy artillery preparation was laid down on the area which seemed to contain the fortress ridge. Two flanking attacks were mounted against the eastern side of the Japanese position as 3/5 inched ahead in the center. In the first assault, Company C of 1/7 destroyed two bunkers before a rising tide of defensive fire overwhelmed its further attempts to advance. The second attack, an enveloping movement made by Companies K and L of 3/7, which General Shepherd had attached to Walt's command, also sputtered to a halt as the Japanese beat back the threat to their rear areas.

Fittingly, the American breakthrough so sorely needed finally came on 3/5's front

[13] The regiment's former executive officer, Lieutenant Colonel William K. Enright, was appointed Assistant D-3 on 7 January as a replacement for Lieutenant Colonel Buse who was slated to take over 3/7.

[14] *Walt ltr.*

[15] *Ibid.*

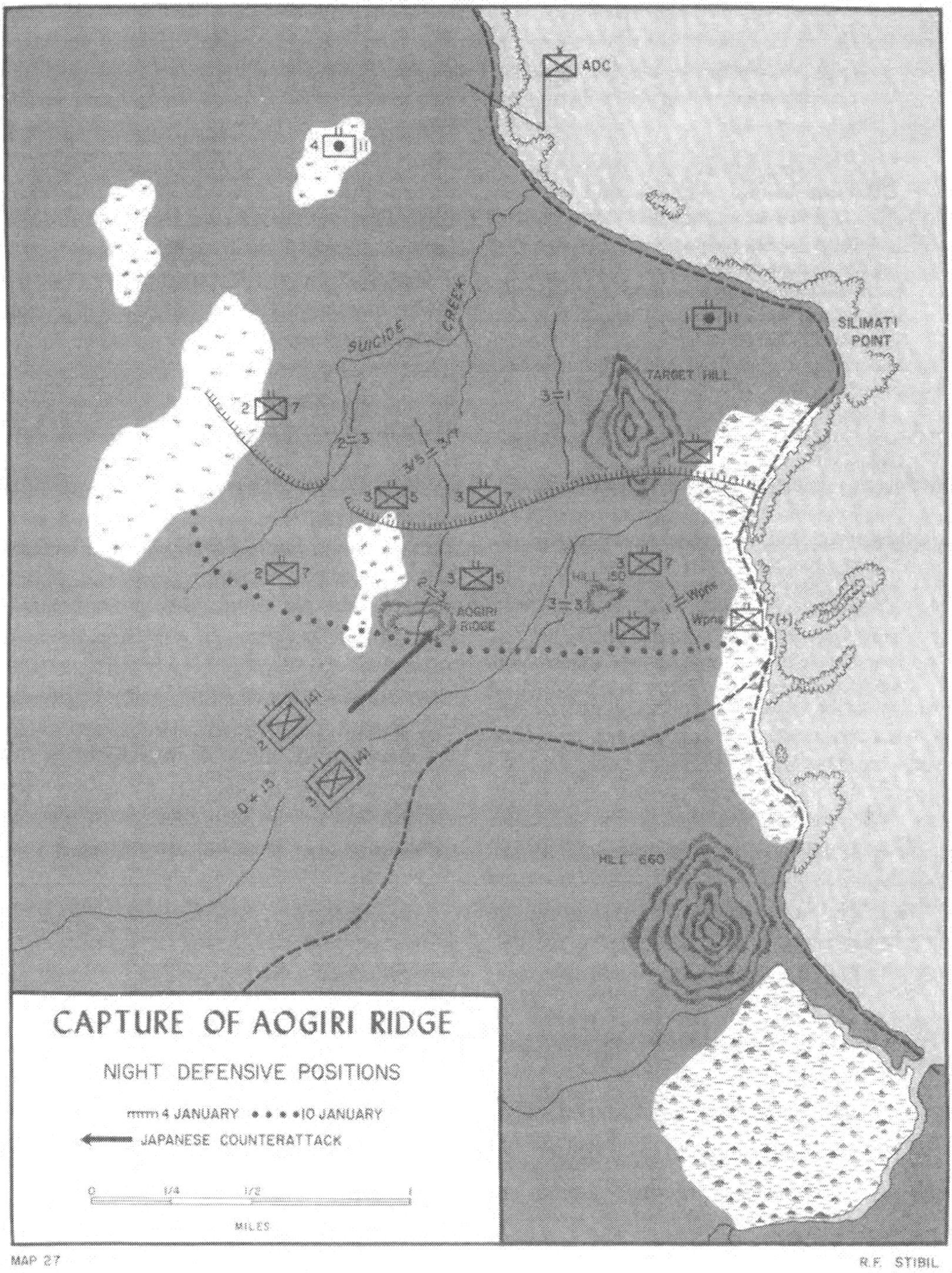
CAPTURE OF AOGIRI RIDGE
NIGHT DEFENSIVE POSITIONS
4 JANUARY
10 JANUARY
JAPANESE COUNTERATTACK
0 1/4 1/2 1
MILES
ADC
SILIMATI POINT
TARGET HILL
SUICIDE CREEK
AOGIRI RIDGE
HILL 150
HILL 660
MAP 27
R.F. STIBIL

just as the day's heavy casualties and rugged going seemed likely to require another withdrawal. The situation is best described in the division special action report:

> The undergrowth was so thick that the men could not see ten yards in front of them. The Jap machine guns had been cleverly concealed among the roots of trees and were well protected by snipers. At dusk the forward elements were almost to the top of one end of the ridge. The situation was desperate. The assault elements had reached the limit of their physical endurance and morale was low. It was a question of whether or not they could hold their hard earned gains. It was then that Lieutenant Colonel Walt's leadership and courage turned the tide of the battle. Calling forward the 37mm gun he put his shoulder to the wheel and with the assistance of a volunteer crew pushed the gun foot by foot up Aogiri Ridge. Every few feet a volley of canister would be fired. As members of the crew were killed or wounded others ran forward to take their places. By superhuman effort the gun was finally manhandled up the steep slope and into position to sweep the ridge. The Marine and Jap lines were only ten yards apart in places. As night came on the Marines dug in where they were.[16]

The wedge-shaped position that Walt's battalion occupied on the forward slope and crest of the ridge was a precarious hold, indeed. On both flanks the Japanese still occupied some of the 37 interconnected bunkers which lined Aogiri's military crest. On the reverse slope, a second line of enemy positions had yet to be taken. The import of the Marine penetration was clear, however, and the Japanese had little choice but to counterattack and hurl back 3/5. Failing this, Aogiri Ridge was surely lost.

Sounds of a pending assault reached the Americans waiting in their foxholes and gun emplacements. Finally, at 0115 on the 10th, the Japanese came screaming up the slope, charging through a driving rain, and the Marines opened fire and cut them down. Three more times the enemy attacked, and Walt's weary men beat back each attempt. The enemy was so close that Walt, in his command post 50 yards behind the most forward positions, could clearly hear the chant that heralded the fourth attack, "Marines you die, prepare to die."[17]

Scant minutes before the Japanese launched a fifth assault, a battalion headquarters detail brought up a resupply of small-arms ammunition. The sorely needed bandoliers and belts were passed down the front lines to men who had used all but their last few rounds. At this juncture, the Marine artillery which had been firing all night was called upon for a maximum effort. Walt directed his forward observer to walk the fire of 1/11 and 4/11 toward 3/5's lines to catch the Japanese as they charged. The artillery officer had to "adjust his fire, not by sight, but by ear, depending always on his ability to pick out the burst of his guns from the tornado of sound about him."[18] The fire was not lifted until 105mm shells were hitting 50

---

[16] *1st MarDiv SAR*, Phase III, p. 10. Actually, the gun was already in position to support the attack and had fired three times before Japanese return fire cut the crew down from nine men to four. Walt called for volunteers to help push the gun forward, and when none were forthcoming, he and his runner crawled up to the gun and put their shoulders to the carriage with the others. Two more shots cleared a tunnel through the growth ahead; a sixth shot silenced an enemy machine gun. Several volunteers then joined the group pushing the gun and there was a general surge forward. *Walt ltr.*

[17] *Walt ltr.*

[18] Luckey, "Cannon, Mud, and Japs," p. 54.

yards from where the Marines crouched in their holes. At the height of this attack, a Japanese major broke through the front lines and almost reached Walt's position before he was killed by fragments from a short round of artillery fire bursting in the trees overhead.[19] The remnants of the enemy force which had started the fifth counterattack wilted and ran in the face of fire from rifles and machine guns that took up the fight where the artillery ceased.

With dawn, the battle was over and not a single Japanese remained alive to defend Aogiri Ridge, soon renamed Walt's Ridge by the division commander. During the night's attacks, most of *3/141*, Colonel Katayama's only strong reserve, had been committed against the Marines; the enemy major who had died leading the fourth attack was apparently the battalion commander. At 0800, the men of 3/5 rose out of their foxholes and walked forward down the ridge, threading their way through scores of bodies sprawled in the awkward poses of those who had died violently. There was no opposition to the advance, and the section of trail behind Aogiri which the Japanese had fought so desperately to hold passed easily into Marine hands.

After the fall of the ridge, only one pocket of resistance remained, the defenses that had held up the units on the right flank of 1/7 through four days of dogged struggle. Elements of the *2d Battalion, 141st Infantry* were fighting to the last man to protect a supply dump that lay along the trail where it passed through the lowland between Aogiri and Hill 150. On 10 January, Lieutenant Colonel Weber's assault units kept probing the enemy position but could make no headway without incurring heavy losses. Even when a platoon of light tanks and two half tracks were finally able to reach the front on the 11th, tank-infantry teams had to fight at close quarters for four hours before the last Japanese soldier died in a futile effort to hold his ground. Once the pocket was wiped out, 1/7 surged forward to straighten out the Marine line which stood poised before the campaign's last major objective, Hill 660.

After the Aogiri Ridge battle, the 1st Marine Division totaled its losses since D-Day as 170 men killed, 6 died of wounds and 4 of other causes, and 636 wounded in action; 588 of the sick and wounded had been evacuated to hospitals on New Guinea. Intelligence officers reported the *Matsuda Force's* losses as 2,402 dead and 11 prisoners; wounded Japanese were estimated to equal the number killed.[20]

HILL 660 [21]

Marine patrols were active all along the front on 12 January trying to fix the limits of the next Japanese defensive position. No enemy troops were encountered along the coastal track short of Hill 660, and none were met at all by scouts who travelled almost a mile beyond Aogiri Ridge along the trail to Nakarop. Behind the screen of patrol activity, the units chosen to make the assault on Hill 660 rested and refitted.

General Shepherd picked the 3d Battalion, 7th Marines for the job of taking

---

[19] *Walt ltr.*

[20] OB Western New Britain, dtd 11Jan44, in ALAMO G-2 Weekly Rept No. 23, dtd 13Jan44, in *ALAMO G-3 Jnl No. 16.*

[21] Unless otherwise noted, the material in this section is derived from: *1st MarDiv SAR*, Phase III, Extension of Beachhead Perimeter and Capture of Hill 660; *1st MarDiv D-3 Jnl-I;* Hough and Crown, *New Britain Campaign.*

the jungle height. The battalion, under its new commander, Lieutenant Colonel Henry W. Buse, Jr., rejoined the companies that had been attached to 3/5 and pulled out of its reserve position to occupy a line of departure to the east of Hill 150. The 1st Battalion, 7th was given orders to keep contact with 3/7 as it advanced to extend the perimeter southward. A thorough mop-up and consolidation of the Aogiri Ridge position was assigned to 3/5, which was to hold where it stood. On the right flank, 2/7, which was occupying ground that had been determined upon as part of the Force Beachhead Line, was directed to dig in and improve its defenses. (See Map 28.[22])

An important role in the pending attack was assigned to a task force organized under Captain Joseph W. Buckley, commanding officer of the 7th Marines Weapons Company. The force was composed of two light tanks, two infantry platoons from 1/7, a 37mm platoon and two 75mm half tracks from Buckley's company, a pioneer platoon with a bulldozer from 2/17, and one of the Army's rocket DUKWs. Buckley's command was to advance down the coastal trail and establish a road block between Hill 660 and Borgen Bay, thus cutting the most favorable escape route for enemy defenders. From captured documents, prisoners of war, and battlefield identifications, the ADC intelligence section had built up a pretty fair picture of what Japanese opposition might be encountered at Hill 660. The strength of the heavily armed roadblock force was considered sufficient for it to hold its own against anything the enemy garrison might try.

The looming hill was too big a target to miss, and bombers, artillery, and mortars all had a field day during the preparatory bombardment. When the Marines moved out at 0800 on the 13th, however, the hill's cover of jungle hid most of the scars of the pounding it had taken. The infantrymen approaching 660 were too experienced to expect that the shells and bombs had done much damage to the Japanese burrowed into the ground beneath the thick foliage.

Lieutenant Colonel Buse's battalion advanced in a column of companies with Company I in the lead. The tanks, which started forward with the assault platoons, were soon left behind bogged in the mud, and the infantry continued without armored support. Company I reached the foot of Hill 660 along its northwest slopes at about 0930, and immediately started through a ravine cluttered with brush and boulders that rimmed the base. As the Marines climbed up the hill proper, they found the slopes so steep that many of them had to sling their rifles and pull themselves upward, seizing holds in the wet undergrowth and clawing their way in the slippery mud underfoot. Suddenly, right in the face of the struggling climbers, the Japanese opened fire with machine guns and rifles from the undergrowth above. The advancing line of skirmishers could do little but fling themselves down and try to work into a position to return the fire.

---

[22] The locations of the front line and boundary lines shown for 12 January on this map, taken from the division periodic report and a similar map done for Hough and Crown, *New Britain Campaign*, have been questioned. According to the former commander of 3/5, the boundary between his battalion and 2/7 was 500 yards east of Aogiri Ridge, and the front line was generally 1,500 yards forward of where it is shown. *Walt ltr.*

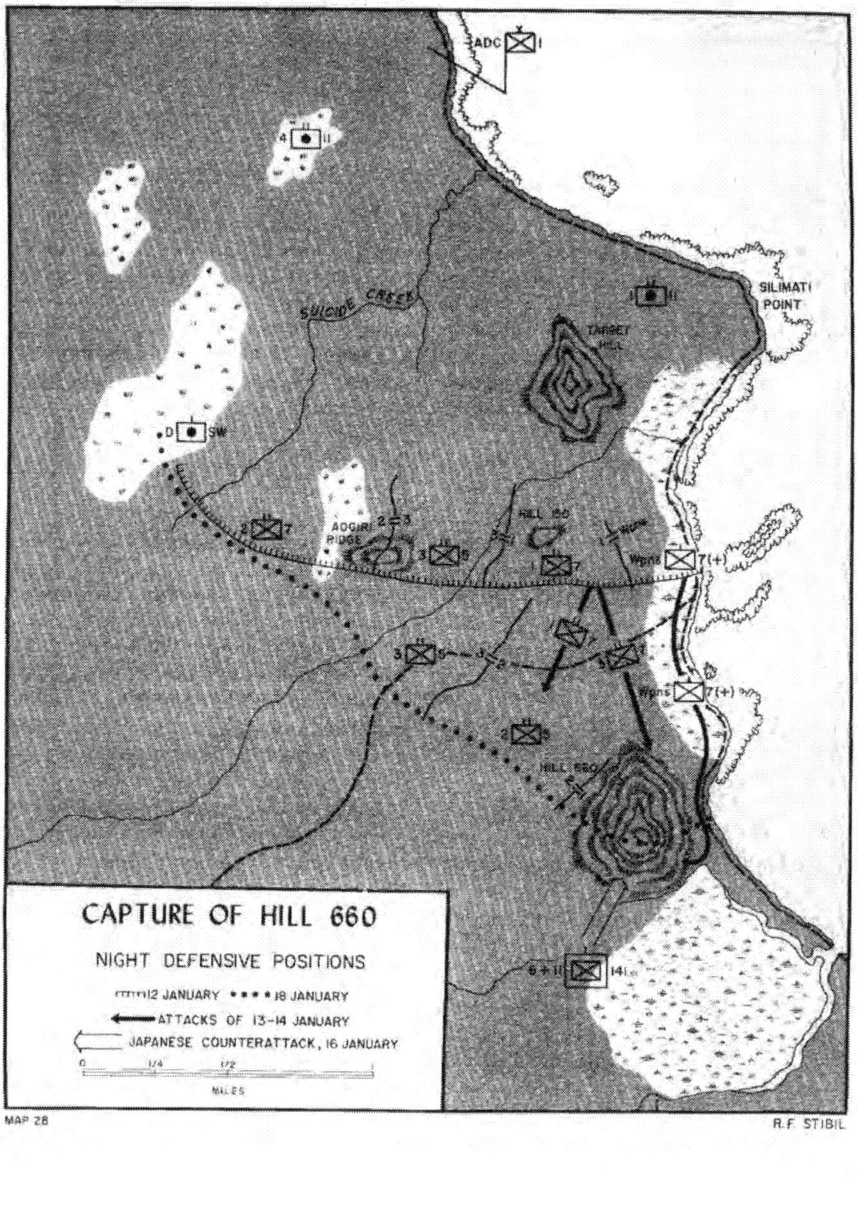

MAP 28

R.F. STIBIL

Lieutenant Colonel Buse sent Company L around to the right of Company I in an effort to outflank the enemy defenders, but all attempts of this sort failed. The Japanese positions stretched around the hill were sited for mutual protection, and Company L was eventually stalled on the lower slopes. One enemy machine gun was able to lay a band of fire behind the Marines on the hill and the two companies were pinned down front and rear. When engineers were finally able to work a light tank forward to the lip of the ravine in midafternoon, its covering fire, together with artillery and mortar support, enabled the stranded Marines to pull back to more defensible night positions.

While 3/7 had not been able to accomplish much during the day's action, Captain Buckley's command was successful in attaining its objective. The pioneers' bulldozer with the task force performed yeoman service helping move the various vehicles through the deepest mud on the coastal track. By 1030, the roadblock was set up, and Buckley had his men digging in for all-around defense on a 100-yard wide stretch of flat land between the bay shore and a swamp that edged the southern face of Hill 660.

During its approach to the objective, Buckley's column was fired on by a nest of automatic weapons located near the crest of the hill. The enemy guns were laid to cover the coastal track and were in defilade positions in relation to the Weapons Company half tracks in the roadblock. Although the Marine 75s could not place direct fire on the troublesome weapons, their high explosive shells hit close enough to drive the Japanese crews away from their pieces whenever it was necessary. With the aid of the half track's suppressive fire, wire crews were able to run a line through to the roadblock from the ADC's command post.

After an uneventful night during which 1/11 peppered Hill 660 with harassing fire, the pack battalion and 4/11 fired a 1,200-round preparation to pave the way for 3/7's attack. Buse's assault elements descended into the base ravine at 0900 and began climbing the hillside, advancing very slowly against determined opposition. As they had on the 13th, the Marines kept working their way around the hill feeling for a soft spot in the Japanese defenses. Tanks were able to follow and deliver supporting fire until they were stopped by two deep ravines that cut the southern slope. Despite the loss of tank support against the enemy machine gun positions, the attacking infantrymen kept moving upward, gradually driving back the Japanese. The ground rose so steeply near the hill top that some assault units had to haul themselves up hand over hand to reach the summit. Fortunately, Marine 60mm mortars were able to silence most of the enemy cannon and machine guns that were clustered in open emplacements on the hill crest before they could do much damage. Riflemen of 3/7 attacking behind the lethal shower of mortar shells quickly scattered the remnants of the defenders in headlong flight down the hillsides. Some of the retreating Japanese ran directly into the sectors of fire of Buckley's roadblock; others took refuge in the swamp. Buse's battalion was clearly in possession of the important ground on Hill 660 as dusk approached.

Two companies of 2/7 moved into line on the right of the 3d Battalion shortly before dark as a wise precaution against counterattack. As the Marines dug in, the heavens opened up and a drenching rain flooded down. All night long the torrent of water fell, soaking the weary men who had climbed and fought through a hot, muggy day. But if it made the life of the victors miserable, the rain also prevented the Japanese from gathering their force for an attack to regain the hill.

At 660's base, the Marines in Captain Buckley's roadblock had a lively night. Small groups of enemy stragglers kept trying to go through or around the American position and were either driven off or killed. Twelve Japanese soldiers were shot down attempting to wade through the offshore waters, and two were killed who ventured too close to the cleared fields of fire of the Marine guns. The enemy activity died away with the coming of daylight and the end of the shrouding rain.

On the 15th, the Marines were treated to a spell of clear weather and sunshine. Most men rested, while combat patrols ranged the brush on the sides of Hill 660 hunting down the Japanese trapped within the American lines. Only a few scattered enemy were caught in the mopping up activity, and there appeared to be no indication that large forces were in the vicinity to threaten the Marine position. As a reasonable precaution, however, additional automatic weapons were brought up to 3/7's front lines and emplaced during the day.

The Japanese were not ready to give up Hill 660 without at least one thrust at the men who had driven them off it. A counterattack force, largely made up of the *6th* and *11th Companies* of the *141st Infantry*, gathered in the swamp south of the hill during the daytime hours of 15 January and then moved forward quietly to the base of the hill well after dark. Finally, at 0630 on the 16th, the enemy troops began a screaming, scrambling charge up the precipitous slopes to get at the Marine positions. The first few soldiers to the top fought at close quarters, but the tremendous volume of defensive small-arms fire drove the Japanese back and pinned them down, as their own fire had once slowed the Marines attempting to take the hill. As soon as he had the enemy cleared away from in front of 3/7's lines, Lieutenant Colonel Buse ordered the rifle companies' 60mm mortar sections to hit the front of the enemy formation while the battalion's 81mm platoon fired on the rear. Then, in a deadly squeeze, the impact areas were moved toward each other. The attack and the attackers died in a welter of mortar explosions.

Patrols on Hill 660 found 110 fresh bodies after the last gasp of the counterattack had faded. Captain Buckley's roadblock crew had wiped out 48 more Japanese in repulsing a weaker effort to overrun their position. Altogether, the three days of action swirling about Hill 660 had cost the enemy over 200 dead and an unknown number of wounded. The cost to the Marines was about 50 men killed and wounded.

The capture of Hill 660 and the repulse of the counterattack to retake it marked the effective end of the Japanese defense of the Cape Gloucester-Borgen Bay area. In the following months, the *Matsuda Force* was to try with increasing desperation to escape from western New Britain, while the BACKHANDER Force kept pressure on the retreating enemy troops.

75MM HALF TRACK *and 37mm gun of Weapons Company, 7th Marines which helped beat back a Japanese counterattack on Hill 660. (USMC 71580)*

JAPANESE FLAGS *are displayed by the weary Marines who captured them, as elements of Combat Team C leave the front lines after 23 days of fighting. (USMC 71602)*

Eastward to Iboki

ARAWE REVISITED[1]

In many respects, Arawe was a sideshow to the main campaign for control of western New Britain. Occasionally, the fighting there was violent, marked by bloody clashes in the enveloping jungle; at other times, days went by with only minor patrol action. General Cunningham had accomplished his major objective when the 112th Cavalry assault troops seized control of Arawe Peninsula. Further operations to clear the Japanese from the area were undertaken primarily to remove a lurking threat to the DIRECTOR Force's position. On the enemy side of the front, Major Komori was determined to hold the Americans back from an objective that they actually did not want—Lupin airfield. (See Map 24.)

Allied press claims of the capture of the grass-choked airstrip, which were broadcast right after Z-Day, considered ground patrolled to be ground controlled. When Komori's *1st Battalion, 81st Infantry* forced the withdrawal of the 112th Cavalry's outposts on 25 December, the Japanese were convinced that they had regained possession of a desirable prize. Komori's primary mission became the denial of the airfield site to the Americans.

The defenses closing off the neck of Arawe Peninsula were the target of repeated small-scale Japanese attacks during the last week of December. The 112th's lines held firm, and the enemy troops reeled back, shaken and hurt, after each unsuccessful effort. The American artillery and mortar fire was particularly galling to the Japanese who were given no rest from punishment even though they broke contact. On the 29th, after eight days of wandering in the jungle, the *1st Battalion, 141st Infantry* reached Komori's positions, and the enemy commander directed the new arrivals to take over the front lines. The depleted companies of *1/81* and the original Merkus garrison were assigned to hold the rear areas of the wide sector from Omoi to the Pulie River which was Komori's defensive responsibility.

With the arrival of *1/141*, the Japanese ceased their attacks on the American positions. Instead, in the jungle about 400-500 yards forward of the 112th's lines, the enemy soldiers constructed a defense in depth, a complex of foxholes, trenches, and weapons emplacements that gave them alternate positions from which to cover approach routes. Patrols of cavalrymen discovered the Japanese were digging in

[1] Unless otherwise noted, the material in this section is derived from: *ALAMO G-3 Jnl;* 112th CavRegt, HistRept 24Nov43–10Feb44, dtd 10Feb44 (WW II RecsDiv, FRC Alex); Co B, 1st TkBn SAR, 9Jan–12May44, dtd 27May44; *17th Div Ops; Komori Diary;* ATIS Item No. 9773, Diary of unidentified platoon commander, 1st MG Co, 1/141, 21Dec43–16Jan44, in ATIS Bul No. 789, dtd 11Mar44 (ACSI Recs, FRC Alex); Miller, *Reduction of Rabaul;* Hough and Crown, *New Britain Campaign.*

on the 1st, but were unable to drive them back. Repeatedly, in following days, the Americans attacked the Japanese, but without success. Shifting frequently from hole to hole, using the concealment offered by a thick mat of undergrowth and the shallow connecting trenches they had dug, the men of the *1/141* were quite successful in holding their ground. On 6 January, General Cunningham told General Krueger that "officers and men participating in these operations report they have not seen a single Japanese and that they are unable to locate machine guns firing on them from a distance of 10 to 20 yards." [2]

Cunningham asked that he be sent reinforcements, noting that artillery and mortar fire seemed to have little effect on the hidden Japanese positions. He stated his belief that to continue attacks "along present lines is to play with the enemy's hand." [3] The ALAMO Force commander was asked for tanks to help root out and destroy the defenses that the American soldiers faced. Krueger took immediate steps to answer the request from Arawe, and a Marine tank platoon was underway from Finschhafen on 9 February, together with a company of the 2d Battalion, 158th Infantry.

The only tank unit available for reinforcement of Cunningham's force was Company B, 1st Tank Battalion, which had been left behind because of the limited operating area for armor at Cape Gloucester when the BACKHANDER Force sailed. When the 1st Marine Division commander was informed of the contemplated commitment of some of his armor reserve—and promised that the tanks would be returned to his control when they were required—he suggested that all of Company B be employed. Rupertus noted that the tank company was the smallest self-sustained unit for combat operations.[4] Accordingly, the remainder of Company B boarded an LCT on the 11th and made a stormy overnight passage through rough seas to Arawe.

From 13 to 15 January, while the 112th Cavalry continued pressuring the Japanese with combat patrols, the Marine tankers worked with the two companies of 2/158 which were to make the principal effort against the enemy position. The infantrymen provided a squad to cover each light tank and rehearsed tactics for the assault, while tank and infantry officers made a thorough reconnaissance of the zone of attack. The plan called for two five-tank platoons, each with an infantry company in support, to advance on a 500-yard-wide front on 16 January. The day's objective was 1,000 yards from the line of departure, and within the intervening distance lay all the maze of defenses that the *1st Battalion, 141st Infantry* had held so doggedly for two weeks.

On the morning of the 16th, a squadron of B–24s dropped 136 1,000-pound bombs on the Japanese defenses, and 20 B–25s followed with a heavy strafing and bombing attack.[5] This aerial preparation, coupled with an intensive artillery and 81mm mortar bombardment, paved the way for the assault. The tanks led off and kept moving forward despite soft ground and

[2] CG, US Forces, APO 323, msg to CG, ALAMO Force, APO 712, dtd 6Jan44, Subj: Ops DIRECTOR TF, in *ALAMO G–3 Jnl No. 15.*

[3] *Ibid.*

[4] *1st MarDiv D–3 Jnl—I,* entries nos. 16 and 17 of 8Jan44.

[5] Craven and Cate, *Guadalcanal to Saipan,* p. 335.

bomb craters which caused several machines to bog down until recovery vehicles could pull them free. Working well together, despite incredibly thick vegetation which practically blinded the tank drivers and commanders, the tank-infantry teams churned and shot their way through the enemy position. When a pocket of resistance developed on the right of the line, a section of the reserve tank platoon and a troop of the 112th quickly finished off the holdouts while the advance continued. By 1600, the predetermined objective had been reached, and Cunningham ordered a withdrawal to the peninsula's main line of resistance. Two tanks, one which had thrown a track on a steep slope and another which was hopelessly bogged down in swampy ground, were destroyed by demolitions to prevent their use as enemy strongpoints when the Americans pulled back.

The attack of 16 January accomplished its objective. Komori ordered the remnants of *1/141* to withdraw to the Lupin vicinity where they could "fight to the glorious end to defend the airfield." [6] The few Japanese who did not get the word to retreat were wiped out on the 17th by flame-throwing tanks and a supporting force of cavalrymen. When the battered enemy troops paused to regroup in positions near the airfield, the *Komori Force's* commander made a head count and found that his two understrength battalions and their supporting units had lost 116 killed in action and 117 wounded in three weeks fighting. In addition, 14 men had died of various illnesses and 80 more were sick enough to be unfit for duty. The sick roll promised to grow, for the Japanese were on short rations and the amount of food to be doled out shrank daily. One ineffective airdrop of supplies was received on New Year's Eve,[7] an event that did more to whet appetites than appease them. Primary reliance was placed on supply by barge from Gasmata and by carrying parties using the trail to Iboki. Neither method was satisfactory; the last barge to get through the gantlet of preying torpedo boats and planes reached the Pulie River mouth on 5 February, and the carriers were unequal to the task of keeping up with consumption. The *Komori Force* slowly starved while it held an objective that the Americans showed little sign of wanting.

Mounting doubts about the utility of the airfield he defended plagued Major Komori. American light planes were flying over his positions, and the Lupin garrison reported that they could hear the takeoffs from Arawe Peninsula. The DIRECTOR Force engineers had built an emergency strip for artillery observation planes on 13 January and, with grading and coral surfacing, it soon came into regular use. By 8 February, the disillusioned Japanese commander was reporting to his superiors that the value of Lupin "is so insignificant that it seems the enemy has no intention of using it." He outlined the increasing difficulty of holding his position with dwindling supplies and concluded that his force would soon be cut off and left "with no alternative but self-destruction." [8]

At first, Komori's broad hints that he be allowed to abandon his untenable defenses were answered by orders that he

[6] *Komori Diary*, entry of 17Jan44.

[7] Each man in 1/141 received 16½ ounces of rice, 5 vitamin pills, and a packet of tobacco from this airdrop. ATIS Item No. 9773, *op. cit.*, entry of 1Jan44.

[8] *Komori Diary*, entry of 8Feb44.

continue to "smash the enemy's project for construction of an airfield." [9] The *Komori Force's* supposed exploits in holding Lupin, recognized by two Imperial citations, formed a bright spot in an otherwise dismal picture of withdrawal and defeat of the Japanese forces in western New Britain. Eventually, the *17th Division* had to face the fact that if it did not give Komori permission to pull out and join the general exodus, he and his men would be isolated and destroyed. On 24 February, Komori's radio crackled out the eagerly awaited retirement order and he lost no time quitting Arawe. Passing on the word to his scattered elements to abandon their positions and head north up the track through Didmop, Komori was soon on his way toward a mid-island trail junction at Upmadung and a rendezvous with the *51st Reconnaissance Regiment* which was to cover the *Matsuda Force's* withdrawal. (See Map 29.)

A month of patrol clashes and ambushes had convinced General Cunningham that it was worthwhile to clear the whole Arawe area of Japanese troops. As *1/141* was abandoning its defensive sector near Lupin, an attack force composed of 2/112 and the tanks of Company B was making final preparations to drive them out. On the 27th, when the American force advanced to the airfield and beyond, they found that their quarry had eluded them.

The bloodless attack saw the birth of a new technique of communication between tank crews and the men they supported. Dissatisfied with the radio links they had with the infantry, particularly the close-in supporting squads, the Marine tankers installed field telephones at the rear of their machines through which the riflemen could contact the tank commanders. The improvement in tank-infantry cooperation was immediate, and the innovation proved to be sound enough to have a permanent part in armored support tactics.

In the several weeks before the Japanese withdrew beyond the Pulie River and gave up the airfield, General Cunningham's force suffered a few scattered casualties in patrol actions. The sum of these added little to the official total for the DIRECTOR Force in the campaign, 118 killed, 352 wounded, and 4 missing in action, which was compiled as of 10 February. That date was declared the end of DEXTERITY Operations by General Krueger. It marked the link-up of Australian troops advancing overland from Sio on the Huon Peninsula with the American task force that had seized Saidor; it also was the day when Marine and Army patrols from the BACKHANDER and DIRECTOR Forces were supposed to have met at Gilnit on the Itni River. This event, which actually took place a few days later than ALAMO Force reported it, signified the completion of the "assigned mission of establishing control over the western tip of New Britain." [10]

SOUTHERN PATROLS [11]

The Gilnit meeting between patrols from the two Allied task forces on New

[9] *Ibid.*, entry of 9Feb44.

[10] *DEXTERITY Rept.*

[11] Unless otherwise noted, the material in this section is derived from: *ALAMO G-3 Jnl; 1st MarDiv SAR*, Phase IV, Extensive Patrolling of Western New Britain-Borgen Bay-Itni River Area and Occupation of Rooke Island; ATIS Item No. 10874, Diary of unidentified member of 51st ReconRegt, 29Dec43–29Mar44, in ATIS Bul No. 939, dtd 20Apr44 (ACSI Recs, FRC Alex), hereafter *51st ReconRegt Diary; 17th Div Ops*; Hough and Crown, *New Britain Campaign*.

Britain was less significant an event than it had appeared it would be during the planning stages of DEXTERITY. Intelligence available before the operation had indicated that only two main routes of withdrawal from Cape Gloucester were available to the Japanese garrison. One of these lay south toward Gilnit and Cape Bushing and the other followed the northern coast. A maze of native trails, most of them narrow and difficult to travel, was known to exist in the jungle waste in the island's interior, but the exact, or even approximate, location of these trails was not known.

Gradually, as the fighting at Cape Gloucester wore on, the weight of evidence accumulating in the hands of the Allies indicated that the northern trail-net was the only practical withdrawal route for the Japanese. The efficiency of the anti-barge campaign, the rugged nature of the terrain along the southern coast, and the presence of DIRECTOR Force at Arawe combined to give the enemy little chance to use the Cape Bushing area as a jump-off point for further movement east by sea or land. Even though the 1st Division became increasingly sure that the Japanese would retreat by northern routes, it could not neglect the possibility that the trails south to the Itni would be used. Native reports that sizeable bodies of enemy troops were in the Gilnit area continued to come in after the *141st Infantry* was identified in the fighting around the BACKHANDER beachhead. (See Map 23.)

The only certain answer to the question of what the Japanese were doing lay in aggressive patrolling. An Army observer attached to the 1st Division during early January noted that the Marines were "patrol conscious" and that "all units are encouraged to exert the maximum effort in patrolling as it is felt this activity is the best means possible for keeping up morale and alertness." [12] This description fitted the actions of the 1st and 5th Marines closely in the period following the capture of the airdrome. While the ADC group drove forward against the enemy troops holding the Borgen Bay defenses, the Marines guarding the newly won airfields sought the elements of the *53d Infantry* that had scattered after the fall of Razorback Hill.

Combat and reconnaissance patrols made a thorough search of the jungle lowland and foothills bordering the airfield perimeter, driving Japanese stragglers before them and securing the ground. The debris left by the enemy in retreat eventually revealed the main track over Mt. Talawe, but progress along its trace was slow and painstaking. Each branching trail, and there were many, had to be checked before the area of patrol effort could be extended. The primary mission of the BACKHANDER troops was the security of the airfields, and there was no inclination to overlook any Japanese group whose attacks might delay construction progress.

Behind the Marine-manned perimeter and the active screen of patrols, Army aviation engineers labored around the clock to build a runway and hardstands on the site of Airfield No. 2. Work on Airfield No. 1 was abandoned almost as soon as it began, when it became apparent that the field that could be built would not be worth the effort necessary to ready it for use. The Japanese had made no attempt to drain their airstrips or to obtain prac-

[12] Col J. F. Bird memo to Deputy CofS, ALAMO For, dtd 9 Jan44, Subj: Rept on BACKHANDER Ops 1–7Jan44, in *ALAMO G–3 Jnl No. 15*.

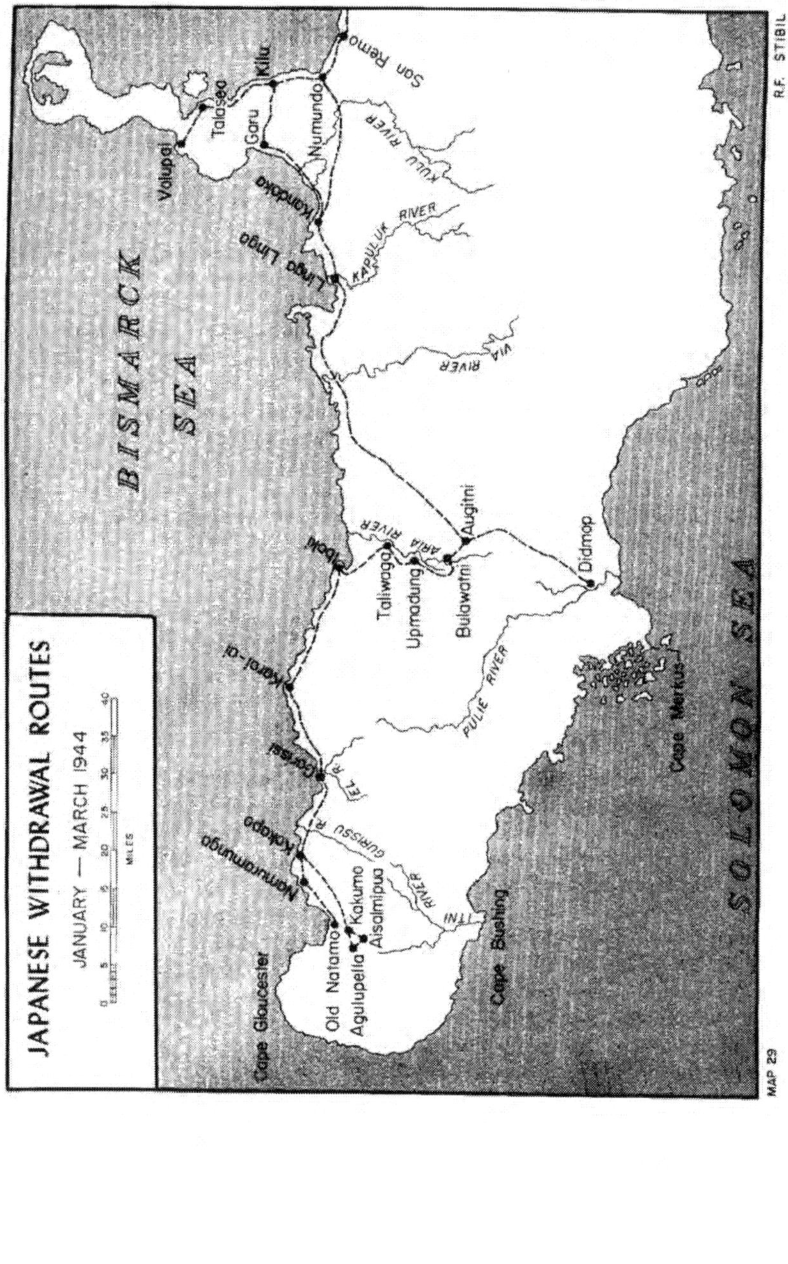
JAPANESE WITHDRAWAL ROUTES
JANUARY — MARCH 1944
MILES
BISMARCK SEA
SOLOMON SEA
Cape Gloucester
Old Natamo
Agulupella
Kakumo
Aisalmipua
Namuramunga
Kokopo
Cape Bushing
INLI
RIVER
GURISSU R
EL R
Karai-ai
Iboki
Taliwaga
Upmadung
Bulawatni
ARIA RIVER
Augitni
Didmop
PULIE RIVER
Cape Merkus
VIA RIVER
KAPULUK RIVER
Linga Linga
Kandoka
KULU RIVER
Numundo
Garu
Volupai
Talasea
Kilu
San Remo
R.F. STIBIL

MAP 29

tical gradients, and, as a result, when the 1913th Aviation Engineer Battalion began work on 3 January it was plagued by drainage problems caused by the heavy rains.[13] Often the engineers' bulldozers and graders appeared to be working in an enormous mud trough, as they sought to find firm ground on which to construct the field. A second aviation engineer battalion began work on the runway on 13 January, and a third came in on the 17th to build the necessary hardstands and roads. The only letup in construction activity occurred when a Japanese bomber made a nuisance raid, and the soldiers, like the Marines on the hills above them, headed for a safe place that would still afford them a view of the awesome antiaircraft barrage put up by the 12th Defense Battalion and the Army's 469th Antiaircraft Artillery (Automatic Weapons) Battalion.

The Japanese had realistically concluded that air raids, and these only in irritant strength, were the only means left to them of hindering work on Cape Gloucester airfield. All during the period that the Marines who had seized the airdrome were conducting local patrols to consolidate the perimeter defenses, elements of the *53d Infantry* were holding blocking positions across the trails that led to Borgen Bay from Mt. Talawe and Sag Sag. General Matsuda had charged the *53d* with the task of defending the inland flank of the Japanese troops battling the Marines for Aogiri Ridge and Hill 660. Since American patrols from the airfield did not venture as far as the enemy trail blocks, and the Japanese showed an unusual lack of offensive spirit, there were no significant contacts between the two forces in the vast area south of Mt. Talawe. When the Marines did come over the mountain, the troops they ran into were not men from the badly mauled *53d Infantry* but elements of the fresh *51st Reconnaissance Regiment*.

Colonel Jiro Sato, the reconnaissance unit's commander, had received orders to evacuate Rooke Island soon after D-Day, but it was not until the night of 7 January that he was able to make his move to New Britain. Then, using landing craft that had been carefully hidden from the eager hunting of Allied planes and torpedo boats, most of Sato's 380-man force sailed undetected across Dampier Strait and up the Itni to Nigol. The regiment's rear echelon made the night crossing on the 15th. After he traveled north and reported to Matsuda's headquarters, Sato was given the mission of holding the western trail approaches to Nakarop and reinforcing the *53d Infantry's* position.

On 12 January, the enemy reconnaissance unit set up in the jungle near Aipati village, not far from the junction of the main government track from Sag Sag to Natamo with the trail leading to the airfield. Thus, on 19 January, when a strong patrol from 3/1 reached this spot, it was an outpost of the *3d Company* of the *51st* which it encountered. The Marines took the Japanese by surprise, killing six and driving off the rest, then scouted around long enough to make sure that they had found the main cross-island track before returning to base. The American discovery of this important trail occasioned the dispatch of strong combat patrols from the airfield, but these started too late to run

[13] OCE, GHQ, AFP, *Airfield and Base Development—Engineers in the Southwest Pacific 1941–1945*, v. VI (Washington: GPO, 1951), pp. 193–194.

into anything but rear-guard action by the Japanese.

By 21 January, the *Matsuda Force* was under orders to withdraw from western New Britain. The *17th Division* commander, General Sakai, had recommended the move about mid-January, arguing that the combination of steadily mounting combat losses and the effective throttling of the sea supply route meant the eventual annihilation of all the troops under Matsuda.[14]

The authorities at Rabaul shared Sakai's pessimistic view of the situation, and General Imamura authorized a withdrawal to the Talasea area.[15] The movement was actually underway before the formal order was issued.

When the *51st Reconnaissance Regiment* took up positions guarding the trail complex south of Mt. Talawe, it was replacing part of the *53d Infantry*, mainly detachments of sick and wounded, who were headed east away from the scene of the fighting. On the 16th, General Sakai added to this movement of "forces not having combat strength" by directing General Matsuda to send all such men to the Iboki-Karai-ai sector and to dispatch elements of his command to "occupy key points along the north coast and protect rear supply lines, making these points bases for counterattacks." [16] The units selected, the *65th Brigade Engineers*, the *31st Road Construction Unit*, and two field hospitals, had the mission of improving the coastal track and preparing casualty convalescent stations at Kokopo, Karai-ai, Iboki, Upmadung, Kandoka, and Numundo. (See Map 29.)

On the 21st, the formal withdrawal order was sent to Matsuda, directing him to disengage his units in contact with the Americans and concentrate in the Iboki area prepared for further movement to Talasea. General Sakai's chief of staff understated the case when he commented that "this withdrawal, under present formations and over existing terrain, will be an extremely difficult one." Ominously, he predicted that if the arrangements to send the sick and wounded to the rear proved "too obvious an obstruction to the efficient execution of the withdrawal, unavoidable instances when wounded and sick must be disposed of are to be expected." [17]

General Matsuda set up a schedule of withdrawal that put the *53d Infantry* and *23d Field Artillery* on the trail first, but only as far as the east bank of the Natamo River, where strong defenses were constructed to cover the retreat of the main body. According to plan, all of the *Matsuda Force* would be across the Natamo by 1 February, using trails that converged on a main track which skirted the immediate coastal region until it reached the vicinity of Kokopo. The existence of this track was known to the Allies, but they had no way of pinpointing its loca-

[14] Docu No. 52399, Statement of Ex-LtGen Yasushi Sakai, in HistDiv, MilIntelSec, GHQ, FEC, Statements of Japanese Officials on WW II, v. III, p. 190 (OCMH).

[15] General Imamura's decision was not made as a direct result of General Sakai's recommendation, but rather was a conclusion he reached independently. Imamura thought that he might be able to supply the *Matsuda Force* at Talasea. *Japanese comments.*

[16] AET No. 2026, ATIS AdvEch No. 2, dtd 18Feb44, 17th Div OpO A No. 82, dtd 16Jan44, in *1DC IntelDocuments.*

[17] ATIS Item No. 10452/e, 17th Div CofS Instns, dtd 21Jan44, based on 17th Div OpO A No. 84, in ATIS Bul No. 883, dtd 4Apr44 (ACSI Recs, FRC Alex).

tion until patrols actually walked along its path.

In recognition of the comparative good shape of the *51st Reconnaissance Regiment*, Matsuda designated Sato's unit as rear guard to cover both stages of the enemy withdrawal, the disengagement from contact in western New Britain and defense of the Japanese rear after most of the men had made the exhausting march to Iboki. The *51st* was to take up positions on the Aria River southeast of Iboki, holding open both the track from the coast which swung inland there and the trail overland to Komori's force at Arawe. On 24 January, Sato moved to Nakarop where he got his withdrawal orders, and, four days later, he was marching east at the tail of the Matsuda column.

The reconnaissance troops, like most Japanese combat units, made a deliberate withdrawal. Covering forces, strong enough to hold off sizeable American patrols, set up in ambush at various stages in the retreat. Usually, when action was joined, the Marines attacked to develop the strength and dispositions of the enemy unit opposing them, then pulled back to gather reinforcements, and came on again to wipe out the blocking force. Quite often when the attack was renewed the Japanese had moved on, and the only sensible course for the following troops was a wary, methodical pursuit.

Most of the Marine patrols reported that the Japanese appeared to be retreating south, an observation explained by the fact that many of the feeder trails in the web that cut the cross-island track led south at first, then east, and finally either north toward Kokopo or south again toward Gilnit. The route that much of the *Matsuda Force* was using in retreat actually led off from the trail to Gilnit at a point below the trail junction at Agulupella. Small wonder then that there was a strong disposition on the part of BACKHANDER Force headquarters to scour the southern region for Japanese troops that might be there. Unquestionably, the victorious Marines had the strength to pursue any reasonable course in clearing western New Britain of the enemy. As it happened, the Japanese withdrawal from the southern part of Itni valley was complete. A patrol of platoon strength could probably have scouted the trail to Gilnit and Cape Bushing with safety, once it was past the area of rear guard action straddling the enemy withdrawal route. Not having the benefit of hindsight, the 1st Marine Division gave the task of checking Gilnit to a composite battalion. (See Map 23.)

The assembly of this battalion came at the end of a week of vigorous patrolling, marked by occasional sharp clashes with the enemy rear guard. Units from all three of the 1st Division's infantry regiments converged on Agulupella, the focal point of patrols coming from Sag Sag, the airfield, and the beachhead. As a result of an exchange of positions immediately following the capture of Hill 660, the 5th Marines with 2/1 attached held the beachhead, and the 7th and 1st Marines occupied the airfield perimeter. Units of the 5th made the initial contact with elements of the enemy's rear guard at Natamo Point on 20 January. Thereafter, the reinforced regiment, operating under the ADC's command, stuck close to the northern shore, driving ahead to close off the coastal track. Its sweep along Borgen Bay became the first leg of an advance that was eventually to find it landing in assault on Willaumez Peninsula.

In the center of the island, the first significant contact with the Japanese guarding Matsuda's rear, elements of the *51st Reconnaissance Regiment*, came on 23 January. A composite company of 1/1 following the trail from the airfield ran into an enemy machine gun and a few protecting riflemen, whose fire held up the patrol temporarily until the advance guard drove them off. Then about 1,500 yards of cautious advance later, the same or a similar small group opened up on the Marines again. In the flurry of return fire two of the Japanese were killed and the rest fled. Holding up for the night in his own ambush position close to Mt. Langila, the patrol commander resumed his advance the next day until his lead elements were pinned down by the fire of at least one enemy rifle platoon reinforced with machine guns. The Japanese positions blended so artfully into the shrouding jungle that the Marines had little at which to aim. Though the patrol's return fire was heavy in volume, it had no apparent effect. Action was broken off, almost miraculously with no Marine casualties, and night defensive positions astride the trail were again occupied. On the 25th, the Marine patrol pulled back to await relief; its ammunition pouches were almost empty and the men were short of food. Half of the rations carried for the patrol by an ANGAU-led group of 40 native porters proved to be spoiled.

Company K of 3/1 made the relief on 26 January and moved out immediately toward the cross-island trail. After an uneventful day's march and a night in perimeter defense, Company K moved unchecked through the area where the Japanese had held off the Marines on the 24th. At the trail junction, the men from K ran into elements of a composite company of the 7th Marines which had landed from LCMs at Sag Sag on the 22d.

Travelling east on the government track and searching the surrounding area thoroughly, the 7th Marines' patrol had reached Aipati on the 24th. The next day, the trail junction was occupied, and the company followed the airfield trail for several hundred yards without finding any sign of the Japanese. On the 26th, this patrol pattern was repeated and in addition a platoon was sent half a mile to the south towards Agulupella; neither unit encountered any enemy. A small reconnaissance patrol, four Marines and three natives, moving cautiously along the main track, was ambushed near Niapaua, where trails forked to Nakarop and Agulupella. One of the scouts was killed before the men were able to slip away.

A 7th Marines patrol from Aipati travelled a mile toward Niapaua on the morning of the 27th without seeing any sign of live Japanese. When Company K of 3/1 reached the cross-island track, it borrowed a machine gun platoon from the 7th's composite company and started east to check the ambush site. Late in the afternoon, the scouts preceding the march formation sighted about 50 of the enemy set up on the far side of a stream that cut across the track. Prudently, the company commander held up for the night, ready to drive ahead in the morning. At 0700, 28 January, the Marines attacked and soon broke through the position spotted the day before. A short distance farther on, Company K ran into a storm of rifle, machine gun, and mortar fire that betrayed the presence of a reinforced company dug in across the track. The Japanese, holding high ground that commanded the Marine position, were able to fend off all attempts to reach them, and the battle lapsed into a stalemated fire fight. After three hours of fruitless exchange, the Marines pulled back out of mortar range, taking 15 casualties with them.

Company K held its ground on the 28th while reinforcements, the rest of the 7th Marines company from Aipati and weapons elements of the 7th sent from the airfield, joined. Major William J. Piper, Jr., executive officer of 3/7, now took command of the combined group. Piper's force found the Japanese position abandoned when it advanced on the 30th and proceeded without hindrance to Niapaua. From there, Piper moved south to Agulupella where he had orders to await the formation of a larger force, designated the Gilnit Group.

Elements of the 5th Marines were also directed to join the enlarged patrol headed for Gilnit. Scouts and combat patrols from 2/1 and 2/5, pressing southwest along the trail behind Aogiri Ridge, encountered Japanese rear guard detachments and drove them off. On the 28th, Company E, 2/5, moved through Magairapua, once Colonel Katayama's headquarters, and then on to Nakarop, where General Matsuda was known to have been located. In both tiny villages and all along the trail between, there were deserted bivouac areas, littered with enemy gear but empty of troops. Matsuda's own quarters was so well camouflaged that its location near the trail was not discovered until several days later.

At Nakarop, Company E was joined by Major Barba's 1/5. Barba had broken off operations along the coast on 28 January and driven through Japanese delaying forces in an effort to reach Niapaua and aid Company K of 3/1 in reducing the trail block it had encountered. On the 29th, a heavily reinforced Company G, 2/5, and a large party of native carriers reached Nakarop from the beachhead. This force, dispatched by the ADC to join the Gilnit Group, plus Barba's battalion with Company E attached, filed into Agulupella on 30 January to unite with Major Piper's command. Lieutenant Colonel Puller was designated by General Rupertus to lead the combined units.

Puller's command, six reinforced rifle companies and headquarters elements, numbered 1,398 Marines, 3 Australian officers, and 150 native carriers. The supply problem posed by its size was staggering and not easily surmounted. So long as the group stayed in the vicinity of Agulupella, native carriers could just barely maintain it by a constant shuttle from the beachhead dumps. The condition of the trails deteriorated steadily as heavy traffic and flooding rains turned the paths into slithery channels of mud. As supply by hand-carry slowed, air drop, both by the division's Piper Cubs and Fifth Air Force B-17s, was instituted to keep Puller's men fed and provided with essential items of equipment.

During the buildup at Agulupella, evidence mounted that the Japanese had not fled south, evidence that soon included the actual withdrawal order dug up in a cache of staff papers found at Matsuda's headquarters, and the 1st Division decided to reduce the size of the Gilnit Group. Accordingly, the 1st Battalion, 5th Marines was detached, and the patrol strength dropped to 384 Marines plus the native carrier group. Puller had units of his command patrolling vigorously throughout the central part of the island while he waited to build up sufficient reserve supplies to make the march to the south. A further complication arose when scouts discovered that a rain-swollen branch of the Itni near Arigilupua, about three miles from Agulupella, was too wide and deep for fording. Puller sent Company K, 3/1, to outpost Arigilupua while a detachment from the 17th Marines bridged the stream.

Captain George P. Hunt, the Company K commander, was given permission by Puller to reconnoiter the trail to the south, while the Gilnit Group was held up by the bridging operation. Hunt selected a small patrol of 11 men from his unit and set out down the well-defined track. With no enemy to stop him and no heavy equipment to slow him down, Hunt reached Nigol on the Itni in one day's march. Bivouacking for the night, the Marines moved about a mile farther to the river bank opposite Gilnit on the next day and then returned to Nigol. The only Japanese sighted in two days was a sick straggler who was sent back to Agulupella.

Puller received Hunt's report of his findings shortly before the main patrol started out for Gilnit. The bridge at Arigilupua was completed on 6 February, and Puller left with about half of his command immediately, meeting Hunt on the trail. Major Piper, who was serving as patrol executive officer, kept a portion of the unit with him and followed Puller, searching all the side trails and bivouac areas encountered. According to plan, the two elements of the patrol leapfrogged each other, exchanging personnel as necessary, with one group always moving ahead on the principal track while the other checked the jungle to either flank. Weapons crewmen, burdened with heavy loads of mortar and machine gun ammunition and parts, found the going particularly rugged. To all patrol members the trip was unforgettable, if only for the monotony of their steady diet of rain and K rations.

The food situation was dictated by the means of resupply. It had been decided to try to keep the patrol subsisted and equipped entirely by air drop, and the division's light planes were handed the task. The little Cubs, whose peak carrying capacity was two cases of K rations, one held on the observer's lap and the other placed on a desk behind him, flew all day long. Often the pilots logged 10–12 hours in the air, landing, refueling, and taking off again, in a regular pattern from dawn to dusk.[18] All drops were made at villages along the route, Arigilupua, Relmen, and Turitei, according to schedule, with special requests filled as they were received.

One of the items asked for by Puller in an urgent dispatch, several hundred bottles of mosquito lotion, raised a few eyebrows at division headquarters, but the request was filled promptly. The Gilnit Group commander's well-known disdain for the luxuries of campaigning caused the wonder, but the explanation was simple and a lesson in jungle existence. As a patrol member later remarked:

> Hell, the colonel knew what he was about. We were always soaked and everything we owned was likewise, and that lotion made the best damn stuff to start a fire with that you ever saw.[19]

---

[18] Capt Richard M. Hunt, "General Rupertus' Improvised Air Force," *Marine Corps Gazette*, v. 33, no. 6 (Jun49), p. 17.

[19] Quoted by LtCol John S. Day in comments on draft of Hough and Crown, *New Britain Campaign*, dtd 7Mar52.

The advance party of the patrol reached the river opposite Gilnit on the 9th, and the main body arrived the following day. Puller sent a small scouting party across the Itni to the patrol objective; again there was no sign of recent enemy occupation. A platoon sent to Cape Bushing encountered no Japanese, but did find ample signs of the area's one-time garrison, the *141st Infantry*. Considerable quantities of weapons, ammunition, and equipment were found in the various enemy camps; surprisingly, there were even some food supplies left behind. Anything that was of use to division intelligence officers, or to ANGAU for distribution to the natives, was set aside; everything else was destroyed.

Puller was ordered to wait in the Gilnit area until he was contacted by the Army patrol dispatched from Arawe. According to the reports the Marines received, the soldiers were held up east of Gilnit by enemy opposition at Attulu Hill, once the *141st Infantry's* command post. Both native and Marine scouts explored the hill and found no Japanese, and Puller radioed division that in his opinion there were no enemy forces in the area to be discovered.

On the 14th, the division made its own contact with Puller when two LCMs with a platoon of 1/7 on board arrived from the airfield. The boats carried some extra supplies for the patrol which was about to set out on its return trip up the track to Agulupella. About midday on 16 February, elements of the Arawe patrol reached Gilnit via the river and met the platoon that Puller had left behind a few hours earlier for that purpose. Its mission fulfilled by the contact with the soldiers, the last Marine unit moved out briskly in the trace of Puller's column, bringing to an effective conclusion 1st Division combat operations in the southern Itni valley.

EASTERN PATROLS [20]

On the maps issued to BACKHANDER troops prior to D-Day, the coastal track that paralleled the shore of Borgen Bay appeared to be the most logical and, in fact, the only practical northern route of withdrawal for the Japanese. By the time enemy resistance collapsed at Hill 660, Marine intelligence officers were reasonably sure that their maps were wrong. Somewhere in the miles of jungle and swamp south and east of the bay were other trails, their existence confirmed by captured papers and the reports of natives and prisoners of war. The names of the villages of Aisalmipua and Kakumo began to crop up as way stations on a frequently used supply route, but the natives could not agree on their location, except to confirm that they lay along a trail from Agulupella to Kokopo. (See Map 29.)

The government track from Sag Sag that ended at the coast a few hundred yards west of Natamo Point was about the only trail that the Allies were sure existed in the Borgen Bay area. They were unaware of the presence of another trail which led to Nakarop from the east side of the point, this one broader, in better shape,

[20] Unless otherwise noted, the material in this section is derived from: *1st MarDiv SAR*, Phase IV, Extensive Patrolling of Western New Britain-Borgen Bay-Itni River Area and Occupation of Rooke Island; 1st MarDiv D-2 Weekly Repts Nos. 5-10, dtd 29Jan-5Mar44; Col William H. Barba comments on draft of Hough and Crown, *New Britain Campaign*, dtd 17Mar52; Maj John S. Stankus comments on draft of Hough and Crown, *New Britain Campaign*, dtd 13Mar52; Hough and Crown, *New Britain Campaign;* OCE, GHQ, AFP, *Amphibian Engineers*.

and partially corduroyed in its wettest stretches. Still a third important track not recorded on pre-invasion maps ran from the village of Old Natamo east of the Natamo River to Kakumo.

In order to effect their retreat without hindrance, the Japanese had to block the trails that led through Agulupella long enough for the main body of the *Matsuda Force* to get clear and started for Kokopo. In the center of the island, the *51st Reconnaissance Regiment* executed this task with seeming ease, and along the coast the same job was performed well by a rear guard made up of elements of the *53d Infantry*, the *39th Field Antiaircraft Battalion*, and the *11th Company* of *3/141*. The vigorous resistance put up by these latter troops had the effect of clouding the issue of what route the *Matsuda Force* had taken. When they had delayed the Marines as long as was necessary to let the main body of the Japanese escape, the defenders just faded away. Helping the rear guard's withdrawal was the natural caution which characterized the patrol operations of veteran troops in the jungle.

At every natural obstacle along the trails that had to be used for any sizable troop movement, there was the threat of ambush. A sudden burst of fire from a single machine gun or a fusillade of shots from a few well-placed riflemen could be the cause of hours of delay. Much of the time the enemy could not be seen and the terrain stalled attempts to outflank his positions, leaving just the few men at the point of a patrol to reply to the fire that swept the trail clear. Under the circumstances, it took steady nerves, quick reactions, and a considerable amount of quiet courage to be a scout and take the lead on a patrol into enemy-held territory. The situation was ideal for a few determined Japanese, probably no more than 300 by the last few days in January, to hold up the advance of thousands.

The coastal track rimming Borgen Bay looped across the base of Natamo Point, making the narrow, jungle-covered spit of land an excellent site for rear guard action. Inland, the several rivers and numerous streams discharging into the bay helped turn the rain-sodden ground into one of the worst stretches of swampland in western New Britain. The Japanese knew that the Americans would have to advance up the coastal corridor and waited for them in prepared positions.

An enemy map of the point, captured on 3 January, showed machine gun emplacements and rifle pits sufficient to hold a reinforced platoon located there. When a patrol started out to check the point on 20 January, there was a strong possibility that the Japanese might be holding it. Still, the shore of Borgen Bay east of Hill 660, was sprinkled with abandoned positions and these might also be vacant. When the patrol, most of Company A, 1/5, led by regimental scouts, reached the estuary that cut into the base of the point, all doubts were dispelled. Automatic weapons fire lashed across the water and forced the Marines to take cover. For two hours, the patrol's supporting mortars and machine guns sought to silence the enemy guns, but to no avail. Finally, because ammunition was getting low, artillery fire was called down to cover the Marines' withdrawal.

On the 21st, the 75mm howitzers of 1/11 fired on the suspected locations of the enemy weapons, and, on the following morning, a reinforced company tried to attack but could make no headway. A platoon that worked its way through the swamp to come up on the east side of the point

PATROL OF MARINES *seeking Japanese troops retreating from western New Britain files across one of the many jungle streams in the Borgen Bay area.* (USMC 72282)

ARMY AMPHIBIAN ENGINEERS *and Marines load Army LCMs at Iboki on 5 March for the overnight run to Talasea.* (USMC 79887)

spotted positions that appeared to hold a company and then pulled back. Ten enemy soldiers were killed and several wounded in the day's exchange of fire, most of these when the patrol, on its first approach, caught the crew of a strongpoint unawares. All through the night, artillery harassed the Japanese on the point, and, at 0910 in the morning, a squadron of A–20s appeared from Finschhafen to bomb and strafe ahead of a two-company attack supported by tanks.

The tanks, mediums of Company A, were transported across the bay by LCM to a beach about a thousand yards from the point and then lumbered up to lend supporting fire to the infantry assault. From offshore, a rocket DUKW laid down a barrage on the Japanese positions, and 1/11 fired in advance of the Marine skirmish line. In a series of short, violent fights, the Japanese were killed or driven from their defenses. In late afternoon, when Natamo Point was securely in American hands, the bodies of 30 enemy soldiers were counted in and around the wrecked and smoking gun pits. Some 15–20 machine guns and two 20mm cannons were destroyed in the day's action. A Japanese 75mm gun located somewhere in the jungle to the southeast of the point started firing late in the afternoon, but 48 rounds in reply from a 155mm seacoast artillery battery located in the beachhead silenced the piece.[21]

About half of the hundred-man garrison of Natamo Point fled the Marines' final attack and escaped down the trail to Agulupella or across the Natamo River to the *53d Infantry's* positions on the eastern bank. In a follow-up advance on the 24th, the 1st Battalion, 5th Marines accounted for 10 more of the enemy while clearing the track as far as the river. The explosion of small-arms, mortar, and artillery fire that greeted any attempt to cross brought an effective halt to the day's advance. For five days, the Natamo River marked the limit of the 5th Marines' advance to the east, as the Japanese held on tenaciously.

On 25 January, an attempt to land a rifle platoon, reinforced with a half track and machine guns, in a coconut grove about 300 yards east of the river was repulsed by heavy Japanese fire. Similarly, a lodgement on the east bank of the river could not be held in the face of overwhelming enemy resistance. Before attempting another frontal assault, patrols from 1/5 followed the river south to find a suitable crossing, but the swamps defeated their efforts. On the 27th, Company B followed the corduroyed trail that led to Nakarop from Natamo Point until a strong Japanese rear guard was encountered about 4,000 yards inland. In the resulting fire fight, the Marines had one man killed and three wounded; the enemy lost 15 soldiers but accomplished his mission of holding up the advance. On the 28th, the 1st Battalion, 5th, less Company B which remained to hold the lines along the Natamo, moved out along this trail with the mission of reaching Nakarop. This battalion, spurred on by reports of the action at Niapaua, made quick work of two rather feeble attempts to delay its progress and bivouacked three and a half miles inland, not far from its objective. But the next day's advance disclosed that the Japanese were gone.

Not only had the enemy disappeared from in front of the Marines working along the trails inland, but he had abandoned his positions guarding the Natamo's

[21] Seacoast Consolidated AR, dtd 23Jan44, in 12th DefBn WarD, Jan44.

west bank, too. Scouts who moved 2,000 yards upstream on the 30th, then forded the river, and came down on the opposite bank found the defenses deserted. On the 31st, a patrol crossed the river mouth and proceeded to Old Natamo; three Japanese who were found in a pillbox near the beach were killed, but no one else was sighted. One of Lieutenant Colonel Puller's patrols reached Kakumo on the same day. The natives told the Marines that the last Japanese had left the village heading east the previous day.

In its advance to the Natamo and movement down the trail to Nakarop, the 1st Battalion, 5th Marines had had 6 men killed and 33 wounded. Estimated enemy dead among the rear-guard detachments was close to 75, and the wounded probably reached a similar figure. In military terms, the cost to the Japanese was negligible for the prize gained, time for the main body of the *Matsuda Force* to get well underway for Iboki, Talasea, and, eventually, Rabaul.

The next obvious step for the 1st Marine Division was aggressive pursuit of the retreating Japanese. Patrols of 2/1, attached to the 5th Marines, followed the coastal track as far as Namuramunga, reaching the village, which was seven miles from Old Natamo, on 2 February. Two other patrols of the 1st Marines' battalion cut directly through the jungle from the coast to find the Agulupella-Kakumo-Kokopo trail and establish conclusively the enemy withdrawal route. Only stragglers, sick and wounded men, were encountered by any of the patrols; the Japanese left behind were too weak from hunger to offer any resistance.[22]

On 3 February, General Rupertus conferred with General Krueger at Finschhafen to discuss further actions by the 1st Marine Division. Both field commanders reacted unfavorably to a directive from GHQ that had just arrived ordering a new Final Beachhead Line in western New Britain, one that would include all of Borgen Bay in the defended perimeter. The feeling among the staff at GHQ was that the new line would prevent the Japanese from returning and shelling the landing beaches supporting the growing airfield. Rupertus and Krueger, knowing that their current problem was not the return of the Japanese but the destruction of the troops that had fled, ignored the message, feeling that it was based on premises no longer valid. Instead, Rupertus was given the go-ahead signal for an immediate pursuit of the *Matsuda Force* as far as Iboki and alerted to Krueger's plan of continuing the advance to the Willaumez Peninsula and beyond.[23]

The 5th Marines got the mission of keeping the pressure on the retreating enemy

[22] During these pursuit operations, a platoon of the 12th Defense Battalion's 155mm guns was positioned within the Yellow Beach perimeter to fire as field artillery in support of 1/11. An air observer spotted "five to eight Japanese barges loading personnel at the eastern tip of Borgen Bay. This information was relayed to an OP on Target Hill and thence to the 155mm guns. These barges were taken under fire at maximum range of 19,200 yards. After adjustment and platoon 10 volleys, the air observer reported 'all barges damaged or sunk with many Japs struggling in the water.'" Col Thomas L. Randall comments on draft manuscript, included with MajGen James M. Masters, Sr., ltr to ACofS, G-3, HQMC, dtd 2Jul62.

[23] ALAMO G-3 note on GHQ directive, dtd 2Feb44, in *ALAMO G-3 Jnl No. 19;* MajGen William H. Rupertus ltrs to LtGen Alexander A. Vandegrift, dtd 4Feb44 and 18Feb44 (Vandegrift Personal Correspondence File, HQMC).

691-360 O—63——27

and, if possible, cutting off and destroying some part of the Japanese force. A simultaneous overland and overwater advance was planned with the help of landing craft of the Army amphibian engineers. The division's Cub planes were to scout ahead of the advancing Marines, spotting suitable landing beaches and keeping tabs on the Japanese. The 5th Marines knew where the enemy was heading and the route he was taking, but the jungle shielded the troops from aerial observers. Under the circumstances, when the Japanese might easily be waiting in ambush for their pursuers, the Marine advance had to be both swift and cautious.

Bad weather was a frequent factor in holding up the LCM-borne phases of the 5th Marines' pursuit. Crashing surf denied the forward beaches to landing craft and placed the burden of catching up on the patrols operating along the coastal track. When the seas and the limited number of LCMs available permitted, large elements of Colonel Selden's unit were able to leap-frog the foot patrols and bite off 10- and 15-mile chunks of the coast at a time. Alternately in the lead, as the 5th moved east, were elements of the attached scout platoon from division tanks, men of the regiment's own intelligence section, and, often, a brace of Army scout dogs and their handlers, loaned for the operation. Kokopo, Gorissi, and Karai-ai were occupied in their turn, and in each village the natives told the same story—the Japanese were still ahead. Prisoners seized along the trail in the mop-up of stragglers confirmed the continued head start.

On the 24th, patrols from sea and land reached Iboki, fully expecting to encounter Japanese resistance at this primary supply base. But there were no enemy defending troops to be found, only sick and starving individuals who had fallen behind. The last cohesive unit of the *Matsuda Force*, the *51st Reconnaissance* rear guard, had passed through the village on the 16th.

Despite his disappointment at missing contact and visiting further destruction on the Japanese, Colonel Selden was able to view his regiment's accomplishment with some pride. On short notice, actually less than a day's warning, the 5th had started its 20-day trek and worked out a successful method of operation that made the best use of the men and transportation available. As Selden later summarized the effort, he had:

> . . . 5,000 men on this jaunt of sixty-odd miles over some of the worst jungle terrain in the world. We kept the Nips on the move by having fresh men out every morning. With few exceptions, men were not called upon to make marches on two successive days. After a one-day hike, they either remained at that camp for three or four days or made the next jump by LCMs.
>
> . . . To have accomplished my march four days prior to the deadline without loss or even having a man wounded was, in our estimation, quite a feat.[24]

ALLIED PROGRESS REPORT [25]

As the 5th Marines moved into a staging area at Iboki Plantation, the final arrangements for continuing the 1st Division's advance along New Britain's north coast were being made. General Krueger intended to make a landing on Willaumez

---

[24] MajGen John T. Selden ltr to HistDiv, HQMC, dtd 7Mar52.

[25] Unless otherwise noted, the material in this section is derived from: *ALAMO G-3 Jnl;* Craven and Cate, *Guadalcanal to Saipan;* Hough and Crown, *New Britain Campaign;* Miller, *Reduction of Rabaul.*

Peninsula, secure it, and drive on for Cape Hoskins and the Japanese airfield there. The landing craft needed for the pending operation would be a mixed and scant force of Navy LCTs and Army LCMs. Most of the amphibious shipping available to Seventh Fleet was tied up in support of landings in the Admiralties.

The success of DEXTERITY was an influential factor in hastening the schedule of operations designed to isolate Rabaul. On 13 February, General MacArthur issued a directive calling for the seizure of Manus in the Admiralties and Kavieng on New Ireland with a probable target date of 1 April. There was strong sentiment at General Kenney's headquarters to slice the delay time and secure the enemy airfields in the Admiralties ahead of the projected D-Day, if the Japanese garrison appeared to be weak. Intensive aerial scouting convinced the Allied Air Forces leader that a reconnaissance in force into the Admiralties could be risked, and he was able to persuade General MacArthur to order the move. A reinforced squadron of the 1st Cavalry Division made the first exploratory landing on Los Negros Island on 29 February and, in a sense, caught a tiger by the tail. The Japanese garrison, much stronger than aerial reconnaissance had indicated, battled fiercely to throw the cavalrymen off the island. General MacArthur made the decision to reinforce the troops ashore rather than withdraw them, and throughout March, American soldiers in overwhelming strength poured into the Admiralties. The capture of two airfields and an excellent deep-water harbor in the islands had the effect of hastening the tempo of operations in the Southwest Pacific and forging an important link in the chain of Allied bases that ringed Rabaul.

So swift was the pace of advance in early 1944, that the strategic importance of Cape Gloucester's airfield shrank steadily while the engineers were still working to get it ready for use. The airfield site at Saidor, its seizure termed by MacArthur a vital "exploitation of the New Britain landings," [26] turned out to be usable by transports and other heavy aircraft several weeks before the runway at Cape Gloucester was ready for regular traffic. Captain Petras landed General Rupertus' plane on Airfield No. 2 on 28 January, and the first Army transport came down on the field on the 31st. Pierced steel planking was laid the whole length of the runway to overcome the effects of heavy rains, but the site was simply a poor one, and a staggering amount of work and materials would have to be devoted in an effort to make Cape Gloucester into a first-class airfield. The changing strategic situation made this task unnecessary. The 35th Fighter Squadron moved into the field on 13 February, while the aviation engineers were still fighting the cape's unsuitable terrain, and the 80th Squadron followed on the 23d. Within a month, recall orders had been issued for both units so that they might be committed in support of MacArthur's advance west along New Guinea's coast toward the Philippines.

Marine operations following the seizure of Cape Gloucester had strong overtones of an aggressive police of the area. The 1st Division's patrols pressing the retreating enemy toward the east made a clean sweep of stragglers at the same time they were trying to find and destroy elements of the main body of the *Matsuda Force*.

---

[26] CinCSWPA msg to CG, ALAMO For, dtd 28Dec43, in *ALAMO G-3 Jnl No. 12*.

The bothersome problem presented by the hulking presence of Rooke Island close inshore to the airfield was taken care of by Company B of the 1st Marines. Landing from LCMs on 12 February, the company patrolled vigorously for a week and confirmed the finding by ALAMO scouts that the Japanese had pulled out. The garrison once considered for Rooke no longer seemed necessary, and the Marines returned to Cape Gloucester on the 20th. By the end of February, New Britain was clear of any effective Japanese force as far east as a line joining Iboki and Arawe.

Talasea and Beyond

RECALL TO RABAUL[1]

The Japanese commanders at Rabaul, General Imamura of *Eighth Area Army* and Admiral Kusaka of *Southeast Area Fleet*, were in an unenviable situation following the loss of Cape Gloucester. They knew that the defensive line toward which the *Matsuda Force* was retreating was untenable. The *17th Division* troops could, and would, undoubtedly, fight doggedly to hold the Allies at bay before Talasea and Cape Hoskins in the north and Gasmata in the south. The enemy's dwindling force of warships and transports could attempt sacrifice runs to keep supplies flowing to the soldiers, and the naval planes of the *Eleventh Air Fleet* could provide weak and sporadic support of the ground action. Not even these few ships and aircraft were to be available, however. The success of the American amphibious assault on Kwajalein prompted the retirement of the *Combined Fleet* from its suddenly vulnerable base at Truk, and the follow-up carrier strike of 16–17 February on Truk decided Admiral Koga to issue recall orders to all Japanese naval aircraft in the Southeast Area.[2]

Enemy interceptors made their last attempt to break up an Allied air attack on Rabaul on 19 February. On the 20th, the fields that ringed Blanche Bay were deserted, "not a single moveable plane remaining"[3] to contest control of the air. The harbor yawned empty too, with the hulks of sunken ships the only reminder of the bustling fleet that had once based there. The Japanese stronghold was forced to rely entirely on its ground garrison for defense. Imamura and Kusaka determined to make that defending force as strong as possible, adding to it every available soldier and sailor on New Britain.

On 23 February, orders to withdraw to Rabaul were received at *17th Division* headquarters at Malalia. General Sakai gladly dropped the plans he had been formulating for holding out against the oncoming Allied troops, for he fully appreciated how isolated and hopeless his fight would have been. In the stead of preparations for a last-ditch defense centered on positions at Cape Hoskins, Sakai began hastily figuring the moves that would get his command back to Rabaul in fighting shape. The *Matsuda Force* was his major problem. The lead section of the weary column of men staggering along towards Talasea was still two weeks' march from the Willaumez Peninsula.

Unless the Allies suddenly broke their pattern of pursuit and surged ahead of the retreating Japanese troops, Sakai could

[1] Unless otherwise noted, the material in this section is derived from: *Eighth Area Army Ops; 17th DivOps; 51st ReconRegt Diary; Komori Diary*; Hough and Crown, *New Britain Campaign*.

[2] *SE Area Nav Ops—III*, pp. 58–63.

[3] *Ibid.*, p. 63.

figure that most of the men en route would reach their objective. Supply dumps located along the withdrawal routes held enough rations to enable the strongest and best-led elements of the *Matsuda Force* to make good their escape. The sick and wounded who fell behind, who lacked the strength to keep up with the main body or even to fend for themselves, were doomed. The kindest fate that might befall them was capture by a Marine patrol. Often the near-naked, emaciated wretches whom the Americans found glassy-eyed and dazed along the trails had not the strength left to survive the trip to the coast. So tangled and rugged was the country through which the enemy columns struggled that scores of stragglers who died a few feet off the track where they had crawled to rest would have lain unnoted but for the unforgettable stench of human remains rotting in the jungle.

The route taken by the defeated Japanese troops after they passed through Iboki headed sharply inland, following the course of the Aria River for 12–14 miles and passing through the native villages of Taliwaga and Upmadung on the west bank and Bulawatni and Augitni on the east. From Augitni, where the trail used by the *Komori Force* to escape Arawe joined, the track headed northeast across mountain slopes and through extensive swamps fed by sluggish, wide, and deep rivers. Hitting the coast at Linga Linga Plantation, a straight-line distance of 35 miles from Iboki, the route crossed the formidable obstacle posed by the Kapaluk River and paralleled the shore to Kandoka at the neck of Willaumez Peninsula. Continuing along the peninsula coast to Garu, the trail then crossed a mountain saddle to the eastern shore at Kilu and turned south to Numundo Plantation. An alternative route from Kandoka to Numundo across the base of the peninsula lay through a 15-mile-wide morass that bulged along the course of the Kulu River. Once the Japanese reached Numundo Plantation, they could follow the coastline trail to the airdrome at Cape Hoskins, to Malalia just beyond, and eventually, with luck, to Rabaul.

The task of keeping the escape route open until the *Matsuda Force* had reached the comparative safety of Malalia was given to two units. The *51st Reconnaissance Regiment* performed the duties of rear guard, insuring the enemy withdrawal from contact, and the *Terunuma Force*, a composite battalion of the *54th Infantry*, held the Talasea area, with orders to defend it against Allied attacks. The delaying actions of Colonel Sato's reconnaissance unit in western New Britain gave General Matsuda the respite he needed to get his command underway to the east. If the *Terunuma Force* carried out its mission equally well, it would hold its positions long enough for the hundreds of survivors of the Cape Gloucester battle to reach the area east of the Willaumez Peninsula. The Japanese commanders considered that they had enough troops in the Cape Hoskins sector to require the Allies to mount a large-scale amphibious operation to take it. And barring such an effort, the *17th Division* was confident that it could pull back to Rabaul with many of its units still in fighting trim. General Sakai estimated that most of his combat troops would reach the stronghold by mid-April and all of the remaining cohesive outfits, including rear-guard detachments, would make it by the middle of May.

Not all the Japanese movement had to be accomplished on foot; there were enough barges available to move a good part of the heavy munitions at Gasmata and Cape Hoskins back to Rabaul. Sick and wounded men, who could not survive a land journey, were given priority in these craft. Some combat units were transported a portion of the way to their goal in overwater jumps from Malalia to Ulamona and then on to Toriu. The first village was a major barge base about 65 miles from Cape Hoskins, and the second, 30 miles farther east, was the terminus of a trail network which led to Rabaul through the mountains of Gazelle Peninsula. The *17th Division* planned that its components would move lightly armed, carrying little reserve ammunition and only a bare subsistence level of rations. If the Allies attempted to cut the retreat route, all available units would concentrate to wipe out the landing force.

Only a few Japanese craft were risked in the dangerous waters west of Willaumez Peninsula, and these were used, on 21 February, to carry *2d Battalion, 53d Infantry* remnants to Volupai, opposite Talasea on Willaumez. The battalion, with reinforcing artillery, was sent on ahead of the main body of the *Matsuda Force* to form the nucleus of a covering force at Ulamona. The barges returned once more to the Aria River before the Marines landed at Iboki and took out General Matsuda, members of his immediate staff, and all the litter patients they could carry. Matsuda was landed at Malalia on the 25th.

Carrying out his orders, Colonel Sato of the *51st Reconnaissance Regiment* saw the last march element of the *Matsuda Force* safely through Upmadung before starting his own unit on the trail. At Augitni, Sato met Major Komori who had brought his troops up from Arawe, and the two groups, both under Sato's command, began moving east on 6 March. On the same date, the leading elements of the *Matsuda Force* column reached the base of Willaumez Peninsula.

VOLUPAI LANDING[1]

The 6th of March was the landing date chosen for the APPEASE Operation—the assault and seizure of the Talasea area of Willaumez Peninsula by the 5th Marines, reinforced. The principal objective included several parts, the government station on the shore of Garua Harbor that gave its title to the whole region, an emergency landing ground nearby, grandly called Talasea Airdrome, and the harbor itself which took its name from Garua Island that formed one of its arms. The landing beach, Red Beach, lay directly across the peninsula from Garua Harbor in the curve of a shallow bay at Volupai. The isthmus connecting the two points, 2½ miles apart, is the narrowest part of the peninsula. (See Map 30.)

Several plantations, one on Garua Island and the others at Volupai on the west coast and Santa Monica, Walindi, and Nu-

[1] Unless otherwise noted, the material in this section is derived from: 1st MarDiv SAR, APPEASE Op, n.d., hereafter *APPEASE SAR;* ATIS Item No. 10443, Talasea Force Personnel Chart, dtd 4Mar44, in ATIS Bul No. 881, dtd 3Apr44 (ACSI Recs FRC, Alex); OCE, GHQ, AFP, *Amphibian Engineers;* Hough and Crown, *New Britain Campaign;* Col Robert Amory, Jr., AUS, and Capt Ruben M. Waterman, AUS, eds., *Surf and Sand, The Saga of the 533d Engineer Boat and Shore Regiment and 1461st Engineer Maintenance Company 1942–1945* (Andover, Mass.: The Andover Press, Ltd., 1947), hereafter Amory and Waterman, *Surf and Sand.*

mundo on the eastern shore, had the only easily traveled ground on Willaumez. The terrain of the rest of the peninsula followed the general pattern of New Britain, mountains and high ground inland covered by rain forest, with foothills and coastal flats choked with swamp forest, secondary growth, and sprawling swamps along the course of the many rain-swollen rivers and streams. Above the isthmus between Volupai and Garua Harbor, the peninsula was little used by the natives or the Japanese because of impassable terrain. Below the narrow neck of land, much of it occupied by Volupai Plantation, there were a number of native villages along the coast and on mountain tablelands. A cluster of four called the Waru villages, about 1,500 yards west of Talasea, and Bitokara village, the same distance northwest of the government station, figured as intermediate objectives in 5th Marines' operation plans.

Red Beach was a 350-yard-wide corridor opening to Volupai Plantation, 400 yards inland; on its northern flank the beach was bordered by a swamp, to the south a cliff loomed over the sand. The cliff was part of the northwest slopes of Little Mt. Worri, a 1,360-foot peak that was overshadowed by 300 feet by its neighbor to the south, Big Mt. Worri. The eastern extension of Big Mt. Worri's ridgeline included Mt. Schleuther (1,130 feet) which dominated Bitokara, Talasea, and the Waru Villages. The trail from Volupai to Bitokara, which was to be the 5th Marines' axis of advance, skirted the base of these heights.

The major obstacle to the proposed landing at Volupai was the reef that extended 3,000 yards out from shore. Obviously impractical as the route for assault waves was the tortuous small-boat passage which wound through the coral formations. To make this narrow waterway safe for supply craft and support troops, the first Marines on Red Beach would have to land from LVTs which could ignore the reef and churn straight on to the beach from the line of departure. The 1st Marine Division would provide the tractors, the Seventh Fleet their transport and escort to the target, and the Army amphibian engineers all the rest of the landing craft needed. An Army officer, Lieutenant Colonel Robert Amory, Jr., would command all shipping during the movement to the objective and the landing.

The understrength company of LCVPs and LCMs that had supported BACKHANDER Force since D-Day could handle some portion of the load in the coming operation, but more engineer boats were needed. As early as 4 February, ALAMO Force alerted the Boat Battalion of the 533d Engineer Boat and Shore Regiment to a probable role in the coming operation. The 533d, a unit of the 3d Engineer Special Brigade, was newly arrived in the forward area and as yet untested in combat. Elements of the boat battalion headquarters, a boat company, a shore company, and a maintenance detachment, all under command of Lieutenant Colonel Amory, were detailed to the job. From Goodenough Island, the engineers and their equipment moved to Finschhafen, and, on 27 February, the advance echelon embarked in its own boats for the 85-mile run to Borgen Bay.

The soldier boatmen made their landfall late at night after a day-long passage through choppy seas, but they got no chance to explore the bivouac area, "in the least atrocious of the various swamps

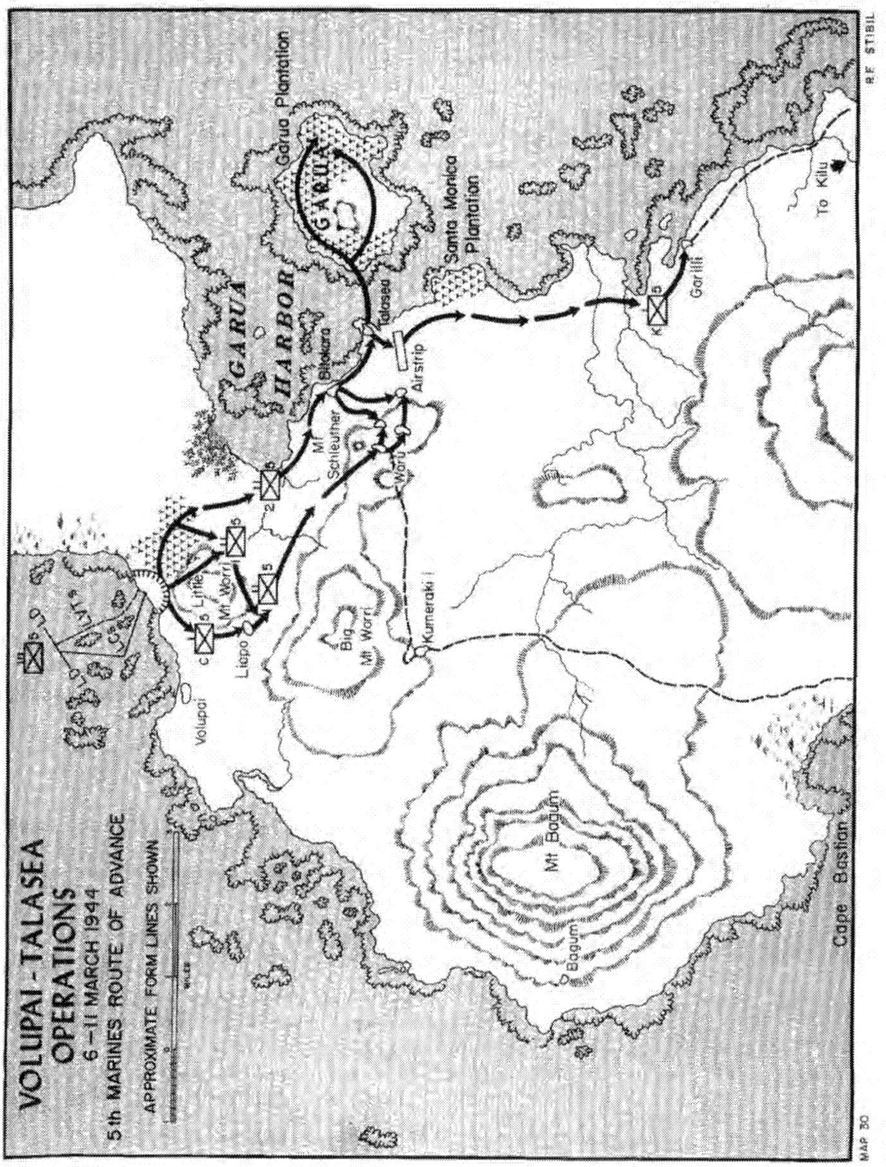
VOLUPAI-TALASEA OPERATIONS
6–11 MARCH 1944
5th MARINES ROUTE OF ADVANCE
APPROXIMATE FORM LINES SHOWN
GARUA HARBOR
GARUA
Garua Plantation
Santa Monica Plantation
Talasea
Bitokara
Airstrip
Mt Schleuther
Waru
Kumeraki
Big Mt Worri
Little Mt Worri
Liapo
Volupai
Mt Bagum
Bagum
Cape Bastian
Garilli
To Kilu
MAP 30
R.F. STIBIL

available," [5] which had been tentatively set aside for them. "Instead," the unit's history relates, "one of the worst 'rat races' of all times was to occupy every minute of every 24 hours for the next week." [6] This period of furious but ordered activity saw the movement of the 5th Marines and all its reinforcing units to Iboki, together with 20-days' supplies for the nearly 5,000 men of the APPEASE task force. Concurrently, the few dozen landing craft available had to be used to transport the 1st Battalion, 1st Marines to Iboki, where it could take over patrol missions from the 5th and be available as a reserve if the APPEASE operation demanded.

While the landing force assembled at its staging point, scouts tried several times to land on Willaumez Peninsula to determine the location and strength of enemy defending forces. Moving at night in torpedo boats, the men were turned back by high seas on one occasion and on another by a sighting of troops moving in the proposed landing area. Finally, early on 3 March, Australian Flight Lieutenant G. H. Rodney Marsland, who had managed Santa Monica plantation before the war, the 1st Division's chief scout, Lieutenant Bradbeer, and two natives landed near Bagum village about nine miles from Volupai. Setting up in the village, the party sent runners out to contact key natives known to Marsland and discover the Japanese dispositions. After nearly 24 hours ashore, the scouts withdrew with some useful information on the location and size of various enemy detachments, but they had surprisingly received no report of the major enemy concentration in the immediate Talasea area.[7]

Defending Willaumez Peninsula was a garrison of 595 men, some 430 of them concentrated in the vicinity of Talasea. A Japanese muster roll, completed at the same time the BACKHANDER scouts were ashore, agreed very closely with the information that was reported to the 5th Marines at Iboki. Volupai had only a rifle platoon and a machine gun squad to defend the beach, 28 men in all, but the bulk of the enemy force was within easy reinforcing distance. The Japanese, all under Captain Kiyomatsu Terunuma, commander of the *1st Battalion, 54th Infantry*, consisted of most of that unit,[8] plus the *7th Company* of the *2d Battalion, 54th*, the *9th Battery, 23d Field Artillery*, a platoon of machine guns, and a platoon of 90mm mortars. Terunuma's orders were to hold his positions north of Kilu and the Walindi Plantation and not to retreat without permission of the *17th Division* commander.

The 5th Marines had no indication that Red Beach was heavily defended; the natives reported that the area was not fortified, and aerial reconnaissance appeared to confirm this intelligence. Support for the APPEASE landing was therefore not overwhelming, but it was adequate for the job at hand. For three days prior to D-Day, Australian Beaufort squadrons based at Kiriwina Island flew bombing and strafing missions against targets in the Talasea and Cape Hoskins vicinity, and, on D-Day, the RAAF planes were to provide cover for the attack flotilla and to

[5] Amory and Waterman, *Surf and Sand*, p. 77.
[6] *Ibid.*, p. 78.

[7] 1stLt John D. Bradbeer Rept on Talasea Recon, dtd 4Mar44, in PhibRecon PtlRepts, Cape Gloucester and Talasea, 11Oct43–9Mar44.
[8] *Japanese comments.*

blast Red Beach ahead of the assault waves. To make up for the absence of naval gunfire support, the 1st Division came up with its own brand of gunboats—medium tanks in LCMs. Four Shermans were added to the platoon of light tanks attached to the 5th Marines to provide the necessary firepower. Tests of the novel means of shelling the beach were made at Iboki to make sure that the tanks could fire from their seagoing gun platforms. The accuracy of the practice firing with 75mm cannon was nothing to boast about,[9] but the makeshift gunboats proved to be practical.

The operation plan of the 5th Marines called for the 1st Battalion, embarked in LVTs, to land in assault and secure a beachhead line which passed through the edge of Volupai plantation. The 2d Battalion, following directly behind in LCMs and LCVPs, was to pass through the 1st's positions and attack up the trail to Bitokara to seize the Talasea area. Two batteries of 75mm pack howitzers of 2/11 were to follow the assault battalions ashore on 6 March to furnish artillery support for the attack across the peninsula. On D plus 1, the 5th's 3d Battalion and reserve elements of the regiment's reinforcing units would move to Red Beach in landing craft which had returned from Willaumez.

Shortly before the APPEASE operation got underway, the 5th Marines got a new commander, Colonel Oliver P. Smith, who had just reported to the division. Colonel Selden stepped up to division chief of staff, replacing Smith who held the position briefly following his arrival on New Britain. A number of experienced senior officers, including Colonel Amor L. Sims, who had been chief of staff, and Colonel Pollock, the D-3, had returned to the States in February to fill key assignments in the continuing build-up of the Marine Corps for the Pacific War. Two of the 5th's battalions also had comparatively new commanders. Major Gordon D. Gayle, who led 2/5, had taken over when Lieutenant Colonel Walt was promoted to regimental executive officer; Lieutenant Colonel Harold O. Deakin, who had the 3d Battalion, assumed command after the battle for Aogiri Ridge was successfully concluded.

The invasion convoy that assembled off Iboki on 5 March for the 57-mile-long run to Volupai included 38 LCMs, 17 LCVPs, 5 LCTs, and 5 MTBs. Each of the Navy's LCTs carried five tractors of Company B, 1st Amphibian Tractor Battalion, and the Marines of 1/5 who would ride them. The torpedo boats were under orders to escort the LCTs, as naval officers were dubious about risking their valuable landing craft in poorly charted waters without adequate communications or proper navigational guides. If anything, the engineer coxswains of Amory's command had more to worry about than the sailors, for their boats were a good bit more thin-skinned than the LCTs, and they would be moving at night through strange waters abounding in coral outcroppings.

Men of the 1st and 2d Battalions, 5th Marines and the regiment's various reinforcing elements began loading their boats at 1300, and, at 2200, the convoy departed, carrying more than 3,000 men and 1,000 tons of equipment. Lieutenant Colonel Amory later categorized his motley convoy as "probably the war's out-

[9] LtCol Rowland L. Hall comments on draft of Hough and Crown, *New Britain Campaign*, dtd 27Mar52.

standing example of overloading small boats," [10] but the movement to the target came off without a major hitch despite heavy rain squalls that struck shortly after midnight and continued for two hours.

There was one mishap en route with potentially serious consequences, but the lack of determined opposition at Red Beach negated its effect. The boat carrying the Army air liaison party attached to the landing force broke down early in the movement, and Major Gayle's boat, which was proceeding independently after a late start, took the crippled craft in tow. Gayle was reluctant to delay his own progress, but considered the liaison group's radios to be of vital importance in contacting supporting air units. As a result of his prudent action, the 2/5 commander arrived off Volupai after his battalion had begun landing, but its executive officer, Major Charles R. Baker, was fully in control.

The convoy of small craft arrived at its appointed place about 3½ miles off the coast at Volupai as dawn was breaking on D-Day. The LCTs closed slowly toward the reef as the Marines looked anxiously skyward for the planes which were supposed to be flashing in to hit the possible enemy positions at Red Beach. None of the RAAF Beauforts appeared, as their fields on Kiriwina were weathered in,[11] and the troops became more and more conscious of their exposure to an unknown enemy waiting on the silent shore. Lieutenant Colonel Amory, from his LCVP at the head of the line of landing craft, radioed Colonel Smith, "Shall we proceed despite air failure," and the landing force commander replied immediately, "Carry on." [12]

At 0825, on Amory's signal, the LCTs lowered their ramps and the LVTs of the first two waves roared into the water and on across the reef. As the tractors started toward the beach, Amory led a boat loaded with navigation buoys and two of the tank gunboats into the coral-free lane that aerial photos showed led to the beach. At the same time, on the opposite (left) flank of the line of departure, the other pair of tanks in LCMs started shoreward, keeping pace with the LVTs for as long as the irregular coral formations would permit. Standing on the bow of his boat, Amory with Flight Lieutenant Marsland at his side, conned the craft through the open water passage while the trailing LCVP dropped buoys to guide the third and succeeding waves of the landing force. The route that had to be used ran 45 degrees to the right of the path the tractors were following over the reefs until it got within 75 yards of the shore, where it "swung sharply to the left to coast six tenths of a mile just barely outside the overhanging trees to the beach at Volupai Plantation." [13]

The tanks opened fire with their machine guns to cover the approach of the

[10] Amory and Waterman, *Sand and Surf*, p. 84. The regiment was fully aware of the overloading but had no alternative. Since there was no follow-up shipping, "resupply could only be accomplished by returning the landing craft used in the assault to Iboki to reload. The Landing Force could, therefore, expect no additional supplies for over 24 hours after landing. The supplies accompanying the Landing Force were increased accordingly." Gen Oliver P. Smith memo to HistBr, G-3, HQMC, dtd 7Jun62.

[11] Craven and Cate, *Guadalcanal to Saipan*, p. 344.

[12] LtCol Robert Amory, Jr., MassNatGrd, ltr to CMC, dtd 25Mar52, including extract from his personal diary for 6Mar44.

[13] *Ibid.*

tractors to the beach, reserving their 75s for any Japanese return fire. The spray of bullets from all the American landing craft, for the LCVPs and the LVTs fired as well, was finally answered by a few scattered shots from the featureless jungle, and then mortar shells began falling amidst the oncoming tractors. At this moment, the assault troops got all the close air support that the 5th Marines received on D-Day, and that from an unexpected source—a Piper Cub circling overhead, with Captain Petras as pilot and Brigadier General David A. D. Ogden, Commanding General, 3d Engineer Special Brigade, as observer. Petras, when he saw the shell bursts, turned his tiny plane in over the tree-tops and started dropping the 25–30 hand grenades he carried on "any of the spots where it looked like there might be some Japs." [14] The results of the impromptu bombing were never checked, but the gallant effort drew considerable praise from the men who witnessed it.

At 0835, the LVTs crawled onto the beach, and the Marines of 1/5's assault platoons clambered over the sides and began advancing cautiously inland. There was little opposition from enemy infantry at first, but mortar rounds continued to fall, with most of them hitting out in the water where the columns of landing boat waves were beginning to thread their way through the buoyed channel. After Lieutenant Colonel Amory waved the tank-LCMs in for a landing, he proceeded out into the narrow passageway to act as control officer to keep the boats from swamping the limited capacity of the beach. After untangling a snarl of landing craft that occurred when the third wave tried to follow the tractors rather than the marked passage, the Army officer spent the morning directing traffic at the corner where the channel turned to parallel the shore.

Major Barba's two assault companies had little difficulty reaching their assigned objective. Company B encountered the only resistance, a small pocket of enemy riflemen it wiped out as it skirted the edge of the swamp that came right up to the northern edge of the Volupai-Bitokara track. Following his orders, Barba established a beachhead perimeter 200 yards inland and dispatched combat patrols to the flanks of his position as he waited for 2/5 to land and pass through his lines.

Immediately after the first tractors touched down on dry land, a reinforced platoon of 1/5 was sent up the slopes of Little Mt. Worri to eliminate an enemy machine gun nest. The existence of the position had been disclosed to the Marsland-Bradbeer scouting party by the natives. When the patrol found the emplacement, which commanded a good field of fire on the beach, it was abandoned. Pushing on through the thick undergrowth, the Marines sighted and engaged a group of enemy soldiers carrying a machine gun down the mountain toward Volupai Plantation.[15] Japanese resistance stiffened appreciably as the 1/5 patrol neared the coconut groves; in the exchange of fire the Marines accounted for a dozen enemy and lost one man killed and another wounded. The patrol leader called for another platoon to help him destroy the position he had developed, but

[14] Maj Theodore A. Petras interview by HistDiv, HQMC, dtd 11Apr50, p. 7.

[15] Col William H. Barba comments on draft of Hough and Crown, *New Britain Campaign*, dtd 24Mar52.

was given orders to hold up where he was until the 2d Battalion arrived on the scene.

The limited area for maneuver directly behind Red Beach and the narrow passage that had to be used to reach the shore slowed unloading appreciably. When Major Gayle reached the seaward end of the channel at 1230, reserve elements of his battalion were still landing, as were the firing batteries of Major Noah P. Wood, Jr.'s 2/11, carried in the LCMs of the last two waves. Company E of 2/5 had already passed through the 1st Battalion's lines at 1100 and run up against the enemy strongpoint located by the 1/5 patrol earlier in the morning.

Three of the medium tanks which had furnished the naval gunfire for the assault landing came up the trail to support Company E in its attack; the fourth tank was bogged down in soft ground at the beach. When the lead Sherman opened fire on the Japanese, it quickly silenced a machine gun that had been holding up the infantry. Then, as the big machine ground ahead on the mud-slickened track, enemy soldiers leaped out of the brush on either flank and attempted to attach magnetic mines to its sides. One man was shot down immediately by covering infantry; the other succeeded in planting his mine and died in its resulting blast, taking with him a Marine who had tried to stop the contact. The mine jammed the tank's turret and momentarily stunned the crew; luckily, the Marines inside also escaped injury from an antitank grenade that hit and penetrated the turret at about the same time. The damaged tank pulled off the trail to let the following armor come through and lead Company E's assault. Later, when the tank attempted to move on up the trail, it exploded a mine that smashed one of its bogie wheels. The presence of land mines resulted in a hurry-up call to division for detectors.

Rooting out the enemy from his trenches and emplacements on the edge of the coconut plantation, the tank-infantry team crushed the opposition and moved ahead with 1/5's 81mm mortars dropping concentrations on any likely obstacle in the way. As Gayle took command of the attack, the Marines had a much clearer idea of what they were going to run up against. At the height of the battle for the enemy position, a map showing the defenses of the Talasea area had been found on the body of a Japanese officer, and, as happened so many times on New Britain, intelligence indoctrination paid off. The document was immediately turned in, not pocketed as a souvenir, and, by 1300, the regimental intelligence section was distributing translated copies.

Once it had passed the Japanese defenses near the beach, the 2d Battalion, with Company E following the track and Company G moving along the mountain slopes on the right flank, made rapid progress through the plantation. At about 1500, five P–39s of the 82d Reconnaissance Squadron at Cape Gloucester flew over the peninsula, but could not locate the Marine front lines, so they dropped their bombs on Cape Hoskins instead.[16] At dusk, Major Gayle ordered his two assault companies to dig in for all-around defense,

[16] 1st MarDiv D–2 Jnl, entry at 1932, 6Mar44. Although this message, which calls the planes P–38s, does not denote the squadron or its location, the operation plan states that four "P–39s" would be on standby at Gloucester. Army Air Forces squadron records at Maxwell Field show that the 82d, which was staging at Cape Gloucester, bombed enemy installations at Cape Hos-

while he set up with his headquarters and reserve in a separate perimeter at the enemy position that had been reduced during the day's fighting. The 2d Battalion, 11th Marines had registered its batteries during the afternoon, and the pack howitzers now fired harassing missions through the night to discourage any counterattacks from forming.

The artillery had taken a beating from the Japanese mortars during the day, as the enemy 90mm shells exploded all over the crowded beach area. Some of the 75s had to set up almost at the water's edge so that they could have an unmasked field of fire. Corpsmen going to the aid of artillerymen who were hit while unloading and moving into position all too often became casualties themselves. During the action on 6 March, the 5th Marines and its supporting elements lost 13 men killed and 71 wounded; 9 of the Marines who died and 29 of those who were wounded were members of 2/11. Fifty of the regiment's seriously wounded men were loaded in an LCM and sent back to Iboki at 1830. The toll of counted enemy dead was 35; if there were Japanese wounded, they were evacuated by the elements of the *Terunuma Force* that had pulled back as 2/5 advanced.

Colonel Smith could count his regiment well established ashore at the end of D-Day. In the next day's attack, he planned to keep pressure on the Japanese, not only along the vital cross-peninsula track, but also in the mountains that overshadowed it. Captain Terunuma, in his turn, was ordering the moves that would stave off the Marine advance and protect the elements of the *Matsuda Force* which were just starting to cross the base of the peninsula.

SECURING TALASEA [17]

To reinforce the shattered remnants of the small Volupai area garrison which had made a hopeless attempt to stem the Marine advance, Terunuma sent his *4th Company*, reinforced with machine guns and mortars, "to check the enemy's attack." [18] At dawn on 7 March, patrols from 2/5 scouting the trail to Bitokara found the Japanese dug in not 50 yards from the Marines' forward foxholes. When Major Gayle's assault company (E) led off the morning's attack, the enemy entrenchments were found deserted. One of the deadly 90mm mortars was captured intact with shellholes from 2/11's fire as close as five feet to its emplacement. Passing swiftly through the abandoned hasty defenses, the 2d Battalion pressed on towards the coast with patrols ranging the foothills that dominated the trail. As Gayle's advance guard neared Mt. Schleuther just before noon, elements of the *4th Company, 54th Infantry* swept the track clear with a deadly concentration of fire. It was soon evident that the Japanese force, which was holding a position on the northwest slopes of the mountain, was too strong to be brushed aside. The pattern of enemy fire, in fact, showed that the Japanese were

---

kins between 1345 and 1430. The 80th Fighter Squadron, part of Gloucester's regular complement, filled in for the missing RAAF planes and dispatched P-38s over Talasea at 1055, 1110, and 1535; none of the Lightnings attacked ground targets. Dr. Robert F. Futrell, USAF HistDiv, ltr to Head, HistBr, G-3, dtd 19Jun62.

[17] Unless otherwise noted, the material in this section is derived from: *APPEASE SAR; 17th Div Ops;* Hough and Crown, *New Britain Campaign.*

[18] ATIS Item No. 10,441, Talasea Garrison Unit OpO No. 42, dtd 7Mar44, in ATIS Bul No. 881, dtd 3Apr44 (ACSI Recs, FRC Alex).

moving to outflank the Marines below and cut them off from the rest of the battalion column.

While Company E built up a strong firing line amid the dripping undergrowth along the trail, Major Gayle sent Company F directly up the steep slopes to hit the enemy troops trying to push west. Moving forward behind artillery and mortar supporting fires, the 2d Battalion Marines beat the Japanese to the dominant ground in the area and drove the losers back with heavy casualties. Coming up on the extreme right of Company F's position, a supporting weapons platoon surprised a machine gun crew setting up, wiped out the luckless enemy, and turned the gun on the retreating Japanese. As the firing died away, the bodies of 40 enemy soldiers testified to the fury of the action. For night security, 2/5 organized a perimeter encompassing its holding on Mt. Schleuther and the track to Bitokara.

Colonel Smith's attack on 7 March was to have been a two-pronged affair with 2/5 moving along the main trail and 1/5 heading into the mountain mass toward the village of Liapo behind Little Mt. Worri and then east to the Waru villages, believed to be the center of enemy resistance. The plan depended upon the 3d Battalion's arrival early on D plus 1 to take over defense of the beachhead. Lieutenant Colonel Deakin had orders to board the landing craft returning from Volupai and make a night passage to Red Beach to be on hand at daybreak to relieve the 1st Battalion. General Rupertus, who was present at Iboki, countermanded the order and directed that no boats leave until after dawn on 7 March. The result of this unexpected change of plans, made to lessen the risk in transit through uncharted waters, was that 3/5 arrived at Volupai late in the afternoon.

After it became clear the reserve would be delayed, a reinforced company of the 1st Battalion was started inland for Liapo to pave the way for the next day's operations. When the trail disappeared in a clutter of secondary growth, the company hacked its way onward on a compass course, but ended the tiring advance some distance short of the target. The approach of darkness prompted the isolated Marine unit to set up in perimeter defense. The night passed without incident.

On 8 March, Major Barba's battalion, moving to the east of Little Mt. Worri, started toward Liapo along two separate paths. Unfortunately, a native guide, dressed in cast-off Japanese clothing, leading a column headed by Company B, was mistaken for the enemy by a similar Company A column. In the brief outburst of fire that followed the unexpected encounter, one man was killed and several others wounded. Shortly afterwards, near Liapo, the battalion found the east-west trail it was seeking and joined the company that had spent the night in the jungle. Although it encountered no Japanese opposition as it moved towards its objective during the remainder of the day's advance, 1/5 found the rugged terrain a formidable obstacle. The men climbed and slid through numerous ravines and beat aside the clinging brush that often obscured the trace of the path they were following. At nightfall, Barba's units set up in defense a few hundred yards short of their goal.

Major Gayle's battalion had a hard time with both enemy and terrain on 8 March, but not until it had seized a sizeable chunk of the regiment's primary objective. A patrol out at daybreak found the last known position of the Japanese manned only by the dead: 12 soldiers, most of them victims of American artillery and mortar fire. Another patrol discovered a sizeable enemy force 500 yards ahead at Bitokara, and Gayle readied a full-strength attack. When the battalion jumped off, with assault platoons converging from the foothills and the track, it found that the Japanese had pulled out again. Bitokara was occupied early in the afternoon, and scouts were dispatched along the shore to Talasea. Again there was plenty of evidence of recent enemy occupation but no opposition. Taking advantage of the situation, Gayle sent Company F to occupy Talasea airdrome nearby.

Scouts who climbed the slopes of Mt. Schleuther, which looked down on Bitokara, soon found where some of the missing Japanese had gone. A well-intrenched enemy force was located on a prominent height that commanded the village, and Gayle made preparations to attack. At 1500, Company E, reinforced with heavy weapons, drove upward against increasing resistance. The fire of a 75mm mountain gun and a 90mm mortar, added to that from rifles and machine guns, stalled the Marines. After an hour's fighting, during which the company sustained 18 casualties, Gayle ordered it to withdraw to Bitokara for the night. When the Japanese began to pepper the village with 75mm rounds, concentrating their shelling on 81mm mortar positions near the battalion command post, Gayle called down artillery on the height. The American howitzers and mortars continued to work over the enemy position all night long.

At 0800 on the 9th, a coordinated attack by companies of both assault battalions was launched to clear Mt. Schleuther and capture the Waru Villages. The artillery and mortar concentrations that preceded the jump-off and the powerful infantry attack that followed hit empty air. One dead soldier and two stragglers were all that was left of the defending force that had fought so hard to hold the hill position on the previous afternoon. The prisoners stated that the main body of the *Terunuma Force* had moved south down the coast on the night of 7 March, leaving a 100-man rear guard to hold off the Marines. This last detachment had taken off in turn after beating back 2/5's attack on the 8th.

Patrols of the 5th Marines searched Garua Island and the entire objective area during the rest of the day and confirmed the fact that the Japanese had departed, leaving their heavy weapons behind. Colonel Smith moved the regimental command post to Bitokara during the afternoon and disposed his 1st Battalion around the Waru villages, the 2d at Talasea and the airdrome, and left the 3d to hold Red Beach. He then informed division that Talasea was secured, and that the 5th Marines would patrol Willaumez Peninsula to clear it of the enemy.

The end of Japanese resistance in the objective area gave the Marines use of an excellent harbor and brought a welcome end to use of Red Beach. Colonel Smith directed that all supply craft would land their cargoes at Talasea from 9 March onward. Marine pioneers and Army shore party engineers were ordered to improve the Volupai-Bitokara track enough to enable them to move all supplies and equipment to Talasea; the job took three days of hard work. The track was deep in mud from the effects of rain and heavy traffic

691-360 O—63——28

as evidenced by three medium tanks trapped in its mire.

Amphibian tractors had again proved to be the only vehicles that could keep up with the Marine advance; 2/5 used the LVTs as mobile dumps to maintain adequate levels of ammunition and rations within effective supporting distance. When the occasion demanded, the tractors were used for casualty evacuation, too. Although the ride back to the beach was a rough one for a wounded man, it was far swifter than the rugged trip that faced him with stretcher bearers struggling through mud.

The cost of the four-day operation to the 5th Marines and its reinforcing units was 17 men killed and 114 wounded. The Japanese lost an estimated 150 men killed and an unknown number wounded. The fighting had been sporadic but sharp, and Captain Terunuma had engaged the Marines just enough to earn the time that the *Matsuda Force* remnants needed to escape. The retirement of the *Terunuma Force* was deliberate; at Garilli, four miles south of Talasea, the Japanese halted and dug in to await the Marines.

MOP-UP PATROLS [19]

When General Sakai issued his withdrawal orders for the various elements of the *17th Division*, he was quite anxious to recover the 1,200-man garrison of Garove Island. Until the Marines landed at Volupai, he was stymied in this wish by *Eighth Area Army's* desire to keep the island in use as a barge relay point. Once the Americans had established themselves on Willaumez Peninsula, Garove was no longer of any value to the Japanese.[20] Resolving to risk some of the few boats and landing craft that he had left to evacuate the garrison, Sakai ordered the *5th Sea Transport Battalion* to load as many men as could crowd aboard the three fishing vessels and the one sampan available and sail for Ulamona. (See Map 31.)

The jam-packed boats, carrying about 700 men, left Garove shortly after midnight on 6 March and reached Ulamona unscathed the next afternoon. On their return voyage that night, the boats were intercepted by American torpedo boats and sunk. Immediately, the commander of the *8th Shipping Engineers*, who was holding three large landing craft in reserve for this purpose at Malalia, sent them directly to the island to bring off the remainder of the garrison. The craft were discovered by torpedo boats and sunk after a running gun battle. Despite these losses, three more landing craft were sent to the island where they picked up 400 men and escaped to Rabaul without encountering any of the deadly torpedo boats or being spotted by Allied aircraft.

The waters around Willaumez Peninsula became increasingly unhealthy for the Japanese as the APPEASE Operation wore on. On the 9th of March, while an LCT carrying supplies to Talasea rounded the northern end of the peninsula, it sighted and shot up four barges lying ill-hidden amidst the overhanging foliage on

[19] Unless otherwise noted, the material in this section is derived from: *APPEASE SAR; 17th Div Ops; 51st ReconRegt Diary; Komori Diary;* Hough and Crown, *New Britain Campaign*.

[20] The 3d Battalion, 1st Marines was alerted for a possible landing on Garove, but scouts who went ashore there on 7 March discovered signs that a considerable garrison was still present and heavily armed. The 1st Division then abandoned the plan to seize the island, figuring the objective not worth its probable cost in casualties.

the shore. On the same day, LCMs ferrying Marine light tanks, which were a more seaworthy load than the Shermans, encountered a Japanese landing craft and drove it ashore with a torrent of 37mm canister and .30-caliber machine gun fire. The terrier-like torpedo boats were the major killers, however, and, after a patrol base was established in Garua Harbor on 26 March, the northern coast of New Britain as far east as Gazelle Peninsula was soon swept clean of enemy craft.

Before the torpedo boats were out in force, however, the Japanese managed to evacuate a considerable number of men from staging points at Malalia and Ulamona. Except for Colonel Sato's rear guard, most elements of the *17th Division* had reached Cape Hoskins by the end of March. Issuing parties, in the carefully laid-out ration depots along the coastal trail, doled out just enough food to keep the men moving east, and then folded up as the units designated for rear guard at Malalia, Ulamona, and Toriu came marching in.

The only good chance that the Marines at Talasea had of blocking the retreat of General Matsuda's depleted command was canceled out by the skillful delaying action of Captain Terunuma. A reconnaissance patrol discovered the Japanese position at Garilli on the 10th, and its destruction was a part of the mission given Company K, 3/5, when it moved out toward Numundo Plantation on the 11th. The company found that the enemy force had abandoned Garilli but was set up along the coastal trail about three miles farther south. The small-scale battle that ensued was the first of many in the next four days, as the Japanese blocked the Marines every few hundred yards and then withdrew before they could be badly hurt. An enemy 75mm gun, dragged along by its crew, helped defend the successive trail blocks and disrupted a number of attempts to rush the Japanese defenses.

Late on the 16th, Company K reached Kilu village and tangled with the *Terunuma Force* for the last time. As the Marines and the Japanese fought, an LCM carrying Lieutenant Colonel Deakin and an 81mm mortar section came through the fringing reef. The enemy artillery piece fired on the landing craft, but failed to score a hit. The arrival of the 81s appeared to turn the trick. As soon as the mortars started firing, Captain Terunuma's men broke contact and faded away.

On the 18th, Company K reached Numundo, and, on the 25th, the whole 2d Battalion outposted the plantation. Five days later, Major Gayle moved his unit to San Remo Plantation about five miles to the southeast and began patrolling west to the Kulu River and east as far as Buluma, a coastal village halfway to Hoskins airdrome. On the peninsula, 1/5 set up a strong ambush at Garu on the west coast. All units patrolled extensively, making repeated visits to native villages, checking the myriad of side trails leading off the main tracks, and actively seeking out the Japanese. Many stragglers were bagged, but only one organized remnant of the *Matsuda Force* was encountered in two months of searching. The task of destroying this unit, what was left of Colonel Sato's reconnaissance regiment, fell to 2/1.

The 1st Marines had the responsibility of maintaining a patrol base at Iboki and spreading a network of outposts and ambushes through the rugged coastal region. In mid-March, Marine and Army patrols both made the trip between Arawe and the north coast, discovering and using the trails that the Japanese had followed. Sick and emaciated enemy fugitives were occasionally found, but the signs all pointed to the fact that those who could walk were now east of this once important boundary.

Only a few engineer boats could be spared from resupply activities for patrol work. These LCMs and LCVPs were used to transport strong units, usually reinforced platoons or small companies, to points like Linga Linga Plantation and Kandoka where the Japanese had maintained ration dumps. The lure of food was irresistible for the starving enemy troops stumbling through the jungle, and the Marines took advantage of the certainty that the Japanese would at least scout the vicinity of places where they had counted on finding rations.

Near Kandoka on 26 March, Colonel Sato's advance party ran head-on into a scout platoon from 2/1. The Japanese caught the Marine unit as it was crossing a stream and cut loose in a blaze of rifle and machine gun fire. For three hours, the Americans were pinned down on the stream banks before the arrival of another 2/1 patrol enabled them to withdraw; one Marine was killed and five wounded. Sato, who had several of his own men wounded in the fight, made the decision to bypass Kandoka and cut through the jungle south of the village. He ordered his men to strip themselves of everything but their weapons and ammunition for the push into the swamps at the base of Willaumez. On the 27th, Marine attempts to locate the Japanese column were unsuccessful, but the days of the *51st Reconnaissance Regiment* were numbered.

While the head of Sato's column was nearing Kandoka, the tail was sighted at Linga Linga by one of the 1st Division's Cubs. The pilot, again the busy Captain Petras, was scouting a suitable landing beach for a large 2/1 patrol. After drawing a map that located the Japanese, he dropped it to the patrol and then guided the engineer coxswains into shore at the rear of the enemy positions. This time the Marines missed contact with Sato's force, but with the aid of the landing craft they were able to move freely along the coast as additional sightings pinned down the location of the enemy unit.

On 30 March, a small Marine patrol sighted Sato's rear guard, 73 men accompanied by the redoubtable enemy colonel himself, who was by this time a litter patient. Major Charles H. Brush, 2/1's executive officer, who was commanding patrol activities in the region, reacted quickly to the report of his scouts. Leaving a trail block force at Kandoka, he plunged into the jungle with the rest of the men available, roughly a reinforced platoon, to engage the Japanese. Shortly before Brush reached the scene, a six-man patrol under Sergeant Frank Chilek had intercepted the Sato column and brought it to a halt by sustained rifle fire. When Brush arrived, Chilek's unit was pulled back by his platoon leader, who had come up with a squad of reinforcements, so that the stronger Marine force would have a clear field of fire. The Japanese were wiped out, and, miraculously, not a Marine was harmed in the brief but furious battle. At least 55 Japanese were killed in one 100-yard stretch of trail, including Colonel Sato, who died sword in hand, cut down by Chilek's patrol.

A few elements of the enemy reconnaissance regiment, those near the head of its column, escaped from the battle on 30 March. Without the inspiration of Sato, however, the survivors fell apart as a unit and tried to make their way eastward as individuals and small groups. Most of these men died in the jungle, victims of malnutrition, disease, and the vicious terrain; others were killed or captured by Marine patrols and outposts.

The 2d Battalion, 5th Marines, operating out of its base at San Remo, accounted for many of these stragglers who blundered into ambushes set on the trails that led to Cape Hoskins. Major Komori, who had led the quixotic defense of an airfield no one wanted, was one of those who met his end in a flurry of fire at a 2/5 outpost.

The major, wracked with malaria, had fallen behind his unit, and, accompanied only by his executive officer and two enlisted men, had tried to struggle onward. On 9 April, a Marine outguard killed Komori and two of his party; the sole survivor was wounded and captured. The death of Major Komori brought to a symbolic end the Allied campaign to secure western New Britain.

Conclusion

RELIEF IN PLACE[1]

Navy and Marine Corps leaders were seriously concerned about the retention of the 1st Marine Division in the Southwest Pacific Area and particularly about its employment in a role that did not take full advantage of its training and experience.[2] Veteran amphibious divisions were scarce throughout the Pacific, and officers of the naval service felt that the shore-to-shore operations which General MacArthur had projected for the remainder of 1944 could be handled well by units that had not made a specialty of amphibious assault. In sharp contrast, the capture of the island targets of Admiral Nimitz' Central Pacific drive demanded trained amphibious divisions. To spearhead its long overwater advances and the unavoidable fierce contests to win secure beachheads, the Navy wanted Marine assault troops.

The allocation of troops to seize various Pacific objectives rested with the Joint Chiefs of Staff. In Washington, the Commandant of the Marine Corps worked through Admiral King to get the 1st Marine Division back under naval command and employed to its full amphibious capability.[3] Neither General Vandegrift nor General Rupertus was convinced that pursuit of the remnants of the Japanese garrison of western New Britain was a task that made the best use of the division.[4] General MacArthur was reluctant to release the Marine unit to Pacific Ocean Areas' command, however, until operations to seize Kavieng and further isolate Rabaul were concluded.

During March, at a series of conferences in Washington attended by key representatives of both MacArthur's and Nimitz' staff, the conflicting points of view regarding the relative strength to be employed in the Central and Southwest Pacific offensives were aired. On the 12th, the Joint Chiefs directed CinCSWPA to complete the isolation of Rabaul with a minimum of forces and to bypass Kavieng, while he made his main thrust west along the New Guinea coast toward the Philippines. CinCPOA was ordered to seize positions in the southern Marianas in June and then to move on to the Palaus in Sep-

[1] Unless otherwise noted, the material in this section is derived from: *GHQ G-3 Jnl*, Mar-Apr-44; Isely and Crowl, *Marines and Amphibious War*.

[2] In a marginal comment to a letter from General Rupertus, informing him that the 1st Marine Division might have to stay on New Britain for a considerable period of time, General Vandegrift noted: "Six months there and it will no longer be a well-trained amphibious division." MajGen William H. Rupertus ltr to LtGen Alexander A. Vandegrift, dtd 4Feb44 (Vandegrift Personal Correspondence File, HQMC).

[3] LtGen Alexander A. Vandegrift ltr to LtGen Holland M. Smith, dtd 15Mar44 (Vandegrift Personal Correspondence File, HQMC).

[4] *Ibid.*; MajGen William H. Rupertus ltr to LtGen Alexander A. Vandegrift, dtd 18Feb44 (Vandegrift Personal Correspondence File, HQMC).

tember. The 1st Marine Division was to be returned to Nimitz' control for employment as an assault division in the Palaus operation.

The JCS left the negotiations regarding the actual redeployment of the 1st Division up to the two senior Pacific commanders. On 31 March, Nimitz radioed MacArthur asking that the division be disengaged as soon as possible and withdrawn to a base in the Solomons. In reply, the general stated that he thought that the division should not be relieved until late June and that when the relief took place it would require extensive use of amphibious equipment since there were no docks at Cape Gloucester. MacArthur indicated that in view of prospective operations in his area such equipment was not available to accomplish the relief.

On 6 April, both Marshall and Nimitz reminded MacArthur of the intended employment of the 1st Marine Division in the Palaus operation, and Nimitz stated that the division would have to be released prior to 1 June in order to "have ample time to prepare for participation in a major amphibious assault."[5] At the same time, Admiral Halsey was asked to determine to what extent his South Pacific Force could support the movement of troops involved. The Pacific Ocean Area's commander pointed out further that the timing and success of operations in the Palaus depended upon the planned use of the 1st Division, and that any delay in the completion of the campaign would "cause corresponding delays in the readiness of the Pacific Fleet"[6] to support MacArthur's operations.

[5] CinCPOA msg to CinCSWPA, dtd 6Apr44, in *GHQ G-3 Jnl*, 9Apr44.

[6] *Ibid.*

The reaction to the messages from Pearl Harbor and Washington was swift. By 8 April, arrangements had been made to relieve the 1st Marine Division with the 40th Infantry Division which was stationed on Guadalcanal. The movement was to be made in two echelons using transports belonging to Halsey's force. To speed the transfer and ease cargo space requirements, the two divisions were directed to exchange in place all equipment that was common to both or could be reasonably substituted therefor. The first elements of the 40th Division to arrive at New Britain would be utilized to relieve the Marines deployed in the Iboki and Talasea areas.

MARINE WITHDRAWAL[7]

Before the Army relief arrived, the 1st Marine Division had begun the inevitable aftermath of a combat operation—a new training cycle. On 17 March, General Rupertus issued a directive to all units at Cape Gloucester outlining a seven-week program of individual and organizational training which laid emphasis on firing practice and tactical exercises from the squad through the regiment. Colonel Smith was ordered to start a similar program for the 5th Marines at Talasea as soon as his situation permitted.

Word of the division's pending departure for the Solomons brought the training schedule to an end, but not before an amphibious reconnaissance school

[7] Unless otherwise noted, the material in this section is derived from: MIS, GHQ, FEC, Ops of the Allied IntelBu, GHQ, SWPA—v. IV, Intel Series, dtd 19Aug48; 40th InfDiv, Hist of BACKHANDER Op 28Apr–27Nov44, n.d. (WW II Recs Div, FRC Alex); *1st MarDiv Mar–May44 War D*; Hough and Crown, *New Britain Campaign*.

graduated a class well-versed in the techniques painstakingly acquired by Lieutenant Bradbeer and the other veteran division scouts. Before the graduates returned to their units, they took part in an actual scouting mission to check the landing beaches and the airdrome on Cape Hoskins. On 13 April, a 16-man patrol landed from LCMs about 5,000 yards west of the enemy airfields and started for the objective. The scouts were split into three parties, one followed the shoreline, another the coastal trail, and the third circled inland.[8] As it approached Cape Hoskins, the center party ran into a Japanese ambush bristling with mortars and machine guns. Despite the enemy fire and a close pursuit, the various elements of the patrol were able to shake loose from contact, get back through the jungle to their rendezvous point, and withdraw without incurring any losses. The next American reconnaissance to Cape Hoskins was made on 7 May, and by the Army, but the Japanese encountered by the Marines, evidently a rear guard, had departed in the general retreat to Rabaul.

Major General Rapp Brush, commanding the 40th Division, flew to Cape Gloucester on 10 April to arrange for the relief in place of the 1st Marine Division by elements of his own unit. The first echelon of the Army division, principally the 185th Infantry and its reinforcing units, reached the cape on 23 April. On the following day, the 1st Marines and detachments from a number of division supporting units boarded the transports that had brought the soldiers and sailed for the South Pacific. The 185th, at the same time, crowded into engineer landing craft at Borgen Bay and sailed for Iboki. Stopping overnight at the plantation, the Army regiment moved on at dawn, leaving behind a platoon to replace a like detachment of the 5th Marines. On the 25th, the soldiers reached Willaumez and the Army commander took over responsibility for the area from Lieutenant Colonel Buse, who had taken command of the 5th Marines when Colonel Smith was promoted and returned to Cape Gloucester to become ADC.[9] The Marine regiment and its attached units boarded the LCMs and LCVPs that had brought their welcome relief force and started back for Cape Gloucester the following evening.

The remainder of the 40th Division arrived on 28 April, and General Rupertus turned over command of the BACKHANDER Force to General Brush. While Captain Petras flew the Marine leader back to the Solomons, the second echelon of the 1st Division loaded its gear and sailed. On 4 May, when the ships that had transported the first echelon returned, the last elements of the division departed. Only one Marine unit, the 12th Defense Battalion, remained at Cape Gloucester, but it too, was relieved later in the month when an antiaircraft artillery group arrived to take its place.[10]

The last group of ships returning to the Solomons joined LSTs carrying Compa-

[8] MajGen Oliver P. Smith ltr to CMC, dtd 31Mar52.

[9] Gen Oliver P. Smith memo to HistBr, G–3, HQMC, dtd 7Jun62.

[10] At one point in the relief of units, General MacArthur intended to retain the 19th Naval Construction Battalion in the Southwest Pacific, but Admiral Nimitz pointed out that the unit, serving as the 3d Battalion, 17th Marines, was organic to the 1st Division. ComSoPac disp to CinCPOA, dtd 14Apr44, and CinCPOA disp to CinCSWPA, dtd 17Apr44, in *GHQ G–3 File*, 18Apr44.

nies A and B, 1st Tank Battalion, and personnel of the division rear echelon who had closed out the Marine supply dumps on New Guinea. Company B had been released from the DIRECTOR Force in mid-April and sent to Finschhafen in anticipation of the 1st Marine Division's withdrawal. Company A, which had the only medium tanks available in the forward area, had been alerted for action in the Admiralties and was actually employed on 22 April as a part of the assault forces at Tanahmerah Bay in the Hollandia operation. A large swamp and a precipitous mountain range immediately behind it prevented the Marine tanks from moving inland, and while the Army infantrymen advanced, the Marines "sat on the beach, fished, and were eventually loaded aboard ship again." [11]

When the 1st Marine Division, reinforced, added up the cost of its four-month campaign on New Britain, the casualty total read 310 men killed or died of wounds and 1,083 wounded in action. The figures could easily have been higher had the operation not been well planned and skillfully executed by veteran troops. When General Rupertus relinquished command of BACKHANDER Force to General Brush, the toll of enemy killed and captured stood at 4,288 and 420, respectively.[12] In postwar calculations, a senior staff officer of *Eighth Area Army* reckoned the Japanese loss in the fighting at Arawe and Cape Gloucester and in the withdrawal to Rabaul at 3,868 killed and died of wounds.[13] It is probable that the actual total of enemy killed lay somewhere between the claim and the recollection.

The 40th Infantry Division had its first clash with the Japanese as soon as it relieved the 1st Division's advance posts. Following his orders from ALAMO Force, General Brush kept pressure on the *17th Division* stragglers and mopped up the few enemy troops that remained alive west of Rabaul. On 7 May, patrols of the 185th Infantry occupied Cape Hoskins airdrome and found the area mined but deserted. In June, a regiment of the 40th relieved the DIRECTOR Force at Arawe, and, in October, units of the division occupied Gasmata. Late in November, the Australian 5th Division relieved the 40th in its positions on New Britain, and the American unit moved west to take part in MacArthur's attack on Leyte.

In its seven months of active patrolling, the 40th Division killed 31 Japanese soldiers and took a mere 18 prisoners, proof enough that the enemy had successfully withdrawn his troops to Rabaul. Close to a thousand enemy were accounted for by natives roaming the jungle that ringed the Japanese stronghold. For self defense, the coastwatchers who manned the observation posts on Gazelle Peninsula and in the jungles to the east had to arm selected natives. An initial air drop of 100 riot guns and ammunition was made on 21 February and proved so worthwhile an idea that it was followed up repeatedly and to such effect that nothing short of all-out enemy retaliation sweeps could have stopped the slaughter. In time, the specter of bushy-haired Melanesians armed with shotgun and knife lying in ambush along every trail put a severe crimp in the aggressiveness of Japanese patrols ranging out from Rabaul.

---

[11] Capt Howard R. Taylor ltr to HistDiv HQMC, dtd 6Jul51.

[12] *1st MarDiv Apr44 WarD*, entry of 28Apr44.

[13] *Eighth Area Army Ops*, p. 197.

CAMPAIGN APPRAISAL[14]

Many serious students of the Pacific War have questioned the selection of Cape Gloucester and Arawe as Allied objectives. In most cases, too little heed has been given to the commander's responsibility to approve operations that are within the reasonable capabilities of his forces. Observers who recognize that the men and munitions available to MacArthur and Nimitz were stretched thinly, argue that the Allies might have made bolder use of limited resources. The conclusion is inescapable that such judgments are based on a knowledge of results.

If, in retrospect, the landing at Saidor seems to have been a wiser move than that at Cape Gloucester, it should not be forgotten that the one was contingent upon the success of the other. If now Arawe's seizure appears to have been a fruitless effort, it did not appear so at the time to many responsible and intelligent men. Away from the pressure of war, it is not hard to see that many of the operations undertaken to reduce Rabaul were unnecessary. At the time, however, DEXTERITY objectives were vital in the opinion of the men who chose them.

In many respects, the seizure of Cape Gloucester was a model amphibious operation. The difficulties overcome in landing a large assault force on an obscure beach with numerous off-lying reefs were formidable. Excellent aerial photography by Allied Air Forces enabled Admiral Barbey's staff to prepare accurate navigational charts for the attack force. A careful plan, with adequate emergency safeguards to insure its execution, provided for essential minesweeping and buoying of boat lanes. Landing craft control procedures were well thought out, and a senior naval officer was made responsible for the safe passage of the craft through the reefs and on to the beaches.[15] Coxswains and wave officers were given panoramic sketches built up from maps and photographs to help them identify beaches as they were seen from boats approaching the shore.

Although there was no hitch in the landing operations at Cape Gloucester, and the Navy and Marine Corps worked together with practiced ease, MacArthur's headquarters realized that the problem of who was in overall control at the time of the landing had been left in the air. Naval amphibious doctrine clearly gave this responsibility to the attack force commander, and at the conclusion of DEXTERITY, GHQ adopted this concept of control for future operations in the Southwest Pacific. Landing force commanders would take charge when their troops were firmly established ashore.[16]

What Admiral Barbey called "the old problem of efficient joint planning"[17] was handled well in the preparations for DEXTERITY. The various staffs—ground, naval, and air—at GHQ and at operating forces levels coordinated their planning activities, and the operations, instructions, and plans that were issued reflected concurrent thinking. Conferences between interested commanders were fre-

[14] Unless otherwise noted, the material in this section is derived from: *DEXTERITY Rept; ALAMO G-3 Jnl; VII PhibFor AR; 1st MarDiv SAR*, all phases and annexes; *APPEASE SAR;* Hough and Crown, *New Britain Campaign*.

[15] *VII PhibFor ComdHist*, p. II-47.

[16] MajGen Stephen J. Chamberlin, USA, memo for CinC, dtd 12Feb44, Subj: OpsInstns for Manus-Kavieng Ops, in *GHQ G-3 Jnl*, 13Feb44.

[17] *VII PhibFor AR*, p. 13.

quent enough to work out solutions to differences regarding objectives, forces available, and timing. The abandonment of Gasmata as a target, the substitution of Arawe, and the diversion of the Gasmata landing force to Saidor were all examples of the flexibility with which changes in the operational situation were met. The 1st Marine Division's strong views on the composition of the BACKHANDER assault force were carefully considered and finally accepted. The decision to cancel parachute troop participation and to strengthen the Marine landing force instead owed a great deal to the Commander, Allied Air Forces and his reluctance to support the air drop as planned.

General Kenney was much more interested in the aerial support that his bombing and attack squadrons could give to DEXTERITY operations than he was in the diversion of transports to a parachute landing. "Gloucesterizing" was an expression that came into use in the Fifth Air Force "to describe the complete obliteration of a target." [18] The word was invented by pilots as a tribute to the thoroughness of the pre-invasion aerial bombardment of Cape Gloucester. Japanese prisoners and captured diaries confirm the devastating effect of the steady hammering by Allied planes. Several hundred enemy troops lost their lives in the bombing and strafing attacks, and most of the permanent installations and fixed defenses near the airdrome were destroyed. Enemy morale skidded downward as anti-barge strikes mounted in intensity and effectiveness with the approach of D-day, and the flow of supplies to the garrison of western New Britain dwindled.

The shortage of fire support ships, and the desire to conceal the chosen landing beaches from the enemy, limited naval gunfire preparations to the morning of the actual assault at both Cape Gloucester and Arawe. The featureless blanket of jungle growth crowding to the water's edge showed few targets that were suited to the flat trajectory of naval guns. Most of the ships' bombardment was confined to area fire which showered the airfield, the hills that broke through the jungle, and the ill-defined beaches. Opportunity targets, such as the anti-boat guns that ripped through the rubber boat formation at Arawe, were sure game for destroyers, but such targets were few.

To bridge the gap between the end of naval gunfire and air bombardment of the beaches and the grounding of the first assault wave, ship-launched rockets were called into play. Both Admirals Barbey and Kinkaid were impressed with the potential of the new weapon, but the lack of opposition to the BACKHANDER and DIRECTOR landings deferred an evaluation of its effect against a stoutly-defended shore. There appeared to be little doubt, however, that the rockets would be a welcome and standard addition to amphibious fire support.

The Yellow Beach assault marked the first time that smoke was used to screen a landing operation in the Southwest Pacific. General Rupertus was not in favor of its employment, arguing that the smoke, dust, and flying debris of the preliminary bombardment was enough to becloud the vision of enemy observers on Target Hill.[19] The Marine general anticipated what happened; the smoke laid by bombers drifted lazily across the land-

[18] Craven and Cate, *Guadalcanal to Saipan*, p. 345.

[19] *Pollock comments.*

ing lanes and obscured the beaches. Fortunately, as Admiral Barbey noted, "landing craft were handled boldly and successfully in it," [20] and the smoke cover was not a serious problem. The lack of Japanese opposition must have had a good bit to do with the VII Amphibious Force commander's belief that the smoke was valuable "in protecting landing craft during the later stages of their approach to the beach." [21] If supporting fires to destroy coast defense guns had been needed, ships' gunners and pilots overhead would have been hard put to locate targets in the thick fog of smoke that hid the coast.

The considerable problems, both logistical and tactical, that were presented by the unexpected swamp behind the Yellow Beaches were met with ingenuity and dispatch. Adapting the dump dispersal and ship unloading plan to the limited stretch of dry ground available, the shore party solved a snarl that might have stalled the entire operation. Mobile loading, which was a key feature of the supply plan for BACKHANDER, worked, but not without considerable difficulty. Much of the trouble that arose in the use of pre-loaded trucks came from the employment of inexperienced and ill-disciplined drivers for a job that demanded skill and individual responsibility. In assessing the operation, ALAMO Force commented that there had been a tendency at Cape Gloucester, common to most amphibious operations, to bring in more motor transport than could be efficiently used. The excess vehicles landed tended to clog the limited road net and delay rather than speed unloading operations.

[20] *VII PhibFor AR*, p. 9.

[21] *Ibid.*

The readiness of Marine pioneers to meet any crisis that cropped up reflected their sound training as the 1st Division's shore party. Rupertus gave the men well-earned praise for meeting the original supply schedules and told Krueger that "I have seen no finer performance of duty on any landing beach by any unit in my career." [22] The contrast with the diligent but slow unloading efforts by improvised and poorly trained shore parties at Tauali and Arawe was marked. Krueger, recognizing that "a highly trained and well equipped shore party is indispensable in any landing operation," used an amphibian engineer shore battalion at Saidor and recommended the use of similar units in any future operations.[23]

The Marine practice of carrying trained replacements into combat as part of the shore party proved itself again at Cape Gloucester. The 300 men that reinforced 2/17 were available as laborers on the beaches and in the dumps at a time when shore party manpower demands were highest. At night, the men joined the pioneers in backing up perimeter defenses and, when the situation demanded, filled in as casualty replacements in hard-hit combat units.

The shore party commander drew attention to the fact that the naval beach party had a good share of the success of unloading operations, noting:

> For the first time a Marine shore party had the benefit of a trained, permanently organized beach party as one of its reinforcing elements. This beach party concept was an innovation of VII Amphibious Force, and its personnel were made avail-

[22] MajGen William H. Rupertus ltr to LtGen Walter A. Krueger, dtd 6Jan44, in *ALAMO G-3 Jnl No. 15.*

[23] *DEXTERITY Rept*, Encl 1, p. 3.

> able several weeks in advance of the landing. They lived and trained with the shore party of which they were a part and were lifted to Cape Gloucester with it. Here the means to control effectively the approach and beaching of landing ships and craft was conclusively demonstrated, and the performance of this beach party fully justified the high praise bestowed by Rupertus.[24]

Although the 1st Marine Division had its own shore party, it did not have an organic unit to provide the services of another element of the amphibian engineers, the boat battalion. In a role particularly well adapted to the shore-to-shore operations of the Southwest Pacific Area, the Army-manned small craft proved themselves a valuable addition to the BACKHANDER Force. In an analysis of their worth, the boat group commander at Talasea noted:

> . . . the First Marine Division maintained actual operational command over a substantial fleet of landing craft. The Army unit manning these was as much a part of the Task Force as any battalion in the division. No longer was it necessary to *request* amphibious lift, it could be *ordered*, and it was, not only for logistical support but for tactical landings and continuous patrolling. The increased mobility, freedom of action, [and] general expedition that this lent to the operations eastward to the San Remo Plantation demonstrated what should have been obvious, that a landing force commander should have as complete control over his boats as he does over his trucks and tanks.[25]

If the 1st Marine Division had continued to serve in the Southwest Pacific, it is probable that boat detachments would have been assigned to its command in future operations. In the Central Pacific, where successive objectives were usually widely separated small islands, operational requirements for amphibious craft were met differently. There, after the assault landing, Navy small boat pools left at the target, together with the landing force's organic LVTs and DUKWs, provided the necessary logistical and tactical support. Still, the practical uses of a boat detachment under direct command were not lost on many Marines, and the 1st Division's D-4 at New Britain voiced his conviction that in "any operation of an amphibious nature wherein a rapid seizure will be followed by an operation involving movement from shore to shore the demand for boat companies will continue to exist." [26]

Important as the engineer boats were to the success of BACKHANDER operations, there was an even greater star performer among the amphibious craft, the LVT. It is difficult to imagine what the fighting at Cape Gloucester would have been like without the support of amphibian tractors. The LVTs took so vital a part in combat operations in the swamp behind the beaches that their accomplishments can not be separated from the achievements of the infantry and supporting artillery. Despite their occasional use as fighting vehicles, the tractors were employed primarily in a logistical support role. Most 1st Division officers were so sold on their usefulness in supply and evacuation that they disapproved of a proposal to put a turret on the Buffalo, agreeing with General Rupertus, who said: "If you put a turret or a canopy on a Buffalo you have simply a light tank, lightly armored and quite slow. You lose

[24] *Ballance ltr.*

[25] LtCol Robert Amory, Jr., MassNatGrd, ltr to CMC, dtd 2Apr52.

[26] BGen William S. Fellers ltr to HistDiv, HQMC, dtd 10Apr52.

the cargo carrying capacity."[27] Regardless of the recommendations of the 1st Division, however, the amphibian tank was already in being and had proved its worth in combat in the Central Pacific. At Peleliu, where the division next landed in assault, LVT(A)s formed the first waves.

The armored vehicle that the 1st Division Marines preferred was the medium tank.[28] The Shermans proved their value repeatedly and repaid many times over the labor that the engineers, pioneers, and Seabees expended to get them through difficult terrain to the front lines. Tank-infantry techniques used in the drive to the airfield so impressed the Army liaison officers with the Marine division that they recommended that they be studied in the U.S. and used in training all units destined for the Southwest Pacific.[29] Marine light tanks served well at Cape Gloucester and Arawe, as they had previously in the South Pacific, but once infantry commanders saw what the mediums could do in the jungle, the cry mounted for more of the heavier-gunned and -armored machines. Light tankmen at Arawe could take credit for pioneering in telephone communication between supporting riflemen and armor, a procedure that became standard throughout the Pacific fighting.

The only unit of BACKHANDER Force to be specially cited for its work at Cape Gloucester was the 11th Marines, which received a Naval Unit Commendation emphasizing the regiments' determination to get into position and fire in support of the assault troops regardless of obstacles.[30] An ALAMO Force observer pinpointed the reason for the high and deserved praise of the regiment when he noted that 1st Division Marines were "very artillery conscious. They claim to have the best artillery in existence and use it effectively at every turn."[31] The 75mm pack howitzer lost ground as a supporting weapon at Cape Gloucester despite its excellent record. Against an enemy that dug in deeply and well on every possible occasion, the heavier, more powerful 105mm howitzer spoke with deadlier effect. To do its best job, however, the 105 needed better ammunition, shells with delay fuzes that would penetrate the jungle cover and blast apart the Japanese bunkers, instead of bursting in the tree canopy or the underbrush.

In one respect, artillery employment at Cape Gloucester did not come up to expectations. The inefficient radios used by the aircraft of the division's squadron of light planes prevented effective artillery spotting. But, if infrequent use was made of the Piper Cubs to direct howitzer fire, there was very little else that the planes and pilots did not do. The range of employment of the makeshift but efficient

[27] Quoted in Col Horace O. Cushman, USA, memo to CofS, ALAMO For, dtd 5Jan44, Subj: Rept of LiaisonO with 1st MarDiv, in *ALAMO G–3 Jnl No. 15*.

[28] *Ibid.*

[29] AG 334 (1Feb44) Rept to TAGO, WD, Washington, D.C., dtd 1Feb44, Subj: Rept of LiaisonOs with the 1st MarDiv in the Op to seize the Gloucester Airdrome area on New Britain, in *ALAMO G–3 Jnl No. 20*.

[30] Curiously, the citation for the 11th Marines includes an incident that describes the employment of a 37mm gun at Aogiri Ridge, a weapon that was manned by its regular crew and men of 3/5, not artillerymen. Col Lewis W. Walt ltr to HistDiv, HQMC, dtd 24Apr52. See Appendix I, Unit Commendations.

[31] Col J. F. Bird, USA, memo to DepCofS, ALAMO For, dtd 9Jan44, Subj: Rept on BACKHANDER Op from 1–7Jan44, in *ALAMO G–3 Jnl No. 15*.

outfit was as wide as the aerial supply of the Gilnit patrol and the impromptu close air support at the Volupai landing. In future operations in the Pacific, the 1st Marine Division would have a regularly constituted observation squadron assigned for operational control, but the "do anything" tradition of the division's first air unit survived.

General Rupertus, writing to the Commandant shortly before DEXTERITY was formally secured, observed: "We have learned much, especially [from] our errors at Guadalcanal, and I feel sure that we have profited by them in this operation." [32] Perhaps the most useful lesson learned was an appreciation of the value of battlefield intelligence. Throughout the fighting in western New Britain, enemy documents were turned in that might have been pocketed or thrown away by troops who were not convinced of their worth. The wealth of material that came back from assault units and intelligence teams closely trailing the advance was systematically and rapidly evaluated by translators with the combat teams and put to use immediately at the appropriate level of command. With the exception of the *51st Reconnaissance Regiment*, which appeared unheralded on the scene, the 1st Division's order of battle officers kept accurate track of the *Matsuda Force* and its state of combat efficiency.

The terrain was the major obstacle to the efficient use of the enemy intelligence that was accumulated. Although the Marines knew early in the fighting approximately where the Japanese headquarters were and the general location of the trails that were being used for troop deployment, the information was of limited use. The jungle shrouded everything. Even when the division's Cubs skimmed the treetops, the pilots and observers could spot little through the green carpet below. The Allied Air Forces photographic planes that did such a fine job establishing the shoreline and fringing reefs of the objective area were far less successful when the runs crossed the interior. The jungle gave up few secrets, even to the most skilled photographic interpreters.

The infamous "damp flat" area back of the Yellow Beaches was known to contain standing water in the rainy season, but the probing cameras did not show the swamp that actually existed. This fact, however, may have been a blessing in disguise, since it was probable that the assault landing would have been switched to other beaches if the situation had been known. The Japanese were completely unprepared for a landing in such an unsuitable place, and what might have been a hotly contested fight for a toehold on the shore never materialized. Because it was a veteran unit, with a well-trained shore party, the 1st Marine Division was able to surmount the miserable terrain and get firmly established before the enemy made a serious attempt to dislodge it.

Certainly, any well-trained, well-led, but untried Allied division could have wrested control of western New Britain from the *Matsuda Force;* the preponderance of strength lay too heavily in the Allies' favor for any other conclusion. Just as surely, the 1st Marine Division did the job faster, better, and at less cost by virtue of its combat experience, its familiarity with the jungle and the Japanese, and its battle-tested unit spirit. The tactics the Marines

[32] MajGen William H. Rupertus ltr to LtGen Alexander A. Vandegrift, dtd 4Feb44 (Vandegrift Personal Correspondence File, HQMC).

used were "book" tactics for jungle warfare; their refinements on basic techniques were those of veterans. Fire discipline at night was excellent, patrolling was careful but aggressive, and weapons were always at hand ready to fire. The enemy's captured guns were expertly manned by Marines and turned against their former owners. Small unit leaders were capable of independent action in brush-choked terrain, where the bitterest fighting was often done at close range with an unseen enemy.

The 1st Marine Division was junglewise and combat-ready when it landed on New Britain. When it left, four months later, its mission accomplished, it was an even more effective team. Ahead lay a summer of intensive training and then combat again, this time at Peleliu, a bloody step closer to Japan.

PART V

Marine Air Against Rabaul

691-360 O—63——29

Target: Rabaul

The overriding objective of the Allied campaign in the Southwest Pacific was, at first, to capture, and, later, to neutralize Rabaul. Each successive advance during 1943 had its worth valued by the miles it chopped off the distance to this enemy stronghold. To a large extent, the key to the objectives and pace of CARTWHEEL operations was this distance, measured in terms of the range of the fighter plane. No step forward was made beyond the effective reach of land-based fighter cover.

The firm establishment of each new Allied position placed a lethal barrier of interceptors closer to Rabaul and its outguard of satellite bases. Equally as important, the forward airfields provided a home for the fighter escorts and dive bombers which joined with long-range bombers to knock out the enemy's airfields. Protected by mounting numbers of Allied planes, many of them manned by Marines, the areas of friendly territory that saw their last hostile aircraft or vessel grew steadily. Japanese admirals learned from bitter experience that their ships could not sail where their planes could not fly.

By carrying the fight to the enemy, Allied air units played a decisive role in reducing Rabaul to impotency. Although this aerial offensive was closely related to the air actions in direct support of CARTWHEEL's amphibious operations, its importance warrants separate accounting.

OBJECTIVE FOLDER[1]

As they fought their way up the Solomons chain and along the enemy coast of eastern New Guinea, few members of Halsey's and MacArthur's naval and ground forces had time to consider any Japanese position but the one to their immediate front. To these men, Rabaul was little more than a worrisome name, the base of the enemy ships and planes that attacked them. Allied pilots and aircrews, however, got to know the Japanese fortress and its defenders intimately. The sky over St. George's Channel, Blanche Bay, and Gazelle Peninsula was the scene of one of the most bitterly fought campaigns of the Pacific War.

To picture Rabaul as it appeared to the men who battled to reach it, to bomb and strafe it, and to get away alive, requires a description of more than the northern tip of Gazelle Peninsula where the town, its harbor, and its airfields were located. To flyers, the approaches were as familiar as

[1] Unless otherwise noted, the material in this section is derived from: AlGeographicalSect, SWPA, Terrain Study No. 22, Area Study of Gazelle Peninsula and Rabaul, dtd 6Oct42 and Terrain Study No. 74, Area Study of Gazelle Peninsula, dtd 3Jan44; MID, WD, Survey of Bismarck Archipelago (S30-675), dtd 5Oct43. Documents not otherwise identified in this part are located in the following files of the Archives, Historical Branch, G-3 Division, Headquarters Marine Corps: Unit Historical Reports; Publications; Aviation; Monograph and Comment.

the objective itself, and a strike directed against Rabaul evoked a parade of impressions—long over-water flights; jungle hills slipping by below; the sight of the target—airfield, ship, or town, sometimes all three; the attack and the violent defense; and then the seemingly longer, weary return over land and sea.

In order to fix Rabaul as an air objective, one should visualize its position in midyear 1943 as the powerful hub of the Japanese airbase system in the Southeast Area. To the west on New Guinea, at Hollandia, Wewak, and Madang, were major airdromes with advance airstrips building on the Huon Peninsula and across Vitiaz Strait at Cape Gloucester. Staging fields in the Admiralty Islands gave enemy pilots a place to set down on the flight from eastern New Guinea to Rabaul. Kavieng's airbase was also a frequent stopover point, not only for planes coming from the west but for those flying south from Truk, home of the *Combined Fleet* and its carriers. Southeast of New Britain in the Solomons lay the important airfields at Buka Passage at one end of Bougainville and at Buin-Kahili and Ballale Island at the other. Forward strips at Vila and Munda in the New Georgia Group marked the limit of Japanese expansion.

Distances in statute miles from Rabaul to the principal bases which guarded it, and to the more important Allied positions from which it was attacked, are as follows:

Truk	795
Guadalcanal	650
Wewak	590
Port Moresby	485
Madang	450
Munda Point	440
Dobodura	390
Lae	385
Admiralties	375
Woodlark	345
Finschhafen	340
Kahili	310
Kiriwina	310
Cape Gloucester	270
Cape Torokina	255
Buka Passage	190
Kavieng	145

The starting point for measuring these distances was a small colonial town that had, in the immediate prewar years, a population of about 850 Europeans, 2,000 Chinese, and 4,000 Melanesians. Quite the most important place in the Australian Mandated Territory of New Guinea, Rabaul was for many years the capital of the mandate. When two volcanic craters near the town erupted in May 1937, the decision was made to shift the government to Lae, but the pace of island life was such that the move had barely begun when the Japanese struck.

The town was located on the shore of Simpson Harbor, the innermost part of Blanche Bay, a hill-encircled expanse six miles long and two and a half wide. One of the finest natural harbors in the Southwest Pacific, the bay is actually the crater of an enormous volcano, with the only breach in its rim the entryway from the sea and St. George's Channel. Two sheer rocks called The Beehives, which rise 174 feet above the water near the entrance to Simpson Harbor, are the only obstacles to navigation within the bay. There is space for at least 20 10,000–15,000-ton vessels, plus a host of smaller craft, to anchor within Simpson's bounds. Separated from this principal anchorage by little Matupi Island is Matupi Harbor, a sheltered stretch guarded on the east and north by a wall of mountains. Just inside the

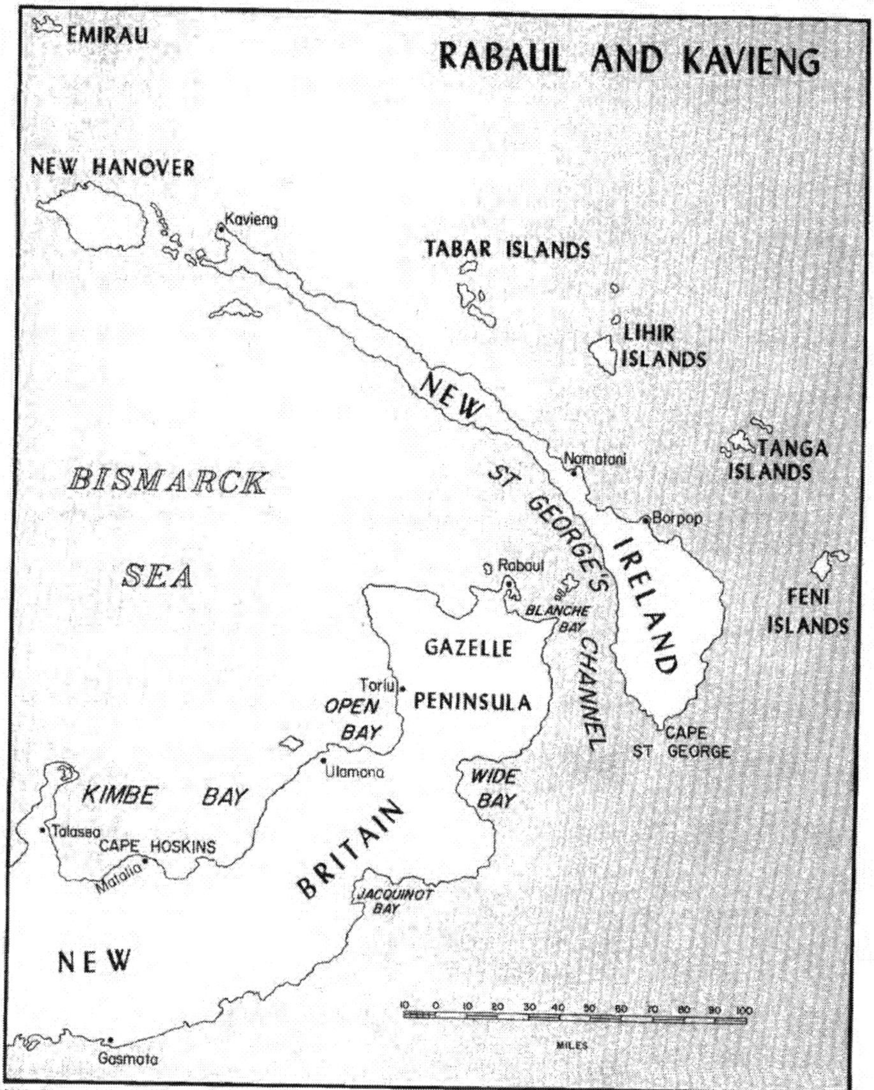

MAP 31 R.F. STIBIL

headlands, Praed Point and Raluana Point, at the entrance to Blanche Bay are two further protected harbors. Both, Escape Bay in the north by Praed Point and Karavia Bay in the south, are less useful, as their waters are too deep for effective anchorages.

Prominent landmarks, as easily recognizable from the air as The Beehives, are the craters that form a part of the hills surrounding Rabaul. Directly east of Matupi Harbor is Mt. Tavurvur (741 feet), which erupted in 1937, and due north is Rabalankaia Crater (640 feet). These two heights give Crater Peninsula its name, but they are overshadowed by the peninsula's mountainous ridge which has three companion peaks, South Daughter (1,620 feet), The Mother (2,247 feet), and North Daughter (1,768 feet). The town of Rabaul nestled between the foothills of North Daughter and the narrow sandy beaches of Simpson Harbor. Across Blanche Bay from Mt. Tavurvur is its partner in the 1937 eruption, Vulcan Crater (740 feet), which juts out from the western shore to form one arm of Karavia Bay.

In the years of peace, the land to the south and east of Blanche Bay was extensively planted in coconuts. The rich volcanic soil there was fertile, and, like most of the northern third of Gazelle Peninsula, the area was relatively flat. Most of the 100 or so plantations on the peninsula were located here, with a large part of them to be found in the region north of the Warangoi and Keravit Rivers. The only other considerable plantation area along the northern coast lay between Cape Lambert, the westernmost point on Gazelle Peninsula, and Ataliklikun Bay, into which the Keravit River emptied. The majority of the 36,000 natives who were estimated to be on the peninsula lived in or near this northern sector.

The rest of Gazelle Peninsula, which is shaped roughly like a square with 50-mile-long sides, is mountainous, smothered by jungle, and inhospitable to the extreme. Two deep bights, Wide Bay on the east coast and Open Bay on the west, set off the peninsula from the rest of New Britain. Access to Rabaul from this part of the island was possible by a coastal track, broken frequently by swamps and rivers, and a web of trails that cut through the rugged interior. For the most part, these routes were hard going and usable only by men on foot.

The wild, inaccessible nature of the central and southern sectors of Gazelle Peninsula made the contrast with the Rabaul area all the more striking. Even before the war, the mandate government had developed a good road net to serve the various villages, plantations, and missions. The Japanese made extensive improvements and expanded the road system to connect with their troop bivouacs and supply dumps. Many of these installations were invisible from the air, hidden in the patches of jungle that interspersed the plantations and farms. The major Japanese construction work, however, was done on airfields, and the five that they expanded or built from scratch became as familiar to Allied aircrews as their own home bases.

Both of the small fields maintained at Rabaul by the Royal Australian Air Force were enlarged and made into major airdromes by the Japanese. Lakunai airfield and its hardstands and revetments occupied all the available ground on a small peninsula that ran out to Matupi Island. A 4,700-foot coral runway, varying in width from 425 to 525 feet, began at Simpson Harbor and ended at Matupi Harbor. The sharply rising slopes of Rabalankaia Crater blocked any extension of the field to the northeast and a small creek was a barrier on the northwest.

The other former RAAF base, Vunakanau airfield, was located at an altitude of 1,000 feet on a plateau about 11 miles southwest of Rabaul. Except for two coconut plantations, the plateau was covered by scrub growth and kunai grass. The ground was quite irregular and laced with deep gullies, and the 5,100 x 750-foot runway the enemy built was the practical limit of expansion. Centered on this grass-covered larger strip was a concrete runway, 4,050 feet long and 140 feet wide. Vunakanau became the largest Japanese airdrome at Rabaul, and its straggling network of dispersal lanes and revetments spread over an area of almost two square miles.

The longest airstrip at Rabaul was constructed at Rapopo on the shore of St. George's Channel about 14 miles southeast of the town and a little over 5 miles west of Cape Gazelle, the northeast corner of Gazelle Peninsula. Designed and built as a bomber field, Rapopo was sliced through the center of a coconut plantation that gave it its name. The clearing for the north-south strip ran 6,900 feet from the sea to a river that effectively barred further extension. A coral-surfaced runway began about 1,600 feet from a low, coastal bluff and occupied the full width of the cleared space.

Well inland from the other airfields, 15 miles southeast of Vunakanau and 8 miles southwest of Rapopo, the Japanese built Tobera airfield. Its runway, 5,300 x 700 feet, with a hard-surfaced central strip 4,800 x 400 feet, was situated on a gently sloping plateau that divided the streams which flowed north to the sea from those which ran south to the Warangoi River. Like most of its companion fields, Tobera was constructed in a plantation area with its dispersal lanes and field installations scattered amidst the coconut trees.

The fifth airfield at Rabaul was located on Ataliklikun Bay just north of the Keravat River and 26 miles southwest of Rabaul. Keravat airfield was plagued by drainage problems and had perhaps the poorest location and the greatest engineering problems. By the end of November 1943, the Japanese had been able to grade and surface a 4,800 x 400-foot runway, but Keravat never became fully operational and saw very limited use as an emergency landing ground.

Caught up and deserted by the rush of events was an auxiliary airfield that was started and never finished on Duke of York Island. The island is the largest of a group of 13 islets that stand in St. George's Channel midway between New Britain and New Ireland. The press for additional airstrips on which to locate and, later, to disperse and protect Rabaul's air garrison was met instead by fields on New Ireland.

There were four operational airfields on that narrow, 220-mile-long island with two, Namatami and Borpop, sited about 50 miles northeast of Rabaul on New Ireland's eastern shore. At Kavieng on the island's northern tip and Panapai close by was an extensive airbase, the largest in the Bismarcks outside of Rabaul's immediate environs. Kavieng and Rabaul had been seized at the same time and grew apace with each other until they both, in turn, were relegated to the backwash of the war by the withdrawal of their aerial defenders.

GARRISON FORCES[2]

In January 1942, Rabaul had a small garrison of about 1,350 men, a reinforced Australian infantry battalion. Kavieng had no defenders at all save a few police boys. The towns themselves and the islands on which they stood were ripe for the taking whenever the Japanese got around to the task. In the enemy's war plans, elements of the *Fourth Fleet* that had seized Guam and Wake made up the *Rabaul Invasion Force*. The assault troops at Rabaul would be the *South Seas Detached Force*, an Army brigade that had landed at Guam, reinforced by two companies of the *Maizuru 2d Special Naval Landing Force*, the victors at Wake.[3] The remainder of the *Maizuru 2d* was detailed to occupy Kavieng.

The *Rabaul Invasion Force* rendezvoused at sea north of the Bismarcks on 19 January, and, on the next day, enemy carrier-based bombers and fighters hit both targets. At Rabaul, the defending air force—five RAAF Wirraway observation planes—was quickly shot out of the sky, and the airstrips were bombed. The carrier planes made diversionary raids on Lae, Salamaua, and Madang on the 21st, and then hit Rabaul again on the 22d, knocking out Australian gun positions on North Daughter and at Praed Point.

After this brief preparation, Japanese transports and supporting vessels sailed into Blanche Bay near midnight on 22 January, and the assault troops began a staggered series of landings shortly thereafter. The enemy soldiers stormed ashore at several points along the western beaches of Simpson Harbor and Karavia Bay, while the naval landing force hit Rabaul and Lakunai airfield. The Australians, spread out in small strongpoints along the shore and on the ridge just inland, fought desperately in the darkness but were gradually overwhelmed and forced to pull back. As daylight broke, the 5,000-man Japanese landing force called down naval gunfire and air support to hammer the retreating defenders. At about 1100, the Australian commander, seeing that further resistance would be fruitless, ordered his men to break contact, split up into small parties, and try to escape.

The Japanese harried the Australian troops relentlessly, using planes and destroyers to support infantry pursuit col-

[2] Unless otherwise noted the material in this section is derived from: ATIS, MIS, GHQ, SCAP, Docu No. 17895 (WDI-46), dtd 9May46, Subj: Full Translation of a Rept on the Japanese Invasion of Rabaul, n.d. (COA, NHD); ATIS, MIS, GHQ, FEC, Japanese Monograph No. 140, Outline of SE Area NavAirOps—Pt IV, dtd Jul49, hereafter *SE Area NavAirOps—IV;* Japanese ResearchDiv, MilHistSec, HQUSAF FE, Japanese Monograph No. 142, Outline of SE Area NavAirOps—Pt V (Dec 43–May 44), n.d., hereafter *SE Area NavAirOps—V; Eighth Area ArmyOps; SE Area NavOps—III;* Statement of LtCol H. H. Carr, CO, 2/22 Bn, AIF, App A to AlGeographicalSect, Terrain Study No. 22, *op. cit.*; USSBS, *Interrogation* No. 446, Cdr Takashi Miyazaki, IJN, II, pp. 413–421, hereafter *Miyazaki Interrogation;* USSBS, *Interrogation* No. 479, Capt Minoru Genda, IJN, dtd 28–29Nov45 (USSBS Recs, National Archives); Masatake Okumiya and Jiro Horikoshi with Martin Caidin, *Zero!* (New York: E. P. Dutton and Co., Inc., 1956), hereafter Okumiya, Horikoshi, and Caidin, *Zero!*; USSBS(Pac), NavAnalysisDiv, Marshalls-Gilberts-New Britain Party, *The Allied Campaign Against Rabaul* (Washington: GPO, 1946), hereafter USSBS, *Campaign Against Rabaul.*

[3] For details of the earlier actions of these Japanese units see Volume I of this series, pp. 75–78, 129–149.

umns. Most of the defenders were eventually trapped and killed or captured on Gazelle Peninsula, but one group of about 250 officers and men stayed a jump ahead of the Japanese, reached Talasea after an exhausting march, and got away safely by boat, landing at Cairns, Queensland, on 28 March. Naturally enough, the fact that they were driven from Rabaul rankled the Australians, but the opportunity for retaliation was still years away.

Flushed with success, the Japanese set about extending their hold throughout the Bismarcks, the Solomons, and eastern New Guinea. Rabaul served as a funnel through which troops, supplies, and equipment poured, at first in a trickle, then in a growing stream until the defeats at Guadalcanal and Buna-Gona checked the two-pronged advance. In the resulting reassessment of their means and objectives, the Japanese reluctantly decided to shift to a holding action in the Solomons in order to concentrate on mounting a sustained offensive on New Guinea. Essential to this enemy decision was the fact that a system of airfields existed between Rabaul and Guadalcanal.

The 650-mile distance from Henderson Field to Vunakanau and Lakunai was a severe handicap to Japanese air operations during the Guadalcanal campaign. The need for intermediate bases was obvious, and enemy engineers carved a succession of airfields from plantations, jungle, and grasslands in the central and northern Solomons during the last few months of 1942. Only Buka, which was operational in October, was completed in time to be of much use in supporting air attacks on the Allied positions on Guadalcanal. Fields at Kahili, Ballale, Vila, and Munda, however, were all in use by the end of February 1943, some as staging and refueling stops and the others as fully operational bases. It was these airfields that furnished Rabaul the shield that the Japanese needed to stave off, blunt, or delay Allied attacks from the South Pacific Area. The task of manning these bases was exclusively the province of the *Eleventh Air Fleet*.

The *Eighth Area Army's* counterpart of Admiral Kusaka's air fleet, the *6th Air Division*, was almost wholly committed to support of operations on New Guinea by the end of 1942. During the bitter fighting in Papua, Japanese air support had been sporadic and sparse, a situation that General Imamura intended to correct. Rapopo Airfield at Rabaul, which became operational in April 1943, was constructed by the Army to accommodate a growing number of planes, and work was begun on a Navy field at Keravat to handle even more. At about this time, the strength of the *6th Air Division* peaked at 300 aircraft of all types. Many of these planes were stationed at Rabaul, but a good part were flying from fields on New Guinea, for Imamura had ordered the *6th* to begin moving to the giant island on 12 April.

On eastern New Guinea, as in the Solomons, airfields closer to the battle scene than those at Rabaul were needed to provide effective air support to Japanese troops. Consequently, two major airbases were developed at Wewak and Madang on the coast northwest of the Huon Peninsula. Despite the heavy use of these fields, the operating efficiency of Army air units dropped steadily in the first part of 1943. The rate at which *6th Air Division* planes were destroyed by Allied pilots and gunners was so great that even an average monthly flow of 50 replacement aircraft could not keep pace with the losses. In July, Tokyo added the *7th Air Division* from the Netherlands East Indies to *Eighth Area Army's* order of battle and followed through by assigning the *Fourth Air Army* to command and coordinate air operations. By the time the air army's headquarters arrived at Rabaul on 6 August, a move of all Army combat aircraft from the Bismarcks was well underway.

In light of the desperate need of the *Eighteenth Army* on New Guinea for air support, *Imperial General Headquarters* had urged General Imamura to leave the air defense of Rabaul entirely to the Navy and concentrate all his air strength in the *Eighteenth's* sector. After discussing the proposed change with Admiral Kusaka, who would acquire sole responsibility for directing air operations at Rabaul, the general ordered the transfer. By the end of August, *Fourth Air Army's* headquarters was established at Wewak, and all Army aircraft, except a handful of reconnaissance and liaison types, were located on New Guinea.

When the last Japanese Army plane lifted from Rapopo's runway, the crucial period of the Allied aerial campaign against Rabaul was still in the offing. The men, the planes, and the units that would fight the enemy's battle were essentially those which had contested the advance of South Pacific forces up the Solomons chain in a year of furious and costly air actions. In that time, Japanese naval air groups were rotated in and out of Rabaul, and were organized and disbanded there with little apparent regard for a fixed table of organization. Two administrative headquarters, the *25th* and *26th Air Flotillas*, operated under *Eleventh Air Fleet* to control the air groups; for tactical purposes, the flight echelons of the flotillas were organized as the *5th* and *6th Air Attack Forces*. Since the accounts of senior surviving air fleet officers, including Admiral Kusaka, differ considerably in detail on enemy strength and organization, only reasonable approximations can be given for any one period.

In September 1943, on the eve of the Allied air offensive against Rabaul, the *Eleventh Air Fleet* mustered about 300 planes and 10,000 men, including perhaps 1,500 flying personnel. Three fighter groups, the *201st Air Group*, the *204th*, and the *253d*, each with a nominal strength of 50 aircraft and 75 pilots, were the core of the interceptor force. One medium bomber unit, the *705th Air Group*, was present, together with elements of at least two more groups, but heavy losses had reduced them all to skeleton proportions of a bomber group's normal strength of 48 planes and 300 crewmen. There was one combined dive bomber-torpedo bomber outfit, the *582d Air Group*, whose strength was 36 attack aircraft and 150 crewmen. Two reconnaissance groups, the *938th* and *958th*, each with 28 float planes and about 100 flying personnel, completed the air fleet's complement of major units. A few flying boats, some transports assigned to each air group, and headquarters and liaison aircraft were also present.

To keep up with the steady drain of combat and operational losses, Tokyo sent 50 replacement aircraft to Rabaul each month. Approximately one-third of these planes were lost in transit, but the remainder, 80 percent of them fighters, reached their destination after a long overwater flight staged through Truk and Kavieng. Land-based naval air units in quiet sectors of the Pacific were drawn upon heavily for planes and pilots and received in exchange battle-fatigued veterans from Rabaul.[4] The drain of Japanese naval planes and personnel from the Netherlands Indies grew so serious toward the end of 1943 that the Army's *7th Air Division* had to be returned to the area to plug the gap.

In every possible way, the Japanese tried to ready Rabaul's air garrison for the certain Allied onslaught. Flight operations from the most exposed forward airstrips in the Solomons were sharply curtailed to conserve aircraft and crews. At all airfields, blast pens and dispersal areas were strengthened and expanded, and antiaircraft guns were disposed in depth to cover approaches. Tobera airfield was rushed to completion to lessen the concentration of aircraft at Vunakanau and Lakunai and to provide space for reinforcements from the *Combined Fleet*. Poised at Truk, two carrier air groups with about 300 planes stood ready to join Kusaka's command when the situation worsened enough to demand their commitment. Although the newest Japanese plane models were to be fed in to Rabaul's air defense as they became available, the overwhelming majority of the planes that would rise to meet the Allied attacks would be from one family of fighters, the Zeros.

ENEMY PLANES AND AIRCREWS[5]

During the first nine months of the war, the Allies tried to identify Japanese aircraft as the enemy did, by the year of initial adoption and type. The calendar the Japanese used was peculiarly their own, with the year 2597 corresponding to 1937, and there were a number of different Type 97s in use, among them an Army fighter, an Army medium bomber, a Navy torpedo bomber, and a Navy flying boat. This was the system that gave rise to the name Zero for the Type O Navy fighter plane that was first employed in 1940 during the fighting in China.

By the time of Pearl Harbor, the Zero had replaced its Type 96 predecessor as the standard Japanese carrier fighter. Based on its performance capabilities, enemy intelligence officers were confident that the plane could gain control of the air over any battle area, and that in aerial combat, "one Zero would be the equal of from two to five enemy [Allied] fighter planes, depending upon the type."[6] This assessment, unfortunately, proved to be too close to the truth for the peace of mind of Allied pilots. In a dogfight, the Zero was at its best; at speeds below 300 miles per hour, it could outmaneuver any plane

[4] USSBS, *Interrogation* No. 360, Capt Hironaka Komoto, IJN, II, p. 288.

[5] Unless otherwise noted, the material in this section is derived from: IntelServ, USAAF, InfoIntelSummary No. 85, Flight Characteristics of the Japanese Zero Fighter Zeke, dtd Mar43, and No. 40-11, Japanese Aircraft and Armament, dtd Mar44, hereafter as part of a series *USAAF InfoIntelSummary* with appropriate number, subject, and date; AirInfoDiv, CNO, Organization and Rank in the Japanese Army and Navy Air Services (Op-35 AID #A2), dtd Aug43; Technical AirIntelCen, NAS, Anacostia, Japanese Aircraft Performance and Characteristics, TAIC Manual No. 1 (OpNav-16-VT#301), dtd Dec44; *Miyazaki Interrogation;* USSBS (Pac), MilAnalysisDiv, *Japanese Air Power* (Washington; GPO, Jul46); WD, *Handbook on Japanese Military Forces, TM 30-480* (Washington, 1Oct44); Okumiya, Horikoshi, and Caidin, *Zero!*

[6] Okumiya, Horikoshi, and Caidin, *Zero!*, p. 60. All material quoted by permission of the publishers, E. P. Dutton & Co., Inc., acting for the copyright holder, Martin Caidin.

that it encountered in 1942. By the end of that year, however, the Zero had officially lost its well-remembered name among the Allies and had become instead, the Zeke.

The name change, part of a new system of enemy aircraft designation, was ordered into effect in the Southwest Pacific in September 1942 and adopted in the South Pacific in December. The Japanese identification method, with all kinds of planes assigned the same type-year, proved too cumbersome for Allied use. In its stead, enemy aircraft were given short, easily pronounced code titles; fighters and floatplanes received masculine names, with feminine names going to bombers, flying boats, and land-based reconnaissance planes. Despite the switch, the name Zero died hard, particularly among Marine pilots and aircrews in Halsey's forces, and it was at least six months before they gave the substitute, Zeke, popular as well as official sanction.[7]

The Zeke was unquestionably the most important enemy plane that fought in the Rabaul aerial campaign. Developed by the Mitsubishi Aircraft Company, the original version of the fighter had two models, one with folding wing tips for carrier use and the other built to operate from land bases. An all-metal, single-engine monoplane, the Zeke had low-set wings tapered to a rounded tip. The pilot sat high in an enclosed cockpit controlling two 7.7mm machine guns synchronized to fire through the propeller and two 20mm cannon fixed in the wings. Performance assets were rapid take-off and high climbing rates, exceptional maneuverability at speeds up to 300 miles per hour, and a total range of 1,580 miles with maximum fuel load and economy speeds. The Zeke's principal liabilities as a combat aircraft, ones it shared with every Japanese military plane, were relatively flimsy construction and a lack of armor protection for pilot, fuel, and oxygen.

Most of the Zekes that defended Rabaul in late 1943 were of a later model than the 1940. The improved planes had the same general appearance but were fitted with racks to carry one 132-pound bomb under each wing and had a more powerful motor that added 15 miles to the former maximum speed of 328 miles per hour at 16,000 feet. Another model of the basic Type O Navy fighter, one with the same engine, armament, and flight performance as the later model Zeke, was the Hamp.[8] Identified at first as a new plane type because of its shorter, blunt-tipped wing, the Hamp was later recognized as a legitimate offspring of the parent Zero. The only other Navy fighter operating out of Rabaul in significant numbers was the Rufe, a slower floatplane version of the Zeke.

The standard enemy land-based naval bomber was the versatile Betty, a 1941 model that was as frequently used on transport, reconnaissance, and torpedo bombing missions as it was for its primary purpose. In the pattern of most enemy medium bombers, the Betty was a twin-engine, mid-wing monoplane with a cigar-shaped fuselage and a transparent nose, cockpit, and tail. Operated by a crew of

[7] Sherrod, *MarAirHist*, p. 135n.

[8] This fighter was at first called the Hap, an unsolicited compliment to General Henry H. (Hap) Arnold, Commanding General, Army Air Forces. Soon after word of it reached Washington, there was an abrupt change in nomenclature. Vern Haugland, *The AAF Against Japan* (New York: Harper & Brothers, 1948), p. 371.

seven to nine men, the plane could carry a maximum bomb load of 2,200 pounds and was armed with four 7.7mm machine guns, all in single mounts, and a 20mm cannon in its tail turret. The Betty was fast, 276 miles per hour at 15,000 feet, and had a range of 2,110 miles at cruising speed with a normal fuel and bomb load. To achieve this relatively high speed and long range, Mitsubishi Aircraft's designers had sacrificed armor and armament. Much of the plane was built of lightweight magnesium, a very inflammable metal, and in the wing roots and body between were poorly protected fuel and oil tanks. "The result was a highly vulnerable aircraft so prone to burst into flames when hit that Japanese aircrews nicknamed it 'Type 1 Lighter.'"[9]

Even more vulnerable to Allied fire than the Betty was the principal dive bomber in the *Eleventh Air Fleet*, the Val. The pilot, who controlled two forward firing 7.7mms in the nose of the monoplane, sat over one unprotected fuel tank and between two others in the wings; the gunner, who manned a flexible-mount 7.7mm in the rear of the cockpit enclosure, was uncomfortably close to the highly explosive oxygen supply. A pair of bomb racks located under each of the plane's distinctive elliptical-shaped wings, and one under the body between the fixed landing gear, enabled the Val to carry one 550- and four 132-pound bombs. The dive bomber's best speed was 254 miles per hour at 13,000 feet, and its normal range at cruising speed with a full bomb load was 1,095 miles. When it flew without escort, the Val was easy game for most Allied fighters.

[9] "Biography of Betty," in *USAAF Infolntel-Summary* No. 44-21, dtd 10Jul44, p. 3.

The slowest of the major plane types at Rabaul, and the one with the poorest performance, was a torpedo bomber, the Kate. The plane was as poorly armed as the Val and was almost as inflammable. The two- to three-man crew all sat in a long, enclosed cockpit atop a slim 33-foot body; the wing span of the monoplane, 50 feet, gave it a foreshortened look. One torpedo at 1,760 pounds was its usual load, although a 1,000-pound bomb or two 250-pound bombs could be carried instead. Since it had a weak engine and its lethal cargo was stowed externally, at emergency speed and its best operating altitude, 8,500 feet, the Kate could only make 222 miles per hour.

Aside from those mentioned, many other Japanese Navy aircraft and an occasional Army plane were encountered and engaged by the Allies in aerial attacks on Rabaul. The Zeke fighter family, however, furnished most of the interceptors and escorts, and the Bettys, Vals, and Kates delivered the dwindling enemy offensive thrusts. A once-numerous fleet of Japanese flying boats, reconnaissance planes, and transports fell away into insignificance by October 1943. The feebly armed and unarmed survivors avoided Allied aircraft like a plague, since they were dead birds if caught.

There was no enemy plane that flew from Rabaul that was not a potential flaming death trap to its crew. To meet the specifications outlined by the Japanese Navy, aircraft designers sacrificed safety to achieve maneuverability in fighters and long range in bombers. Heightening the losses suffered by these highly vulnerable planes was the plummeting level of skill of their flying and maintenance personnel.

CAPTURED JAPANESE ZERO, *showing U.S. plane markings, aloft over the San Diego area during test flights.* (*USN 80–G–11475*)

JAPANESE VAL DIVE BOMBERS *are shown armed and ready to take off on a bombing mission in film captured early in 1943.* (*USN 80–G–345598*)

By 1943, the problem of keeping aircraft in forward areas in good operational condition and adequately manned had become acute. The senior staff officer of the *25th Air Flotilla* during the critical period of the battle for Rabaul recalled:

> In the beginning of the war, during 1942, if 100% of the planes were available for an attack one day, the next day 80% would be available, on the third day 50%. In 1943, at any one time, only 50% of the planes were ever available, and on the next day following an all-out operation only 30% would be available. By the end of 1943, only 40% at any one time would be serviceable. In 1942, the low availability was due to lack of supply; from 1943 on, it was due to lack of skill on the part of maintenance personnel and faulty manufacturing methods. Inspection of the aircraft and spare parts, prior to their delivery to Rabaul, was inadequate, and there were many poorly constructed and weak parts discovered. The Japanese tried to increase production so fast that proper examination was impossible.[10]

Japanese naval aviation had begun the war with 2,120 aircraft of all types, including trainers. In April 1943, after 16 months of heavy fighting, the total strength stood at 2,980, which meant that the manufacturers had been able to do little more than keep pace with combat and operational losses. In the succeeding year, the production rate nearly doubled, but losses soared also; there were 6,598 planes on hand in April 1944, but the standard of construction had deteriorated badly.[11]

Even more serious than the sag in the quality of naval aircraft maintenance and production was the steady attrition of experienced flight personnel. The pilots who began the war averaged 800 hours of flying time, and many of them had combat experience in China. Relatively few of these men survived until the end of 1943; a great many died at Coral Sea and Midway and in air battles over Guadalcanal. Others crashed trying to stretch the limited range of Vals and Kates to cover the long stretch between Rabaul and Guadalcanal. The replacements, pilots and aircrews alike, could not hope to match the worth of the men whose places they took.

Two years of flight training and practice had been the prewar requisite to make a man a qualified naval pilot or "observer" (bombardiers, navigators, and gunners). In 1941, the training time was cut about in half. Pilots spent about 60 hours in primary and intermediate trainers, observers spent 44, both in a six-month period. Flight training in combat types, spread over a four-to-six month period, was 100 hours for pilots and 60 hours for observers. Thereafter, 150 hours of tactical flight training was programmed for men in the units to which they were assigned. At Rabaul, however, this phase was spent in combat, and those few who survived 150 hours could count themselves as living on borrowed time.

The majority of flying personnel in the Japanese Navy were warrant officers, petty officers, and naval ratings. Regular and reserve officers selected for pilot and observer training were intended for command billets; they were few in number, and, as combat flight leaders, their losses were disproportionately great. In the Rabaul area by the fall of 1943, a representative Betty unit with 11 planes had only one officer among 23 pilots and one among 38 observers, while all of the 39

---

[10] *Miyazaki Interrogation*, p. 418.

[11] USSBS, *Interrogation* No. 414, Cdr J. Fukamizu, IJN, II, p. 374, Anx B.

radiomen and mechanics were enlisted men.[12]

An experienced Japanese combat air commander, operations officer at Buin-Kahili during September 1943, characterized these aircrews as personifying:

> . . . Japan's people on the battlefield, for they came from every walk of life. Some of them carried the names of well-known families; some non-commissioned officers were simple laborers. Some were the only sons of their parents. While we maintained strict military discipline on the ground, with proper observance for rank, class, and age, those differences no longer existed when a crew's plane lifted its wheels from the ground.
>
> The enemy cared little about the groups which constituted our aircrews and there existed no discrimination on the part of the pilot who caught our planes in his sights! Our air crews were closely knit, for it mattered not one whit whether an enlisted man or an officer manned the machine guns or cannon. The effect was exactly the same. Unfortunately this feeling of solidarity of our aircrews was unique in the Japanese military organization.[13]

Fighter pilot or bomber crewman, the Japanese naval flyer who fought at Rabaul was aware that he was waging a losing battle. The plane he flew was a torch, waiting only an incendiary bullet to set it alight. The gaping holes in his unit left by the death of veterans were filled by young, inexperienced replacements, more a liability than an asset in combat air operations. Despite the handicaps under which he fought—out-numbered, out-gunned, and out-flown—the enemy flyers fought tenaciously right up to the day when Rabaul was abandoned to its ground defenders.

[12] ComSoPac, Weekly AirCIntelRepts, 7Feb43–27May44, Rept of 3–9Oct43, hereafter *SoPac ACI Repts.*

[13] Okumiya, Horikoshi, and Caidin, *Zero!*, p. 294.

Approach March

COMAIRSOLS[1]

Even a cursory study of the organizational structure of air command in the South Pacific can produce a headache for the orderly mind. Many officers held two or three billets concurrently, in units of their own service as well as elements of SoPac task forces. The resulting maze of administrative and command channels might appear unworkable, but it functioned smoothly as a result of Admiral Halsey's emphasis on the principle of unity of command. He "insisted that each commander of a task force must have full authority over all components of his force, regardless of service or nationality."[2] Under this tenet, Commander, Aircraft, Solomons (ComAirSols), directed the combat operations of all land-based air in the Solomons during CARTWHEEL.

Rear Admiral Charles P. Mason was the first officer to hold the title ComAirSols; he assumed command on 15 February 1943 at Guadalcanal. Actually, Mason took over a going concern, as he relieved Brigadier General Francis P. Mulcahy, who had controlled all aircraft stationed at the island during the final phase of its defense. Mulcahy, who became Mason's chief of staff, was also Commanding General, 2d Marine Aircraft Wing. The fact that a general headed the staff of an admiral is perhaps the best indication of the multiservice nature of AirSols operations. Since Mason brought only a few officers

---

[1] Unless otherwise noted, the material in this section is derived from: *ComSoPac Apr–Oct43 WarDs;* HistDiv, AC/AS Intel, Data pertaining to the ThirteenthAF in the Campaigns of the Lower, Central, and Northern Solomons, 29Mar42–27Sep44 (File 750–01, AF Archives, Maxwell AFB), hereafter *ThirteenthAF Data;* StrikeComd, AirSols WarDs, 4Apr43–1Jun44, hereafter *StrikeComd WarDs;* Col William O. Brice interview by HistDiv, HQMC, dtd 30Jan45, filed in folder, ComAirSols Repts—Orders—Plans, 1943–44; AvnHistUnit, OP–519B, DCNO(Air), *The Navy's Air War, A Mission Completed,* Lt A. R. Buchanan, USNR, ed. (New York and London: Harper and Brothers [1946]), hereafter Buchanan, *Navy's Air War;* Craven and Cate, *Guadalcanal to Saipan;* SqnLdr J. M. S. Ross, RNZAF, *Royal New Zealand Air Force—Official History of New Zealand in the Second World War 1939–45* (Wellington: War History Branch, Department of Internal Affairs, 1955), hereafter Ross, *RNZAF;* Sherrod, *MarAirHist.*

[2] ComThirdFlt ltr to CominCh, dtd 3Sep44, Subj: Narrative Account—SoPac Campaign (COA, NHD), p. 4. A former chief of staff of AirSols recalls that the command chain was so confused in the beginning that "a Navy squadron commander, land based on Guadalcanal, could not prescribe the hours that the air crews taking care of the planes would work. These hours were prescribed by the CASU [Carrier Aircraft Service Unit] commander whose chain of command ran through a dubious chain of island commanders back to Admiral Halsey. When this was brought to the amazed attention of Admiral Halsey, he immediately issued orders that any air personnel under the operational control of ComAirSols would be under his direct command." LtGen Field Harris ltr to ACofS, G–3, HQMC, dtd 22Oct62.

with him to help run the new command with its enlarged scope of activity, he kept Mulcahy's veteran staff. Experience, not rank, seniority, or service, determined the assignments.

Vice Admiral Aubrey W. Fitch, as Commander, Aircraft, South Pacific (ComAirSoPac), was Admiral Mason's immediate superior. The senior officer retained two areas of flight operations under his direct control; sea search by long range Navy patrol planes and Army bombers, and transport operations by South Pacific Combat Air Transport Command (SCAT). Throughout its long and useful life (November 1942–February 1945), SCAT's complement of Marine and Army transports was headed by MAG–25's commanding officer. SCAT's operations area moved northward with the fighting during 1943, and by August's end, all regularly scheduled flights in SoPac's rear areas were being handled by the Naval Air Transport Service (NATS).[3]

Admiral Fitch, in addition to his immediate concern with the far-ranging sea search and transport operations, coordinated the multitude of air combat and support activities within the whole of Halsey's extensive command area. In administrative and logistical matters, there was a headquarters at Pearl Harbor above ComAirSoPac. Air Force, Pacific Fleet (AirPac) controlled allocation and distribution of Navy and Marine planes, materiel, and aviation personnel throughout the Pacific and was responsible for advance training and combat readiness of squadrons.

Subordinate to ComAirPac, and charged with responsibility for Marine aviation's role in his sphere, was Major General Ross E. Rowell, commanding Marine Aircraft Wings, Pacific (MAWPac). In his training, administrative, and supply capacities, Rowell dealt with a comparable headquarters within Admiral Fitch's command, Marine Aircraft, South Pacific (MASP). With Admiral Halsey's approval, MASP was established on a tentative basis on 21 April 1943 to coordinate the administrative and logistical workload of the 1st and 2d MAWs. For almost a year, until 3 December, when the Commandant of the Marine Corps was finally successful in convincing Admiral King that a separate headquarters was necessary, the 1st Wing commander headed MASP also, using officers and men from the wing headquarters and service squadrons to handle the additional duties. Throughout the period when it was operating without a T/O sanctioned by CominCh, MASP was under Major General Ralph J. Mitchell.

Neither Mitchell's 1st Wing nor Mulcahy's 2d functioned as tactical or operational commands. In common with the higher air headquarters of other American services and that of the Royal New Zealand Air Force (RNZAF) in the South Pacific, the Marine wings and their descending chain of groups and squadrons were primarily concerned with the host of collateral duties necessary to get planes in the air, armed, fueled, and manned for a combat mission. ComAirSols, and the various operational task forces he set up, planned and controlled all sorties against the enemy in the combat area.[4]

[3] ComSoPac Serial 01369, dtd 16Aug43, Subj: Opns of SCAT, in SCAT Statistics and Correspondence, 1942–1944.

[4] The subdivision of the South Pacific into combat, forward, and rear areas is succinctly described in Ross, *RNZAF*, p. 135, as: "the Combat Area in which the Allied forces were in actual contact with the enemy; the Forward

The Army counterpart of MASP was the Thirteenth Air Force which came into being on 14 December 1942.[5] Throughout most of the Guadalcanal campaign, the Army Air Forces units fighting in the South Pacific were nominally part of the Seventh Air Force based in Hawaii. Actually, most of the administrative and logistical support of the AAF squadrons and groups was channeled through the headquarters of Major General Millard F. Harmon, Halsey's senior Army commander and a veteran pilot himself. Harmon was vitally interested in achieving closer control and coordination of these units and strongly urged Washington to authorize formation of a new air force. Adding impetus to his request was the general's feeling, shared at AAF headquarters, that the Navy was not utilizing Army aircraft, particularly heavy bombers, to their fullest combat potential.

While General Harmon "had no intention of capsizing the accepted principle of unity of command," he was interested in "gaining for the AAF full operational control of its own aircraft."[6] He wanted to insure that AAF views on proper employment of its planes and personnel were fully considered. He argued that "no one can build up a force, train it, dispose it, and supply it and be held responsible for its operational effectiveness without some direct contact and influence on operational control."[7] Both Admirals Halsey and Fitch supported General Harmon's request for a separate SoPac command of AAF units, and General Marshall, agreeing, designated them the Thirteenth Air Force. By 13 January, organizational work was far enough along so that headquarters squadrons for the force and its subordinate XIII Fighter and Bomber Commands could be activated. The Thirteenth's commander, Brigadier General Nathan F. Twining, and his staff set up for work close to Admiral Fitch's headquarters on Espiritu Santo.

For much the same reason that the Thirteenth located near ComAirSoPac—to have a strong voice in the employment of its aircraft—the RNZAF assigned a senior liaison officer to Fitch's staff. On 10 March 1943, after the New Zealand War Cabinet had decided to deploy most of the country's operational squadrons in the South Pacific's forward area, a suitable command echelon, No. 1 (Islands) Group under Group Captain Sidney Wallingford, was activated to administer the RNZAF units. At the time, one New Zealand bomber-reconnaissance squadron was flying from Guadalcanal and another from Espiritu Santo, and two fighter squadrons were getting ready to move up from rear area bases. As the RNZAF strength gradually built up during 1943, the New Zealanders took an increasingly prominent role in the drive to isolate Rabaul.

Navy planes, other than those flying from carriers, were administered by Commander, Fleet Aircraft, Noumea, an echelon on a par with MASP, Thirteenth Air

---

Area which, although not in contact with the enemy, might be liable to attack, and which was organized for defense and for supporting operations in the Combat Area; and the Rear Area. As the campaign moved north, so did the boundaries of the respective areas."

[5] Dr. Robert F. Futrell, USAF HistDiv, ltr to Head, HistBr, G-3, HQMC, dtd 6Nov62.

[6] Craven and Cate, *Guadalcanal to Saipan*, p. 71.

[7] MajGen Millard F. Harmon, USA, ltr to Gen Henry H. Arnold, USA, dtd 25Nov42, quoted in *Ibid.*, p. 70.

Force, and No. 1 (Islands) Group. Rear Admiral Marc A. Mitscher had the command during the last days of the Guadalcanal campaign and kept it until 4 April 1943, when Admiral Halsey designated him Admiral Mason's relief as ComAirSols. Like Mason, Mitscher brought relatively few staff officers with him and melded them easily into the existing command setup. Another Marine, Brigadier General Field Harris, became AirSols chief of staff to replace General Mulcahy, who went to New Zealand for a short, well-deserved rest before returning to the combat area and his next tactical assignment as ComAir New Georgia.

By the time Mitscher assumed command, AirSols had shaken down into the organizational pattern it was to follow throughout the air offensive against Rabaul—three major functional task forces: fighters; medium and heavy bombers; and light bombers and reconnaissance planes. Each command had its beginning with the mixed bag of aircraft and pilots, crewmen and mechanics, that had defended Guadalcanal as the Cactus Air Force, taking its name from the island's code name. In the urgent haste of getting anyone and everything that could fly and fight to Henderson Field, niceties of squadron and group organization and concerns with service of origin were often forgotten and usually ignored. The Marine command echelons that were on the island controlled all aircraft that were sent up from the rear area and employed them according to function and performance. General Mulcahy was the first island air commander to bring in a full wing operating staff and the first to have enough planes and personnel to warrant its employment.

In the course of the air battles over Guadalcanal and its surrounding seas, two task forces evolved, one composed of fighters and the other of everything else that would fly. Until 16 October 1942, when MAG–14 relieved MAG–23 as the administrative and maintenance agency at Henderson Field, Cactus Air Force was too small to worry about intermediate echelons of tactical command. The 1st MAW commander, General Geiger, and a small operations staff directly controlled all missions. Senior Marine fighter pilots, first Lieutenant Colonel William J. Wallace, then Lieutenant Colonel Harold W. Bauer, acted as fighter commanders, and, in like manner, the most experienced pilots of other aircraft types, regardless of service, helped plan and lead strikes. When most of MAG–23's surviving pilots and aircrews were pulled out of Guadalcanal in October for a rest and a training assignment in the States, Cactus Air Force had grown to a size and complexity that precluded Geiger's direct supervision of all flights.

On the arrival of MAG–14, its commander, Lieutenant Colonel Albert D. Cooley, was named to head an Air Search and Attack Command which would control all bombing, reconnaissance, and rescue operations. Direction of fighter activity, still largely an informal tactical arrangement, remained with Lieutenant Colonel Bauer. After Bauer was reported missing in action on 15 November, Lieutenant Colonel Samuel S. Jack took over as fighter commander. On 28 December, General Mulcahy, now heading Cactus Air, established a Fighter Command and confirmed Jack as its head. When Colonel William O. Brice relieved Lieutenant Colonel Cooley as Commanding Officer of

MAG-14 on 19 December, he also assumed command of Air Search and Attack.

In April, at the time Admiral Mitscher took over AirSols, MAG-14, in its turn, was due for a rest from combat; Lieutenant Colonel Edward J. Pugh's[8] MAG-12 was in line to make the relief as Guadalcanal's top Marine administrative and logistical echelon. Mitscher decided to make MAG-12's commander responsible for running Fighter Command, and brought Marine Colonel Christian F. Schilt up from Admiral Fitch's staff to head a smaller but more easily controlled Air Search and Attack Command. Under Schilt, in what was soon known as Search and Strike Command and, by mid-summer, simply as Strike Command, were all dive and torpedo bombers and short-range reconnaissance planes. The aircraft types assigned to Strike Command insured that it would be primarily composed of Navy and Marine air crews, with a substantial leaven of New Zealanders.

At the same time the new Strike Command was formed, with its headquarters and most of its strength at Henderson Field, the medium and heavy bombers that had served under Cooley and Brice were concentrated under a separate task force at Carney Field near Koli Point. To head this Bomber Command, Admiral Mitscher approved the appointment of the Army's Colonel William A. Matheny. By reason of its assigned aircraft and personnel, Bomber Command was almost wholly an AAF organization, and its commander concurrently led XIII Bomber Command.

During the Allied approach to New Georgia and the first month of operations ashore, Admiral Mitscher continued to command AirSols. On 25 July, Admiral Halsey initiated a practise of rotating the top tactical air command among the various services, and Mitscher was relieved by the Thirteenth Air Force's commander, General Twining. Holding to the joint service nature of AirSols, Twining chose a Navy captain, Charles F. Coe, as his chief of staff and continued the assignments of many Navy and Marine officers who had been a part of Mitscher's command organization.[9] Twining's AirSols bomber chief was still Colonel Matheny, but Fighter Command went to the XIII Fighter Command's Brigadier General Dean C. Strother and Strike Command to Marine Lieutenant Colonel David F. O'Neill.

On their detachment, Admiral Mitscher and General Harris sent a message to Air Sols personnel addressed "to the best air force we know and the one best known to the Japs."[10] The organization they praised was clearly in the ascendency, already a good deal stronger than the *Eleventh Air Fleet* was or could hope to be. Although some of this Allied strength lay in increased allotments of planes and men, even more stemmed from a complete

[8] HqSq-14 Muster Roll, Apr43 (Unit Diary-Sect, RecsBr, PersDept, HQMC).

[9] "General Twining also chose a Marine aviator, Colonel William G. Manley, as his operations officer. Further, RNZAF Air Commodore Sidney Wallingford and his staff were attached to the AirSols staff (by direction of ComAirSoPac, I believe) for operational training, administrative, and logistical liaison with the RNZAF units operating directly under the operational control of AirSols task unit commanders. Thus the AirSols staff was both a joint and a combined air staff, composed of Army, Navy, Marine, and RNZAF officers." VAdm Charles F. Coe ltr to ACofS, G-3, HQMC, dtd 9Oct62.

[10] Quoted in Theodore Taylor, *The Magnificent Mitscher* (New York: W. W. Norton Co., Inc., 1954), p. 161.

reversal of form between opposing fighter aircraft. The fighter plane called the turn in the advance toward Rabaul, and the day of the Zeke had long passed. In its stead stood the Corsair, the Hellcat, and the Lightning.

ALLIED PLANES AND AIRCREWS [11]

One of the more significant events in the history of the air war in the Pacific was a crash landing on 3 June 1942 from which the plane emerged virtually intact. The pilot, a Japanese petty officer, was less fortunate and broke his neck. The plane, a Zero, had had its fuel line punctured by antiaircraft fire during a raid on the U.S. naval base at Dutch Harbor. When the luckless Japanese pilot was unable to get back to his carrier, the *Ryujo*, he attempted a landing on an isolated Aleutian island. Five weeks later, an American scouting party found the plane upside down in a marsh, its pilot dead in the cockpit.

The invaluable find, a new aircraft on its first combat mission, had been built at the Mitsubishi plant only four months before it went down. Returned to the States with careful haste, the plane was completely disassembled by engineers and technicians and rebuilt in its original undamaged condition ready for flight test. At San Diego, in the last months of 1942, the Zeke was skillfully flown against major American fighter aircraft to measure comparative performance and to fathom the Japanese plane's weaknesses. The findings were revealing and reinforced the combat experience of Allied pilots; in essence, they boiled down to one warning: "Never attempt to dog fight Zeke." [12]

While the tests revealed that the enemy fighter could out-maneuver any of its opponents at speeds below 300 miles per hour, they also confirmed defects cited in combat pilots' reports from the Pacific. The Zeke had comparatively poor diving ability, gave sluggish response to controls at high speed, and performed best at medium and low altitudes. The lack of armor for the pilot and the inflammable fuel supply both emphasized the experience of the leading Marine ace at Guadalcanal, Captain Joseph J. Foss, who stated: "If you hit a Zero at the base of its wing, it's just POW! and it disintegrates." [13] The response to these findings was twofold, to accelerate production of new American fighters that could clearly outclass the Zeke, and to emphasize aerial

[11] Unless otherwise noted, the material in this section is derived from: *USAAF InfoIntel Summary* No. 85, Flight Characteristics of the Japanese Zero Fighter Zeke, dtd Mar43; "Flight Characteristics of the Japanese Type Zero Mk II Fighter Hap," in *USAAF InfoIntelSummary* No. 43–45, dtd 30Sep43; Technical AirIntelCen, NAS, Anacostia, Representative Enemy and Allied Aircraft: Comparative Performance and Statistics, TAIC Manual No. 2 (OpNav-16-V #T302), dtd Oct44; Buchanan, *Navy's Air War*; Ross, *RNZAF*; Wesley Frank Craven and James Lea Cate, eds., *Men and Planes—The Army Air Forces in World War II*, vol. 6 (Chicago: University of Chicago Press, 1955); Craven and Cate, *Guadalcanal to Saipan*; Okumiya, Horikoshi, and Caidin, *Zero!*; Sherrod, *MarAirHist*; Ray Wagner, *American Combat Planes* (Garden City, N.Y.: Hanover House, 1960).

[12] *USAAF InfoIntelSummary* No. 85, *op. cit.*, p. 1.

[13] Quoted in "Lessons Learned in Combat with Jap Pilots," in *USAAF InfoIntelSummary* No. 40–43, dtd 10Aug43, p. 3.

combat tactics that took full advantage of the Japanese plane's limitations.

The more important Allied fighters that met the enemy attack as part of Cactus Air Force were the F4F (Grumman Wildcat) flown by the Navy and Marine Corps, and the Army's P-38 (Lockheed Lightning), P-39 (Bell Airacobra), and P-40 (Curtiss Warhawk). After the fighting on Guadalcanal ended, two new American planes began to make their appearance; one, the new standard Navy fighter, the F6F (Grumman Hellcat), and the other—the plane that was to become synonymous with Marine air for the next decade—the F4U (Chance-Vought Corsair). Like all military aircraft, these planes underwent constant modification and improvement, and the various models that fought the Japanese carried a steadily changing array of identifying numbers and letters. In general, it should be remembered that each new version of a basic plane type could do a little more than its predecessor, fly a bit faster, climb higher, or carry a greater pay load or heavier armament.

The system used by the Navy to designate its aircraft gave a letter to denote function, followed by the number of that type made by a particular company, then gave the manufacturer's code and any model numbers and letters: *e.g.*, F4U-1C, the third version (C) of the first model of the fourth fighter (F) manufactured by Chance-Vought (U).[14] The Army Air Forces used a letter function symbol with a number to indicate sequence within a type; letters appended to the number indicated the model: *e.g.*, P-38H, the eighth model (H) of the thirty-eighth fighter (P) accepted by the AAF.[15] While Allied pilots and aircrews were vitally interested in the improved performance indicated by the modification symbols, the basic designations were in more common usage and were employed interchangeably with the colorful names chosen by the manufacturers or the service concerned.

The Wildcat, a stubby, mid-wing monoplane, was the mainstay of Navy and Marine fighter strength for the first 18 months of the war. Slow, when measured against its opponents, the F4F could make about 320 miles per hour at its best altitude, 19,400 feet. With a maximum fuel load, the plane had a total range of 1,100 miles, well under the Zeke's capability; its normal combat range was 770 miles. The Wildcat was sturdily built and was equipped with self-sealing fuel tanks and armor for its vitals so that it could absorb terrific punishment. As one Marine pilot noted, "a Zero can't take two seconds' fire from a Grumman, and a Grumman can sometimes take as high as fifteen minutes' fire from a Zero."[16] As it could take it, the American carrier fighter could also dish it out, and the destructive impact of the fire of its six .50 caliber wing guns blasted hundreds of enemy planes to pieces.

Grumman's successor to the F4F, its production accelerated by the menace of the Zeke, was the F6F Hellcat, which had

[14] *Marine Corps Aircraft 1913-1960—Marine Corps Historical Reference Series No. 20* (Washington: HistBr, G-3 Div, HQMC, 1961) p. 22.

[15] Army aircraft functional symbols were: A (Attack), B (Bombardment), C (Cargo), F (Photographic), L (Liaison), P (Pursuit), and T (Training). The Navy used: B (Bomber), F (Fighter), J (Utility), N (Training), O (Observation), P (Patrol), R (Transport), S (Scout), and T (Torpedo).

[16] Quoted in Sherrod, *MarAirHist*, p. 83.

greater speed, increased range (but still not as much as the Japanese fighter), and improved maneuverability. In high compliment to its performance, the Japanese considered it to be "the only aircraft that could acquit itself with distinction in a fighter-*vs.*-fighter dogfight."[17] In appearance, the Hellcat resembled its predecessor, having the same thick-bodied fuselage and square-tipped wings with a cockpit canopy set high over the fuel tanks between the wing roots. The armament was the same, but the ammunition load was greater, and the F6F was even better protected from enemy fire. The plane could make 375 miles per hour at 17,500 feet, had a climbing rate of 3,500 feet a minute, and a service ceiling of 37,300 feet.

Developed simultaneously with the F6F, the F4U had poor downward visibility (corrected in later models) and a relatively high landing speed, both attributes that made it unattractive as a carrier fighter. While the Navy was hesitant about using the Corsair, the Marines were enthusiastic. The distinctive-looking, gull-winged monoplane was produced in such quantity that all Marine fighter squadrons in the Pacific were equipped with it by July 1943. The powerful Corsair drew a high rating when flown against the captured Zeke in the San Diego tests, with the findings: "Zeke is far inferior to the F4U-1 in level and diving speeds at all altitudes. It is inferior in climb at sea level, and inferior above 20,000 feet . . . Zeke cannot stay with the F4U in high speed climbs."[18] In combat, the disparity of performance proved equally wide; the Japanese called the Corsair "the first single-engine American fighter seriously to challenge the Zero."[19] The F4U's top speed was 417 miles per hour at 20,000 feet; it had a normal range of 1,015 miles with a maximum double that. Armed like the F6F with six wing-mounted .50s, and protected by armor and self-sealing tanks, the Corsair was deadly when flown by an experienced pilot.

Tactics developed to counter the Zeke's maneuverability capitalized on the uniformly high diving performance of American planes, and the mutual protection of two-plane sections fighting as a team and keeping each other's tail clear of enemy attackers. One plane that was singularly proficient in the high speed diving engagement was the AAF's P-38. The two-engined fighter with its distinguishing twin tail booms was designed for high altitude interception and clearly outclassed the Zeke above 20,000 feet, where it could hit maximum speeds just over 400 miles per hour. After making the initial mistake of trying to fight the Zeke on its own terms, Lightning pilots soon learned to fly high out of reach and dive to the attack, firing a nose concentration of four .50s and a 20mm cannon. The plunging dive, launched at the attackers' initiative, carried through Japanese formations and away at speeds that left little chance of being tagged by pursuers. The P-38 was capable of performing a wide variety of tasks and was particularly good as a reconnaissance and photographic plane, since it had a range of 1,500 miles with full tanks and was almost invulnerable to air attack so long as it flew above the Zeke's service ceiling.

[17] Okumiya, Horikoshi, and Caidin, *Zero!*, p. 222.

[18] *USAAF InfoIntelSummary* No. 85, *op. cit.*, p. 5.

[19] Okumiya, Horikoshi, and Caidin, *Zero!*, p. 221.

The Army's utility fighters were the P–39 and P–40, which went through continual redesign and improvement and fought throughout the war, although in gradually decreasing numbers. Both low-wing monoplanes carried the same engine, one that limited effective operations to heights below 15,000 feet. The Airacobra's engine was mounted behind the pilot and the Warhawk's was in the nose; the engine airscoop immediately behind the P–39's cockpit enclosure and the P–40's deep-throated intake under its engine gave each aircraft one of its primary identifying characteristics. Neither plane was particularly fast, the Airacobra could hit 368 miles per hour at maximum efficiency and the Warhawk 350, but both aircraft could out-dive and pull away from the Zeke at lower altitudes. Beyond that accomplishment, Allied pilots (Commonwealth air forces used the P–40 extensively, calling it the Kittyhawk) relied on superior flying skill and wingman protection when jumped by Japanese fighters. The two planes proved to be particularly suited for low-level ground support as strafers and fighter-bombers and saw most use in the latter part of the war in that role. The P–39 delivered a heavy punch with a 37mm gun in its nose firing through a hollow propeller shaft and two .30 and four .50 caliber machine guns in its wings; the P–40 carried the common American fighter armament of six .50s.

When the Lightnings and Corsairs came into common use, the pattern for the AirSols offensive deployed each type at the altitude where it performed best. A typical large-scale raid late in 1943 with bombers at 20,000 feet would have P–39s or P–40s furnishing low cover and P–38s flying at about 30–34,000 feet; between the bombers and the Lightnings would be F4Us in staggered layers of four to eight planes weaving over an area two to four miles wide. No matter where the Japanese attacked, they had to penetrate a screen of fighters operating at maximum efficiency and be ever wary of the escorts waiting to dive on them from above.

One of the mainstays of naval aviation in World War II, the dive bomber, found little favor with the AAF. The Navy's SBD–3, the Douglas Dauntless, was tried out as the A–24 in New Guinea in 1942 and won a verdict of "too slow, too limited in range, and too vulnerable to enemy fighters" from Army pilots.[29] The Army's further development of light bombers tended, thereafter, to concentrate on fighters equipped as bombers. While recognizing the faults of the SBD and working to replace it with a better aircraft, the Navy found it effective as a carrier-borne attack plane, and the Marines were sold on its accuracy against both shipping and point targets ashore. The Dauntless, a single-engine low-wing monoplane with a thick body and a narrow upswept tail, carried a crew of two, a pilot and a radioman-gunner. For defense, the gunner manned a pair of flexible mounted .30s firing to the rear from the cockpit enclosure, and the pilot controlled two .50s fixed in the nose. The dive bomber had a range of 1,345 miles with a 1,000-pound bomb load and 1,580 miles when used as a scout; its best speed was 250 miles per hour at 16,000 feet. Since, like all American combat aircraft, the SBD carried protective armor and self-sealing tanks, it was not nearly as vulnerable to Japanese fighters as was the Val, its enemy counterpart, to Allied hunters.

---

[29] Craven and Cate, *Men and Planes, op. cit.*, p. 198.

Unlike the hapless Kate, the American Navy's standard torpedo bomber throughout most of the war was a relatively high performance aircraft. The TBF (Grumman Avenger) had a top speed of 271 miles per hour at 12,000 feet and made only a few miles less when it was carrying its internally stowed torpedo. Fat-bodied, with a long canopied cockpit ending in a power-operated turret for a .50 caliber machine gun, the TBF looked a little like the Wildcat from below. More than one Japanese pilot weak on plane recognition discovered to his sorrow that the difference between the two Grummans included a ventral-mounted .30 caliber machine gun manned by the bombardier. To round off the plane's armament, the pilot at first had a .30 mounted in the engine cowling and, in later modifications, a pair of .50s mounted in the wings. The Avenger's combat range with a 1,760-pound torpedo was 1,215 miles.

In the early stages of the war, the Navy relied on its flying boats for planes that could deliver a heavier bomb load than the carriers' SBDs and TBFs. These patrol bombers, the PBY (Consolidated Catalina), PB2Y (Consolidated Coronado), and PBM (Martin Mariner), were excellent for sea search and anti-submarine work and invaluable in rescuing downed flyers; properly fitted for the job, they made effective cargo and personnel transports. The PBYs, when equipped with radar for night reconnaissance and bombing, were justly famed as the Black Cats, that made darkness a misery for outlying Japanese garrisons and the vessels that tried to supply them. All the flying boats, however, were slow and prime game for enemy fighters and antiaircraft gunners. As a result, in areas where Japanese planes swarmed, better armed and protected Army heavy bombers had to be used for reconnaissance missions, a fact that bothered AAF commanders who felt that their planes should be employed in their primary bombardment function. Eventually, as more aircraft were manufactured, the Navy procured land bombers, and the majority of its patrol planes in the latter stages of the war were land-based.

When the Navy did get four-engine land bombers, it took the AAF's B–24 (Consolidated Liberator) in both a twin-tail (PB4Y) and single-tail (PB4Y–2) version. After 1942, the Liberator gradually succeeded the B–17 (Boeing Flying Fortress) in the South Pacific campaign against Rabaul. The Fortress, aptly named for its guns and armor, could fight its way through to a target and home again, but its practical combat range was less than 800 miles when fully loaded and its bomb capacity was relatively small. General Harmon wanted the B–24 for Halsey's command because it could carry a larger bomb load over a longer distance and still hold its own with Japanese interceptors.

While the Liberator was not quite as strong defensively as the Fortress, it carried ten .50 caliber machine guns in flexible single mounts or paired in power turrets, and its 10-man crew could put up a whale of a battle. With a range of 2,850 miles carrying a 2,500-pound bomb load and 2,000 miles with 8,000 pounds, a speed of 287 miles per hour at 26,700 feet, and a service ceiling of 32,600 feet, the B–24 was also a formidable offensive weapon. One experienced Japanese fighter commander who fought in the Solomons termed the B–17 and the B–24, "the most difficult" aircraft for Zekes to shoot down.[21]

The AAF was pre-eminent in the medium bombardment field, and two of its bombers, much alike in performance, were used extensively in the Pacific—the B–25 (North American Mitchell) and B–26 (Martin Marauder). Both were twin-engine monoplanes with the same top speed, 285 miles per hour, and a bomb capacity that crept steadily upward during the fighting to reach 4,000 pounds carried over a 1,200-mile range by 1945. Medium bombers specialized in strafing and low-level bombing runs, and, as a result, both planes were flying arsenals with their six-man crews firing as many as 12 .50 caliber machine guns, and, in the B–25's case, often a 75mm nose cannon to boot. The Marauder, a sleek high-wing, needle-like aircraft, was plagued with troubles when it was first introduced and won a reputation as a difficult plane to fly and fight. In contrast, the Mitchell, a twin-tail, mid-wing plane that looked a lot like the Liberator, was a pilots' favorite. It was the B–25, rechristened the PBJ by the Navy, that the Marines adopted and used extensively during the last year of the war.

The Navy and Army used many of the same planes in another category, transports. The majority of the aircraft that were employed were military versions of one prewar commercial model, the Douglas twin-engine DC–3, which could carry a cargo payload of as much as 10,000 pounds or a 6,500-pound passenger load. The Army called this plane the C–47 (Skytrain) and the Navy dubbed it the R4D, but by any name it was the workhorse of the air, dependable and employed everywhere. The four-engine Douglas DC–4, the Army's C–54 (Skymaster), saw limited use by the Navy as the R5D, but, as the larger plane was in limited supply, in its stead the Coronado and Mariner were successfully adapted to haul cargo and passengers. Marine transport squadrons used the R4D, which, unarmed and unarmored, flew at considerable risk in the combat areas of the Pacific.

One aircraft problem, shared by all the services, and never adequately solved until late in the war, was the development of an effective night fighter. Although conventional fighters working with ground searchlights were occasionally able to down night intruders, the score was not impressive. What was needed was a fast plane equipped with radar and capable of reaching high altitudes that could work with ground controllers to find and destroy enemy attackers. For their first night fighter, the Marines reluctantly chose the twin-tail PV–1 (Vega Ventura), which was the best aircraft they could get for the job in October 1942 when the first VMF(N) squadron began forming. The plane had a rated service ceiling of 26,300 feet and a practical one well below that, and the fact that many interceptions would occur above 25,000 feet was well recognized. The Ventura, used by the Navy as a patrol plane, was a twin-engine mid-wing monoplane that could perform adequately as a low-altitude medium bomber; in its night fighter version, the plane carried radar and six .50 caliber machine guns in its nose. The men who crewed the night fighters were highly

[21] LCdr Mitsugu Kofukada, IJN, quoted in Okumiya, Horikoshi, and Caidin, *Zero!*, p. 226.

trained,[22] a description that fitted all of the Allied pilots and aircrews who were fighting in the Pacific at the time of the air offensive against Rabaul.

When the war started, American service pilots, particularly the men in command billets, were veterans of hundreds of hours of flying in all types of aircraft. Fledgling pilots and aircrewmen underwent an extensive training program before they ever joined a squadron, and kept on learning after they reported for duty. With wartime expansion, many easier-paced schedules of prewar years were discarded, but the concept of extensive ground and flight schooling was retained. In many instances in the early part of the war, when American aircraft were no better than on a par with their opponents and often no real match at all, pilot skill was all that could be relied upon. A continuous stream of experienced flying personnel returned to the States from the active war theaters to instruct the men in training and pass on life-saving tips of aerial combat. In the case of Marine trainees, who had only one adversary to get to know, all indoctrination was concentrated on beating the Japanese.

After 1942, most naval pilots were the products of a training system which included pre-flight school for basic instruction and physical conditioning, followed by three months of primary training about equally divided between ground and flight school. Next phase in the program was intermediate training, 14 weeks at Pensacola or Corpus Christi mainly spent flying, at the end of which successful students were designated naval aviators (officers) or naval aviation pilots (enlisted men), the latter group a very small percentage of the whole. At this point, Marine pilots went to Cherry Point or El Toro to begin at least two months of operational training in high performance aircraft of the type they would fly in combat, and Navy pilots reported to naval commands for similar instruction.

The Army Air Forces pilot training program was closely akin to the system used by the Navy with a primary indoctrination course, then basic flying school, followed by advanced school, and completed with transition training to handle combat aircraft. After transition, a new Army pilot, like his Navy and Marine counterpart, had 140–150 hours flying time behind him and the expectation that he would add many more before he met the enemy. The requirements for aircrewmen and mechanics of all the services were met in a manner similar to pilot training: multi-stage courses, tailored to job requirements, concluded with practice work on combat aircraft before assignment to operational units.

Once they had joined a combat squadron, new Allied flying personnel could count on the fact that they would not be expended by unbroken action. Unlike most Japanese flyers, who had to fight until exhaustion hastened their end in battle, Allied pilots and aircrews were given regular respites from the intense strain of combat flying. In Halsey's area, after a Marine squadron fought for four to six weeks under ComAirSols, it moved to the rear area while combat crews were given a week's leave in Sydney or Auckland, and then, after two weeks to a month spent training and absorbing replacements at Efate or Espiritu Santo, the squadron

[22] Sherrod, *MarAirHist*, devotes a chapter (pp. 158–169) to the development, training, and employment of Marine night fighter squadrons which gives an interesting picture of the problems overcome.

went back into action. The benefit of such a program, common to all Allied air units once the first desperate days of understrength, shortage-plagued fighting were over, was incalculable. Although it gave rise to envious and often ribald comment from ground troops, the system of combat air crew rotation to rest centers undeniably saved lives. While it was impossible to give every combat veteran in the South and Southwest Pacific a vacation from war with a taste of civilization and a temperate climate thrown in, it was feasible for flying personnel. The privilege paid off, as it was intended to, in increased operational efficiency and prolonged combat employment of veteran squadrons.

NORTHWEST FROM HENDERSON FIELD[23]

In reconstructing the course of aerial operations during CARTWHEEL, the historian is necessarily struck with the wide disparity between claimed and admitted losses by both sides.[24] Overclaiming was a common fault, and contemporary public accounts as well as many memoirs based on such material are poor sources of relative scores. A reasonably accurate picture of the results of air action can be established, however, by using Allied official reports for Allied losses and captured documents, helped out by postwar assessments, for the toll of damage to the Japanese.

Some of the inflated statistics published by the enemy can be traced to a losing side's natural eagerness to accept the most glowing pilots' victory reports and to an equal reluctance to release news of plummeting strength. Allied commands had less excuse for exaggerated totals, since concerted efforts were made to cross-check claim and counterclaim in order to keep accurate tallys.[25] Most AirSols flyers prided themselves on asking credit for

[23] Unless otherwise noted, the material in this section is derived from: ComAirPac Analysis of AirOps, Central, South, and SoWesPac, Oct43, dtd 22Nov43, and Nov43, dtd 24Dec43, hereafter *ComAirPac Analysis* with appropriate months; *SoPac ACI Repts; StrikeComd WarDs; ThirteenthAF Data; SE Area NavOps—III; SE Area NavAirOps—IV;* [BuDocks] *Building the Navy's Bases in World War II—History of the Bureau of Yards and Docks and the Civil Engineer Corps*, v. II (Washington, 1947); Buchanan, *Navy's Air War;* Craven and Cate, *Guadalcanal to Saipan;* Morison, *Breaking the Bismarcks Barrier;* Ross, *RNZAF;* Sherrod, *MarAirHist.*

[24] The difficulty of reconciling opposing figures is well illustrated by a ComAirSoPac comparison of Japanese and Allied claims and admissions of losses during four air battles of mid-1943 (*SoPac ACI Rept*, 10–16Oct43, p. 3):

Date	Japanese loss reports		Allied loss reports	
	Own	Allied	Own	Japanese
1 Apr 43	9	57	6	16
6 Jun 43	9	41	7	23
12 Jun 43	7	24	6	26
16 Aug 43	17	27	3	27

[25] The conclusion of the Army's historian of the CARTWHEEL campaign regarding both sides' claims of damage to ships and planes is: "First, Japanese claims were wildly exaggerated whereas American claims were merely exaggerated. Second, Japanese commanders apparently took the claims seriously, so that nonexistent victories often served as the basis for decision. On the other hand, American commanders, taking human frailty into account, evaluated and usually scaled down claims so that decisions were normally based on more realistic estimates of damage." Miller, *Reduction of Rabaul*, p. 232.

nothing but sure kills and observed hits, yet the nature of air warfare is such that a hasty backward glance from a swiftly maneuvering plane was often all the confirmation possible of a claim. Under the circumstances, all manner of targets were "destroyed" several times over. Nowhere was this tendency more pronounced than in air combat, for, as the historian of Marine aviation in the Pacific has cogently observed:

> Nothing is more difficult than an accurate count during an air battle in which several dozen planes are involved; it is very easy for two pilots to claim the same plane at which both are shooting. The smoking plane may get back to its base; it may not even have been actually smoking.[26]

The flashing complexity of a single aerial affray illustrates the difficulty of reconstructing a history containing a succession of such combats. The snarling tangle of interceptors and escorts is, however, only a part of the story, although it is often the part that seizes the imagination and overbalances many popular narrations. A review of air operations lends itself all too easily to a style of telling that places the individual in the forefront, somtimes to the neglect of the group effort. Certainly the highlighted pilot ace and the sharpshooting bomber crew were invaluable, and there is no disposition to downgrade their vital skills and example here, but the larger framework in which they acted will be the theme of this account.

From a Marine aviator's viewpoint, and indeed from that of many other AirSols flyers, 12 March 1943 was the start of a new chapter in the air war against the Japanese. The day marked the debut of the Corsair as a combat plane, as Major William E. Gise led VMF-124's flight echelon up from Espiritu Santo to Henderson Field. There was work for the gull-winged fighters immediately as 12 of the pilots, with only a hasty briefing on Solomons topography, flew escort for a rescue mission to Vella Lavella. Next day, the F4Us made the 600-mile round trip to Bougainville as part of the escort for B-24s attacking shipping at Buin. A similar mission on the 14th ran into about 50 Zekes over Kahili, and the meeting was not a happy one for AirSols. One Corsair was shot down and another lost in a collision with an enemy fighter. Japanese naval pilots also accounted for two of the P-40s flying low, two of the heavy bombers, and the whole top cover, four P-38s. The total enemy loss was three Zekes.

Fortunately, this inauspicious beginning was not a portent of the Corsair's future performance. The Marine pilots were new to the plane, new to combat, and had far less operational flying time, 20 hours on the average,[27] than was the case with men who arrived later in the year as replacements and reinforcements.[28] It took a little while for the F4U and the

[26] Sherrod, *MarAirHist*, p. 201.

[27] Air Technical AnalysisDiv, CNO, Interview with 1stLt Kenneth Walsh, USMC (OpNav-35 #E17), dtd 23Nov43, in MASP Survival and Interviews folder.

[28] Speaking of these later pilots from all the services, the Strike Command operations officer commented: "The efforts of operational training in the various training commands have paid a high dividend. Young pilots who haven't flown much can be given a mission that two years before the war wouldn't have been given to a division of squadron commanders." AirIntelGru, DivNavIntel, CNO, Interview of LCdr H. H. Larsen, USN (OpNav-16-V #E31), dtd 27Feb44, pp. 1-2.

men who flew it to get completely shaken down in combat, but when Admiral Yamamoto launched his *I Go* attacks in early April, the Corsairs were ready and able to meet the best pilots and planes the Japanese could send up. Confidence in the F4U grew as its record of victories mounted, and pilots could say as one veteran did; "The Corsair was a sweet-flying baby if ever I flew one. No longer would we have to fight the Nip's fight, for we could make our own rules." [29] Respected but unregretted, the Marines' Wildcats swiftly passed from the scene, and, by 2 July, all eight of the fighter squadrons under MASP were flying Corsairs.[30]

One of the greatest assets of the F4U was its range; unlike the F4F, the swifter fighter could cover the distance from Guadalcanal to southern Bougainville and return, carrying fuel to spare for air combat. Since it flew best at the altitude where Zekes were wont to intercept, the Corsair eased the lot of the Warhawks and Lightnings, letting each type fly at a height where it was on a par with or superior to the enemy fighter. With adequate escort available, daylight raids by Liberators and Fortresses on targets at Ballale, Buin, and Kahili increased. Fighter sweeps into the northern Solomons were flown regularly.

Japanese airfields closer to Guadalcanal, Munda, and the little-used liaison strip at Vila, were not neglected, however, while the heavy bombers and long-legged fighters ranged beyond the New Georgia Group. Strike Command sent a steady procession of SBDs and TBFs to New Georgia, accompanied by AirSol's usual varied collection of fighters, to keep the enemy runways bomb-cratered and their defending gun crews fearful. Despite the pounding it took, the Japanese kept Munda in use as an emergency strip, and its threat was constant. Any letup in the Allied air attacks and Rabaul's 300-plane garrison could begin staging raids through Munda to hit the swelling complex of fields on Guadalcanal.

Without auxiliary tanks, Navy and Marine dive bombers could not join in attacks on Bougainville targets and return with safety, but torpedo bombers could make the trip and did. The TBFs were used primarily on night harassing missions, hitting shipping and airfield installations by flare light. Enemy attempts at interception, using day fighters and searchlights to locate targets, were even less successful than similar Allied attempts.

Aside from their more common employment as bombers, the Avengers were occasionally used for another type of mission, offensive aerial mining, with results hard to assess. On the night of 20 March, Major John W. Sapp led 42 TBFs from his own VMSB-143 [31] and three Navy squadrons up to Bougainville to mine the waters off Buin-Kahili. While 18 Army heavy bombers dropped clusters of fragmentation bombs on shore targets and attracted the attention of searchlights and antiaircraft, the TBFs slipped down to 1,500 feet and parachuted a pattern of 1,600-pound magnetic mines into the enemy harbor. None of the Avengers was hit, and the entire raiding force got back

---

[29] Col Gregory Boyington, *Baa! Baa! Black Sheep* (New York: G. P. Putnam's Sons, 1958), p. 129.

[30] VMF-213 got its new F4Us on 11 March, VMF-121 on 15 April, VMF-112 and -221 on 19 May, VMF-122 on 16 June, VMF-214 on 19 June, and VMF-123 on 2 July.

[31] VMSB-143 was redesignated VMTB-143 on 31 May 1943.

safely. On the following night, 40 torpedo bombers and 21 B–17s and B–24s mounted another mining strike to the same area; again the Japanese went scoreless.

Careful study by the Navy indicates that this mine plant probably claimed two merchantmen and damaged a destroyer, but the results of mining in poorly charted enemy waters can never be completely known. Admiral Halsey was pleased enough with the reported damage to order a resumption of aerial mining in May, and on the 19th, 30 TBFs from VMSB–143 and VT–11, with a supporting flight of six heavy bombers carrying 100-pound fragmentation bombs, sortied for Buin-Kahili. This time enemy antiaircraft ignored the relatively light diversionary attack and concentrated search lights and guns on the TBFs as they parachuted their mines. The Navy and the Marine squadron each lost two planes to the hail of defending fire. On 20 May, four Liberators and four Fortresses with a mixed load of 100- and 300-pound bombs, accompanied 30 mine-laden Avengers to the Shortlands. Surprised by the Allied attack, the Japanese engaged the bombers and devoted little fire to the mining planes; all TBFs returned to base after laying their deadly cargo. The Avenger crews felt themselves lucky to have escaped whole, as the enemy fire was heavy and the mined area was close inshore.

A final mission of the mining program, the target again Buin-Kahili waters, was mounted on 23 May. About midnight on the 22d, while the main striking force was taking off from Guadalcanal, five B–24s hit Kahili's airstrip and defenses, breaking off their attack when a flight of 14 B–17s arrived to hit shore defenses during the mining run. Of 26 TBFs employed, only 20 carried mines, while two Navy and four Marine planes each had a load of four 500-pound bombs. Two of these Marine Avengers served as prowlers, unsuccessfully seeking enemy shipping during the attack, and the remaining bomb-loaded torpedo planes attacked searchlights and antiaircraft positions on offshore islands. The bombing was effective; enemy fire was erratic and probing lights were knocked out almost as soon as they flashed on. No AirSols planes were downed, and all returned without mishap, helped along the way by the flares that a RNZAF Hudson (Lockheed PBO) dropped near Vella Lavella as navigational aids.[32]

One of the mines of this series was credited with causing damage to the enemy light cruiser *Yubari* on 5 July, but otherwise nothing definite was learned of the mission's success. TBFs were not used for mine laying again until after the Bougainville landing, but Strike Command had learned that aerial mining in constricted and heavily defended waters required effective supporting and diversionary attacks. Many Avenger pilots were convinced that, without such support, losses among mine-laying planes would be prohibitive.

The more usual run of Allied air raids on Buin-Kahili and the Shortlands stepped up appreciably after the Seabee-constructed airfields in the Russells opened for business. The advance echelon of Lieutenant Colonel Raymond E. Hopper's MAG–21 landed on Banika on 14 March, the rest of the group arrived on 4 April, and the first of the island's two fields was unofficially christened on the 13th, two days before its completion, when a dam-

[32] CO, StrikeComd, AirSols ltr to ComAirSols, dtd 29May43, Subj: Mine laying in Kahili and Shortland Island Areas.

aged P–38 made an emergency landing. By the time both airstrips were in full operation in late June, MAG–21's three fighter squadrons were being employed primarily as escorts for bombers with interception scrambles limited to the intrusion of an occasional snooper picked up on radar.

Following the enemy's unsuccessful *I Go* attacks of early April, Japanese fighters and bombers steered clear of Guadalcanal in daylight for several weeks. Then on 25 April, a force of 16 Bettys and 20–25 Zekes was spotted southeast of New Georgia by a flight of four Corsairs led by Major Monfurd K. Peyton. The Marine planes, all from VMF–213, were returning to base from a strafing mission at Vila. When the F4Us circled to intercept the bomber formation, they were jumped by enemy fighters, but bore in despite the odds. Five Zekes were gunned down in the resulting affray and two Corsairs and one pilot were lost, but the entire Japanese attack formation was turned back.

While daylight raids were scarce, enemy night attacks on Guadalcanal and Banika, sometimes in formations as large as eight bombers, were frequent. The physical damage done on such visits was meager, but the wear and tear on nerves and tempers was great, and many a fervent wish for an effective night fighter was voiced by troops chased into trenches and dugouts by "Condition Red" alerts. A squadron of the AAF's first night fighters, P–70s, which began operating from Guadalcanal in March was generally ineffective, as the plane could not operate at the heights where enemy bombers flew. Lightnings practiced in night work easily reach the required altitude and occasionally flamed an unwary raider caught in the glare of probing searchlights, but a lack of airborne radar limited the P–38s' effectiveness. Not until late fall, when the first Navy and Marine night fighter squadrons began operating in the South Pacific did the Allies achieve control of the skies over their positions at night as well as in daytime. The dawn-to-dusk mastery of the air by AirSols interceptors was conclusively demonstrated in the bloody repulse of the series of raids which the Japanese launched against Guadalcanal between 7 and 16 June 1943.

Reinforced by 58 fighters and 49 bombers that the *Combined Fleet* transferred from Truk to Rabaul on 10 May, the *Eleventh Air Fleet* sought to check the Allies' aggressive air attacks by hitting at the ultimate source of AirSols offensive strength, its fighters. On 7 June, Admiral Kusaka sent approximately 80 Zekes, a number of them new Hamp models with bombs carried under the wings, flying toward the Russells and spoiling for a fight. Warned by coastwatchers, Fighter Command obliged the enemy naval pilots by sending up more game than they wanted, 104 interceptors, with about half deployed over the shipping at Guadalcanal and the rest stacked in layers between the Russells and New Georgia. For about an hour and a half, Japanese and Allied fighters tangled in a blinding rain storm all over a 50-mile-long battle zone. Finally, after the defenders shot down 23 Zekes, and antiaircraft guns on Banika accounted for a 24th,[33] the raid was turned

[33] Postwar research by Japanese military historians indicates that nine carrier fighters failed to return from this attack and five were heavily damaged. Chief, WarHistOff, DefAgency of Japan, ltr to Head, HistBr, G–3, HQMC, dtd 29Mar63, hereafter *Japanese Air Comments*.

691-360 O—63——31

back short of Henderson Field. Allied losses were seven fighters in combat, with all pilots recovered but one, and two planes crashed as a result of the foul weather.

On 12 June, Admiral Kusaka again tried a fighter sweep with about 50 Zekes and experienced the same dismal failure. Intercepted north and west of the Russells by 49 Allied fighters of the 90 scrambled, the Japanese attacking force lost half its strength before it turned back.[34] Five American fighters were downed and one RNZAF P–40; four of the pilots survived to be picked up by rescue amphibians. Coastwatchers reported Japanese bombers had come south past Bougainville during the day, but none showed up in the lower Solomons when the Zekes failed to clear a path.

Despite its heavy losses, a month's allotment of replacement aircraft in two days of combat, the *Eleventh Air Fleet* staged a third big attack on 16 June. Prompted by sightings of large numbers of ships moving into the waters off Guadalcanal during the build-up for the TOENAILS operation, Admiral Kusaka this time sent at least 24 dive bombers along with 70 of his fighters. Amply forewarned by coastwatchers, and vectored into position by New Zealand ground intercept radar, AirSols fighters virtually destroyed the raiding force. Seventy-four of the 104 planes sent aloft by Fighter Command made contact, and no two accounts agree on the exact total of damage, but one thing is certain, the relative score was incredibly high in favor of the defenders. AirSols pilots originally claimed 49 Zekes and 32 Vals; ship and ground antiaircraft fire added 17 planes to that count. Six Allied fighters were destroyed and five pilots were lost. The few bombers that got through to Guadalcanal before they were shot out of the sky damaged one cargo ship sufficiently to force it ashore and set an LST afire. Enemy records are curiously blank regarding this raid; there is no doubt, however, that the number of planes that got back to Rabaul was woefully low. One lucky survivor who returned with tales of substantial Allied shipping losses found no witness to substantiate or dispute his fable.[35]

The *Eleventh Air Fleet* had no time to lick its wounds and recover. Less than a week after the 16 June attack, Marine raiders landed at Segi, heralding the launching of the drive to seize Munda airfield. Reacting to the grave threat posed by Allied seizure of bases in the New Georgia Group, Kusaka threw every plane he had against the attacking forces. To give his subordinate badly needed reinforcements, Admiral Koga ordered the air groups of the *2d Carrier Division* at Truk forward to Kahili. The commitment of 150 additional Zekes, Vals, and Kates to the Solomons air battles, a move that crippled the offensive power of the *Combined Fleet*, precipitated violent air action, but had little overall effect on the outcome of the campaign. The balance of air power was now so overwhelmingly

[34] According to recent Japanese research, "77 Zero fighters took off to engage in the aerial combat on June 12, of which 3 turned back to their base without getting to their destination. Thus, the Japanese lost 6 Zero Fighters and one fighter made an emergency landing." *Ibid.*

[35] Cited in Morison, *Breaking the Bismarcks Barrier*, p. 140. The Japanese War History Office indicates that 13 Vals and 12 Zekes failed to return from this attack. *Japanese Air Comments.*

with AirSols that the final result could not be doubted.

The imbalance was found not only in relative quantity and quality of aircraft, but also in what the enemy *6th Attack Force* commander called the "world of difference between the ability of the Japanese and Americans to construct air bases in the combat theaters." [36] While taking judicious note that most Japanese forward airbases had been built and maintained by "primitive manpower," in contrast to those that seemed to be the product of "mass mechanical invasion on jungle, coral, and rock," the enemy officer made an even more significant assessment, recalling:

> One of the major points which has too often been overlooked in an evaluation of fighting power, but which determined to a large extent the efficiency of air units, was that of hygienic installations. Japanese engineers paid scant attention to this problem, dismissing the pressing matter of mosquito protection by simply rigging mosquito nets in personnel quarters. Sanitary facilities were basically crude and ineffective; certainly they contributed nothing to the morale of ground and air crews.
>
> The Americans, by contrast, swept clean vast areas surrounding their ground installations with advanced mechanical aids. Through exhaustive disinfecting operations, they banished flies and mosquitos from their airbases and paid similar attention to every phase of sanitation and disease.
>
> Some may consider this a prosaic matter, but it was vital to the men forced to live on desert islands and in the midst of jungles swarming with disease and insect life. The inevitable outcome of such neglect was a tremendous difference in the health of the American and Japanese personnel who were assigned to these forward air facilities.[37]

Wracked by disease, starved for proper foods, living in wretched squalor, with AirSols night intruders banishing sleep, Japanese flyers at Kahili were literally wearied to the point where they were often victims of their own poor reactions in combat. The living conditions of mechanics and armorers were considerably worse than those of flight crews, and the numbed senses of maintenance personnel working through the night to patch damaged planes unwittingly caused the deaths of many flyers. Topping the bitter cup of enemy naval aviators was the knowledge that they had slight chance to live if their planes went down any distance from a Japanese base. A gross wastage of veteran pilots and crews occurred because the Japanese had no air-sea rescue apparatus comparable to the extensive Allied setup. The *2d Carrier Division*'s operations officer believed that "naval commanders were so afraid of the possible sacrifices which might be the consequences of attempting to rescue our crews which were shot down that often we abandoned on the open sea those men whom we could obviously have saved." [38] The fault was not entirely with commanders either, as the Japanese staff officer further noted that "our own combat men, the flying mates of

---

[36] LCdr Mitsugu Kofukuda, IJN, quoted in Okumiya, Horikosi, and Caidin, *Zero!*, p. 229.

[37] *Ibid.*, p. 230. Commander Kofukuda's comments prompted a former commanding officer of MAG-24 to call attention to the equally high quality of the aircraft maintenance effort which complemented the know-how of airbase construction, and in particular "to the training, leadership, and ingenuity of the Marine ground crews who kept a high percentage of aircraft in operation, to the naval aviation supply system that got the goods to them, and to the designers and manufacturers who produced special handling equipment to reduce the manpower required and above all to speed accomplishment of the tasks." Col Lewis H. Delano ltr to CMC, dtd 27Nov62.

[38] Okumiya, Horikoshi, and Caidin, *Zero!*, p. 312.

the same men who were shot down and adrift at sea, would not, even under orders, take any unnecessary chances to save their lives." [39]

If Japanese flyers "accepted their abandonment stoically," [40] there was no need for such resignation on the part of any Allied aircrewman who survived a crash or bailed out from a plunging wreck. In the vicinity of home fields, small amphibians were quick to the scene of any water landings, racing crash boats for the rescue honors. Hudsons and Venturas were stationed on the return routes of Allied air attack forces to spot downed planes and protect and keep in sight crew survivors. Flying boats, nicknamed Dumbos after a popular cartoon character, a flying elephant, made the pickup under the cover of a heavy fighter escort. Many men who swam or paddled ashore on the various islands owed their lives and freedom to friendly natives who cared for the injured and got the flyers back to the nearest coastwatcher, often after near-incredible adventures dodging enemy searching parties. In the Bougainville vicinity, where the Melanesians were less well disposed toward the colonial government, downed aviators were sometimes turned over to the Japanese, but the coastwatchers were usually able to call upon AirSols for a bombing and strafing mission against any village that actively supported the enemy. The harsh punishment, and the reason for it, were not lost on the offenders.

Bougainville and its offshore islands were by no means neglected during TOENAILS, even though most of the AirSols effort was in direct support of New Georgia operations. Dauntless dive bombers, helped along by 55-gallon belly tanks to increase their range, began joining Buin-Kahili strikes in early June, and they continued to hit such targets when their presence was not more urgently needed by ComAir New Georgia. Mitchells made their first appearance in Bomber Command's array in June, and the medium bombers too had a hand in the reduction of Bougainville installations when General Mulcahy did not put in a call for their support against enemy forces on New Georgia. Most of the missions flown against targets in the northern Solomons hampered Japanese efforts to support their beleaguered troops in the central Solomons.

One such strike, larger than most but still representative of many others, was mounted on 17 July, after aerial reconnaissance had disclosed that a large concentration of shipping lay off Buin. Led in by seven B-24s which bombed from high altitude, an attack force of 37 SBDs and 35 TBFs covered by 114 fighters dove on the enemy vessels with the Corsairs of the escort keeping close company. Zekes rising from Kahili's runways to intercept were shot down by the zooming F4Us almost before the enemy pilots knew what hit them. Surprise seemed to be complete, and AirSols flyers claimed 47 Zekes and five floatplanes, with 41 of the 52 credited to pilots of the four participating Marine fighter squadrons.[41] Excited Avenger and Dauntless crews were sure that they had sunk four destroyers and an oiler; postwar assessment gave

[39] *Ibid.*

[40] *Ibid.*

[41] *Japanese Air Comments.* In this instance, as in others previously cited, the loss figure supplied by the Japanese War History Office—13 Zekes—appears to be too low in relation to the carefully checked contemporary credits to Allied flyers.

them the destroyer *Hatsuyuki* and damage to three others. The Japanese got slim pickings for their heavy losses: one SBD, one TBF, two P–38s, and one F4U.

In case the Japanese did not absorb the lesson that a 192-plane strike taught on the 17th, another equally heavy attack on the same area was made on the 18th, again with considerable damage to enemy shipping. Then, on the evening of the 19th, a Black Cat spotted and trailed an enemy task force near Choiseul, giving the lead to Strike Command which sent six Avengers up from Henderson Field, each carrying a 2,000-pound bomb. Dropping their loads from masthead height, the TBFs sank the destroyer *Yugure* and put a hole in the side of the heavy cruiser *Kumano*. A further attack during darkness by five more Avengers and eight Mitchells failed to score, but another destroyer, the *Kiyoami*, was sent to the bottom after daylight on the 20th by skip-bombing B–25s. Two days later, a shipping strike of 12 B–24s, 16 SBDs, and 18 TBFs covered by 122 fighters caught the seaplane tender *Nisshin* off Bougainville's south coast and sent it to the bottom with all the 24 medium tanks and most of the 600-odd troops it carried.

The ceaseless attacks on targets in the northern Solomons, while the fighting on New Georgia coursed its slow way to an end, left no doubt in Japanese minds of the general area of the next major Allied objective. When Munda airfield finally fell to the XIV Corps in early August, the enemy's only valid reason for continuing the fight in the central Solomons was to win time to strengthen Bougainville defenses.

In August and September, Seabees worked feverishly on the fields at Segi, Munda, the small island of Ondonga six miles northwest of Munda Point, and Barakoma on the east coast of Vella Lavella. As these Allied airbases came into heavy use, the forward fields of the Japanese became untenable. Munda had been rendered impotent by continued strikes mounted from Guadalcanal and Banika, and now Ballale and Kahili were emptied of planes by similar relentless attacks. Japanese auxiliary airstrips on Bougainville at Kara near Kahili, at Tenekau and Kieta on the northeast coast, and at Bonis on the Buka Passage were never finished or were knocked out of action almost as soon as they came into use.

In mid-October, headquarters of Strike Command, Fighter Command, and AirSols all moved to New Georgia, keeping pace with the shorter-ranged aircraft that were crowding into the expanding airdromes on the newly won islands. Bomber Command's Liberators continued to fly from Carney and Koli Point Fields on Guadalcanal, and its Mitchells were based in the Russells. The B–24s and PB4Ys made Buka their special target, and Japanese ships and barges drew a good share of the attention of the heavily gunned B–25s. To handle the enemy bases in southern Bougainville, Strike Command sent about a hundred planes a day in the last two weeks of October to bomb and strafe runways, defending antiaircraft, and whatever else seemed a profitable mark.

Since the SBD–TBF attack formations had abundant fighter cover, most opposition came from the enemy guns ringing the airfields. The tactics developed by Strike Command to deal with antiaircraft fire were calculated to give the Japanese gunners nightmares. As Lieutenant Colonel O'Neill's operations officer, Lieutenant Commander Harold H. Larsen, USN, outlined the procedure, the strike setup against Ballale, Kahili, and Kara was:

> . . . to have the dive bombers go down and hit the guns, with as many diving simultaneously as possible. Torpedo planes came down and hit the field with a lot of variations, due to the fact that the Japs soon caught on that the torpedo planes would hit the field and they would come out of their holes after the dive bombers went away and wallop the torpedo planes as they pulled out. So we had little sneakers arranged here and there—some dive bombers would lay up in the air until the SBDs had all gone over, and then come down and hit some of the Japs who got sassy; or they would wait until after the torpedo planes had finished their attacks and come down; or a group of four to six torpedo planes would come down in the center of the torpedo plane attack on the field and hit any guns that happened to open fire.[42]

The air offensive against the remaining Japanese positions in the Solomons, was so extensive in nature by the time of the Bougainville operation that local airbase commanders, or air operations officers as they were usually designated, acted as deputies for ComAirSols in tactical command of all aircraft assigned to their fields. Through local headquarters of the type commands, Fighter, Strike, and Bomber, directions were issued for various missions, with joint operations coordinated by the AirSols operations officer. On the eve of the Empress Augusta Bay landings, local tactical air control had been passed to Commander Air Guadalcanal, except for heavy bomber sorties which were handled by the Air Operations Officer, Koli Point, and to local commands at Banika, Segi, Ondonga, and Barakoma. Fighter and Strike Commands directly controlled all missions originated from Munda's fields.[43]

Perhaps the best way of showing how much the precursor Cactus Air Force of 1942 had grown in a year of steady reinforcement, aircraft improvement, and operational success is to outline AirSols strength at the start of the amphibious campaign in the northern Solomons:

Munda	
VF(N)-75	6 F4U-2*
12th Fighter Sqn	25 P-39
VC-24	24 SBD
VC-38	9 SBD
VC-40	9 SBD
VMSB-144	24 SBD
VMSB-234	10 SBD
VMSB-244	24 SBD
VC-38	9 TBF
VC-40	9 TBF
VMTB-143	10 TBF
VMTB-232	20 TBF
17th Photo Sqn	3 F5A+
Barakoma	
VMF-212	20 F4U
VMF-215	20 F4U
VMF-221	20 F4U
Ondonga	
70th Fighter Sqn	25 P-39
VF-17	36 F4U
No. 15 RNZAF Sqn	21 P-40
No. 17 RNZAF Sqn	21 P-40
Segi	
VF-33	24 F6F
VF-38	12 F6F
VF-40	12 F6F
Russells	
VMF-211	20 F4U
VMF(N)-531	5 PV-1
VB-138	12 PV-1
VB-140	15 PV-1
70th Bomb Sqn	16 B-25
75th Bomb Sqn	16 B-25
390th Bomb Sqn	16 B-25

[42] Larsen interview, *op. cit.*, p. 1.

[43] ComAirSols OPlan No. T1-43, dtd 21Oct43, in ComAirSoPac Correspondence; OpOs and Plans folder.

Guadalcanal

44th Fighter Sqn	25 P-38
Reserve (AAF)	10 P-40
Reserve (AAF)	10 P-39
VB-102	15 PB4Y
VB-104	12 PB4Y
5th & 307th Bomb Groups	48 B-24,4 SB-24*
No. 3 RNZAF Sqn	15 PV-1
VP-23	12 PBY5
VP-54	6 PBY5A
VP-71	15 PBY5
VD-1	7 PB4Y (Photo)
17th Photo Sqn	3 F5A*
VS-54	14 SBD3
VS-64	8 OS2U3*
VS-68	8 OS2U3*
SCAT	21 C-47/R4D [44]

The Japanese considered that the seizure of a foothold at Torokina and the construction of airfields there was the move that "decided the fate of Rabaul." [45] Once Marines were ashore on Bougainville, and Seabees and engineers were at work with bulldozer and grader, the neutralization of the airfields on Gazelle Peninsula was inevitable. Before the Japanese pulled out their air garrison, however, four months of heavy air attacks, begun by SWPA Allied Air Forces, intensified by South and Central Pacific carrier planes, and finished by AirSols, were necessary.

[44] *Ibid.*, Annex A. Aircraft not previously identified in the text marked * are: F4U-2, the night fighter version of the Corsair; F5A, the photo-reconnaissance version of the P-38; OS2U3, the Chance-Vought Kingfisher, a single float scout plane; SB-24, a radar-equipped Liberator developed for night bombing. Listed under Guadalcanal are planes actually based at Florida Island which came under control of Commander Air Guadalcanal. Although this operation plan showed two P-38s as being attached to VMF(N)-531, the former commanding officer says that the squadron controlled only its own PV-1s. BGen Frank H. Schwable ltr to HistBr, G-3, HQMC, dtd 7Nov62.

[45] *SE Area NavAirOps*—IV, p. 20.

Knockout by Torokina

SOUTHWEST PACIFIC AND CARRIER AIR SUPPORT [1]

On 23 February 1942, a month after its fall to Japanese landing forces, Rabaul was bombed by six B–17s of the Fifth Air Force. This attack, mounted from Townsville, Australia, was the first of a series of raids by small groups of Allied heavy bombers on the enemy base. Between March and August, SWPA planes dropped an average of 130 tons of bombs a month on targets at Rabaul. On the 7th of August, 18 Flying Fortresses took off from Port Moresby, climbed over the 13,000-foot barrier of the Owen Stanley Mountains, and flew across the Solomon Sea to strike Vunakanau Field in support of the Marines landing at Guadalcanal. Frequently thereafter, American fortresses and Australian Catalinas bombed the airfields and crowded harbors at Rabaul in order to harass the Japanese and inflict as much damage as possible.

Only a relatively small number of Allied planes had enough range to participate in Rabaul raids, and those few were husbanded carefully by scheduling most strikes at night. Before any really sizeable daylight air attack could be launched, bases closer to the enemy stronghold had to be taken to serve as home fields and staging points for fighter escorts and light and medium bombers. Consequently, the interest of MacArthur's planners in acquiring airfields on the eastern slopes of the Owen Stanleys and on Woodlark and Kiriwina was fully as great as the eagerness of Halsey's staff to move into the New Georgias. Both area commanders wanted a clearer shot at Rabaul with longer times over target and more protection for bombers.

In part, the heavy losses of Japanese naval aircraft in the spring of 1943 during the Bismarck Sea battle, the *I Go*

---

[1] Unless otherwise noted, the material in this section is derived from: *ComAirPac Oct–Nov43 Analyses*; CTF 38 Rept of First Rabaul Strike—5Nov43, dtd 8Dec43; CTF 38 Rept of Second Strike on Shipping in the Rabaul Area—11Nov 43, dtd 8Dec43; CTG 50.3 AR of attack on enemy ships at Rabaul and subsequent enemy aircraft raid on TG 50.3, dtd 9Dec43 (all in COA, NHD); *SE Area NavOps—III; SE Area NavAirOps—IV*; Lt Roger Pineau, USNR, "Summary of Enemy Air Raids on Rabaul," 12Oct43–29Feb44, n.d., compiled for the Morison naval history project from Japanese documents; Maj Harris G. Warren, USAAF, "The Fifth Air Force in the Conquest of the Bismarck Archipelago, November 1943–March 1944," dtd Jan46 (AAF HistStudy No. 43, USAF Archives, Maxwell AFB), hereafter Warren, "FifthAF in the Bismarcks;" Craven and Cate, *Guadalcanal to Saipan*; Halsey and Bryan, *Admiral Halsey's Story;* George C. Kenney, *General Kenney Reports, A Personal History of the Pacific War* (New York: Duell, Sloan and Pearce, 1949), hereafter Kenney, *Reports*; Morison, *Breaking the Bismarcks Barrier*; George Odgers, *Air War Against Japan 1943–1945—Australia in the War of 1939–1945 (Air)* (Canberra: Australian War Memorial, 1957), hereafter Odgers, *RAAF Against Japan*; USSBS, *Campaign Against Rabaul*.

fiasco, and costly attacks on Guadalcanal in June opened the way for CARTWHEEL advances. During the summer, while AirSols planes beat off enemy aircraft attacking the New Georgia beachheads and raided the northern Solomons in their turn, General Kenney's Allied Air Forces concentrated on cutting down Japanese Army air strength on New Guinea. Rabaul, in a sense, had a breathing spell, but it was only a lull before a devastating storm broke.

The first telling blow of the air offensive that eventually neutralized Rabaul was struck on 12 October 1943. On that date, in the first of a series of raids planned in support of the pending Bougainville operation, Allied Air Forces mounted the largest strike of the war against Rabaul. General Kenney later stated that every SWPA plane "that was in commission, and that could fly that far, was on the raid"[2]—87 B–24s, 114 B–25s, 12 RAAF Beaufighters, 125 P–38s, and 11 weather and reconnaissance planes. Operational accidents and mechanical failures on the long haul from takeoff to target forced 50 of the fighters and bombers to turn back, but the successive attacking waves had strength to spare to overwhelm the 32 Zekes that rose to intercept.

The Mitchells came first, speeding low over the waters of St. George's Channel to avoid discovery by Japanese coastwatchers and radar. At the mouth of the Warangoi River, the nine squadrons of B–25s and their cover of P–38s roared inland just above the jungle to strafe enemy planes at Vunakanau and Rapopo and to leave a deadly litter of 20-pound parachute fragmentation (parafrag) bombs in their trace. This initial attack surprised the Japanese, and there was little effective opposition to the Americans. The RAAF Beaufighters coming in behind the Mitchells were not so lucky. Delayed in their takeoff from Dobodura by the cloud of dust raised by the B–25s, the Australian light bombers missed rendezvous with their escort over Kiriwina and had to fight their way through Zekes to complete their mission of strafing Tobera. After the Beaufighters completed their attack on the airfield, the Liberator squadrons, each plane carrying six 1,000-pound bombs, struck at shipping in Simpson Harbor. Happily claiming a staggering total of damage—one B–24 squadron reported 48 hits for 48 bombs dropped—the big bombers got back to Port Moresby after losing only two of their number. The total of Allied planes shot down during the day's action was five.

While the Japanese lost nothing like the "extremely optimistic"[3] figures for ship and plane losses estimated from the original claims of returning aircrews, the actual destruction wrought was significant. One 6,000-ton transport and several smaller ships were sunk, three destroyers and a bevy of small craft were damaged. Japanese records indicate that two of their interceptors were downed[4] and 45 planes destroyed or damaged on the ground. Allied aerial photographs indicated a much higher figure for enemy aircraft losses, although one smaller than the 177 of the first excited claims.

---

[2] Kenney, *Reports*, p. 313.

[3] Craven and Cate, *Guadalcanal to Saipan*, p. 321.

[4] Recent Japanese research indicates that four, rather than two fighters were lost, but disclaims the damage to the destroyers. *Japanese Air Comments.*

JAPANESE ANTIAIRCRAFT CREWS *abandon their 75mm guns at Rabaul during strafing runs by Army Air Forces B–25s.* *(USAF A–26636AC)*

PARAFRAG BOMBS *drop toward Japanese Bettys in revetments at Vunakanau field during a B–25 attack in October 1943.* *(USAF 25899AC)*

On 13 October, a heavy schedule of follow-up attacks was launched with an early dawn raid on shipping in Simpson Harbor by a squadron of RAAF Beaufort torpedo bombers. Visibility was poor during the Australians' attack, and the weather changed for the worse soon after, forcing 70 heavy bombers and 100 fighters already en route to turn back 150 miles from their objective. Continuing bad weather put off the next large strike until the 18th, and then only part of the attacking force got through, 54 B-25s that flashed in under a 200-foot ceiling of storm clouds to hit Tobera, Rapopo, and shipping in the harbor. Again Allied damage claims were unusually high and admitted Japanese losses questionably low. Kenney's flyers told of shooting down 10-12 planes and destroying 41 more on the ground and of sinking a small freighter and a corvette; Japanese records admit the loss of three interceptors and a 100-ton submarine chaser, while claiming nine B-25s against the actual loss of three.

The disparity of claim and counterclaim continues through the reports of 100-plane SWPA raids on 23, 24, and 25 October. More than 175 enemy aircraft were reported as destroyed or heavily damaged in these attacks, while only five Allied fighters and bombers were shot down. The Japanese admitted loss or damage to about two-fifths of the number of planes claimed, and, in their turn, decided that they had made 36 sure kills in the same three days of air battles. Regardless of Allied exaggerations, the actual Japanese losses were high, and the combat effectiveness of the *Eleventh Air Fleet* plummeted.

On 29 October, General Whitehead's 1st Air Task Force, which was the controlling headquarters for the SWPA attacks on Rabaul, sent 46 Liberators with an escort of 57 Lightnings against Vunakanau, where they reported destroying 9 planes on the ground and 16 in the air. Massive attacks planned for the next two days to support the Cape Torokina landings were grounded by unfavorable weather reports, and the dreary picture looked the same on 2 November. Two reconnaissance planes discovered, however, that the sky was clearing over the target, and that Rabaul's harbor was jammed with ships and its airfields held 237 planes of all types. The planned raid was quickly rescheduled.

Eighty B-25s with 80 P-38s were en route to the tempting target by 1100. Two squadrons of Lightnings led the way in a fighter sweep of the harbor, and were closely followed by four squadrons of Mitchells which strafed antiaircraft positions ringing Simpson's shore. This suppressive attack opened the path for the rest of the Mitchells, 41 in all, to hit shipping from a new approach route by swinging in over Crater Peninsula, North Daughter, and Rabaul town. Attacking through a frantic swarm of enemy fighters, the B-25s dropped to mast-top height to skip-bomb and strafe in the scramble of wildly dodging ships. Cruisers and destroyers fired their big guns into the water to send up towering columns of spray in the path of the attacking planes, while antiaircraft batteries fired without letup.

The damage claims that came out of this hotly-contested fight were high as usual, but the actual destruction was high, too. Two merchant ships and a mine sweeper were sunk, and a 10,000-ton oiler plus a number of smaller ships were hit. The Japanese admitted losing 20 planes, and the Allied Air Forces had eight B-25s and nine P-38s shot down. In General Kenney's opinion, the Japanese planes his attack force encountered on 2 November "put up the toughest fight the Fifth Air Force encountered in the whole war." [5]

The increased savagery of the air battles over Rabaul was easily accounted for—reinforcements had arrived. Admiral Koga of the *Combined Fleet* had launched Operation *Ro* and sent 173 planes of the *1st Carrier Squadron, Third Fleet* to reinforce the 200-odd aircraft that Admiral Kusaka still had in his *Eleventh Air Fleet*. Koga's move was a desperate one, a gamble that immobilized his carriers at Truk while an all-out attempt was made to check the Allied advance into the northern Solomons. Operation *Ro*'s start was put off from mid-October to the end of the month when the *Combined Fleet* commander sortied from Truk with his main body, expecting to crush a U.S. invasion attempt in the Marshalls. A week's fruitless stay in Eniwetok's spacious lagoon convinced Koga that his intelligence was faulty, and after the enemy force returned to its base in the Carolines, the *1st Air Squadron* began staging into Rabaul's fields through Kavieng. As a result of the delay occasioned by the false alarm, the Japanese carrier aircraft reached the New Britain stronghold just as the Bougainville operation got underway.

Immediately caught in a swirl of air battles over Cape Torokina and Rabaul, many of the *Third Fleet's* Zekes, Kates, and Vals and the harried survivors of Kusaka's air groups fell victim to the guns of AirSols rampaging fighters, Kenney's raiding groups, and American carrier planes. On 5 November, for the first time in the war, U.S. carriers launched a strike against Rabaul.[6]

As soon as Admiral Koga learned of the American landing at Bougainville, he determined to reinforce the *Eighth Fleet* ships at Rabaul. Early on 4 November, AirSols Liberators on patrol over the Bismarck Sea sighted and attacked two enemy convoys, one a part of this reinforcement effort. Two oilers and two transports were damaged. About noon, a B–24 spotted 19 Japanese vessels, including 6 heavy cruisers, headed for the northern entrance to St. George's Channel. As soon as the patrol plane reported its find, Admiral Halsey determined to attack the enemy ships. He meant to stave off the probability of another night sea battle off Cape Torokina, one which Admiral Merrill's battered cruisers and destroyers, then refitting at Guadalcanal, could not possibly win. The threat posed by the Japanese heavy cruisers, Halsey considered, "was the most desperate emergency that confronted me in my entire term as COMSOPAC." [7]

Although ComSoPac expected that its "air groups would be cut to pieces," [8] he ordered the carrier task force (TF 38), which had supported the Bougainville landings, to attack the concentration of shipping at Rabaul. As he later dramatically stated his motive, "we could not let the men at Torokina be wiped out while

[5] Kenney, *Reports*, p. 319. The most recent breakdown of Japanese losses indicate that 4 Zekes were shot down and that 11 other planes were destroyed or heavily damaged on the ground. *Japanese Air Comments*.

[6] A carrier task force built around the *Lexington* was scheduled to hit Rabaul on 21 February 1942, but it was discovered and attacked while it was still 250 miles east of New Britain. Although the Japanese planes were beaten off, it was decided that the invaluable carrier should not be risked once surprise was lost, and the raid was called off.

[7] Halsey and Bryan, *Admiral Halsey's Story*, p. 181.

[8] *Ibid*.

we stood by and wrung our hands." [9] The carriers, *Saratoga* and *Princeton*, and their escorts were refueling near Rennell Island when the attack order was received on the evening of the 4th. Streaking north at 27 knots, the task force reached its launch position, a point 57 miles southwest of Cape Torokina, at 0900 the next morning. Rear Admiral Frederick C. Sherman, commanding TF 38, ordered into the attack virtually every plane—52 Hellcats, 23 Avengers, and 22 Dauntlesses—that his carriers could fly off. Combat air patrol over the task force was flown by AirSols Navy F6Fs operating from Barakoma.

Halsey's target priority was cruisers first, destroyers second, and Sherman's orders to his strike leader were "not to spread his attacks too thin over too many targets, but to concentrate sufficient forces to do serious damage to as many ships, particularly cruisers, as possible." [10] Two hours after take-off, the American planes flying over St. George's Channel sighted their objective, a cluster of 40–50 vessels in Simpson Harbor. The carrier bombers turned to the attack, roaring across Crater Peninsula, as the Hellcats stuck close overhead to ward off some 70 enemy fighters that had risen to intercept. As the dive bombers maneuvered to attack, the TBFs slipped down low to make their torpedo runs. The SBDs struck first in screaming dives that concentrated on the eight heavy cruisers in the violently dodging covey of warships and auxiliaries below them. As soon as the Dauntlesses had released their bombs, the Avengers cut in among the Japanese ships like wolves in a sheep herd—only these sheep could fight back. The antiaircraft fire was fierce and unceasing: one cruiser was so plagued by TBFs that it fired its main battery guns at them. Speeding through the tempest of flak and smoke, the carrier bombers rendezvoused and headed for home. The Hellcat escort, which had kept formation above the harbor during the attack, now closed the rear of the SBDs and TBFs and fought off the Zekes that tried to follow, refusing to be drawn off into individual dog fights.

Amazingly, the strike group returned with relatively small losses: five F6Fs, four TBFs, and one SBD were missing. Twenty Hellcats, nine Avengers, and eight Dauntlesses were damaged, about one out of five seriously. What was TF 38's score against the Japanese? The returning bomber crews figured they had made certain or very probable hits on six heavy cruisers, two light cruisers, and four destroyers. Twenty-five enemy planes were claimed as shot down in combat, and another 25 were listed as probable kills. While the Japanese admitted losing only an improbable four planes in postwar assessments, they confirmed the heavy damage to the warships. No ships were sunk, but four heavy cruisers were crippled, three of them severely, and two light cruisers and two destroyers were also hit. Most of the destruction was caused by the SBDs; only two American torpedoes found a mark. Whatever the exact toll of damage, the raid can only be considered an unqualified success, since it accomplished its purpose. As a result of his costly lesson in air superiority, Admiral Koga decided not to risk his heavy cruisers in an attack on the Torokina beachhead and ordered his ships back to Truk.

[9] *Ibid.*

[10] CTF 38 Rept—5Nov43, *op. cit.*, p. 2.

At noon on 5 November, according to plan, a follow-up raid by SWPA Liberators reached Rabaul, found the airfields deserted, and bombed the town instead. The P–38 escort for the B–24s saw a number of enemy fighters, but the Japanese pilots steered clear of the American formations. Most of Admiral Kusaka's aircraft were out looking for Sherman's carriers, and, at 1255, the *Saratoga* and *Princeton* were sighted as they were recovering their last planes. By the time a flight of 18 torpedo-laden Kates arrived on the scene at dusk, the carriers were long gone. Attacked instead were an LCT, an LCI gunboat, and a PT boat proceeding from Torokina to the Treasurys. The little ships weathered a torpedo attack, even shot down one of the bombers, and limped back to Torokina still afloat. The returning enemy aircrews claimed that they had blown up and sunk a large carrier, set a medium carrier ablaze which later sank, and sunk two heavy cruisers and a light cruiser or destroyer!

By crediting the wild lies of the Kate crews, the Japanese fostered a comforting belief that they had come out far ahead in 5 November's air battles with the American naval planes. The self-delusion could not have lasted more than a week. A second carrier strike, stronger than the first, was in the offing, with destructive raids by SWPA planes spanning the interlude. Kenney's bombers, 26 B–24s with an escort of 60 P–38s, hit Rapopo on 7 November, dropping 167 1,000-pound bombs on its runways and dispersal areas. Five of the Lightnings were lost in running battles with Japanese interceptors, and returning AAF pilots claimed to have shot down 22 of the enemy Zekes. Between the 7th and the 11th, the frustrating barrier of storms that so often screened Gazelle Peninsula from Allied raiders caused General Whitehead's headquarters to cancel or divert to other targets several more large-scale daylight attacks mounted from New Guinea bases. RAAF Beauforts and American Liberators continued to get through at night, but not in any sizeable numbers. The Rapopo strike of 7 November proved to be the last daylight raid on Rabaul carried out by General Kenney's flyers.

On 11 November, Admiral Halsey scheduled a heavy carrier attack against shipping in Simpson Harbor. He asked that land-based bombers from the Southwest Pacific Area hit Rabaul's airfields and ordered ComAirSols to send a powerful strike group of Thirteenth Air Force B–24s to bomb enemy vessels as they tried to flee the attack of the Navy's dive and torpedo bombers. On this occasion, the planes of three more carriers, the *Essex*, *Independence*, and *Bunker Hill* of Rear Admiral Alfred E. Montgomery's TG 50.3, were available to reinforce Admiral Sherman's air groups. The attack plan called for the SWPA bombers to hit first, with the *Saratoga-Princeton* planes coming in next, followed by those from Montgomery's carriers. The AirSols Liberators were to arrive on the scene and make their bombing runs as the SBDs and TBFs flushed the Japanese ships from their anchorages.

Bad weather caused the attack plans to miscarry in part; only 13 of General Kenney's B–24s were able to break through the night's storm front and reach their target, Lakunai field. Admiral Sherman's task force arrived at its launch position southeast of the Green Islands at 0530 and began flying off its strike group immediately. Again TF 38 made a maximum effort, sending up 36 F6Fs, 23 SBDs, and 15 TBFs, while AirSols Navy fighters flew combat air patrol over the carriers, landing on the flight decks for fuel and servicing as necessary. At Rabaul, dense cloud cover obscured most of the harbor, and the carrier planes sighted only a cruiser and four destroyers through the openings below them. These they attacked as the enemy ships, guns blazing, scurried for the protection of a rain squall. Several bomb and torpedo hits were claimed as certain or probable, but poor visibility prevented any sure assessment. Japanese air opposition was light, and the carrier planes returned to their ships after losing two planes in the attack, the same number that they claimed of the enemy.

A little over an hour after Admiral Sherman's carriers began launching, TG 50.3 flew off the first of its strike group from a position west of Bougainville. Each of Admiral Montgomery's carriers kept eight Hellcats on board to reinforce the AirSols combat air patrol, and about 165 planes in all were dispatched, 23 of them brand-new SB2Cs (Curtiss-Wright Helldivers), a heavier-armed and faster replacement for the SBD. Like the TF 38 raiders before them, the second group of carrier planes found cloud cover heavy over Simpson Harbor and shipping elusive. Japanese interceptors, alerted by the earlier attack, were aloft and waiting, and the American planes had to fight their way in to their targets and out again. Enemy antiaircraft fire or the guns of Zekes accounted for seven fighters, three dive bombers, and three torpedo bombers. In payment for these losses, the carrier aircrews claimed two destroyers sunk, several other warships damaged, and 35 Japanese planes downed. As the naval pilots headed for home, 42 AirSols B–24s attacked on schedule, but the results of high-level bombing through fleeting cloud gaps at dodging targets went unobserved.

Admiral Halsey had directed that, if possible, a second strike be mounted by both carrier forces, but Admiral Sherman was forced to withdraw his ships to the south as soon as his air group returned. The escorting destroyers were low on fuel, as zero wind conditions had forced the task force to operate at continuous full speed to launch and land planes of the striking force and the AirSols cover. The next morning, when TF 38 was well away from the threat of enemy air attack, the carriers fueled the destroyers for the run back to Espiritu Santo.

After his planes returned from their first strike on Rabaul, Admiral Montgomery was ready to launch a second attack. The Japanese, who followed the American planes back to their carriers, had different ideas. The enemy reconnaissance planes reported the task group's location, and, at noon, Admiral Kusaka sent out a strike group of 67 Zekes, 27 Vals, 14 Kates, and a small flight of Bettys. Marine Corsairs of VMF–212 and –221 had taken their turn on station over the task group earlier in the day, but when the Japanese approached, the combat air patrol was Corsairs and Hellcats from two shore-based Navy squadrons. Radar on the *Independence* detected the first enemy plane 115 miles away, and when the striking force was 80 miles out, the combat air patrol was vectored to intercept: at 40 miles the enemy was sighted and attacked. In the running fight that ensued, the American fighters were credited with shooting down 15 planes.

At 1355, as the Japanese launched their first dive bombing attack, they flew right into the midst of 64 Hellcats and 23 Avengers which had been launched to take part in TG 50.3's second attack on Rabaul. In a wild, confused battle all over the sky, punctuated by heavy and accurate antiaircraft fire, the Americans so harassed the enemy that attacks on the carriers were uncoordinated and not pressed home with resolution. After the last Japanese plane drew off, AirSols fighters landed on the carriers for fuel and servicing before heading back to Barakoma. By the time the land-based planes had cleared the flight decks, it was too late to complete a second strike on Rabaul before dark. Admiral Montgomery cancelled the attack, recovered aircraft, and retired.

The cloud cover at Rabaul had helped keep down Japanese naval losses. One destroyer was sunk, a light cruiser and a destroyer were badly damaged, and three other warships were torn up by strafing bombers. If the ship losses were light, considering the weight of the American attacks, the enemy plane losses were not. Admiral Kusaka lost 17 Vals, 14 Kates, 8 Zekes, and several Bettys in the day's battles, nothing like the 111 planes the carrier aircrews claimed, but still a crippling toll.

On 12 November, as the American carriers were withdrawing, unharmed except in the imaginations of enemy pilots, Admiral Koga ordered the *Third Fleet* planes at Rabaul back to Truk. Although the Emperor issued an Imperial Rescript praising the results of the *Ro* Operation—the bogus damage claims were truly impressive—Japanese plane losses "had put the carrier air force in a position where further combat would rob it of even a skeleton force around which to rebuild."[11]

[11] *SE Area NavOps—III*, p. 25.

In less than two weeks of furious action, the *1st Air Squadron* had lost 43 of its 82 Zekes, 38 of 45 Vals, 34 of 40 Kates, and all 6 of its reconnaissance planes. In feeble replacement for the carrier aircraft, Admiral Koga diverted 26 Vals from the Marshalls air garrison to Rabaul.

With his carrier plane reinforcements gone, and his own *Eleventh Air Fleet's* strength down to less than 200 planes of all types and states of repair, Admiral Kusaka could do little to interfere with Bougainville operations. Except for small-scale night harassing attacks on the beachhead, enemy air attacks virtually ceased after mid-November. At the same time, Allied air strikes on Rabaul also fell off drastically in size and number. Australian Beauforts were the only aircraft to attack the enemy fortress for a month, as AirSols planes concentrated on patrol and close support missions at Cape Torokina, and the Allied Air Forces hit targets on New Guinea and western New Britain. During the comparative lull, both sides were preparing for the final phase of the battle for control of the air over Rabaul, the AirSols assault mounted from fields on Bougainville.

FIGHTER SWEEPS AND ESCORTS[12]

With the Bougainville beachhead well established and a new phase of the CARTWHEEL campaign pending, AirSols again had a change in commanders. In

[12] Unless otherwise noted, the material in this section is derived from: *ComAirPac Dec43 Analysis; SoPac ACI Repts;* IntelSect, SoPacFor, The Air Assault on Rabaul 17Dec43–19Feb44, n.d., hereafter *Air Assault on Rabaul;* CTG 37.2 ARs of First, Second, and Third Kavieng Strikes, 25Dec43–4Jan44, dtd 1Feb44 (COA, NHD);

keeping with Admiral Halsey's policy of rotating the top job among the participating services, General Twining's relief was a Marine. On 20 November 1943, General Mitchell took over as ComAirSols, retaining his positions as Commanding General, MASP and 1st MAW. With Mitchell's advent, there was no change in the heads of major tactical commands and little in the staff of AirSols. Colonel O'Neill and General Matheny, both recently promoted, continued to lead the Strike and Bomber Commands, respectively, while Marine Colonel William O. Brice, who had taken over on 24 October, ran Fighter Command. Brigadier General Field Harris, as ComAirNorSols, was slated to direct air operations originating from Bougainville once its fields were open for use.

The principal employment of AirSols squadrons in late November and early December was the support of operations in the northern Solomons. Only a small portion of the missions flown were in direct support of the troops at Cape Torokina; the majority of strikes were sent against Japanese bases elsewhere on Bougainville or on neighboring islands. Sea traffic between Rabaul and the enemy garrisons in the Solomons thinned to an insignificant trickle, as virtually every plane and pilot in AirSols had a hand in a successful and continual barge hunt.

Not only were the waters around Bougainville unsafe for the Japanese, but the air over the sea was equally unhealthy. Bettys on reconnaissance south of New Britain were shot down with such regularity that their patrols had to be curtailed drastically and halted altogether eventually. The vulnerable bombers had once carried two or three pilots in a crew of eight men, now only one pilot was risked in a cut-down crew of five or six men. Although Admiral Kusaka asked for additional medium bombers for patrol missions, he was turned down; none were available for the Southeast Area.[13]

With his search sector south of Rabaul closed to all except night-flying scouts and an occasional lucky daylight reconnaissance pilot who escaped the eager AirSols hunters, the Japanese air commander had to rely on radar and coastwatchers for warning of Allied raids. The *Eleventh Air Fleet* had 11 radar sets in the Rabaul-Kavieng area with a maximum interception range of 90 miles, and a like number of smaller sets stripped from aircraft which could pick up planes at 72 miles. All these stations were in operation during the height of the air battles over Rabaul.

Gazelle Peninsula was guarded from every approach. If an AirSols strike group took a course that brought it near northern Bougainville, radar on Buka could pick up the planes and give Rabaul 50–60 minutes warning. If the raiders swung west over the Solomon Sea to come in from the south, radar on the peninsula's east coast spotted them and provided a half an hour's notice of impending attack. Similarly, enemy interceptors had a 30-

ComAirNorSols FtrComd MissionRepts, 17Dec43–26Jan44, hereafter *FtrComd Missions*; IntelSect, MASP, FMF, The Combat Strategy and Tactics of Maj Gregory Boyington, USMCR, dtd 19Jan44, reprinted by AirIntelGru, DivNavIntel, CNO, as OpNav-16-V #S42, dtd 15Feb44; AirIntelGru, DivNavIntel, CNO, Interview of Lt Joseph E. Butler, USNR (Op Nav-16-V #E17), dtd 22May44 (COA, NHD), hereafter *Butler Interview*; *SE Area NavOps—III*; *SE Area NavAirOps—V*; Sherrod, *MarAirHist*; Ross, *RNZAF*; USSBS, *Campaign Against Rabaul*.

[13] *SE Area NavAirOps—IV*, p. 42.

minute alert if Allied planes appeared on the screens of any of the five sets at Cape St. George. When the attackers skirted wide around New Ireland's southern tip and then roared west across the narrow island for Rabaul, radar near Borpop airfield gave 20 minutes alarm. The coverage to the north and west of Rabaul was equally effective and thorough, and the chances for surprise were slim.

The radar sets available to SoPac forces were superior to those used by the Japanese, and the disparity carried over to techniques of radar employment. Not only were enemy planes picked up farther away from their targets, they were also frequently set up for a kill by ground control intercept (GCI) radar working with night fighters. In contrast to the enemy, who by choice or force of circumstance included their aircraft sets in Rabaul's early warning system, AirSols made extensive use of airborne radar for night bombing and interception.

Defending the Bougainville beachhead, the Venturas of VMF(N)-531 and Corsairs of VF(N)-75 proved themselves efficient night fighters. It took a little while before American ground commanders were willing to silence antiaircraft guns in favor of interceptors closing on enemy raiders, but when Kates and Bettys began to flame out of the sky with regularity, the night fighters won enthusiastic acceptance. Unaccountably, Japanese hecklers flew low enough for the Marine Venturas to intercept effectively, and the record of the two night fighter squadrons was about the same despite the Corsair's superior flight performance. The number of planes shot down by means of radar interception was not large—6 were claimed by VF(N)-75 during 4 months in the combat area and 12 by VMF(N)-531 during 10 months—but the effect was all that could be desired. The Japanese quickly grew chary of risking their planes in areas protected by the GCI-night fighter teams.

During the period when the Navy and Marine night fighters were winning their spurs over Cape Torokina, tension was mounting throughout AirSols command as the plans for the pending assault on Rabaul took shape. The progress of the Seabees working on the airfields within the IMAC perimeter was avidly followed not only at Allied headquarters but in the squadrons themselves. AirSols veterans were already familiar with the tactics that General Mitchell would employ to knock out the enemy base; they had worked effectively in neutralizing Bougainville's airfields and would do so again. The pressure would be constant, destructive, and varied in nature.

The fighter plane was the key to the successful prosecution of the AirSols offensive. As escorts, the fighters made large-scale bombing raids feasible, particularly by SBDs and TBFs, which were much more vulnerable to enemy attack than the heavily-armed B-24s and B-25s. Operating independently of bombers, fighter formations could range at will over Japanese airfields, challenging enemy interceptors to fight. This tactic, the fighter sweep, was honed to a fine edge at Kahili, where the marauding squadrons based at Munda, Ondonga, Segi, and Barakoma made steady inroads on enemy strength during missions calculated to clear the sky of Japanese Zekes and Hamps.

For the individual Allied pilot, the risk entailed in taking part in a mission intended to force air combat was considerable. For the Japanese pilot who met the attack, the risk was much greater and the chance of survival poorer. Even the latest model Zekes were no match for the Corsairs and Hellcats which now predominated among the AirSols fighters, while the Warhawks (Kittyhawks) and Airacobras, and the Lightnings particularly, could hold their own in combat, and, at the proper altitudes, could outperform the Japanese planes. The enemy naval pilots were engaged in a losing battle, and most of them knew it, but they fought on despite a strong sense of impending doom.[14]

Apprehensively, the Japanese awaited the completion of the first Allied airfield on Bougainville, knowing it marked the beginning of the SoPac attack on Rabaul. On 9 December, ground crews of VMF-212 and -215 landed in the IMAC beachhead and moved to the Torokina fighter field, where the 71st Naval Construction Battalion was putting the rough finish on its work. The next day, 17 Corsairs of VMF-216 christened the runway for operational use; they were followed in by six SBDs and four SCAT transports with additional personnel and equipment. In a week's time, after extensive preparations were made to fuel and service the hundreds of planes that would stage through Torokina, General Mitchell was ready to launch the first fighter sweep against Rabaul. As sweep leader, ComAirSols choose Major Gregory Boyington, commanding officer of VMF-214, a veteran fighter pilot with 20 enemy planes to his credit, six of them shot down over China during his service as a member of the American Volunteer Group.

At first light on 17 December, a powerful fighter force took off from New Georgia airfields for Bougainville. After a fueling stop at Torokina, where the pilots received a final briefing on the mission, Boyington in the lead plane of the sweep was again airborne at 0830. In the next 40 minutes, 30 more Marine Corsairs, 23 RNZAF Kittyhawks, and 23 Navy Hellcats joined up over the beachhead and fitted themselves into a ladder-like attack formation.[15] When the Allied fighters arrived over their objective at 1005, only one lonely Rufe floatplane was sighted in the air, but the P-40s flying low in the lead spotted about 40 enemy planes taking off from Lakunai and swept down to intercept. Two Zekes were shot down as they were climbing from the runway, and three more enemy planes were claimed by the New Zealanders in the resulting battle. Other Japanese fighters, 70 in all, took off during the 40 minutes that the Allied planes circled over the harbor, town, and airfields, but few enemy pilots showed any inclination to climb up to the 25,000-30,000-foot heights where the Corsairs and Hellcats awaited them. Boyington, using a radio channel that he knew the Japanese monitored, taunted the enemy to come up and fight but only got an unrewarding response, "Come on down, sucker," for his efforts.[16]

In addition to the RNZAF bag of five Zekes, a Navy pilot of VF-33 claimed one of the enemy fighters, and a Marine of Boyington's squadron flamed the unlucky Rufe which had greeted the sweep's arrival. Seven planes, however, were slim pickings, especially when three Kittyhawks were downed, and only one RNZAF pilot was recovered. The Japanese, preoccupied with their air attacks on the

[14] Okumiya, Horikoshi, and Caidin, *Zero!*, p. 302*ff*.

[15] Actually, 81 planes took off for the sweep but 5 turned back for mechanical reasons.

[16] *FtrComd Missions*, 17Dec43.

Arawe landing force, were little disposed to tangle with such a formidable fighter force so long as the Allied planes did not attack ground targets. On the whole, the results of the first fighter sweep were disappointing.

A curious aspect of the mission on the 17th was that 27 Tonys, a Japanese Army fighter with a distinctive appearance not at all like that of the Zeke family, were sighted. In the following weeks of attacks, hundreds of reports of Tonys were made by AirSols pilots and aircrews who claimed to have engaged and shot down many of the planes. Japanese records, however, agree that the only Army planes at Rabaul in this period were reconnaissance types, and these were present in small numbers. Apparently the reports were the result of a consistent mistake in identification, though what Zeke model got credit for being a Tony is hard to visualize. Whatever their type, the enemy interceptors soon got over the shyness they displayed on the 17th.

General Mitchell's attack plan called for a continuous round of strikes against Rabaul following the opening fighter sweep, but bad weather turned back the raid planned for 18 December. Secondary targets were hit instead, or, as AirSols intelligence officers phrased it: "Rabaul's Japs were blessed and Bougainville's damned when weather prevented rendezvous of the large Liberator strike with fighter escort and alternative targets on Bougainville were taken." [17] On the 19th, 16 B–24s broke through the weather front and attacked shipping in Simpson Harbor and Rabaul town as clouds obscured the primary targets, the airfields. Evidently, the presence of the big bombers was what was needed to overcome Japanese reluctance to close with Allied fighters. The 50-plane escort was hotly engaged during withdrawal and made a modest claim of having shot down four Zekes; the enemy admitted the loss of five planes, making it almost a unique occasion in the history of such claims. Japanese pilots in their turn got a more usual score, two Corsairs instead of the eight fighters they asked credit for.

AirSols second fighter sweep over Rabaul on 23 December got markedly better results than the first. In response to Major Boyington's conviction that he had had too many planes to control effectively on the 17th, the sweep force was held to 48 fighters. Taking advantage of the Japanese eagerness to intercept and break up bombing raids, Boyington's fighters were scheduled to hit soon after 18 B–24s with a 46-plane escort attacked Rabaul's airfields. When the sweep group arrived on the scene, 25 minutes after the bombing attack began, about 40 Zekes were chasing the retiring bomber formation. Over Cape St. George, the Allied fighters tore into the enemy interceptors and had a field day, claiming 30 Zekes for a loss of three F4Us. Since the F6Fs of the bomber escort were credited with shooting down three planes while losing one of their number, and the bomber crews added their own claim for 6 enemy fighters, the day's total score was 39. In their postwar reconstruction of this air battle, the Japanese recalled losing 6 fighters and accounting for 5 B–24s and 19 fighters.

The discrepancy in figures was duplicated on 24 December when an AirSols attack in the pattern of the previous day was mounted. This time the Liberators concentrated on Vunakanau and the

[17] ComAirSols IntelSummary, 18–19Dec43.

escorts, 16 P-38s and 32 F4Us, shot down 6 Zekes; the trailing fighter sweep, composed of 24 P-40s and 22 F6Fs, accounted for 14 enemy fighters, while losing 7 of their own number over the target. Two RNZAF Venturas on rescue and patrol duty over St. George's Channel during the strike added at least 2 Zekes to raise the day's claims to 22. The Japanese version of the action saw 58 Allied planes go down as only 6 Zekes were lost.

After the combined attack on Christmas Eve, General Mitchell switched back to a week of separately mounted bombing strikes and fighter sweeps. Liberators with heavy escorts struck the airfields on the 25th and 30th, and 49- and 45-plane sweeps were over Rabaul on the 27th and 28th. These forays cost AirSols nine planes, and, in addition, a B-25 that was shot down by antiaircraft fire while it was attacking the radar station near the lighthouse at the tip of Cape St. George. A Dumbo landed and rescued the Mitchell crew from close inshore despite fire from machine guns and artillery. Allied pilots and aircrews claimed to have shot down 74 enemy planes over Rabaul between Christmas and New Year's Day.[18]

The machine guns of enemy fighters were only one means, although the principal one, by which Rabaul was defended. Several times during the late December strikes against the Japanese base, AirSols pilots reported that enemy planes were trying to break up attacking formations and destroy aircraft by air-to-air bombing. Most of these bombs were incendiaries, generally of 70-pound size with a bursting charge of picric acid and an explosive load of about 200 phosphorus-filled steel pellets.[19] Zekes, flying above and head on to Allied aircraft, released these bombs so that they would explode in the path of the targeted planes. The incidence of such attacks increased sharply in the new year, and a number of planes were damaged by the spectacular phosphorus fireballs, although actual losses charged to such air-to-air bombing were slight.

Far more dangerous to the attackers were the Japanese antiaircraft guns ringing any worthwile target at Rabaul. At least 260 guns, ranging in size from 13mm machine guns to 12.7cm cannon, were manned by enemy Army and Navy crews throughout the whole of the AirSols attack. Fortunately for the Allied flyers, the only fire control radar the Japanese had was the first such piece manufactured in Japan; it had many mechanical defects and was ineffective. Enemy range and height finders were not too accurate either, and Zekes flying alongside bomber formations were used to radio altitude and speed data to the guns. Communications difficulties marred the usefulness of this makeshift system.[20]

AirSols had available a limited number of Liberators. While these bombers could release their loads from heights above the reach of the heaviest Japanese antiaircraft guns, the destructive effect of such high altitude bombing did not approach the saturation level that the AirSols offensive required. As soon as the Piva bomber field at Torokina was operational, a new stage in the attack on Rabaul would

[18] ComAirSols IntelSummary, 25-31Dec43.

[19] MilAnalysisDiv, USSBS(Pac), *Japanese Air Weapons and Tactics* (Washington: GPO, Jan47), p. 47.

[20] USSBS, *Interrogation* No. 224, Cdr Yasumi Doi, IJN, I, p. 209.

begin, and Strike Command would commit its SBDs and TBFs. When the field on Stirling Island was ready for use, Bomber Command would add B-25s to the assault. The enemy would be hit by bombers and fighters from high, low, and medium altitudes, from every direction possible, and around the clock.

Until General Mitchell was able to unleash the full offensive power of his command, Admiral Sherman's carriers were called upon to heighten the effect of the blows that AirSols could deliver. Intelligence that the Japanese had heavy troop reinforcements en route to the Bismarcks and a strong reserve of aircraft nesting at Kavieng prompted ComSoPac to order an attack on the New Ireland base set for Christmas morning. Since SWPA forces were poised to launch the Cape Gloucester operation at this time, the carrier strike might well disrupt the Japanese aerial counterattack that was sure to be mounted when word of the landings reached Admiral Kusaka.

Before dawn on the 25th, Sherman's carrier task group, composed of the *Bunker Hill*, *Monterey*, and six destroyers, started launching aircraft from a position 150 miles northeast of Kavieng. At first light, the planes—31 F6Fs, 28 TBFs, and 27 SB2Cs—joined up and headed for the enemy base, their primary target being shipping in the harbor. Air opposition was negligible; most of Kavieng's fighters had moved to Rabaul on the 24th, decoyed there by an American cruiser-destroyer bombardment of the Buka-Bonis area (heretofore a usual prelude to a SoPac amphibious landing). With the defending Zekes gone, escort Hellcats were able to stick to the attack plan and precede the light bombers in a strafing run to suppress ships' antiaircraft fire during the bombing. Only a few ships were present, and skip-bombing TBF's sunk one of these, a 5,000-ton freighter. Another medium-sized freighter was damaged, and a 500-ton mine sweeper was driven on the rocks.[21]

The strike group was back on board its carriers by 1045, with only one TBF missing. The combat air patrol shot down three enemy bombers during retirement, and ships' antiaircraft got two more, when Bettys tried a torpedo attack. Instead of heading back for port, the carriers stayed at sea on ComSoPac's orders, waiting for a chance to catch the Japanese reinforcement convoys.

On 1 January, a second strike was launched against Kavieng when search planes reported enemy warships near the harbor. This time, about 30 Zekes were present to add their power to the intense antiaircraft fire of 2 light cruisers and 2 destroyers. The Japanese fighters dropped phosphorus bombs on the SB2Cs as they dove to the attack, but without effect. Although the American aircrews were certain that they had hit their targets repeatedly, the actual damage to the skillfully handled ships was slight. The Zekes and ships' guns combined to down two Hellcats and a Helldiver: the carrier aircrews' claim was 14 planes, twice the admitted Japanese losses.

A third attack was launched on 4 January in an attempt to sink a cruiser force reported as being just north of Kavieng. The strike group found the cruisers actually were large destroyers and attacked the radically maneuvering ships with little luck. Reefs prevented effective torpedo

[21] *Japanese Air Comments.*

runs, and the torpedoes that were dropped were set to run too low to hit destroyers; the only damage was caused by strafing.[22] One F6F was shot down by the Zekes that harried the attack formation like wolves; three enemy fighters were downed. The combat air patrol accounted for one Betty scouting the carriers and wiped out a ferry group, a bomber and six Zekes, heading for Kavieng. According to ComSoPac's orders, no further strikes were sent against the enemy ships on the 4th; destroyers were not considered worth the risks of a second attack.

Admiral Sherman's task group retired after the third strike on Kavieng unmolested by the *Eleventh Air Fleet*. After their withdrawal, the carriers and destroyers refueled and headed for the Central Pacific to take part in the Marshalls operation. Although the Rabaul-centered cordon of enemy bases had experienced its last carrier air raid, the cessation attracted little notice among garrison members. Instead of an occasional unpleasant taste of ship-based dive and torpedo bombers, the Japanese were now to be force-fed a steady diet of SBD–TBF attacks mounted from Bougainville.

PIVA PUNCH[23]

By the year's end, the Seabees had the northernmost of the two Piva airstrips they were working on, Piva Uncle, ready for use as a staging base for light bombers. Full-scale operations from the field, however, required an additional week of logistic preparations, so that repeated 50–100 plane missions could be mounted. While fuel and supply dumps and servicing facilities were expanded to handle the planned strikes, the pace of attack against Rabaul never slackened. If anything, it increased.

In order that the Japanese garrison would get no respite from the daily round of Liberator raids and fighter sweeps during the first week in January, three squadrons of RAAF Beauforts bombed the enemy airfields at night. Instead of hitting their targets in mass formations, the Kiriwina-based Australians made single plane attacks in succession, a harassing tactic used to good advantage by both sides throughout the fighting in the South Pacific. The last of these Beaufort missions was flown against Lakunai and Tobera on the night of 7–8 January; thereafter, the task of hitting Rabaul targets at night was handled by SoPac aircraft. The area of operations of the RAAF planes on Kiriwina was restricted to central New Britain east of Arawe and west of Wide Bay, an Allied Air Forces decision that disappointed the Australians who preferred a more decisive role in the fighting.

Emphasizing the fact that the air battles were not all one-sided in favor of the attackers was the loss of Major Boyington on 3 January during a fighter sweep over Rabaul. Before the sweep leader disappeared, he was seen to shoot down his 26th enemy plane, a feat that ranked him with Major Joseph J. Foss as the lead-

---

[22] *Ibid.*

[23] Unless otherwise noted, the material in this section is derived from: *ComAirPac Jan–Feb44 Analyses*; ComAirSolPac Daily IntelBuls, Jan–Feb44; *SoPac ACI Repts; Air Assault on Rabaul; StrikeComd WarDs; FtrComd Missions; Butler Interview; SE Area NavOps—III; SE Area NavAirOps—V*; Craven and Cate, *Guadalcanal to Saipan*; Morison, *Breaking the Bismarcks Barrier*; Odgers, *RAAF Against Japan*; Okumiya, Horikoshi, and Caidin, *Zero!*; Ross, *RNZAF*; Sherrod, *MarAirHist*.

ing Marine Corps ace of the war.[24] Boyington parachuted from his crippled Corsair into St. George's Channel and was picked up by an enemy submarine and taken to Rabaul; eventually, he reached Japan and spent the rest of the war in a prison camp.

The day on which Major Boyington was downed was one of the few in early January when the weather observations of the attacking squadrons could read: "Visibility unlimited, clear over target, ceiling 30,000 feet."[25] Much more frequently, Rabaul was partially or wholly protected by rain squalls and heavy cloud cover and returning flyers reported: "Heavy solid front on New Ireland coast extending south from St. George Cape," or "Heavy overcast to 8,000 feet, built up in thick layers."[26] The weather was so changeable and often varied so much over the course of a day that even reports of clear skies by reconnaissance planes might no longer hold true two or three hours later when a strike group reached its objective.

On 5 January, an unbroken wall of clouds over St. George's Channel prevented the first land-based SBD–TBF strike on Rabaul from even reaching its target. The 26 Dauntlesses and 21 Avengers that took part returned to their home fields on New Georgia, but not before the dive bombers attacked enemy troop concentrations on Bougainville. On the 7th, a similar light bomber group staged through Piva Uncle, picked up its fighter escort over Bougainville, and headed for Tobera field. Again the weather was poor and the primary target was closed in. Rapopo field was visible, but the strike was briefed for another secondary target, the radar installation near the Cape St. George lighthouse, which was attacked instead. Over Rabaul, antiaircraft fire was heavy and interceptors were numerous and aggressive, but the 72-plane escort was a match for the Zekes. Twelve enemy planes were claimed against a loss of three F6Fs, two in a mid-air collision, and two SBDs which crashed on the way back to base.

Finally, on 9 January, the Navy and Marine bombers were able to get their first good shot at Rabaul targets. Twenty-three SBDs and 16 TBFs flew up from Munda, fueled at Piva Uncle, and winged toward Tobera, home field for most of the Japanese fighters. The enemy had ample warning of the raid, and about 40 interceptors were airborne when the AirSols planes arrived on the scene. The Zekes did not close until the bombers nosed over to make their dives, and then the escort, 62 fighters, beat off the Japanese handily. The SBDs struck first, concentrating their half-ton bombs on defending gun positions; the TBFs followed, hitting the runway and apron with 2,000 pounders. Antiaircraft fire was light over the field but heavy on the retirement course which followed the Warangoi River to its mouth. Thirteen Japanese planes were claimed by the escort which lost a Hellcat and two Kittyhawks; the strike group had six aircraft damaged that made it back to Piva Uncle. One badly shot-up Marine Avenger, its gunner dead at his post, had to ditch off Torokina; the pilot was rescued uninjured.

[24] Actually, before Boyington went down himself, he finished off two more planes, making his an unequalled score among Marine pilots in World War II. His total of 28 victories includes the six planes he shot down as a member of the American Volunteer Group.

[25] *FtrComd Mission Repts*, 3Jan44.

[26] *Ibid.*, 5Jan44 and 6Jan44.

MECHANICS OF VMF–211 *repair a Corsair which returned from a strike with its wings and fuselage full of holes from antiaircraft shell fragments.* (*USMC 77816*)

Once the light bombers had scored, they came back again and again to hit Rabaul with increasing strength and effect. The Japanese reported that they met each attack in early January with all their remaining fighters, but that they could not check the raids in spite of what they termed "the remarkable results obtained each time"[27] by their interceptors. In general, the enemy fighter pilots hit the Allied formations as they were heading in for Rabaul, attempted to penetrate the screen of escorts and get at the bombers, and often followed the SBDs and TBFs into their dives, through the defending antiaircraft fire, and all the way out to the rendezvous area before sheering away. The job of fending off the Japanese planes that broke through the high, medium, and low cover fell to the RNZAF Kittyhawks. Flying close cover for the American bombers, the New Zealanders shared the bombing runs, braved the storm of short-range flak, and guarded the tails of the SBDs and TBFs as they streaked for home.

Despite the diversity of plane types and pilots flying in AirSols attack formations, air discipline was tight; the cardinal rule observed by fighters and bombers alike was to stay joined up and fight together. By the time of the Rabaul battle, the Japanese were following the same course and fought in two- and three-plane sections that stuck together fairly well. Occasionally, since aircraft on both sides used voice radio in the medium high frequency range, there was an exchange of insults calculated to goad the incautious pilot into reckless action. The Strike Command operations officer noted that "we would call them and tell them, 'We're coming, you little so-and-sos, you'd better get ready,' and they would tell us what they were going to do to us when we got there."[28] This ruse had little effect; Japanese actions showed that they were husbanding their dwindling strength, employing their fighters to best advantage in the face of mounting odds.

During the first few weeks of determined assault on Rabaul's airdromes and the planes they harbored, AirSols fighters and bombers made few attacks on the shipping in Simpson Harbor. Allied intelligence officers kept close tab on the number and type of ships present, however, through the reports of returning pilots and aircrewmen and the findings of frequent reconnaissance missions. General Mitchell had no intention of letting the enemy continue to reinforce and resupply Rabaul with impunity. He intended, in fact, to use his planes to choke off all significant contact by sea between the Japanese stronghold and its sources of supply, just as AirSols had already shredded the lifeline between Rabaul and its satellite bases in the Solomons.

Just before dawn on 14 January, after a steady procession of Liberators and Mitchells had made the night hellish for Rabaul's garrison, six TBFs attempted a raid on the shipping in Simpson Harbor. Foul weather forced these Avengers to turn back, but later in the day the enemy ships were attacked by 16 more in company with 37 SBDs. The primary target of the bombers, Lakunai airfield, was closed in, and the strike group was alerted as it flew east across New Ireland to hit its secondary target, the shipping. About 30 enemy fighters intercepted the Allied

[27] *SE Area NavAirOps—V*, p. 18.

[28] *Butler Interview*, p. 2.

formation while it was still 40–50 miles from its mark; over Blanche Bay the number of defending Zekes doubled.

The escort of American and New Zealand fighters beat off most of the enemy attackers, giving the SBDs a clear shot at a harbor full of scurrying ships. Nosing over into steep dives at about 8,000 feet, the Dauntless pilots aimed their planes at the biggest ships, trying to drop their half-ton bombs right down the stacks. The TBFs followed the dive bombers in swift, shallow approaches that brought them down to masthead height where they tried to slam 2,000-pound bombs against the sides of their elusive targets.

Nine direct hits on seven cargo vessels plus hits on two destroyers were claimed, as well as 20 damaging near misses. All but two of the direct hits were credited to the torpedo bombers. During the running battle with enemy interceptors, the escort and the bombers reported they shot down 29 planes and probably got 16 more; the corresponding AirSols loss was 8 fighters, with 4 pilots recovered, 2 SBDs, and a TBF. The admitted Japanese loss was 3 planes, plus bomb damage to a destroyer and an oiler; their claimed bag of Allied aircraft was 65.

Strike Command's attacks on shipping continued, the results improved, and, by the end of January, seven merchantmen and an oiler had been sunk and three more ships badly damaged. The TBFs, which proved to be more effective in shipping attacks than the SBDs, used 4- to 5-second delay fuses on 1-ton bombs, came in very fast 30–40 feet off the water, headed directly for the targeted ships, dropped their bombs when close aboard, and streaked directly over the ships for the rendezvous point. The pilots said it was pretty hard to miss, but no one was anxious to stick around to check results. The Strike Command operations officer said that he tried:

> . . . to get the pilots to slow down and join up as soon as possible but coming in like that, being shot at continuously, and with phosphorous bombs dropped on them consistently by the enemy fighter planes, they just couldn't do it. I did persuade one squadron commander, one day, to slow down and try to wait for the other men, and he said he eased back on the throttle just a little and he was the last man out.[29]

As the tempo of AirSols attacks in the new year increased, and their strength and effectiveness grew apace, Allied pilots noted a definite falling off in the number of Japanese planes rising to intercept. For a time in early January, it appeared that "there was a question as to whether reinforcements for the Bismarcks sinkhole would be forthcoming." [30] Admiral Koga and his staff, however, were convinced by the intensity of the AirSols attack that the next major Allied move would be made in the Southwest rather than the Central Pacific. Consequently, the *Combined Fleet* commander decided to commit the *2d Air Squadron* to Rabaul, denuding the carriers *Runyo*, *Hiyo*, and *Ryuho* of their air groups to add 69 fighters, 36 dive bombers, and 23 torpedo bombers to the *Eleventh Air Fleet*. When the carrier planes arrived, surviving flying personnel of the battered *26th Air Flotilla*, which had fought at Rabaul since the Guadalcanal landings, were withdrawn to Truk to refit and reorganize.

---

[29] *Butler Interview*, p. 2.

[30] *Air Assault on Rabaul*, p. 7.

Many of the air flotilla's veteran pilots and crewmen, despite the knowledge that they would be free of the hopeless fight at Rabaul, were unhappy to be going. The men felt there was an implication of failure in their relief, and that the Navy expected the *2d Air Squadron* to accomplish a task that the *26th Flotilla* had found impossible—stemming the AirSols attack. In truth, however, all that was expected of the new planes and pilots was that they would preserve "as long as possible the strategic position to which the Truk advance base is the key." [31]

By the end of January, Rabaul was set up for a rain of knockout blows, a trip-hammer series of strikes that would batter it into impotence. In 873 daylight sorties during the month, AirSols bombers had dropped over 775 tons of bombs on airfields and shipping. The steady pounding had put each of the enemy fields out of action for varying periods of time, although the Japanese managed to keep at least one runway open for their interceptors. The shipping losses had been so severe that the enemy commanders knew that their principal supply line must soon be cut off.[32] The month's toll of defending aircraft, 120 according to Admiral Kusaka's postwar recollection,[33] was more than half of the planes available before the *2d Air Squadron* reinforcements arrived.

In view of the 1,850 escort and sweep sorties that Allied fighters flew over Rabaul in January, the combat loss of 65 planes was relatively low. Only 19 bombers of all types were shot down by Zekes or antiaircraft. In reckoning their tally of enemy planes for the month, AirSols pilots and gunners arrived at a figure of 503. Japanese aircrews claimed an even higher number of victories, 618 planes. Regardless of what the actual relative score was, one fact is certain, AirSols' combat losses had no dampening effect on the intensity of its offensive; the damage inflicted on the Japanese, however, was telling.

The *2d Air Squadron* pilots, many of them young men new to combat, aged quickly as they fought against increasingly unfavorable odds. The carrier group's operations officer, in recalling the atmosphere of the time, evoked a picture of real desperation:

> The days passed in a blur. Every day we sent the Zeros up on frantic interception flights. The young and inexperienced student pilots had become battle-hardened veterans, their faces showing the sudden realization of death all about them. Not for a moment did the Americans ease their relentless pressure. Day and night the bombers came to pound Rabaul, to smash at the airfield and shipping in the harbor, while the fighters screamed low in daring strafing passes, shooting up anything they considered a worthwhile target. . . .
>
> It was obvious that so long as we continued the battle in its present fashion, the Americans would grind us under.[34]

The Japanese could make no effective response to the relentless AirSols attacks. There were not enough enemy fighters on

[31] *SE Area NavOps*—III, p. 62.

[32] CinCPac–CinCPOA Item No. 11,955, Diary of MajGen Masatake Kimihara, ACofS, Eighth AreaA, 1Jan–9Jun44, entry of 6Feb44, in CinCPac–CinCPOA Translations No. 1, dtd 31Oct44.

[33] USSBS, *Campaign Against Rabaul*, p. 50. *SE Area NavAirOps*—V, p. 19–a, lists losses for the month as 46.

[34] Okumiya, Horikoshi, and Caidin, *Zero!*, p. 309.

the scene even to slow down the pace, and those that were available were outclassed by newer and better Allied aircraft. The only other means of defense available to the Japanese, antiaircraft fire, was well countered by repeated light bomber strikes. As they had before at Munda and Kahili, Strike Command's SBDs concentrated on gun positions protecting the airfields, while TBFs ploughed up the runways. Complementing the SBD–TBF attacks, B-25 strafers laid a destructive spread of parafrag bombs throughout the airdrome areas, aiming particularly at aircraft on the ground. The attack pattern was varied enough to keep enemy gun crews harassed and apprehensive.

Perhaps the most discouraging aspect of the stepped-up Allied offensive to the Japanese was the order that had to be issued to the *2d Air Squadron* fighter pilots "to attack or defend yourselves only when the battle circumstances appear particularly favorable to you."[35] This official admission of Allied superiority put a severe crimp in enemy aircrew morale, even though the imbalance had been obvious for weeks. The massive AirSols attack formations, aggregating 200 planes a day by early February, were too strong to stop or turn aside.

Frequent bad weather was Rabaul's only sure defense, but its shielding effect often did not extend to alternate targets, particularly the airfields at Namatami and Borpop and the radar at Cape St. George on New Ireland.[36] These enemy installations were frequently attacked, since they directly supported Rabaul; the net effect of damage done to them was a reduction in the strength of the key Japanese base.

Everything began to turn sour for the Japanese in February, as concurrent operations in both the Southwest and Central Pacific made the Rabaul airbases untenable. Admiral Halsey was ready to move a New Zealand landing force into the Green Islands, only 115 miles from Rabaul, on the 15th. To support this operation and also to provide cover for a pending carrier attack on Truk, General Kenney's Fifth Air Force bombers began a series of large-scale raids on Kavieng on the 11th.[37] In the offing at the month's end was a SoWesPac move into the Admiralties to seize the enemy airfields there and cut off the Bismarcks from New Guinea.[38]

The net was closing on Rabaul, and the Japanese knew it. Yet neither the Kavieng air raids, the Green Islands landing, nor the enemy-anticipated invasion of the Admiralties was the deciding factor in the Japanese decision to pull all combat aircraft out of Rabaul.

Credit for forcing that move went to Central Pacific task forces under Vice Admiral Raymond A. Spruance which struck Truk on 17 and 18 February. Planes from nine carriers hit airfield installations and shipping in the atoll's anchorage in a two-day spree that saw at least 70 enemy planes destroyed in the air

[35] *Ibid.*

[36] The Marine Corps' third leading ace in World War II, First Lieutenant Robert M. Hanson of VMF-215, who shot down 25 planes, was killed by antiaircraft fire during a strafing run at Cape St. George on 3 February.

[37] Craven and Cate, *Guadalcanal to Saipan*, p. 355.

[38] Details on the Allied seizure of the Green Islands and the Admiralties will be covered in the following chapter.

LOADED FOR A RABAUL STRIKE, *Marine TBFs roll down a taxiway toward the Piva bomber runway on 17 February 1943.* (USMC 81362)

HEADED FOR VUNAKANAU, *a formation of Marine SBDs with bombs suspended from their wings takes part in the April strikes on Rabaul.* (USMC 81436)

and on the ground,[39] and more than 200,000 tons of merchant shipping sunk. Two enemy cruisers, four destroyers, and a subchaser were sent to the bottom by a combination of air and surface attacks.[40]

Admiral Nimitz, in ordering the strike on Truk, had hoped to catch the entire *Combined Fleet* in its lair. Photographs taken by two Marine PB4Ys of VMD–254 on 4 February had shown the enemy fleet to be present. The planes had taken off from the newly-built airfield on Stirling Island and flown unescorted the 1,000 miles to the Japanese bastion, taken their pictures, and returned after 12 hours in the air to land at the Piva bomber strip on Bougainville. Unfortunately, the Japanese had spotted one of the planes, although they were unable to intercept or bring it down.[41] The sight of the four-engined land bomber overhead, and the realization of all it portended with American forces firmly established in the Gilberts and Marshalls, was enough to convince Admiral Koga the time had come to pull back from his exposed position. Accordingly, the enemy commander ordered the *Combined Fleet* to weigh anchor and head for home waters. The bulk of the ships left on 10 February; most of the vessels caught by the American carrier attack a week later were auxiliaries and escorts delayed in sailing by the weather.

Many of the Japanese planes shot up on the ground at Truk were replacements meant for Rabaul. Their loss in the Carolines emphasized the futility of further aerial defense of the New Britain base. A good portion of the strategic value of Rabaul to the Japanese lay in its usefulness as a shield for Truk against attack. The twisted wreckage of the Zekes littering the atoll's airfields, the missing ships vanished beneath the waves of the anchorage, and the towering columns of smoke rising from gutted supply dumps gave ample evidence that Truk was vulnerable—and that Rabaul's role in its defense was ended.

On 17 February, as soon as news of the American carrier strike reached Admiral Koga, he dispatched orders to Admiral Kusaka to send all serviceable aircraft at Rabaul to Truk. The Carolines' base was only a way-station now, and the ultimate destination of these planes was airfields on the new Japanese defensive perimeter running from Western New Guinea through the Palaus, the Marianas, and the Volcano-Bonins.

On the night of 17–18 February, American destroyers lent insulting but unwitting emphasis to the enemy decision to strip Rabaul of its remaining offensive power. A bombardment group of five destroyers steamed through St. George's Channel with a PB4Y overhead to spot its targets and fired 3,868 rounds of 5-inch at enemy installations at Praed Point and in Rabaul town. At about the same time, a similar group of destroyers shelled Kavieng. Although Japanese coast defense guns replied in both instances, they

[39] RAdm Samuel Eliot Morison, *Aleutians, Gilberts and Marshalls—History of United States Naval Operations in World War II*, v. VII (Boston: Little, Brown and Co., 1950), p. 320. Admiral Morison's history usually is the best informed source for assessment of Japanese losses. American claims of losses inflicted at the time of the attack ranged from 128 to more than 200 planes.

[40] NavHistDiv, Off of CNO, ND, *United States Naval Chronology, World War II* (Washington, 1955), pp. 77–78.

[41] NavAnalysisDiv, USSBS(Pac), *The Reduction of Truk* (Washington: GPO, Feb47), p. 4.

hit nothing, and the raiders retired unscathed.[42] There was no aerial pursuit.

What last-gasp resistance there was from the *2d Air Squadron* was offered on the morning of 19 February when 36 Zekes rose to intercept a 139-plane attack formation, centered on 71 Strike Command bombers. The SBDs and TBFs which hit Lakunai included six Avengers of VMTB-143 armed with both 500-pound bombs and 5-inch rockets, the latter a relatively new air weapon which proved effective against point targets. When the AirSols escort and the Japanese interceptors tangled, the respective claims were 23 planes shot down by American pilots[43] and 31 by Japanese. A B-24 strike group which followed the light bombers to attack Tobera and Lakunai was also intercepted; the Liberator crews claimed three Zekes. AirSols actual loss was one Corsair with 10 planes damaged; the Japanese loss appears to have been eight fighters.[44]

On 20 February, the only aircraft remaining in Rabaul were about 30 damaged fighters, a few Navy utility types, and 4 Army reconnaissance planes.[45] An attempt was made to evacuate some of the invaluable veteran ground crews in two of the last merchant vessels to visit Rabaul, but AirSols bombers sank the vessels on the 21st. The next day, American destroyers cruising the waters off New Ireland sank the rescue tug that picked up the survivors. The loss of the maintenance crews, whose skills represented the experience of many years, was equally as damaging to the Japanese naval air arm as the loss of veteran air crews in combat with Rabaul's attackers.

Although the men of AirSols had no way of knowing it, they had won the air battle of Rabaul; the Japanese would never come back. No exact figure for the number of enemy planes that escaped to Truk can be established, but a consensus of the recollections of key officers of the *Eleventh Air Fleet* would indicate no more than 70 got out, although one source says 120 made it.[46] Even this larger figure makes little difference in the overall assessment of the results of the AirSols offensive.

In the two months that passed between the first Allied fighter sweep over Rabaul on 17 December and the last opposed bombing raid of 19 February, the Japanese lost at least 250 planes, and very probably more. Again the records are incomplete and contradictory, but only as to the number of planes involved, not as to the fact of defeat and withdrawal. Emphasizing the difficulty of assessing the claims of air warfare is the first box score compiled for the 17 December–19 February battle by intelligence officers at ComAirSoPac—151 Allied planes lost in destroying 789 Japanese planes.[47] The comparable Japanese claim for the same period—

[42] ComFAirWingOne Rept of Night Missions, 17–18Feb44, dtd 27Feb44 (COA, NHD).

[43] New Zealand P-40s escorting TBFs hitting Vunakanau on 13 February shot down two Zekes bringing the total RNZAF score in the Pacific to 99. Much to the disappointment of the New Zealanders, they never got the century.

[44] Pineau, "Summary of Enemy Air Raids on Rabaul," *op. cit.;* ComAirSols Bombers and Fighters Mission Rept, 19Feb44.

[45] MilAnalysisDiv, USSBS(Pac), Answer to Questionnaire No. 5, Bismarck and Solomon Islands, 7Dec41–19Feb44, prepared under LtCol Sakuyuki Takagi, IJA (USSBS Recs, National Archives).

[46] *Miyazaki Interrogation*, p. 414.

[47] ComAirSoPac IntelBul, dtd 22Feb44.

this from a postwar history of their operations by the Japanese—was that they lost 142 of their own planes in shooting down 1,045 Allied aircraft.[48]

The withdrawal of the Japanese defending aircraft on 19 February 1944 did not signify the end of the Allied air offensive against Rabaul. Far from it. The aerial attack went on—and on—and on. It continued in an unceasing round until the end of the war, and hundreds of combat aircraft, many of them Marine planes, took part in the frustrating campaign to neutralize the enemy base. The Japanese fought back, at least the antiaircraft gunners did, whenever they were offered a target; the rest of the huge enemy garrison was immobilized, dug in and waiting for an amphibious assault that never came.

[48] *SE Area NavOps—III*, p. 59.

The story of those 18 months of hazardous but largely routine aerial attacks on Rabaul should be considered apart from the few weeks of intensive air battles that ended the offensive threat of the key enemy base. In that short span of fighting, as in the preceding months of methodical advance that made it possible, no one service can claim to have had the pre-eminent part. In a very real sense, Admiral Halsey's South Pacific Forces, and in particular, Aircraft, Solomons, were joint commands. The admiral once rather pithily recalled:

> Whenever I hear blather about inter-service friction, I like to recall that our Army, Navy, and Marine airmen in the Solomons fought with equal enthusiasm and excellence under rear admirals, then under a major general of the Army, and finally under a major general of Marines.[49]

[49] Halsey and Bryan, *Admiral Halsey's Story*, p. 186.

PART VI

Conclusion

Encirclement

In the late summer of 1943, while the Joint Chiefs were deciding to neutralize Rabaul rather than capture it, General MacArthur's staff was preparing plans for the operations which would follow Bougainville and Cape Gloucester and complete the encirclement of the key New Britain base. With a tentative target date of 1 March 1944, MacArthur intended to seize Kavieng, using SoPac forces, and the Admiralties, employing his own SWPA troops, planes, and ships. The establishment of Allied airfields at Finschhafen and Cape Gloucester meant that the Admiralties' landings could be covered adequately by land-based fighters, but Kavieng operations required carrier air support. Even the boost in range given SoPac fighters by airfields at Cape Torokina would not be enough to provide effective escorts and combat air patrols over Kavieng.

Once the Central Pacific offensive got underway with operations in the Gilberts, it appeared that mounting demands on the Pacific Fleet's shipping resources would serve to put off D-Day at Kavieng until about 1 May 1944.[1] Faced with the possibility that there would be "a six months interval between major South Pacific operations" which might "kill the momentum of the South Pacific drive," Admiral Halsey consulted General MacArthur, who gave "his unqualified approval" to the scheme for an intermediate operation "which would keep the offensive rolling, provide another useful base, and keep the pressure on the enemy." [2]

As Halsey ordered his staff to prepare the plans for the seizure of the Green Islands, the intermediate target he had selected, he also directed them to study the possibility of seizing Emirau Island in the St. Matthias Group as an alternative to Kavieng. ComSoPac felt that the time was ripe for another bypass operation, one that would achieve the same objective as the proposed large-scale Kavieng assault, but at much less cost. The admiral argued vigorously for his point of view at Pearl Harbor in late December, and in Washington in January, during a short leave he spent in the States.[3]

Although General MacArthur indicated on 20 December that the possession of airfields at either Kavieng or Emirau would accomplish his mission of choking off access to Rabaul,[4] he was soon firm again in his belief that the New Ireland base would have to be captured. This was the stand that SWPA representatives took at a coordinating conference held on 27 January

[1] ComSoPac 1st end, dtd 29Apr44, to ComIIIPhibFor Rept of the Seizure and Occupation of Green Islands, 15Feb–15Mar44, dtd 16Apr44.

[2] *Ibid.*

[3] Halsey and Bryan, *Admiral Halsey's Story*, pp. 186–188.

[4] SJC [MajGen Samuel J. Chamberlin] memo for jnl, dtd 21Dec43, Subj: Conference at GHQ, 20Dec43, in *GHQ G–3 Jnl.*

at Pearl Harbor, where preparations went ahead for a simultaneous assault on Kavieng and the Admiralties. The new tentative target date was 1 April, a month and a half after SoPac forces were slated to secure the Green Islands.

GREEN AND ADMIRALTY ISLANDS LANDINGS[5]

Before the Green Islands was chosen as the next SoPac objective after Bougainville, several other prospective targets were considered and rejected. A proposal to seize a foothold in the Tanga Islands, 35 miles east of New Ireland, was turned down because the operation could not be effectively covered by land-based fighters. Similarly, the capture of enemy airstrips at Borpop or Namatami was discarded because carrier support as well as a large landing force, would be required to handle Japanese resistance.[6] Nissan, the largest of the Green Islands, was not only close enough to Torokina for AirSols fighter support, but also was weakly defended. (See Map 32.)

Located 37 miles northwest of Buka and 55 miles east of New Ireland, Nissan is an oval-shaped atoll 8 miles long with room on its narrow, flat main island for a couple of air strips. With Rabaul only 115 miles away and Kavieng about a hundred miles farther off, the Allied objective was clearly vulnerable to enemy counterattack once it was taken. By 15 February, however, the swing of fortune against the Japanese made that risk readily acceptable. In fact, Admiral Halsey reported that the campaign to neutralize Rabaul's air strength "had succeeded beyond our fondest hopes."[7]

Although aerial photographs and the scanty terrain intelligence available regarding the Green Islands indicated that Nissan was suitable for airfield development, nothing sure was known. A 24-hour reconnaissance in force, to be launched close enough to D-Day to prevent undue warning and consequent reinforcement of the garrison, was decided upon to obtain detailed information. New Zealand infantrymen of the 30th Battalion made up the main body of the 330-man scouting party; they were reinforced by American Navy specialists who would conduct the necessary harbor, beach, and airfield surveys.

The landing force loaded on board APDs at Vella Lavella at midday on 29 January, rehearsed the operation that evening, and got underway for the target at dawn. Escorted and screened by destroyers, the high-speed transports hove to off Nissan just after midnight and started debarking troops immediately. The column of LCVPs was led into the atoll lagoon by two PT boats that had sounded a clear passage during a previous reconnaissance mission. By 0100, all troops were ashore near the proposed airfield site;

[5] Unless otherwise noted, the material in this section is derived from: *GHQ G-3 Jnl;* ComIIIPhibFor Rept of the Seizure and Occupation of GreenIs, 15Feb–15Mar44, dtd 16Apr44 (COA, NHD); CO StrikeComdGreen Rept, dtd 21May44 (COA, NHD); ComDesron 45 (CTG 31.8) AR, 28Jan–1Feb44, dtd 10Feb44 (COA, NHD); *SE Area NavOps—III;* Halsey and Bryan, *Admiral Halsey's Story;* Miller *Reduction of Rabaul;* Morison, *Breaking the Bismarcks Barrier;* Warren, "Fifth AF in the Bismarcks."

[6] ComIIIPhibFor Rept of the Seizure and Occupation of GreenIs, 15Feb–15Mar44, dtd 24Mar44 (COA, NHD), cancelled by ComIIIPhibFor Rept, dtd 16Apr44, *op. cit.*

[7] ComSoPac 1st end, dtd 29Apr44, to ComIII PhibFor Rept, dtd 16Apr44, *op. cit.*

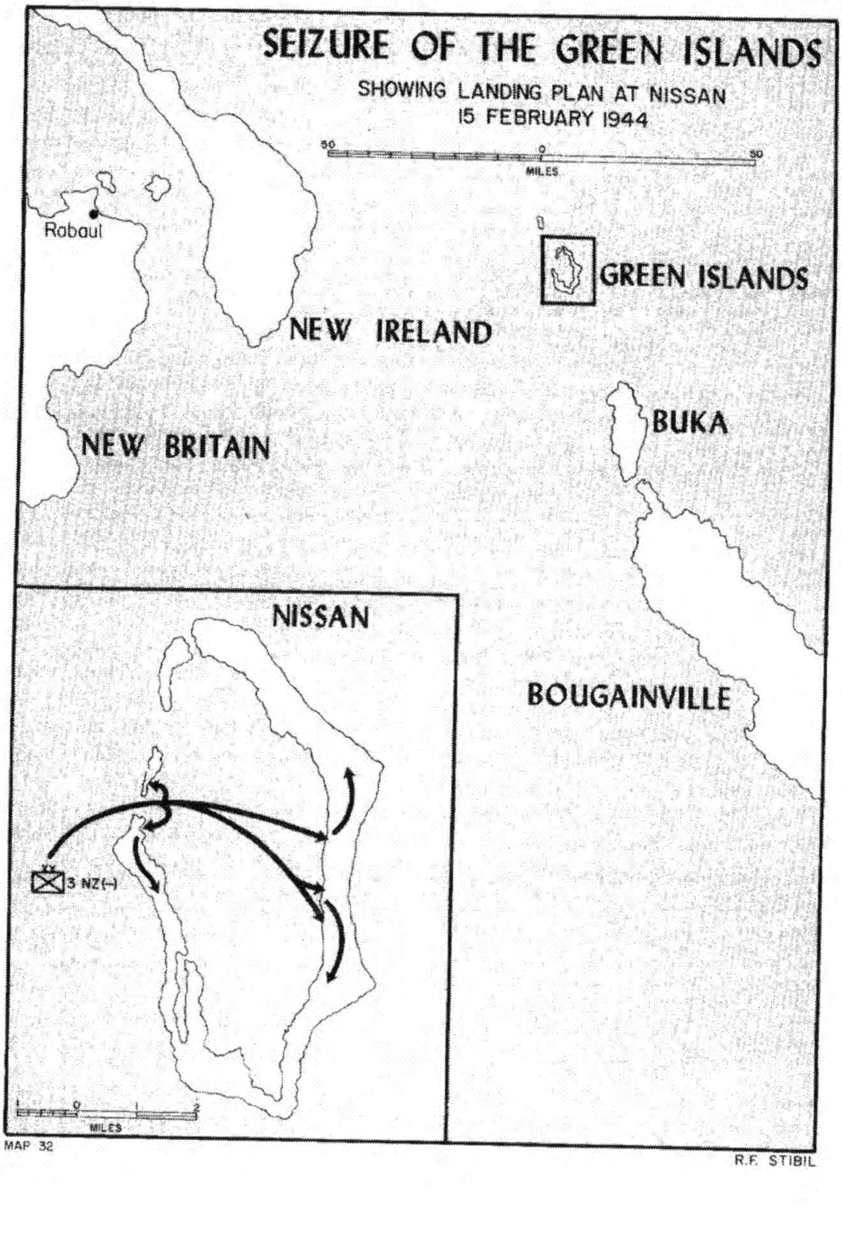
SEIZURE OF THE GREEN ISLANDS
SHOWING LANDING PLAN AT NISSAN
15 FEBRUARY 1944
MILES
Rabaul
GREEN ISLANDS
NEW IRELAND
BUKA
NEW BRITAIN
NISSAN
BOUGAINVILLE
3 NZ(-)
MILES
MAP 32
R.F. STIBIL

there was no sign of enemy opposition. Once it was obvious that the landing was safely effected, the transports and escorts shoved off for the Treasurys in order to be well away from Nissan by daylight.

The atoll reconnaissance started when dawn broke, and the findings were very encouraging. Preliminary estimates that Nissan could accommodate a fair-sized airbase and that its lagoon and beaches could handle landing ships proved accurate, and, despite evidence that about 100 Japanese occupied the islands, there was only one clash with the defenders. In an exchange of fire with a well-hidden machine gun, an LCVP-borne scouting party lost three men and had seven others wounded. Showing that Rabaul was well aware of the raid, seven Zekes appeared during the afternoon to strafe and bomb the landing craft; one sailor was killed and two were wounded.

Right on schedule, at 0010 on 1 February, the APDs and their escorts arrived in the transport area off Nissan. After breasting a choppy sea, the landing craft and the troops were all back on board ship by 0145. On the return voyage, the escorts added a bonus prize to the successful mission when the destroyers *Guest* and *Hudson* sank a Japanese submarine, the *I-171*, with a barrage of depth charges.

Since the reconnaissance of Nissan confirmed earlier estimates of its value as an objective, Admiral Halsey's operation plan for its capture, issued on 24 January, went unchanged. Admiral Wilkinson, as Commander Task Force 31, was directed to seize the Green Islands using Major General H. E. Barrowclough's 3d New Zealand Division (less the 8th Brigade) as the landing force. ComAirSols would provide reconnaissance and air cover, and as in previous SoPac operations, a commander and staff to control air activities at the objective. Brigadier General Field Harris, well experienced in this type of assignment after similar service at Bougainville, was designated ComAirGreen. To cover the landings, two cruiser-destroyer task forces would range the waters north, east, and south of the islands, while a tightly echeloned procession of APDs, LCIs, and LSTs ran in toward the target from the west, unloaded, and got clear as soon as possible. The possibility of a Japanese surface attack could not be discounted with Truk presumably still the main *Combined Fleet* base, and an aerial counterattack from Rabaul and Kavieng was not only possible but probable.

The pending assault did not catch the Japanese unawares, but the incessant AirSols strikes on *Eleventh Air Fleet* bases at Rabaul, coupled with RAAF and Fifth Air Force attacks on Kavieng, gave the enemy no chance for effective countermeasures. The original garrison of the Green Islands, 12 naval lookouts and 60 soldiers who operated a barge relay station, fled to the nearby Feni Islands on 1 February after briefly engaging the Allied reconnaissance force. About a third of these men returned to Nissan on the 5th to reinforce a small naval guard detachment that had been sent by submarine from Rabaul after word of the Allied landing was received. The combined garrison stood at 102 men on 14 February, when Japanese scout planes reported that a large convoy of transports, screened by cruisers and destroyers, was headed north from the waters off Bougainville's west coast.

Japanese aircraft harassed the oncoming ships throughout the moonlit night approach, but managed to score on only one target, the light cruiser *St. Louis*, which steamed on despite damage and casualties from one bomb hit and three near misses. The amphibious shipping reached its destination unscathed, and, at 0620, the first wave of New Zealander assault troops from the APDs was boated on the line of departure. In order to spare the atoll's natives, there was no preliminary or covering fire as the LCVPs raced shoreward. AirSols planes were overhead, however, ready to pounce on any Japanese resistance that showed, destroyers had their guns trained ashore, and LCI gunboats shepherded the landing craft to the beaches.

All landings were unopposed, the first at 0655 on small islets at the entrance to the lagoon and those immediately following which were made near the prospective airfield site. About 15 Japanese dive bombers attempted to hit the transports at about this time, but a fury of antiaircraft fire from every available gun caused them to sheer off after some ineffectual bombing. The AirSols combat air patrol, all from VMF-212, claimed six of the attacking planes; the Japanese admitted losses of four Bettys, two Kates, six Vals, and a Rufe during both the night heckling and the unsuccessful thrust at the landing ships.

The New Zealanders sent patrols out as soon as the landing force was firmly set up ashore, but these encountered only slight resistance. The operation proceeded smoothly and without encountering any unforeseen snags. As soon as the APDs discharged the assault troops, they picked up an escort and headed south, while 12 LCIs beached on Nissan and quickly unloaded. At 0835, an hour before the LCIs left, 7 LSTs, each loaded with 500 tons of vehicles and bulk cargo, entered and crossed the lagoon and nosed into shore. When the LSTs retired at 1730, Admiral Wilkinson in his flagship and the remainder of TF 31's ships accompanied them, leaving behind 6 LCTs to serve the budding base.

On D-Day, 5,800 men had been landed—to stay. Although there were almost 100,000 Japanese troops located close by on the Gazelle Peninsula and New Ireland, they were held at bay by superior Allied air and naval strength. The situation of the Japanese units located south of the newest SoPac outpost was "hopeless" in General MacArthur's view, and he reported to the JCS that the successful landing "rings the curtain down on [the] Solomons campaign." [8]

Despite the overwhelming odds against them, the defenders of Nissan Atoll fought tenaciously against the New Zealanders, killing 10 and wounding 21 of the 3d Division's men in 5 days of mopping-up action.[9] The last pocket of resistance was not wiped out until the 19th when the Japanese remnant sent Admiral Kusaka the message: "We are charging the enemy and beginning radio silence." [10]

What little help Rabaul could offer its doomed outguard on Nissan was confined to night bombing, and even that proved costly and futile for the Japanese. The attacking planes lost three of their number to VMF(N)-531 Venturas vectored to their targets by one of the squadron's GCI teams. Even the nuisance value of night raiders was lost when the *Combined Fleet* ordered all flyable aircraft out of Rabaul

[8] MacArthur disps to Marshall, dtd 14Feb44 and 15Feb44, in *GHQ G-3 Jnl.*

[9] Gillespie, *New Zealand History*, p. 188.

[10] *SE Area NavOps-III*, p. 62.

on the 19th. With the departure of defending Zekes, the New Britain base lay wide open to AirSols attack, particularly to strikes that could be mounted or staged from fields in the Green Islands. Even more important, Kavieng, where the Japanese still had some planes, was within easy reach of fighters and light bombers dispatched by ComAirGreen.

Seabee units outdid themselves and surpassed all base development goals. The fighter field was able to handle its first emergency landing on 4 March, the date on which Admiral Wilkinson passed command of the Green Islands to General Barrowclough. Three days later, AirSols fighters staged through Green, as Nissan was usually called, to attack Kavieng. Completion of the bomber field was scheduled for 1 April, but the first group of light bombers, 36 SBDs and 24 TBFs from Piva, was able to stage for a strike and hit Kavieng on 16 March. On the 19th, VMSB-243, VMTB-134, and part of VB-98 were detached from Strike Command, Piva and shifted to Green and General Harris' command.

The light bombers did not get to settle in at their home field for a while though, but shared instead the fighter strip with the Corsairs of VMF-114 and -212. The Thirteenth Air Force pre-empted the bomber field when its B-24s landed on Green en route to strikes on Truk. Until hardstands for the Liberators were completed on 15 April, the Marine and Navy bombers competed with Seabee construction equipment for room. "Frequently trucks hauling coral would be sandwiched between sections of planes taxiing and often [an] entire strike [group] would inch by fighters parked along the ends of the taxiway. Each TBF would have to taxi with folded wings and unfold them only when in position along the strip." [11]

The temporary crowding served a useful purpose, however; it made maximum use of Green's airfields at a time when many of the missions flown by AirSols planes helped isolate the newest Bismarck's battleground, the Admiralties. The seizure of these islands, 200 miles from Wewak and 260 miles from Kavieng, snapped the last link between General Imamura and his *Eighth Area Army* troops fighting on New Guinea.

In terms of their eventual usefulness, the Admiralties far outshone any other strategic objective that was seized during the operations against Rabaul. Seeadler Harbor, contained in the hook-like embrace of the two main islands, Manus and Los Negros, is, if anything, as fine as Rabaul's harbor and well able to handle warships and auxiliaries of all sizes. In 1942, at Lorengau village on Manus, the largest island, the Japanese had built an airfield and followed up in 1943 by constructing another at Momote Plantation on Los Negros. Both fields were used as staging points for traffic between Rabaul and New Guinea. (See Map 33.)

When Allied advances on the Huon Peninsula and in the Solomons threatened the Bismarcks area, Admiral Kusaka and General Imamura both ordered more of their troops into the Admiralties. A naval garrison unit from New Ireland was able to get through to Lorengau in early December, but ships carrying Army reinforcements from Japan and the Palaus were either sunk by American submarines or turned back by the threat of their torpedoes. In late January, Imamura dispatched one infantry battalion from

[11] CO StrikeComdGreen Rept, *op. cit.*, p. 4.

SEABEE EQUIPMENT *is unloaded from an LST at Nissan Island on D-Day of the Green Islands operation.* (*USMC 77990*)

FIRST WAVE ASHORE *on Los Negros, troopers of the 1st Cavalry Division, advance toward Momote airfield.* (*SC 187412*)

Kavieng and another from Rabaul on board destroyers that reached Seeadler despite harassment by Allied aircraft. These soldiers, together with those of a transport regiment and the naval contingent already present, made up a formidable defense force of about 4,400 men.

Realizing that he was charged with defending a prize that the Allies could ill afford to ignore, the Japanese commander in the Admiralties decided that deception was one of his most effective weapons. When General Kenney's planes attacked Lorengau and Momote airfields, the enemy leader ordered his men not to fire back. He told them, in fact, not to show themselves at all in daylight. His ruse had the desired effect; reconnaissance planes could spot few traces of enemy activity. On 23 February, three B-25s "cruised over Manus and Los Negros for ninety minutes at minimum altitude without having a shot fired at them or seeing any signs of activity either on the airdromes or along the beaches." [12] To American Generals Kenney and Whitehead, the situation seemed ripe for a reconnaissance in force, one that might open the way for an early occupation of the Admiralties and the consequent upgrading of the target dates for all later operations.

General MacArthur, impressed by the promise of a quick seizure of an important objective, accepted General Kenney's proposal that a small force carried on destroyers and APDs land on Los Negros and seize Momote airfield, repair it, and hold it ready for reinforcement by air if it proved necessary. In case Japanese resistance proved too stiff, the reconnaissance force could be withdrawn by sea. If, on the other hand, the enemy garrison was weak, the original landing force would be strong enough to hold its own and open the way for reinforcing echelons.

Acting on General MacArthur's orders, issued on 24 February, the 1st Cavalry Division (Major General Innis P. Swift) organized a task force of about 1,000 men, most of them from its 1st Brigade, to make the initial landing. If all went well, follow-up echelons would bring in more cavalrymen plus Seabees and other supporting troops to mop up the Japanese and begin base construction. Commanding the reconnaissance force, its backbone the 2d Squadron, 5th Cavalry, was Brigadier General William C. Chase. The SWPA's veteran amphibious force commander, Admiral Barbey, was responsible for the conduct of the operation. General MacArthur decided that both he and Admiral Kinkaid would accompany the attack group that transported Chase's troops in order to evaluate at first hand the results of the reconnaissance.

On 27 February, two days before D-Day, a small party of ALAMO scouts landed on Los Negros about a mile south of Momote; they reported the jungle there to be a bivouac area alive with enemy troops. The scouts' finding was too inconclusive to bring about any change in the size of Chase's force, but the information did result in the detail of a cruiser and two destroyers to blanket the bivouac area with naval gunfire when the landing was attempted. The rest of the covering force, another cruiser and two more destroyers, was assigned Lorengau and Seeadler Harbor as an area of coverage. Nine destroyers of the attack group, each transporting about 57 troopers, were assigned fire support areas which would di-

[12] Craven and Cate, *Guadalcanal to Saipan*, p. 559.

rectly cover the landing attempt. Three APDs, each with 170 men on board, would land the first three waves of assault troops.

The chosen landing area, a beach near the airfield, could be reached only through a 50-yard-wide opening in the fringing reef that closed the narrow entrance to a small harbor on the eastern shore of Los Negros. The site seemed so improbable for a landing that the Japanese concentrated most of their strength to meet an attack from the Seeadler Harbor side of the island.

On 27 February, the 1st Cavalry Division troops boarded ship at Oro Bay, and the attack group moved out to rendezvous with its escort off Cape Sudest. Through a heavy overcast on the morning of the 29th, the American ships approached the Admiralties and deployed to bombardment and debarkation stations off the coast of Los Negros. At 0728, three B–24s bombed Momote, but poor visibility cancelled out most of the rest of the preparatory air strikes. Cruisers and destroyers began shelling the island at 0740 and continued firing as the troops in LCVPs crossed the line of departure, 3,700 yards out, and headed for the beach. Fifteen minutes later, as the first wave passed through the harbor channel, enemy machine guns on the headlands opened fire on the boats while heavier guns took on the cruisers and destroyers. Counterbattery fire was prompt and effective; the Japanese guns fell silent. At 0810, a star shell fired from the cruiser *Phoenix* signalled the end of naval gunfire and brought in three B–25s, all that had reached the target in the foul flying weather, to strafe and bomb the gun positions on the headlands.

The first troops were on the beach at 0817 and moving inland; the few Japanese defenders in the vicinity pulled back in precipitous haste. Enemy gun crews manning the weapons interdicting the entrance channel began firing again when naval gunfire lifted. Destroyers pounded the gun positions immediately and drove the crews to cover, a pattern of action that was repeated throughout the morning. The American landings continued despite the Japanese fire, and by 1250, General Chase's entire command had landed. The cost of the operation thus far was two soldiers killed and three wounded, a toll doubled by casualties among the LCVP crews. Five enemy dead were counted.

The cavalrymen advanced across the airfield during the afternoon, but pulled back to man a tight 1,500-yard-long perimeter anchored on the beach for night defense. General MacArthur and Admiral Kinkaid went ashore about 1600, conferred with General Chase, and heard the reports that the cavalrymen had run across signs of a considerable number of enemy troops. After assessing the available intelligence, and viewing the situation personally, the SWPA commander ordered General Chase to stay put and hold his position at the airfield's eastern edge. As soon as the senior commanders were back on board ship, orders were dispatched to send up more troops and supplies to reinforce the embattled soldiers. Two destroyers remained offshore to furnish call-fire support when the rest of the task group departed at 1729.

The first of the counterattacks that the cavalrymen expected, and had prepared for as best they could with their limited means, came that night. The Japanese on Los Negros, who outnumbered the Americans handily, did not take advantage of their strength and made no headway in a series of small-scale attacks that sometimes penetrated the perimeter but never seriously threatened the integrity of the position. With daylight, American patrols pushed out from their lines until they ran into heavy enemy resistance, then pulled back to let the destroyers and the force's two 75mm howitzers fire on the Japanese. Aircraft made nine supply drops for the cavalrymen during the day, and, toward evening, Fifth Air Force planes bombed the enemy positions, despite the ineffectual attacks of several Japanese Army fighters which showed up from Wewak. There was an unsuccessful assault on the cavalrymen's lines at dusk, and another night of infiltration attempts that ended with a two-day count of 147 Japanese dead within American lines.

By dawn of 2 March, the Japanese had lost their chance to drive out the reconnaissance force, for the first reinforcement echelon, 1,500 more troopers and 428 Seabees, stood offshore. An American destroyer and two minesweepers of the landing ship escort attempted to force the entrance of Seeadler Harbor, but uncovered a hornet's nest of coast defense guns which forced them to sheer off. Warned away from Seeadler for the time being, the amphibious craft landed their troops and cargo on the beaches guarded by Chase's men.

Once the fresh troops were ashore, General Chase attacked and seized the airfield against surprisingly light resistance. There was ample evidence, however, that the Japanese were readying an all-out attack. It came on the night of 3–4 March, a night of furious fighting that saw 61 Americans killed and 244 wounded, with 9 of the dead and 38 of the wounded Seabees, who backed up the cavalry's lines. At the focal point of the attack, 167 enemy soldiers fell; hundreds more fell all along the perimeter.

The remainder of the 1st Cavalry Division joined its advance forces in the Admiralties during the following week. The Japanese on Los Negros were either killed or driven in retreat to Manus. Air and ship bombardment eliminated the enemy guns that had shielded the Seeadler entrance, and on 9 March, the 2d Cavalry Brigade entered the harbor and landed on Los Negros. The cavalry division commander, General Swift, now planned the seizure of Lorengau airdrome and the capture of Manus.

On 15 March, following a series of actions that cleared the small islands fringing the harbor of the enemy, the 2d Brigade landed on Manus and fought its way to the airfield. Even though the main objective was quickly secured, the big island was far from won. It was two months before the combat phase of the operation was ended and the last organized resistance in the Admiralties faded. The count of Japanese dead reached 3,280, 75 men were captured, and another 1,100 were estimated to have died and been buried by their own comrades. The 1st Cavalry Division lost 326 troopers and had 1,189 of its men wounded in the protracted and bitter fighting.

While the battle raged, the naval construction battalions and the Army engineers turned to on the airfields and naval base projected for the islands. Momote was operational by 7 March, and, on the 9th, a squadron of Australian Kittyhawks from Kiriwina moved in as part of the garrison. The RAAF planes, soon reinforced, flew cover for B-25 bombers at first and then began to fly bombing and

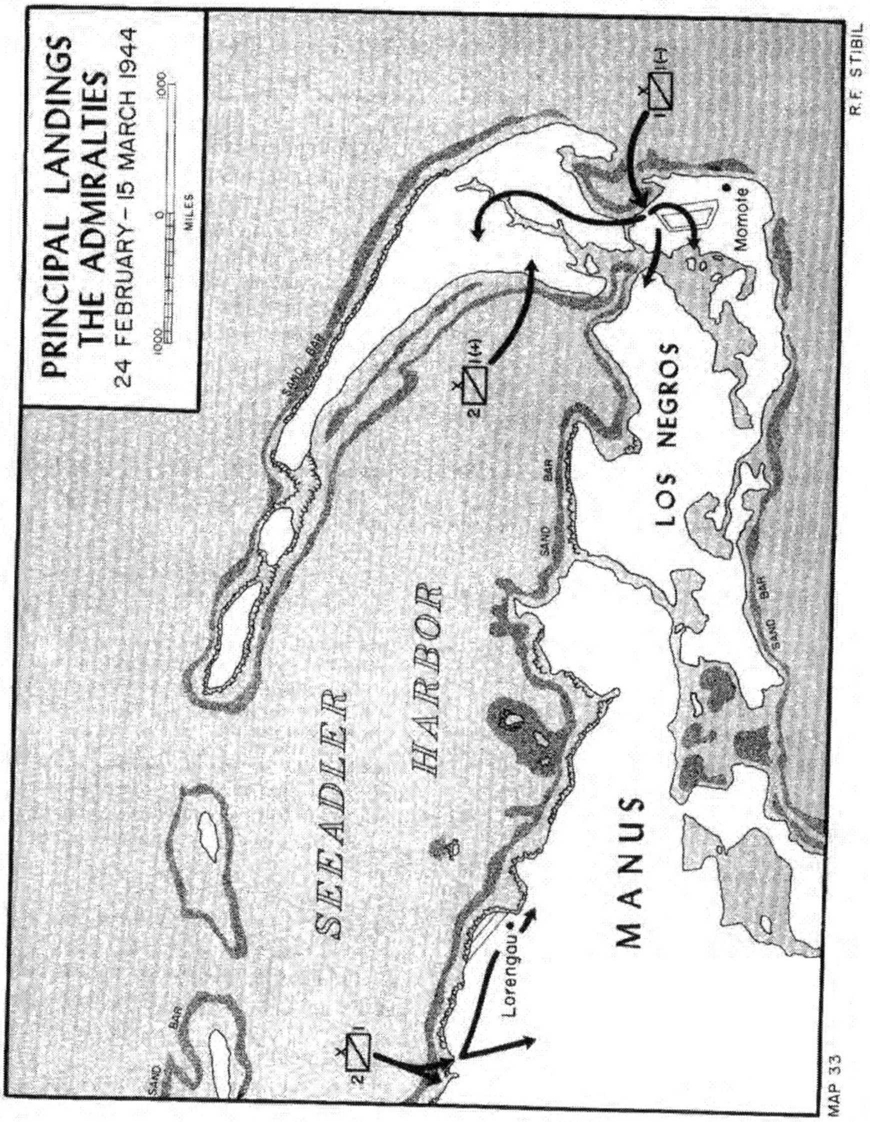
PRINCIPAL LANDINGS
THE ADMIRALTIES
24 FEBRUARY – 15 MARCH 1944
1000 0 1000
MILES
SEEADLER
HARBOR
MANUS
LOS NEGROS
Lorengau
Momote
SAND BAR
R.F. STIBIL

MAP 33

strafing strikes of their own in support of the cavalrymen's offensive. On 16 March, the Australian squadrons were given the primary mission of protecting all Allied shipping in the vicinity of the Admiralties.[13]

When Lorengau airfield proved unsuitable for extensive development, the engineers and Seabees shifted their tools and machines to Mokerang Plantation on Los Negros, about 7,000 yards northwest of Momote on the Seeadler shore. The new field was operational by 21 April. The naval base, including two landing strips for carrier aircraft on outlying islands, flourished. Manus, as the whole base complex was generally known, grew to be as important in staging and supporting Allied operations during 1944 as Guadalcanal had been in 1943 and Espiritu Santo in 1942.

EMIRAU: THE LAST LINK[14]

At one time in the planning for the operations that would follow Bougainville and Cape Gloucester, General MacArthur and his staff had considered it necessary to make a landing at Hansa Bay between the Japanese *Eighteenth Army's* bases at Madang and Wewak. The early and successful move into the Admiralties, from which planes could easily interdict both enemy positions, crystallized opinion against the Hansa Bay venture. In its stead, on 5 March, MacArthur proposed to the JCS that he completely bypass the Madang-Wewak area and take a long step forward in his advance toward the Philippines by seizing Hollandia in Netherlands New Guinea. In the same message, the general reaffirmed his conviction that Kavieng had to be taken to insure the complete neutralization of Rabaul.

In Washington, where the need for taking Kavieng had been seriously questioned, considerable weight was obviously given to Admiral Halsey's opinion, voiced in person in January, that "the geography of the area begged for another bypass,"[15] and that:

> . . . the seizure of an airfield site in the vicinity of the St. Matthias Group appeared to be a quick, cheap operation which would insure the complete neutralization of Kavieng and complete the isolation of Rabaul and the Bismarcks in general. Furthermore, the Carolines would be brought just that much nearer as a target for our own aerial operation.[16]

The fact that Admiral Nimitz joined in recommending that the Kavieng operation be dropped in favor of the much less expensive seizure of Emirau may have been decisive.

On 12 March, the JCS issued a new directive for future operations in the Pacific, cancelling Kavieng and Hansa Bay and ordering the capture of Hollandia and Emirau, the latter as soon as possible. General MacArthur immediately issued

[13] Odgers, *RAAF Against Japan*, pp. 174–180.

[14] Unless otherwise noted, the material in this section is derived from: *GHQ G–3 Jnl;* CTF 31 Rept of the Seizure and Occupation of Emirau Island, 20Mar–7Apr44, dtd 16Apr44; CG Emirau LdgFor Rept of Ops, 15Mar–9Apr, dtd 20Apr44; *1st MAW Feb–Apr44 WarDs*; Kenneth W. Condit and Edwin T. Turnbladh, *Hold High the Torch, A History of the 4th Marines* (Washington: HistBr, G–3 Div, HQMC, 1960); Halsey and Bryan, *Admiral Halsey's Story*; Miller, *Reduction of Rabaul*; Morison, *Breaking the Bismarcks Barrier;* Rentz *Bougainville and the Northern Solomons.*

[15] Halsey and Bryan, *Admiral Halsey's Story*, p. 188.

[16] ComSoPac 1st end, dtd 1May44, to CTF 31 Rept, dtd 16Apr44, *op. cit.*

orders halting preparations for the Kavieng attack, then only 18 days away, and directed Admiral Halsey to seize Emirau instead, using a minimum of ground combat forces. In his turn, ComSoPac ordered his amphibious force commander, Admiral Wilkinson, to take the new objective by 20 March and recommended that the 4th Marines be used as the landing force. The message from Halsey at Noumea to Wilkinson at Guadalcanal was received early on the morning of 15 March when loading had already started for Kavieng.

The I Marine Amphibious Corps, composed of the 3d Marine Division and 40th Infantry Division, had been the chosen landing force for Kavieng. For that operation, the 3d Division was reinforced by the 4th Marines, and the regiment was ready to load out when word was received of the change in plans. Fortunately, the headquarters of III Amphibious Force, IMAC, the 4th Marines, and the transport group which was to carry the troops to the target were close together and planning got underway immediately. General Geiger noted that the several staffs "had only about six or eight hours to work up the Emirau plans"[17] that had resulted from Admiral Halsey's earlier interest in the island as a SoPac objective. Late in the afternoon of the 15th, the admiral flew in to Guadalcanal from New Caledonia and quickly approved the concept of operations that had been developed.

Commodore Lawrence F. Reifsnider was to command the amphibious operation, with Brigadier General Alfred H. Noble, ADC of the 3d Marine Division, in command of the landing force. General Noble, who was also slated to become the island's first commander, had a small staff made up of IMAC and 3d Division personnel. An air command unit for Emirau, under Marine Colonel William L. McKittrick, was formed from the larger headquarters that had been organized to control air operations at Kavieng.

The 4th Marines, commanded by Lieutenant Colonel Alan Shapley, was the newly formed successor to the regiment captured on Corregidor.[18] It was activated on 1 February from former raider units, after the Commandant decided there was no longer enough need to justify the existence of battalions specially raised for hit-and-run tactics.[19] On the 22d, the Commandant directed General Geiger to reinforce the regiment by the addition of a pack howitzer battalion, engineer, medical, tank, and motor transport companies, and reconnaissance, ordnance, war dog, and service and supply platoons. Only the tank and medical companies had been added by the date the regiment sailed for Emirau. For the landing operation, the 3d Division provided amphibian tractor and pioneer companies and motor transport and ordnance platoons; the 14th Defense Battalion furnished a composite automatic weapons battery.

The pace of preparations for Emirau was so swift that it put a temporary crimp

[17] MajGen Roy S. Geiger ltr to LtGen Alexander A. Vandegrift, dtd 25Mar44 (Vandegrift Personal Correspondence File, HQMC).

[18] The story of the 4th Marines' defense of Corregidor's beaches it told in Part IV of Volume I of this series.

[19] The Headquarters and Service Company of the 1st Raider Regiment and the 1st, 3d, and 4th Raider Battalions become the Headquarters and Service Company and the 1st, 2d, and 3d Battalions of the 4th Marines. The 2d Raider Battalion formed the regimental weapons company.

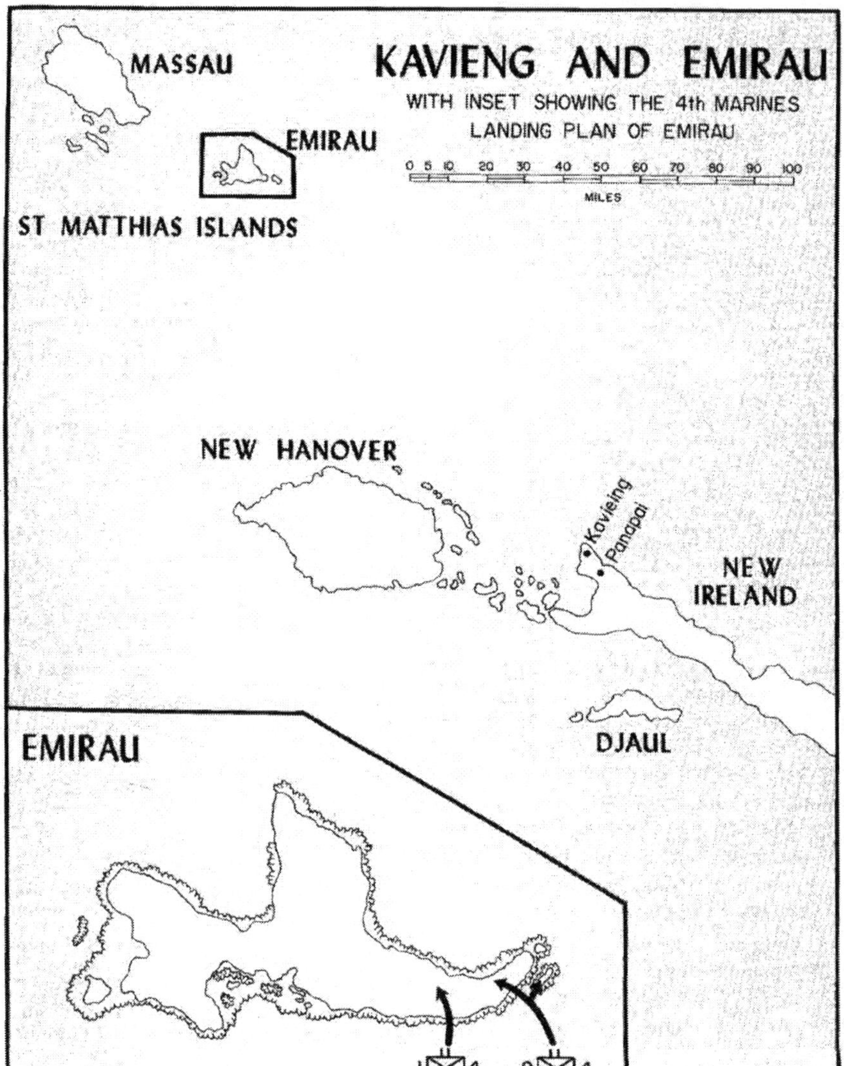
KAVIENG AND EMIRAU
WITH INSET SHOWING THE 4th MARINES LANDING PLAN OF EMIRAU
0 5 10 20 30 40 50 60 70 80 90 100
MILES
MASSAU
EMIRAU
ST MATTHIAS ISLANDS
NEW HANOVER
Kavieng
Panapai
NEW IRELAND
DJAUL
EMIRAU
1 4
2 4
3 4
R.F. STIBIL

MAP 34

in JCS plans for the employment of the Marines released by the cancellation of Kavieng. On 14 March, MacArthur received and passed on to Halsey for compliance, a JCS directive that the 3d Marine Division, the 4th Marines, and the 9th and 14th Defense Battalions were to be released to the control of CinCPOA immediately. By the time the admiral received this order, it was too late to replace the 4th Marines and still meet Emirau's D-Day of 20 March; some platoons of the 14th Defense Battalion were already on board ship. Consequently, ComSoPac outlined the situation to Admiral Nimitz and promised to release all units required for future operations as soon as possible. In view of the circumstances, Admiral Nimitz concurred in the temporary transfer of troops for use at Emirau.

The target for the operation is an irregularly shaped island eight miles long, hilly and heavily wooded. It lies in the southeastern portion of the St. Matthias Group, about 25 miles from Massau, the other principal island. Situated 90 miles northwest of Kavieng, Emirau was considered suitable for development as a base for fighters, bombers, and torpedo boats. All intelligence indicated that the Japanese had not occupied the islands in any appreciable strength, and a photo reconnaissance mission flown by VD-1 on 16 March revealed no trace of enemy activity or installations. (See Map 34.)

Even though little opposition was expected, detailed provisions for strong air and naval gunfire support were a part of the Emirau operation plan. The naval bombardment group that was to have shelled Kavieng under the cancelled plan was ordered to hit the town and its airfields anyway as insurance against interference from the enemy. While it was anticipated that there would be no need for preliminary bombardment of Emirau before the landing, two destroyers of the escort were prepared to deliver call fire, and planes from two supporting escort carriers were to be overhead, ready to strafe and bomb as necessary. The cruisers and other destroyers of the escort would take station to screen the landing. Should a Japanese surface threat materialize, the 4 battleships and 15 destroyers pounding Kavieng on D-Day would be available as a weighty back-up power. On Green, planes of VMSB-243, VMTB-134, and VB-98 were on standby for possible employment, reinforcing the carrier planes.

The formidable support preparations for Emirau made the unopposed amphibious operation seem anticlimactic. The loading, movement, and landing of General Noble's force was conducted in an aura of orderly haste. New shipping assignments, necessitated by the change in plans, forced the 4th Marines to sort and redistribute all the supplies in its beach-side dumps during the night of 15 March. Loading began the next morning and continued through the 17th when the troops went on board ship, and Commodore Reifsnider's attack group sailed from Guadalcanal.

The ships left in two echelons, grouped by cruising speed and destined to rendezvous on D minus 1. The Marines of the two assault battalions, 1/4 and 2/4, were on board nine APDs; the remainder of the landing force traveled to the target on three LSDs and an APA. One LSD transported the 66 LVTs that would land the assault waves over Emirau's fringing reef, another carried three LCTs, two of them loaded with tanks, and the third had three

LCTs on board bearing radar and antiaircraft guns.

At 0605 on 20 March, the attack group arrived in the transport area, the LSD launched her LVTs, and the assault troops transferred to the tractors using the APDs' boats which were supplemented by those from the APA. Then, as the men of the reserve battalion, 3/4, scrambled down the nets into boats to be ready for employment wherever needed, Corsairs of VMF–218 flashed by overhead to make a last minute check of the island for signs of the Japanese. Right on schedule, the assault waves crossed the island's encircling reef and went ashore on two beaches about 1,000 yards apart near the eastern end of the island, while a detachment of 2/4 secured a small islet that sheltered the easternmost beach. Soon after the assault troops landed, the 3d Battalion's boats grounded on the reef, and the reserve waded ashore through knee-deep water. During these landings, a few shots were fired in return to supposed enemy opposition, but subsequent investigation showed that there were no Japanese on the island.

Had there been opposition, one hitch in the landing plan could have been fateful. The tanks were not launched in time for the assault, since their LSD's flooding mechanism was only partially operative. A fleet tug with the escort was able to drag the loaded LCTs out through the stern gate by means of a towline. Although by this time the success of landing was assured, the tanks were run ashore anyway both as insurance and for training.

Supplies began coming in about 1100, first from the APDs and then from the APA, with the LCTs helping the ships' boats to unload. By nightfall, 844 tons of bulk cargo had been landed in addition to the weapons and equipment that went ashore in the assault. All the ships sailed just after sunset, leaving General Noble's force of 3,727 men to hold the island and prepare for follow-up echelons.

Emirau's natives told the Marines that only a handful of Japanese had been on the island, and they had left about two months before the landing. Intelligence indicated that there were enemy fuel and ration dumps on Massau and a radio station on a nearby island. On 23 March, destroyers shelled the areas where the reported installations lay and, according to later native reports, succeeded in damaging the dumps and radio enough to cause the Japanese to finish the job and try to escape to Kavieng. On the 27th, a destroyer intercepted a large canoe carrying enemy troops about 40 miles south of Massau; the Japanese opened up with rifles and machine guns, and the ship's return fire destroyed them all. This episode furnished the last and only vestige of enemy resistance in the St. Matthias Group.

The first supply echelon reached Emirau on 25 March, bringing with it the men and equipment of a battalion of the 25th Naval Construction Regiment. The Seabees and the supplies landed over beaches and dumps that had been prepared by the 4th Marines. Five days later, three more naval construction battalions arrived to turn to on the air base and light naval facilities. An MTB squadron began patrolling on the 26th while its base was being readied. Sites for two 7,000-foot bomber strips and a field 5,000-feet-long for fighters were located and surveyed before the month's end. On 31 March, heavy construction on the airfields began.

In view of the island's projected role as an important air base, General Noble's relief as island commander was a naval aviator, Major General James T. Moore, who had been Commanding General, 1st Marine Aircraft Wing, since 1 February. General Moore, with advance elements of the wing headquarters squadron, arrived on Emirau on 7 April, following by two days the forward echelon of MAG–12. Relief for the 4th Marines by the island's garrison, the 147th Infantry Regiment, took place on 11 April, and the Marines left on the same ships that brought the Army unit. At noon on the 12th, acting on Admiral Wilkinson's orders as operation commander, General Moore formally assumed command of all ground forces on Emirau.

Throughout April, airfield construction continued at a steady but rapid tempo in order to ready the island for full use in the interdiction of Japan's Bismarcks bases. The first emergency landing was made on the 14th when a Navy fighter came down on one of the bomber strips. On the 29th, SCAT transports began operating regularly from the new fields, and, on the 2d of May, the first squadron of the MAG–12 garrison, VMF–115, arrived and sent up its initial combat air patrol. In the next two weeks, several more Marine fighter, dive, and torpedo bomber squadrons moved up from Bougainville and Green. By mid-May, Emirau was an operating partner in the ring of SWPA and AirSols bases that throttled Rabaul and Kavieng.

THE MILK RUNS BEGIN [20]

The prime target at the hub of the encircling Allied airfields, Rabaul, had no respite from attack even while the SWPA forces were seizing the Admiralties and SoPac troops were securing Emirau. If anything, the aerial offensive against the enemy base intensified, since the absence of interceptors permitted both a systematic program of destruction and the employment of fighters as bombers.

First on the list of objectives to be eliminated was Rabaul proper. The town was divided into 14 target areas which, in turn, were further subdivided into two or three parts; each was methodically wiped out. Two weeks after the opening attack of 28 February, the center of the town was gutted, and most strikes, thereafter, were aimed at more widely spaced structures on the outskirts. By 20 April, only 122 of the 1,400 buildings that had once comprised Rabaul were still standing and these were "so scattered that it was no longer a paying proposition to try to make it a 100 percent job." [21]

Weeks before the AirSols staff reached this conclusion, the task of reducing the town to rubble and charred timbers was left pretty much to the fighter-bombers, while the B–24s, B–25s, SBDs, and TBFs concentrated on the two largest enemy supply dumps, one about two miles west of Rapopo and the other on the peninsula's north coast three miles west of Rabaul. Bomber pilots found that 500-

---

[20] Unless otherwise noted, the material in this section is derived from: *SoPac ACI Repts;* Off of NavAirCIntel, ComSoPac, The Reduction of Rabaul, 19Feb–15May44, dtd 8Jun44, hereafter *Reduction of Rabaul;* ComAirNorSols WarD, 15–30Jun44, hereafter *ComAirNorSols WarD* with appropriate months; MASP Correspondence on Ops and TacEmpl of Units, 1943–44; *Thirteenth-AF Data;* AirSols Ftr-StrikeComd WarD, 15Mar–1Jun44, dtd 15Jul44; *1st MAW Mar–Jun44 WarDs; Eighth Area ArmyOps; SE Area NavAirOps—V;* Craven and Cate, *Guadalcanal to Saipan;* Ross, *RNZAF;* Sherrod, *MarAirHist.* USSBS, *Allied Campaign Against Rabaul.*

[21] *Reduction of Rabaul,* p. 4.

pound bombs containing clusters of 128 smaller incendiary bombs were more effective than high explosive in laying waste to these sprawling areas of storage tents, sheds, and ammunition piles.

Over the course of three months, during which the major destruction of aboveground installations was accomplished, an average of 85 tons of bombs a day was dropped on Rabaul targets. The attack was a team effort, done in part by all the plane types assigned to AirSols command. The two Liberator groups of the Thirteenth Air Force provided a normal daily effort of 24 planes until 23 March, when all heavy bombers were diverted to attacks on Truk and other targets in the Carolines. The Thirteenth's B–25 group also sent up an average of 24 planes a day. The strength of Marine and Navy light bombers varied during the period, but generally there were three SBD and three TBF squadrons at Piva and three SBD and one TBF squadron at Green; in all about 160–170 planes were available, with a third to a half that number in daily use. Even when the Japanese attacked the Torokina perimeter in March, and much of the air support of the defending Army troops was furnished by Piva-based light bombers, there was little letup in the relentless attack on Rabaul. The SBD–TBF squadrons at Green increased their efforts, which were supplemented daily by the attacks of 48–60 fighters equipped to operate as bombers.

Once the Zekes disappeared from the sky over the Bismarcks, AirSols had a surplus of fighter planes. Consequently, all Army P–38s, P–39s, and P–40s, and RNZAF P–40s, were fitted with bomb racks, after which they began making regular bombing attacks. At first, the usual loading was one 500-pound bomb for the Airacobras, Warhawks, and Kittyhawks and two for the Lightnings, but before long, the single-engine planes were frequently carrying one half-ton bomb apiece and the P–38s, two. Except for some bombing trial runs by Corsairs against targets on New Ireland, AirSols, Navy and Marine fighters in this period confined their attacks on ground targets to strafing runs. Later in the year, all fighter aircraft habitually carried bombs.[22]

The pattern of attacks was truly "clock-round," giving the enemy no rest, with the nighttime segment of heckling raids dominated by Mitchells. Army B–25s drew the job at first, but with the entry into action of VMB–413 in mid-March, the task gradually was given over to Marine PBJs. The Marine squadron, the first of five equipped with Mitchells to serve in the South and Southwest Pacific, proved particularly adept at night operations as well as the more normal daylight raids. General Matheny, the veteran bomber commander, specially commended the unit for its development of "the dangerous, tiresome mission of night heckling against the enemy bases to the highest perfection it has attained in the fourteen months I have been working under ComAirSols."[23]

The object of the heckling missions was to have at least one plane over the target all night long. For the enemy troops below, the routine that was developed must

[22] Much of the pioneering work in perfecting the Corsair as a fighter-bomber was done by Marine squadrons operating against enemy islands in the Marshalls. Their story will be covered in the fourth volume of this series.

[23] Quoted in VMB–413 Hist, 1Mar43–1Jul45, dtd 15 Jul45, p. 5.

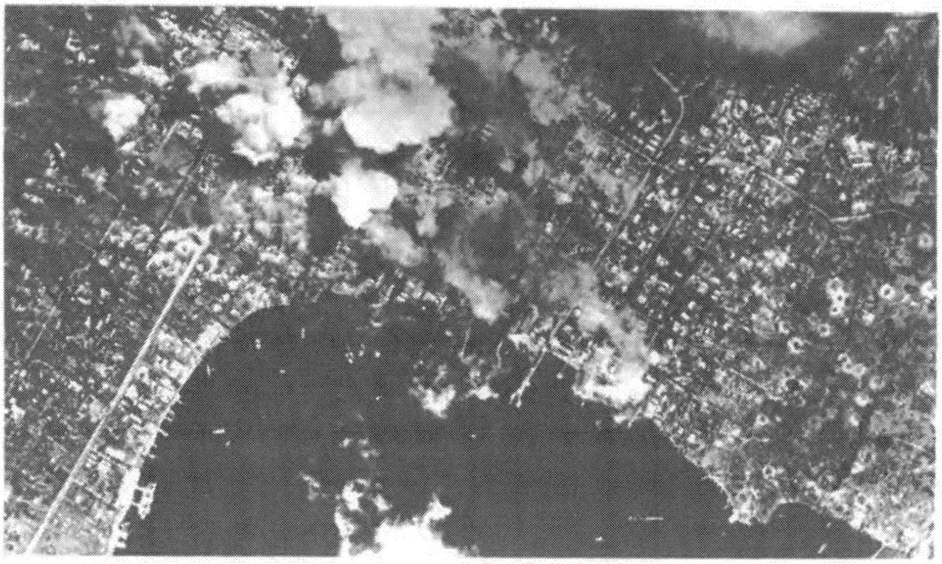

TOWN OF RABAUL *shows the effect of area saturation bombing in this photograph tcken from a Navy SBD during an attack on 22 March 1944.* (*USN 80–G–220342*)

CORSAIRS AT EMIRAU *in position along the taxiway to the new airfield which was operational less than two months after the landing.* (*USMC 81362*)

have been nerve-wracking. At dusk, the first PBJ:

> . . . appeared over Rabaul just as the Japanese began their evening meal. It dropped several bombs and retired. Minutes later, it came in again, hundreds of feet lower. More bombs dropped and it circled away. This pattern was repeated until, on its last run, the plane strafed the area.
>
> As the sound of its motor died away, the Japanese heard the second plane coming in on schedule to repeat the maddening process which went on night after night.[24]

The enemy troops that were subjected to the mass air raids of spring 1944 were surprisingly better off than aerial observers could tell. Spurred on by the punishing attacks which scored heavily against major targets, the Japanese dispersed a substantial portion of their supplies out of sight under cover of the jungle. Even more significant was the fact that every man not bedridden or wounded labored to dig caves and tunnels to shelter the troops and materiel needed to fight should the Rabaul area be invaded. By the end of May, enough supplies were underground to insure that the Japanese could make a prolonged defense. The digging-in process kept up until the end of the war, making Rabaul a fortress in fact as well as name.

The responsibility for the defense of the Rabaul area was a dual one, with Japanese Army and Navy troops holding separate sectors. The battered town and the mountainous peninsula east of it, from Praed Point to the northern cape, was defended by elements of the *Southeast Area Fleet.* Other naval troops, primarily antiaircraft artillery and air base units converted to infantry, held positions in the vicinity of Vunakanau and Tobera airfields. *Eighth Area Army* defended the rest of Gazelle Peninsula north of the Keravat and Warangoi Rivers. General Imamura, deeply imbued with the offensive spirit of Japanese military tradition, prepared battle plans which would meet an invasion attempt, wherever it occurred, with vigorous counterattacks. If all else failed, he felt that "the members of the whole army should commit the suicide attack."[25] Admiral Kusaka believed it was his primary duty to keep "his forces safe as long as possible and planned to hold on and destroy the enemy fighting strength"[26] by a tenacious defense of the elaborate fortifications the Navy had constructed in the hills back of Rabaul. Despite the difference in philosophy of ultimate employment, however, officers and men of both services worked together well, readying themselves to meet an attack that never came.

The growing desperation of the Japanese position in the Bismarcks was borne home to General Imamura by an order which the area army's assistant chief of staff characterized as "a cruel, heartless, unreasonable measure."[27] On 25 March, the units on New Guinea which had been under Imamura's command, the *Fourth Air* and *Eighteenth Armies*, were transferred by *Imperial General Headquarters* to the control of the *Second Area Army* defending western New Guinea. Since by this time the only contact the *Eighth Area Army* had with its erstwhile troops was by radio, the transfer was a practical move,

---

[24] John A. DeChant. *Devilbirds, The Story of United States Marine Corps Aviation in World War II* (New York: Harper and Brothers, 1947), p. 140.

[25] *Eighth Area ArmyOps*, p. 191.

[26] *Ibid.*

[27] Entry of 15Mar44 in CinCPac–CinCPOA Item No. 11,955, Diary of MajGen Masataka Kimihara, 1Jan–9Jun44, in CinCPac–CinCPOA Translations No. 1, dtd 31Oct44.

however disheartening its effect may have been on the army staff at Rabaul. As if to soothe the sense of isolation and loss, Tokyo directed General Imamura to defend Rabaul as a foothold from which future offensive operations would be launched.

The emptiness of this promise of future Japanese offensives was emphasized by the changes Rabaul's impotence wrought in the dispositions of Allied forces in the South and Southwest Pacific. From the start of the New Georgia operation, most of the combat troops, planes, and ships assigned to Admiral Halsey's command had operated in the SWPA under General MacArthur's strategic direction. On the 25th of March, the JCS issued a directive that outlined a redisposition of forces to take effect on 15 June, by which the bulk of SoPac strength was assigned to MacArthur's operational control for the advance to the Philippines. CinCSWPA would get the Army's XIV Corps Headquarters and Corps Troops, plus six infantry divisions. Added to the Seventh Fleet were 3 cruisers, 27 destroyers, 30 submarines, 18 destroyer escorts, an amphibious command ship, an attack transport, an attack cargo ship, 5 APDs, 40 LSTs, and 60 LCIs. The Thirteenth Air Force was also to be transferred, but with instructions that its squadrons would support Pacific Ocean Areas' operations as required.

Marine ground forces in the South Pacific were assigned to Admiral Nimitz' command, as CinCPOA, to take part in the Central Pacific drive. The majority of Marine air units, however, were detailed to the SWPA as the core strength of the aerial blockade of bypassed enemy positions in the Solomons and Bismarcks.

Under the assignment of forces first worked out by JCS planners, the Royal New Zealand Air Force units, which had played such an important role in the AirSols campaign against Rabaul, were relegated to the SoPac garrison. This decision was unacceptable to the New Zealand Government which wanted its forces to continue their active role in the Pacific fighting. The end result of representations by the New Zealand Minister in Washington was the allocation of seven squadrons—four fighter, two medium bomber, and one flying boat—to the SWPA and seven squadrons of the same types to the South Pacific. Since, by this time in 1944, all RNZAF units were either equipped or in the process of being equipped with U.S. Navy planes, an overriding factor in squadron assignment was the ease of maintenance and resupply in areas that would be manned primarily by U.S. Navy and Marine units. Under the plans developed, the deployment of the RNZAF to assigned SWPA bases, Bougainville, Los Negros, Emirau, and Green, would not be completed until late in the year.

Many of the units that officially became part of General MacArthur's command in June were already under his operational control two months earlier. By mid-April, the 13th Air Task Group, comprising heavy bombers of the Thirteenth Air Force under Major General St. Clair Street, was operating against the Palaus and Carolines to protect the flank of the Hollandia task forces. One heavy bombardment group of Street's command moved from Munda and Guadalcanal to Momote field on Los Negros on 20 April. The other group of the Thirteenth's B–24s followed soon after, and both bombed enemy bases that threatened MacArthur's further moves up the New Guinea coast and Nimitz' thrust into the Marianas.

Since he was to have a second American air force operating under his headquarters, General Kenney recommended and had approved the formation of a new command, Far East Air Forces (FEAF), whose principal components would be the Fifth and Thirteenth Air Forces. In addition to heading FEAF, Kenney remained Commanding General, Allied Air Forces, and, as such, commanded all other air units assigned to the Southwest Pacific Area, including those that had been a part of AirSols.

On 15 June 1944, all military responsibility for the area and the Allied units west of 159° East Longitude and south of the Equator passed to General MacArthur. Coincident with this change, Admiral Halsey relinquished his command of the South Pacific Area to his deputy of eight months standing, Vice Admiral John H. Newton, and went to sea as Commander, Third Fleet. The AirSols units became part of a new organization, Aircraft, Northern Solomons, with an initial strength of 40 flying squadrons, 23 of them Marine. The seven Thirteenth Air Force squadrons that were included in AirNorSols were under orders to join FEAF, and eight of the Navy's and RNZAF's were headed for SoPac garrison duty.[28]

Reflecting the preponderance of Marine elements assigned to AirNorSols was the appointment of General Mitchell as its commander. Mitchell, who had turned over leadership of AirSols to the Army Air Forces' Major General Hubert R. Harmon on 15 March,[29] had shortly thereafter relieved Admiral Fitch as ComAirSoPac. Throughout this period, the Marine general continued to head MASP also, but with his assumption of duties as ComAirNorSols, he designated Brigadier General Claude A. Larkin to succeed him in the South Pacific command and preside over its dissolution. According to plans for the future employment of its component wings, the 1st in the Southwest and the 2d in the Central Pacific, there was no longer any need for MASP. Completing the picture of Marine air command changes, General Mitchell took over the 1st Wing at the same time he became ComAirNorSols and established the headquarters of both organizations on Bougainville.

The command reorganization of 15 June 1944 marked the end of an important phase of the Pacific fighting, one which saw the onetime scene of violent battle action gradually become a staging and training center for combat on other fronts. In fitting tribute to the men who drove the Japanese back from the Solomons and Bismarcks, Admiral Halsey sent a characteristic farewell message to all ships and bases when he departed Noumea, saying: "'Well done' to my victorious all-services South Pacific fighting team. You have met, measured, and mowed down the best the enemy had on land and sea and in the air."[30]

[28] ComAirNorSols OPlan No. 1, dtd 6Jun44, Anx A, EstDistr of Air Units, ComAirNorSols.

[29] Two other Marine officers served briefly as ComAirSols. Brigadier General Field Harris, who relieved General Harmon on 20 April, and Major General James T. Moore, who relieved General Harris on 31 May. *StrikeComd WarD*, 15Mar–15Jun44.

[30] Entry of 15Jun44 in *ComAirNorSols Jun44 WarD*.

THE MITCHELLS REMAIN[51]

Controlling most of the strikes against Japanese targets when AirNorSols was activated was Fighter-Strike Command, headed by Colonel Frank H. Schwable. The title of this headquarters, the successor to AirSols Strike Command, reflected the shift in emphasis of fighter missions from air combat to strafing and bombing in company with SBDs and TBFs. The life of the new command was short, however, for General Mitchell abolished all separate type commands on 21 August when he centralized direction of tactical air operations under his own headquarters with Colonel Schwable as operations officer. Responsibility for controlling the aircraft assigned to various missions at each AirNorSols base remained with the area air commander who was also, in most cases, a Marine air group commander. Marines of the group headquarters doubled as members of the air commander's staff, serving together with representatives of other Allied and American units flying from the particular base.

Logistic support of the AirNorSols squadrons, except for those in the Admiralties, was to be the responsibility of ComSoPac until December when agencies of the Seventh Fleet could take over. For Mitchell's Marine units this function, once channeled through MASP, was made the responsibiltiy of Marine Air Depot Squadron 1, which remained in the South Pacific to handle the 1st Wing's personnel and supply needs. A similar role as a rear echelon for 2d Wing units staging to Central Pacific bases was performed by MAG-11's service squadron. All men and equipment that had been part of MASP were distributed to other units or returned to the States. General Larkin, who decommissioned MASP on 31 July, wrote its informal but apt epitaph in a letter to General Rowell at MAWPac:

> Certainly hate to see this command go under but it has outlived its usefulness, and it is always good news when units can be done away with rather than having to form new ones. At least it is an indication that we are doing okay with the war in this area by reducing and going forward.[52]

The inevitable result of the continuous Allied advance was that fields that had once bustled with combat air activity—at Noumea, Efate, Espiritu Santo, Guadalcanal, Banika, and Munda—were relegated to limited use or closed down. Newer, fully developed bases like Green and Emirau carried the burden of the attack against Rabaul and Kavieng, while most of the strikes aimed at the thousands of Japanese troops still active in the northern Solomons were mounted from the Piva fields. The more profitable enemy targets, however, those that could be reached only by heavy bombers and the few short-range planes that could crowd onto advanced airstrips, were hit less frequently than the bypassed positions. In July, as an example, SWPA land-based air forces flew over

[51] Unless otherwise noted, the material in this section is derived from: *ComAirNorSols Jun–Dec44 WarDs; 1st MAW Jun44–Aug45 WarDs;* ComAirEmirau WarDs, Jan–Aug45; MAG-61 WarDs, Jul–Aug45; *Eighth Area ArmyOps;* Maj Charles W. Boggs, Jr., *Marine Aviation in the Philippines* (Washington: HistDiv, HQMC, 1951); Odgers, *RAAF Against Japan;* Ross, *RNZAF;* Sherrod, *MarAirHist;* USSBS, *Allied Campaign Against Rabaul.*

[52] BGen Claude A. Larkin ltr to MajGen Ross E. Rowell, dtd 31Jul44, in MASP correspondence folder on Formation, Organization, and Disbandment. General Larkin, on 4 August, became Deputy Commander, AirNorSols.

8,000 sorties against targets in bypassed areas but only 3,000 against targets in the forward areas.[33] There was a constant danger that bombing attacks against blockaded enemy forces would degenerate into what ComAirPac called "mere weight lifting,"[34] an ineffective use of the air weapon, which was perhaps the most powerful available to Allied commanders.

General Kenney was well aware of the fact that some of his most effective aviation units, the veteran Marine squadrons, were tied down in the Solomons and Bismarcks. He intended to employ them in the seizure of Mindanao, and, to make the Marine units available, he determined to replace them with RNZAF and RAAF squadrons. In like manner, General MacArthur planned to relieve the American infantry divisions on Bougainville and New Britain with Australian troops. There was a strong current of opinion at MacArthur's headquarters that further operations in British and Australian territories and mandates should be undertaken by Commonwealth forces. On 12 July, CinCSWPA confirmed this concept in a letter to the Australian commander, General Blamey, stating:

> A redistribution of Allied forces in the SWPA is necessitated by the advance to the Philippine Islands. Exclusive of the Admiralties, it is desired that Australian forces assume the responsibility for the continued neutralization of the enemy in the Australian and British territory and mandates in the SWPA by the following dates:
>
> Northern Solomons-Green Island-Emirau—1 Oct 44
>
> Australian New Guinea—1 Nov. 44
>
> New Britain—1 Nov 44[35]

[33] *ComAirPac Jul44 Analysis*, p. 11.

[34] *ComAirPac Mar44 Analysis*, p. 19.

[35] Quoted in Odgers, *RAAF Against Japan*, p. 292.

General Blamey ordered the Australian II Corps to relieve the American XIV Corps on Bougainville, the 40th Infantry Division on New Britain, and the garrisons on Emirau, Green, Stirling, and New Georgia. The 6th Australian Division was designated to replace the American XI Corps in eastern New Guinea. Not content with holding defensive perimeters, the Australians intended to seek out and destroy the Japanese wherever this would be done without jeopardizing Allied positions.

Since the Australians planned an active campaign with a limited number of troops—two brigades on New Britain and four on Bougainville—plentiful and effective close air support was a necessity. Some of it would be provided by RAAF reconnaissance and direct support aircraft operating under control of Australian ground force commanders, but most of the planes would come from ComAirNorSols. According to plans, a New Zealand Air Task Force under Group Captain Geoffrey N. Roberts, was to take over control of air operations from AirNorSols when the 1st Marine Aircraft Wing moved to the Philippines.

One big hitch in this plan for RNZAF replacement of the 1st Wing occurred when the first target in the American return to the Philippines was shifted northward from Mindanao to Leyte. This change cancelled the wing's prospective employment, for, as General Kenney later explained: ". . . the movement forward of any air units in the Southwest Pacific depended upon the location of the unit under consideration, the availability of shipping, and the availability of airdromes in the forward zone." [36] In the perennially tight shipping situation, the distance of the wing from the Philippines acted against its employment as a unit. As a consequence, the changeover date from AirNorSols to RNZAF command, originally projected for 1 November 1944, was repeatedly delayed and did not take place until 15 July 1945. In the interim, the 1st Wing's operational strength was pared down to its transport and medium bomber groups; all fighter and dive bomber squadrons were transferred piecemeal to the Philippines. [37]

About 20 September, wing headquarters received the first word that its seven dive bomber squadrons would be employed in the Luzon campaign. For the most effective combat control, it was decided to employ two air groups, one of four squadrons and the other of three, and on 1 October, by transfers and joinings, MAG–24 became an all-SBD outfit (VMSB–133, –236, –241, and –341). A new Headquarters, MAG–32, was sent out from Hawaii to command the wing's remaining Dauntless squadrons (VMSB–142, –224, and –243). Air intelligence officers with experience in close air support techniques as practiced in the Marshalls and Marianas reported from the Central Pacific to assist in training the SBD crews. General Mitchell issued a training directive which indicated that the dive bombers would "be employed almost exclusively as close support for Army ground forces in an advancing situation" and that their basic mission would "be largely confined to clearing obstacles immediately in front of friendly troops." [38] Army units worked closely with the Marine squadrons during the training to formulate realistic problems of troop support. Whenever Japanese antiaircraft concentrations were light, the SBD pilots practiced their air support techniques during the regular routine of strikes on enemy targets.

The monotonous pattern of attacks on the same targets, day after day, went on regardless of the pending deployment of various wing units. One virtue of the situation was that many Marine pilots and aircrewmen got their first taste of combat flying during these months of strikes against bypassed objectives. Once, whole squadrons had been sent back to the States after completing a combat tour, but now, only the individual veterans returned and the squadrons remained, kept up to strength by replacements. The flying, gunnery, and bombing experience gained while hitting Rabaul and Kavieng and tackling the Japanese positions in the northern Solomons was invaluable. Although combat and operational casualties were low, there was enough opposition from enemy gunners, enough danger from the treacherous weather, to make pilots handle any AirNorSols mission with prudence. The urge for more violent action was always present, however, and the flyers were cautioned repeatedly in orders against "jousting with A/A [antiaircraft] positions in any area at any time." [39]

[36] Gen George C. Kenney ltr to Maj Charles W. Boggs, Jr., dtd 27Oct50, quoted in Boggs, *Marine Aviation, op. cit.*, p. 8.

[37] By the end of August 1944, all torpedo bomber squadrons in the 1st Wing had been transferred to the States or to other wings for employment in the Central Pacific.

[38] ComAirNorSols TrngMemo No. 1, dtd 21Oct44, in *ComAirNorSols Oct44 WarD.*

[39] ComAirSols OpMemo No. 37, dtd 5Apr44, Subj: Regs on Bombing and Strafing, in *StrikeComd WarDs;* ComAirNorSols OpInstrn No. 13–44, dtd 12Aug44, Subj: Bombing and Strafing Regs, in *ComAirNorSols Aug44 WarD.*

About the only variety that flyers had in what the 1st Wing's history called a "deadly routine of combat air patrol, milk run bombing, and night heckling" was experimenting with new weapons and techniques.[40] Incendiaries of different types were tested against Japanese installations, and bomb loadings were varied to measure destructive effect. This diversity brought no letup in the weight of the attack delivered against major enemy objectives until the end of the summer when, as the air operations commander at Piva noted, "practically all of the good targets in these areas had been destroyed."[41] As the Japanese went underground to find cover in the faceless jungle, the number of obvious targets steadily lessened, and many AirNorSols strikes blasted and burned area targets in a systematic destruction pattern much like that which leveled the town of Rabaul. Even the gardens that the Japanese troops planted to supplement their rations were sprayed with oil in hope that the crops would wither and die.

During September and October, one spectacular new air weapon, a drone bomb, was tested against Japanese targets in the AirNorSols area. The drones, specially built planes capable of carrying a 2,000-pound bomb, were radio controlled by torpedo bombers of a special naval test unit. Synchronized television screens in drone and control planes enabled the controllers to view what was ahead of the drones and to crash them against point targets. After test attacks on a ship hulk beached at Guadalcanal, the test unit moved up to Stirling and Green and made 47 sorties in conjunction with F4Us, SBDs, and PBJs. The results were inconclusive. Two of the pilotless bombs were lost en route to targets because of radio interference, mechanical defects caused five crashes, Japanese antiaircraft shot down three, and five drones had television failures and could not locate a target. Of those drones that did attack, 18 hit their objective and 11 missed or near missed. ComAirNorSols concluded that there was a future for this weapon, but that it needed more development work and better aircraft for drones. Evidently, the Chief of Naval Operations, who in August had turned down a request to use SBDs as drones, agreed with General Mitchell's evaluation. Since the "better aircraft" were needed elsewhere, the test unit was decommissioned shortly after completing its last strike on 26 October.[42]

Vastly more effective than the imaginative drone bombs were the attacks by more workaday aircraft. The Corsairs, in particular, expanded their usefulness through regular bombing missions, since there was little call for them in their role as interceptors. It was this aspect of the Corsairs' capabilities, however, that brought about their employment in the Philippines.

Fighter planes were badly needed at Leyte where Third Fleet carriers had stayed a month beyond the time of their scheduled departure for a strike on Japan in order to fly cover for amphibious shipping. Two of the Seventh Fleet's escort carriers had been lost in the Battle of Leyte Gulf and four more had been dam-

[40] 1st MAW Hist Jul41–Jun46, n.d., p. 13.

[41] ComAirPiva WarD, 15Jun–30Nov44, p. 21.

[42] STAG One Rept of SvcTest on Drones Attacking Beached Hulk on Cape Esperence, GuadalcanalIs, dtd 6Aug44; ComAirNorSols Rept on Ops of STAG One Det in NorSols Area, dtd 30Oct44.

aged, so that Admiral Kinkaid was desperately short of planes for air defense. Fifth Air Force P–38s, based at a muddy, inadequate forward airstrip at Tacloban, had their hands full defending the immediate beachhead area and could do little to augment the shipping protection afforded by carrier aircraft. Greatly increasing the seriousness of the air picture was the advent of the *kamikazes*, the Japanese suicide pilots who crashed their bomb-laden planes against shipping targets.

Admiral Halsey, who was anxious to free his carriers for the attack on the Japanese home islands, saw a solution to his problem in the Marine Corsairs of the 1st Wing. He reminded General MacArthur that these fighters were available and had proved themselves repeatedly when they flew under Halsey's command.[43] They were capable of reinforcing the Army Air Forces' planes in interceptor and ground support roles and would be a welcome addition to the air cover of the Seventh Fleet's ships. Deciding quickly to employ the Marine planes, MacArthur ordered them brought forward. On 30 November, General Mitchell received a directive from Allied Air Forces to transfer four of the 1st Wing's Corsair squadrons to operational control of the Fifth Air Force on Leyte. The planes were to arrive at Tacloban by 3 December.

As soon as the order was received, MAG–12 (VMF–115, –211, –218, and –313) was alerted for the move and ceased all combat operations under ComAirNorSols. With Marine PBJs as navigational escorts, the flight echelons of group headquarters and the fighter squadrons arrived in the Philippines on schedule, after covering 1,957 miles from Emirau via Hollandia and Peleliu. A shuttle service by R4Ds of MAG–25, supplemented by C–47s of the Fifth Air Force, carried essential maintenance men and material forward to insure that the Corsairs kept flying. On 5 December, MAG–12 pilots flew their initial combat patrols in the Philippines and shot down the first of a long string of enemy planes.[44]

On 7 December, a week after the first Marine fighters were ordered to Leyte, MAG–14 and the remaining four Corsair squadrons in the 1st Wing were put on 48-hours notice for a forward movement. This time the destination was an airfield yet to be built on Samar, and the move was not so precipitate as that of MAG–12. The first squadron of MAG–14 to fly in from Green, VMO–251,[45] arrived on Samar on 2 January. The forward echelons of the group headquarters and service squadrons, and of VMF–212, –222, and –223, had arrived by 24 January. Again PBJs guided and escorted the Corsairs and, stripped of most of their guns and extra weight, helped transport key personnel and priority equipment from Green

[43] Halsey and Bryan, *Admiral Halsey's Story*, p. 231.

[44] The honor of shooting down the first enemy plane in the Philippines credited to a Marine went to a pilot of VMF(N)–541, who bagged an Oscar before dawn on the 5th. The squadron, equipped with a night fighter version of the F6F, had arrived at Tacloban on 3 December, exchanging places with an Army P–61 night fighter squadron. MacArthur had asked Nimitz for the temporary loan of the faster Hellcats to improve dawn and dusk interception at Leyte. CinCSWPA disp to CinCPOA, dtd 26Nov44, quoted in Sherrod, *MarAirHist*, p. 275.

[45] VMO–251, originally an observation squadron, was re-equipped as a fighter squadron in the summer of 1942, but was not redesignated VMF–251 until 31 January 1945.

LEYTE INVASION FLEET *assembled in Seeadler Harbor in the Admiralties symbolizes the move forward from the Solomons and Bismarcks.* (SC 283167)

MARINE MITCHELLS *fly over Crater Peninsula during one of the ceaseless round of suppressive attacks that neutralized Rabaul.* (USMC 114743)

and Piva in addition. Marine and Army transports planes carried the bulk of the men and gear of the forward echelon.

The ground echelons of the Marine fighter groups were not able to begin moving to the Philippines until February when shipping priorities eased after the Luzon landings. In contrast, a good part of the ground echelons of MAG–24 and –32 squadrons preceded their planes to Luzon and helped establish a field at Mangaldan near Lingayen Gulf. Flights of SBDs began arriving from the Bismarcks on 25 January, and, by the end of the month, all seven Dauntless squadrons were operational.[46]

The withdrawal of Marine fighters and dive bombers from operational control of ComAirNorSols placed the burden of maintaining the aerial blockade of the bypassed Japanese on RNZAF Corsairs and Venturas and MAG–61's Mitchells. The New Zealanders smoothly took on all fighter-bomber commitments that the Marines had handled; there was no break in the unremitting pattern of harassing attacks and watchdog patrols. RNZAF Corsairs also flew close and direct support strikes for the Australian infantrymen on Bougainville, working closely with the RAAF tactical reconnaissance aircraft attached to the II Corps. The Mitchells and Venturas also flew ground support missions for the Australians, but spent most of their time making bombing attacks on Rabaul, Kavieng, and the other principal Japanese bases.

The Venturas, which were not fitted with a bombsight suitable for medium-level (9,500–13,000 feet) drops until April 1945, relied on Marine PBJs as strike leaders in this type of mission. When the Mitchell released its bombs, the accompanying RNZAF bombers dropped theirs also. The resulting concentration of hits was particularly effective against the larger targets found at Rabaul, where most of the medium-level bombing was done. Low-level attacks by both the Mitchell and Ventura squadrons were aimed primarily at targets that were not so well protected by antiaircraft as those at Rabaul.

Only one squadron of MAG–61's Mitchells was freed from the frustrating round of policing missions in the Bismarcks and Solomons. On 3 March, on orders from Allied Air Forces, VMB–611 was transferred to MAG–32 with orders to move forward from Emirau to the Philippines. By the end of the month, 611's PBJs were operating from fields on Mindanao. The four bombing squadrons remaining in MAG–61, VMB–413, –423, –433, and –443, served the last months of the war at Emirau and Green. Orders to deploy to the Philippines were finally received just prior to the end of the fighting.

As if to signify the near completion of the aerial campaign that had begun at Guadalcanal almost three years before, General Mitchell relinquished command of AirNorSols and the 1st Marine Aircraft Wing on 3 June 1945 and returned to the States two days later. A little over a month after the general's departure, the long-awaited transfer of control of air operations to the RNZAF finally took place. On 15 July, Air Commodore Roberts assumed command from Marine Brigadier General Lewie G. Merritt and AirNorSols was dissolved. General Merritt's 1st Wing now came under the

[46] The story of Marine participation in the Philippines liberation campaign will be covered in the fourth volume of this series.

orders of the New Zealand Air Task Force.[47]

By the time Air Commodore Roberts took over the direction of air operations, the primary mission of most combat aircraft in his command was support of the Australian ground forces of II Corps. On Bougainville, in a nine-months-long offensive, the Australians had pushed the Japanese back in all directions from the Torokina perimeter and were driving on the enemy positions at Buin, Numa-Numa, and Bonis. On New Britain, the Australians, operating from a base camp at Jacquinot Bay on the southern coast, kept aggressive patrols forward in the Open Bay-Wide Bay region of Gazelle Peninsula, sealing off the Japanese at Rabaul from the rest of the island. In March, when an airfield was opened at Jacquinot, RAAF planes and, later, RNZAF Corsairs and Venturas, flew ground support missions and attacked the enemy at Rabaul. (See Map 31.)

On 3 August, General Kenney directed General Merritt to move the headquarters of the 1st Wing and MAG-61 to the Philippines. Six days later, Marine planes flew their last bombing mission against Rabaul. Six PBJs from VMB-413, six from VMB-423, five from VMB-443, and one from group headquarters took part; an RNZAF Catalina went along as rescue Dumbo. Each Mitchell carried eight 250-pound bombs which were dropped through heavy cloud cover with unobserved results; the targets were storage and bivouac areas near Rabaul and Vunakanau.

When the fighting ended on 14 August, some Mitchells had already flown to the Philippines, the remainder made the trip by the 19th. The wing's command post shifted from Bougainville to Zamboanga on Mindanao on 15 August. Ahead of the Marine squadrons lay months of hectic peacetime employment in North China as part of the American occupation forces. Behind the flyers and ground crews was a solid, lasting record of achievement in every task of aerial combat and blockade that had been asked of them.

---

[47] General Mitchell's relief had been Major General Louis E. Woods, but Woods held command only one day before he was ordered to Okinawa to take over Tactical Air Force, Tenth Army. General Merritt took command of AirNorSols and the 1st Wing on 10 June.

Appraisal

EPILOGUE [1]

Until the Japanese Emperor issued his rescript directing his forces to lay down their arms, troops of the *Eighth Area Army* and *Southeast Area Fleet* were still full of fight. On Bougainville, they were locked in desperate struggle with units of the Australian II Corps; on New Britain and New Ireland they were ready for battle but frustrated by lack of an opponent. Had the Allied seizure of Rabaul been necessary, the operation certainly would have been a bloody one.

Despite the steady pounding that Allied aircraft gave the enemy base—20,967 tons of bombs dropped in 29,354 sorties (over half of them flown by Marine planes) [2]—the Japanese had plenty of guns left with which to fight. According to postwar interrogations of officers of the garrison, only 93 out of a total of 367 antiaircraft guns were destroyed, 1 of 43 coast defense guns, and none of the thousands of infantry supporting weapons, ranging in size from light machine guns to 150mm howitzers. Since ground and beach defenses were seldom subjected to air attack, the high survival rate of the guns is not unusual. Even if they had been primary targets, however, many would have escaped destruction in the jungle or the caves where they were hidden.

By the war's end, the Japanese had built or improved more than 350 miles of tunnels and caves, where they had stored all their essential supplies and equipment. These stocks were sufficient to support the garrison well beyond the time when it surrendered. Ironically, it was the efficiency of the Allied naval and air blockade that was responsible for the favorable enemy logistic situation. In large part, Rabaul's troops subsisted on rations, dressed in uniforms, and used equipment that had been intended for garrisons cut off in the northern Solomons and eastern New Guinea.

Wherever supplies were short, the Japanese improvised. Issue rations were supplemented by extensive gardens, devoted primarily to cassava and sweet potato plants. Factories were set up which turned out black powder and sulfuric acid for explosives, manufactured flame throwers and mortars, and fabricated enough antitank mines to arm each man with one. Over 30,000 bombs were fused and planted as antipersonnel mines. The Japanese at Rabaul were prepared to do battle, and many of them, after 18 months of constant aerial attacks, were even anxious to meet

[1] Unless otherwise noted, the material in this section is derived from: *Eighth Area ArmyOps; Hattori, Complete War History;* USSBS, *Campaign Against Rabaul.*

[2] Of this total, U.S. Army Air Forces planes dropped 11,037 tons in 7,490 sorties; U.S. Navy planes, 1,458 in 4,608; British Commonwealth planes, 947 in 2,538; and U.S. Marine Corps planes, 7,142 in 14,718. Table No. 1 in USSBS, *Campaign Against Rabaul*, p. 263.

an opponent that they could come to grips with.

Fortunately, the encounter never took place. The Allied casualty list of an amphibious assault at Rabaul would have been as lengthy and grim as any of the Pacific War. When the order came for the Japanese to cease fighting, *Eighth Area Army* had about 57,000 men and *Southeast Area Fleet* about 34,000 on Gazelle Peninsula, with an additional 7,700 Army and 5,000 Navy troops a night's barge trip away on New Ireland.[3] These men, as part of the amazing display of national discipline evident throughout the Pacific, accepted the Emperor's surrender order without incident.

On 6 September 1945, General Imamura and Admiral Kusaka boarded HMS *Glory*, standing off Rabaul, and surrendered the forces of the *Eighth Area Army* and the *Southeast Area Fleet* to General Vernon A. H. Sturdee, commanding the Australian First Army. Two days later, at Torokina, the Japanese who had fought so tenaciously on Bougainville formally capitulated to the Australian II Corps' commander, Lieutenant General Sir Stanley Savige. At each ceremony, Air Commodore Roberts, RNZAF, was present as New Zealand's senior representative.[4] In a larger sense, he represented also his predecessors, ComAirNorSols and ComAirSols, and the thousands of Allied flyers who had a part in neutralizing Rabaul's offensive power.

SUMMARY

At times in the first eight months of the war, it appeared that the tidal wave of Japanese expansion would never ebb. Yet, like its natural counterpart, the enemy wave washed to a halt, and then receded. Guadalcanal and Papua were the Japanese high water marks in the southern Pacific.

The naval battles off Guadalcanal, virtually a standoff as far as ships' losses were concerned, hurt the Japanese far more than the Allies. Confronted by ample evidence of America's superior productive capacity, the enemy could ill afford to trade ship for ship. Once the Cactus Air Force won control of the skies of the southern Solomons from the Zekes, the Japanese realized they faced unacceptable shipping losses if they continued the fight for Guadalcanal. The resulting evacuation of enemy troops from the key island foreshadowed other retreats and defeats certain to come.

Less than a month after the threat posed by the planes at Henderson Field forced the Japanese to pull out of Guadalcanal, a smashing victory won by land-based Allied aircraft crippled enemy efforts to hold positions on the opposite flank of the Solomon Sea. The heavy transport losses in the Battle of the Bismarck Sea ended large-scale reinforcement of the *Eighteenth Army* fighting in northeast New Guinea. Although the Japanese fought just as hard as before to hold what they had, they fought with fewer men, fewer weapons, and less food and supplies.

[3] Strength figures for the Japanese garrison at the time of its surrender vary considerably because some sources include Korean labor troops and civilians attached to the military, while others omit them wholly or in part. A postwar compilation, issued in June 1950, by the Japanese Demobilization Bureau, arrived at these figures for Japanese only: Bismarcks—57,530 Army, 30,854 Navy; Solomons—12,330 Army, 16,729 Navy; Eastern New Guinea—12,100 Army, 1,200 Navy. Cited in Hattori, *Complete War History*, IV, p. 464.

[4] Ross, *RNZAF*, p. 311.

When the successful capture and defense of Guadalcanal and the simultaneous seizure of the Buna-Gona area of New Guinea wrote "finish" to the Japanese advance, the stage was set for a coordinated Allied offensive aimed at the enemy strategic citadel, Rabaul. General MacArthur's ELKTON plans, as revised in Washington in the light of forces available to the South and Southwest Pacific, formed the basis for the JCS CARTWHEEL directive of 28 March 1943. Under its provisions, a series of intermediate objectives were to be taken before the culminating assault on Rabaul. The common determinant for the selection of these objectives was their utility as air bases.

The seizure of the Russell Islands by Admiral Halsey's forces on 21 February, though not a part of the ELKTON concept, was, in spirit at least, the opening move of the drive on Rabaul. The airdrome that was constructed on Banika housed fighters and medium bombers which supported CARTWHEEL operations in the central and northern Solomons. The advance to the boundary of the South Pacific Area was characteristic of Admiral Halsey's infectious determination to maintain the initiative over the Japanese. He was equally anxious to get on with his first operation under CARTWHEEL, the seizure of New Georgia, but had to agree to several delays of D-Day in order to coordinate his attacks with those of Southwest Pacific forces. The joint landing date finally agreed upon was 30 June; the simultaneous targets were the Trobriands, New Georgia, and Nassau Bay near Salamaua on New Guinea.

The Japanese threat to Segi brought Marines to New Georgia nine days ahead of schedule, and the lack of enemy opposition enabled Army shore parties to land on Woodlark and Kiriwina a week before the garrison arrived. Otherwise, the main landings went ahead as planned. Four months of determined fighting were necessary before the successive Allied objectives on New Guinea, Salamaua, the Markham Valley, and Lae were captured. In the smaller compass of the New Georgia Group, the defeat of the Japanese took equally as long.

New Georgia was far from the best-managed or best-fought campaign of the Pacific War. It was, however, a time of learning for the Allied leaders and men involved, even though the learning process was prolonged and painful. The troops that finally broke out of the jungle to take Munda airfield were combat-wise, and their commanders had learned to make more realistic estimates of the time and men necessary to root the Japanese out of heavily defended objectives. Once Munda was in Allied hands, the enemy situation deteriorated. The rest of the island group was taken with increasing skill and spirit, with each assault demonstrating a greater familiarity with the tools and techniques of amphibious operations and the demands of jungle warfare. The Japanese finally gave way before the persistent pressure and evacuated their surviving forces from Kolombangara to fight again another day.

By the end of a summer of fighting marked by a gradual increase in Allied strength, it was apparent that the outer perimeter of Japanese island defenses soon would collapse. On 30 September, *Imperial General Headquarters* ordered the commanders of these vulnerable positions to do their utmost to hold out as long as possible. Time was needed for the construction of a cordon of defenses along a line arcing from the Marianas through the Palaus and western New Guinea to the

Philippines. General Imamura and Admiral Kusaka responded to the directive, which, in effect, conceded the eventual loss of Rabaul, by reinforcing Army and Navy garrisons in the northern Solomons, the Bismarcks, and eastern New Guinea. Both enemy commanders retained a large portion of their troops and materiel on New Britain and New Ireland, however, in the belief that a showdown battle for possession of Rabaul was inevitable.

The conviction of the Japanese leaders, that Rabaul would have to be taken, was shared in Brisbane, but not so freely accepted in Noumea, Pearl Harbor, or Washington. What ComSoPac, CinCPac, and JCS planners envisioned instead was the possibility that Rabaul could be bypassed and its strength neutralized by an aerial blockade mounted from bases within fighter range. Although General MacArthur opposed this concept, it won acceptance from the Combined Chiefs of Staff at the Quebec Conference in August and became a part of Allied strategy. The large number of troops, ships, and planes that would have been necessary to capture Rabaul were allotted instead to other operations of the two-pronged drive on Japan. To ensure the isolation of the enemy fortress, Kavieng and the Admiralties were designated as targets for SWPA forces in addition to the remaining CARTWHEEL objectives in eastern New Guinea, western New Britain, and southern Bougainville.

In the fall of 1943, Australian and American forces steadily drove the Japanese back from coastal and inland positions on the Huon Peninsula. The Allied goal was the seizure and security of airfields from which planes could support operations on both sides of Vitiaz Strait. Once Nadzab and Finschhafen were operational, air superiority over the strait and adjacent areas was assured.

The SoPac operation parallel to that on the Huon Peninsula entailed the seizure of a foothold on Bougainville. While the Marine parachutists' diversionary raid on Choiseul and the New Zealanders' capture of the Treasurys were part of the overall campaign, the main event was the IMAC landing at Cape Torokina on 1 November. By shunning the areas where the Japanese were concentrated, and hitting instead a lightly defended objective that required extensive base development, Admiral Halsey drastically reduced his assault casualty lists and avoided the prolonged battle to seize a major fortified position that had characterized the New Georgia operation.

The Japanese, recognizing the grave threat to Rabaul, reacted violently and swiftly to the Bougainville landing. AirSols and carrier planes, Third Fleet ships, and the dogged fighting of Marines and soldiers holding the beachhead beat back all attacks. Within the protection of the perimeter, Seabees and engineers overcame formidable natural obstacles to construct Torokina and Piva airfields and make them ready for their essential role in the reduction of Rabaul. By mid-December, ComAirSols was able to launch a sustained aerial attack designed to wipe out every vestige of the enemy's offensive power.

By design, the start of the massive AirSols assault on Rabaul coincided with the opening phase of the last CARTWHEEL operation, the seizure of western New Britain. One of the enigmas of the ensuing campaign was that the Japanese paid an inordinate amount of attention to the preliminary landing at Arawe, although they were well aware of its limited strategic value. The tiny peninsula seemed to have a special attraction for pilots of the *Eleventh Air Fleet*, even after the 1st Marine Division's landings at Cape Gloucester made the Allied main objective obvious. The Japanese never had a chance to mount any telling air attacks on the new beachheads, however. They were too busy trying to defend Rabaul.

On 2 January 1944, three days after the Marines at Cape Gloucester seized their airdrome objective, other SWPA forces sailed through Vitiaz Strait, now secure on both shores, and landed at Saidor. The seizure of an enemy position on the New Guinea coast west of the Huon Peninsula was a giant stride forward on the way to the Philippines. Before the next scheduled amphibious operation was launched, the strategic situation was changed drastically by the evacuation of Japanese aircraft from Rabaul.

Credit for forcing the enemy withdrawal belongs to many Allied commands, but to none in so large a measure as to Aircraft, Solomons. The American and New Zealand pilots, aircrewmen, and ground personnel who fought as part of AirSols the long way up the Solomons chain made Rabaul a yawning grave for Japanese naval aviation. The final two months of incessant attacks, made possible by possession of the Bougainville airfields, disintegrated the defending air fleet. Although the order to pull out was precipitated by the devastating American carrier raid on Truk, the end of Japanese air operations at Rabaul was already certain.

The seizure of the Green Islands, just before the *Combined Fleet* ordered all serviceable aircraft withdrawn from New Britain, emphasized the steady worsening of the Japanese situation. Only a feeble attempt was made to punish an Allied amphibious force making a landing within easy range of any plane based at Rabaul. Once fields at Green were operational, it was inevitable that fighters and light bombers based there would own the skies over Gazelle Peninsula and southern New Ireland. Possession of Green also meant that ComAirSols could begin a systematic program of attacks on Kavieng, one of the two staging bases through which aerial reinforcements still could reach Rabaul.

As long as the Japanese had airfields at Kavieng and in the Admiralties, Allied leaders felt that Rabaul's air garrison might be rebuilt. The cost of such a risk-laden move appeared to be prohibitive, but there was no guarantee that future events might not make it appear worthwhile to *Imperial General Headquarters*. If the two positions were taken or neutralized, however, nothing but a trickle of enemy long-range aircraft would get through. The isolation of Rabaul would be complete.

The enemy avenue of approach from the Admiralities was blocked on 29 February, when a small Army reconnaissance force, outnumbered but not outmatched, was able to seize a beachhead on Los Negros. The Japanese garrison, cut off from all outside help, fought doggedly but hopelessly until it was wiped out. The fighting did little to impede the building progress of a base that was destined to play a major part in the advance to the Palaus and the Philippines. Seeadler Harbor proved to be everything in the way of an advance naval base that Rabaul's Simpson Harbor might have been, with the added virtues of a more favorable location and a cheaper price. Most missions flown from the airfields constructed on Los Negros supported the drive west along the New Guinea coast or struck enemy bases in the Carolines. There was little call for the squadrons in the Admiralties to hit Rabaul or Kavieng. The capture of Emirau on 20 March 1944 sealed the fate of both enemy bases.

The decision to bypass Kavieng in favor of Emirau, like the earlier decision to bypass Rabaul, was made in Washington. In both cases, the consensus of JCS opinion, reinforced by the recommendations of Admirals Nimitz and Halsey, overweighed General MacArthur's belief that the major enemy bases constituted such a threat that they would have to be taken. By using Marines and ships that were ready to take part in the Kavieng operation, Admiral Halsey was able to effect a swift and bloodless occupation of Emirau. Planes flying from the airbase that was soon built on the island pounded Kavieng until the war's end, and took their turn, as well, in the raids flown against Rabaul.

Even though the taking of Emirau meant that the enemy's last chance of reinforcing the Southeast Area was gone, there was no thought of surrender on the part of General Imamura or Admiral Kusaka. Instead, the Japanese commanders kept their men keyed up, ready to fight a battle that never took place. Most of the Allied leaders and men who took part in the campaign against Rabaul passed on to more active fronts, and those who remained had the thankless task of keeping the Japanese beaten down.

For the most part, except where the Australians kept the ground campaign alive, what was left of the war in the Solomons and Bismarcks was a deadly boring routine for pilots and aircrews. Marine and RNZAF squadrons drew the majority of the unwanted assignments of maintaining the aerial blockade, and they did their job well. In light of postwar analyses of the destruction wrought by air attacks on the bypassed Japanese bases, it appears that much of the bombing effort was wasted, once the enemy was forced to go underground in order to survive. In fact, it now seems plausible to believe that the Japanese could have been contained just as well by using fewer planes and men.

The evaluation of any military campaign breeds such second guessing. Benefiting from knowledge of the situation of both sides at a given moment, it is easy to decide that certain operations were unnecessary and that others should have been conducted differently than they were. The men who planned and fought the battles, however, did so without the enlightenment provided by hindsight. They learned, instead, from the mistakes that they unwittingly made in the process of becoming veteran fighters.

Of all the lessons that were absorbed during the successful campaign to isolate and neutralize Rabaul, none was more important than the absolute necessity for interservice and inter-Allied cooperation. Few commands in the Pacific war evidenced such wholehearted subordination of self-interest as the South Pacific Forces who won their way from Guadalcanal to Emirau. Admiral Nimitz saw this spirit as "a guiding directive to all armed services of the United States, now and in the future." [5] There can be no more fitting memorial to the bitter fighting and sacrifice of the Rabaul campaign than its ample proof that the separate services meshed together well as one fighting team.

---

[5] CinCPac-CinCPOA 1st End, dtd 15Sep44, to ComThirdFlt ltr to ComInCh, dtd 3Sep44, Subj: SoPac Campaign, Narrative account (COA, NHD).

Bibliographical Notes

This history is based principally upon official Marine Corps records: the reports, diaries, journals, orders, etc., of the units and commands involved in the operations described. Records of the other armed services have been consulted where they are pertinent. On matters pertaining to activities at high strategic levels, the authors have drawn on the records of the Joint Chiefs of Staff.

In order to cover gaps and inadequacies that occur in the sources consulted, extensive use has been made of the knowledge of key participants in the actions described. These men have been generous with their time in answering specific and general queries, in making themselves available for interviews, and in commenting on draft manuscripts. The military historical offices of the other services, of the New Zealand Government, and of the Japanese Government have read and commented upon those draft chapters bearing upon the activities of their own units.

Because this volume deals with the whole of the Allied campaign to neutralize Rabaul, many of the records used relate to more than one of the component operations. Such sources have been fully cited in the text and are discussed in relation to the particular operation where they have the most pertinency. All records cited, except as otherwise noted, are on file at, or obtainable through, the Archives of the Historical Branch, G-3 Division, Headquarters, U.S. Marine Corps.

A number of published works of general interest have been consulted frequently in the writing of this volume. The more important of these are listed below.

Books

Wesley Frank Craven and James Lee Cate, eds. *The Pacific: Guadalcanal to Saipan—August 1942 to July 1944—The Army Air Forces in World War II*, v. 4. Chicago: University of Chicago Press, 1950. This is the Air Force's official history for the period of the Rabaul neutralization campaign. Well documented, the book is a reliable source for the actions of Fifth Air Force and Thirteenth Air Force units and the attitudes and decisions of their commanders.

FAdm William F. Halsey and LCdr J. Bryan, III. *Admiral Halsey's Story*. New York: McGraw-Hill Book Company, Inc., 1947. This popular treatment of one of the most spectacular figures of the Pacific war presents a fascinating and useful picture of South Pacific command planning and decisions.

John Miller, Jr. *CARTWHEEL: The Reduction of Rabaul—The War in the Pacific—United States Army in World War II*. Washington: Office of the Chief of Military History, Department of the Army, 1959. A basic military source work, this volume of the Army's official history presents a comprehensive view of the CARTWHEEL campaign with particularly good coverage on the planning aspects.

Samuel Eliot Morison. *Breaking the Bismarcks Barrier—History of United States Naval Operations in World War II*, v. VI. Boston: Little, Brown and Company, 1950. Rear Admiral Morison's history was written with every cooperation from the Navy and can be considered its official history, even though the author disclaims this evaluation. Morison is at his best in describing action at sea and in analyzing Japanese moves and motives.

Robert Sherrod. *History of Marine Corps Aviation in World War II*. Washington: Combat Forces Press, 1952. This is a highly readable account of Marine air activities which was written with substantial Marine Corps research support; its text includes the results of many interviews and eyewitness accounts no longer available for study.

United States Strategic Bombing Survey (Pacific), Naval Analysis Division. *The Campaigns of the Pacific War*. Washington: Government Printing Office, 1946. This report attempts to present the broad picture of the war from the

Japanese viewpoint through brief descriptions of the various campaigns, but, unfortunately, it was prepared too soon after the event to gain deep perspective. The text contains many inaccuracies. The book is of great value, however, in presenting translations of many enemy documents that reveal Japanese wartime thinking.

United States Strategic Bombing Survey (Pacific), Naval Analysis Division. *Interrogations of Japanese Officials.* 2 vols. Washington: Government Printing Office, 1946. This is a companion report to *Campaigns* (above) and similarly of value in telling the Japanese side of the story.

PART I

STRATEGIC SITUATION—SPRING 1943

Official Documents

The JCS records, especially those of the Pacific Military Conference in March 1943, were particularly helpful in developing the course of the ELKTON plans as they fared in Washington. The transcripts and summaries give considerable useful background information on the state of U.S. and Allied forces in the South and Southwest Pacific. The war diaries of Commander, South Pacific furnish an excellent chronological narrative with emphasis on important messages sent and received.

Intelligence surveys by various higher headquarters were used extensively to build a picture of the state of Allied knowledge of enemy troops and terrain. In the case of the Russell Islands operation, action reports and war diaries of the units concerned furnished the narrative base. The main sources for the status report on the FMF were a study of Marine Corps ground training in World War II prepared in the Historical Branch and a history of FMFPac prepared at Pearl Harbor about 1951.

Japanese Sources

In the years immediately following the end of the war, former Japanese officials working under the auspices of General MacArthur's headquarters prepared a series of monographs detailing Japanese actions in many Pacific and Asian campaigns and at the various headquarters in the home islands. In the middle 50s, a number of these original studies were revised and expanded, again by knowledgeable Japanese. The monographs vary considerably in their value, but, on the whole, they are honestly presented and useful in gaining an insight into Japanese actions. The Office of the Chief of Military History, Department of the Army, which has a complete file of these studies, has prepared an annotated guide and index, *Guide to Japanese Monographs and Japanese Studies on Manchuria 1945–1960* (Washington, 1961), which is an excellent aid in evaluating the individual items.

Among the several Japanese monographs of the series that were used with this part, No. 45, the 382-page history of the *Imperial General Headquarters, Army Section*, was particularly helpful. It provides an overall view of the progress of the war as seen from Tokyo and contains appendices of Army orders. The operations record of the *Seventeenth Army* (No. 35 of the series) is valuable for its development of the Army's early actions in the central and northern Solomons campaigns. Similarly, the Japanese account (No. 99 of the series) of Southeast Area naval operations from February through October 1943 gives the Navy's view of the beginnings of joint defensive measures.

Books

Cdr Eric A. Feldt, RAN. *The Coastwatchers.* Melbourne and New York: Oxford University Press, 1946. This is a personalized account of the coastwatchers by one of their leaders which gives a good picture of individual exploits and of the overall contribution of these valiant men to the success of operations in the South and Southwest Pacific Areas.

Richard M. Leighton and Robert W. Coakley. *Global Logistics and Strategy, 1940–1943—The War Department—United States Army in World War II.* Washington: Office of the Chief of Military History, Department of the Army, 1955. This book is an excellent, objective examination of the background of Allied action in the early years of the war.

John Miller, Jr. *Guadalcanal: The First Offensive—The War in the Pacific—United States Army in World War II.* Washington: Historical Division, Department of the Army, 1949. This work is one of the first Army official histories written; it is also one of the best, and gives adequate, objective coverage to Marine actions in the first offensive of the war.

Samuel Milner. *Victory in Papua—The War in the Pacific—United States Army in World*

War II. Washington: Office of the Chief of Military History, Department of the Army, 1957. This is the basic source for any narrative of the fighting in New Guinea that paralleled the action on Guadalcanal.

PART II

TOENAILS OPERATION

Official Records

Discussions of the operations at New Georgia are based on the records of the units concerned. Included in the documents are special action reports, war diaries, and informal combat reports of the tactical units involved as well as the journals and special reports of the various staff sections. It must be remembered that the New Georgia operation was conducted by a composite force of Navy, Marine Corps, and Army units and that few arrangements for submission of action reports had been made. Accordingly, the various units reported either to the next senior echelon or to their own service, whichever they deemed proper. As a result, the reports of some Army units are in Marine Corps archives and *vice versa.* In general, however, most reports of tactical units are held by the service concerned. It must also be remembered that the desirability of maintaining official records was not fully recognized at this point of the war and that most commanders were naturally more interested in accomplishment of the combat mission than they were in keeping records. Consequently, most existing records are incomplete. The exceptions are the post-operation reports of the New Georgia Occupation Force (XIV Corps) and the 37th Infantry Division. These records are invaluable for a comprehensive account of the drawn-out Munda campaign.

One great assistance to the study of the New Georgia operation was the mid-1943 order by the Marine Corps which directed the preparation and submission of war diaries by tactical units. This resulted in the preparation of a number of organizational histories and post-operation reports which filled several large gaps in the general account of the campaign.

At the conclusion of the war, the Historical Section of the South Pacific Base Command prepared a manuscript of the history of the New Georgia campaign. This account includes a large number of well-drawn maps. This manuscript, held by the Office of the Chief of Military History, Department of the Army, is helpful as a guide to obscure documents and memoranda which might not otherwise be encountered by researchers. The base command's manuscript forms the basis for many of the later histories of the New Georgia fighting.

Other official records which were informative included the combat narratives published during 1944 by the Office of Naval Intelligence. Two of these once-classified booklets used in this portion of the book were No. IX: *Bombardments of Munda and Vila-Stanmore, January–May, 1943* and No. X: *Operations in the New Georgia Area, 21June–5August 1943.* Taken from action reports of the commands and ships involved, these narrative accounts were helpful in synthesizing naval actions and coordinating the Navy's contributions to the combat action ashore.

Unofficial Sources

During the writing of the Marine Corps monograph on the New Georgia campaign, Major John N. Rentz of the Historical Division obtained a number of written comments from participants of all services, and these letters and memoranda, together with a number of personal interviews, form the basis for many of the personal recollections which augment the operational reports of the tactical units. Certain key individuals, who also commented on the draft of this book, helped clarify command problems encountered during the fighting. Valuable, in addition, were a number of articles and vignettes by combat correspondents in the *Marine Corps Gazette* and *Leatherneck* magazine of late 1943 and early 1944. These unofficial sources are helpful in filling in the background to combat operations.

Japanese Sources

Japanese records used in this account, in addition to the three monographs mentioned previously, were obtained mainly from captured documents interpreted by South Pacific Forces during the campaign and may be procured from either the Naval History Division or Marine Corps Historical Branch archives. A fourth monograph used in this account, No. 34 of the series held by the Office of the Chief of Military History, was the account of *Seventeenth Army* operations from May 1942 to January 1943, which provides useful background information on units that were engaged during the New Georgia fighting.

Books and Periodicals

A number of biographies and memoirs of ranking officers were consulted for information for this part of the book, but the most informative was Admiral Halsey's. Other published sources from which information was obtained include:

Oliver A. Gillespie. *The Pacific—The Official History of New Zealand in the Second World War, 1939–1945.* Wellington: War History Branch, Department of Internal Affairs, 1952. This is a useful study which describes the course of employment of New Zealand forces as seen from the New Zealand viewpoint.

Col Samuel B. Griffith, II. "Corry's Boys," *Marine Corps Gazette*, v. 36, no. 3 (Mar52), and "Action at Enogai," *Marine Corps Gazette*, v. 38, no. 3 (Mar54). These are personal experience stories by the former commanding officer of the 1st Raider Battalion during the fighting on New Georgia.

Jeter A. Isley and Philip A. Crowl. *The U.S. Marines and Amphibious War.* Princeton: Princeton University Press, 1951. This book deals more with the development of amphibious doctrine and equipment than with operational history. The authors, however, have a number of pertinent conclusions relative to the campaign.

Maj John N. Rentz. *Marines in the Central Solomons.* Washington: Historical Branch, HQMC, 1952. This monograph forms the basis for this account. It is well written and contains considerable detail of Marine Corps small unit activities in the New Georgia Group.

Col Joseph E. Zimmer. *The History of the 43d Infantry Division, 1941–1945.* Baton Rouge, La.: Army and Navy Publishing Company, 1947. This is a perceptive history of the unit that did most of the fighting on New Georgia by the former commanding officer of 1/169.

PART III

NORTHERN SOLOMONS OPERATIONS

Official Records

In contrast to the Guadalcanal and New Georgia operations, the Northern Solomons campaign is fully documented from its earliest planning stages through the completion of the fighting. Most of the material in this section of the book was derived from the records of the tactical units and staff sections which participated in the three landings which comprised the Northern Solomons venture—the Treasurys, Choiseul, and Empress Augusta Bay. The III Amphibious Force war diaries for the months of October and November and the action report prepared after the Cape Torokina landings are valuable for information on the Navy's participation in the planning and execution of these operations. These documents are held by the Classified Operational Archives, Naval History Division.

The most informative account of the entire Northern Solomons campaign from its inception to its conclusion, however, is contained in the action report of I Marine Amphibious Corps. This account, in three parts, provides a day-by-day narrative of the three operations as well as a discussion of the planning difficulties, logistics preparations, and administrative problems of the campaign. Included are a number of overlays and maps plus special reports by various staff sections and tactical units. Also valuable are the separate administrative, intelligence, operational, and supply and evacuation journals of the corps which accompany the overall report.

The 3d Marine Division, which made the initial landings at Cape Torokina, provided a complete resume of the entire operation in the combat report written after the division's return to Guadalcanal. In addition to a narrative account of the campaign, the combat report includes a special report by each staff section of the division and action reports by each of the tactical units of the division as well as attached units. The three records—III Amphibious Force, I Marine Amphibious Corps, and the 3d Marine Division—provide a complete and comprehensive assessment of the entire campaign.

A contemporary account of the Bougainville operation, written prior to the end of the war by the Historical Section, Headquarters, Marine Corps, was of great assistance in outlining the campaign. This mimeographed study uses the above-mentioned records as the basis for the narrative. It is well written and quite descriptive in a number of instances. Equally as useful in maintaining the thread of action in the whole campaign was the Third Fleet Narrative Report prepared in the late summer of 1944.

Another once-classified account of the Solomon Islands campaign prepared by the Office of Naval Intelligence was also of value. This booklet, No. XII—*The Bougainville Landing and the Bat-*

tle of Empress Augusta Bay, 27 October–2 November 1943, was published in 1945. It describes the naval battles which were part of the Northern Solomons campaign.

Unofficial Documents

The comments and interviews obtained by Major Rentz in the writing of the monograph on Bougainville were also helpful in the preparation of this book. A number of the staff officers of IMAC as well as the 3d Marine Division submitted lengthy comments regarding the planning, preparations, and execution of the campaign, and all of these were of great value in filling in several gaps in the records. The various accounts were consulted and compared so that an accurate presentation could be made. As might be expected, recollections of one event may start a chain reaction which results in further recollections and remembrances. All of these were helpful, although not all could be used.

An account on the development of naval gunfire support during this period was also informative. This manuscript, "Naval Gunfire Support in the Solomon Islands Campaign," was written by Colonel Frederick P. Henderson in 1954 and traces the growth of fire support by naval vessels through the various South Pacific operations. It was especially valuable in regard to the Bougainville operation which was the proving ground for many gunfire support theories developed as a result of experience gained in earlier actions.

Among the comments received in regard to the draft of this book, those of Lieutenant General Edward A. Craig, Major General Victor A. Krulak, and Colonel Robert T. Vance were particularly helpful. General Craig was able to add considerably to the story of the 9th Marines, in particular during the Piva Trail battle. General Krulak's suggested corrections and additions to the narrative of the Choiseul raid were carefully based on contemporary records and clarified a number of points on which there had been conflicting or incomplete information. Colonel Vance's comments and sketch maps helped fix many details of the action of the 3d Parachute Battalion.

Japanese Sources

The intelligence journals and reports of various IMAC headquarters contain numerous partial translations which give a running picture of the Japanese situation. In addition to those Japanese monographs of the series previously mentioned, No. 100, covering the activities of Southeast Area naval forces from October 1943 to February 1944, was consulted frequently. It contains a daily operations log of naval air activities and is more concerned with naval aviation than other naval forces.

Books

The following books, in addition to those already mentioned, were used extensively in the preparation of the Bougainville chapters.

1st Lts Robert A. Aurthur and Kenneth Cohlmia. *The Third Marine Division*. LtCol Robert T. Vance, ed. Washington: Infantry Journal Press, 1948. This volume includes a colorful description of all the combat operations of the division in World War II.

John Monks, Jr. *A Ribbon and a Star, The Third Marines at Bougainville*. Illustrated by John Falter. New York: Henry Holt and Company, 1945. Although this book deals principally with the 3d Marines, it is, undoubtedly, the favorite of every Marine who fought at Bougainville because of its descriptive passages and sketches. The author and illustrator have captured the feeling of combat and the island.

Maj John N. Rentz. *Bougainville and the Northern Solomons*. Washington: Historical Section, Division of Information, HQMC, 1948. This official monograph contains a highly informative account of the entire campaign with great detail on the combat at Cape Torokina. Especially helpful was an outstanding descriptive appendix on the Northern Solomons islands.

PART IV

THE NEW BRITAIN CAMPAIGN

Official Documents

By far the most useful records of the CARTWHEEL operations on New Britain are the daily journals and message files of General Headquarters, SWPA and of ALAMO Force. These voluminous documents include memoranda of staff conversations, orders, plans, special reports, and just about every conceivable type of message bearing on military operations. They must be searched carefully, however, as documents that bear on a common topic are occasionally filed together out of chronological order. Like General Krueger's DEXTERITY Operation report, which provides a good summary of New Britain actions, these reports are available from the World War II Records Division, Federal Records Center, Alexandria, Va.

The 1st Marine Division's action report for the Cape Gloucester operation, which was prepared in large part by one of the authors of the later campaign monograph, is well written and often exciting reading. The narrative, organized around phases of the fighting, is sometimes shaky on details, but subsequent comment by participants in the actions described clarified many points. The corrected narrative was the basis of the monographic account. The division's Talasea action report is not as complete, relatively speaking, as that covering Cape Gloucester, but it furnishes an adequate basis for a narrative when supplemented by contemporary documents of other commands.

The plans, orders, and reports of naval elements of Admiral Barbey's amphibious forces are particularly good for the earlier part of the campaign. The basic report and historical account of General Cunningham's command, supplemented by the messages contained in the ALAMO G-3 File, give a clear picture of the situation at Arawe. On the whole, the documentation of the operations in western New Britain is excellent at the higher levels and complete enough at lower echelons to insure that careful research will produce a reliable account.

Unofficial Documents

The letters and interviews resulting from the preparation of the New Britain campaign monograph are unusually complete and detailed. The comments, based on draft narratives and questions circulated by the Marine Corps Historical Branch, were used extensively in the writing of that narrative and have been consulted often in the preparation of this shorter account. Frequently, different aspects of the comments have been emphasized in this book.

Through the generosity of General Vandegrift, his personal correspondence when he was Commandant was made available for Historical Branch use. The letters that he received from General Rupertus are valuable in following the course of the preparations for the operation, the fighting itself, and the various aspects of the 1st Division's employment in the Southwest Pacific Area. Extracts from this correspondence, together with copies of some of the letters, are available in the Marine Corps Historical Branch Archives for use by qualified researchers.

Among the letters received in comment upon the draft narrative of this part, those from the other service historical agencies have been very effective in clarifying some of the language used and pointing the way to a more accurate account. General Shepherd and Admiral Barbey, who provided the most useful critical readings of the draft chapters, elaborated on their comments in later conversation with the author. Admiral Barbey's comprehensive remarks on the organization and philosophy of employment of amphibious forces in the SWPA were valuable in analyzing the separate development of amphibious techniques in the Central and Southwest Pacific.

Japanese Sources

The Allied Translation and Intelligence Section of General MacArthur's headquarters maintained forward echelons with the 1st Marine Division on Cape Gloucester which screened Japanese documents as they were picked up. Working closely with the language personnel of the division's own intelligence section, these ATIS translators were partially responsible for the effective flow of enemy intelligence to combat troops. The later full translation of such Japanese material in ATIS bulletins and other publications made the reconstruction of the actions of the *Matsuda Force* relatively easy. There is a wealth of Japanese material available from the Cape Gloucester operation, and credit for its recovery can be traced directly to the indoctrination the troops received in the importance of turning in any documents they found.

Two further Japanese monographs of the series held at the Office of the Chief of Military History were used extensively with this part. They are complementary, one (No. 127) deals with the operations of the *Eighth Area Army* and the other (No. 128) covers the activities of the *17th Division*. Together, the two studies give a good picture of operations in western New Britain as seen from Rabaul.

A manuscript translation of the book put out by the Matsu Publishing Company in Tokyo in 1955, *Dai Toa Senso Zenshi* [The Complete History of the Greater East Asia War], was made available by the Office of the Chief of Military History. This excellent study, written by Takushiro Hattori, who was a ranking staff officer during the war and an historian afterwards, was very helpful in understanding Japanese actions during the fighting on New Britain. The book contains enough detail, based in part upon the studies for the Japanese monographs mentioned above, to be a useful strategic review for every major campaign in the war.

The War History Office, Defense Agency of Japan, very kindly consented to read the draft manuscripts of the Marine Corps operational history and began its welcome review with this part. The task, which involved a considerable amount of translation and research, was time consuming but worthwhile. The comments received, while not voluminous, have been excellent and have helped to clarify several heretofore moot points.

Books and Periodicals

Col Robert Amory, Jr., AUS, and Capt Ruben M. Waterman, AUS, eds. *Surf and Sand, The Saga of the 533d Engineer Boat and Shore Regiment and the 1461st Engineer Maintenance Company 1942–1945.* Andover, Mass.: The Andover Press, Ltd., 1947. This is the unit history of the Army amphibian engineers who were attached to the 1st Marine Division on New Britain.

General Headquarters, Army Forces, Pacific, Office of the Chief Engineer. *Amphibian Engineer Operations—Engineers in the Southwest Pacific 1941–1945,* v. IV. Washington: Government Printing Office, 1950. Although a good source for the activities of the Army small boat units that supported the Marines, this work contains some minor inaccuracies.

LtCol Frank O. Hough and Maj John A. Crown. *The Campaign on New Britain.* Washington: Historical Branch, HQMC, 1952. The basic source for the narrative of Marine actions on New Britain, this monograph contains numerous quotes from the draft chapter comments of participants. Among the several informative appendices is an outstanding one on the vegetation of the island and its effect on military operations, prepared by Captain Levi T. Burcham.

LtCol Robert B. Luckey. "Cannon, Mud, and Japs," *Marine Corps Gazette,* v. 28, no. 10 (Oct44). This is an interesting and very readable account of the employment of artillery at Cape Gloucester by the former executive officer of the 11th Marines.

George McMillan. *The Old Breed: A History of the First Marine Division in World War II.* Washington: Infantry Journal Press, 1949. This unit history, which concerns itself more with the spirit of the 1st Division than with a recital of details of its combat actions, is generally accorded to be one of the finest books of its type written after the war.

United States Strategic Bombing Survey (Pacific), Military Analysis Division. *Employment of Forces Under the Southwest Pacific Command.* Washington: Government Printing Office, Feb47. Based closely upon studies prepared by historians with General MacArthur's headquarters, this booklet is a useful summary of actions in the SWPA.

PART V

MARINE AIR AGAINST RABAUL

Official Documents

The terrain studies of Rabaul prepared by various intelligence agencies were an important factor in understanding Rabaul as a target complex. The South Pacific air combat intelligence reports provided the best running account of air action and a good picture of the steady deterioration of Japanese airfield and aircraft strength. The archives of the Marine Corps Historical Branch contain enough material on various South Pacific air commands, including the all-important Strike Command, to develop a good picture of air action. There are voluminous Marine squadron and group reports of varying quality which can be exploited for a more detailed story than space allowed in this book.

The USAF Historical Archives at the Air University, Maxwell Field, Alabama, furnished the reports of Fifth and Thirteenth Air Force actions which supplement the material available in Navy and Marine records. Since ComAirSols was always a joint command, its activities lend themselves to treatment as an integrated whole. It is difficult to separate Marine air's contributions to the reduction of Rabaul from those of other services and our Allies. In order to present a balanced picture of the situation, this part was written with the joint aspect of the air offensive always in mind.

The sections concerning characteristics of major Japanese and Allied combat aircraft were taken primarily from Army Air Force and Navy intelligence publications. These booklets, plus published interviews with pilots and operations officers with experience in the South Pacific area, provide a good means for assessing relative plane performance. Material on Japanese air crew training and experience levels was also found in intelligence reports as well as in the publications of the United States Strategic Bombing Survey.

Unofficial Documents

There is no body of letters and interviews in the Marine Corps Historical Branch archives relating to the air campaign against Rabaul as there is in the case of other campaigns which have been covered by monographic studies. Although there are a few pertinent letters among the papers acquired from the Sherrod aviation history project, these have limited value to a history of broad scope. Comments on the draft of this part from key commanders and staff officers, from the historical agencies of other services, particularly that of the Air Force, and from the New Zealand War History Office have been a useful check on the coverage and treatment of the aerial campaign.

Japanese Sources

Two more Japanese monographs of the highly useful series prepared for General MacArthur's headquarters were consulted frequently in the writing of this part. Both cover the activities of naval air during the period when Admiral Kusaka's *Eleventh Air Fleet*, with reinforcements from the *Combined Fleet's* carrier air groups, defended Rabaul. Monograph No. 140, Southeast Area Naval Air Operations (July–November 1943) is written in journal form with missions, claims, and losses featured and little discussion of combat operations. No. 142 which covers naval air operations from December 1943 to May 1944, provides a general review of the period when the Japanese lost the air battle over Rabaul. Included as an appendix to this last study is an analysis by a former staff officer of the *25th Air Flotilla* of Japanese air operations in the Southeast Area throughout the Allied advance on Rabaul and its subsequent isolation.

The difficult problem of assessing Japanese aircraft losses was eased considerably by the careful analysis of the draft manuscript made by the War History Office of the Defense Agency of Japan. The Japanese comments have been utilized as appropriate throughout the finished narrative.

Books

The fourth volume of the official history of the Army Air Forces, edited by Craven and Cate, and Sherrod's history of Marine Corps aviation have been the most important source works used for this part. In addition to these two books, both already cited as overall sources for this volume, the following were referred to frequently:

Wesley Frank Craven and James Lea Cate, eds. *Men and Planes—The Army Air Forces in World War II*, v. 6. Chicago: University of Chicago Press, 1955. An excellent volume in this basic reference series, this book provides considerable information on the aircraft used by the Army Air Forces and the training of its aircrews.

Deputy Chief of Naval Operations (Air), Aviation History Unit, OP-519B. *The Navy's Air War, A Mission Accomplished*. Lt A. R. Buchanan, USNR, ed. New York and London: Harper and Brothers [1946]. A summary of naval aviation's contribution to the war, this book is useful because of its information on aircrew training and aircraft development.

George C. Kenney. *General Kenney Reports, A Personal History of the Pacific War*. New York: Duell, Sloan and Pearce, 1949. An interesting memoir that sheds some light on command decisions in the SWPA, this work has the fault, however, of relying on the damage statistics and claims of the time written about rather than those which have been proved more accurate by later research.

George Odgers. *Air War Against Japan 1943–1945—Australia in the War of 1939–1945 (Air)*. Canberra: Australian War Memorial, 1957. This work is the prime source for information about the activities of the RAAF in the Southwest Pacific.

Masatake Okumiya and Jiro Horikoshi with Martin Caidin. *Zero!* New York: E. P. Dutton and Co., Inc., 1956. A fascinating book written by the designer of the Zero and an experienced Japanese naval pilot with the help of a veteran American writer on aviation matters. This account provides an exciting and informative history of the most formidable fighter used by the Japanese during the war.

SqnLdr J. M. S. Ross, RNZAF. *Royal New Zealand Air Force—Official History of New Zealand in the Second World War.* Wellington: War History Branch, Department of Internal Affairs, 1955. An objective and useful study of the RNZAF actions in the South and Southwest Pacific, this work merits close scrutiny.

United States Strategic Bombing Survey (Pacific), Naval Analysis Division. Marshalls-Gilberts-New Britain Party. *The Allied Campaign Against Rabaul.* Washington: Government Printing Office, 1946. More than half of this study is taken up with appendices, which include extensive interviews with General Imamura, Admiral Kusaka, and principal subordinates. The narrative is particularly good in its summation of the effects of the Allied air campaign on the Japanese stronghold.

PART VI

CONCLUSION

Official Documents

The basic sources for the narrative of the seizure of the Green Islands and Emirau were the action reports of the III Amphibious Force. The account of fighting in the Admiralties was based upon the description in the official Army history. The story of the aerial attacks that obliterated the town of Rabaul and destroyed the supplies that the Japanese were unable to disperse or move underground is well covered in the SoPac study, *The Reduction of Rabaul*, which covers the period 19 February–15 May 1944.

The narrative of the 18 months of Allied aerial attacks on Rabaul and Kavieng, which followed the Japanese evacuation of all flyable aircraft from the bastion, was found in the reports and war diaries of ComAirSols and ComAirNorSols. Once Marine Mitchells bore the brunt of the 1st Marine Aircraft Wing's interdiction attacks, the reports of MAG–61 and ComAirEmirau became the basic sources.

In summing up the period covered by this volume, the most useful documents were the action reports prepared by principal commands for each operation covered and the narrative account of Third Fleet activities prepared by Admiral Halsey's staff and submitted to CinCPOA in September 1944. Much of the material already cited was reviewed again before the last chapter was written.

Unofficial Documents

Many of the senior officers who commented upon pertinent draft parts of this volume made significant observations on the course of the war in the South and Southwest Pacific Areas. These comments were carefully considered and, in many cases, are incorporated in the evaluations made in the summary chapter.

Japanese Sources

No one can read the monographs prepared by Japanese historians for the use of American military forces or follow the comments that they made on the draft of this volume without tremendous respect for their honesty and lack of subterfuge. The study made by Takushiro Hattori, previously mentioned, reflects this objective and analytical approach throughout its pages. The manuscript translation of Hattori's work, together with material derived from Japanese sources in the relevant volumes of the United States Strategic Bombing Survey's works, have been used to review the Japanese part in the Rabaul campaign.

Books

The basic published sources that underlie the narrative of this book were used again in preparing the concluding chapters. In addition to these volumes, listed in the opening section of these notes, the following were of particular use:

Maj Charles W. Boggs, Jr. *Marine Aviation in the Philippines.* Washington: Historical Division, HQMC, 1951. This official monograph was useful in developing the story of the deployment of 1st Wing squadrons from the Bismarcks and Solomons to the Philippines.

Kenneth W. Condit and Edwin T. Turnbladh. *Hold High the Torch, A History of the 4th Marines.* Washington: Historical Branch, G–3 Division, HQMC, 1960. This regimental history provided a useful source of information regarding the employment of the newly formed 4th Marines in the Emirau operation.

691–360 O—63——36

Guide to Abbreviations

A-20 — Army twin-engine attack plane, the Douglas Havoc
AA — Antiaircraft
AAF — Army Air Forces
ABC — American-British-Canadian
AC/AS — Assistant Chief of Staff, Air Staff
ACI — Air Combat Intelligence
ACofS — Assistant Chief of Staff
ACSI — Assistant Chief of Staff, Intelligence (Army)
Actg — Acting
ADC — Assistant Division Commander
Adm — Admiral
Admin — Administrative
AET — Advance Echelon Translation
AF — Air Force
AFB — Air Force Base
AFP — Army Forces, Pacific
AID — Air Information Division
AIF — Australian Imperial Forces
Air — Aircraft; Air Forces
AKA — Cargo ship, attack
Al — Allied
Alex — Alexandria
An — Annual
ANGAU — Australia-New Guinea Administrative Unit
Anx — Annex
APA — Transport, attack
APc — Transport, coastal (small)
APD — Transport, high speed
App — Appendix
AR — Action Report
Arty — Artillery
ATIS — Allied Translation and Intelligence Service
Avn — Aviation
B-17 — Army four-engine bomber, the Boeing Flying Fortress
B-24 — Army four-engine bomber, the Consolidated Liberator
B-25 — Army twin-engine bomber, the North American Mitchell
B-26 — Army twin-engine bomber, the Martin Marauder
BAR — Browning Automatic Rifle
Bd — Board
BGen — Brigadier General
Bn — Battalion
Br — Branch
Brig — Brigade
Bu — Bureau
Bul — Bulletin
C — Combat
C-2 — Corps Intelligence Office(r)
C-47 — Army twin-engine transport, the Douglas Skytrain.
C-54 — Army four-engine transport, the Douglas Skymaster.
Capt — Captain
Cav — Cavalry
Cbt — Combat
CCS — Combined Chiefs of Staff
Cdr — Commander
CEC — Civil Engineer Corps
Cen — Center
CG — Commanding General
CIC — Combat Intelligence Center
CinC — Commander in Chief
Cir — Circular
CMC — Commandant of the Marine Corps
CNO — Chief of Naval Operations
CO — Commanding Officer
Co — Company
CofS — Chief of Staff
Col — Colonel
Com — Commander
Comd — Command
CominCh — Commander in Chief, U.S. Fleet.
CP — Command Post
Cpl — Corporal
CSNLF — Combined Special Naval Landing Force
CT — Combat Team
CTF — Commander Task Force
CTG — Commander Task Group
CWO — Chief Warrant Officer

Curr .... Current
D-2 .... Division Intelligence Office(r)
D-3 .... Division Operations and Training Office(r)
DA .... Department of the Army
DD .... Destroyer
Def .... Defense
Dep .... Depot; Deputy
Dept .... Department
DesDiv .... Destroyer Division
DesRon .... Destroyer Squadron
Det .... Detachment
Dir .... Director
Disp .... Dispatch
Distr .... Distribution
Div .... Division
Docu .... Document
DUKW .... Amphibious truck
Ech .... Echelon
Empl .... Employment
End .... Endorsement (Navy)
ESB .... Engineer Shore Brigade
Est .... Estimate
Evac .... Evacuation
FAdm .... Fleet Admiral
FEC .... Far East Command
FEAF .... Far East Air Forces
F5A .... Army photo plane version of the P-38
F4F .... Navy-Marine single-engine fighter, the Grumman Wildcat
F4U .... Navy-Marine single-engine fighter, the Chance-Vought Corsair
F6F .... Navy-Marine single-engine fighter, the Grumman Hellcat
Flt .... Fleet
FMF .... Fleet Marine Force
FO .... Field Order
For .... Force
FRC .... Federal Record Center
Fwd .... Forward
G-2 .... Division (or larger unit) Intelligence Officer(r)
G-3 .... Division (or larger unit) Operations and Training Office(r)
Gen .... General
GHQ .... General Headquarters
GO .... General Order
GPO .... Government Printing Office
Grd .... Guard
Gru .... Group
Hist .... History; Historical
Hq .... Headquarters
HQMC .... Headquarters, United States Marine Corps
ICPOA .... Intelligence Center, Pacific Ocean Areas
IGHQ .... Imperial General Headquarters
IJA .... Imperial Japanese Army
IJN .... Imperial Japanese Navy
Ind .... Indorsement (Army)
Inf .... Infantry
Info .... Information
Instn .... Instruction
Intel .... Intelligence
Is .... Island(s)
Jap .... Japanese
JCS .... Joint Chiefs of Staff
JICPOA .... Joint Intelligence Center, Pacific Ocean Areas
JSP .... Joint Staff Planners
JSSC .... Joint Strategic Survey Committee
KIA .... Killed in Action
LCdr .... Lieutenant Commander
LCI .... Landing Craft, Infantry
LCM .... Landing Craft, Medium
LCP .... Landing Craft, Personnel
LCP(R) .... Landing Craft, Personnel (Ramp)
LCT .... Landing Craft, Tank
LCVP .... Landing Craft, Vehicle and Personnel
Ldg .... Landing
LSD .... Landing Ship, Dock
LST .... Landing Ship, Tank
LT .... Landing Team
Lt .... Lieutenant
Ltr .... Letter
LVT .... Landing Vehicle, Tracked
LVT(A) .... Landing Vehicle, Tracked (Armored)
MAC .... Marine Amphibious Corps
Maj .... Major
MAG .... Marine Aircraft Group
Mar .... Marine(s)
MASP .... Marine Aircraft, South Pacific
MAW .... Marine Aircraft Wing(s)
MBDAG .... Marine Base Defense Aircraft Group
MBDAW .... Marine Base Defense Aircraft Wing
MC .... Medical Corps
Med .... Medical
Memo .... Memorandum

MG.......... Marine Gunner
MIA.......... Missing in Action
MID.......... Military Intelligence Division
MIS.......... Military Intelligence Service
Misc.......... Miscellaneous
MLR.......... Main Line of Resistance
Mm.......... Millimeter
MS.......... Manuscript
Msg.......... Message
MTB.......... Motor Torpedo Boat
NATS.......... Naval Air Transport Service
Nav.......... Navy; Naval
NCB.......... Naval Construction Battalion
ND.......... Navy Department
NGOF.......... New Guinea Occupation Force
NHD.......... Naval History Division
NLF.......... Northern Landing Force
NLG.......... Northern Landing Group
Nor.......... Northern
NZ.......... New Zealand
O.......... Order
OB.......... Order of Battle
Obj.......... Objective
OCMH.......... Office of the Chief of Military History
Off.......... Office
ONI.......... Office of Naval Intelligence
Op.......... Operation
OPlan.......... Operation Plan
Org.......... Organizational
OS2U.......... Navy single-engine float plane, the Chance-Vought Kingfisher
P-38.......... Army twin-engine fighter, the Lockheed Lightning
P-39.......... Army single-engine fighter, the Bell Airacobra
P-40.......... Army single-engine fighter, the Curtiss Warhawk
P-61.......... Army twin-engine night fighter, the Northrop Black Widow
P-70.......... Army night fighter version of the A-20
Pac.......... Pacific
Para.......... Parachute
PB4Y.......... Navy-Marine four-engine bomber, the Consolidated Liberator
PBJ.......... Navy-Marine twin-engine bomber, the North American Mitchell
PBM.......... Navy twin-engine seaplane, the Martin Mariner
PBO.......... Navy twin-engine bomber, the Lockheed Hudson
PB2Y.......... Navy twin-engine seaplane, the Consolidated Coronado
Phib.......... Amphibious; Amphibious Forces
Plat.......... Platoon
PMC.......... Pacific Military Conference
POA.......... Pacific Ocean Areas
POW.......... Prisoner of War
PV.......... Navy-Marine twin-engine bomber and night fighter, the Vega Ventura
R-2.......... Regimental Intelligence Office(r)
R-3.......... Regimental Operations Office(r)
RAAF.......... Royal Australian Air Force
Rad.......... Radio
RAdm.......... Rear Admiral
RAN.......... Royal Australian Navy
RCT.......... Regimental Combat Team
Rdr.......... Raider
Recon.......... Reconnaissance
Recs.......... Records
Regs.......... Regulations
Regt.......... Regiment
R5D.......... Navy-Marine four-engine transport, the Douglas Skymaster
R4D.......... Navy-Marine twin-engine transport, the Douglas Skytrain
RN.......... Royal Navy
RNZAF.......... Royal New Zealand Air Force
SAR.......... Special Action Report
SBD.......... Navy-Marine single-engine dive bomber, the Douglas Dauntless
SB-24.......... Army night bombing version of the B-24
SB2C.......... Navy-Marine single-engine dive bomber, the Curtiss-Wright Helldiver
SC.......... Submarine Chaser
SCAP.......... Supreme Commander Allied Powers
SCAT.......... South Pacific Combat Air Transport Command
SE.......... Southeast
Sec.......... Section; Secretary
Serv.......... Service
Sgt.......... Sergeant
Sit.......... Situation

SMS .... Marine Service Squadron
SNLF .... Special Naval Landing Force
So .... South
Sols .... Solomons
Sqn .... Squadron
STAG .... Special Task Air Group
Stf .... Staff
Strat .... Strategic
Subj .... Subject
SWPA .... Southwest Pacific Area
Tac .... Tactical
TAGO .... The Adjutant General's Office
TAIC .... Technical Air Intelligence Center
TBF .... Navy-Marine single-engine torpedo bomber, the Grumman Avenger
TBX .... Medium-powered field radio
Tele .... Telegram
TF .... Task Force
TG .... Task Group
Tg .... Telegraph
Tk .... Tank
TM .... Technical Manual
TNT .... Trinitro-toluol, a high explosive
T/O .... Table of Organization
Trac .... Tractor
Trans .... Transport; Translation
Trng .... Training
TSgt .... Technical Sergeant
U .... Unit
USA .... United States Army
USAF .... United States Air Force
USAFFE .... United States Army Forces in the Far East
USAFISPA .... United States Army Forces in the South Pacific Area
USASOS .... United States Army Service of Supply
USMC .... United States Marine Corps
USN .... United States Navy
USNR .... United States Naval Reserve
USSBS .... United States Strategic Bombing Survey
VAdm .... Vice Admiral
VB .... Navy dive bomber squadron
VC .... Navy composite squadron
VD .... Navy photographic squadron
VF .... Navy fighter squadron
VF(N) .... Navy night fighter squadron
VMB .... Marine bomber squadron
VMD .... Marine photographic squadron
VMF .... Marine fighter squadron
VMF(N) .... Marine night fighter squadron
VMO .... Marine observation squadron
VMSB .... Marine scout bomber squadron
VMTB .... Marine torpedo bomber squadron
WarD .... War Diary
WD .... War Department
WIA .... Wounded in Action
WW II .... World War II
YMS .... Harbor mine sweeper

Military Map Symbols

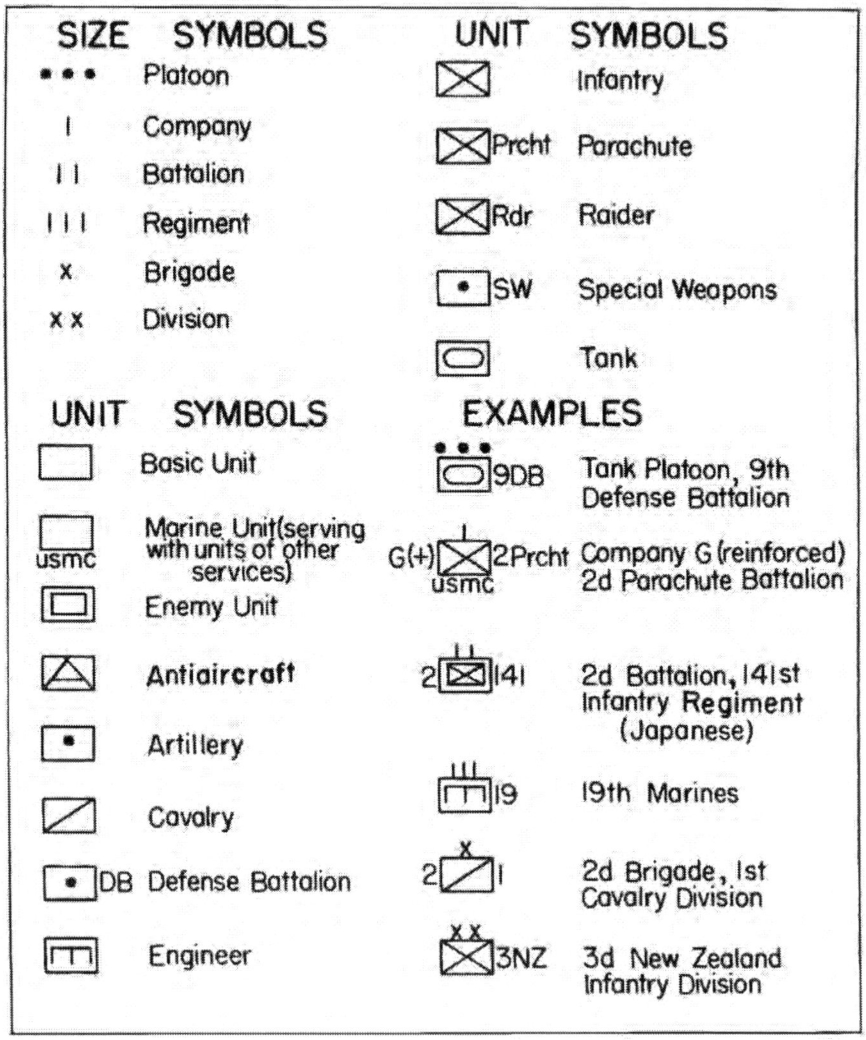

Chronology

The following listing of events is limited to those coming within the scope of this book, and those forecasting events to be treated in the volumes to follow.

1943

23Jan	Casablanca Conference approves ELKTON plan for operations against Rabaul.
23–24Jan	Kolombangara bombarded by cruiser-destroyer and carrier group.
7–8Feb	Japanese destroyers successfully evacuate 13,000 troops from Guadalcanal.
9Feb	Organized resistance on Guadalcanal ends.
12Feb	Gen MacArthur issues ELKTON I plan.
15Feb	ComAirSols, a joint air command, established on Guadalcanal.
21Feb	Russell Islands seized by 43d InfDiv troops reinforced by Marines.
2–5Mar	Battle of Bismarck Sea; US and Australian aircraft bomb Japanese destroyers and troop transports en route to Lae, New Guinea.
6Mar	U.S. naval force bombards Vila-Munda area.
	First Japanese air raids on Russells.
29Mar	CARTWHEEL directive issued by JCS.
7Apr	FAdm Isoruku Yamamoto begins "I" Operation, designed to drive Allies out of Solomons and New Guinea.
15Apr	First of Russell Islands' air strips operational.
18Apr	Adm Yamamoto is killed when his plane is shot down by P–38s.
26Apr	Gen MacArthur issues ELKTON III, superseding previous ELKTON plans.
2May	Japanese commanders at Rabaul create *Southeast Detached Force* for the defense of the central Solomons.
25May	TRIDENT Conference in Washington ends; CCS decide to seize Marshalls and to move against Japanese outer defenses.
3Jun	Adm Halsey issues orders for assault on New Georgia Islands.
5Jun	First long-range daylight raid by Marine SBDs and TBFs on ships in Kahili-Buin waters.
16Jun	New Georgia Occupation Force FO #1 issued; sets D-Day as 30Jun.
21Jun	One-half of 4th RdrBn lands at Segi Point; begins operations in eastern New Georgia.
22–23Jun	Army units begin Trobriand Islands invasion with landing on Woodlark Island.
28Jun	4th RdrBn meets first resistance of New Georgia campaign while approaching Viru Harbor.
30Jun	Army troops, reinforced by Marine elements, land on Vangunu and Rendova. Army troops, reinforced by 12th DefBn, seize Kiriwina.
1Jul	Viru Harbor seized. 9th DefBn shells Munda from Rendova.
3Jul	Southern Landing Group lands on Zanana Beach.

5Jul	Northern Landing Group lands at Rice Anchorage. Cruiser-destroyer force bombards Vila, Kolombangara, and Bairoko Harbor.
5-6Jul	Battle of Kula Gulf; U.S. naval task force engages 10 Japanese destroyers carrying reinforcements and supplies to Kolombangara. Some troops land during battle.
8Jul	MajGen William H. Rupertus relieves MajGen Alexander A. Vandegrift as CG, 1st MarDiv.
10Jul	Northern Landing Group attacks and takes Enogai.
11Jul	Adm Halsey issues directive for an attack on an unannounced position in the Bougainville area; Gen Vandegrift, CG, IMAC, selected to head invasion force. Segi Point air strip become operational.
11-12Jul	Cruiser-destroyer force bombards Munda.
13Jul	2d naval Battle of Kolombangara marks end of Japanese attempts to resupply and reinforce their New Georgia garrison by destroyer.
20Jul	Northern Landing Group launches unsuccessful attack on Bairoko Harbor; falls back to Enogai under cover of one of heaviest air strikes of central Solomons campaign. Marine land-based aircraft attack Japanese shipping south of Choiseul; two enemy destroyers sunk.
5Aug	Munda airfield, main objective of central Solomons campaign, falls.
6-7Aug	Three Japanese destroyers sunk, one damaged, in Battle of Vella Gulf.
8-9Aug	Main body of *Southeast Detached Force* moves to Kolombangara.
9Aug	Northern and Southern Landing Groups of New Georgia Occupation Force establish contact.
11Aug	Adm Halsey issues orders for Vella Lavella invasion.
13Aug	Japanese *Imperial Headquarters* issues Navy Staff Directive No. 267, authorizing abandonment of central Solomons after delaying actions.
14Aug	Marine aircraft begin operations from Munda airfield as ComAir New Georgia opens command post there.
15Aug	Vella Lavella invaded in force; 4th DefBn included in invasion group.
24Aug	QUADRANT Conference in Quebec ends; CCS decide to attack Japan along both central and southwest Pacific routes.
25Aug	New Georgia campaign ends as Bairoko Harbor is seized without opposition.
27Aug	Marines and Seabees occupy Nukufetau, Ellice Islands; Arundel Island occupied by Army troops.
28Aug	Forward echelon of 7th DefBn occupies Nanumea, Ellice Islands.
29Aug	1st RdrRegt withdraws from New Georgia operation.
31Aug	1st MarDiv alerted for movement from Melbourne to advance staging area.
1Sep	ComAirNorSols formed at Espiritu Santo under command of BGen Field Harris in preparation for northern Solomons offensive.
4Sep	V Amphibious Corps (VAC) formed under command of MajGen Holland M. Smith. Australian troops land near Lae, New Guinea.
11Sep	CinCSWPA requests Adm Halsey to strike in northern Solomons in accordance with JCS directives.

15Sep.......... MajGen Charles D. Barrett relieves Gen Vandegrift as CG, IMAC.

16Sep.......... Army troops fighting on Arundel Island reinforced by three platoons of Marine defense battalion tanks.

17Sep.......... 3d NZ Div lands on Vella Lavella relieving Army landing force.

19Sep.......... 1st MarDiv combat teams begin departure from Melbourne.

20–21Sep.......... MajGen Sasaki withdraws last Japanese survivors from Arundel as island is declared secure by Allied forces.

22Sep.......... Adm Halsey issues warning order for northern Solomons invasion of Treasury Islands and Empress Augusta Bay area of Bougainville. Gen MacArthur issues orders for DEXTERITY. Australian troops land at Finschhafen.

24Sep.......... ALAMO scouts begin reconnaissance of Cape Gloucester area.

25Sep.......... Forward echelon of IMAC Corps Troops land on Vella Lavella.

27Sep.......... IMAC issues instructions to 3d MarDiv for Bougainville operation. ComAirSols planes begin operations from Barakoma air strip.

28Sep.......... Japanese begin withdrawal from Kolombangara.

1Oct.......... Adm Halsey informs Gen MacArthur of decision to invade Bougainville on 1Nov and is promised maximum air assistance from SWPA air units. Low-level reconnaissance flights made over Cape Torokina region with ground officers acting as observers.

2–3Oct.......... Japanese complete safe withdrawal of some 9,400 troops from Kolombangara. Australian troops capture Finschhafen.

6Oct.......... Action in central Solomons ends as Army units make unopposed landing on Kolombangara.

6–7Oct.......... Battle of Vella Lavella; nine Japanese destroyers evacuating troops from Vella Lavella attacked by U.S. naval force.

8Oct.......... Gen Vandegrift reassumes command of IMAC upon death of Gen Barrett.

9Oct.......... 3d NZ Div declares Vella Lavella secure.

15Oct.......... IMAC issues OpO #1 directing 3d MarDiv to seize Cape Torokina. Beginning of intensified preinvasion air bombardments of Bougainville by Allied aircraft.

22Oct.......... IMAC directs 2d ParaBn to land on Choiseul, night of 27–28Oct, to conduct diversionary raid.

27Oct.......... Marine advance party lands at Atsinima Bay, north of Karuma River on Bougainville, to prepare for assault. 8th NZ Brig lands on Treasury Islands.

28Oct.......... 2d ParaBn lands on Choiseul.

31Oct.......... Bougainville invasion groups head for target area after rendezvous west of Guadalcanal.

1Nov.......... IMAC lands at Cape Torokina with 3d and 9th Marines and 2d RdrRegt in assault. First successful night air interception in Pacific by VMF(N)-531 aircraft.

1–2Nov.......... Battle of Empress Augusta Bay; U.S. fleet turns back Japanese naval attempt to counteract landing on Cape Torokina.

4Nov.......... 2d ParaBn withdraws from Choiseul.

5Nov.......... First carrier-based air strike at Rabaul.

6Nov........ Elements of 21st Marines arrive to reinforce Bougainville beachhead.

7Nov........ Japanese counterattack Bougainville beachhead by landing troops near Laruma River.

8Nov........ Battle of Koromokina Lagoon ends as the Japanese landing force is defeated by elements of 3d, 9th, and 21st Marines. First elements of 37th InfDiv arrive at Bougainville. MajGen Roy S. Geiger assumes command of IMAC as Gen Vandegrift is ordered home to become 18th Commandant of the Marine Corps.

11Nov........ Additional elements of 21st Marines arrive on Bougainville.

13Nov........ Pre-invasion bombardment of western New Britain targets begins.

17Nov........ Japanese aircraft attack convoy carrying Marine reinforcements to Bougainville; APD *McKean* is sunk with loss of some personnel from 21st Marines.

19Nov........ Battle of Piva Forks begins as final elements of 37th InfDiv arrive.

20Nov........ MajGen Ralph H. Mitchell assumes command of AirSols.

25Nov........ Carrier-based aircraft bomb Kavieng, New Ireland. Battle of Cape St. George concludes series of night naval engagements of the Solomons campaigns.

26Nov........ 3d MarDiv advances to Piva River line, having defeated Japanese in Battle of Piva Forks.

29Nov........ 1st ParaBn conducts Koiari Beach raid on Bougainville.

30Nov........ EUREKA Conference at Teheran ends. Stalin agrees to commit Russian forces against Japan after Germany is defeated. CG, ALAMO Force issues Field Order #5 for Arawe and Cape Gloucester operations.

7Dec........ SEXTANT Conference at Cairo concludes; CCS set up timetable for offensive against Japan.

15Dec........ Operation DEXTERITY begins with invasion of Arawe. BACKHANDER force conducts final rehearsal at Cape Sudest, New Guinea.

17Dec........ First AirSols fighter sweep over Rabaul from Bougainville air strips.

26Dec........ 1st MarDiv lands on Cape Gloucester at Silimati Point and Tauali.

28Dec........ Relief of 3d MarDiv begins as Americal Division takes responsibility for eastern sector of Bougainville beachhead.

30Dec........ 1st Marines secures Cape Gloucester airfield.

1944

1Jan........ BGen Lemuel C. Shepherd, Jr., issues first ADC attack order for the drive on Hill 660.

2Jan........ Task Force MICHAELMAS lands at Saidor, New Guinea.

11Jan........ Aogiri Ridge taken, renamed Walt's Ridge.

13Jan........ CinCPac-CinCPOA GRANITE plan issued; outlines tentative operation plans CATCHPOLE (Marshalls) and FORAGER (Marianas).

16Jan........ 3d MarDiv completes withdrawal from Bougainville. Hill 660 taken by 3/7 on Cape Gloucester.

31Jan........ Marines and Army troops land on Kwajalein and Majuro Atolls.

1Feb........ Marshalls invasion continues as Marines land on Roi and Namur Islands.

2Feb........ Roi and Namur secured.

7Feb........ Kwajalein Atoll secured.

10Feb	Operation DEXTERITY declared at an end by CG, ALAMO Force.
12Feb	Elements of the 1st Marines land on Rooke Island.
16Feb	Western New Britain secured as Gilnit patrol group meets Army patrol from Arawe on Itni River.
18Feb	Marines and Army units land on Eniwetok Atoll.
19Feb	Rabaul installations attacked heavily by Marine, Navy, and Army aircraft; after this date, the enemy abandons air defense of Rabaul.
21Feb	3/5 lands at Karai-ai.
25Feb	2/5 lands at Iboki Plantation.
29Feb	1st Cavalry Division lands in the Admiralties.
6Mar	5th Marines, Reinforced, land at Volupai Plantation for the Talasea operation.
8Mar	Japanese begin attack on 37th InfDiv sector on Bougainville.
9Mar	Talasea declared secure.
18Mar	Plans for Hollandia invasion issued jointly by CinCSWPA and CinCPac-CinCPOA.
20Mar	4th Marines seize Emirau.
24Mar	Last intensive enemy attack on Bougainville beachhead.
25Mar	IMAC, Corps Troops, 1st and 3d MarDivs transferred to CinCPOA operational control.
27Mar	Japanese begin withdrawal from Empress Augusta Bay area.
1Apr	Army troops begin advance up Numa Numa Trail on Bougainville.
22Apr	Army forces land at Hollandia. 1st MarDiv patrol clashes with enemy on Cape Gloucester for last time.
23Apr	1st MarDiv turns responsibility for New Britain over to 40th InfDiv.
15Jun	CinCSWPA assumes command of all forces west of longitude 159° East; South Pacific campaign against Japanese virtually ended.
21Jun	3d DefBn, which landed 1Nov43 on Bougainville and was last FMF ground unit in active SoPac area, withdrawn to Guadalcanal.
27Nov	40th InfDiv relieved on New Britain by Australian 5th Division.
3Dec	First Marine air units assigned to Philippines campaign at Leyte.
1945	
11Jan	Marine Air Groups, Dagupan (1st ProvMAW) organized at Luzon.
15Aug	1st MAW Hq moved to Philippines.
6Sep	Japanese surrender Rabaul to the Australians.

Fleet Marine Force Status—30 April 1943 [1]

Unit and location	Strength			
	USMC		USN	
	Off	Enl	Off	Enl
Outside U.S.A.				
South Pacific Area				
New Caledonia				
Special & Service Battalion, IMAC	134	975	8	23
Regulating Station (Transients), IMAC	43	1,046	24	59
1st Corps Motor Transport Battalion (less Company C)	21	416	1	9
1st Corps Medical Battalion	1	109	38	348
1st Corps Naval Construction Battalion			20	791
IMAC Barrage Balloon Group	43	910	2	8
1st Marine Raider Battalion	35	873	3	19
1st Marine Parachute Regiment (less 4th Battalion)	99	1,928	13	66
1st Marine Depot Company*	3	110		
14th Replacement Battalion*	39	1,199	10	111
1st Separate Wire Platoon	1	45		
1st Base Depot	44	700	1	21
4th Base Depot (w/Company C, 1st Corps Motor Transport Battalion)	48	831		
Marine Air Depot Squadron-1, 1st MAW	12	308	5	16
Marine Air Base Squadron-1, 1st MAW	10	340		
Headquarters Squadron-25, MAG-25	31	223	17	90
Service Squadron-25, MAG-25	11	302		
Marine Utility Squadron-152, MAG-25	46	223		
Marine Utility Squadron-153, MAG-25	20	220		
Marine Utility Squadron-253, MAG-25	57	226		
Area Sub-Total	698	10,993	142	1,561
New Zealand				
2d Marine Division	884	16,727	115	1,525
155mm Howitzer Battalion, IMAC	30	572	1	14
2d Antitank Battalion	31	698	1	12
3d Defense Battalion	57	1,024	4	22
4th Defense Battalion	54	1,096	5	19
16th Replacement Battalion*	30	1,018	19	108
2d Base Depot	15	295	1	14

See footnote at end of table.

Unit and location	Strength			
	USMC		USN	
	Off	Enl	Off	Enl
New Zealand—Continued				
3d Marine Division (less 3d Marines, Reinf)	650	12, 303	95	1, 172
1st Aviation Engineer Battalion	29	615	3	13
3d Base Depot	11	226	1	9
Headquarters Squadron-2, 2d MAW	49	409	6	13
Headquarters Squadron-14, MAG-14	23	382	5	7
Service Squadron-14, MAG-14	11	244		10
Area Sub-Total	1, 874	35, 609	256	2, 938
Guadalcanal-Tulagi				
9th Defense Battalion	48	1, 071	3	16
14th Defense Battalion	38	772	3	22
2d Aviation Engineer Battalion	36	615	1	9
Marine Fighter Squadron-123, MAG-11	33	254	1	-------
Marine Fighter Squadron-124, MAG-12	27	259	1	7
Marine Scout-Bomber Squadron-132, MAG-11	34	257	1	4
Marine Scout-Bomber Squadron-143, MAG-12	34	328	1	2
Marine Scout-Bomber Squadron-233, MAG-11	31	233	1	4
Area Sub-Total	281	6, 247	12	64
Florida Island				
11th Defense Battalion	47	1, 088	4	22
Russell Islands				
10th Defense Battalion	46	1, 070	4	22
Espiritu Santo				
1st Marine Raider Regiment (less 1st Battalion)	108	2, 514	14	79
Headquarters Squadron-1, 1st MAW	72	506	15	11
Marine Air Repair & Salvage Squadron-1, 1st MAW	7	181	-------	7
Marine Photographic Squadron-154, 1st MAW	29	415	2	10
Headquarters Squadron-11, MAG-11	31	500	11	6
Service Squadron-11, MAG-11	17	333	-------	7
Marine Fighter Squadron-112, MAG-11	35	250	1	8
Marine Fighter Squadron-121, MAG-11	52	250	1	4
Marine Fighter Squadron-213, MAG-11	29	206	1	7
Marine Scout-Bomber Squadron-131, MAG-11	28	223	-------	-------
Marine Scout-Bomber Squadron-141, MAG-11	8	270	1	5
Marine Fighter Squadron-122, MAG-12	33	226	1	7

See footnote at end of table.

Unit and location	Strength			
	USMC		USN	
	Off	Enl	Off	Enl
Espiritu Santo—Continued				
Marine Scout-Bomber Squadron-142, MAG-12	32	267	1	8
Marine Scout-Bomber Squadron-234, MAG-12	27	259	1	6
Headquarters Squadron-21, MAG-21	17	267	5	12
Service Squadron-21, MAG-21	8	172		
Marine Fighter Squadron-214, MAG-21	27	233		
Marine Fighter Squadron-221, MAG-21	32	205		
Area Sub-Total	592	7, 277	54	177
Efate				
Marine Scout-Bomber Squadron-144, MAG-11	38	268	1	
Headquarters Squadron-12, MAG-12	27	341	7	38
Service Squadron-12, MAG-12	11	224		
Area Sub-Total	76	833	8	38
Funafuti				
5th Defense Battalion (Reinf)	47	1, 046	34	279
Samoa Islands				
Headquarters, Defense Force	24	108	7	4
Signal Company, Defense Force	7	172		
Base Depot, Fleet Marine Force	48	686	2	15
13th Replacement Battalion	20	992	5	38
15th Replacement Battalion	29	990	5	60
4th Garrison Replacement Detachment	8	298		
3d Marines (Reinf)	214	4, 503	23	447
2d Defense Battalion (Reinf)	133	1, 820	19	148
3d Marine Brigade (including 22d Marines)	230	3, 731	25	187
Headquarters Squadron-13, MAG-13	34	267	7	18
Service Squadron-13, MAG-13	10	194		
Marine Fighter Squadron-111, MAG-13	26	159	1	8
Marine Fighter Squadron-441, MAG-13	30	239	1	8
Marine Scout-Bomber Squadron-151, MAG-13	40	281	1	8
Marine Scout-Bomber Squadron-241, MAG-13	31	221		
Area Sub-Total	884	14, 661	96	941

See footnote at end of table.

Unit and location	Strength			
	USMC		USN	
	Off	Enl	Off	Enl
Wallis Island				
8th Defense Battalion (Reinf)	115	2, 171	37	720
Southwest Pacific Area				
Australia				
1st Marine Division	956	16, 825	121	1, 709
7th Replacement Battalion*	22	957	5	39
Area Sub-Total	978	17, 782	126	1, 748
Central Pacific Area				
Oahu				
Headquarters, Marine Forces, 14th Naval District	32	121	6	18
12th Defense Battalion	47	1, 089	4	22
Headquarters Squadron, MAWPac	14	50	1	
Headquarters Squadron-4, 4th MBDAW	27	235	15	
Marine Utility Squadron-252, 4th MBDAW	24	187	1	
Headquarters Squadron-24, MAG-24	22	280	8	23
Service Squadron-24, MAG-24	8	194		
Marine Fighter Squadron-222, MAG-24	38	228	1	7
Marine Scout-Bomber Squadron-235, MAG-24	39	285	1	8
Marine Scout-Bomber Squadron-236, MAG-24	20	249	1	8
Area Sub-Total	271	2, 918	38	86
Midway				
6th Defense Battalion	84	2, 075	3	31
Headquarters Squadron-22, MAG-22	9	111	2	3
Service Squadron-22, MAG-22	2	85		
Marine Fighter Squadron-215, MAG-22	30	225	1	8
Marine Scout-Bomber Squadron-244, MAG-22	31	236	1	8
Area Sub-Total	156	2, 732	7	50
Palmyra				
1st Defense Battalion	57	1, 195	3	28
Marine Fighter Squadron-211, 4th MBDAW	41	296	2	6
Area Sub-Total	98	1, 491	5	34

See footnote at end of table.

Unit and location	Strength USMC Off	Strength USMC Enl	Strength USN Off	Strength USN Enl
Johnston				
16th Defense Battalion	46	888	1	12
Marine Scout-Bomber Squadron-243, 4th MBDAW	16	137	1	7
Area Sub-Total	62	1, 025	2	19
Guantanamo Bay, Cuba				
13th Defense Battalion	68	1, 232	4	16
St. Thomas, Virgin Islands				
Marine Scouting Squadron-3	26	88		
West Coast, U.S.A.				
Camp Elliott				
Headquarters, Fleet Marine Force, San Diego Area	20	51	2	
Headquarters Company, Amphibious Corps, Pacific Fleet	38	232	8	16
Service Company, Amphibious Corps, Pacific Fleet	4	102		
Signal Company, Amphibious Corps, Pacific Fleet	28	332		
Reconnaissance Company, Amphibious Corps, Pacific Fleet	7	98		2
Amphibious Tractor Detachment, Amphibious Corps, Pacific Fleet	1	25		1
1st Corps Signal Construction Company	6	199		
1st Corps Signal Operation Company	4	198		
17th Replacement Battalion	10	145	15	77
18th Replacement Battalion	4	57	6	
Training Center, Camp Elliott	407	6, 578	59	478
Base Depot, Fleet Marine Force	31	398	3	23
Company B, 4th Parachute Battalion	4	157		
Area Sub-Total	564	8, 572	93	597
Camp Pendleton				
24th Marines (Reinf)	187	3, 680	14	91
1st Corps Tank Battalion (Medium)	30	807	1	9
Training Center, Camp Pendleton	187	2, 655	22	510
Area Sub-Total	404	7, 142	37	610

See footnote at end of table.

Unit and location	Strength			
	USMC		USN	
	Off	Enl	Off	Enl
Camp Dunlap				
2d Airdrome Battalion	53	853	3	27
Camp Gillespie				
Parachute Training School	21	590	1	11
San Diego				
Headquarters Squadron, Service Group, Marine Fleet Air, West Coast	46	212	11	28
Supply Squadron-5	17	235		
Air Regulating Squadron-1	7	632		4
Air Regulating Squadron-2	6	660		178
Air Regulating Squadron-3	111	128	2	3
Air Regulating Squadron-4	7	554		
Area Sub-Total	194	3,011	14	224
Kearney				
Headquarters Squadron-15, MAG-15	35	231	5	17
Service Squadron-15, MAG-15	11	274		
Marine Observation Squadron-155, MAG-15	25	269		
Marine Observation Squadron-251, MAG-15	7	206		7
Marine Photographic Squadron-254, MAG-15	25	353	2	8
Marine Utility Squadron-353, MAG-15	43	338		
Area Sub-Total	146	1,671	7	32
El Toro				
Headquarters Squadron-23, MAG-23	15	128	9	15
Service Squadron-23, MAG-23	7	140		
Marine Fighter Squadron-223, MAG-23	5	141		
Marine Fighter Squadron-224, MAG-23	6	150		
Marine Scout-Bomber Squadron-231, MAG-23	13	169		
Marine Scout-Bomber Squadron-232, MAG-23	42	226		
Headquarters Squadron-41, MBDAG-41	23	228	5	14
Service Squadron-41, MBDAG-41	13	173		
Marine Fighter Squadron-113, MBDAG-41	10	243		
Marine Fighter Squadron-212, MBDAG-41	36	240	1	8
Area Sub-Total	170	1,838	15	27

See footnote at end of table.

Unit and location	Strength			
	USMC		USN	
	Off	Enl	Off	Enl
Santa Barbara				
Headquarters Squadron-42, MBDAG-42	13	151	6	18
Service Squadron-42, MBDAG-42	12	191		
Marine Fighter Squadron-422, MBDAG-42	10	116		
Area Sub-Total	35	458	6	18
El Centro				
Headquarters Squadron-43, MBDAG-43	26	565	4	12
Service Squadron-43, MBDAG-43	8	122		
Marine Fighter Squadron-216, MBDAG-43	6	90		
Area Sub-Total	40	777	4	12
Mojave				
Headquarters Squadron-44, MBDAG-44	24	128	5	13
Service Squadron-44, MBDAG-44	9	117		
Marine Fighter Squadron-225, MBDAG-44	6	63		
Area Sub-Total	39	308	5	13
East Coast, U.S.A.				
New River				
Headquarters Battalion, Training Center	64	1,092	19	80
School Battalion, Training Center	165	679	35	54
Signal Battalion, Training Center	70	2,826	1	56
Quartermaster Battalion, Training Center	72	676		
Engineer Battalion, Training Center	116	906	22	1,087
Artillery Battalion, Training Center	130	680	1	6
Parachute Battalion, Training Center	19	555	2	30
Rifle Range Battalion, Training Center	12	312		
Infantry Battalion, Training Center	22	561		
Barrage Balloon Activities	16	113	9	370
23d Marines (Reinf)	260	3,177	23	129
25th Marines (Reinf)	202	2,724	10	124
1st Airdrome Battalion	46	1,030	3	21
51st Composite Defense Battalion	20	511	3	22
19th Replacement Battalion	31	1,068	5	61

See footnote at end of table.

Unit and location	Strength			
	USMC		USN	
	Off	Enl	Off	Enl
New River—Continued				
Company A, 4th Parachute Battalion	13	265		
2d Marine Depot Company	3	110		
3d Marine Depot Company	3	110		
Marine Scout-Bomber Squadron-341, MAG-34	14	146		
Area Sub-Total	1, 278	17, 541	133	2, 040
Norfolk				
Base Depot, Fleet Marine Force	24	207		4
Cherry Point				
Headquarters Squadron-3, 3d MAW	27	171	8	10
Marine Bomber Squadron-413, 3d MAW	2	16		
Headquarters Squadron-31, MAG-31	8	37	3	34
Service Squadron-31, MAG-31	4	70		
Headquarters Squadron-32, MAG-32	7	109	3	32
Service Squadron-32, MAG-32	6	72		
Headquarters Squadron-33, MAG-33	10	77	4	41
Service Squadron-33, MAG-33	5	55		
Headquarters Squadron-34, MAG-34	8	87	3	27
Service Squadron-34, MAG-34	6	80		
Headquarters Squadron-35, MAG-35	10	27		
Service Squadron-35, MAG-35	5	62		
Marine Observation Squadron-351, MAG-35	1	28		
Marine Utility Squadron-352, MAG-35	10	50		
Headquarters Squadron-53, MAG-53	16	35	1	
Service Squadron-53, MAG-53	6	83		
Marine Night Fighter Squadron-531, MAG-53	13	159		
Marine Night Fighter Squadron-532, MAG-53	4	57		
Area Sub-Total	148	1, 275	22	144
Oak Grove				
Marine Fighter Squadron-321, MAG-32	14	121		
Atlantic				
Marine Scout-Bomber Squadron-331, MAG-33	7	117	1	2

See footnote at end of table.

Unit and location	Strength			
	USMC		USN	
	Off	Enl	Off	Enl
Parris Island				
Marine Fighter Squadron-311, MAG-31	12	126		
Total Fleet Marine Force (Ground) Overseas	4, 812	91, 745	698	8, 306
Total Fleet Marine Force (Air) Overseas	1, 507	14, 060	131	409
Total Fleet Marine Force (Ground) in U.S.A.	2, 329	34, 759	267	3, 289
Total Fleet Marine Force (Air) in U.S.A.	819	9, 258	73	471
Total Fleet Marine Force Overseas	6, 319	105, 802	829	8, 715
Total Fleet Marine Force in U.S.A.	3, 148	44, 017	340	3, 760
Total Fleet Marine Force	9, 467	149, 822	1, 169	12, 475

[1] Strength figures and unit designations and locations were abstracted from the FMF Status Reports, Ground and Air, for April 1943 prepared by the M-3 Section, Headquarters Marine Corps. Units en route to the indicated areas are designated by an asterisk *.

Table of Organization E–100 Marine Division—15 April 1943 [1]

Unit	USMC		USN		Totals	
	Off	Enl	Off	Enl	Off	Enl
Division Headquarters	(51)	(139)	(8)	(13)	(59)	(152)
Special Troops	157	2, 098	14	46	173	2, 144
Headquarters Battalion	(83)	(736)	(11)	(21)	(94)	(757)
Headquarters Company	(60)	(328)	(11)	(21)	(71)	(349)
Signal Company	(17)	(313)			(17)	(313)
Military Police Company	(6)	(95)			(6)	(95)
Special Weapons Battalion	(38)	(703)	(2)	(14)	(40)	(717)
Headquarters & Service Battery	(13)	(73)	(2)	(14)	(15)	(87)
Antiaircraft Battery	(7)	(300)			(7)	(300)
3 Antitank Batteries (each)	(6)	(110)			(6)	(110)
Tank Battalion	(36)	(659)	(1)	(11)	(37)	(670)
Headquarters & Service Company	(11)	(64)	(1)	(9)	(12)	(73)
3 Light Tank Companies (each)	(6)	(155)			(6)	(155)
Scout Company	(7)	(130)		(2)	(7)	(132)
Service Troops	78	1, 682	42	398	120	2, 080
Service Battalion	(27)	(614)	(2)	(18)	(29)	(632)
Headquarters Company	(8)	(43)	(2)	(10)	(10)	(52)
Service & Supply Company	(13)	(433)		(9)	(13)	(442)
Ordnance Company	(6)	(138)			(6)	(138)
Motor Transport Battalion	(28)	(489)	(1)	(9)	(29)	(498)
Headquarters & Service Company	(13)	(165)	(1)	(9)	(14)	(174)
3 Transport Companies (each)	(5)	(108)			(5)	(108)
Amphibian Tractor Battalion	(22)	(454)	(1)	(9)	(23)	(463)
Headquarters & Service Company	(7)	(64)	(1)	(9)	(8)	(73)
3 Tractor Companies (each)	(5)	(130)			(5)	(130)
Medical Battalion	(1)	(125)	(38)	(362)	(39)	(487)
Headquarters & Service Company	(1)	(5)	(3)	(12)	(4)	(17)
5 Medical Companies (each)		(24)	(7)	(70)	(7)	(94)
Engineer Regiment	74	1, 548	35	860	109	2, 408
Headquarters & Service Company	(22)	(256)	(5)	(7)	(27)	(263)
Engineer Battalion	(21)	(614)	(1)	(9)	(22)	(623)
Headquarters Company	(6)	(44)	(1)	(9)	(7)	(53)
3 Engineer Companies (each)	(5)	(190)			(5)	(190)

See footnote at end of table.

Unit	USMC		USN		Totals	
	Off	Enl	Off	Enl	Off	Enl
Engineer Regiment—Continued						
Pioneer Battalion	(31)	(678)	(3)	(32)	(34)	(710)
Headquarters Company	(7)	(78)	(3)	(32)	(10)	(110)
3 Pioneer Companies (each)	(8)	(200)			(8)	(200)
Naval Construction Battalion			26	812	26	812
Headquarters Company			(14)	(134)	(14)	(134)
3 Construction Companies (each)			(4)	(226)	(4)	(226)
Artillery Regiment	188	2, 956	9	54	197	3, 010
Headquarters & Service Battery	(20)	(171)	(4)	(9)	(24)	(180)
3 Pack Howitzer Battalions (each)	(34)	(561)	(1)	(9)	(35)	(570)
Headquarters & Service Battery	(13)	(129)	(1)	(9)	(14)	(138)
3 Pack Howitzer Batteries (each)	(7)	(144)			(7)	(144)
2 Howitzer Battalions (each)	(33)	(551)	(1)	(9)	(34)	(560)
Headquarters & Service Battery	(15)	(125)	(1)	(9)	(16)	(134)
3 Howitzer Batteries (each)	(6)	(142)			(6)	(142)
3 Infantry Regiments (each)	137	2, 984	11	110	148	3, 094
Headquarters & Service Company	(21)	(146)	(5)	(14)	(26)	(160)
Weapons Company	(8)	(189)			(8)	(189)
3 Infantry Battalions (each)	(36)	(883)	(2)	(32)	(38)	(915)
Headquarters Company	(10)	(93)	(2)	(32)	(12)	(125)
Weapons Company	(8)	(220)			(8)	(220)
3 Rifle Companies (each)	(6)	(190)			(6)	(190)
Division Totals	908	17, 236	133	1, 688	1, 041	18, 924

[1] All unit strength figures enclosed in parentheses are included in the strength totals of parent units.

MAJOR WEAPONS AND TRANSPORTATION—MARINE DIVISION

Weapons	Number	Transportation	Number
Carbine, .30 cal., M-1	11,074	Ambulance:	
Flamethrower, portable	24	¼-ton, 4 x 4	48
Gun:		½-ton, 4 x 4	11
37mm, antitank	54	Car, 5-passenger	3
40mm, antiaircraft	16	Motorcycle	12
75mm, antitank, self-propelled	12	Station wagon, 4 x 4	12
Gun, machine:		Tractor:	
.30 cal., M1919A4	682	amphibian	100
.30 cal., M1917A1	108	miscellaneous	73
.50 cal., M2	343	Trailer:	
Gun, submachine, .45 cal	78	¼-ton, cargo	92
Howitzer:		½-ton, dump	20
75mm pack	36	1-ton, cargo	125
105mm	24	1-ton, water	81
Launcher, rocket, antitank, M-1	243	miscellaneous	123
Mortar:		Truck:	
60mm	81	¼-ton, 4 x 4	375
81mm	81	¼-ton, 4 x 4, with radio	134
Pistol, .45 cal	299	1-ton, 4 x 4, cargo	268
Rifle, .30 cal., M-1	8,030	1-ton, 4 x 4, with radio	22
Rifle, Browning, automatic	558	2½-ton, 4 x 4, cargo	48
Shotgun, 12 gauge	306	2½-ton, 6 x 6, cargo	198
Tank, light, with armament	54	2½-ton, 6 x 6, dump	51
Tank, light, recovery	3	miscellaneous	51

Marine Task Organization and Command List[1]

MARINE GROUND UNITS

A. SEIZURE OF THE RUSSELLS (21 February–20 June 1943)

3d Marine Raider Battalion
(21Feb–20Mar43)

CO........ Col Harry B. Liversedge (to 15 Mar43)
LtCol Samuel B. Yeaton (from 15Mar)

10th Defense Battalion
(24Feb–20Jun43)

CO........ Col Robert E. Blake

Detachment, 11th Defense Battalion
(21Feb–28Mar43)

CO........ Maj Joseph L. Winecoff

B. NEW GEORGIA OPERATION (20 June–16 October 1943)[2]

Forward Echelon, IMAC Corps Troops
(25Sep–16Oct43)

CO........ Maj Donald M. Schmuck

Headquarters, 1st Marine Parachute Regiment
(8–16Oct43)

CO........ LtCol Robert H. Williams
ExO........ Maj Jackson B. Butterfield (actg)
R-3........ Maj Walter S. Osipoff

1st Parachute Battalion

CO........ Maj Richard Fagan

2d Parachute Battalion
(1Sep–10Oct43)

CO........ LtCol Victor H. Krulak

3d Parachute Battalion

CO........ Maj Robert T. Vance

Headquarters, 1st Marine Raider Regiment
(5Jul–28Aug43)

CO........ Col Harry B. Liversedge
ExO........ (None shown for the period)
R-3........ LtCol Joseph P. McCaffery

1st Raider Battalion

CO........ LtCol Samuel B. Griffith, II

4th Raider Battalion
(21Jun–11Jul; 18Jul–28Aug43)

CO........ LtCol Michael S. Currin

4th Defense Battalion
(15Aug–16Oct43)

CO........ Col Harold S. Fassett

9th Defense Battalion
(20Jun–31Aug43)

CO........ LtCol William J. Scheyer

C. TREASURY-BOUGAINVILLE OPERATION AND CONSOLIDATION OF NORTHERN SOLOMONS (28 October 1943–15 June 1944)[3]

I Marine Amphibious Corps
(1–21Dec43)

CG........ MajGen Roy S. Geiger

[1] Unless otherwise noted, names, positions held, organization titles, and periods of service were taken from the muster rolls of the units concerned, held in the Diary Unit, Files Section, Records Branch, Personnel Department, Headquarters Marine Corps. Units are listed only for those periods, indicated by the dates below parent unit designation, for which they are entitled to campaign participation credit. This information is derived from muster rolls and U.S. Bureau of Naval Personnel, *Navy and Marine Corps Awards Manual*—NAVPERS 15,790 (Rev. 1953) with changes (Washington, 1953–1958). The muster rolls have been the final authority when there is a conflict in dates of unit entitlement within the overall campaign period as cited by the Awards Manual. In the case of Marine air units, many of which participated in the campaigns as flight or advance echelons only, the unit commander who was actually in the combat area is shown where muster rolls reveal this information. In order to conserve space, only units of battalion and squadron size, or larger, and sizeable separate detachments are listed for each operation, although smaller organizations may have participated also.

[2] Includes: New Georgia-Rendova-Vangunu Occupation, 20 Jun–31Aug43; Vella Lavella Occupation, 15Aug–16Oct43.

[3] Includes Choiseul Island Diversion, 28Oct–4Nov43; Occupation and Defense of Cape Torokina, 1Nov–15Dec43; Consolidation of Northern Solomons, 15Dec43–15Jun44.

CofS........ BGen Alfred H. Noble (to 18 Dec43)
BGen Oscar R. Cauldwell (from 18Dec)
C-1........ LtCol Joseph D. Burger
C-2........ LtCol William F. Coleman
C-3........ LtCol Edward W. Snedeker
C-4........ LtCol Frederick L. Wieseman

2d 155mm Artillery Battalion (Provisional)

(18Nov-21Dec43)

CO........ LtCol Joe C. McHaney

Corps Transportation Company, 1st Corps Motor Transport Battalion

(8Nov-25Dec43)

CO........ Maj Franklin H. Hayner

1st Corps Signal Battalion

(6Nov-21Dec43)

CO........ LtCol Frederick A. Ramsey, Jr.

Headquarters, 1st Marine Parachute Regiment

(4Dec43-12Jan44)

CO........ LtCol Robert H. Williams
ExO........ Maj Jackson B. Butterfield (actg)
R-3........ Maj Walter S. Osipoff

1st Parachute Battalion

(23Nov43-12Jan44)

CO........ Maj Richard Fagan (to 11Jan44)
Maj Robert C. McDonough (from 11Jan)

2d Parachute Battalion

(28Oct-4Nov43)

CO........ LtCol Victor H. Krulak

3d Parachute Battalion

CO........ Maj Robert T. Vance (to 10Dec43)
Maj Harry L. Torgerson (from 10Dec)

2d Marine Raider Regiment (Provisional)

(1Nov43-12Jan44)

CO........ LtCol Alan Shapley
ExO........ LtCol Joseph W. McCaffery (KIA 1Nov)
Capt Oscar F. Peatross (from 1Nov)
R-3........ Capt Oscar F. Peatross

2d Raider Battalion

CO........ Maj Richard T. Washburn (from 1Nov)

3d Raider Battalion

CO........ LtCol Fred D. Beans

3d Defense Battalion

(1Nov43-21Jun44)

CO........ LtCol Edward H. Forney

Branch No. 3, 4th Base Depot

(1Nov43-16Jan44)

CO........ Col Kenneth A. Inman

3d Marine Division

(1Nov-21Dec43)

CG........ MajGen Allen H. Turnage
ADC........ BGen Oscar R. Cauldwell
CofS........ Col Robert E. Blake
D-1........ LtCol Chevey S. White
D-2........ LtCol Howard J. Turton
D-3........ LtCol James D. Snedeker (to 12Nov43)
Col Walter A. Wachtler (12Nov-16Dec)
LtCol Alpha L. Bowser (from 17Dec)
D-4........ Col William C. Hall

Division Headquarters and Service Battalion

CO........ LtCol Samuel D. Puller (to 14Nov-43)
(None shown for 14Nov)
LtCol Hartnoll D. Withers (15-30Nov)
LtCol Samuel D. Puller (1-16Dec)
(None indicated after 16Dec)

Division Special and Service Troops

CO........ Col Walter A. Wachtler (to 12Nov-43)
LtCol James D. Snedeker (12-14-Nov)
LtCol Hartnoll D. Withers (15Nov-9Dec)
(None shown for 10Dec)
LtCol James M. Smith (from 11-Dec)

3d Amphibian Tractor Battalion

CO........ Maj Sylvester L. Stephan (to 5Dec-43; 8-9Dec; from 17Dec)
Maj Ervin F. Wann (6-7Dec; 10-16Dec)

3d Medical Battalion

(1Nov-25Dec43)

Co........ Cdr Gordon M. Bruce (MC)

3d Motor Transport Battalion

(1Nov–25Dec43)

CO........ Maj Stewart W. Purdy

3d Service Battalion

CO........ LtCol Ion M. Bethel

3d Special Weapons Battalion

(17Nov–25Dec43)

CO........ LtCol Durant S. Buchanan

3d Tank Battalion

(1Nov–25Dec43)

CO........ LtCol Hartnoll J. Withers (to 15Nov43; from 10Dec)
(None shown for period 16Nov–9Dec)

3d Marines

(1Nov–25Dec43)

CO........ Col George W. McHenry (to 17 Dec43)
Col Walter A. Wachtler (from 17 Dec)
ExO........ LtCol George O. Van Orden
R–3........ Maj Sidney S. McMath (to 21Dec)
(None shown from 22Dec)

1st Battalion, 3d Marines

CO........ Maj Leonard M. Mason (WIA 1Nov43)
Maj John D. Brody (2–18Nov)
Maj Charles J. Bailey, Jr. (from 19Nov)

2d Battalion, 3d Marines

CO........ LtCol Hector de Zayas

3d Battalion, 3d Marines

CO........ LtCol Ralph M. King

9th Marines

(1Nov–28Dec43)

CO........ Col Edward A. Craig
ExO........ LtCol James A. Stuart
R–3........ LtCol Ralph L. Houser

1st Battalion, 9th Marines

CO........ LtCol Jaime Sabater (to 19Nov43)
LtCol Carey A. Randall (from 19 Nov)

2d Battalion, 9th Marines

CO........ LtCol Robert E. Cushman, Jr.

3d Battalion, 9th Marines

CO........ LtCol Walter Asmuth, Jr.

12th Marines

(1Nov43–1Jan44)

CO........ Col John B. Wilson
ExO........ LtCol John S. Letcher
R–3........ LtCol William T. Fairbourn

1st Battalion, 12th Marines

CO........ LtCol Raymond F. Crist, Jr.

2d Battalion, 12th Marines

(6Nov43–1Jan44)

CO........ LtCol Donald M. Weller

3d Battalion, 12th Marines

CO........ LtCol Jack Tabor

4th Battalion, 12th Marines

CO........ LtCol Bernard H. Kirk

19th Marines

(1Nov43–1Jan44)

CO........ Col Robert M. Montague (to 7Dec43)
LtCol Robert E. Fojt (from 7Dec)
ExO........ LtCol Robert E. Fojt (to 7Dec)
Maj William V. D. Jewett (from 7Dec)
R–3........ Capt Minetree Folkes, Jr.

1st Battalion, 19th Marines (Engineers)

CO........ Maj Ralph W. Bohne

2d Battalion, 19th Marines (Pioneers)

CO........ LtCol Harold B. West (to 7Dec43)
Maj Halstead Ellison (from 7Dec)

21st Marines

(6Nov43–9Jan44)

CO........ Col Evans O. Ames
ExO........ LtCol Arthur H. Butler
R–3........ Maj James W. Tinsley

1st Battalion, 21st Marines

CO........ LtCol Ernest W. Fry, Jr.

2d Battalion, 21st Marines

CO........ LtCol Eustace R. Smoak

3d Battalion, 21st Marines

(17Nov43–9Jan44)

CO........ LtCol Archie V. Gerard

D. NEW BRITAIN CAMPAIGN AND TALASEA OPERATION (26 December 1943–25 April 1944) [4]

[4] Includes: Cape Gloucester landing and operations, 26Dec43–1Mar44; Talasea landing and operations, 5Mar–25Apr44.

Headquarters, 1st Marine Division

(26Dec43–1Mar44)

CG........ MajGen William H. Rupertus
ADC...... BGen Lemuel C. Shepherd, Jr.
CofS...... Col Amor L. Sims (to 4Feb44)
Col Oliver P. Smith (4–29Feb)
Col John T. Selden (from 1Mar)
D-1....... Maj Elmer W. Myers
D-2....... LtCol Edmund J. Buckley (to 24Feb44)
Col Harold D. Harris (from 24Feb)
D-3....... Col Edwin A. Pollock (to 30Jan44)
LtCol William K. Enright (from 30Jan)
D-4....... Col William S. Fellers

Assistant Division Commander Group (to February 1944)

ADC...... BGen Lemuel C. Shepherd, Jr.
CofS...... Col Herman H. Hanneken
ADC-2.... Capt Gene E. Gregg
ADC-3.... Maj John S. Day
ADC-4.... Capt Robert T. Crawford

Division Headquarters and Service Battalion

(1Jan–1Mar44)

CO........ LtCol Frank R. Worthington

Division Special Troops

CO........ Col Herman H. Hanneken
(None shown after 20Feb44)

Provisional Air Liaison Unit [5]

CO........ Capt James Harris
1stLt Richard M. Hunt

1st Tank Battalion

CO........ LtCol Charles G. Meints

1st Medical Battalion

CO........ Capt Everett B. Keck (MC) (to 28Feb44)
Cdr Stanley P. Wallin (MC) (from 28Feb)

1st Marines

CO........ Col William J. Whaling, Jr. (to 29Feb44)
Col Lewis B. Puller (from 29Feb)
ExO....... LtCol Harold D. Harris (to 24Feb44)
(None shown after 24Feb)
R-3....... Maj Martin F. Rockmore (to 7Jan44)
Capt John N. Rentz [6] (7–20Jan)
Capt Arthur Larson (from 21Jan)

1st Battalion, 1st Marines

CO........ LtCol Walker A. Reaves

2d Battalion, 1st Marines

CO........ LtCol James M. Masters, Jr. (to 10Feb44)
Maj Charles H. Brush, Jr. (from 10Feb)

3d Battalion, 1st Marines

CO........ LtCol Joseph F. Hankins

5th Marines

(29Dec43–1Mar44; 5Mar–25Apr44)

CO........ Col John T. Selden (to 29Feb44)
Col Oliver P. Smith (1Mar–9Apr)
LtCol Henry W. Buse, Jr. (from 10Apr)
ExO....... LtCol William K. Enright (to 6Jan44)
LtCol Lewis W. Walt (6–8Jan)
Maj Harry S. Connor (9–12Jan)
LtCol Lewis W. Walt (13–31Jan)
(None shown for 1–8Feb)
LtCol. Odell M. Conoley (9–20Feb)
LtCol Henry W. Buse, Jr. (21Feb–9Apr)
Maj Harry S. Connor (from 10Apr)
R-3....... Maj Gordon D. Gayle (to 6Jan44)
Maj Harry S. Connor (from 6Jan)

1st Battalion, 5th Marines

CO........ LtCol William H. Barba

2d Battalion, 5th Marines

CO........ LtCol Lewis W. Walt (to 6Jan44)
Maj Gordon D. Gayle (from 6Jan)

3d Battalion, 5th Marines

(30Dec43–1Mar44; 5Mar–25Apr44)

CO........ LtCol David S. McDougal (WIA 7Jan44).

---

[5] This unit did not have an official T/O nor was it listed in the division's muster rolls. It existed, however. The only mention made anywhere of its commanding officers is to be found in Captain Richard M. Hunt, "General Rupertus' Improvised Air Force," *Marine Corps Gazette*, v. 33, no. 6 (Jun49), although the inclusive dates of command and relief are not indicated.

[6] No R-3 is shown in the muster rolls for the period 7–20Jan. Capt Rentz, who was Assistant R-3 at this time, was on active duty at HQMC when Hough and Crown's *New Britain Campaign* was written and has been listed as the R-3 in that monograph.

CO........ Maj Joseph S. Skoczylas (WIA 7Jan).
LtCol Lewis B. Puller (7–8Jan)
LtCol Lewis W. Walt (9–12Jan)
LtCol Harold O. Deakin (13Jan–10Apr).
Maj Walter McIlhenny (from 11Apr)

7th Marines

CO........ Col Julian N. Frisbie (to 22Feb44)
Col Herman H. Hanneken (from 22Feb).
ExO........ LtCol Lewis B. Puller (to 23Feb44)
(None shown after 23Feb)
R-3........ Maj Victor H. Streit

1st Battalion, 7th Marines

CO........ LtCol John E. Weber

2d Battalion, 7th Marines

CO........ LtCol Odell M. Conoley (to 8Feb44)
Maj Charles S. Nichols, Jr. (8–14Feb).
LtCol John W. Scott, Jr. (from 15Feb).

3d Battalion, 7th Marines

CO........ LtCol William R. Williams (to 4Jan44).
LtCol Lewis B. Puller (4–5Jan)
LtCol Henry W. Buse, Jr. (6Jan–20 Feb).
Maj William J. Piper, Jr. (from 21 Feb).

11th Marines

CO........ Col Robert H. Pepper (to 31Jan44)
Col William H. Harrison (from 31Jan)
ExO........ LtCol Robert B. Luckey (to 15 Feb44)
(None shown for 15–16Feb)
LtCol Thomas B. Hughes (from 17Feb)
R-3........ Maj Louis A. Ennis (to 16Feb44)
(None shown for 16–21Feb)
Maj Elliott Wilson (from 22Feb)

1st Battalion, 11th Marines

CO........ LtCol Lewis J. Fields

2d Battalion, 11th Marines

(26Dec43–1Mar44; 5Mar–25Apr44)

CO........ LtCol Noah P. Wood, Jr.

3d Battalion, 11th Marines

(19Feb–1Mar44)

CO........ LtCol Forest C. Thompson

4th Battalion, 11th Marines

CO........ LtCol Thomas B. Hughes (to 17Feb44)
LtCol Louis A. Ennis (from 17Feb)

5th Battalion, 11th Marines

(30Dec43–1Mar44)

CO........ LtCol Charles M. Nees

17th Marines

CO........ Col Harold E. Rosecrans (to 19Feb44)
Col Francis I. Fenton (from 19Feb)
ExO........ LtCol Robert G. Ballance (from 22Feb)
(None shown prior to this date[7])
R-3........ Maj John P. McGuinness (to 22Feb)
Maj Levi A. Smith, Jr. (from 22Feb)

1st Battalion, 17th Marines (Engineers)

CO........ LtCol Henry H. Crockett

2d Battalion, 17th Marines (Pioneers)

CO........ LtCol Robert G. Ballance[7] (to 22Feb44)
Maj Austin S. Igleheart, Jr. (from 22Feb)

12th Defense Battalion

CO........ Col William H. Harrison (to 31 Jan44)
LtCol Merlyn D. Holmes (from 31Jan)

E. EMIRAU LANDING AND OCCUPATION (20 March 1944–12 April 1944)

I Marine Amphibious Corps Task Group A[8]

(20Mar–12Apr44)

Force Commander.... BGen Alfred H. Noble
CofS............ Col Gale T. Cummings
F-1............ Maj Ormond R. Simpson
F-2............ LtCol Sidney S. Wade
F-3............ LtCol George O. Van Orden
F-4............ LtCol Leonard M. Mason

[7] Although Lieutenant Colonel Ballance is shown in the muster rolls as the Commanding Officer, 2/17, he served as the regimental executive officer in the period 26Dec43–22Feb44. According to Ballance, Major Levi A. Smith, Jr., served as 2/17's commander during this same period. *Ballance ltr.*

[8] Extracted from Emirau Landing Force Journal (Emirau Area OpFiles A10-1, 2, 3, and 4, HistBr, HQMC).

4th Marines, Reinforced
(20Mar–12Apr44)

CO.................. LtCol Alan Shapley
ExO.................. LtCol Samuel D. Puller
R-3.................. Maj Orville V. Bergren

1st Battalion, 4th Marines

CO.................. LtCol Charles L. Banks

2d Battalion, 4th Marines

CO.................. Maj John S. Messer

3d Battalion, 4th Marines

CO.................. Maj Ira J. Irwin

4th Pack Howitzer Battalion
(20Mar–12Apr44)

CO.................. Maj Robert H. Armstrong

14th Defense Battalion
(20–25Mar44)

CO.................. LtCol William F. Parks

MARINE AIR UNITS

Headquarters and Detachments, 1st Marine Aircraft Wing

(A—9Feb–20Jun43; 22May44–15Mar45)[9]
(B—21Jun43–1May44)
(C—27Aug–15Dec43)

CG.............. MajGen Roy S. Geiger (to 20Apr43)
MajGen Ralph J. Mitchell (21Apr43–31Jan44)
MajGen James T. Moore (1Feb–14Jun)
MajGen Ralph J. Mitchell (from 15Jun)

AWC.............. BGen James T. Moore (to 1Jul43)
BGen Claude A. Larkin (from 3Aug)

CofS.............. BGen Louis E. Woods (to 27May43)
BGen James T. Moore (28 May–1Sep)
Col Clayton C. Jerome (1Sep 43–31Jan44)

CofS.............. Col William L. McKittrick (1Feb–16Jun)
Col. Stanley E. Ridderhof (from 17Jun)

W-1.............. LtCol Thomas C. Ennis (to 21Oct43)
Col William B. Steiner 22Oct43–31Jan44)
Capt Howard H. Parker (1Feb–1Sep)
LtCol Carl L. Jolly (2Sep44–11Mar45)
Maj Walter N. Gibson (from 12Mar)

W-2.............. LtCol John C. Munn (to 26Mar43)
Capt David B. Decker (27 Mar–26Nov)
Capt Peter Folger (26Nov43–19Jun44)
Capt Frank E. Walter (20Jun 43 1Oct)
Capt William H. Powell (2Oct–9Dec)
Capt Harlow P. Rothert (from 10Dec)

W-3.............. Col Christian F. Schilt (to 19Mar43)
LtCol Joe A. Smoak (19 Mar–10Apr)
LtCol Paul A. Moret[10] (21 Apr–7Jun)
Col Marion A. Dawson (8Jun43–13Jan44)
Col William B. Steiner (14–31Jan)
Col Alexander W. Kreisler, Jr. (1Feb–15Aug)
Col Frank W. Schwable (16 Aug–31Oct)
Col Edward A. Montgomery (1Nov44–6Mar45)
Col Charles J. Schlapkohl (from 7Mar)

W-4.............. LtCol Albert D. Cooley (to 21Mar43)
(None shown for 21–31Mar)
Col Herbert B. Becker (31Mar 43–8Jun44)
Col Zebulon C. Hopkins (9Jun–25Sep)

---

[9] Under each unit listed there will appear a letter designation for each operation in which the unit participated, and dates of involvement. Following are the campaigns and dates of entitlement:

A. Consolidation of the Solomons	8Feb43–15Mar45
B. New Georgia Operation	20Jun–16Oct43
C. Bismarck Archipelago Operation	25Jun43–1May44
D. Treasury-Bougainville Operation	27Oct–15Dec43

[10] Killed in plane crash 7Jun43.

W-4 ---- LtCol Otto E. Bartoe (from 26Sep)
CO, HqSqn-1 ---- Capt Herman J. Jesse (to 15Feb43)
Capt Carlos Martinez (15Feb-20Aug)
Maj John T. Rooney (21Aug-9Oct)
LtCol Eugene B. Diboll (10 Oct-31Dec)
Maj Loren P. Kesler (1Jan-4Feb44)
Capt James C. White, Jr. (5Feb-19Jun)
Capt Walter E. Sallee (20Jun-12Sep)
Maj Walter N. Gibson (13 Sep-8Oct)
Capt Robert W. Baile (from 9Oct)

Headquarters and Forward Echelon, 2d Marine Aircraft Wing

(A—9Feb-20Apr; 29Jun-16Oct43)

CG ---- BGen Francis P. Mulcahy
CofS ---- Col Walter G. Farrell (to 25Aug43)
Col Elmer H. Salzman (from 25Aug)
W-1 ---- 1stLt Robert G. Coddington
W-2 ---- Col Elmer H. Salzman (to 25Aug43)
LtCol Etheridge C. Best (from 25Aug)
W-3 ---- LtCol William C. Lemly (to 25Apr43)
LtCol Etheridge C. Best (25 Apr-24Aug)
LtCol Eugene F. Syms (from 25Aug)
W-4 ---- LtCol Franklin G. Cowie
CO, HqSqn-2 ---- Maj William K. Snyder

Forward Echelon, Marine Aircraft Group 14

(A—9Feb-3Apr43; 27Oct43-15Jan45)
(B—20Aug-16Oct43)

CO ---- Col William O. Brice (to 16Mar44)
LtCol Roger T. Carleson (16-Mar-25Sep)
Col Zebulon C. Hopkins (from 26Sep)
ExO ---- LtCol Perry O. Parmelee (to 17Dec43)
LtCol Joe A. Smoak (18Dec43-31Jan44)
LtCol Roger T. Carleson (1-Feb-15Mar)
Maj Floyd E. Beard, Jr. (16Mar-28Nov)
(None designated after 28Nov)
GruOpsO ---- Maj Clyde T. Mattison (to 14Jul43)
Maj Arthur R. Stacy (15Jul-26Nov)
(None shown for 27Nov)
Maj Floyd E. Beard, Jr. (28 Nov43-15Mar44)
Maj Walter J. Carr, Jr. (16 Mar-?Jun)
Maj Floyd E. Beard, Jr. (?Jun-28Nov)
(None designated after 28Nov)
CO, HqSqn-14 ---- Capt Stanley M. Adams (to 25Nov43)
Capt Arnold Borden (25Nov43-5Jun44)
Maj Donald S. Bush (6Jun-7Dec)
Capt Robert M. Crooks (from 8Dec)
CO, SMS-14 ---- Maj Arthur R. Stacy (to 13Jul43)
Maj Kenneth H. Black (13Jul-13Dec)
Capt Walter A. Johnson (14Dec43-2Oct44)
Capt Droel H. Looney (from 3Oct)

Marine Aircraft Group 21

(A—13Mar-20Jun43)

CO ---- LtCol Raymond E. Hopper (to 17May43)
LtCol Nathaniel S. Clifford (actg) (from 17May)
ExO ---- LtCol Nathaniel S. Clifford (to 17May)
(None shown after 17May)
GruOpsO ---- Capt Charles W. Somers, Jr. (to 10May43)
Maj Wilfred H. Stiles (10May-1Jun)
(None designated 2-9Jun)

GruOpsO — Maj George F. Britt (from 10Jun)
CO, HqSqn-21 — Maj Joseph T. Cain
CO, SMS-21 — LtCol Robert M. Haynes (to 12May43)
Maj Douglas J. Peacher (from 12May)

Forward Echelon, Marine Aircraft Group 24

(D—15Dec43)
(A—16Dec43-30Apr44)

CO — Col William L. McKittrick (to 20Feb44)
LtCol Lewis H. Delano, Jr. (from 20Feb)
ExO — LtCol Roger T. Carleson (to 1Jan44)
LtCol Lewis H. Delano, Jr. (1Jan-19Feb)
LtCol Robert W. Clark (from 20Feb)
GruOpsO — LtCol Lewis H. Delano, Jr. (to 19Feb)
Maj Max J. Volcansek, Jr. (19Feb-26Apr)
(None shown after 26Apr)
CO, HqSqn-24 — Capt Alan Limburg (actg) (to 26Jan44)
Maj Lawrence L. Jacobs (from 26Jan)
CO, SMS-24 — LtCol Robert W. Clark (to 20Feb44)
Capt Watt S. Ober (from 20Feb)

Marine Aircraft Group 25

(A—9Feb-20Jun43; 27Oct43-15Mar45)
(B—21Jun-16Oct43)

CO — Col Perry K. Smith (to 10Jul43)
Col Wyman F. Marshall (10Jul-15Dec)
Col Allen C. Koonce (16-31Dec)
Col William A. Willis (1Jan-24Jul44)
Col Allen C. Koonce (25Jul44-12Feb45)
Col Harold C. Major (from 13Feb)
ExO — Col Wyman F. Marshall (to 14Feb43)
LtCol Elmore W. Seeds (14Feb-31May)
LtCol William K. Lanman, Jr. (1-3Jun)
LtCol Federick E. Leek (4Jun-20Aug)
Col William A. Willis (21Aug-27Sep)
LtCol John P. Coursey (28Sep-14Nov)
(None shown 15-24Nov)
LtCol William H. Klenke, Jr. (25Nov44-9Feb45)
Col Warren E. Sweetser, Jr. (from 10Feb)
GruOpsO — Col Wyman F. Marshall (to 5Apr43)
LtCol Harry F. Van Liew (5Apr-4Jul)
LtCol Elmore W. Seeds (5Jul-10Oct)
LtCol Harry F. Van Liew (11Oct43-2Aug44)
LtCol Harry H. Bullock (3Aug-26Sep)
LtCol Theodore W. Sanford, Jr. (27Sep44-4Mar45)
LtCol William H. Klenke, Jr. (from 5Mar)
CO, HqSqn-25 — Capt Dave J. Woodward, Jr. (to 9Sep43)
Maj Jonathan W. Dyer (9Sep43-15Jan44)
Maj Thomas M. Heard (16Jan-16Jun)
Capt LeRoy M. James (17Jun-20Oct)
Maj Theodore E. Beal (21Oct-9Nov)
Maj Charles J. Prall (from 10Nov)
CO, SMS-25 — Maj Ralph R. Yeamans (to 21May43)
Maj Jack A. Church (21May-15Jul)
Maj Ralph R. Yeamans (16Jul-16Nov)
Maj Jack A. Church (17Nov43-9Mar44)
LtCol Millard T. Shepard (10Mar-13Nov)
LtCol Albert S. Munsch (from 14Nov)

Flight Echelon, Marine Aircraft Group 61

(A—18Jul44–15Mar45)

CO	Col Perry K. Smith
ExO	LtCol Frederick B. Winfree
GruOpsO	LtCol Stewart W. Ralston (to 17Aug44)
	Maj Peter V. Metcalf (17Aug–8Nov)
	LtCol Stewart W. Ralston (from 8Nov)
CO, HqSqn–61	Maj Peter V. Metcalf (to 17Aug44)
	Capt Claude A. Wharton (from 17Aug)
CO, SMS–61	Maj Jack W. Julian (to 1Jan45)
	LtCol Roswell B. Burchard, Jr. (from 1Jan)

Marine Fighter Squadron 112

(A—9Feb–20Jun43)

CO	Maj Paul J. Fontana (to 27Mar43)
	Maj Robert B. Fraser (from 27Mar)

Marine Fighter Squadron 114

(C—28Mar–4May44)

CO	Maj Robert F. Stout

Marine Fighter Squadron 115

(C—19Apr–1May44)
(A—2May–30Nov44)

CO	Maj Joseph J. Foss (to 20Sep44)
	Maj John H. King, Jr. (from 20Sep)

Flight Echelon, Marine Fighter Squadron 121

(A—4Mar–20Jun43)
(B—21Jun–22Jul43)

CO	Maj Roy L. Vroome (to 14May43)
	Capt Robert E. Bruce (from 14May)

Flight Echelon, Marine Fighter Squadron 123

(A—9Feb–20Jun43)
(B—15Aug–18Sep43)

CO	Maj Edward W. Johnston (to 19Apr43)
	Maj Richard M. Baker (from 19Apr)

Flight Echelon, Marine Fighter Squadron 124

(A—12Feb–1Jun43)
(B—21Jun–6Sep43)

CO	Maj William E. Gise (MIA 13-May43)
	Capt Cecil B. Brewer (13May–25Jun)
	Maj William H. Pace (26Jun–13Jul)
	LtCol William A. Millington (from 14Jul)

Flight Echelon, Marine Scout-Bomber Squadron 131

(A—9Feb–2May43)

CO	Capt Jens C. Aggerbeck, Jr. (to 15Mar43)
	Capt George E. Dooley (from 15Mar)

Flight Echelon, Marine Scout-Bomber Squadron 132

(A—8Feb–29Mar43)
(B—22Jun–1Aug43)

CO	Maj Louis B. Robertshaw (to 27May43)
	Maj Russell D. Rupp (from 27-May)

Marine Scout-Bomber Squadron 133

(A—24Aug–11Dec44)

CO	Maj Lee A. Christoffersen

Flight Echelon, Marine Torpedo-Bomber Squadron 134

(D—26Nov–27Dec43)
(C—17Feb–1May44)

CO	LtCol Alben C. Robertson

Flight Echelon, Marine Scout-Bomber Squadron 141

(A—9Feb–19Jun43)
(B—20Jun–3Sep43)

CO	Capt Claude A. Carlson, Jr. (to 8Mar43)
	1stLt Oscar J. Camp, Jr. (8 Mar–1Apr)
	Maj Howard F. Bowker, Jr. (2Apr–14May)
	Capt Middleton P. Barrow (15 May–25Aug)
	1stLt John E. Lepke (from 26 Aug)

Flight Echelon, Marine Torpedo-Bomber Squadron 142

(A—9Feb–26Apr44; 19Sep–19Dec44)

CO .... Maj Robert H. Richard (to 9-Jun44)
Capt Hoyle R. Barr (from 18-Jul)

Marine Scout-Bomber Squadron 143

(A—15Feb–20Jun43)

CO .... Maj John W. Sapp, Jr. (to 14May43)
Capt Warren G. Mollenkamp (14May–7Jun)
MG Alvie D. Godwin (actg) (8–20Jun)

Ground Echelon, VMTB-143 [11]
(B—20Jul–29Aug43)
Capt Timothy A. Moynihan

Forward Echelon, VMTB-143
(D—27Oct–30Nov43)
(C—19Jan–3Mar44)
Capt Timothy A. Moynihan (to 13Nov43)
1stLt William O. Cain (13–28Nov)
Capt Henry W. Hise (from 29Nov)

Marine Torpedo-Bomber Squadron 144

(A—9Feb–20Jun43)

CO .... Capt Roscoe W. Nelson (to 20Apr43)
Maj Frank E. Hollar (from 20Apr)

Flight Echelon, VMTB-144
(B—21Jun–1Aug43)
(D—27Oct–22Nov43)
Maj Frank E. Hollar

Flight Echelon, Marine Utility Squadron 152

(A—9Feb–19Jun43; 4Aug44–15Mar45)
(B—20Jun–5Aug43)

CO .... Maj Elmore W. Seeds (to 13Feb43)
Maj Dwight M. Guillotte (from 13Feb)
LtCol Albert W. Munsch (to 14Nov44)
LtCol John P. Coursey (from 14Nov)

[11] VMSB designation changed as of 31May43.

Detachment, Flight Echelon, Marine Utility Squadron 153

(A—8Apr–19Jun43; Jun44–15Mar45)
(B—20Jun–18Aug43)
(C—10–15Dec43)

CO .... Maj William K. Lanman, Jr. (to 1Jun43)
LtCol Elmore W. Seeds (1Jun–4Jul)
Maj Robert B. Bell (5Jul–4Nov)
Maj Freeman W. Williams (5Nov43–22May44)
Maj Theodore W. Sanford, Jr. (23May–29Jul)
LtCol Harold F. Brown (from 30Jul)

Detachment, Flight Echelon, Marine Photographic Squadron 154

(A—9Feb–20Jun43)
(B—21Jun–16Oct43)

CO .... LtCol Elliot E. Bard

Flight Echelon, Marine Fighter Squadron 211

(D—17Oct–22Nov43)
(C—30Dec43–1Feb44)
(A—2Jun–30Nov44)

CO .... Maj Robert A. Harvey (to 26Jan44)
Maj Thomas V. Murto, Jr. (26Jan–5May)
Maj Thaddeus P. Wojcik (6May–18Oct)
Maj Stanislaus J. Witomski (from 19Oct)

Flight Echelon, Marine Fighter Squadron 212

(D—27Oct–27Nov43)
(C—7Jan–17Feb; 19Feb–30Apr44)
(A—7Jun–8Dec44)

CO .... Maj Stewart B. O'Neil (to 31Dec43)
Maj Hugh M. Elwood (1Jan–23Apr44)
Maj Wilbur A. Free (24Apr–8May)
Maj Boyd C. McElhany, Jr. (9May–18Nov)
Maj Quinton R. Johns (from 18Nov)

Marine Fighter Squadron 213
(A—3Apr–20Jun43)

CO............ Maj Wade H. Britt, Jr. (to 13Apr43)
Flight Echelon, VMF–213
(B—21Jun–29Jul; 5Sep–16Oct43)
Maj Gregory J. Weissenberger (13Apr–22Aug43)
Maj James R. Anderson (from 22Aug)

Flight Echelon, Marine Fighter Squadron 214
(A—10Mar–14May43)
(B—22Jul–2Sep43)
(C—14Sep–20Oct43)
(D—28Nov–15Dec43)
(C—16Dec43–6Jan44)

CO............ Maj George F. Britt (to 9Jun43)
Maj Henry A. Ellis, Jr. (9Jun–11Jul)
Maj William H. Pace (12Jul–7Aug)
Capt John R. Burnett (8Aug–6Sep)
Maj Gregory Boyington (7Sep-43–3Jan44, MIA)
Capt Lawrence H. Howe (from 4Jan)

Marine Fighter Squadron 215
(B—25Jul–6Sep43)

CO............ Maj James L. Neefus (to 30Sep44)
Ground Echelon, VMF–215
(D—27Oct–27Nov43)
LtCol Herbert H. Williamson
Flight Echelon, VMF–215
(A—7Jan–7May44)
(C—22Apr–1May44)
Maj Robert G. Owens, Jr. (to 28Feb44)
Maj James K. Dill (from 28Feb)

Flight Echelon, Marine Fighter Squadron 216
(D—23Nov–15Dec43)
(C—5Feb–28Mar44)

CO............ Maj Rivers J. Morrell, Jr. (to 22Jan44)
Maj Benjamin S. Hargrave, Jr. (from 22Jan)

Flight Echelon, Marine Fighter Squadron 217
(A—28Jan–17Mar44)
(C—28Jan–17Mar44)

CO............ Maj Max R. Read, Jr.

Flight Echelon, Marine Fighter Squadron 218
(C—3Feb–16Mar44)
(A—30Apr–6Jun; 23Jul–30Nov44)

CO............ Maj Horace A. Pehl (to 28-Sep44)
Maj Robert T. Kingsbury (from 28Sep)

Marine Fighter Squadron 221
(A—17Mar–10May43)
(B—27Jun–24Aug43)
(D—27Oct–19Nov43)

CO............ Capt Robert R. Burns (to 1Jun43)
Maj Monfurd K. Peyton (1-Jun–16Aug)
Capt John S. Payne (17–24Aug)
Maj Nathan T. Post, Jr. (25-Aug–11Oct)
Maj Edwin S. Roberts, Jr. (from 12Oct)

Flight Echelon, Marine Fighter Squadron 222
(B—5Sep–15Oct43)
(D—19Nov–15Dec43)
(C—16–23Dec43)
(A—3Feb–19Mar; 8May–16Jun; 5Aug–8Dec44)

CO............ Capt Max J. Volcansek, Jr. (to 5Nov43)
Maj Alfred N. Gordon (5Nov-43–4Apr44)
Maj Roy T. Spurlock (from 5Apr)

Flight Echelon, Marine Fighter Squadron 223
(D—28Nov–15Dec43)
(C—16Dec43–8Jan44)

CO............ Maj Marion E. Carl
VMF–223
(A—17Feb44–11Jan45)
Maj Robert P. Keller (to 3-Jul44)
Maj David Drucker (3Jul–13Oct)
Maj Robert F. Flaherty (from 14Oct)

Flight Echelon, Marine Torpedo-Bomber Squadron 232

(B—23Sep–16Oct43)
(D—27Oct–15Dec43)
(C—20Jan–1May44)
(A—2May–19Jun44)

CO............ Maj Rolland F. Smith (to 26Apr44)
Maj Menard Doswell III (from 26Apr)

Flight Echelon, Marine Scout-Bomber Squadron 233 [12]

(A—13Mar–5Apr43)
(B—13Aug–21Sep43)
(D—2Nov–11Dec43)
(C—3Jan–10Mar44)

CO............ Capt Elmer L. Gilbert, Jr. (to 1May43)
Maj Claude J. Carlson, Jr. (1–25May)
Maj William J. O'Neill (26May–3Sep)
Maj Royce W. Coln (from 4Sep)

Flight Echelon, Marine Scout-Bomber Squadron 234

(A—9–28Feb; 12Apr–20Sep43)
(B—4Aug–7Sep43)

CO............ Maj William D. Roberson (to 5Apr43)
Maj Otis V. Calhoun, Jr. (5Apr–30Sep)

Detachment, Flight Echelon, VMSB-234
(D—27Oct–25Nov43)
Maj Harold B. Penne (1–26Oct)
Capt Edward J. Montagne, Jr. (from 27Oct)

Flight Echelon, Marine Scout-Bomber Squadron 235

(B—4Sep–16Oct43)
(D—27Nov–15Dec43)
(C—16–31Dec43)
(A—23Mar–6May; 8Jun–13Sep44)

CO............ Capt Everett E. Munn (to 10Feb44)
Maj Glenn L. Todd (10Feb–17May) [13]
Capt Edward C. Willard (18May–15Aug)
Maj James A. Feeley, Jr. (from 16Aug)

Flight Echelon, Marine Scout-Bomber Squadron 236

(B—7Sep–16Oct43)
(D—27Nov–15Dec43)
(C—16Dec43–7Feb44)
(A—28Apr–6Jun; 1 Aug–22Nov44)

CO............ Maj Floyd E. Beard, Jr. (to 10Nov43)
Maj William A. Cloman, Jr. (10Nov43–12Jun44)
Maj Edward R. Polgrean (13Jun–13Oct)
Capt Glen H. Schluckbier (14–30Oct)
Maj James A. Feeley, Jr. (from 31Oct)

Flight Echelon, Marine Scout-Bomber Squadron 241

(C—9Feb–17Mar44)
(A—4May–11Jun; 31Jul–20Sep44)

CO............ Maj James A. Feeley, Jr. (to 12Aug44)
Maj James C. Lindsay (from 12Aug)

Marine Torpedo-Bomber Squadron 242

(C—29Feb–25Apr44)

CO............ Maj William W. Dean

Flight Echelon, Marine Scout-Bomber Squadron 243

(D—20Nov–15Dec43)
(A—16–27Dec43; 16Jun–23Dec44)
(C—17Mar–27Apr44)

CO............ Maj Thomas J. Ahern (to 3Oct44)
Maj Joseph W. Kean, Jr. (from 13Oct)

Flight Echelon, Marine Scout-Bomber Squadron 244

(B—18Oct–29Nov43)
(C—10Feb–22Mar44)
(A—17May–24Jun; 31Jul–13Nov44)

CO............ Maj Robert J. Johnson (to 25Jan44)

[12] VMSB-233 was redesignated VMTB-233 on 22May43.

[13] Muster Rolls show him as CO until 17May, but departing for US 7May.

CO.......... Maj Harry W. Reed (25Jan-17Apr)
Capt Richard Belyea (18Apr-1Jul)
Maj Frank R. Porter, Jr. (from 2Jul)

Flight Echelon, Marine Observation Squadron 251
(A—9Feb-11May43; 18Jun-30Dec44)

CO.......... Maj Joseph N. Renner (to 13Mar43)
Capt Claude H. Welch (13-Mar-14May)
Maj Carl M. Longley (4Jun-31Oct)
Capt Robert W. Teller (1-5-Nov)
Maj William C. Humberd (from 6Nov)

Detachment, Flight Echelon, Marine Utility Squadron 253
(B—20Jun-31Aug43)

CO.......... LtCol Henry C. Lane

Flight Echelon, Marine Photographic Squadron 254
(A—12Dec43-30Sep44)

CO.......... Maj Edwin P. Pennebaker

Flight Echelon, Marine Fighter Squadron 313
(A—25Sep-30Nov44)

CO.......... Maj Joseph H. McGlothlin, Jr.

Flight Echelon, Marine Fighter Squadron 321
(C—5Dec43-27Jan44; 17Mar-24Apr44)

CO.......... Maj Edmund F. Overend

Flight Echelon, Marine Scout-Bomber Squadron 341
(C—1Jan-10Feb; 6Apr-1May44)
(A—2May-30Nov44)

CO.......... Maj George J. Waldie, Jr. (to 24Jan44)
Maj James T. McDaniels (24-Jan-19May)
Maj Walter D. Persons (20-May-14Aug)
Maj Christopher F. Irwin, Jr. (from 15Aug)

Ground Echelon, VMSB-341
(C—20Mar-1May44)
Maj James T. McDaniels

Flight Echelon, Marine Bomber Squadron 413
(C—15Mar-1May44)
(A—2May44-15Mar45)

CO.......... LtCol Andrew B. Galatian, Jr. (to 14Aug44)
LtCol Stewart W. Ralston (14Aug-7Nov)
LtCol Roswell B. Burchard, Jr. (8Nov44-1Jan45)
LtCol Robert B. Cox (from 2Jan)

Marine Bomber Squadron 423
(A—13May44-15Mar45)

CO.......... LtCol John L. Winston (to 19Jul44)
LtCol Norman J. Anderson (from 19Jul)

Marine Bomber Squadron 433
(A—21Jul44-15Mar45)

CO.......... Maj John G. Adams

Marine Bomber Squadron 443
(A—27Aug44-15Mar45)

CO.......... LtCol Dwight M. Guillotte

Advance Echelon, Marine Night-Fighter Squadron 531
(B—12Sep-16Oct43)

CO.......... Col Frank H. Schwable

Rear Echelon, VMF(N)-531
(B—10-16Oct43)
(D—27Oct-15Dec43)
(C—16Dec43-1May44)
(A—2May-15Jul44)

CO.......... Col Frank H. Schwable (to 18Feb44)
LtCol John D. Harshberger (18Feb-13May)
Capt James H. Wehmer (from 14May)

Marine Bomber Squadron 611
(C—15Dec43-1May44)

CO.......... LtCol George A. Sarles

Flight Echelon, VMB-611
(A—17Nov-23Dec44; 11Feb-9Mar45)
LtCol George A. Sarles

Marine Casualties [1]

Location and date	KIA		DOW		WIA		MIAPD		POW [2]		TOTAL	
	Officer	Enlisted	Officer	Enlisted	Officer	Enlisted	Officer	Enlisted	Officer	Enlisted	Officer	Enlisted
Marines												
New Georgia [3] (20Jun–16Oct43)	8	145	0	10	31	384	1	57	0	0	40	596
Bougainville [4] (28Oct43–15Jun44)	18	334	7	81	77	1,172	6	286	0	0	108	1,873
Cape Gloucester (26Dec43–1Mar44)	19	245	1	49	40	775	0	124	0	0	60	1,193
Talasea	2	10	0	16	8	125	0	9	0	0	10	160
Aviation [5]	92	104	1	15	108	114	232	339	17	5	452	577
Sea-duty	1	19	0	5	8	87	3	58	0	0	12	169
Miscellaneous [6]	0	1	1	0	5	1	0	0	0	0	6	2
Total Marines	140	858	10	176	277	2,658	242	873	17	5	688	4,570
Naval Medical Personnel Organic to Marine Units [1]												
New Georgia	0	3	0	0	0	11	0	0	0	0	0	14
Bougainville	1	8	1	2	1	29	0	0	0	0	2	39
Cape Gloucester	1	8	1	2	1	29	0	0	0	0	2	39
Talasea	0	0	0	0	0	4	0	0	0	0	0	4
Marine Aviation	1	2	0	0	0	2	0	0	0	0	1	4
Total Navy	2	21	1	4	2	86	0	0	0	0	5	111
Grand Total	142	879	11	180	279	2,744	242	287	17	5	693	4,681

[1] These final Marine casualty figures were compiled from records furnished by Statistics Unit, Personnel Accounting Section, Records Branch, Personnel Department, HQMC. They are audited to include 26 August 1952. Naval casualties were taken from NavMed P-5021, *The History of the Medical Department of the Navy in World War II*, 2 vols (Washington: Government Printing Office, 1953), II, pp. 1-84. The key to the abbreviations used at the head of columns in the table follows: KIA, Killed in Action; DOW, Died of Wounds; WIA, Wounded in Action; MIAPD, Missing in Action, Presumed Dead; POW, Prisoner of War. Because of the casualty reporting method used during World War II, a substantial number of DOW figures are also included in the WIA column.

[2] Included are 4 officers who died while POWs, and 2 who escaped.

[3] Includes: Rendova, Arundel, Vella Lavella, Enogai, and Vangunu operations.

[4] Includes: Choiseul operation and consolidation of Northern Solomons.

[5] Includes: All operations in Solomons-New Britain area during period 9Feb43-15Mar45.

[6] Includes: Arawe, Russell Islands, and Treasury Islands operations.

Unit Commendations

THE SECRETARY OF THE NAVY,
Washington.

The President of the United States takes pleasure in presenting the PRESIDENTIAL UNIT CITATION to the

MARINE FIGHTING SQUADRON TWO HUNDRED FOURTEEN

for service as set forth in the following

CITATION:

"For extraordinary heroism in action against enemy Japanese forces at Guadalcanal, April 7, 1943; Munda, July 17 to August 30, 1943; Northern Solomons, September 16 to October 19, 1943; and Vella Lavella and Torokina, December 17, 1943, to January 6, 1944. The first squadron to strafe Kahili, the first to operate from Munda while the field was under heavy enemy artillery fire, and the first to lead a fighter sweep on Rabaul, Marine Fighting Squadron TWO HUNDRED FOURTEEN executed bomber escort missions, strafing attacks, search sweeps and patrol missions. Superbly serviced and maintained by its ground crews despite enemy shellfire and nightly bombing attacks, this unit destroyed or damaged 273 Japanese aircraft during these campaigns and, in some of the most bitterly contested air combats on record, contributed substantially to the establishment of an aerial beachhead over Rabaul and paved the way for Allied bombers to destroy Japanese shipping, supply dumps and shore installations. Frequently outnumbered but never outfought, Marine Fighting Squadron TWO HUNDRED FOURTEEN achieved an outstanding combat record which reflects the highest credit upon its skilled pilots, air and ground crews and the United States Naval Service."

For the President.

JAMES FORRESTAL,
Secretary of the Navy.

THE SECRETARY OF THE NAVY,
Washington.

The Secretary of the Navy takes pleasure in commending the

SOUTH PACIFIC COMBAT AIR TRANSPORT COMMAND

consisting of

Marine Aircraft Group TWENTY FIVE
Marine Headquarters Squadron TWENTY FIVE
Marine Service Squadron TWENTY FIVE
Marine Transport Squadron ONE HUNDRED FIFTY TWO
Marine Transport Squadron ONE HUNDRED FIFTY THREE
Marine Transport Squadron TWO HUNDRED FIFTY THREE
403rd Troop Carrier Group and the 801st Evacuation Hospital of the Thirteenth Troop Carrier Squadron, United States Army Forces

for service as follows:

"For outstanding heroism in support of military operations in the forward areas of the South Pacific from December 10, 1942, to July 15, 1944. Flying unarmed, land-based planes without escort despite dangers from Japanese land, sea and air forces, treacherous tropical storms and mechanical failures at sea far from base, the South Pacific Combat Air Transport Command delivered bombs, ammunition, gasoline and vital supplies to combat troops in close and direct contact with the enemy. Frequently taken under fire by hostile antiaircraft guns and fighters while airborne, and by Japanese artillery and Naval gunfire while on the ground at advanced fields, the pilots, aircrewmen and ground echelons served with courage, skill and daring in maintaining uninterrupted support of our forces in the forward areas and contributed essentially to the rout of the Japanese from strategically important bases in the South Pacific. This gallant record of achievement reflects the highest credit upon the South Pacific Combat Air Transport Command and the United States Naval Service."

All personnel attached to and serving with the South Pacific Combat Air Transport Command from December 10, 1942, to July 15, 1944, are hereby authorized to wear the NAVY UNIT COMMENDATION Ribbon.

JOHN L. SULLIVAN,
Secretary of the Navy.

THE SECRETARY OF THE NAVY,
Washington.

The Secretary of the Navy takes pleasure in commending the

MARINE FIGHTING SQUADRON TWO HUNDRED TWENTY ONE

for service as follows:

"For outstanding heroism in action against enemy Japanese forces in the Solomon Islands Area from March 17 to November 17, 1943. Operating with courage and determination in the face of adverse weather, difficult living conditions and inadequate equipment, Marine Fighting Squadron TWO HUNDRED TWENTY ONE carried out daily effective strikes against Munda Airfield and the Kahili Area, in addition to participating in major defensive operations against superior Japanese forces over the Russell Islands and repulsing an attack on our surface vessels in the waters surrounding Tulagi. Relentless in seeking out the enemy, these fighter pilots intercepted a large striking force of Japanese twin-engined bombers attempting to attack our landing forces in Blanche Channel, blasted sixteen of the hostile bombers from the sky and contributed in large measure to the complete annihilation of the striking force and to the success of the Rendova operation. The first squadron to operate from the advanced base at Vella Lavella, Marine Fighting Squadron TWO HUNDRED TWENTY ONE fiercely countered the enemy's aerial attacks and, by completely destroying an entire Japanese squadron refueling on Kara Airfield, aided materially in insuring the success of landings on Treasury Island and Bougainville, at Empress Augusta Bay. By their constant vigilance, aggressiveness and devotion to duty in the face of grave peril, the pilots and crews of this gallant squadron were instrumental in denying to the enemy the strategic Solomon Islands Area, achieving a distinguished combat record in keeping with the highest traditions of the United States Naval Service."

All personnel attached to and serving with Marine Fighting Squadron TWO HUNDRED TWENTY ONE during the period from March 17 to November 17, 1943, are authorized to wear the NAVY UNIT COMMENDATION Ribbon.

JOHN L. SULLIVAN,
Secretary of the Navy.

THE SECRETARY OF THE NAVY,
Washington.

The Secretary of the Navy takes pleasure in commending

MARINE FIGHTING SQUADRON TWO FIFTEEN

for service as follows:

"For outstanding heroism in action against enemy Japanese forces in the Solomon Islands and Bismarck Archipelago Areas from July 24, 1943, to February 15, 1944. Undaunted in the face of hostile fighter opposition and intense antiaircraft fire, Marine Fighter Squadron TWO FIFTEEN carried out numerous patrols and fighter sweeps and escorted many bombing attacks against Japanese shipping, airfields and shore installations. Individually heroic and aggressive, the gallant pilots of this fighting squadron shot down 137 enemy planes, probably destroyed 45 others and accounted for 27 on the ground, an exceptional combat record attesting the superb teamwork of the daring flight echelon and the resourceful, tireless and skilled ground echelon which serviced and maintained the planes despite daily hostile shellfire and nightly bombing attacks. The destruction and damage inflicted on the enemy by Marine Fighting Squadron TWO FIFTEEN contributed substantially to the successful completion of the New Georgia, Bougainville and Rabaul Campaigns and reflect the highest credit upon the United States Naval Service."

All personnel attached to the flight and ground echelons of Marine Fighting Squadron TWO FIFTEEN are hereby authorized to wear the NAVY UNIT COMMENDATION Ribbon.

JAMES FORRESTAL,
Secretary of the Navy.

THE SECRETARY OF THE NAVY,
Washington.

The Secretary of the Navy takes pleasure in commending the

THIRD MARINES, THIRD MARINE DIVISION

for service as follows:

"For outstanding heroism in action against enemy Japanese forces during the invasion, seizure, occupation and defense of Empress Augusta Bay Beachhead, Bougainville, Solomon Islands, from November 1 to December 22, 1943. In action against the enemy for the first time, the THIRD Marines landed on an extremely wide front in the face of perilous surf and beach conditions and through flanking fire of hostile machine guns, anti-boat guns, mortars, small arms and artillery from heavily entrenched positions on Cape Torokina and Puruata Island. Pressing forward through almost impenetrable jungle and swampy terrain, this Regiment completely reduced the intricate system of mutually supporting Japanese pillboxes, bunkers, fire trenches and foxholes which constituted the Cape Torokina defense, and secured its portion of the objective by evening of D-Day. Shifted to the left flank of the beachhead, the THIRD Marines smashed a Japanese counter-landing and drove steadily forward despite difficulties of terrain, supply and communication and, developing the main enemy position in a meeting engagement on the Numa Numa Trail, completely wiped out the Japanese 23rd Infantry. In continuous action as a front line regiment for a total of fifty-two consecutive days, the gallant men and officers of the THIRD Marines, by their skill in jungle warfare and their aggressive fighting spirit, contributed greatly to the success of the campaign and enhanced the highest traditions of the United States Naval Service."

All personnel attached to and serving with the THIRD Marines at Bougainville from November 1 to December 22, 1943, are authorized to wear the NAVY UNIT COMMENDATION Ribbon.

JAMES FORRESTAL,
Secretary of the Navy.

THE SECRETARY OF THE NAVY,
Washington.

The Secretary of the Navy takes pleasure in commending the

TWELFTH MARINES, THIRD MARINE DIVISION

for service as follows:

"For outstanding heroism in action against enemy Japanese forces in the Empress Augusta Bay Beachhead, Bougainville, Solomon Islands, from November 1, 1943, to January 12, 1944; and in the invasion and seizure of Guam, Marianas, July 21 to August 10, 1944. Divided for landing into small elements dispersed over 5000 yards of beach at Empress Augusta Bay, the TWELFTH Marines overcame perilous surf and beach conditions and an almost impenetrable wall of jungle and swampy terrain to land their pack howitzers, initial ammunition and equipment by hand, to occupy firing positions, emplace guns, set up all control facilities and deliver effective fire in support of the THIRD Marine Division beachhead by afternoon of D-Day. In action for 73 days while under continual Japanese air attacks, the TWELFTH Marines aided in smashing an enemy counterattack on November 7–8, silenced all hostile fire in the Battle of Cocoanut Grove on November 13, and delivered continuous effective fire in defense of the vital beachhead position. At Guam, they landed in the face of enemy mortar and artillery fire through treacherous surf and, despite extreme difficulties of communication, supply and transportation, and the necessity of shifting from one type of fire to another, rendered valuable fire support in night and day harassing fires, counterbattery fires and defensive barrages, including the disruption of an organized counterattack by seven Japanese battalions on the night of July 26–27. By their individual heroic actions and their skilled teamwork, the officers and men of the TWELFTH Marines served with courage and distinction during the THIRD Marine Division's missions to secure the Empress Augusta Bay Beachhead and to aid in the recapture of Guam, thereby enhancing the finest traditions of the United States Naval Service."

All personnel attached to and serving with the TWELFTH Marines during these periods are hereby authorized to wear the NAVY UNIT COMMENDATION Ribbon.

JAMES FORRESTAL,
Secretary of the Navy.

THE SECRETARY OF THE NAVY,
Washington.

The Secretary of the Navy takes pleasure in commending the

III AMPHIBIOUS CORPS SIGNAL BATTALION

for service as set forth in the following

CITATION:

"For extremely meritorious service in support of military operations, while attached to the I Marine Amphibious Corps during the amphibious assault on Bougainville, and attached to the III Amphibious Corps during operations at Guam, Palau and Okinawa, during the period from November 1, 1943, to June 21, 1945. The first American Signal Battalion to engage in amphibious landings in the Pacific Ocean Areas, the III Amphibious Corps Signal Battalion pioneered and developed techniques and procedures without benefit of established precedent, operating with limited and inadequate equipment, particularly in the earlier phase of these offensive actions, and providing its own security while participating in jungle fighting, atoll invasions and occupation of large island masses. Becoming rapidly experienced in guerrilla warfare and the handling of swiftly changing situations, this valiant group of men successfully surmounted the most difficult conditions of terrain and weather as well as unfamiliar technical problems and, working tirelessly without consideration for safety, comfort or convenience, provided the Corps with uninterrupted ship-shore and bivouac communication service continuously throughout this period. This splendid record of achievement, made possible only by the combined efforts, loyalty and courageous devotion to duty of each individual, was a decisive factor in the success of the hazardous Bougainville, Guam, Palau and Okinawa Campaigns and reflects the highest credit upon the III Amphibious Corps Signal Battalion and the United States Naval Service."

All personnel attached to the III Amphibious Corps Signal Battalion who actually participated in one or more of the Bougainville, Guam, Palau and Okinawa operations are hereby authorized to wear the NAVY UNIT COMMENDATION Ribbon.

JAMES FORRESTAL,
Secretary of the Navy.

THE SECRETARY OF THE NAVY,
Washington.

The Secretary of the Navy takes pleasure in commending the

NINTH MARINE DEFENSE BATTALION

for service as follows:

"For outstanding heroism in action against enemy Japanese forces at Guadalcanal, November 30, 1942, to May 20, 1943; Rendova-New Georgia Area, June 30 to November 7, 1943; and at Guam, Marianas, July 21 to August 20, 1944. One of the first units of its kind to operate in the South Pacific Area, the NINTH Defense Battalion established strong seacoast and beach positions which destroyed 12 hostile planes attempting to bomb Guadalcanal, and further engaged in extensive patrolling activities. In a 21-day-and-night training period prior to the Rendova-New Georgia assault, this group calibrated and learned to handle new weapons and readily effected the conversion from a seacoast unit to a unit capable of executing field artillery missions. Joining Army Artillery units, special groups of this battalion aided in launching an attack which drove the enemy from the beaches, downed 13 of a 16-bomber plane formation during the first night ashore and denied the use of the Munda airfield to the Japanese. The NINTH Defense Battalion aided in spearheading the attack of the Army Corps operating on New Georgia and, despite heavy losses, remained in action until the enemy was routed from the island. Elements of the Battalion landed at Guam under intense fire, established beach defenses, installed antiaircraft guns and later, contributed to the rescue of civilians and to the capture or destruction of thousands of Japanese. By their skill, courage and aggressive fighting spirit, the officers and men of the NINTH Defense Battalion upheld the highest traditions of the United States Naval Service."

All personnel attached to and serving with the NINTH Defense Battalion during the above-mentioned periods are authorized to wear the NAVY UNIT COMMENDATION Ribbon.

JOHN L. SULLIVAN,
Secretary of the Navy.

THE SECRETARY OF THE NAVY,
Washington.

The Secretary of the Navy takes pleasure in commending the

ELEVENTH MARINE REGIMENT

for service as follows:

"For outstanding heroism while serving with the FIRST Marine Division in action against enemy Japanese forces at Cape Gloucester, New Britain, from December 26, 1943, to April 30, 1944. Tortured by tropical insects, torrential rain and never-ending sniper fire, the Eleventh Marine Regiment slashed through dense jungle and through mud which mired artillery pieces and prevented movement except by man-handling. Refusing to be stopped by any and all obstacles, officers and men worked as an indomitable team under raking enemy fire, fighting their way over twisted, covered trails to provide heavy-weapons fire for the assault infantry troop. With fire from a half-ton field gun, they tore a swatch through the jungle screening a strategic ridge and, in the midst of hand-to-hand fighting with a stubbornly resisting enemy, inched forward up the 40-degree slope to place the field piece on the commanding crest. There they guarded it through the night against the fury of repeated banzai attacks until, in the rain-drenched blackness of early dawn, they stopped the charging Japanese with relentless artillery fire and insured the security of this dominating position. Their fortitude, determination and courageous fighting spirit in the face of almost insurmountable odds throughout this campaign reflect the highest credit upon the Eleventh Marine Regiment and the United States Naval Service."

All personnel attached to and serving with the Eleventh Marines at Gloucester Bay from December 26, 1943, to April 30, 1944, are authorized to wear the NAVY UNIT COMMENDATION Ribbon.

JOHN L. SULLIVAN,
Secretary of the Navy.

Index

    fighter sweeps, 486, 488, 490–491, 493, 499, 502

U. S. GOVERNMENT PRINTING OFFICE : 1964 O - 691-360

Made in the USA
Coppell, TX
05 November 2026

11292664R00352